# FLUVIAL SEDIMENTOLOGY VI

SPECIAL PUBLICATION NUMBER 28 OF THE
INTERNATIONAL ASSOCIATION OF SEDIMENTOLOGISTS

# Fluvial Sedimentology VI

EDITED BY

## N.D. SMITH AND J. ROGERS

*b*

**Blackwell
Science**

© 1999 The International Association of
Sedimentologists and published for them by
Blackwell Science Ltd
Editorial Offices:
Osney Mead, Oxford OX2 0EL
25 John Street, London WC1N 2BL
23 Ainslie Place, Edinburgh EH3 6AJ
350 Main Street, Malden
   MA 02148 5018, USA
54 University Street, Carlton
   Victoria 3053, Australia
10, rue Casimir Delavigne
   75006 Paris, France

Other Editorial Offices:
Blackwell Wissenschafts-Verlag GmbH
Kurfürstendamm 57
10707 Berlin, Germany

Blackwell Science KK
MG Kodenmacho Building
7–10 Kodenmacho Nihombashi
Chuo-ku, Tokyo 104, Japan

First published 1999

Set by Graphicraft Limited, Hong Kong
Printed and bound in Great Britain at the Alden Press,
Oxford and Northampton

The Blackwell Science logo is a
trade mark of Blackwell Science Ltd,
registered at the United Kingdom
Trade Marks Registry

DISTRIBUTORS

Marston Book Services Ltd
PO Box 269
Abingdon, Oxon OX14 4YN
(*Orders*: Tel: 01235 465500
          Fax: 01235 465555)

USA
Blackwell Science, Inc.
Commerce Place
350 Main Street
Malden, MA 02148 5018
(*Orders*: Tel: 800 759 6102
          781 388 8250
Fax: 781 388 8255)

Canada
Login Brothers Book Company
324 Saulteaux Crescent
Winnipeg, Manitoba R3J 3T2
(*Orders*: Tel: 204 837-2987)

Australia
Blackwell Science Pty Ltd
54 University Street
Carlton, Victoria 3053
(*Orders*: Tel: 3 9347 0300
          Fax: 3 9347 5001)

A catalogue record for this title
is available from the British Library

ISBN 0-632-05354-2

Library of Congress
Cataloging-in-publication Data

Fluvial sedimentology VI / edited by N.D. Smith and J. Rogers.
        p.       cm. — (Special publication number 28 of the
   International Association of Sedimentologists)
     ISBN 0-632-05354-2
     1. Sedimentology—Congresses.
     2. River sediments—Congresses.
   I. Smith, Norman D. (Norman Dwight)
   II. Rogers, J. (John), 1944–.   III. International Conference on
   Fluvial Sedimentology (6th: 1998: Cape Town, South Africa)
   IV. Series: Special publication . . . of the International
   Association of Sedimentologists; no. 28.
   QE471.2.F57   1999
   551.3′53—dc21                                  98-53191
                                                  CIP

For further information on
Blackwell Science, visit our website:
www.blackwell-science.com

# Contents

# Preface

This volume is a product of the Sixth International Conference on Fluvial Sedimentology, held 22–26 September 1997 in Cape Town, South Africa. The conference was attended by approximately 300 delegates from 34 countries, and 236 scientific presentations were given, either orally or as posters. This volume contains 31 papers by authors from 13 countries. Twenty-nine of these papers developed directly from the Cape Town conference, and two were submitted independently of the meeting. Their contents represent a wide variety of fluvial subjects, as has become customary for the proceedings volumes arising from these conferences (see references).

The International Conference on Fluvial Sedimentology (ICFS) has been held every four years since its inception in Calgary in 1977. It operates without regular sponsorship or institutional affiliation. Each conference is organized by a volunteer host selected by delegates at the previous meeting, which has so far resulted in a widespread geographical distribution of conference venues — Calgary (Canada), Keele (UK), Fort Collins (USA), Sitges (Spain), Brisbane (Australia), and Cape Town (South Africa). The next conference is scheduled for 2001 at the University of Nebraska in Lincoln, USA.

The scope of the ICFS meetings has grown steadily over their 20-year existence: from a 3-day, single-session, one-field-trip meeting in Calgary to the 5-day, multiple-session, multiple-field-trip event in Cape Town. An additional feature of the Cape Town conference, and one that affects the contents of this volume, was that for the first time, several special symposia were introduced into the meeting. One of these symposia, 'Avulsion — origins, processes and depositional responses', resulted in seven papers in this volume appearing under the heading of 'Avulsion: Modern and Ancient'. A symposium entitled 'Placer formation through time' produced several papers, which will appear in a special issue of *Economic Geology* (1999, Vol. 94 no. 5), marking the first time that at least one placer-related paper has not appeared in an ICFS proceedings volume. Four papers were developed from a third symposium, 'Textural variability and grain-size differentiation in gravel-bed rivers', now published in the *Journal of Sedimentary Research* (1999, Vol. 69, no. 1).

A word about the title of this volume is in order. Because of the loose organization of the ICFS enterprise, the editors of the previous five proceedings volumes chose their own titles, with the result that each previous volume is named substantially differently from the others despite a substantial similarity in their contents (Calgary: Miall, 1978, '*Fluvial Sedimentology*'; Keele: Collinson & Lewin, 1983, '*Modern and Ancient Fluvial Systems*'; Fort Collins: Ethridge *et al.*, 1987, '*Recent Developments in Fluvial Sedimentology*'; Sitges: Marza & Puigdefabregas, 1993, '*Alluvial Sedimentation*'; Brisbane: Fielding, 1993, '*Current Research in Fluvial Sedimentology*'. Given the success of the Cape Town meeting and the current plans for a seventh ICFS in Lincoln in 2001, it would appear that these conferences and their resulting proceedings volumes may be settling in for a long run. It therefore would seem prudent to attempt to achieve some uniformity in the volume titles, if for no other reason than the prospect that unique titles in the future will become increasingly difficult to invent. The title of this volume, '*Fluvial Sedimentology VI*', was selected because it duplicates the title of the initial volume (Miall, 1978) as well as the conference title, and furthermore it identifies the position of the volume in the whole series.

We thank the authors for their cooperation with deadlines and their willingness to make requested modifications to their manuscripts with firm upper lips. We are especially grateful to the numerous reviewers whose comments and recommendations were invaluable in bringing this volume forth. They are: J.F. Aitken, J. Alexander, A. Arche, P. Ashmore, P.J. Ashworth, W.J. Autin, J.L. Best, M.D. Blum, R.G. Brakenridge, C.S. Bristow, D.W. Burbanks, M. Church, E. Cotter, R.W. Dalrymple, P. DeCelles, J. Diemer, G.B. Doyle, K.A. Eriksson, P.G. Eriksson, F.G. Ethridge, W.K. Fletcher, L. Frostick, S.L. Gabel, D. Germanowski, M.L. Goedhart, B. Gomez, M.J. Guccione, D.J. Harbor, E.J. Hickin, J. Isbell, H.M. Jol, L.S. Jones, A.D. Knighton, R.A. Kostaschuk, M.J. Kraus, M.F. Lapointe, D.A. Leckie, G.H. Mack, S.D. Mackey, F. Magilligan, S.B. Marriott, S.J. McLelland, L.A.K. Mertes, J.R. Miller, G.S. Morozova, G.C. Nadon, G.C. Nanson, J. Peakall, M. Pérez-Arlucea, J.E. Pizzuto, R.L. Slingerland, D.G. Smith, L.C. Smith, I.G. Stanistreet, T.E. Törnqvist, B.R. Turner, P. Whiting, B.J. Willis, M.C. Wizevich, K.J. Woolfe and M.J. Zaleha. Many thanks to them all.

## REFERENCES

COLLINSON, J.D. & LEWIN, J. (Eds) (1983) *Modern and Ancient Fluvial Systems.* Spec. Publs int. Ass. Sediment., No. 6, 575 pp. Blackwell Scientific Publications, Oxford.

ETHRIDGE, F.G., FLORES, R.M. & HARVEY, M.D. (Eds) (1987) *Recent Developments in Fluvial Sedimentology.* Spec. Publ. Soc. econ. Paleont. Miner., Tulsa, **39**, 389 pp.

FIELDING, C.R. (Ed.) (1993) *Current Research in Fluvial Sedimentology. Sediment. Geol.,* **85** (1–4), 1–656.

MARZO, M. and PUIGDEFABREGAS, C. (Eds) (1993) *Alluvial Sedimentation.* Spec. Publs int. Ass. Sediment., No. 17, 640 pp. Blackwell Scientific Publications, Oxford.

MIALL, A.D. (Ed.) (1978) *Fluvial Sedimentology.* Mem. Can. Soc. petrol. Geol., Calgary, **5**, 859 pp.

NORMAN D. SMITH
*Lincoln, Nebraska, USA*

JOHN ROGERS
Cape Town, South Africa

# Sediment Transport and Bedforms

*Spec. Publs int. Ass. Sediment.* (1999) **28**, 3–13

# Turbulent sand suspension over dunes

R. KOSTASCHUK* *and* P. VILLARD†

*\*Department of Geography, University of Guelph, Guelph, Ontario, N1G 2W1, Canada*
*(Email: rkostasc@uoguelph.ca); and*
*†Division of Science and Technology, Department of Geography, Tamaki Campus, University of Auckland,*
*Private Bag 29019, Auckland, New Zealand*

## ABSTRACT

Processes of sand suspension in rivers are poorly understood, but flume studies suggest that large, turbulent, coherent flow structures generated over dunes may play an important role. This study links field and flume studies by visualizing sand suspension over large dunes in the Fraser Estuary, Canada. An acoustic profiler provides nearly instantaneous 'snapshots' of strongly intermittent suspension events, and velocity and sand-concentration profiles are used to create time- and space-averaged contour maps of velocity and sand suspension. The acoustic images show suspension structures originating at crests and at lower stoss sides of dunes. The contour maps reveal turbulent structures, characterized by low velocity and high turbulence intensity, at the dune trough, the lower stoss side, the crest, and above the lee side downstream of the crest. High sand concentration and flux on the contour maps occur in a zone that originates close to the bed on the stoss side of the dune and extends downstream of the dune crest. A secondary zone occurs further above the bed. The acoustic images and contour maps show that sand is suspended in coherent flow structures characterized by low mean velocity and high turbulence intensity. These structures are interpreted as upwelling ejections generated at crests and lower stoss sides of dunes. The results of this study improve our understanding of sand suspension processes and have important implications for sand transport in large rivers.

## INTRODUCTION

### Background and purpose

Turbulence is a fundamental characteristic of alluvial flows and is linked to many geomorphological processes and features, such as sediment transport (Drake *et al.*, 1988; Rood & Hickin, 1989; Thorne *et al.*, 1989; Lapointe, 1992; Babakaiff & Hickin, 1996), bedform development (Jackson, 1976; Best, 1992; Muller & Gyr, 1996a), and the origin of stream meanders (Levi, 1991; Rhoads & Welford, 1991; Yalin, 1992). Recent laboratory and field evidence suggests that large, turbulent, coherent flow structures, generated by dunes in unidirectional water flows, are fundamental agents of sediment suspension (for detailed reviews see Best (1993) and Nezu & Nakagawa (1994) ). The experimental approaches used in laboratory and field studies are very different, however, and the data are often not directly comparable. Many laboratory studies in small flumes, for example, use flow visualization and high-frequency vertical and horizontal velocity measurements, techniques not easily extrapolated to large dunes in the field. Muller & Gyr's (1986, 1996b) laboratory flow visualizations showed that coherent flow structures occur as vortices that originate at the dune crest and develop above and within the lee-side flow-separation zone. These structures can interact with the bed at the point of flow reattachment on the next dune downstream and may propagate to the surface. Strong potential for sediment suspension is suggested from detailed velocity measurements over fixed bedforms (e.g., Nezu & Nakagawa, 1989; Nelson *et al.*, 1993; McLean *et al.*, 1994, 1996; Bennett & Best, 1995, 1996), which reveal high levels of turbulence at the point of reattachment of flow separation and along the shear layer originating at the crest. The mobile-bed flume experiments of Iseya & Ikeda (1986) support the hypothesis that bed sediment is suspended in coherent structures generated at reattachment.

Studies of river flows have often identified 'boils', circular patches of upwelling fluid at the flow surface (e.g., Matthes, 1947; Jackson, 1976; Rood & Hickin, 1989; Babakaiff & Hickin, 1996). Boils usually contain higher suspended bed-material concentrations than the ambient flow (Rood & Hickin, 1989), implying that they are linked to sediment-entraining events at the bed. Matthes (1947) proposed that boils are the surface manifestation of coherent flow structures that he called 'kolks', an interpretation later supported by Jackson (1976). Jackson speculated that kolks originated in the lee-side flow-separation zone, although he had no data to support this. Kostaschuk & Church (1993) and Kostaschuk & Villard (1996a) used acoustic flow visualizations to identify coherent flow structures containing high suspended-sand concentrations.

This research expands on earlier work by Kostaschuk & Church (1993) and Kostaschuk & Villard (1996a) and aims to link field and flume studies by visualizing sand suspension over dunes in the Fraser River estuary, Canada. Dunes in the Fraser are long, low bedforms without lee-side flow reversals (Kostaschuk & Villard, 1996b) and represent many dunes in large rivers dominated by sand transport in suspension (e.g., Smith & McLean, 1977; Ashworth et al., 1996). We use an acoustic profiler to provide images of suspension structures being generated over dunes and velocity and sand-concentration profiles to examine the effect of these structures on mean-flow and sand-suspension patterns. We show that sediment suspension is linked to coherent flow structures, typified by low mean velocity and high turbulent intensity, which are interpreted as upwelling fluid ejections.

## Dunes in the Fraser Estuary

The Fraser River has a mean annual discharge of 3400 m$^3$ s$^{-1}$, with maximum flows over 11 000 m$^3$ s$^{-1}$, and discharges into the Strait of Georgia, a mesotidal marine basin on the west coast of Canada (Fig. 1). The Main Channel of the Fraser Estuary has a sand bed that experiences maximum sediment transport during low tides when river discharge is high (Kostaschuk et al., 1989a,b). Dunes in the estuary range from 0.1 m to 4 m in height and 2 m to more than 100 m in length (Kostaschuk et al., 1989a). The largest dunes have a curved, concave-downstream planform geometry with circular depressions on crests and associated pits in troughs (Kostaschuk & MacDonald, 1988). Dunes migrate around low tide when sandy bed material is in transport (Kostaschuk et al., 1989a,b; Kostaschuk & Church, 1993; Kostaschuk & Villard, 1996b).

Kostaschuk & Villard (1996b) describe two main dune types in the Fraser Estuary. Symmetric dunes have stoss and lee sides of similar length, mean stoss and lee slope angles < 8°, and rounded crests. Lee slopes of symmetric dunes reach maximum angles of 11–18° near the dune trough. Asymmetric dunes have superimposed small dunes on stoss sides, long stoss sides with slopes < 3°, and short, straight lee slopes up to 19°. Kostaschuk & Villard (1996b) conducted a detailed examination of velocity profiles and found no evidence for lee-side flow reversal in any of their measurements. Kostaschuk & Villard (1996b) speculate that symmetric dunes are equilibrium features, but asymmetric dunes are transitional forms between symmetric dunes and smaller dunes adjusted to lower flow velocity.

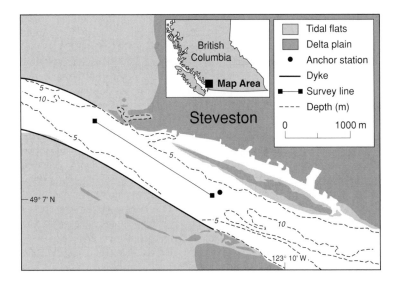

**Fig. 1.** Study area showing locations of anchor station (1989) and survey line (1990).

## METHODS

A Datasonics DFT 25 kHz acoustic profiler (e.g., Kostaschuk & Church, 1993) was used in June 1990 to visualize suspended sand structures over large dunes in the Fraser Estuary. The profiler transducer was mounted on a launch that ran along a survey line in the centre of the estuary channel (Fig. 1). Data were collected during the 2–3 h period surrounding low tide when flow is steady, bed sediment is in suspension, and boils are apparent at the surface. A run usually required 10–15 min to complete and several replicate runs were made during each survey period. Profile lines varied from 465 m to 1690 m in length. Measurements of flow velocity and other conditions were not possible during these surveys because the vessel was travelling in the channel.

Measurements of velocity and suspended sediment were taken over seven symmetric dunes in June and July 1989 during the 2–3 h period around low tide. Kostaschuk & Villard (1996b) provide a detailed analysis of velocity and concentration profiles over three of these dunes — this study also includes unpublished data from four additional dunes. Velocity and suspended bed-material concentration profiles were measured from a launch anchored at a single station (Fig. 1). A 200-kHz echosounder was used to measure water depth and to position the vessel over each dune. Velocity was measured with a Marsh McBirney Model 527 speed-direction electromagnetic current meter with a sampling radius of 0.2 m and a listed accuracy of $\pm 1$ cm s$^{-1}$. A pump sampler was used to obtain suspended sediment samples. Velocity was typically measured at 10 points above the bed for each profile, and suspended sediment was measured at five points. Measurements started at 0.5 m above the bed and finished at 8.5 or 10 m above the bed, depending on flow depth. Velocity was sampled at 1 Hz for a 90 s period at each point, and the pump sample usually required around 30 s to complete. Particle-size analysis was used to separate the wash load concentration (particles < 0.125 mm: McLean & Church, 1986) from the sand concentration in suspended-sediment samples.

Point velocity and sand concentration were normalized, so that data from individual dunes could be combined and used to construct contour maps of point horizontal mean velocity, point horizontal turbulence intensity, point suspended-sand concentration and point suspended-sand flux. The vertical position of each point in the flow column ($y$) was measured relative to the trough of each dune, then divided by dune height ($H$) to normalize it. The horizontal position of each point ($x$) was measured relative to the crest of each dune and divided by the dune stoss-side length ($L_s$) for normalization. Each mean velocity measurement ($u$) at a point in the flow was normalized by the average near-surface velocity for that dune ($\bar{u}_s$). Normalized turbulence intensity is the standard deviation of the point velocity ($s$) divided by the mean point velocity. Point suspended-sand concentration ($c$) was normalized by the average near-surface suspended-sand concentration for each dune ($\bar{c}_s$). Point sand flux ($g$, where $g = uc$) was normalized by the average near-surface sand flux for each dune ($\bar{g}_s$).

A total of 240 point velocity and 98 point sand-concentration measurements were obtained (Fig. 2). All of the

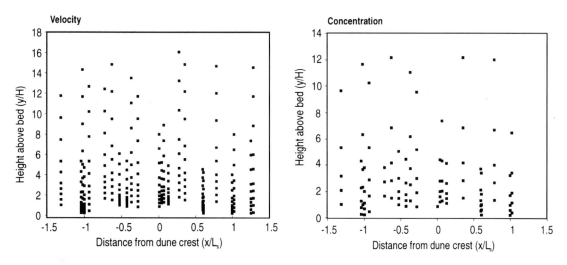

**Fig. 2.** Sampling locations for normalized velocity and concentration measurements. See text for explanation of symbols and normalization procedures.

observations were used for the contouring procedures, but the maps focus on the lower and central portion of the flow where the data are concentrated. Linear kriging (e.g., Davis, 1986), a simple moving-average interpolation method, was used to produce the sampling grid for the contour maps. Kriging estimates points on an evenly spaced grid by calculating a weighted average from a group of the nearest observations. We selected this procedure over alternative trend-based methods because it avoids extreme estimates when data are sparse. A disadvantage of kriging is that some resolution is lost in the filtering process. As the contour maps are intended to provide an aggregate and general representation of flow and sediment dynamics, this is not a serious limitation.

## RESULTS

River discharge, cross-sectional velocity and flow depth were lower during the 1989 field season compared with 1990, but tidal height was similar (Table 1). The lower cross-sectional velocity in 1989 reflects the lower river discharge. Flow depth is less in 1989 because surveys were conducted along the edge of the navigation channel, compared with acoustic profiles collected in the deeper mid-channel area in 1990, and discharge was lower. Mean dune length, height and symmetry ratio are all smaller in 1989 compared with 1990, but dune steepness and lee-side slopes are larger (Table 2). The 1990 dunes have stoss slopes, maximum lee slopes, and symmetry ratios characteristic of symmetric dunes. Some 1990 dunes also have smaller superimposed dunes on the stoss side (Fig. 3), a morphology that Kostaschuk & Villard (1996b) describe as characteristic of asymmetric dunes.

Acoustic flow visualization revealed distinct sand-suspension events on all surveys, but not all dunes actively generated acoustic events. Replicate profile lines showed that individual dunes were consistently active

over the survey period on each date, and in one case even on consecutive days. Acoustic structures usually originated on the lower stoss sides of dunes, but some occurred directly above dune crests (Fig. 3). The upward-facing surface of many structures 'bends' in a downstream direction, rising at angles from 12 to 36° near the bed and decreasing to 5–11° above. Some structures extend more than 10 m above the bed, but most are confined to the lower 6–8 m of flow.

Figure 4 illustrates the general patterns of mean velocity, turbulence intensity, sand concentration and sand flux with height above the bed. Mean velocity increases nonlinearly with height above the bed, whereas turbulence intensity, sand concentration and sand flux all decrease.

**Table 1.** Flow conditions in 1989 and 1990. $Q$ is mean daily discharge from the Water Survey of Canada for Fraser River at Hope (130 km upstream of the study site), $T$ is low tide height from the Canadian Tide and Current Tables for Point Atkinson in Strait of Georgia (30 km north of the study site), $U_p$ is the mean cross-sectional channel velocity at Steveston predicted from the Fraser River Mathematical Model, $h$ is mean flow depth over each dune in 1989 and for the acoustic survey profile in the centre of the channel in 1990.

| Survey | $Q\,(\mathrm{m^3\,s^{-1}})$ | $T\,(\mathrm{m})$ | $U_p\,(\mathrm{m\,s^{-1}})$ | $h\,(\mathrm{m})$ |
|---|---|---|---|---|
| 16/06/89 | 6780 | 0.9 | 1.3 | 11.0 |
| 19/06/89 | 6790 | 1.1 | 1.5 | 10.3 |
| 20/06/89 | 6480 | 0.4 | 1.4 | 10.4 |
| 21/06/89 | 5960 | 0.4 | 1.4 | 10.0 |
| 01/07/89 | 4980 | 0.3 | 1.6 | 10.3 |
| 02/07/89 | 4720 | 0.2 | 1.6 | 12.0 |
| 03/07/89 | 4510 | 0.2 | 1.6 | 11.0 |
| 20/06/90 | 8770 | 0.2 | 1.9 | 10.9 |
| 21/06/90 | 8590 | 0.1 | 1.8 | 12.0 |
| 24/06/90 | 8620 | 0.2 | 1.8 | 11.3 |
| 27/06/90 | 8820 | 1.4 | 1.7 | 11.6 |
| 1989 mean | 5750 | 0.5 | 1.5 | 10.7 |
| 1990 mean | 8700 | 0.5 | 1.8 | 11.5 |

**Table 2.** Dune morphology in 1989 and 1990. Dunes from 1989 are those over which velocity and suspended-sediment data were gathered, $1990_e$ are active dunes with acoustic events, $1990_i$ are inactive dunes without acoustic events, $p$ is the independent $t$-test probability comparing the morphology of active and inactive dunes groups in 1990, $n$ is the number of dunes, $L$ is mean dune length, $H$ is mean dune height, $H/L$ is mean dune steepness, $\alpha$ is mean lee-side slope angle, $\alpha_{max}$ is the mean value of the steepest lee-side slope angle, $S$ is the mean symmetry ratio (stoss-side length divided by lee-side length). Values in parentheses for 1989 are maxima and minima.

| Year | $n$ | $L\,(\mathrm{m})$ | $H\,(\mathrm{m})$ | $H/L$ | $\alpha\,(^0)$ | $\alpha_{max}\,(^0)$ |
|---|---|---|---|---|---|---|
| 1989 | 7 | 27 (18–34) | 1.33 (0.68–1.93) | 0.049 (0.031–0.072) | 7.4 (6.8–8.3) | 11.1 (8.9–18) |
| $1990_e$ | 15 | 64 | 1.95 | 0.031 | 6.6 | 10.2 |
| $1990_i$ | 15 | 74 | 2.20 | 0.030 | 6.4 | 12.8 |
| $p$ | 15 | 0.44 | 0.25 | 0.76 | 0.87 | 0.71 |

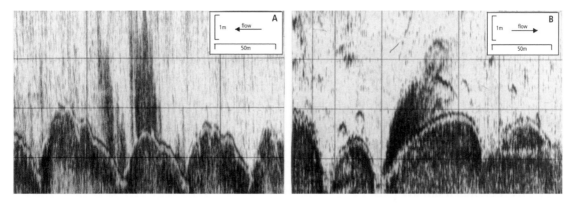

**Fig. 3.** Acoustic flow visualization along the survey line (Fig. 1): A, 27 June 1990; B, 21 June 1990.

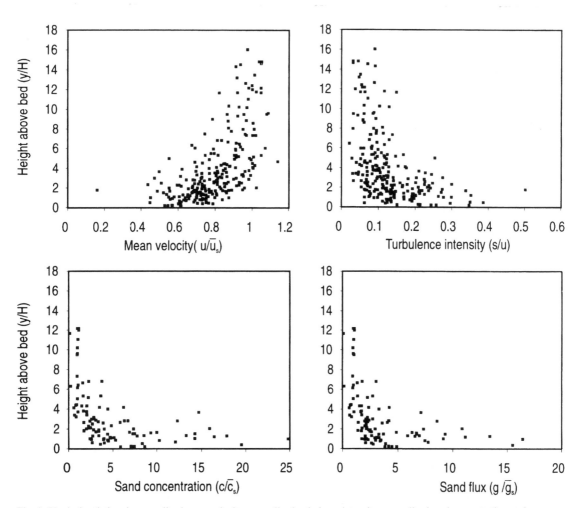

**Fig. 4.** Vertical variations in normalized mean velocity, normalized turbulence intensity, normalized sand concentration, and normalized sand flux. Sample positions are shown on Fig. 2. See text for explanation of symbols and normalization procedures.

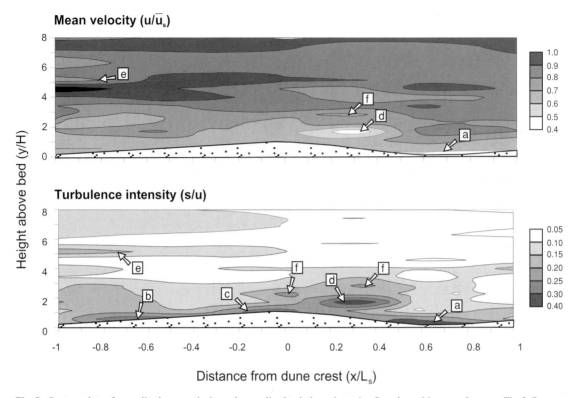

**Fig. 5.** Contour plots of normalized mean velocity and normalized turbulence intensity. Sample positions are shown on Fig. 2. See text for explanation of symbols and normalization and contouring procedures.

Spatial variations in these patterns are illustrated on the contour maps (Figs 5 & 6). The upper portion of the flow ($> 4y/H$) is a reasonably homogeneous region of high velocity, low turbulence intensity, low sand concentration and low sand flux, although there are localized zones of higher mean velocity and turbulence intensity above the lower stoss side. The lower region of the flow ($< 4y/H$) has lower velocity, higher turbulence intensity and higher sand concentrations and flux.

Well-defined zones of high turbulence intensity occur near the bed in the dune trough (a on Fig. 5), on the lower stoss side (b on Fig. 5) and at the dune crest (c on Fig. 5). Similar zones occur further above the bed downstream of the dune crest (d on Fig. 5) and the lower stoss side (e on Fig. 5). A poorly defined zone of moderate turbulence intensity also exists higher in the flow above the crest (f on Fig. 5). Some zones of high turbulence intensity are also characterized by low mean velocity (a, d, e and f on Fig. 5). All velocities measured on the lee side are directed downstream, with no evidence for lee-side flow reversal.

Discrete zones of high sand concentration and flux occur in the lower flow region. The largest zone begins close to the bed on the lower stoss side and extends slightly downstream of the crest (a on Fig. 6). Smaller zones of high concentration and flux extend downstream of the crest above the bed (b on Fig. 6) and above the bed over the upstream trough (c on Fig. 6). A zone of high concentration also occurs in the downstream trough (d on Fig. 6).

## DISCUSSION

The smaller dunes in 1989 compared with 1990 (Table 2) reflect the shallower flows and lower flow velocities in 1989 (Table 1). Lower dune steepness (height/length), lower lee-side angles and higher symmetry ratios for dunes in 1990 may be the result of higher suspended-sand transport rates produced by higher flow velocity. Kostaschuk & Villard (1996b) found that high flow velocity caused 'rounding' of the dune crest and

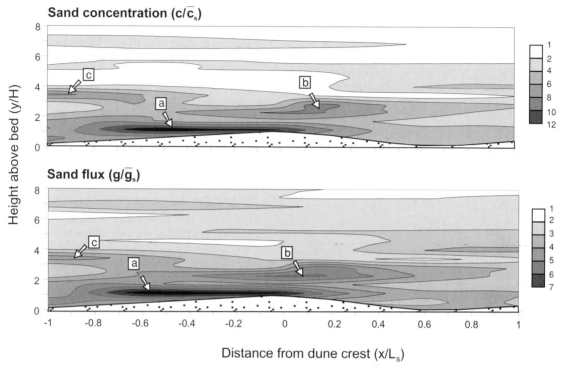

**Fig. 6.** Contour plots of normalized sand concentration and normalized sand flux. Sample positions are shown on Fig. 2. See text for explanation of symbols and normalization and contouring procedures.

enhanced deposition of suspended sediment in the dune trough, both of which would contribute to lower steepness and gentler, but longer, lee-side slopes in 1990. They also found that these processes cause lower symmetry ratios, but our data show lower ratios in 1989. We have no physical explanation for this, but it could be biased by the small sample size in 1989. Some dunes in 1990 have superimposed dunes similar to the asymmetric dunes described by Kostaschuk & Villard (1996b), even though all other morphological characteristics resemble symmetric dunes. This suggests that a few 1990 dunes may be in the process of transition to asymmetric forms.

Acoustic flow visualization showed that some dunes were consistently generating suspension structures and others were not. We examined possible morphological controls on event generation by using independent *t*-tests to compare characteristics of acoustically active and inactive dunes (Table 2). The tests showed no significant difference in any of the morphological variables between the two groups, indicating that two-dimensional dune morphology does not influence suspension activity. It is possible that planform characteristics of Fraser dunes,

such as crest depressions and trough pits (Kostaschuk & MacDonald, 1988), may serve as foci for sediment suspension, but we have no way to test this for our data set.

It is well-known that velocity and sand concentration can vary considerably over time periods of several minutes (e.g., Lapointe, 1992) and our sampling intervals of 90 s for velocity and around 30 s for concentration certainly do not capture all temporal variability inherent in the Fraser data. In addition, the maps of concentration and flux are based on a lower sampling density than velocity. These sampling considerations suggest that the contour maps are useful as general representations of flow and sediment fields over dunes, but detailed interpretations of patterns must be treated with some caution. Additional caution must be used in comparing acoustic images with patterns on the contour maps because of the very different temporal and spatial scales of the data. The contour maps represent time-averaged and dune-averaged velocity and suspended-sediment concentration. In contrast, the acoustic images are nearly instantaneous 'snapshots' of strongly intermittent suspension events. The measurements used for the contour maps, however, certainly capture and

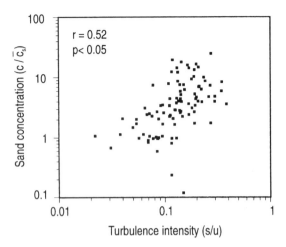

**Fig. 7.** Scatter plot of sand concentration versus turbulence intensity. See text for explanation of symbols and normalization procedures.

'average' the instantaneous structures illustrated on the acoustic images.

The contour maps (Figs 5 & 6) show that discrete zones of low velocity and high turbulence intensity generally correspond with zones of high sand concentration and flux, an appearance supported by a moderate but highly significant correlation between turbulence intensity and sand concentration (Fig. 7). This correlation indicates that sand is held in suspension by events associated with high turbulence intensity in the flow. The largest zone of high sand concentration and flux on the contour maps (a on Fig. 6) occurs close to the bed on the stoss side of the dune and extends above the lee downstream of the crest. High turbulence intensity near the bed, however, occurs in zones in the trough (a on Fig. 5), the lower stoss side (b on Fig. 5), the crest (c on Fig. 5), and above the lee side downstream of the crest (d on Fig. 5). More data were used in construction of the velocity contour maps compared with the sediment maps, and the sand sampling pattern may not be sufficiently dense to reveal different zones. It is also possible that sand eroded on the lower stoss side and transported in suspension over the stoss obscures sand-suspension sites further downstream. Although it is difficult to locate the origin of the portion of the sand-suspension zone that extends beyond the dune crest on the contour map (a on Fig. 6), it is probably the product of suspension structures apparent on the lower stoss sides of dunes on acoustic images (Fig. 3). The zone of high turbulence intensity on the lower stoss side (b on Fig. 5) is probably responsible for the erosion and suspension of the sand.

The secondary zone of high concentration and flux above the bed on the contour map (b on Fig. 6) most likely is associated with the zone of high turbulence intensity situated on the crests of the dunes on the contour map (c on Fig. 5) and on the acoustic images (Fig. 3). The zone of high sand concentration and flux above the upstream trough (c on Fig. 6) is probably from dunes located upstream. Low mean velocity and high turbulence intensity in the dune trough (a on Fig. 5) are related to high sand concentration (d on Fig. 6), but not high flux. High sand concentration in the trough results from sediment settling from above that is kept in suspension by high turbulence intensities. The absence of significant sand flux in the trough is the result of low horizontal velocities.

Figure 8 is a preliminary, conceptual model for flow and turbulent suspended-sediment structures over large, low-angle, unseparated-flow dunes dominated by sediment transport in suspension. This model is speculative in nature and is designed to serve as a guide for further investigation. Flume experiments and field measurements of dunes with flow separation (e.g., McLean & Smith, 1986; Nelson *et al.*, 1993; Bennett & Best, 1995) show that the mean flow consists of an outer flow, a wake region that expands downstream of the crest, a lee-side separation zone of flow reversal and an internal boundary layer. Our contour maps of mean velocity and turbulence reveal a region of outer flow characterized by high velocity and low turbulence intensity (Fig. 8A), but over these low-angle dunes the separation zone is replaced by a lee-side deceleration zone of downstream flow (Fig. 8A). The upper boundary of the wake region is indicated by the zone of high turbulence intensity that extends downstream of the crest on our contour maps (Fig. 8A). An internal boundary layer probably exists on the stoss side downstream of the reattachment point (Fig. 8A), but our measurements are too far from the bed to define it.

Bennett & Best (1995) found high horizontal turbulence intensity in dune troughs, a pattern similar to our contour plots, but did not find the zone of high horizontal turbulence intensity at the dune crest. Their experiments, however, showed that turbulent 'sweep' events (inrushes of fluid toward the bed) were concentrated near the dune crest and at reattachment. They also found that turbulent 'ejection' events (upwellings of fluid toward the surface) occurred along the shear layer between the wake and outer flow regions. More recently, Bennett & Best (1996) suggested that ejection events also originate at reattachment, where sweeps are concentrated. Reattachment ejection structures could originate from large vortices that develop in the wake region (Muller & Gyr, 1986, 1996b), or instabilities that originate along the shear layer above the deceleration zone (Itakura & Kishi,

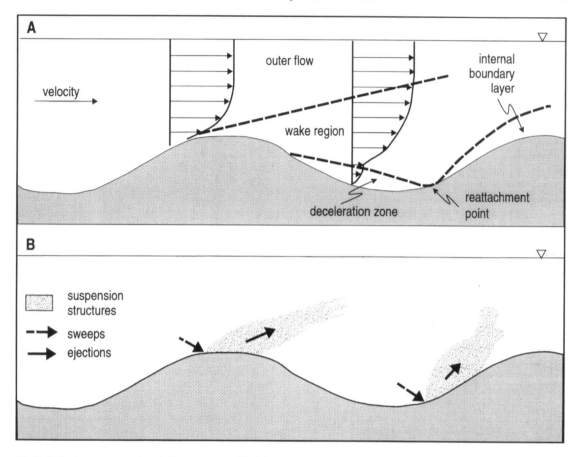

**Fig. 8.** Preliminary conceptual model for flow (A: modified from McLean & Smith, 1986), and suspension structures (B: proposed sweep positions from Bennett & Best, 1996) for large dunes dominated by sand transport in suspension. See text for explanation. Not to scale.

1980). These flume experiments show that ideal conditions for generation of upwelling, 'ejection-like' suspension structures exist at dune crests and at reattachment. Large turbulent sweep structures over dunes could erode bed sediment (e.g., Best, 1992) that is then suspended in the ejection structures (Fig. 8B). Our acoustic data (Fig. 3) provide dramatic images of these events in the Fraser River. Kostaschuk & Church (1993) found that acoustic suspension structures in the Fraser River have strong positive upward vertical velocities and horizontal velocities lower than the mean—features characteristic of ejections.

Most large rivers have fine-grained beds in their lower reaches, and sand transport is overwhelmingly in suspension (e.g., Kostaschuk & Ilersich, 1995). Acoustic time series (Kostaschuk & Church, 1993; Kostaschuk & Villard, 1996b) show that coherent suspension structures occupy most of the flow and are likely the dominant mechanisms of sediment transport in these systems. Although many formulae are available for sediment transport in fluvial systems, most rely on time-averaged flow properties and do not attempt to account for the fundamental role of coherent flow structures (Nezu & Nakagawa, 1994). Recently, Hogg *et al.* (1994, 1996) have proposed models to simulate the transport of bedload by sweeps and suspension transport by ejection structures. We believe that the model of Hogg *et al.* (1994) should be developed to provide a more complete simulation of sand suspension by coherent flow structures. Such a model is crucial to understanding sedimentary processes and to reliable prediction of sediment transport rates, especially in rivers dominated by sediment transport in suspension. Our understanding of the fluid and sediment dynamics of turbulent suspension structures remains poor, however, and will improve only with combined field and flume investigations.

## SUMMARY

An acoustic profiler provides nearly instantaneous 'snapshots' of strongly intermittent suspension events, and velocity and sand-concentration profiles are used to create time- and spatially averaged contour maps of flow and sand-suspension structures over dunes in the Fraser River. Acoustic flow visualization shows suspension structures originating at both dune crests and at the lower stoss sides of dunes. The contour maps reveal turbulent structures, characterized by low velocity and high turbulence intensity, in the dune trough, the lower stoss side, the crest and above the lee side downstream of the crest. High sand concentration and flux on the contour maps occur in a zone that originates close to the bed on the stoss side of the dune and extends downstream of the dune crest, and in a zone further above the bed on the stoss side. The acoustic images and contour maps show that sand is suspended in coherent flow structures characterized by a low mean velocity and high turbulence intensity. These structures are interpreted as upwelling ejections generated at dune crests and on the lower stoss sides of dunes.

## ACKNOWLEDGEMENTS

Special thanks to Mike Church and John Luternauer for their ongoing support. Financial support was provided by the Natural Sciences and Engineering Research Council of Canada and the Geological Survey of Canada. Sherry Ilersich helped in the field and Ray Sanderson piloted the survey launch. Mike Church reviewed an early version of the manuscript and made many helpful comments.

## REFERENCES

Ashworth, P.J., Best, J.L. & Roden, J. (1996) *Bedforms and Bar Dynamics in the Main Rivers of Bangladesh.* River Survey Project FAP 24, Special Report No. 9, University of Leeds, Leeds, UK.

Babakaiff, C.S. & Hickin, E.J. (1996) Coherent flow structures in Squamish River estuary, British Columbia, Canada. In: *Coherent Flow Structures in Open Channels* (Eds Ashworth, P.J., Bennett, S.J., Best, J.L. & McLelland, S.J.), pp. 321–342. John Wiley & Sons, Chichester.

Bennett, S.J. & Best, J.L. (1995) Mean flow and turbulence structure over fixed, two-dimensional dunes: implications for sediment transport and bedform stability. *Sedimentology*, **42**, 491–513.

Bennett, S.J. & Best, J.L. (1996) Mean flow and turbulence structure over fixed ripples and the ripple–dune transition. In: *Coherent Flow Structures in Open Channels* (Eds Ashworth, P.J., Bennett, S.J., Best, J.L. & McLelland, S.J.), pp. 281–304. John Wiley & Sons, Chichester.

Best, J.L. (1992) On the entrainment of sediment and initiation of bed defects: insights from recent developments within turbulent boundary layer research. *Sedimentology*, **39**, 797–811.

Best, J.L. (1993) On the interactions between turbulent flow structure, sediment transport and bedform development: some considerations from recent experimental research. In: *Turbulence: Perspectives on Flow and Sediment Transport* (Eds Clifford, N.J., French, J.R. & Hardisty, J.), pp. 61–93. John Wiley & Sons, Chichester.

Davis, J.C. (1986) *Statistics and Data Analysis in Geology.* John Wiley & Sons, Chichester.

Drake, T.G., Shreve, L.R., Dietrich, W.E., Whiting, P.J. & Leopold, L.B. (1988) Bedload transport of fine gravel observed by motion picture photography. *J. Fluid Mech.*, **192**, 193–218.

Hogg, A.J., Huppert, H.E. & Soulsby, R.L. (1994) The dynamics of particle-laden fluid elements. In: *Sediment Transport Mechanisms in Coastal Environments* (Eds Belorgey, M., Rajaona, R.D. & Sleath, J.F.A.), pp. 64–78. Proceedings of Euromech 310.

Hogg, A.J., Dade, W.B., Huppert, H.E. & Soulsby, R.L. (1996) A model of an impinging jet on a granular bed, with application to turbulent, event-driven bedload transport. In: *Coherent Flow Structures in Open Channels* (Eds Ashworth, P.J., Bennett, S.J., Best, J.L. & McLelland, S.J.), pp. 101–124. John Wiley & Sons, Chichester.

Iseya, F. & Ikeda, H. (1986) Effect of dune development on sediment suspension under unsteady flow conditions. In: *Proceedings of the 30th Japanese Conference on Hydraulics: Tokyo, Japan*, pp. 505–510. Japanese Society of Civil Engineers, Tokyo. (In Japanese.)

Itakura, T. & Kishi, T. (1980) Open channel flow with suspended sediment on sand waves. In: *Proceedings of the Third International Symposium on Stochastic Hydraulics* (Eds Kikkawa, H. & Iwasa, Y.), pp. 599–609. Japanese Society of Civil Engineers, Tokyo.

Jackson, R.G. (1976) Sedimentological and fluid-dynamic implications of the turbulent bursting phenomenon in geophysical flows. *J. Fluid Mech.*, **77**, 531–560.

Kostaschuk, R.A. & Church, M.A. (1993) Macroturbulence generated by dunes: Fraser River, Canada. *Sediment. Geol.*, **85**, 25–37.

Kostaschuk, R.A. & Ilersich, S.A. (1995) Dune geometry and sediment transport: Fraser River, British Columbia. In: *River Geomorphology* (Ed. Hickin, E.J.), pp. 19–36. John Wiley & Sons, Chichester.

Kostaschuk, R.A. & MacDonald, G.M. (1988) Multitrack surveying of large bedforms. *Geo-Mar. Lett.*, **8**, 57–62.

Kostaschuk, R.A. & Villard, P.V. (1996a) Turbulent sand suspension events: Fraser River, Canada. In: *Coherent Flow Structures in Open Channels* (Eds Ashworth, P.J., Bennett, S.J., Best, J.L. & McLelland, S.J.), pp. 305–319. John Wiley & Sons, Chichester.

Kostaschuk, R.A. & Villard, P.V. (1996b) Flow and sediment transport over large subaqueous dunes: Fraser River, Canada. *Sedimentology*, **43**, 849–863.

Kostaschuk, R.A., Church, M.A. & Luternauer, J.L. (1989a) Bedforms, bed-material and bed load transport in a salt-wedge estuary — Fraser River, British Columbia. *Can. J. Earth Sci.*, **26**, 1440–1452.

KOSTASCHUK, R.A., LUTERNAUER, J.L. & CHURCH, M.A. (1989b) Suspended sediment hysteresis in a salt-wedge estuary, Fraser River, Canada. *Mar. Geol.*, **87**, 273–285.

LAPOINTE, M.F. (1992) Burst-like sediment suspension events in a sand bed river. *Earth Surf. Process. Landf.*, **17**, 253–270.

LEVI, E. (1991) Vortices in hydraulics. *J. Hydraul. Eng.*, **117**, 399–413.

MATTHES, G.H. (1947) Macroturbulence in natural stream flow. *Trans. Am. geophys. Union*, **28**, 255–262.

MCLEAN, D.G. & CHURCH, M.A. (1986) *A Re-examination of Sediment Transport Observations in the Lower Fraser River.* Environment Canada, Sediment Survey Section, Report IWD-HQ-WRB-SS-86-6, Ottawa.

MCLEAN, S.R. & SMITH, J.D. (1986) A model for flow over two-dimensional bedforms. *J. Hydraul. Eng.*, **112**, 300–317.

MCLEAN, S.R., NELSON, J.M. & WOLFE, S.R. (1994) Turbulence structure over two-dimensional bedforms: implications for sediment transport. *J. geophys. Res.*, **99**, 12729–12747.

MCLEAN, S.R., NELSON, J.M. & SHREVE, R.L. (1996) Flow–sediment interactions in separating flows over bedforms. In: *Coherent Flow Structures in Open Channels* (Eds Ashworth, P.J., Bennett, S.J., Best, J.L. & McLelland, S.J.), pp. 203–226. John Wiley & Sons, Chichester.

MULLER, A. & GYR, A. (1986) On the vortex formation in the mixing layer behind dunes. *J. Hydraul. Res.*, **24**, 359–375.

MULLER, A. & GYR, A. (1996a) The role of coherent structures in developing bedforms during sediment transport. In: *Coherent Flow Structures in Open Channels* (Eds Ashworth, P.J.,

Bennett, S.J., Best, J.L. & McLelland, S.J.), pp. 227–236. John Wiley & Sons, Chichester.

MULLER, A. & GYR, A. (1996b) Geometrical analysis of the feedback between flow, bedforms and sediment transport. In: *Coherent Flow Structures in Open Channels* (Eds Ashworth, P.J., Bennett, S.J., Best, J.L. & McLelland, S.J.), pp. 237–248. John Wiley & Sons, Chichester.

NELSON, J.M., MCLEAN, S.R. & WOLFE, S.R. (1993) Mean flow and turbulence fields over two-dimensional bedforms. *Water Resour. Res.*, **29**, 3935–3953.

NEZU, I. & NAKAGAWA, H. (1989) Turbulent structure of backward-facing step flow and coherent vortex shedding from reattachment in open channel flows. In: *Turbulent Shear Flows 6*, pp. 313–337. Springer-Verlag, Berlin.

NEZU, I. & NAKAGAWA, H. (1994) *Turbulence in Open-Channel Flows.* A.A. Balkema, Rotterdam.

RHOADS, B.L. & WELFORD, M.R. (1991) Initiation of river meanders. *Progr. phys. Geogr.*, **15**, 127–157.

ROOD, K.M. & HICKIN, E.J. (1989) Suspended sediment concentration and calibre in relation to surface flow structure in Squamish River estuary, southwestern British Columbia. *Can. J. Earth Sci.*, **26**, 2172–2176.

SMITH, J.D. & MCLEAN, S.R. (1977). Spatially-averaged flow over a wavy surface. *J. geophys. Res.*, **82**, 1735–1746.

THORNE, P.D., WILLIAMS, J.L. & HEATHERSHAW, A.D. (1989) *In situ* acoustic measurement of marine gravel threshold and transport. *Sedimentology*, **36**, 61–74.

YALIN, M.S. (1992) *River Mechanics.* Pergamon Press, New York.

*Spec. Publs int. Ass. Sediment.* (1999) **28**, 15–32

# Dune growth, decay and migration rates during a large-magnitude flood at a sand and mixed sand–gravel bed in the Dutch Rhine river system

W. B. M. TEN BRINKE\*, A. W. E. WILBERS† *and* C. WESSELING†

\**National Institute for Inland Water Management and Waste Water Treatment (RIZA), P.O. Box 9072, 6800 ED Arnhem, The Netherlands; and*
†*Department of Physical Geography, Utrecht University, P.O. Box 80115, 3508 TC Utrecht, The Netherlands*

## ABSTRACT

During February–March 1997, measurements were carried out on the growth and decay of bedforms in the Rivers Rhine and Waal throughout an entire flood. During the rise, peak and fall of the flood, detailed echosoundings were made of two sections: one section upstream where the Rhine divides into the Waal and the Pannerdensch Kanaal and the river bed is a mixture of sand and gravel, and one section in the sandy part some 30 km downstream. These echosoundings were made on a daily basis. The soundings in the upstream section were made using a single-beam echosounder. In the downstream section both a single- and a multibeam echosounder were used. From this time series of echosoundings, bedform dynamics throughout the flood were quantified. These results aim to quantify bedload sediment transport from the day-to-day migration of bedforms daily (dune tracking).

The results show very clearly the growth, decay and migration rates of dunes on the river bed. These dunes covered most of the bed for a couple of days before and after peak discharge at the sand–gravel-bed section, and for the entire period of echosoundings at the sand-bed section. The dunes in the sandy part of the river were much longer and higher than those upstream. Dunes kept growing even after the water level dropped; daily average dune height and dune length in the sand-bed section were up to 1.2 m and 52–59 m, respectively, with smaller dunes of some 50 cm height and 15 m length superposed on the large ones. Dunes at the sand–gravel-bed section were generally 20–60 cm high and 10–15 m long. At the sand-bed section dune patterns were very similar from day to day, enabling the use of correlation techniques to quantify dune migration rates. At the sand–gravel-bed section, however, the migration rate of the dunes was too fast for the same dunes to be recognized from day to day. Bedload sediment transport at the sand-bed section, like the dune properties, showed hysteresis when plotted versus discharge. Bedload transport was highest before the discharge peak.

## INTRODUCTION

In a densely populated country such as The Netherlands, rivers serve many purposes, such as shipping, fresh water for agriculture and drinking water, habitat for aquatic life, and safety against flooding. Serving these, sometimes conflicting, purposes at the same time requires a thorough understanding of natural and human-induced morphological processes. These processes are related to the behaviour of alluvial sand and gravel beds in a country largely made up of fluvial deposits in the delta of the Rhine and Meuse rivers. Clearly, knowledge of bedform dynamics and sediment transport is an important issue in The Netherlands. The River Rhine–Waal is by far the largest river in The Netherlands. This river is part of the

Rhine river system (Fig. 1). In fact, the Rhine–Waal is the most important river in western Europe. Its shipping density is highest of all the inland waterways of the world. Over centuries the character of this river, as all Dutch rivers, has been strongly influenced by humans. Human impact includes dredging and excavation work, groynes and, locally, rip-rap on the river banks. The present Dutch Rhine riparian landscape is characterized mainly by pastures, separated from the main channel by groynes (Fig. 2). The present policy of the Dutch authorities is to return riverine pastures to natural riparian zones without compromising the river's other functions. Engineering work, such as the excavation of secondary channels, therefore,

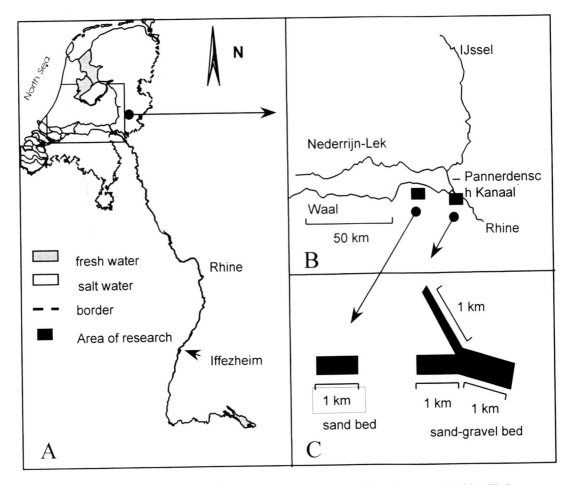

**Fig. 1.** The River Rhine and its tributaries in The Netherlands. The areas of research in the Dutch Rhine and the River Waal are indicated.

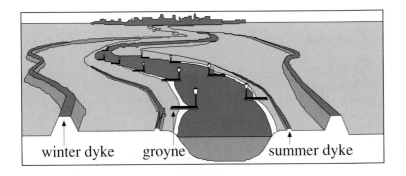

**Fig. 2.** Schematic cross-section of a Dutch river floodplain.

should be designed such that bed-level changes in the main channel are limited. Numerical morphological models calibrated with field data, therefore, are needed. These field data relate to bed-level behaviour and sediment transport at relatively high discharges. A campaign on bed-level behaviour and sediment transport was carried out during a flood in February–March 1997. The outcome of this research is presented in this paper.

The measurements aim at:

**1** quantifying bedload sediment transport from the migration rates and properties of the dunes (dune tracking);
**2** quantifying the temporal and spatial variability of bed roughness during floods.

Studies predicting the hydraulic roughness based on dune dimensions (and grain size) will be part of an other project.

## DUNE PROPAGATION AND BEDLOAD TRANSPORT

### Bedload and suspended load in the Dutch Rhine river system

Over the past decades, sediment transport in the Dutch Rhine system has been studied by several researchers in different ways. The sediment is transported as suspended load and bedload and generally is referred to as sand-sized because gravel is only a small part of the transported load (Table 1).

The oldest data available are from Van Til (1956). He carried out several *in situ* measurements of suspended and bedload sediment transport in all branches of the Rhine for a large range of discharges, and used these data to test sediment transport formulae. He calculated the sediment transport for an average yearly discharge curve and for the large-magnitude flood of 1926 (Fig. 3). In the 1980s and early 1990s again a series of *in situ* sediment transport measurements was carried out, also for different discharges and addressing both suspended load and bedload. From these measurements, relationships between discharge and sediment transport were determined, and from these relationships and the yearly discharge curves the yearly sediment transport was calculated for the period 1980–1995 (Kleinhans, 1996). Van Dreumel (1995) produced a sediment budget for the downstream, estuarine

**Table 1.** Bed material sediment transport in the Waal according to several studies. The data of Van Til and Kleinhans refer to the upstream and those of Van Dreumel to the downstream part of the Waal. The data of Dröge refer to a position 10 km upstream of the Waal.

|  | Conditions | Reference | Bedload | Suspended load | Total (× 1000 m³ yr⁻¹) |
|---|---|---|---|---|---|
| 1901–1950 | Average year | Van Til (1956) | 175 | 110 | 285 |
| 1926 | Year with flood | Van Til (1956) | 265 | 245 | 510 |
| 1980–1995 | Average year | Kleinhans (1996) | 287 | 368 | 655 |
| 1988 | Year with flood | Kleinhans (1996) | 319 | 535 | 854 |
| 1995 | Year with flood | Kleinhans (1996) | 314 | 518 | 832 |
| 1982–1992 | Average year | Van Dreumel (1995) |  |  | 700* |
|  | Average year | Dröge (1992) |  |  | 412† |

\* A dry density of 1500 kg m⁻³ was assumed.
† Data refer to particles > 200 μm.

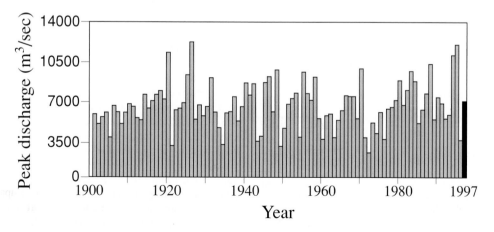

**Fig. 3.** The highest Rhine discharges for each year since 1900 and the peak discharge of March 1997 (in black).

part of the Rhine–Meuse river system, based on a detailed study of soundings, dredging works and sediment transport measurements in this area over the period 1982–1992. From this budget the yearly sand output from the River Waal was calculated. Dröge (1992) quantified the sediment transport output volume of the German Niederrhein from transport measurements and echosoundings in this area. This material is input into the Dutch Rhine–Waal.

Table 1 shows that average yearly transport of bed material in the Waal is roughly 500 000 $m^3$ $yr^{-1}$. The sediment transport during a large-magnitude flood can also be estimated from the data in Table 1 by subtracting the data for an average year from the data for a year with a flood, as provided by both Van Til and Kleinhans. Thus for the Waal, sediment transport volumes of 225 000, 199 000 and 177 000 $m^3$ $flood^{-1}$ can be calculated for the floods of 1926, 1988 and 1995 (Fig. 3), respectively.

## The principles of dune tracking

Among the first to use the propagation of bedforms to quantify bedload sediment transport were Engel and Lau (1980). Following Havinga (1982) a formulation for bedload transport from the migration of dunes can be derived. Continuity of mass implies

$$\frac{\delta q_b}{\delta x} + \frac{\delta z}{\delta t} = 0 \tag{1}$$

where $q_b$ is bedload ($m^2$ $s^{-1}$), $x$ is horizontal distance (m), $z$ is bed level (m) and $t$ is time (s). Assuming undisturbed propagation of the dunes at a celerity $c = \Delta x / \Delta t$

$$z(x, t + \Delta t) = z(x - c\Delta t, t) \tag{2}$$

and thus

$$z(x,t) + \frac{\delta z}{\delta t}\Delta t = z(x,t) - \frac{\delta z}{\delta x}\Delta x \tag{3}$$

resulting in

$$\frac{\delta z}{\delta t} = -c\frac{\delta z}{\delta x} \tag{4}$$

Combining eqns (1) and (4) leads to

$$\frac{\delta q_b}{\delta x} - c\frac{\delta z}{\delta x} = 0 \tag{5}$$

Integration yields

$$\int_0^x \frac{\delta q_b}{\delta x}\,dx = c\int_0^x \frac{\delta z}{\delta x}\,dx \tag{6}$$

Hence

$$q_b(x) = c(z(x) - z_0) \tag{7}$$

where $z_0$ is the height of the dune where $q_b(x) = 0$. The average bedload over one dune wavelength $\lambda$ results from the integration of eqn (7)

$$q_b = \frac{1}{\lambda}\int_0^\lambda q_b(x)\,dx = \frac{c}{\lambda}\int_0^\lambda (z(x) - z_0)\,dx \tag{8}$$

When dunes are large and gentle, the flow follows the bed topography and $z_0$ is the level of the dune trough. Then eqn (8) becomes

$$q_b = cfH \tag{9}$$

where $H$ is dune height (= dune crest – dune trough) (m) and $f$ is a dimensionless form factor.

The form factor $f$ is defined as $f = V/H\lambda$, where $V$ is dune volume per running metre. This factor $f$ is 0.5 in the case of triangular dunes. Generally, however, the dunes are too steep for the flow to follow the bed topography. The flow separates at the dune crest and reattaches at a point on the stoss side of the next dune. The sediment particles below this point of reattachment, therefore, are not transported. Thus, sediment transport takes place over a depth of $(1 - \alpha)H$ instead of $H$, where $\alpha$ is the part of the dune height below the point of reattachment. Adding this term to eqn (9) results in

$$q_b = f(1 - \alpha)cH = \beta cH \tag{10}$$

where $\beta$ is the so-called bedload discharge coefficient. This coefficient is to be determined by comparing the bedload transport according to eqn (10) with the sediment transport quantified from measurements with bedload samplers. In that way, $\beta$ combines the effect of the form factor, the point of reattachment, and additional effects resulting from suspended particles settling in the dune troughs, being suspended at the dune front, or jumping from crest to crest and therefore not taking part in the bedload calculated from the migration of the dunes. This bedload discharge coefficient has been determined by several researchers from flume experiments, field conditions and a combination of both (Table 2).

The value of $\beta$ according to data from Havinga & 't Hoen (1986) is relatively high. They compared the bedload transport from dune tracking with the volumes of sand deposited in a trench dug in the River IJssel (Fig. 1). During a 2-week period bedload calculations were made by subtracting the soundings for two consecutive days throughout these 2 weeks. Remarkably, most of their values for $\alpha$ in eqn (10) were negative, implying that the point of zero transport is below the dune trough, which is physically impossible. In a discussion between Havinga (1983) and Engel & Lau (1983) about these results, Engel & Lau attributed these deviating results to the settling of

**Table 2.** Published bedload discharge coefficients ($\beta$).

| Reference | Conditions | $\beta$ |
|---|---|---|
| Simons *et al.* (1965; in Van den Berg, 1987) | Flume experiments | 0.5 |
| Engel & Lau (1980, 1981) | Flume experiments | 0.33 |
| Havinga (1982) | IJssel River | 0.6 |
| Havinga & 't Hoen (1986) | IJssel River | 0.99 |
| Hansen (1966; in Van den Berg, 1987) | Skive-Karup River | 0.49 |
| Jinchi (1992) | ShenShui River | 0.53 |
| Mardjikoen (1966; in Van den Berg, 1987) | Flume experiments + Hii River | $0.6 \pm 0.21$ |

sediment from suspension in the trench, which results in an overestimation of the bedload transport into the trench. On the other hand, part of the bedload transport is not in the form of the displacement of dunes but results from sediment jumping from crest to crest. If this part of the bedload is significant, a comparison of bedload transport based on dune tracking and siltation rate would result in negative values for $\alpha$. These values should not be interpreted as the point of zero transport being lower than the dune trough. It merely indicates that the depth of bedload transport is higher than the dune height.

Most of the values for $\beta$ in the literature are between 0.5 and 0.6. Pending more information about the process of bedload transport in Dutch rivers, in this paper a bedload discharge coefficient $\beta$ for the River Waal of 0.55 is assumed.

## STUDY LOCATION

The River Rhine originates in the Alps and flows through Switzerland and Germany to The Netherlands (Fig. 1A). The Rhine river system in The Netherlands consists of a set of three distributaries originating from the River Rhine at two bifurcations, just after the Rhine has passed the Dutch–German border (Fig. 1B). These three distributaries are the Waal, Nederrijn-Lek and IJssel. These branches have embankments constructed along their entire length. The discharge ratio between these distributaries is approximately 6:2:1.

The average discharge of the Rhine near the Dutch–German border is 2300 m$^3$ s$^{-1}$, stemming from both rain and snowmelt. In December 1993 to January 1994 and January to February 1995 the river Rhine experienced maximum discharges of 11 000 and 12 000 m$^3$ s$^{-1}$, respectively, among the highest Rhine discharges ever recorded (Fig. 3). With respect to these discharges the high discharge of 1997 with a peak value of 7000 m$^3$ s$^{-1}$, which is the subject of this paper, was relatively modest.

A peak discharge of $\geq$ 7000 m$^3$ s$^{-1}$ is experienced on average every 4 yr.

The branches of the Dutch Rhine are generally sand bed rivers with median particle sizes for bed material of about 0.5–4 mm. In 1995 the upper 5–10 cm of the bed of the River Rhine–Waal in The Netherlands was sampled at three positions across the river with cross-sections at regular distances of 1 km down its length, and the grain-size composition was determined by nested sieving (Fig. 4). The sediment of the bed is mainly sand for most of the river length downstream of the Dutch–German border. Upstream, near the border, the bed is a mixture of sand and gravel. The average water depth of the Rhine–Waal is in the order of 5 m. The length and width of the study reach are 104 km and 260–340 m, respectively. Bed gradient is $1.1 \times 10^{-4}$. The research was carried out at two locations on the Rhine–Waal: at the bifurcation of the Rhine into the Waal and Pannerdensch Kanaal (sand–gravel bed), and some 30 km downstream in the Waal (sand bed) (Fig. 1c). At both locations the river is relatively straight (sinuosity = 1.1).

## METHODS AND ANALYSES

### Single-beam and multibeam echosounding

Near the bifurcation of the Rhine into the Pannerdensch Kanaal and the Waal echosoundings were made connecting 1-km subsections in each of these rivers (Fig. 1C). These subsections were sounded along 1 km tracks parallel to the banks. These tracks were distributed over the entire wetted cross-section between the groynes, 10 m apart for the Waal subsection and 20 m apart for the other two subsections. The survey vessel was equipped with a digital reading single-beam echosounder (ATLAS DESO 25) and a two-dimensional horizontal positioning system (differential global positioning system – DGPS), controlled by a desk-top computer. Deviations from the

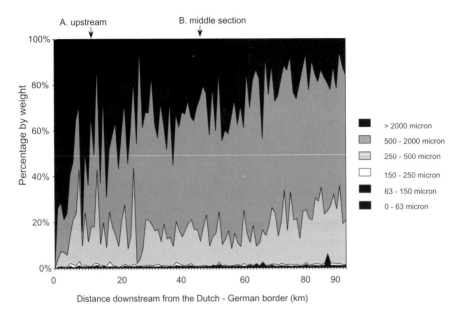

**Fig. 4.** The grain-size distribution of the bed of the Dutch Rhine and the River Waal, farther downstream, versus distance along the river, based on three samples across the river at each 1-km section.

programmed course are computed and presented to the helmsman of the ship to enable him to correct the course immediately. The actual position, together with the related water depth data, are stored on tape and processed on a main frame computer at the office.

The sand-bed section halfway down the Waal was sounded along 1-km tracks parallel to the banks with two different types of echosounders. From one vessel a similar single-beam echosounder was used as for the section near the bifurcation. These tracks were 10 m apart and covered the wetted cross-section between the groynes. From another vessel a multibeam echosounder was used. This multibeam SIMRAD echosounder consists of a set of 128 echosounders underneath the centrepoint of the vessel that scan the bed topography in a line perpendicular to the boat's track. Generally, the results of the outer eight sensors (four on each side of the vessel) are not considered for reasons of accuracy. Each sensor transmits a 95 kHz signal with a beam width of 1.5°. The beam width of all sensors combined is 150°. This way the river bed is scanned over a width of five times the water depth. The entire bed area in between the groynes was sounded by sailing 14 tracks, including ample overlap. All movements of the vessels (pitch, roll, heath, heading) are corrected automatically, as are the influences of salinity and temperature on sound velocity. Positioning is by means of DGPS. The combined accuracy of the determination of

position and elevation is such that bed topography can be determined with an absolute accuracy of 20 cm. Considering the high density of data (6–7 depth values $m^{-2}$), the average bed level per square metre can be calculated far more accurately.

**Dune tracking using GIS-based software**

In order to perform the calculations of the bedload transport by making use of dune propagation, several dune characteristics have to be measured. Because of the large amount of data that was obtained by echosounding, this could be done only through a computer program. The DT2D (dune tracking in two dimensions) program performs most of the calculations without interaction with the user. The user is prompted to give only some of the thresholds that are needed during the calculations and during dune fitting.

Data (from single-beam and/or multibeam echosounders) are loaded into an internal data base. Then the program calculates the dune migration distance for every echosounding track by using a cross-correlation technique. The match for the echosoundings of a certain track at time $T_1$ and $T_2$ for which the maximum cross-correlation is calculated is the dune migration distance from $T_1$ to $T_2$. The dune migration rate is this distance divided by $\Delta T = T_1 - T_2$. Cross-correlation is calculated from

$$r_j = \frac{n\sum\limits_{i=1}^{n} Z_{1i}Z_{2i} - \sum\limits_{i=1}^{n} Z_{1i}\sum\limits_{i=1}^{n} Z_{2i}}{\sqrt{\left(n\sum\limits_{i=1}^{n} Z_{1i}^2 - \left[n\sum\limits_{i=1}^{n} Z_{1i}\right]^2\right) \times \left(n\sum\limits_{i=1}^{n} Z_{2i}^2 - \left[n\sum\limits_{i=1}^{n} Z_{2i}\right]^2\right)}}$$

(11)

where $r_j$ is the cross-correlation for matching position $j$, $n$ the number of points, $Z_{1i}$ the $z$ value in the track at $T = T_1$ and $Z_{2i}$ the $z$ value in the track at $T = T_2$. The program also calculates a digital elevation model (DEM) using an inverse distance technique.

After these first calculations the program shows the position of all the tracks in a map window and the depth profiles of one or two of these tracks in another window. The depth profile of a specific track at a specific time can be selected from the tracks on the map. The user can locate dunes on the tracks interactively. This is done by giving the lower limit of the length of bedforms that may be considered dunes, which is the smallest distance between two successive troughs of a bedform. Additionally, a filter may be used to smooth outliers that are not considered to be dunes.

When the dunes are located the program determines the dune length, height, volume and slope of the stoss- and lee-side for each dune. Next these dune characteristics are averaged for each track and exported as ASCII files for further calculations in a spreadsheet.

## RESULTS

### Dune pattern and planform

On 27 February the Rhine discharge exceeded 3500 m$^3$ s$^{-1}$ and the campaign of daily echosoundings of the selected transects in the Rhine river system was started. On this day the upstream section was already partly covered with dunes (Fig. 5). In Fig. 5 the variations in bed topography

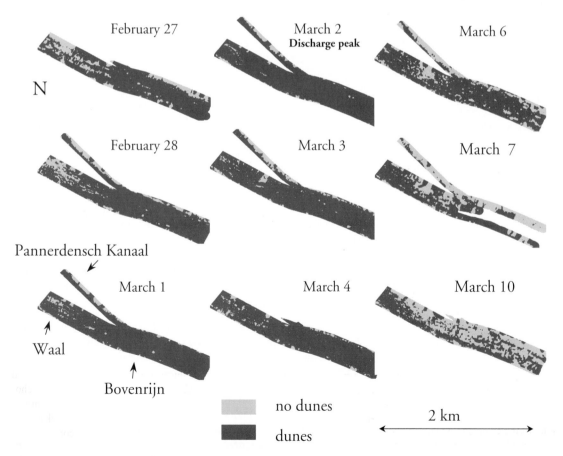

**Fig. 5.** The presence of dunes in the upstream sand–gravel-bed section of the Dutch Rhine, the River Waal and the Pannerdensch Kanaal during the rise and fall of the flood of February–March 1997.

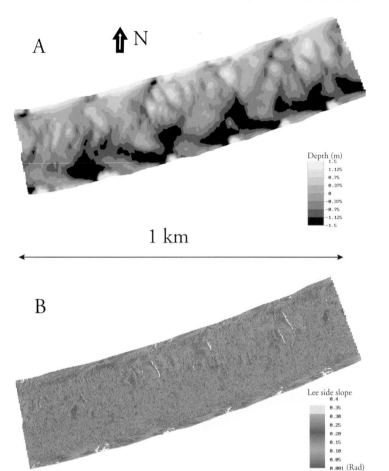

**Fig. 6.** The pattern of dunes in the middle sand-bed section of the Waal during the rise of the flood of February–March 1997, visualized as a depth map with respect to Dutch Ordnance Datum (A), and by presenting the slope of the bed (B).

are considered to be dunes when the ratio between bedform length and bedform height is less than 70. When discharge increased towards its peak on 2 March the area covered with dunes increased, with an almost complete coverage at the time of peak discharge. Just a few days after the discharge had started to decline the bed flattened. A week after the peak discharge the pattern of dunes had almost completely disappeared.

For the downstream section somewhat different results were obtained. Again, the bed was already almost completely covered with a pattern of dunes on the first day of echosounding (27 February; Fig. 6) but this pattern did not change throughout the campaign. After 2 weeks the pattern of dunes was similar to the pattern at the beginning of the campaign. Owing to the use of a multibeam echosounder in this area this pattern can be displayed in great detail, showing large dune crests perpendicular to

the bank and smaller dunes superposed on these large dunes. Although the large dunes covered the entire cross-section of the bed, they were highest and steepest in the northern half of the bed.

### Dune height, length and steepness

For the sand–gravel bed section the dune geometry changed remarkably going from the Rhine into the Waal and Pannerdensch Kanaal. Figure 7 shows the average dune properties in the subsections of the Rhine and Waal. For each day an average value for the length, height and steepness of the dunes was calculated for the entire 1-km subsections. The dunes in the Pannerdensch Kanaal were comparable to the ones in the Waal. Dune height and dune length decreased where the Rhine divides into the Waal and Pannerdensch Kanaal, from average values up to

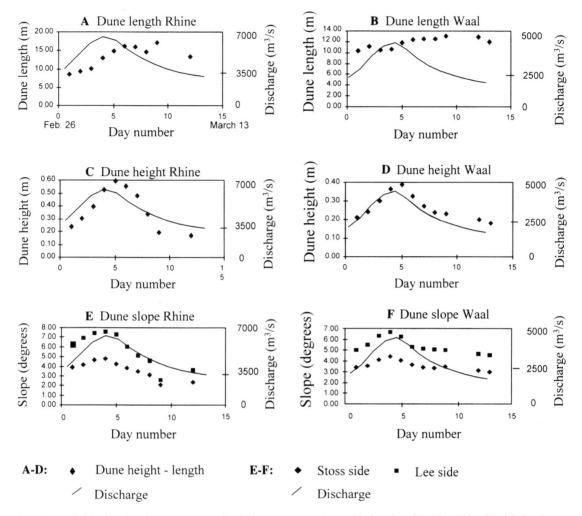

**Fig. 7.** Dune height, length and steepness versus time in the upstream sand–gravel-bed section of the River Rhine–Waal during the flood of February–March 1997.

60 cm height and 15 m length to average values up to 40 cm height and 13 m length. The dunes were largest after the peak discharge. Dune steepness closely followed the variation in discharge. Dune steepness was highest at peak discharge and was up to 4–5° for the stoss side and up to 7–8° for the lee side. The hystereses between the dune properties and the discharge curve are shown in Fig. 8, where these properties are plotted versus discharge. The hystereses are strongest for the Rhine subsection, with anticlockwise loops for dune length (Fig. 8A) and dune height (Fig. 8C) and clockwise loops for dune steepness (Fig. 8E). The hysteresis for dune steepness is small or negligible for the Waal subsection (Fig. 8F).

For the sand-bed section a distinction is made between large dunes and small dunes, the latter being the small bedforms superposed on the larger ones. An average value for the length, height and steepness of the dunes was calculated for each day for the entire 1-km section, for both the large and the small dunes. The results are presented in Fig. 9 together with the discharge curve. Remarkably, the length of the large dunes was highest during the rise of the flood and decreased at peak flood and during the fall of the flood (Fig. 9A). The length of the small dunes, however, was highest during the fall of the flood (Fig. 9B). The height of both the large and small dunes increased on the falling limb of the discharge curve

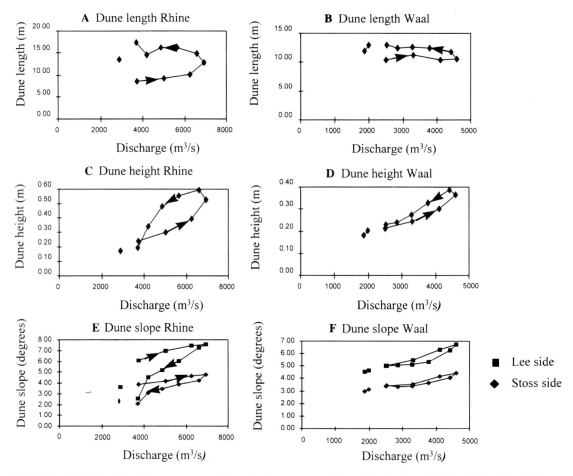

**Fig. 8.** Dune height, length and steepness versus discharge in the upstream sand–gravel-bed section of the River Rhine–Waal during the flood of February–March 1997.

(Fig. 9C&D). Dune steepness was relatively constant throughout the 2-week period (Fig. 9E&F). The smallest dunes were steepest. Generally, lee sides were about twice as steep as the stoss sides.

The hysteresis of dune growth and decline with changes in discharge is shown in Fig. 10, where dune properties are plotted versus discharge. For the large dunes the direction of the loop for dune length (Fig. 10A) is contrary to the direction of this loop for the smaller dunes (Fig. 10B) and the dunes of Rhine (Fig. 8A) and Waal (Fig. 8B), and also contrary to the direction of the loop for dune height (Fig. 10C&D). For dune steepness the hysteresis is anticlockwise for the large dunes (Fig. 10E) and negligible for the small dunes (Fig. 10F).

**Dune migration**

For the sand-bed section the downstream displacement of the dunes can be visualized very clearly by plotting the planform of the dune crests, derived from the multibeam echosoundings, in successive time steps. In Fig. 11 this reconstruction is shown for the entire period for which multibeam data are available. The black areas at regular distances along the margins of the plots are groynes. The dune planform is visualized by plotting the steepest part of these dunes. The position of one of these dunes is indicated by an arrow. Although the dunes could be traced very easy from day to day, the planform gradually changed, resulting in dune crests on day 12 that were completely different from those on day 1.

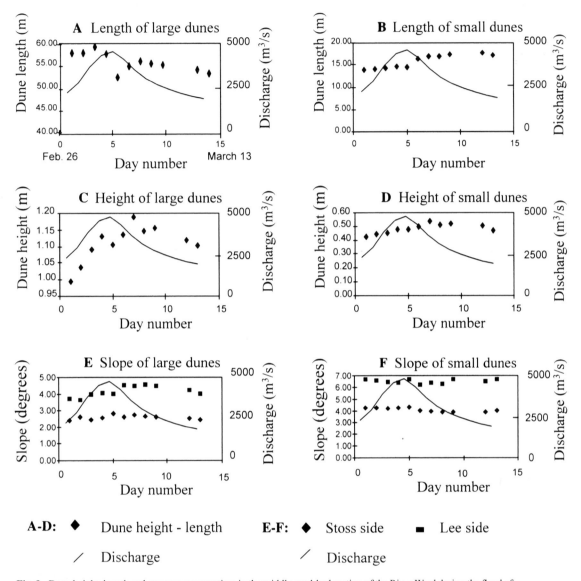

**Fig. 9.** Dune height, length and steepness versus time in the middle sand-bed section of the River Waal during the flood of February–March 1997.

The migration of the dunes resulted from erosion of the stoss sides and deposition on the lee sides. This is visualized by subtracting the echosoundings of two successive days (Fig. 12). The pattern of erosion and sedimentation is quite regular, following dune crestlines orientated perpendicular to the river banks. A plot such as Fig. 12 was obtained for all successive days of echosounding. The migration rate of the dunes was not constant in time.

When plotted versus discharge the hysteresis in the migration rate is obvious (Fig. 13). In the sand–gravel-bed section the migration rate and the change of dune properties appeared too fast to successfully apply our techniques for dune tracking. Probably, the displacement of the sediment itself was not so much faster than in the sand-bed section, but with the wavelength of the dunes being so much smaller, the displacement of the sediment

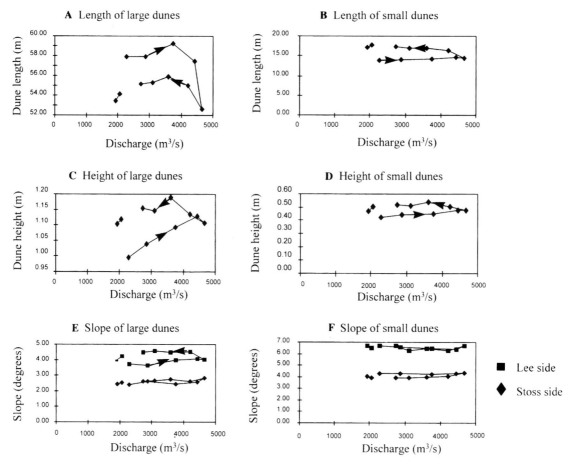

**Fig. 10.** Dune height, length and steepness versus discharge in the middle sand-bed section of the River Waal during the flood of February–March 1997.

in number of wavelengths per day was too fast in the sand–gravel-bed section to be resolved on daily data. Generally, the same dunes could not be traced from day to day by cross-correlation. For only a limited number of dunes, however, the dunes could be recognized from day to day visually. For these dunes, the migration rate was quantified, and these were plotted versus dune length together with the results for the sand-bed section (Fig. 14). Although based on two clusters far apart, the results in the figure follow an exponentially reducing migration rate versus dune length.

**Bedload transport**

For the sand-bed section, cross-section-averaged bedload sediment transport is shown in Fig. 15. Bedload transport

was highest on the rising limb of the discharge curve, resulting in a clockwise hysteresis of bedload versus discharge.

The single-beam echosoundings were made along 25 1-km transects, 10 m apart. For each of these transects the bedload transport based on dune tracking was quantified and averaged over the entire period of measurements. The results are shown in Fig. 16D, together with the results on the time-averaged dune properties (Fig. 16A–C). The remarkably high values for dune height near the northern and southern river banks reflect scour holes downstream of the groynes. As these scour holes do not migrate, these features do not influence the quantification of bedload transport. Bedload transport variability across the river is quite small. Bedload is largest in the northern half of the cross-section, which is a very gentle inside bend.

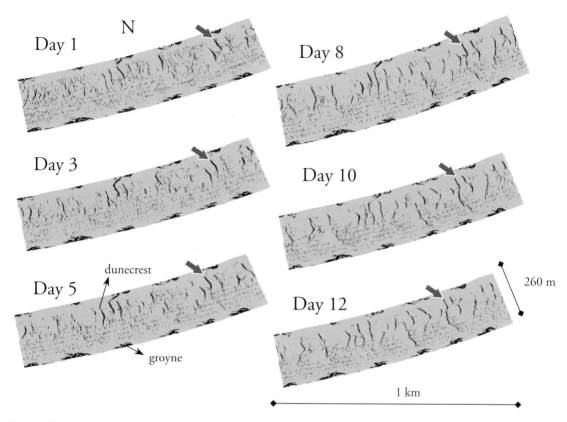

N

Day 1

Day 8

Day 3

Day 10

dunecrest

Day 5

Day 10

Day 12

groyne

260 m

1 km

**Fig. 11.** The downstream displacement of the dune crests in the middle sand-bed section of the River Waal in successive time steps.

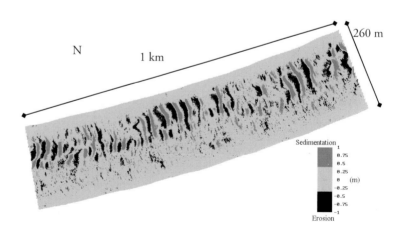

**Fig. 12.** The difference of two successive multibeam echosoundings (time step is 24 h) in the middle sand-bed section, showing erosion on the stoss sides and deposition on the lee sides.

N       1 km       260 m

Sedimentation
1
0.75
0.5
0.25
0     (m)
-0.25
-0.5
-0.75
-1
Erosion

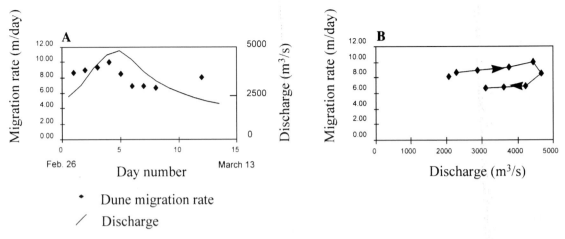

Fig. 13. Dune migration rate versus time (A) and discharge (B) in the middle sand-bed section of the River Waal during the flood of February–March 1997.

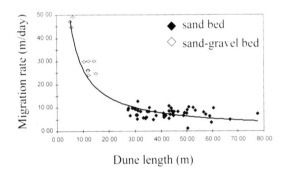

Fig. 14. Dune migration rate versus dune length for the dunes in the sand–gravel bed and the sand bed during the flood of February–March 1997.

## DISCUSSION

### Dune growth and decline in the Dutch rivers

The results presented in this paper show that even relatively small floods induce bedforms that cover the bed of the entire wetted cross-section. Considering the short duration of the flood, these bedforms have dimensions that are quite large when compared with the dimensions of the river. The use of a multibeam echosounder is a most powerful tool to tackle the temporal and spatial variability of these bedforms.

There is a strong difference in the dune properties and the way the dunes responded to the discharge curve for

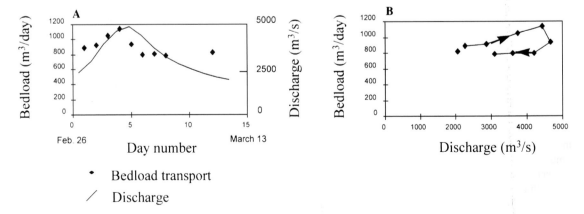

Fig. 15. Bedload sediment transport versus time (A) and discharge (B) in the middle section of the River Waal during the flood of 1997.

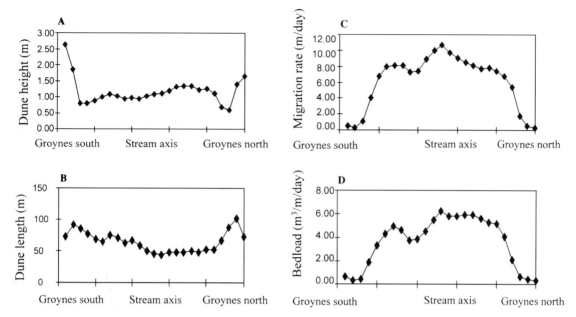

**Fig. 16.** Average dune height (A), length (B), migration rate (C), and bedload sediment transport (D) across the middle section in the River Waal during the flood of 1997.

the sand–gravel-bed section and the sand-bed section. In the sand–gravel-bed section the bed responded more closely to the discharge curve. Four days before the peak discharge some parts of the bed were still flat. At peak discharge the entire bed of the wetted cross-section was covered with dunes, whereas 1 week later the bed was flat again. The response of bedforms to discharge variations depends on the ratio between dune volume and bedload sediment transport. The smaller this ratio the faster bedforms will respond. Clearly, this is the case for the sand–gravel-bed section when compared with the sand-bed section. The fast response of the sand–gravel bedforms to the hydrodynamic forcing seems to be reflected also in the changing dune properties going from the Rhine into the Waal and Pannerdensch Kanaal. Unfortunately no information on changes in flow conditions near the bed is available to clarify this coupling between hydro- and morphodynamics.

In the sand-bed section, dunes were present during all days that echosoundings were made. Remarkably, dunes in this section were relatively long (and low) on the rising limb of the discharge curve and became shorter (and higher) with time. Laboratory experiments have shown that with increasing discharge, bedform length increased whereas bedform height remained more or less constant (Termes, 1986; in Julien & Klaassen, 1995). Results for the 1984 flood of the River Meuse in The Netherlands

showed that for a given discharge, both the dune height and wavelength were larger under falling discharge than increasing discharge. Starting with a flat bed, one would expect dunes to grow in all dimensions, as was observed in the sand–gravel-bed section. In the sand-bed section, however, relatively long bedforms were present at relatively low discharges on the rising limb of the discharge curve. The presence of these bedforms at such an early stage of the flood points at either dunes originating rapidly once shear stress increases or dunes being present at relatively modest discharges (such as before the flood). Earlier investigations of dunes in the River Rhine (Van Urk, 1982) showed a time lag in dune development (dune height) responding to the discharge curve of approximately 4 days. It was also shown that the time lag of the dune heights was much smaller than the time lag of the dune lengths. Results presented here confirm the shorter time lag of dune height when compared with dune length. In fact, it is very difficult to determine this time lag for dune length because dune length did not really increase and decrease in time responding to the discharge curve. In the sand–gravel-bed section, dune length remained relatively large until the end of the campaign. It looks as though dunes disappear after peak discharge by decreasing in height, whereas dune length stays the same until the bed has become flat again. During this flood dune height responded to the discharge curve with a time lag of only 1 (sand–gravel-bed section) to 2 days (sand-bed section), which is much shorter than earlier results by Van Urk (1982). Probably these time lags depend strongly on the size and duration of the peak discharge.

Whether or not time lag for dune growth depends on the form and amplitude of the discharge curve, the dune properties themselves do. Julien & Klaassen (1995) presented results on dune properties of the Rivers Rhine and Meuse in The Netherlands during the large-magnitude flood of 1988 (Fig. 3). They determined average values for dune length and height for the Dutch Rhine near the Dutch–German border of typically 20 and 1.5 m, respectively. These dunes also refer to the sand–gravel-bed section, where the dunes in 1997 were up to 15 m long and 60 cm high. Julien & Klaassen related dune wavelength $\lambda$ and dune height $\Delta$ to flow depth and median grain size and found, as a first approximation, $\lambda \cong 6.5$ h and $\Delta \cong 2.5$ $h^{0.7}$ $d_{50}^{0.3}$. Following this relationship, average dune length and height should be 69 m and 1.78 m for the sand–gravel-bed section ($h = 10.6$ m, $d_{50} = 1.3$ mm) and 53 m and 1.26 m for the sand-bed section ($h = 8.1$ m, $d_{50} = 0.77$ mm). Clearly, for the dunes in the sand–gravel-bed section this is not the case. This probably results from the fact that the peak discharge was relatively small and lasted for a couple of days only. It takes time for dunes to grow, especially when the bed material is coarse grained. Most likely, a high discharge that lasts for at least a week is needed for the large dunes to be formed. In the winter of 1993–1994 and in the early spring of 1995 two of the largest Rhine floods ever recorded occurred (Fig. 3). These floods lasted for about a month and resulted in dune properties that were well in line with the results of Julien & Klaassen (1995).

### Dune steepness and implications for flow separation

Dune-length and dune-height variation versus discharge showed anticlockwise hysteresis except for the large dunes in the sand-bed section. In this sand-bed section, the hystereses of dune length (clockwise) and dune height (anticlockwise) strengthen one another as far as dune steepness is concerned, which results in a slight anticlockwise hysteresis for dune steepness versus discharge. Changes in dune length and height seem to be equally important in determining changes in dune steepness. In the Rhine subsection, dune height and dune steepness show similar variations in time but opposing hystereses owing to the combined effect of variations in height and length.

Kostaschuk & Villard (1996) studied dunes in the lower reach of the Fraser River, a sand-bed river with discharge conditions comparable to the River Rhine

($Q_{\text{mean annual, Fraser}} = 3400$ $m^3$ $s^{-1}$, $Q_{\text{mean annual, Rhine}} = 2300$ $m^3$ $s^{-1}$; $Q_{\text{max, Fraser}} = 11\,000$ $m^3$ $s^{-1}$, $Q_{\text{max, Rhine}} = 11\,000–12\,000$ $m^3$ $s^{-1}$). They found symmetric dunes with equal stoss- and lee-side slopes $< 8°$ and asymmetric dunes with stoss-side slopes $< 3°$ and lee-side slopes up to $19°$. In both sections of the River Rhine dunes were asymmetric with lee-side slopes $< 8°$, being about twice as steep as the stoss sides. One might expect the formation of dunes to depend upon both the value of the discharge peak and the duration of the flood. Julien & Klaassen (1995), however, also found relatively low slopes for the dunes in the Dutch Rhine during the larger, longer lasting flood of 1988: height:length ratio = 0.02–0.04. In 1997 this ratio was 0.03–0.04 for the dunes in the sand–gravel-bed section and 0.02–0.03 for the dunes in the sand-bed section. Possibly, a strong interaction with the suspended load plays a role. During the echosoundings of 1997 both suspended and bedload sediment transport were measured in detail near the sand–gravel-bed section (Kleinhans, 1997). These results showed that at least 60% of the total sediment transport (suspended + bedload) was suspended load. Kostaschuk & Villard (1996) indicated that the lower slopes of the lee sides of their symmetric dunes, especially, resulted from the settling of suspended sediment in the dune troughs. This may seem to be an explanation for the relatively low slopes of the dunes in the Rhine distributaries. The theory seems unlikely, however, considering the constant difference between the slopes of lee and stoss sides. Possibly, the definition of dune slope in the dunetrack software is inappropriate. This slope is defined as the slope of the line connecting the highest (top) and lowest (trough) bed levels. Probably, the steepest tangent to the lee-side slope of the dunes provides angles more in line with the results of Kostaschuk & Villard (1996). It is not clear yet whether dunes are steep enough for the flow to separate at or near the crests, and thus whether the dunes contribute to the bed roughness. According to Dyer (1986) dunes are likely to have separated flow at height:length ratios $> 0.067$.

### Sediment transport

For the data of the sand-bed section, dune tracking was applied succesfully. The dunes could be recognized on a day-to-day basis very well. In fact, in future campaigns a time lag of 48 h instead of 24 h may be used without compromising the success of dune tracking as bedload is calculated from the displacement of the larger dunes, which do not change that much from day to day. In cases where the data are to be used for research on bed roughness, however, the echosoundings should be made on a daily basis because bed roughness depends on the smaller dunes where flow separation is more likely. For the data on the sand–gravel-bed section, dune tracking could not be used because these dunes migrated more than one wavelength per day and cross-correlation could not be used to track the same dunes from day to day. In addition, deformation of dunes is a greater problem when dunes are small (Gabel, 1993). In future campaigns the sand–gravel-bed section must be sounded twice a day. The hystereses of the dune properties when plotted versus discharge are reflected in the bedload sediment transport. Bedload transport is highest during the rise in discharge. This is in line with the theory that the current associated with a certain discharge loses more energy as a result of friction on the falling limb, the dunes being higher and steeper, so less energy is available for moving sediment particles (Richards, 1982). Similar results were shown by Kuhnle (1992).

The great advantage of using dune tracking for the determination of bedload transport is the possibility to quantify the variability of this transport across the river in great detail in a short period of time. Generally, this period is short enough to assume constant discharge conditions. Quantifying this spatial variability across the river from Helley Smith type bedload samplers takes a couple of days owing to the great number of samples that have to be taken to account for random (inaccuracies in the instrumentation) and systematic (along and across dune) variations. During the flood of 1997, as part of the same project, suspended and bedload sediment transport was also determined directly by using an acoustic device for suspended sediments and a Helley Smith sampler for the bedload. This was done in the 1-km part of the Waal in the sand–gravel-bed section. It took 3.5 days to complete one set of measurements across the river. Naturally, discharge conditions changed such that sediment transport varied owing to a combination of spatial and temporal variability of processes that are hard to unravel. These so-called direct sediment transport measurements, however, have to be carried out in combination with dune tracking because dune tracking also has a major disadvantage, namely that only a certain part of the sediment transport is quantified. A large part of the total load is suspended load or bedload that moves across the bed without taking part in the displacement of bedforms. From the direct sediment transport measurements the total sediment transport in the 1-km part of the Waal was quantified for a period of 8 days (two complete sets of measurements) covering the discharge peak and the falling limb. Total transport was 32 000 m$^3$, of which 12 800 m$^3$ was bedload. Bedload resulting from the migration of dunes must be less than 12 800 m$^3$. Unfortunately, this cannot be checked in this section. In the sand-bed section, bedload according to dune tracking during the same period was 7300 m$^3$. It seems that bedload in the form of migrating dunes is a small part of the total load. Also, the choice of bedload discharge coefficient may be too small. A good quantification of this coefficient, therefore, is of utmost importance.

## CONCLUSIONS

The results show very clearly the growth, decline and migration rates of dunes in the distributaries of the Dutch Rhine. These dunes covered most of the bed for a couple of days before and after peak discharge at the sand–gravel-bed section and for the entire period of echosoundings at the sand-bed section. The dunes in the sandy part of the river were much longer and higher than those upstream. Dunes still kept on growing once the water level dropped: dunes in the sand-bed section were on average up to 1.2 m high and 52–59 m long, with smaller dunes of some 50 cm height and 15 m length superimposed on the large ones. Dunes at the sand–gravel-bed section were generally 20–60 cm high and 10–15 m long. When compared with data from the literature, the dunes during the flood of 1997 were relatively small. This probably is the result of the short duration of the peak discharge. At the sand-bed section, dune patterns were very similar from day to day, enabling the use of correlation techniques to quantify dune migration rates. At the sand–gravel-bed section, however, the migration rate of the dunes was too fast for the same dunes to be recognized from day to day.

The time lag of dune growth responding to the discharge curve was relatively small, especially for the sand–gravel-bed section. This time lag was in the order of 1 day for the sand–gravel-bed section to 2 days for the sand-bed section. The hystereses of dune properties versus discharge were generally anticlockwise.

Remarkably, dune length versus discharge showed a clockwise hysteresis for the large dunes in the sand-bed section. From the variation of dune length with time, it may be inferred that either dunes in the sand-bed section originate rapidly once shear stress increases, or dunes are present at relatively modest discharges. Changes in dune height and dune length seem to be equally important in determining changes in dune steepness. The combination of the hystereses of the dune properties results in a clockwise hysteresis for bedload sediment transport at the sand-bed section: bedload transport was highest before the discharge peak. Bedload transport quantified through dune tracking is probably a small part (< 40%) of the total load. A good quantification of the bedload discharge coefficient, therefore, is of utmost importance.

## ACKNOWLEDGEMENTS

This research was supported financially by the Directorate Eastern Netherlands (DON) and the Head Office of Rijkswaterstaat. The authors wish to thank A. Wolters, M. Schropp (RIZA) and E.S.P. Smit (DON) for their cooperation in organizing the campaign, the crew of the vessels *Krayenhoff, Beyerinck, Conrad, Onderzoeker* (DON) and *Wijtvliet* (Directorate Zeeland) as well as their supervisors for carrying out the echosoundings, J.H. van den Berg and M.G. Kleinhans (Utrecht University) for their assistance during the analyses, E. van Velzen, J.P.G. van de Kamer, J. Alexander, S.L. Gabel and N.D. Smith for their comments on an earlier draft of the manuscript, and finally A.G. de Boer for laying the foundation of some of the principles of the dunetrack software during his practical work at RIZA.

## REFERENCES

DRÖGE, B. (1992) Geschiebehaushalt des Rheins unterhalb Iffezheim. *Proceedings, 5th IAHR Symposium on River Sedimentation*, April, Karlsruhe.

DYER, K.R. (1986) *Coastal and Estuarine Sediment Dynamics*. John Wiley & Sons, Chichester.

ENGEL, P. & LAU, Y.L. (1980) Computation of bedload using bathymetric data. *Proc. Am. Soc. civ. Eng. J. Hydraul. Div.*, **106**, 369–380.

ENGEL, P. & LAU, Y.L. (1981) Bedload discharge coefficient. *Proc. Am. Soc. civ. Eng. J. Hydraul. Div.*, **107**, 1445–1454.

ENGEL, P. & LAU, Y.L. (1983) Bedload discharge coefficient. Closure to discussion. *Proc. Am. Soc. civ. Eng. J. Hydraul. Div.*, **109**, 161–162.

GABEL, S.L. (1993) Geometry and kinematics of dunes during steady and unsteady flows in the Calamus River, Nebraska, USA. *Sedimentology*, **40**, 237–269.

HANSEN, E. (1966) Bedload investigations in Shive-Karup river. *Technical University of Denmark, Bull. no. 12*, 1–8.

HAVINGA, H. (1982) *Bedload Determination by Dune Tracking*.

Technical report ISO/TC 113/SC 6 N 155, Proceedings International Standards Organization/Technical Committee, 113, New Delhi.

HAVINGA, H. (1983) Bedload discharge coefficient. Discussion. *Proc. Am. Soc. civ. Eng. J. Hydraul. Div.*, **109**, 157–160.

HAVINGA, H. & 'T HOEN, J.P.F. (1986) *Prototype Testing of the Dunetrack Method*. Report National Institute for Inland Water Management and Waste Water Treatment RIZA (The Netherlands), 86.52. (Text in Dutch.)

JINCHI, H. (1992) Application of sandwave measurements in calculating bedload discharge. *Proceedings Symposium Erosion and Sediment Transport Monitoring Programmes in River Basins, Oslo*. Publication 210, pp. 63–70. International Association of Hydrological Sciences, Wallingford.

JULIEN, P.Y. & KLAASSEN, G.J. (1995) Sand-dune geometry of large rivers during floods. *J. Hydraul. Eng.*, **121**, 657–663.

KLEINHANS, M.G. (1996) *Sediment Transport in the Dutch Rhine Branches*. Report Utrecht University ICG 96/7. (Text in Dutch.)

KLEINHANS, M.G. (1997) *Sediment Transport in the River Waal during the Flood of 1997*. Report Utrecht University ICG 97/7. (Text in Dutch.)

KOSTASCHUK, R. & VILLARD, P. (1996) Flow and sediment transport over large subaqueous dunes: Fraser River, Canada. *Sedimentology*, **43**, 849–863.

KUHNLE, R.A. (1992) Bedload transport during rising and falling stages on two small streams. *Earth Surf. Process. Landf.*, **17**, 191–197.

MARDJIKOEN, P. (1966) Some characteristics of sandwaves in open channels with movable bed. *Delft Hydraulics Laboratory; Report R. 7*, Delft, the Netherlands, 17.

RICHARDS, K. (1982) *Rivers: Form and Process in Alluvial Channels*. Methuen, London.

SIMONS, D.B., RICHARDSON, E.V. & NORDIN, C.F. (1965) Unsteady movement of ripples and dunes related to bedload transport. *Proceedings of IAHR Congress, Leningrad*, **3.29**, 1–8.

VAN DEN BERG, J.H. (1987) Bedform migration and bedload transport in some rivers and tidal environments. *Sedimentology*, **34**, 681–698.

VAN DREUMEL, P.F. (1995) *Mud and Sand Budget of the Dutch Delta Area*. Report Rijkswaterstaat, Dordrecht. (Text in Dutch.)

VAN TIL, K. (1956) Calculations for the bedload in view of the canalization of the Lower Rhine–Lek. *De Ingenieur*, **41**, 19–29. (Text in Dutch.)

VAN URK, A. (1982) Bedforms in relation to hydraulic roughness and unsteady flow in the Rhine branches (The Netherlands). In: *Mechanics of Sediment Transport* (Eds Mutlu Sumer, B. & Muller, A.), pp. 151–157. A.A. Balkema, Rotterdam.

*Spec. Publs int. Ass. Sediment.* (1999) **28**, 33–41

# Bedforms of the middle reaches of the Tay Estuary, Scotland

S. F. K. WEWETZER* *and* R. W. DUCK

*Department of Geography, University of Dundee, Dundee DD1 4HN, Scotland*
*(* Email: S.F.K.Wewetzer@dundee.ac.uk)*

## ABSTRACT

The distribution of bedforms in part of the middle reaches of the macrotidal Tay Estuary, Scotland, has, for the first time, been surveyed systematically using side-scan sonar. The main bedform types recorded were dunes of various sizes and morphologies, with small wavelength dunes occupying most of the channel areas. Dunes of three wavelength classes (small, 0.6–5.0 m; medium, 5.0–10.0 m; large, 10.0–100.0 m) were recorded on Middle Bank, a major sand bank that divides the study area into the main Navigation Channel to the south and Queen's Road Channel to the north. Medium height dunes (0.25–0.50 m) are the dominant dune height class in the study area, characterizing Middle Bank as well as most channel areas. Dune dimensions measured from sonographs were examined in terms of intercorrelations of wavelength, height and corresponding water depth. Although some researchers have found significant correlations between these variables in flume experiments and in field studies of intertidal environments, dune height and wavelength were not correlated or weakly correlated with water depth in this study. It is suggested that the relationships between these parameters established previously are not generally applicable in estuarine environments.

## INTRODUCTION AND SETTING

This study was undertaken with two principal aims. The first was to survey the hitherto unknown spatial distribution of subtidal and intertidal bedform types in part of a major, macrotidal estuary using side-scan sonar. The second was to investigate relationships among the geometrical parameters of bedforms and water depth.

The study area, the Tay Estuary, is the northern of two major estuaries on the east coast of Scotland. The estuary, which receives its freshwater from a 6500 km$^2$ catchment area, has a complex origin resulting from a combination of geological constraints, Pleistocene glaciation, river erosion and sea-level change. The major draining systems are the Rivers Tay and Earn, which contribute the greatest volume of freshwater of any river basin in the UK, with a long-term mean discharge of about 180 m$^3$ s$^{-1}$. The marine input depends mainly on the tidal cycle. The tidal range is between 4 and 6 m (macrotidal) with a tidal reach of 50 km.

Two multipier bridges, the Tay Railway Bridge and the Tay Road Bridge, cause obstructions to the free flow of the tidal waters and geographically define the field study area (Fig. 1). Previous studies have demonstrated that this section of the middle reaches of the Tay Estuary is highly dynamic and is dominated by sand, migrating sandbanks and migrating channels (Buller & McManus, 1975).

In the reach between the Railway and Road Bridges, Middle Bank (Fig. 1) is the largest intertidal sand bank and acts as a natural divide between the Queen's Road Channel to the north and the main Navigation Channel to the south. It is the largest positive relief feature in this sector, being over 1 km long and up to 200 m in width.

The dominant bottom sediment type found between the two bridges is slightly gravelly sand (nomenclature of Folk, 1974), which covers most of the middle region of the study area, including Middle Bank (Buller & McManus, 1975; Wewetzer, 1997). The deep areas, such as the main Navigation Channel, are lined with coarser fractions incorporating pebbles and gravelly sand together with mussels (mainly *Mytilus edulis*).

## EQUIPMENT AND METHODS

A systematic side-scan sonar survey of the study area was carried out between August 1993 and March 1995 in order to determine the distribution of bedforms. Full

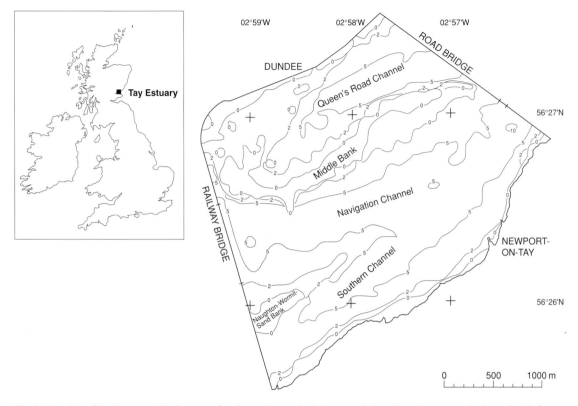

**Fig. 1.** Location of Tay Estuary and bathymetry of study area (water depths in metres below Chart Datum – equivalent to level of Lowest Astronomical Tide).

spatial coverage of the area was achieved. During the initial part of the field programme a dual-channel Klein Hydroscan, Model 401, operating at a frequency of 400 kHz and recording sonographs on wet paper, was employed. After February 1994, a dual-channel Waverley Sonar 3000, operating at a frequency of 100 kHz coupled to a thermal linescan recorder, was used. Position fixing was achieved by means of a Magellan NAV 1000 PLUS GPS receiver. Although Magellan (1990) suggests an accuracy of 25 m or better, the accuracy achieved during the field-work was of the order of ±8 m. All field work was carried out from the research vessel *Mya* of the University of Dundee, Tay Estuary Research Centre (TERC), Newport-on-Tay, Scotland, during the final stage of the flood tide, at slack water and the beginning stage of the ebb tide. This work forms part of a wider study of the middle reaches of the Tay Estuary being undertaken by the authors. Thus, additional data needed to support and verify sonograph interpretation, the details of which are not reported in this paper, were provided by vertical beam echo-sounding and bottom sediment sampling.

The sedimentary bedforms recorded by side-scan sonar were analysed in terms of their dimensions and classified on the basis of the scheme of Ashley (1990), according to which all the bedforms identified in the study area were various types of dunes. The two most important descriptors, according to Ashley's (1990) classification, are dune wavelength ($L$) and dune height ($H$). The methods for measuring these from sonographs are described below, followed by the observations recorded for the study area. In order to establish any intercorrelations between $L$, $H$ and corresponding water depth ($d$), following studies by Allen (1970), Yalin (1977), Dalrymple *et al.* (1978), Zarillo (1982), Flemming (1988), Berné *et al.* (1993) and others, the dunes identified from sonographs were analysed statistically for:

**1** the whole study area;

**2** the study area divided into ebb and flood tidal subsets;

**3** the study area divided into intertidal and subtidal areas.

Dune wavelength, or crest-to-crest spacing, is the distance between the crests of adjacent dunes. Owing to the geometry of the acoustic pulse travelling through water

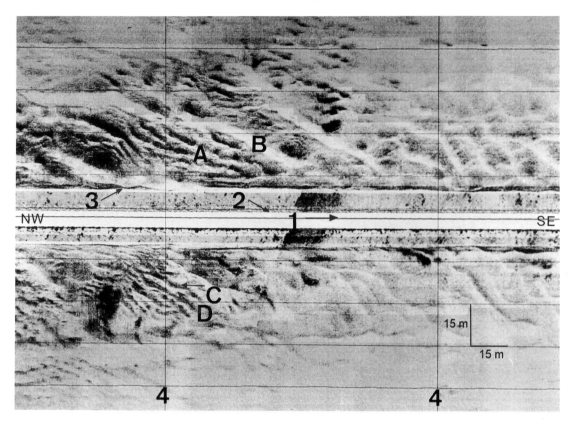

**Fig. 2.** Example of a sonograph recorded using the Waverley Sonar 3000 (100 kHz) from the southern edge of Middle Bank, middle reaches of the Tay Estuary. 1, survey line; 2, first surface return; 3, first bottom return; 4, GPS fixes; A, platform dunes (wavelength 15–25 m) with superimposed dunes; B, acoustic shadow of A; C, small wavelength dune (3–5 m) with no superimposed dunes; D, acoustic shadow of C.

and its reflection, $L$ is generally easy to measure, especially when a relatively large dune field is recorded. Commonly, one channel of the sonograph allows a better interpretation when giving a good reflection from the steep lee side facing the outgoing pulse (appearing dark on the sonograph) followed by its acoustic shadow appearing as a blank area on the trace (Fig. 2). The other channel will record the gentler stoss slope appearing in a lighter tone on the sonograph, which does not always produce a distinct acoustic shadow. Slopes facing away from the transducer give only light reflections and may not be detectable. Therefore, dune fields observed on both channels were easily detected, although small individual dunes may have been overlooked.

Dune height is the vertical distance between the trough and the crest of a dune. It is not possible to measure the height of dunes ($H$) directly from sonographs. One has to measure the acoustic shadow produced by an object or

bedform ($l_s$), the height of the tow fish above the sea bed ($h_f$) as well as the slant range from the tow fish to the target ($r_s$). With the aid of these three measurements, the height of a dune may be calculated as follows (Klein, 1985),

$$H = \frac{h_f \times l_s}{r_s + l_s} \quad (1)$$

As not all of the measurements can be made with exact precision, owing to the resolution of the sonograph print-out and the sometimes indistinct acoustic shadow, the results of dune height determination should be regarded as estimates.

## DUNE WAVELENGTH

The morphological classification of dunes after Ashley (1990) divides crest-to-crest spacing into four classes:

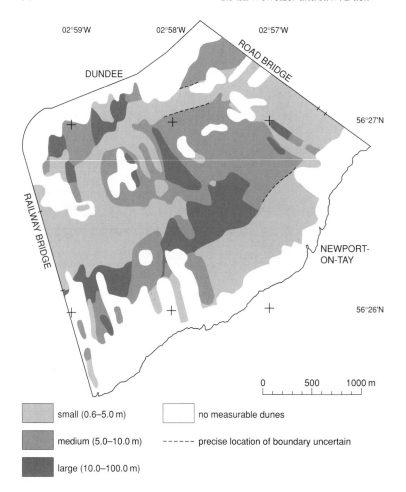

small (0.6–5.0 m)

medium (5.0–10.0 m)

large (10.0–100.0 m)

no measurable dunes

- - - - - precise location of boundary uncertain

**Fig. 3.** Dune wavelength during flood and slack tidal conditions as interpreted from side-scan sonographs. Classification after Ashley (1990).

small, 0.6–5.0 m; medium, 5.0–10.0 m; large, 10.0–100.0 m; very large, over 100.0 m. No dunes of very large wavelength were identified in the study area. The dominant wavelength of dunes observed was of the smallest class (0.6–5.0 m), which occupy the deep areas of the Southern Channel (Fig. 1), the main Navigation Channel and Queen's Road Channel. The dunes on Middle Bank are mainly of medium wavelengths, except in the centre area where large wavelength dunes occur just east of a region of small wavelength dunes. The shallower regions along the northern shore of the study area reveal a mixture of small, medium and large wavelength dunes (Fig. 3). Over most of the study area dunes of two sizes are present. These consist of platform dunes (Ashley, 1990) with smaller superimposed dunes. In the computation of relationships between dune dimensions and water depth only the platform dunes are included.

## DUNE HEIGHT

Ashley (1990) divides dune height into four classes: small dunes, 0.05–0.25 m; medium, 0.25–0.50 m; large, 0.50–3.00 m; very large, over 3.00 m (note that the height values have been modified from those given by Ashley ((1990) according to Dalrymple & Rhodes (1995)). No very large dune heights were calculated from acoustic shadows recorded on the sonographs of the study area. The majority of the dunes, including the channels and Middle Bank, are of medium height as also confirmed by echo-sounding data. A major section of dunes of large height stretches from the Road Bridge towards the centre of the estuary. This covers the middle sections of the main Navigation Channel. A few small patches of dunes of large height are also found in the southwestern corner

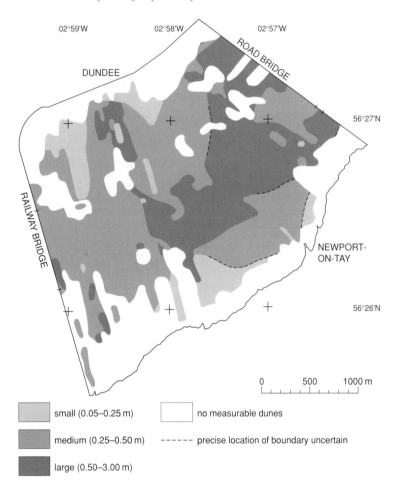

02°59'W  02°58'W  02°57'W

ROAD BRIDGE

DUNDEE

56°27'N

RAILWAY BRIDGE

NEWPORT-
ON-TAY

56°26'N

0    500    1000 m

**Fig. 4.** Dune height during flood and slack tidal conditions as interpreted from side-scan sonographs. Classification after Ashley (1990).

| | small (0.05–0.25 m) | | no measurable dunes |
| | medium (0.25–0.50 m) | ----- | precise location of boundary uncertain |
| | large (0.50–3.00 m) | | |

close to the Railway Bridge. Dune heights along the shorelines where the waters are shallow belong to the smallest class. The centre section revealed a remarkably abrupt transition from small-height dunes in the northern and southern shore areas to large heights towards the middle of the estuary (Fig. 4).

## GEOMETRICAL PARAMETERS OF DUNES

Many researchers have investigated the relationships between dune wavelength and height and their occurrence at certain water depths. The present state of knowledge of intertidal bedforms is far greater than that of subtidal bedforms, which are difficult to observe visually. Zarillo (1982), Goedheer & Misdorp (1985), Harris & Collins (1985) and Berné *et al.* (1993) have conducted research on subtidal dune morphology in tidal environments using side-scan sonar. Dalrymple *et al.* (1978), Dalrymple (1984), Allen *et al.* (1994), and many others have studied intertidal bedforms. All of these authors investigated the relationships between dune morphology and water depth.

In this section, dune morphologies recorded in the study area of the Tay Estuary by side-scan sonar are examined in terms of intercorrelations of wavelength, height and water depth. Table 1 presents the results of statistical analyses using the coefficient of determination ($r^2$) for linear regression analysis. Analysis for non-linear regression, such as power and logarithmic functions, was also carried out, but $r^2$ values were close to zero so were not investigated further.

A commonly accepted (Ashley, 1990; Dalrymple & Rhodes, 1995; Kostaschuk *et al.*, 1995; and others) relationship between dune height ($H$) and crest-to-crest

**Table 1.** Coefficients of determination ($r^2$) computed for linear regression analysis of geometrical parameters of dunes and water depth; (s), significant correlation at the 0.01 level.

(a)  All dunes

|          | Wavelength | Height |
|----------|------------|--------|
| Height   | 0.321 (s)  |        |
| Water depth | 0.000   | 0.064  |

(b)  Dunes recorded during flood- and *ebb-* tidal conditions

| Flood/*ebb* | Wavelength | Height | Water depth |
|-------------|------------|--------|-------------|
| Wavelength  |            | *0.365* (s) | *0.000* |
| Height      | 0.258 (s)  |        | *0.091* |
| Water depth | 0.000      | 0.071 (s) |         |

(c)  Intertidal and *subtidal* dunes

| Intertidal/*subtidal* | Wavelength | Height | Water depth |
|-----------------------|------------|--------|-------------|
| Wavelength            |            | *0.332* (s) | *0.002* |
| Height                | 0.007      |        | *0.046* |
| Water depth           | 0.028      | 0.377 (s) |         |

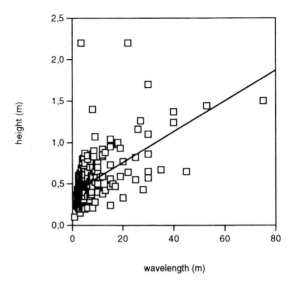

**Fig. 5.** Plot of subtidal dune wavelength versus dune height as measured from sonographs ($H = 0.019L + 0.390$; $r^2 = 0.332$).

spacing ($L$) for subtidal dunes was determined by Flemming (1988) as

$$H = 0.0677\,L^{0.8098} \tag{2}$$

whereas Zarillo (1982) found that maximum dune wavelengths were of the order of $2\pi H$ or

$$H = \frac{L}{6.283} \tag{3}$$

in a tide-dominated estuary in Georgia. Dalrymple *et al.* (1978) found varying relationships between dune heights and wavelengths in intertidal environments depending on the type of dune measured (see Table 3).

The relationships computed in this study between $H$ and $L$ (Table 2) for the whole data set and the subtidal and the intertidal data (Figs 5 & 6) sets revealed three different equations, none of which are similar to those determined by Flemming (1988) or Zarillo (1982). A further

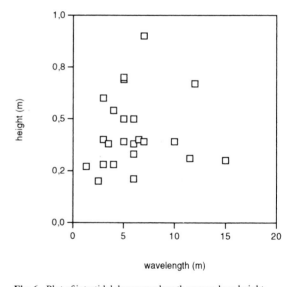

**Fig. 6.** Plot of intertidal dune wavelength versus dune height as measured from sonographs (insignificant relationship, see Table 1).

**Table 2.** Summary of equations of relationships between dune dimensions (height, $H$ and wavelength, $L$) and water depth ($d$) recorded by side-scan sonar in the middle reaches of the Tay Estuary.

|                     | $H$–$L$             | $H$–$d$             | $L$–$d$             |
|---------------------|---------------------|---------------------|---------------------|
| Total data set      | $H = 0.019L + 0.375$ | $H = 0.030d + 0.220$ | $L = 0.029d + 8.167$ |
| Intertidal data set | $H = 0.004L + 0.408$ | $H = 0.061d + 0.207$ |                     |
| Subtidal data set   | $H = 0.019L + 0.390$ | $H = 0.028d + 0.250$ |                     |

**Table 3.** Intertidal bedform size and morphological characteristics (after Dalrymple *et al.*, 1978).

| Dune type | Wavelength (m) | Height (m) | Characteristic |
|---|---|---|---|
| Type 1: small to medium, simple, two-dimensional dune | 0.1–25.0 | 0.05–0.50 | Straight to smoothly sinuous<br>Lack scour pits<br>Height constant along crestline<br>Wavelengths and heights poorly correlated:<br>$H = 0.0947L^{0.346}$ ($r^2 = 0.213$ for $n = 70$) |
| Type 2: small to medium, simple, three-dimensional dune | 0.05–14.00 | 0.05–0.70 | Sinuous to lunate<br>Scour pits<br>Height variable along crestline<br>Wavelengths and height well correlated<br>$H = 0.0865L^{0.787}$ ($r^2 = 0.621$ for $n = 255$) |
| Type 3: large to very large, compound dune | 10.0–215.0 | 0.15–3.40 | Straight to sinuous<br>Lack scour pits<br>Height constant along crestline<br>Wavelengths and heights moderately correlated<br>$H = 0.0635L^{0.733}$ ($r^2 = 0.626$ for $n = 58$) |

division of the intertidal data set into the three types of intertidal dunes as described by Dalrymple *et al.* (1978) (Table 3) was made. Type 1 dunes (small to medium, simple, two-dimensional dune) revealed no correlation, but only two dune groups of this type were recorded. The $r^2$ value computed for dunes of type 2 (small to medium, simple, three-dimensional dune) was $r^2 = 0.000$, although 19 dune groups of this type were recorded in the intertidal environment. For the large-scale dune types (type 3), only one group was recorded and no further analysis was possible.

Measurements of dune height and wavelength in previous studies have shown a positive relationship with water depth (Dalrymple & Rhodes, 1995). Allen (1970) suggested the formula

$$H = 0.086d^{1.19} \tag{4}$$

for the relationship between dune height ($H$) and water depth ($d$). Yalin (1977), however, suggests that dune height should be approximately 17% of the water depth

$$H = 0.167d \tag{5}$$

based on a combination of theory and empirical observations, and wavelength ($L$) and water depth ($d$) are related to each other as

$$L = 6d \tag{6}$$

In common with the findings of Bokuniewicz *et al.* (1977), no significant correlation between dune wavelength and water depth was found in this study (Table 1, Figs 7 & 8). It is questionable how closely related these two parameters are in natural environments. Water depth changes with the tidal state, especially in estuaries, and

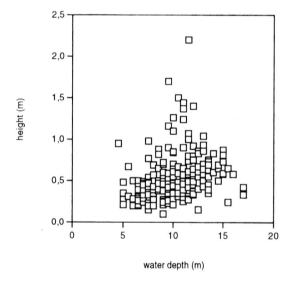

**Fig. 7.** Plot of water depth versus dune height as measured from side-scan sonographs for the total data set (insignificant relationship, see Table 1).

varies in the study area by 4 to 6 m according to the tidal state. For the following computations the water depth was derived on the basis of the first bottom and first surface return on the sonographs (Fig. 2). The first bottom return is a good indication of towfish height above the sea-bed and the first surface return is a good indication of towfish depth below the water surface (Fish & Carr, 1990). The combination of the two reveals the water depth at the time of the survey (generally close to high water in this study).

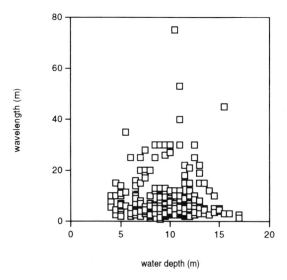

**Fig. 8.** Plot of water depth versus dune wavelength as measured from side-scan sonographs for the total data set (insignificant relationship, see Table 1).

Analysis of the data collected suggests different relationships between $H$ and $d$ for the three data sets (Table 2). None of the computed equations is similar to those of Yalin (1977). A relationship between $L$ and $d$ (Table 2) was computed for all dunes recorded by side-scan sonar but $r^2$ is insignificant for all three data sets.

## DISCUSSION AND CONCLUSIONS

The variety of relationships computed between dune height and wavelength, dune height and water depth, and dune wavelength and water depth suggest that these parameters do not just depend on one another but are part of a far more complex system. Many authors have tried to establish relationships between the parameters examined above. Dalrymple *et al.* (1978), Zarillo (1982) and Flemming (1988) established positive relationships between dune height and dune wavelength but with different formulae. Flemming's (1988) formula has been accepted and used in studies where dune height could not be measured but had to be calculated for sediment transport predictions (Kostaschuk *et al.*, 1995). Yalin (1964, 1977, 1987), similarly to Allen (1968), used a combination of theory and empirical observations to establish relationships between dune height and water depth. By contrast, Goedheer & Misdorp (1985) investigated dunes solely in natural environments and were unable to establish a relationship between dune height and water depth,

but recorded the highest dunes in their area of study, Oosterschelde (southwest Netherlands), in the shallowest waters. Goedheer & Misdorp (1985), however, studied only a small part of the Oosterschelde, namely one of the subtidal channels, the Schaar van Colijnsplaat. Similar to the Tay Estuary, a semidiurnal tide occurs in this subtidal channel, but the tidal amplitude ranges from about 2.3 m on a neap tide to only about 3.0 m on a spring tide, compared with the larger tidal range in the Tay Estuary. Bokuniewicz *et al.* (1977) also found no correlation between dune height and water depth in studies of bedforms in the eastern Long Island Sound, USA. Although semidiurnal tides also occur in the Long Island Sound, this estuary is far larger than the Tay Estuary, being 150 km long and up to 40 km wide compared with 50 km in length and up to 4 km wide.

The results of the regression analysis carried out in this study reveal that the relationships between dune height and wavelength, dune height and water depth, and dune wavelength and water depth vary according to spatial (intertidal versus subtidal) and temporal (ebb tidal versus flood tidal) data subdivision. Moreover, this study highlights the fact that the relationships established previously between these parameters are not generally applicable in estuarine environments. Further information is required, in particular (cf. Dalrymple & Rhodes, 1995) on the influence of additional variables such as flow strength, sediment textural characteristics (grain size, shape and sorting) and sediment availability, before more generally applicable relationships can be derived.

## ACKNOWLEDGEMENTS

We would like to thank Ian Lorimer, Tay Estuary Research Centre (TERC), for his invaluable assistance during field-work. This study was part of a project funded by the Commission of the European Communities (Human Capital and Mobility Fund), which is gratefully acknowledged.

## REFERENCES

ALLEN, J.R.L. (1968) *Current Ripples – their Relation to Patterns of Water and Sediment Motion.* North-Holland Publishing Company, Amsterdam.

ALLEN, J.R.L. (1970) *Physical Processes of Sedimentation.* Elsevier, New York.

ALLEN, J.R.L., FRIEND, P.F., LLOYD, A. & WELLS, H. (1994) Morphodynamics of intertidal dunes: a year-long study at Lifeboat Station Bank, Wells-Next-The-Sea, Eastern England. *Philos. Trans. R. Soc. London, Ser. A,* **347**, 291–345.

ASHLEY, G.M. (1990) Classification of large-scale subaqueous bedforms: a new look at an old problem. *J. sediment. Petrol.*, **60**, 160–172.

BERNÉ, S., CASTAING, P., LE DREZEN, E. & LERICOLAIS, G. (1993) Morphology, internal structure and reversal asymmetry of large subtidal dunes in the entrance of the Gironde Estuary (France). *J. sediment. Petrol.*, **63**, 780–793.

BOKUNIEWICZ, H.J., GORDON, R.B. & KASTENS, K.A. (1977) Form and migration of sandwaves in a large estuary, Long Island Sound. *Mar. Geol.*, **24**, 185–199.

BULLER, A.T. & MCMANUS, J. (1975) Sediments of the Tay Estuary, Part I: Bottom sediments of the upper and upper middle reaches. *Proc. R. Soc. Edinburgh, Ser. B*, **75**, 41–64.

DALRYMPLE, R.W. (1984) Morphology and internal structure of sandwaves in the Bay of Fundy. *Sedimentology*, **31**, 365–382.

DALRYMPLE, R.W. & RHODES, R.N. (1995) Estuarine dunes and bars. In: *Geomorphology and Sedimentology of Estuaries* (Ed. Perillo, G.M.E.), pp. 359–422. Elsevier, Amsterdam.

DALRYMPLE, R.W., KNIGHT, R.J. & LAMBIASE, J.J. (1978) Bedforms and their hydraulic stability relationships in a tidal environment, Bay of Fundy, Canada. *Nature*, **275**, 100–104.

FISH, J.P. & CARR, H.A. (1990) *Sound Underwater Images: a Guide to the Generation and Interpretation of Side Scan Sonar Data*. Lower Cape Publishing, Orleans, WA.

FLEMMING, B.W. (1988) Zur Klassifikation subaquatischer, strömungstransversaler Transportkörper. *Bochumer Geol. Geotechn. Arb.*, **29**, 44–47.

FOLK, R.L. (1974) *Petrology of Sedimentary Rocks*. Hemphill Publishing Company, Austin, TX.

GOEDHEER, G.J. & MISDORP, R. (1985) Spatial variability and variations in bedload transport direction in a subtidal channel as indicated by sonographs. *Earth Surf. Process. Landf.*, **10**, 375–386.

HARRIS, P.T. & COLLINS, M.B. (1985) Bedform distributions and sediment transport paths in the Bristol Channel and Severn Estuary, U.K. *Mar. Geol.*, **62**, 153–166.

KLEIN (1985) *Side-scan Sonar Record Interpretation*. Klein Associates, Salem, NH.

KOSTASCHUK, R.A., LUTERNAUER, J.L., BARRIE, J.V., LEBLOND, P.H. & VON DEICHMANN, L.W. (1995) Sediment transport by tidal currents and implications for slope stability: Fraser River delta, British Columbia. *Can. J. Earth Sci.*, **32**, 852–859.

MAGELLAN (1990). *Magellan NAV 1000 PLUS User Guide*. Monrovia, CA.

WEWETZER, S.F.K. (1997) *Bedforms and sediment transport in the middle Tay Estuary, Scotland: a side-scan sonar investigation*. Unpublished PhD thesis, University of St. Andrews.

YALIN, M.S. (1964) Geometrical properties of sand waves. *Proc. Am. Soc. civ. Eng.*, **90**, 105–119.

YALIN, M.S. (1977) *Mechanics of Sediment Transport*. Pergamon Press, Toronto.

YALIN, M.S. (1987) On the formation mechanism of dunes and ripples. *Euromech Colloq. Proc.*, 261.

ZARILLO, G.A. (1982) Stability of bedforms in a tidal environment. *Mar. Geol.*, **48**, 337–351.

*Spec. Publs int. Ass. Sediment.* (1990) **28**, 43–57

# Flow structure and transport of sand-grade suspended sediment around an evolving braid bar, Jamuna River, Bangladesh

S. J. McLELLAND\*,†,‡, P. J. ASHWORTH\*, J. L. BEST†, J. RODEN†,\* and G. J. KLAASSEN§

\**School of Geography, University of Leeds, Leeds, LS2 9JT, UK;*
†*School of Earth Sciences, University of Leeds, Leeds, LS2 9JT, UK; and*
§*WL/Delft Hydraulics, P.O. Box 177, 2600 MH Delft, Netherlands, and IHE Delft, P.O. Box 3015, 2601 DA Delft, Netherlands*

## ABSTRACT

Detailed vertical profiles of time-averaged flow velocities and sediment concentration were taken during three periods of mid-channel bar development in the Jamuna River, Bangladesh. Bar growth was initiated downstream from a major flow convergence and generated a bar 4 km long and 1 km wide in a channel up to 15 m deep. Flow velocities and the concentrations of sand-grade suspended sediment were quantified using an acoustic Doppler current profiler (ADCP). Bed morphology was measured using echo-sounding and all positions were located using a differential global positioning system (DGPS).

These data reveal no evidence for channel-scale, coherent helical flow cells in either distributary around the braid bar. Instead, the structure of flow is dominated by a simpler flow divergence over the bar head, flow convergence at the bar tail and flow that is usually parallel to the thalwegs in each distributary. During the later stages of bar growth, flow is directed over the bar top from one distributary towards the other as the bar begins to adopt a more asymmetrical morphology. In addition, large sand dunes migrate up the bar stoss side, producing an accretionary dune front at the bar head. These dunes are strongly linked to high suspended bed-sediment concentrations as flow shallows on to the bar top. A shadow of low suspended bed-sediment concentration is located in the bar lee during the early stages of bar growth, this also being a region of small sand dunes.

The lack of coherent secondary flows, around large kilometre-scale bars, may be explained through the large width-to-depth ratio of these channels, the low curvature of the anabranches, the complexity of flow over the bar top as it interacts with flow in the anabranches and the significant influence of large-scale dune-bedform roughness. These factors suggest that current models for the processes of mid-channel bar creation, growth and preservation, derived from studies of smaller rivers, require substantial revision before application to kilometre-scale sand-braid bars.

## INTRODUCTION

Mid-channel bars may be defined broadly as accumulations of sediment that are not attached to the adjacent channel banks at all water levels, and which possess a width that scales with bankfull channel width (Parker, 1976; Yalin, 1992), but a height that is independent of bar-top water depth. Mid-channel bars are ubiquitous in braided rivers and their development is coupled

intimately with local flow structure (e.g. Ashworth, 1996; Richardson *et al.*, 1996), sediment dispersal (e.g. Ashworth *et al.*, 1992a; Laronne & Duncan, 1992; Lane *et al.*, 1995) and channel change (e.g. Ashmore, 1991; Thorne *et al.*, 1993; Goff & Ashmore, 1994). Furthermore, mid-channel bars may form the predominant channel-scale depositional element within braided alluvium (e.g. Cant & Walker, 1978; Bridge *et al.*, 1986; Bridge, 1993; Bristow, 1993) and thus dictate the geometry of facies associations and internal channel heterogeneity.

‡ Present address: Department of Geography, University of Exeter, Exeter EX4 4RJ, UK. (Email: S.J.McLelland@exeter.ac.uk)

Much of the previous work on mid-channel bars has largely concerned the morphology of braid-bar growth (e.g. Leopold & Wolman, 1957; Smith, 1974; Hein & Walker, 1977; Ashmore, 1982, 1991; Rundle, 1985; Fujita, 1989; Ferguson *et al.*, 1992). More recently, Bridge & Gabel (1992) described two-dimensional flow structure around a mid-channel island, and Whiting & Dietrich (1991) and Whiting (1997) described flow processes at the head of a mid-channel bar at flow stages less than 45% bankfull depth. The channel morphology at both these study sites, however, was static and the dynamics of bar evolution could not be studied. Richardson *et al.* (1996) describe the three-dimensional flow structure around a large compound braid bar in the Jamuna River, Bangladesh, which had undergone several stages of growth and migration (using a separately collected data set as part of the River Survey Project (Flood Action Plan 24) to that detailed here; see Delft Hydraulics *et al.*, 1996b), but were not concerned with the initial stages of braid-bar formation and its influence on the local flow structure. Ashworth (1996) used a Froude-scale mobile-bed model of a mid-channel bar to quantify the change in surface flow speed and direction as the bar became emergent, but this study had a limited spatial distribution of two-dimensional velocity data and there are no comparable field studies.

Recent work has suggested that bifurcating flow around a mid-channel braid bar can be represented by flow in two meander bends (Ashworth *et al.*, 1992b; Bridge, 1993) and therefore that the flow structure over and around a braid bar may consist of a channel-wide helical cell in each anabranch, which transports sediment in towards the mid-channel bar from the outer bank (cf. Ashworth *et al.*, 1992b, fig. 25.3, p. 504). Richardson *et al.* (1996, fig. 25.2, p. 523) suggest a more complicated model of flow structure around a braid bar, with skew-induced helical cells confined to each anabranch thalweg, radial outward flow occurring in the shallow flow over the bar top and small, contrarotating secondary flow cells being present at the outer bank. Difficulties exist, however, when assessing the contribution of secondary flow to the bar growth process because there is no independent method of orientating the flow vectors in strongly divergent and bifurcating flow (McLelland *et al.*, 1994).

The importance of secondary flow for bedload sorting, over and around mid-channel bars, has been highlighted by Ashworth *et al.* (1992b), who suggested that the movement of fluid and sediment inward towards the bar tail may explain the process of bar-tail accretion and downstream bar migration (see Ashworth, 1996, fig. 1, p. 105). In sand-bed rivers, sediment conveyance is dominated by bedform migration and suspended, sediment transport.

Although the dynamics of suspended sediment over sand bedforms has been studied in a number of different environments (e.g. Jackson, 1976; Lapointe 1993; Bennett & Best, 1995; Kostaschuk & Villard, 1996), little work has been undertaken on the spatial distribution of suspended sediment around mid-channel bars. This paucity of data probably results from the fact that most studies of braid-bar dynamics have been confined to gravel-bed rivers, where suspended sediment is often considered to be a minor contributor to bar growth. Additionally, conventional direct-sampling techniques are time-consuming and cannot rapidly yield comprehensive suspended-sediment data sets throughout the whole flow depth and around an entire bar.

This paper reports on a unique, high-resolution, spatially intensive bathymetric and flow survey data set obtained recently under the River Survey Project (Flood Action Plan 24) in Bangladesh (Delft Hydraulics *et al.*, 1996b). The results describe the principal characteristics of flow structure and the spatial distribution of sediment transport, measured simultaneously around a kilometre-scale sand bar during three stages of mid-channel bar evolution and allow examination of the opportunities for development of channel-scale, coherent, helical flow cells around large sand-bed braid bars.

## FIELD SITE

The Jamuna River (the Bengali name for the Brahmaputra) rises in the Tibetan Plateau and flows south for 220 km through Bangladesh before joining the Ganges to form the Padma River, which then converges with the Meghna and discharges into the Bay of Bengal (Fig. 1). The Jamuna is one of the largest and most dynamic sand-bed braided rivers in the world (Schumm & Winkley, 1994), with peak discharges over $1 \times 10^5$ m$^3$ s$^{-1}$, annual suspended loads up to $725 \times 10^6$ t, maximum bank-erosion rates of 1 km yr$^{-1}$ and bed scour of up to 50 m (Coleman, 1969; Klaassen & Vermeer, 1988; Hossain, 1993; Thorne *et al.*, 1993; Best & Ashworth, 1997). The Jamuna braidbelt is approximately 10 km wide and exhibits characteristics of both a braided and an ana-stomosing channel-pattern, with large (> 10 km long), vegetated, metastable 'permanent' bars or islands that are inundated at bankfull stage and 'temporary', mobile braid bars (< 5 km long), the elevation of which corresponds to the dominant discharge, calculated as $3.8 \times 10^4$ m$^3$ s$^{-1}$ by Thorne *et al.* (1993).

Braided reaches of the Jamuna contain many different types and size of bars (Coleman, 1969; Bristow, 1987; Klaassen & Masselink, 1992; Klaassen *et al.*, 1993;

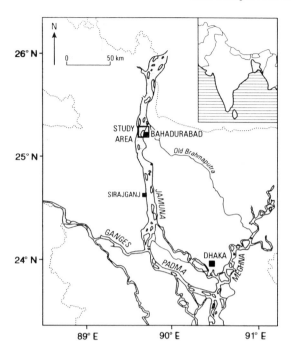

**Fig. 1.** Location map showing the major rivers of Bangladesh and the study area just north of Bahadurabad on the Jamuna River.

Peters, 1993), with one of the most common being mid-channel bars. These bars are often associated with the confluence–diffluence unit (cf. McLelland *et al.*, 1996), where flow convergence leads to bed scour and subsequent sediment deposition results in flow divergence and bar growth (Carson & Griffiths, 1987; Ferguson *et al.*, 1992; Ferguson, 1993; Klaassen *et al.*, 1993; Thorne *et al.*, 1993; Mosselman *et al.*, 1995). Mid-channel bars can also develop by enlargement of the inner chute of an accreting point bar, which isolates a bar within the channel (Richardson *et al.*, 1996). In the Jamuna, mid-channel bars have planform morphologies similar to the unit or longitudinal bars described for gravel-bed rivers by Smith (1974), Bluck (1979) and Church & Jones (1982) and often possess steep cutbanks, resulting from basal erosion and bank collapse (Richardson *et al.*, 1996). Over 40% of the channel bed and submerged bar-tops are covered by dune bedforms at any flow stage, although this spatial coverage may rise to 100% during the highest discharges (Delft Hydraulics *et al.*, 1996a). Dune morphology is often three-dimensional, with crest heights of up to 6 m at high flow and dune lee-face angles that range from true slipfaces (*c*. 35°) to low-angle lee faces of 3–4°.

A study reach 8 km long and 5 km wide, which was downstream of a major flow convergence between the first order 'west' and 'east' channels (Fig. 2A&B) near

Bahadurabad (Fig. 1), was selected for intensive bathymetric and flow surveys (see below). Surveys were undertaken between November 1993 and March 1996 in order to follow a period of mid-channel bar initiation, growth and emergence. The flow regime of the Jamuna is strongly seasonal as a result of monsoonal rains, with low flow between February and April followed by a rapid increase to a flood peak in late July and August (Fig. 3). Mean annual peak discharge calculated from the 27-yr daily discharge record near the study reach at Bahadurabad (Fig. 2A) is 69 000 m$^3$ s$^{-1}$, which is about 30% greater than the estimated bankfull discharge (Delft Hydraulics & DHI, 1996a). Water-surface slope in the study reach is stage-dependent and ranges between 5.5 and $9.1 \times 10^{-5}$. Average bed and suspended-sediment grain size are 0.14 mm (fine sand) and 0.042 mm (coarse silt) respectively (Delft Hydraulics & DHI, 1996b).

## DATA ACQUISITION

Bathymetric and flow data were acquired during a series of ship surveys that were undertaken between November 1993 and February 1996. In this paper, data are presented from three survey periods: August 1994, March 1995 and September 1995. These surveys consisted of a series of parallel cross-sectional lines, spaced 250 m apart and approximately perpendicular to the main-channel thalweg (058°, Fig. 2B). Additional downstream lines were surveyed in the anabranch-channels and over the bar top during high-flow stages. Ship location was fixed using a differential global positioning system (DGPS), which operates to within ±2 m in mobile mode with a ground reference station.

Bed topography was measured using a SIMRAD dual frequency (30 and 210 kHz) echo-sounder with digital data recorded at 1-s intervals, which represents a ground spacing of approximately 1.5 m at a constant ship speed. Bed-height data have a resolution of ±0.05 m. All bed heights were reduced to a common datum, termed standard low water level (SLW), which is determined from long-term stage measurements at Bahadurabad. In March 1995, ship survey data were supplemented by a land survey using a portable DGPS and a total station electronic-distance meter. All topographic data were contoured from a 25 m grid spacing using a data-visualization program (Spyglass Transform®), with kriging used to interpolate unmeasured grid data. Kriging utilized a spherical distribution model and a normalized variance of zero was applied to all surveyed grid data. Between 18 500 (March 1995) and 51 400 (September 1995) individual survey points were used to produce the bathymetric maps of the

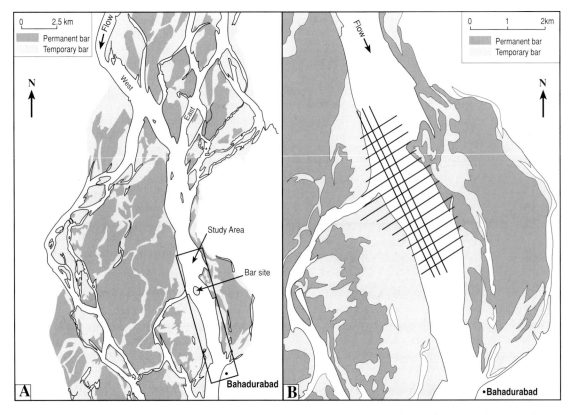

**Fig. 2.** (A) Location of the survey area and position of study bar in the main east channel. 'Permanent' bars or islands are vegetated and normally stable over several years. 'Temporary' bar is sediment within less stable braid bars that may be reworked during each monsoon. (B) Location and orientation of ship survey lines of August 1994. Note that different survey lines were used during each survey period, depending on the flow stage and location of the study bar (see Table 1 for further information).

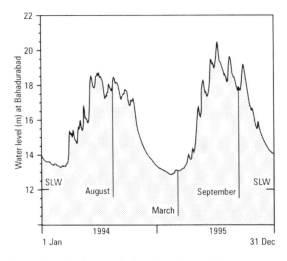

**Fig. 3.** Stage hydrograph for 1994–1995 recorded at Bahadurabad. The three survey periods are superimposed on the hydrograph (see Table 1 for further information). Standard Low Water level (SLW) is 12.03 m at Bahadurabad (see text for further details).

study site. Bank edges were defined using the mean water level for each survey period.

Flow data were collected in water depths of greater than 2 m using an RD Instruments four-beam, 300 kHz acoustic Doppler current profiler (ADCP). Beams are oriented at 20° from the vertical and consist of monochromatic sound pulses that are transmitted in a downward direction. At any depth, the radial speed of water can be determined for each beam axis from the Doppler shift of the sound frequency reflected back from acoustic reflectors such as sediment particles. Using a geometrical calculation, these four radial-beam velocities yield three perpendicular components of flow velocity; downstream, vertical and cross-stream. The ADCP measures a complete velocity profile every 1 s and outputs an ensemble average every 6 s for an array of 0.5 m vertical bins. Measurement bins start 2.16 m below the water surface and extend down to approximately 6% of the flow depth from the channel bed. Standard errors of the ADCP velocity components and flow direction measurements are ±4.6% and ±1% respectively (Sarker, 1996), assuming that the flow is

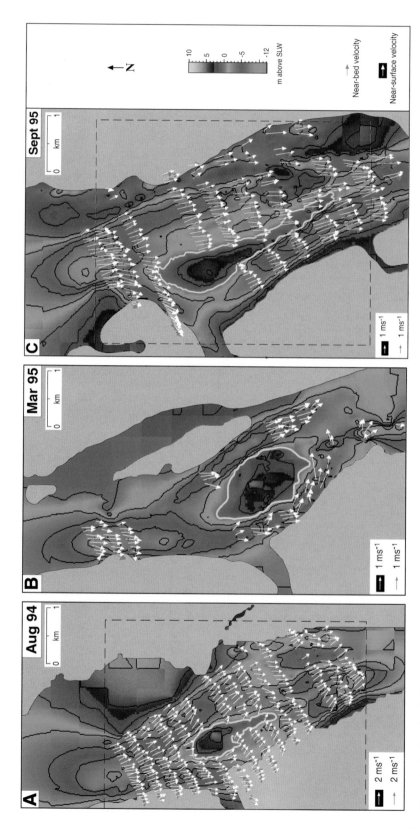

**Plate 1** A–C. Near-bed (red) and near-surface (white) flow vectors superimposed on bathymetric maps of the study area during the three survey periods. Near-surface vectors represent flow at a depth of 2.66 m whilst near-bed vectors represent flow at the second deepest measurement bin (see text for explanation of measurement bins). Vectors are shown at cross-stream intervals of approximately 50–100 m, where the flow depth is greater than 3.66 m. Note the difference in scale for vector arrows on each diagram. Banklines are constructed for 0 m SLW measured at Bahadurabad (see Fig. 2), however local flow conditions may exceed this bank elevation. Study bar outlines are highlighted by a yellow contour at 0 m SLW. See text for details of bathymetric map construction. The northerly co-ordinates are aligned for all three maps, but they each begin at different easterly co-ordinates. Boxes shown by the dashed lines in (a) and (c) delimit the area shown in the corresponding diagrams (a) and (c) of Figs 5 and 7. Plate 1B and Figs 5B and 7B show the same survey area between A and C).

homogeneous within each measurement bin. Any error in the spatial distribution of the DGPS data is considerably smaller than the grid scale on which the data are analysed and does not add any error to the velocity vectors because the flow direction is calculated independently.

As well as measuring flow velocity, the ADCP also records the intensity of the back-scattered signal, which is expressed in decibels (dB) and adjusted for the attenuation of sound through the water column (for a water temperature of 4°C). Back-scatter intensity reflects the concentration of particles in the measurement volume and is dependent on the acoustic properties of both the fluid and the sediment particles (Thorne *et al.*, 1991). Acoustic attenuation varies with water temperature, although at the low operating frequency of the ADCP (300 kHz) there is only a 2% variation in back-scatter value for a depth variation of 5–10 m at the average water temperature in the Jamuna (25°C, Sarker, 1996). As it is not possible to calculate the acoustic properties of the sediment particles, the strength of the back-scatter signal has been calibrated against known sampled suspended-sediment concentrations. Sarker (1996) describes the calibration techniques used by the River Survey Project (FAP 24), which used point-integrated sampling with a pump-bottle sampler to measure the suspended sediment concentration at a known flow depth and compared this with the ADCP back-scatter intensity. The concentration of the sand fraction in the suspended sediment ($c$), is given by the relationship:

$$\sqrt{c} = -68.0 + 1.04i, \quad r^2 = 0.79, \quad \text{sample size} = 73 \quad (1)$$

where $i$ is the signal intensity (Sarker, 1996). The silt fraction is ignored in this calibration because particles in this size range have a negligible effect on the back-scatter signal. It should be noted, however, that although 74% of the bed material in the study reach is sand, silt can constitute between 15 and 98% of the total suspended-sediment concentration (Sarker, 1996). A minimum grain size cannot be defined to give a threshold for ADCP response, although back-scatter intensity diminishes rapidly for smaller grain sizes. Hence, eqn (1) is not resolved for backscatter-signal intensities of less than 70 dB, which yield concentrations less than $\sim 25$ mg l$^{-1}$. Processed back-scatter data therefore represent concentrations of sand-grade suspended sediment that have been calculated for each flow-measurement bin and the depth-integrated values plotted by contouring data on to a 25-m grid using Spyglass Transform®. Kriging was again utilized to interpolate unmeasured grid data using a spherical distribution model and a normalized variance of zero was applied to all measured suspended-sediment data in the grid (which exceeded 1500 measurement points for each survey period).

## RESULTS

Figure 4 shows the size and position of the study bar at three different flow stages (August 1994, March 1995 and September 1995), which illustrate three key stages in the growth and development of the study bar. Plate 1A–C (facing p. 46) shows the near-bed (red, 0.5 m above the deepest measurement bin) and near-surface (white, 2.66 m below the water surface) flow vectors superimposed on the bed topography for the same three periods, and Fig. 5A–C shows the corresponding depth-averaged flow vectors ($V$) superimposed on the pattern of suspended-sand sediment transport where $V = \sum_{1}^{n} V_k / n$, and $V_k$ is the flow

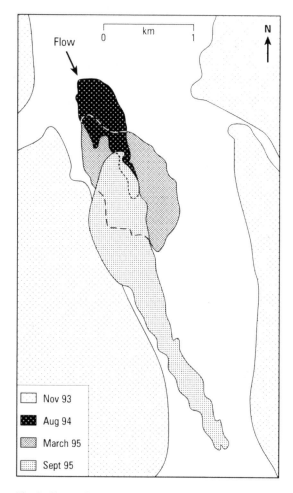

**Fig. 4.** Changes in the study bar position and morphology for the three study periods. The area shown as November 1993 indicates the channel and bank edges before the initial development of the study bar. The bar margin is defined by the 0 m relative to SLW contour.

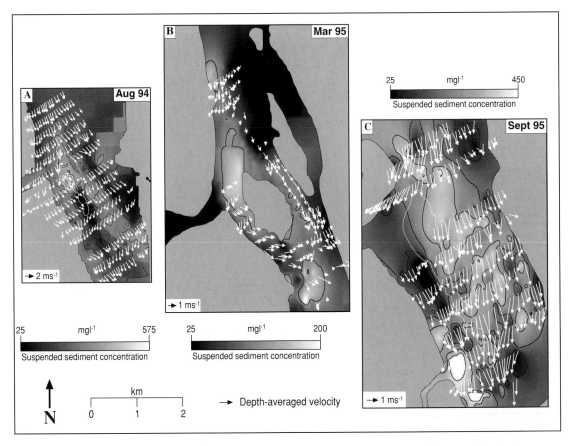

**Fig. 5A–C.** Depth-averaged flow vectors superimposed on contour maps of suspended sand-grade sediment concentration for the study area during the three survey periods. See text for calculation of suspended sediment concentration and details of map construction. Contour lines are shown at intervals of 100 mg l$^{-1}$. Flow vectors are shown at cross-stream intervals of approximately 50–100 m, where the flow depth is greater than 2.16 m. Note the difference in both the range of the suspended sediment concentration scale bar and the vector arrow length on each diagram. Banklines are constructed for 0 m SLW measured at Bahadurabad (see Fig. 2), however, local flow conditions may exceed this bank elevation. Study bar outlines in (A) and (C) are defined by 0 m SLW contour (see Fig. 4) and shown as a white outline. The northerly co-ordinates are aligned for all three maps, but they each begin at different easterly co-ordinates. Corresponding maps in Figs 5A–C and 7A–C are identical in size.

vector in measurement bin *k* and *n* is the number of measurement bins. For clarity, Plate 1A–C & Fig. 5A–C show only velocity profile data at a cross-stream spacing of 50–100 m, which represents less than 20% of the total data set where measurements are available at a spacing of approximately 10 m.

**Bar evolution during the study period**

Mid-channel bar development was initiated by a widening of the main channel during the monsoon floods of 1994 and sediment deposition in the channel centre, forming a central bar core that grew by downstream

accretion (Delft Hydraulics *et al.*, 1996a). At close to the maximum annual flood discharge (Fig. 3), flow measurements in August 1994 (Plate 1A & Fig. 5A) document flow around a symmetrical, narrow mid-channel bar (~ 0.5 km wide, ~ 1.5 km long and a maximum height of +2.9 SLW, Fig. 4). Following the initial bar growth, bar widening occurred during the recession limb of the monsoon flood hydrograph. Flow measurements were obtained in March 1995 (Plate 1B & Fig. 5B), after lateral accretion on the bar margins created a broad bar-top platform (~ 1 km wide, Fig. 4) with two bar-tail limbs extending downstream from the bar nucleus, and incision within both anabranch channels produced bed erosion of

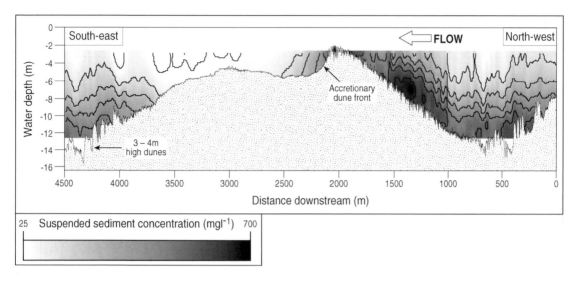

**Fig. 6.** Longitudinal survey line over the centre of the bar top during August 1994 showing the variation of suspended-sediment concentration through the flow depth. SLW is at −5.83 m depth on the vertical scale. Note the suspended-sediment concentration measurements begin at 2.66 m and are interpolated down to the deepest flow measurement. Contour lines are shown at an interval of 150 mg l$^{-1}$.

up to 7 m (Plate 1B). Rising discharge at the beginning of the 1995–1996 monsoon season led to erosion of the bar top and infilling of the eastern anabranch. These channel changes, however, were reversed at the peak of the monsoon flood in July 1995, when the bar narrowed slightly and bar-tail aggradation led to an extension in bar length by ~ 2 km (Fig. 4). The peak flows in 1995 represented the 20-yr flood, and flow stage in September 1995 was similar to that measured in August 1994 (Fig. 3). Significant channel change occurred during this late high flow (Plate 1C & Fig. 5C), with reactivation of the eastern anabranch supplying sediment to promote further downstream bar-tail growth (~ 800 m) and lateral accretion into the western anabranch (~ 200 m). A full description of bar evolution and morphology is presented in Delft Hydraulics *et al.* (1996a). The results below focus on the key attributes of flow structure and suspended-sediment distribution at these three stages of bar growth.

## Flow structure and spatial distribution of sand-grade suspended sediment

*August 1994*

The bed topography (Plate 1A) shows deep bed scour (~ 10 m below SLW) upstream and downstream from the study bar, suggesting that flow is concentrated at these nodes. Flow clearly diverges over the bar head (Plate 1A & Fig. 5A) and converges around the bar tail. The flow divergence is not initiated until close to the bar head, with near-bed flow divergence beginning slightly further upstream than divergence at the water surface. At the bar head (Plate 1A), deflection of flow towards the western anabranch occurs as the flow shallows on to the bar top (14–2 m water depth, see Fig. 6), whereas significant deflection of flow away from the bar in the eastern anabranch does not occur until flow is parallel with the broadest section of the bar top. Both near-bed and near-surface velocities are generally lower in the western anabranch than those in the eastern anabranch (Plate 1A), even though approximately 60% of the total discharge flows through the western anabranch (Table 1). As the bar becomes narrower downstream, vectors in the western anabranch begin to be directed towards the bar and show convergence with flow from the eastern anabranch at the end of the bar tail. In contrast, flow in the eastern anabranch does not converge towards the west until beyond the end of the bar tail, where flow becomes concentrated and accelerates towards the deepest part of the confluence scour.

The flow vectors in Plate 1A reveal no consistent pattern of deviation between near-bed and near-surface flow vectors and therefore suggest that large-scale helical secondary flow cells are not present across the entire channel width. For clarity, Plate 1A shows only velocity profile data at a spacing of 50–100 m, but even if all profiles are

**Table 1.** Source of flow and morphological survey data. Survey times are shown on the discharge record in Fig. 3. The change in water level during each survey is small (typically ±0.05 m in 24 h). Standard low water (SLW) at Bahadurabad is 12.03 m (i.e. mean flow in August 1994 was 5.83 m above SLW). For more details, see text.

| Survey period | Water level (m) | Number of survey lines | West anabranch discharge (m³ s⁻¹) | East anabranch discharge (m³ s⁻¹) |
|---|---|---|---|---|
| 10–14 August 1994 | 17.86 | 19 | 9 150 | 6 650 |
| 7–8, 13 March 1995 | 13.18 | 21 | 1 610 | 1 210 |
| 12–16 September 1995 | 17.86 | 27 | 5 000 | 14 000 |

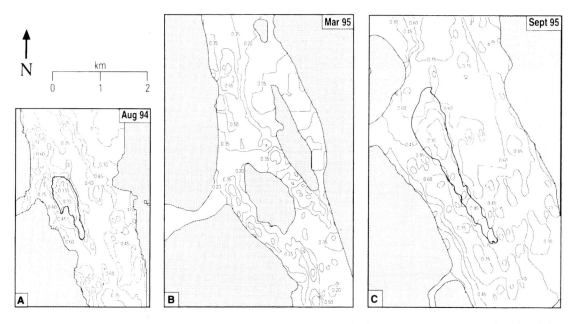

**Fig. 7A–C.** Contour maps of near-bed flow acceleration for the study area during the three survey periods. Contours show the relative height above the bed of the maximum flow velocity. The interval of contour lines is the same for each diagram. Banklines enclosing the stippled area and study bar outlines (shown by thick black lines) are defined by 0 m SLW contour (see Fig. 4). The northerly co-ordinates are aligned for all three maps, but they each begin at different easterly co-ordinates. Corresponding maps in Figs 5A–C and 7A–C are identical in size.

included (at approximately 10 m spacing) there is no *consistent* deviation in near-bed and near-surface flow direction. There is, however, an oscillation in the direction of both near-bed and near-surface flow vectors in the western anabranch and to a lesser extent in the eastern anabranch. At the bar head, flow vectors show only a small difference between near-bed and near-surface velocities, which suggests convective flow acceleration owing to the reduction in flow depth at the bar head (cf. Whiting & Dietrich, 1991; Whiting 1997). The spatial distribution of the relative height above the bed of the maximum flow velocity (Fig. 7A) demonstrates the marked change in the velocity distribution as flow shallows on to the bar head. The reduction in height of the

maximum velocity indicates an acceleration of near-bed flow relative to the near-surface water, which may be a result of either near-bed flow acceleration or lateral flow divergence at the water surface. There is no significant spatial redistribution of flow momentum in the eastern anabranch channel but, as flow shallows and converges towards the bar tail in the western anabranch, there is a reduction in the relative height of the maximum velocity.

The concentrations of suspended sand are similar in both anabranch channels and show little downstream variation, ranging from ~ 200 mg l⁻¹ near the channel margins to a maximum of ~ 350 mg l⁻¹ at the thalweg. As flow shallows on to the bar head there is a rapid increase in suspended-sediment concentration up to 575 mg l⁻¹

(Fig. 5A) while on the bar lee-side and bar top there is a zone of very low suspended-sediment concentration (< 100 mg l$^{-1}$). This shadow of low sand-grade suspended-sediment concentrations is as broad as the bar itself and extends over 2 km downstream from the bar head, beyond the bar tail and into the confluence scour zone. A longitudinal survey of the bed profile (Fig. 6) indicates that the peak in suspended-sediment concentration coincides both with dunes migrating up the bar stoss-side and a 3-m high accretionary dune front at the bar head. The accretionary front is generated as dunes, which have migrated on to the bar head and stall as flow depth decreases on to the bar. Maximum dune height reduces from 4 m in the thalweg to 0.5 m on the bar top. Downstream from the bar head, bedforms reduce in size to 0.2–0.3 m or less, which is mirrored by the substantial decrease in flow velocities and suspended-sediment concentration over the bar platform (Plate 1A, Figs 5A & 6). These data reveal the significant role of dunes in generating sediment suspension, which has been recorded in many other studies (see Jackson, 1976; Kostaschuk & Villard, 1996).

*March 1995*

During the falling stage and low-flow period of 1994–1995 (Plate 1B) the position of the scour holes, both upstream from the bar head and downstream from the bar tail, remained static and a similar depth to that measured in August 1994 (Plate 1A). Bed scour in the anabranch channels, however, accompanied a widening of the braid bar, with maximum scour of the anabranch channels adjacent to the broadest section of the bar (Plate 1B) and greatest erosion in the eastern anabranch. These changes in bed morphology were associated with only minor variations in the distribution of discharge between the western and eastern anabranches (Table 1). Although limited flow data are available around the bar head at low flow (Plate 1B & Fig. 5B), the flow vectors demonstrate a marked flow convergence around the bar tail. Deflection of the flow vectors away from the anabranch channel centre-line is greater than during the higher flow stage in August 1994 (Plate 1A & Fig. 5A), particularly in the western anabranch channel. Furthermore, Table 2 illustrates that there is greater deviation between near-surface and near-bed flow vectors compared with the flow pattern in August 1994. Despite the increased deflection of flow around the bar, however, there is still no evidence of a consistent direction for the deviation between near-surface and near-bed flow directions (Plate 1A) that would indicate the presence of a channel-scale helicoidal flow pattern, as suggested in models of flow structure

**Table 2.** Difference between near-surface and near-bed flow vectors for each survey period. The mean deviation is calculated from the absolute difference between near-surface and near-bed flow vectors and therefore does not indicate the direction of the deviation. The range given for the mean deviation is one standard deviation of the distribution and the maximum deviation is calculated from the 99th percentile of each distribution.

| Survey period | Mean deviation | Maximum deviation |
|---|---|---|
| 10–14 August 1994 | 7.3 ± 7.7° | 27.7° |
| 7–8, 13 March 1995 | 11.7 ± 8.9° | 40.6° |
| 12–16 September 1995 | 6.3 ± 5.6° | 25.2° |

around mid-channel bars (Ashworth *et al.*, 1992b; Bridge, 1993; Richardson *et al.*, 1996).

The spatial distribution of the height above the channel bed of the maximum flow velocity illustrates pronounced acceleration of near-bed relative to near-surface velocities as flow shallows on to the bar head (Fig. 7B). In contrast with the pattern in August 1994 (Fig. 7A), however, the maximum velocity remains closer to the bed throughout the study reach, which suggests that convective flow acceleration is more important than at the higher flow stage (August 1994). This difference suggests that relative changes in topographic relief enhance the acceleration of near-bed flow at low stage, as also suggested by Whiting (1997). In contrast to Whiting (1997), however, the magnitude of convective acceleration is greater at the higher flow stage, which may be attributed to flow shallowing rapidly on to the bar top, whereas at low flow, the bar is emergent and this rapid change in flow depth does not occur.

The range of sand-grade suspended-sediment concentration measured during March 1995 (Fig. 5B) is approximately one-third of that recorded in August 1994 (Fig. 5A), although it is noticeable that the maximum suspended-sediment concentrations are found at the bar head at both flow stages (~ 175 mg l$^{-1}$ in March 1995). Suspended-sediment concentration in March 1995 is relatively high close to the bar margin in both anabranch channels (~ 100 mg l$^{-1}$), but decreases to almost zero at the west bank of the western anabranch. In addition, there is a reduction in the suspended-sediment load of the western anabranch compared with the eastern anabranch, which probably corresponds to the increase in bed scour of the eastern anabranch. At the bar tail, a core of high suspended-sediment concentration is carried round from the anabranch channels, but downstream from the bar tail, flow with a low suspended-sediment concentration moves from the west bank towards the channel centre-line, following the converging eastern anabranch flow.

Downstream from the bar, a marked increase in suspended-sediment concentration occurs just downstream from the scour hole (Plate 1B) associated with the bar-tail flow convergence and possibly associated with the presence of larger dunes in these deeper channels.

*September 1995*

Flow data from September 1995 (Plate 1C & Fig. 6C) clearly illustrate a strong divergence of flow towards the western anabranch at the bar head. Similar to the situation in August 1994, however, there is less deflection of flow into the eastern anabranch, because the thalweg upstream from the mid-channel bar is aligned almost parallel with the eastern anabranch. Immediately downstream from the widest part of the braid bar there is marked deceleration of flow close to the bar margin, particularly in the western anabranch (Plate 1C). Downstream from the bar head there is limited deflection of flow vectors from the alignment of both anabranch thalwegs, although some flow convergence occurs around the bar tail, particularly in the eastern anabranch. In comparison with August 1994 there is an increase in the width of the eastern anabranch, which now conveys approximately 75% of the total discharge (Table 1).

Except at the entrance to the western anabranch, the majority of near-bed and near-surface flow vectors are aligned parallel to the thalweg (Plate 1C), with a negligible difference between the near-bed and near-surface vectors (Table 2) in response to the weak channel curvature effects on the flow. At the bar head, limited near-bed flow acceleration occurs upstream of the flat bar-top platform (Fig. 7C). This contrasts with the situation in the previous two surveys, where convective acceleration occurred as the flow shallowed at the bar head. Additionally, the minimum depth of the maximum flow velocity is only 30% of the total flow depth during the September 1995 survey (Fig. 7C), compared with 15 and 20% for the August 1994 and March 1995 surveys, respectively. It is also noticeable that the western anabranch shows a marked increase in the height of the maximum velocity, which may be associated with flow from the eastern anabranch crossing the bar top into the western anabranch (Plate 1C & Fig. 5C).

As shown in the previous two survey periods (Fig. 5A & B) there is an increase in sand-grade suspended-sediment concentration at the bar head in September 1995 (Fig. 5C). However, the range of suspended-sediment concentration is less than that measured in August 1994 (Fig. 5A), although the flow stage for both survey periods is almost identical (Fig. 3 & Table 1). Sediment concentrations in the western anabranch are very low (< 50 mg l$^{-1}$), suggesting that this anabranch is being starved of sediment, as also is revealed by the bed-morphology contours, which show a blocking of the entrance to the western anabranch (Fig. 4C) and a reduction in the proportion of discharge flowing through this channel (Table 1). In marked contrast to the previous two surveys, suspended sediment is transported over the bar top as flow begins to move from east to west over the bar top, with no zone of low suspended-sediment concentration in the bar lee.

## DISCUSSION

The results presented here document, for the first time, flow structure and the spatial distribution of sand-grade suspended sediment at several stages of braid-bar growth, at the kilometre scale, in a river with a high width-to-depth ratio. Within a period of several months, at the peak of the 1994 flood hydrograph, a mid-channel braid bar 4 km long, 1 km wide and up to 15 m high was formed downstream from a flow convergence. To understand and predict why these bars evolve, it is necessary to elucidate the processes of bar initiation, maintenance and growth.

### Bar evolution and bed sediment transport

Planform changes upstream from the study reach revealed significant bank erosion during early 1994, which yielded large quantities of sediment to the channel upstream of the study reach (Delft Hydraulics *et al.*, 1996a). Deposition of this sediment, which has been mobilized both as sand dunes and high concentrations of suspended load, may be initiated by a reduction in sediment-transport capacity downstream of the constriction. It has been reasoned above that the initial bar accumulation may have been generated by the deceleration and stacking of dunes within the study reach. Ground-penetrating radar studies of the bar subsurface (Delft Hydraulics *et al.*, 1996a) confirm that the base of the bar is dominated by large (up to 6 m high) dune foresets, which attest to dune stacking. Once the initial bar core has been generated, however, it is then necessary to consider which processes are responsible for continued bar growth.

Initial development of the bar in August 1994 (Fig. 4 & Plate 1A) suggests that bedforms have a major role in controlling bar aggradation through the migration and amalgamation ('stacking') of dunes at the bar head (Fig. 6). The formation of the accretionary dune front (Fig. 6), which separates dunefields upstream from a relatively flat or rippled bar top downstream, may be explained by the temporal fluctuations in discharge at the peak of the flood hydrograph (Fig. 3), which produce a variation in water height over the bar top of 0.5–2.0 m. These water-depth fluctuations may be sufficient to permit dunes to migrate on to the bar top, where they then stall as a result of declining flow velocities or shear stress.

The presence of high suspended-sediment concentrations around the bar margins during March 1995 (Fig. 5B), occurs in conjunction with the reduction in flow velocities in the same region, and a decrease in suspended-sediment concentration towards the bar tail, suggesting that deposition occurs along the bar margin. This pattern of deposition may explain the substantial increase in bar width at this time (Fig. 4). This mode of bar growth through lateral accretion has been documented previously by Bristow (1987).

The reduction in suspended-sediment concentration noted during September 1995 (Fig. 5B) is explained by the reduction in sediment supply to the study reach, which was probably higher in August 1994 and March 1995 owing to 600 m of bank erosion over a strip of east bank extending from 5 to 10 km upstream of the study bar (Delft Hydraulics *et al.*, 1996a). In marked contrast to the situation in August 1994 when the bar top was in a shadow of very low suspended-sediment concentrations (Figs 5A & 6), there is significant lateral transport of sediment across the bar top during September 1995 (Fig. 5C). This transformation in the pattern of sediment transport may be explained by the lateral convergence of flow from the western to the eastern anabranch combined with the reduced role of the accretionary dune front at the bar head. In combination, these mechanisms reflect the diminishing importance of the western anabranch and the translation of the bar towards a diagonal bar form similar to that described by Lewin (1976). This process of channel change is similar to the scour-pool enlargement, oblique riffle growth and extension and general development of a sinuous thalweg described for smaller gravel-bed braided rivers by Ferguson & Werritty (1983, fig. 3, p. 185).

## Controls on the development of channel-scale secondary flows

Much previous work on the three-dimensional flow structure in open channels has been undertaken in single-thread meandering channels (e.g. Hey & Thorne, 1975; Bridge & Jarvis, 1976; Bathurst *et al.*, 1979; Dietrich & Smith, 1983) and has shown that the flow field is predominantly controlled by the resultant forces arising from channel-planform curvature and lateral gradients in bed topography (e.g. Dietrich, 1987). In general, channel

curvature induces radial forces and lateral variations in water-surface elevation, which skew the downstream velocity and boundary shear stress fields, leading to the development of a helicoidal secondary flow pattern. In addition, significant bed topography (usually an inner point bar) causes flow shoaling and convective accelerations across the channel as the flow is steered around the point bar rather than directly over it (Ikeda, 1984; Dietrich, 1987). At all three flow stages in this study, the flow structure around the mid-channel bar indicates no evidence for coherent, channel-scale, secondary flow cells but instead a much simpler large-scale pattern of flow divergence over the bar head, convergence at the bar tail and usually thalweg-parallel flow in the anabranches (Plate 1A–C). The absence of channel-scale helical flow cells in the anabranches contradicts the findings of Richardson *et al.* (1996) and is different from that found for flow within meander bends. There are a number of important issues, however, that govern the presence and strength of secondary flow in braided rivers.

First, laboratory and theoretical work have shown that the magnitude of skew-induced cross-channel flow is dependent on the channel width-to-depth ratio (Dietrich, 1987; Yalin, 1992), with deep, narrow channels dominated by skew-induced helicoidal flow and wide, shallow channels having topographically forced flow with convective acceleration. The average width-to-depth ratio for both anabranches in the Jamuna study reach was usually in excess of 100. Yalin (1992) suggests that it is difficult to envisage the existence of any cross-channel circulation of flow (i.e. helicoidal secondary flow) in channels with a width-to-depth ratio greater than 100, although convective accelerations will be present in any geometry of a sinuous channel (Yalin, 1992, pp. 192–193). Dietrich (1987) also supports this view and suggests that if a large, relatively flat bar-top surface is present (as in the Jamuna study reach), then the cross-stream pressure gradient will be weak on this surface, possibly leading to topographically forced, radially outward flow over the bar top into the anabranches.

A second, related issue in explaining the lack of channel-scale coherent helical flow is the relatively small planform curvature of the two anabranches in the Jamuna study reach. Mid-channel bar morphology and anabranch bed topography (i.e. scour pools and shallows) are strongly influenced by the ratio of the radius of curvature to channel width (Friedkin, 1945), with the magnitude of curvature-induced secondary circulation being suppressed in straighter channels. The two anabranches in the Jamuna study reach at all three stages of bar development are relatively straight (particularly in September

1995, see Plate 1C), reducing the opportunity for cross-channel skewing of flow. The combination of straight anabranches with a large width-to-depth ratio and a broad, flat, mid-channel bar top, minimizes the opportunity for secondary flow generation and instead the flow is dominated by simple diverging and converging flow throughout the water depth, induced by topographically forced convective acceleration.

A third reason for the absence of channel-scale helical flow cells is illustrated by recent research concerning the interaction of overbank flows with flow in the adjacent meandering channel (e.g. Naish & Sellin, 1996; Willetts & Rameshwaran, 1996). This work has demonstrated the significant changes to the secondary flow within the meandering channel that may be generated by complex flow interactions in compound channels. Indeed, Naish & Sellin (1996) illustrate that overbank flow entering the inner bank of a curved channel may alter both the size and intensity of secondary circulation, and if the overbank flow is sufficiently deep, create an additional helical flow cell with an opposite sense of rotation. The implications of this work for the present mid-channel braid bar are important at high stages when flow over the bar top will interact with the near-surface flow in the opposite anabranch channel. Such over-bar flow was particularly important during September 1995 when a strong flow from the east and into the west anabranch may have disrupted the generation of simple, single cell, helicoidal flow in each of the anabranches.

A fourth factor influencing the generation and coherence of channel-scale secondary flows within open channels is the role of bedform roughness. Dunes present in the thalwegs and bar margins of the present study reach ranged from several decimetres to 4 m in height (Delft Hydraulics *et al.*, 1996a). Previous work has shown that dune roughness may influence the flow structure through:

1 generation of macroturbulence, which, especially during high flow, reaches the water surface as vigorous 'boils' (Coleman, 1969);

2 generation of transverse vorticity through bedform obliquity with respect to mean flow (Dietrich, 1987). Observations at high-flow stage in 1994 and 1995 revealed strong, discrete upwellings of flow at the water surface, which reached several tens of metres in diameter with a mean periodicity of 25 s and vertical velocities up to 0.3 m s$^{-1}$ (Delft Hydraulics *et al.*, 1996a). This macroturbulence originates as Kelvin–Helmholtz instabilities on the lee side of steep dunes (lee-side angles greater than ~ 12°) with flow separation (Müller & Gyr, 1986; Bennett & Best, 1995), or in zones of flow expansion over low-angle dunes, which probably do not possess permanent flow-separation zones (e.g. Kostaschuk & Villard,

1996). Such frequent dune-related macroturbulence may counteract inward secondary currents in some parts of the bend (Dietrich, 1987) by modifying both local velocity and Reynolds-stress distributions (e.g. Bennett & Best, 1995). Dietrich & Smith (1983) and Dietrich (1987) have demonstrated that flow-separation zones created by three-dimensional dunes alter near-bed flow patterns and sediment-transport paths in sand-bed meanders. Inwardly moving troughwise, near-bed secondary flows, generated in the low-pressure separation zone, situated in the lee of oblique dune crests may temporarily overcome the topographically induced outward flows that may occur over the bar top (Dietrich, 1987). Hence, the complexity of the flow structure in the anabranch thalwegs and bar margins will be increased where dune bedforms are prevalent and dunes may be expected to either mask, or possibly deform, any larger channel-scale helical flows, particularly near the channel bed.

## SUMMARY

This paper presents a detailed study of the evolution of a kilometre-scale sand bar in the Jamuna River, Bangladesh, using simultaneous measurements of flow, suspended-sediment transport and bed topography. Three key stages in the development of the bar under study are documented. Initially a symmetrical mid-channel bar, 4 km long, 1 km wide and up to 15 m high, developed downstream of a flow convergence. Following initial bar growth, bar widening occurred, together with incision of both anabranch channels. In the final stages, the bar narrowed and 1.5 km of bar-tail extension produced an asymmetrical bar and channel morphology. High-resolution acoustic Doppler current profiler (ADCP) data illustrate the salient characteristics of both flow structure and the spatial distribution of sand-grade suspended-sediment transport during this bar evolution. The principal conclusions are:

1 there is no evidence for channel-scale, coherent helical flow cells in either anabranch channel around the braid bar or over the bar top at any flow stage;

2 topographically forced convective accelerations, generated by the combination of straight anabranches, large channel width-to-depth ratios and a broad, flat, mid-channel bar top, minimize the opportunity for the generation of secondary flows – instead, flow structure is dominated by a simpler divergence of flow over the bar head, flow convergence at the bar tail and flow that usually is parallel to the thalwegs in each distributary;

3 the majority of the bar is composed of large sand dunes, which are linked to high concentrations of suspended

sand-grade sediment on the bar stoss-side – such large-scale bedform roughness may alter near-bed secondary flow patterns and cause local disruption of any coherent channel-scale helical-flow structure;

4 transport of suspended sand-grade sediment, downstream from the bar head, changes progressively during the development of the bar, with initially, a shadow of low suspended-sediment concentration located in the lee of a series of stacked dunes on the bar head – bar widening is associated with high suspended-sediment concentration adjacent to the bar margin and bar extension accompanies increased suspended-sediment transport over the bar top.

These results suggest that existing models for mid-channel braid-bar creation, growth and stability derived from studies of smaller rivers, which rely on the presence of channel-scale, helical flow cells, require substantial revision before they can be applied to the dynamics of large sand braid-bars.

## ACKNOWLEDGEMENTS

We gratefully acknowledge the logistic and financial support of Delft Hydraulics and the Danish Hydraulics Institute as main contractors of the River Survey Project (Flood Action Plan 24), the Flood Plan Co-ordinating Organisation (FPCO) as client and the European Union as funding agency for the River Survey Project (FAP24). In particular, we thank them for permission to publish this collaborative data set. J.J. Peters offered valuable advice on the FAP24 research programme. Funding was also provided by NERC grant GR9/02034 and a NERC studentship to JR (GT4/94/176/G). SJM is grateful for the award of a Research Fellowship from Fitzwilliam College, Cambridge, while undertaking some of the data analysis associated with this paper. Presentation of this paper at the International Conference on Fluvial Sedimentology in Cape Town was made possible through financial assistance from Fitzwilliam College and the Department of Geography, University of Cambridge to SJM and the Royal Society to PJA and JLB. The manuscript was greatly improved by the thorough and constructive review comments of Mike Church and Peter Whiting.

## REFERENCES

ASHMORE, P.E. (1982) Laboratory modelling of gravel braided stream morphology. *Earth Surf. Proc. Landf.*, **7**, 201–225.

ASHMORE, P.E. (1991) How do gravel bed rivers braid? *Can. J. Earth Sci.*, **28**, 326–341.

ASHWORTH, P.J. (1996) Mid-channel bar growth and its relationship to local flow strength and direction. *Earth Surf. Process. Landf.*, **21**, 103–123.

ASHWORTH, P.J., FERGUSON, R.I., ASHMORE, P.E., PAOLA, C., POWELL, D.M. & PRESTEGAARD, K.L. (1992a) Measurements in a braided river chute and lobe: 2. Sorting of bedload during entrainment, transport and deposition. *Water Resour. Res.*, **28**, 1887–1896.

ASHWORTH, P.J., FERGUSON, R.I. & POWELL, D.M. (1992b) Bedload transport and sorting in braided channels. In: *Dynamics of Gravel Bed Rivers* (Eds Billi, P., Hey, R.D., Thorne, C.R. and Tacconi, P.), pp. 497–513. John Wiley & Sons, Chichester.

BATHURST, J.C., THORNE, C.R. & HEY, R.D. (1979) Secondary flow and shear stress at river bends. *Proc. Am. Soc. civ. Eng., J. Hydraul. Div.*, **105**, 1277–1295.

BENNETT, S.J. & BEST, J.L. (1995) Mean flow and turbulence structure over fixed, two-dimensional dunes: implications for sediment transport and bedform stability. *Sedimentology*, **42**, 491–514.

BEST, J.L. & ASHWORTH, P.J. (1997) Scour in large braided rivers and the recognition of sequence stratigraphic boundaries. *Nature*, **387**, 275–277.

BLUCK, B.J. (1979) Structure of coarse grained braided stream alluvium. *Trans R. Soc. Edinburgh*, **70**, 181–221.

BRIDGE, J.S. (1993) The interaction between channel geometry, water flow, sediment transport and deposition in braided rivers. In: *Braided Rivers* (Ed. Best, J.L. & Bristow, C.S.), Spec. Publ. geol. Soc. London, No. 75, pp. 13–71. Geological Society of London, Bath.

BRIDGE, J.S. & GABEL, S.L. (1992) Flow and sediment dynamics in a low sinuosity braided river: Calamus River, Nebraska Sandhills. *Sedimentology*, **39**, 125–142.

BRIDGE, J.S. & JARVIS, J. (1976) Flow and sedimentary processes in the meandering River South Esk, Glen Clova, Scotland. *Earth Surf. Process.*, **1**, 303–336.

BRIDGE, J.S., SMITH, N.D., TRENT, F., GABEL, S.L. & BERNSTEIN, P. (1986) Sedimentology and morphology of a low sinuosity river; Calamus River, Nebraska Sand Hills. *Sedimentology*, **33**, 851–870.

BRISTOW, C.S. (1987) Brahmaputra River: channel migration and deposition. In: *Recent Developments in Fluvial Sedimentology* (Eds Ethridge, F.G., Flores, R.M. & Harvey, M.D.), Spec. Publ. Soc. econ. Paleont. Miner., Tulsa, **39**, 63–74.

BRISTOW, C.S. (1993) Sedimentary structures exposed in bar tops in the Brahmaputra River, Bangladesh. In: *Braided Rivers* (Ed. Best, J.L. & Bristow, C.S.), Spec. Publ. geol. Soc. London, No. 75, pp. 277–289. Geological Society of London, Bath.

CANT, D.J. & WALKER, R.G. (1978) Fluvial Processes and facies sequences in the sandy braided South Saskatchewan River, Canada. *Sedimentology*, **25**, 624–648.

CARSON, M.A. & GRIFFITHS, G.A. (1987) Bedload transport in gravel channels. *J. Hydrol. (N. Z.)*, **26**, 1–151.

CHURCH, M. & JONES, D. (1982) Channel bars in gravel-bed rivers. In: *Gravel-bed Rivers* (Eds Hey, R.D., Bathurst, J.C. & Thorne, C.R.), pp. 291–324. John Wiley & Sons, Chichester.

COLEMAN, J.M. (1969) Brahmaputra River: channel processes and sedimentation. *Sediment. Geol.*, **3**, 129–239.

DELFT HYDRAULICS & DHI (1996a) *Floodplain Levels and Bankfull Discharge*. FAP24 River Survey Project, Special Study Report 6 (prepared for FPCO), Dhaka, Bangladesh, 20 pp.

DELFT HYDRAULICS & DHI (1996b) *Physical Properties of River Sediments*. FAP24 River Survey Project, Special Study Report 14 (prepared for FRCO), Dhaka, Bangladesh, 39 pp.

DELFT HYDRAULICS, DHI & UNIVERSITY OF LEEDS (1996a) *Bedform and Bar Dynamics in the Main Rivers of Bangladesh*. FAP24 River Survey Project, Special Study Report 9 (prepared for FPCO), Dhaka, Bangladesh, 107 pp.

DELFT HYDRAULICS, DHI & UNIVERSITY OF NOTTINGHAM (1996b) *Study of Secondary Currents and Morphological Evolution in a Bifurcated Channel*. FAP24 River Survey Project, Special Study Report 16 (prepared for FRCO), Dhaka, Bangladesh, 107 pp.

DIETRICH, W.E. (1987) Mechanics of flow and sediment transport in river bends. In: *River Channels: Environment and Process* (Ed. Richards, K.S.), pp. 179–227. Special Publication Series No. 18, The Institute of British Geographers.

DIETRICH, W.E. & SMITH, J.D. (1983) Influence of the point bar on flow through curved channels. *Water Resour. Res.*, **19**, 1173–1192.

FERGUSON, R.I. (1993) Understanding braiding processes in gravel-bed rivers: progress and unsolved problems. In: *Braided Rivers* (Ed. Best, J.L. & Bristow, C.S.), Spec. Publ. geol. Soc. London, No. 75, pp. 73–87. Geological Society of London, Bath.

FERGUSON, R.I. & WERRITTY, A. (1983) Bar development and channel change in the gravelly River Feshie, Scotland. In: *Modern and Ancient Fluvial Systems* (Eds Collinson, J.D. & Lewin, J.), Spec. Publs int. Ass. Sediment., No. 6, pp. 181–193. Blackwell Scientific Publications, Oxford.

FERGUSON, R.I., ASHMORE, P.E., ASHWORTH, P.J., PAOLA, C. & PRESTEGAARD, K.L. (1992) Measurements in a braided chute and lobe 1. Flow pattern, sediment transport and channel change. *Water Resour. Res.*, **28**, 1877–1886.

FRIEDKIN, J.F. (1945) *A Laboratory Study of the Meandering of Alluvial Rivers*. U.S. Waterways Experimental Station, Vicksburg, MS, 40 pp.

FUJITA, Y. (1989) Bar and channel formation in braided streams. In: *River Meandering* (Eds Ikeda, S. & Parker, G.). *Water Resour. Monogr., Am. geophys. Union*, **12**, 417–462.

GOFF, J.R. & ASHMORE, P.E. (1994) Gravel transport and morphological change in braided Sunwapta River, Alberta, Canada. *Earth Surf. Process. Landf.*, **19**, 195–212.

HEIN, F.J. & WALKER, R.G. (1977) Bar evolution and development of stratification in the gravelly braided Kicking Horse River, British Columbia. *Can. J. Earth Sci.*, **14**, 562–570.

HEY, R.D. & THORNE, C.R. (1975) Secondary flows in river channels. *Area*, **7**, 191–195.

HOSSAIN, M.M. (1993) Economic effects of riverbank erosion: some evidence from Bangladesh. *Disasters*, **17**, 25–32.

IKEDA, S. (1984) Flow and bed topography in channels with alternate bars. In: *River Meandering: Proceedings of the Conference Rivers '83* (Ed. Elliott, C.M.), pp. 733–746. American Society of Civil Engineers.

JACKSON, R.G. (1976) Sedimentological and fluid-dynamic implications of the turbulence bursting phenomenon in geophysical flow. *J. Fluid Mech.*, **77**, 531–560.

KLAASSEN, G.J. & MASSELINK, G. (1992) Planform changes of a braided river with fine sand as bed and bank material. In: *5th International Symposium on River Sedimentation*, Karlsruhe, Germany, 6–10 April, pp. 459–471.

KLAASSEN, G.J. & VERMEER, K. (1988) Confluence scour in large braided rivers with fine bed material. *Proceedings of the International Conference on Fluvial Hydraulics*, Supplementary Volume, Budapest, Kultúra, pp. 395–408.

KLAASSEN, G.J., MOSSELMAN, E. & BRÜHL, H. (1993) On the prediction of planform changes of braided sand-bed rivers. In: *Advances in Hydroscience and Engineering* (Ed. Wang, S.S.Y.), pp. 134–146. Proceedings of the International Conference on Hydroscience and Engineering, Washington, DC. University of Mississippi, Oxford, MS, 7–11 June.

KOSTASCHUK, R. & VILLARD, P. (1996) Turbulent sand suspension events: Fraser River, Canada. In: *Coherent Flow Structures in Open Channels* (Eds Ashworth, P.J., Bennett, S.J., Best, J.L. & McLelland, S.J.), pp. 305–319. John Wiley & Sons, Chichester.

LAPOINTE, M. (1993) Monitoring alluvial sand suspension by eddy correlation. *Earth Surf. Process. Landf.*, **18**, 157–175.

LANE, S.N., RICHARDS, K.S. & CHANDLER, J.H. (1995) Within-reach spatial patterns of process and channel adjustment. In: *River Geomorphology* (Ed. Hickin, E.J.), pp. 105–130. John Wiley & Sons, Chichester.

LARONNE, J.B. & DUNCAN, M.J. (1992) Bedload transport paths and gravel bar formation. In: *Dynamics of Gravel Bed Rivers* (Eds Billi, P., Hey, R.D., Thorne, C.R. & Tacconi, P.), pp. 177–205. John Wiley & Sons, Chichester.

LEOPOLD, L.B. & WOLMAN, M.G. (1957) River channel patterns – braided, meandering and straight. *U.S. geol. Surv. Prof. Pap.*, **282B**, 85.

LEWIN, J. (1976) Initiation of bedforms and meanders in coarse-grained sediment. *Geol. Soc. Am. Bull.*, **87**, 281–285.

McLELLAND, S.J., ASHWORTH, P.J. & BEST, J.L. (1994) *Velocity Correction Methods in Open Channels: Description, Critique and Recommendations*. Consultancy Report for FAP24, Dhaka, Bangladesh, 17 pp.

McLELLAND, S.J., ASHWORTH, P.J. & BEST, J.L. (1996) The origin and downstream development of coherent flow structures at channel junctions. In: *Coherent Flow Structures in Open Channels* (Eds Ashworth, P.J., Bennett, S.J., Best, J.L. & McLelland, S.J.), pp. 459–490. John Wiley & Sons, Chichester.

MOSSELMAN, E., HUISINK, M., KOOMEN, E. & SEIJMONSBERGEN, A.C. (1995) Morphological changes in a large braided sand-bed river. In: *River Geomorphology* (Ed. Hickin, E.J.), pp. 235–247. John Wiley & Sons, Chichester.

MÜLLER, A. & GYR, A. (1986) On the vortex formation in the mixing layer behind dunes. *J. Hydraulic Res.*, **24**, 359–375.

NAISH, C. & SELLIN, R.H.J. (1996) Flow structure in a large-scale model of a doubly meandering compound river channel. In: *Coherent Flow Structures in Open Channels* (Eds Ashworth, P.J., Bennett, S.J., Best, J.L. & McLelland, S.J.), pp. 631–654. John Wiley & Sons, Chichester.

PARKER, G. (1976) On the cause and characteristic scales of meandering and braiding in rivers. *J. Fluid Mech.*, **76**, 457–480.

PETERS, J.J. (1993) Morphological studies and data needs. *Proceedings International Workshop on Morphological Behaviour of Major Rivers in Bangladesh*, FAP24, Dhaka, Bangladesh, 12 pp.

RICHARDSON, W.R.R., THORNE, C.R. & MAHMOOD, S. (1996) Secondary currents and channel changes around a braid bar in the Brahmaputra River, Bangladesh. In: *Coherent Flow Structures in Open Channels* (Eds Ashworth, P.J., Bennett,

S.J., Best, J.L. & McLelland, S.J.), pp. 520–543. John Wiley & Sons, Chichester.

RUNDLE, A. (1985) Bar morphology and the formation of multiple channels: the Rakaia, New Zealand. *Z. Geomorphol. Suppl.*, **35**, 15–37.

SARKER, M.H. (1996) *Morphological processes in the Jamuna River*, Unpublished MSc Thesis, International Institute for Infrastructural Hydraulic and Environmental Engineering, Delft, The Netherlands, 175 pp.

SCHUMM, S.A. & WINKLEY, B.R. (1994) The character of large alluvial rivers. In: *The Variability of Large Alluvial Rivers* (Eds Schumm, S.A. & Winkley, B.R), pp. 1–13. American Society of Civil Engineers, New York.

SMITH, N.D. (1974) Sedimentology and bar formation in the upper Kicking Horse River, a braided outwash stream. *J. Geol.*, **82**, 205–223.

THORNE, C.R., RUSSELL, A.P.G. & ALAM, M.K. (1993) Planform channel evolution of the Brahmaputra River, Bangladesh. In: *Braided Rivers* (Eds Best, J.L. & Bristow, C.S.), Spec. Publ. geol. Soc. London, No. 75, pp. 257–276. Geological Society of London, Bath.

THORNE, P.D., VINCENT, C.E., HARDCASTLE, P.J., REHMAN, S. & PEARSON, N. (1991) Measuring suspended sediment concentration using acoustic backscattering devices. *Mar. Geol.*, **98**, 7–16.

WHITING, P.J. (1997) The effect of stage on flow and components of the local force balance. *Earth Surf. Process. Landf.*, **22**, 517–530.

WHITING, P.J. & DIETRICH, W.E. (1991) Convective accelerations and boundary shear stress over a channel bar. *Water Resour. Res.*, **27**, 783–796.

WILLETTS, B.B. & RAMESHWARAN, P. (1996) Meandering overbank flow structures. In: *Coherent Flow Structures in Open Channels* (Eds Ashworth, P.J., Bennett, S.J., Best, J.L. & McLelland, S.J.), pp. 609–629. John Wiley & Sons, Chichester.

YALIN, M.S. (1992) *River Mechanics*. Pergamon Press, Oxford, 219 pp.

# Modern Fluvial Environments

*Spec. Publs int. Ass. Sediment.* (1999) **28**, 61–69

# Effective discharge for overbank sedimentation on an embanked floodplain along the River Waal, The Netherlands

N. E. M. ASSELMAN*

*The Netherlands Centre for Geo-Ecological Research, Department of Physical Geography, Utrecht University, PO Box 80.115, 3508 TC Utrecht, The Netherlands*

## ABSTRACT

As little is known about the effectiveness of floods of different magnitude on floodplain sedimentation, the aim of this study was to determine the effective discharge for overbank sedimentation in The Netherlands. The study was carried out on a grass-covered, embanked, floodplain section along the River Waal, the main distributary of the River Rhine in The Netherlands. The floodplain section consists of two parts. One part is protected from low floods by a minor river dyke, whereas the other part lies adjacent to the main river channel. In the latter section, sediment accumulation was measured during a series of floods using sediment traps made of artificial grass. The measurements provide an indication of the importance of floods of different magnitude on the deposition of suspended sediment. A sedimentation model was applied to compute the long-term effective discharge for sedimentation on both floodplain sections.

The sedimentation measurements showed an increase in sediment accumulation with flood magnitude. The increase was stronger for sand than for silt- and clay-size material. Sedimentation did not increase proportionally with sediment transport rates, however, suggesting that trapping of fine-grained suspended sediment becomes less efficient at very high discharge. The model results confirm this hypothesis. On the low floodplain section, most sediment is deposited at a river discharge of about 5500 $m^3 s^{-1}$ or less, i.e. a discharge at which the floodplain is just inundated and which occurs, on average, about 1.3 times a year. On the floodplain section bordered by a minor river dyke, a slightly higher effective discharge for sedimentation was found (up to 7000 $m^3 s^{-1}$). This discharge level occurs about once every 2 yr. It is concluded, therefore, that the effective discharge for floodplain sedimentation in The Netherlands is a moderate discharge, during which the floodplain is just inundated. Such discharges occur relatively frequently.

## INTRODUCTION

Infrequent and extreme events, such as large floods, tsunamis and large landslides, have an important effect on the development of landforms. As stated by Wolman & Miller (1960), however, a more accurate picture of the overall effectiveness of various geomorphological processes should include not only the rare extreme events, but also events of moderate intensity that occur much more frequently. They illustrate the magnitude–frequency problem by wondering who would be most effective in cutting down a forest: a dwarf, a man or a giant? The dwarf makes little progress at a time, but works almost

24 h a day. The man is strong and works hard, but sometimes he takes a day off. The giant is extremely strong, but also very lazy.

Since 1960, long records of water discharge and suspended-sediment concentrations have become available for many large rivers throughout the world. From these records, the effective discharge for suspended-sediment transport, i.e. the discharge at which the largest part of the annual suspended-sediment load is transported, can be determined easily (e.g. Andrews, 1980; Webb & Walling, 1982, 1984; Asselman, 1997). Less is known about the effective discharge for overbank sedimentation. Reconstructions of past floodplain sedimentation rates often show the combined effect of floods of different magnitude, averaged over several years (Alexander &

* Present address: WL/Delft Hydraulics, PO Box 177, 2600 MH Delft, The Netherlands. (Email: nathalie.asselman@wldelft.nl)

Prior, 1971; Costa, 1975; Trimble, 1983; Brakenridge, 1984; Knox, 1989; Walling & Bradley, 1989; Walling *et al.*, 1992; He & Walling, 1996), whereas most studies on contemporary rates of floodplain sedimentation are carried out during one single event only (Kesel *et al.*, 1974; Mansikkaniemi, 1985; Gretener & Strömquist, 1987; Walling & Bradley, 1989; Heusch *et al.*, 1993; Brunet *et al.*, 1994; Asselman & Middelkoop, 1995; Gomez *et al.*, 1995). This complicates the assessment of the relative importance of events of different magnitude on average overbank sedimentation rates.

The aim of this study, therefore, was to determine the importance of different discharge stages on overbank sedimentation on an embanked floodplain section in The Netherlands. Sediment accumulation was measured during a series of floods in order to obtain an indication of the importance of floods of different magnitude on deposition of suspended sediment. A sedimentation model was applied to compute the long-term effective discharge for floodplain sedimentation.

## STUDY AREA

The study was carried out on an embanked grass-covered floodplain section along the River Waal, the main distributary of the River Rhine in The Netherlands (Fig. 1A). The mean Rhine discharge near Lobith (German–Dutch border) is 2200 m$^3$ s$^{-1}$, of which the Waal transports about two-thirds. Peak discharges usually range between 5000 and 10 000 m$^3$ s$^{-1}$. Suspended-sediment concentrations near Lobith vary between 30 mg l$^{-1}$ during periods of low discharge and 200 mg l$^{-1}$ during floods (Rijkswaterstaat, 1992).

The floodplain section studied consists of two parts: the Variksche Plaat (VP) and the Stiftsche Uiterwaard (SU). Sedimentation was measured during a series of floods at VP. At SU, sedimentation was measured during a single event only. Both floodplain sections have an irregular relief, each featuring a residual channel (Fig. 1B). They are grass-covered and land use is restricted to summer grazing. The VP directly borders the main channel, whereas the SU is protected from low

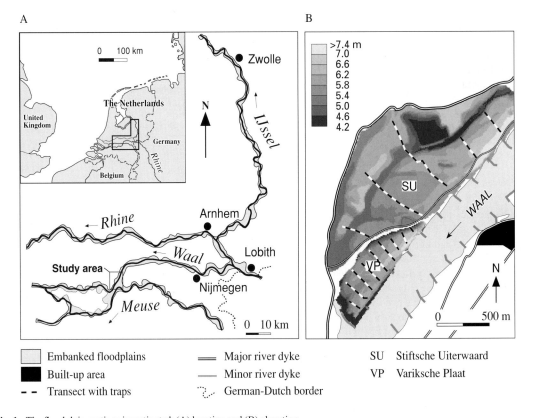

**Fig. 1.** The floodplain sections investigated: (A) location and (B) elevation.

floods by a minor river dyke. Inundation of the lower parts of VP starts at relatively low discharge (less than 4000 m³ s⁻¹) by water flowing through the downstream end of the abandoned channel. Current velocities are low and current directions are opposite to the mean primary flow direction in the main channel. At higher discharges, between about 5000 and 6500 m³ s⁻¹ in the River Rhine at Lobith, the entire VP is inundated, and water flows in a direction more or less parallel to the flow in the main channel. Inundation of the area behind the minor river dyke (SU) starts at discharges exceeding about 6000–6500 m³ s⁻¹.

## SEDIMENT ACCUMULATION DURING FLOODS OF DIFFERENT MAGNITUDE

### Data collection and preparation

Sediment traps of artificial grass were used to collect sediment deposited by overbank flow. The artificial grass mats measured $50 \times 50$ cm² and consisted of plastic blades 1.5 cm long fixed to a pliable (plastic) base. Similar traps have been used and tested by Mansikkaniemi (1985). A few days before inundation the sediment traps were placed on the floodplain sections. The traps were placed in a semi-regular grid consisting of several transects perpendicular to the main channel, and a few clusters. Sample spacing in the transects was about 50 m, with some adjustment for the local relief. The spacing between transects was about 120 m at VP, and 250 m at SU (Fig. 1). In total about 140 traps were used. After emergence, the traps were taken to the laboratory where the sediment was removed, collected, dried and weighed. Also, grain-size analyses were undertaken. The pipette method was used for silt and clay-size material, whereas sand was sieved. A more detailed description of the sedimentation measurements is given by Asselman & Middelkoop (1995).

Sediment transport was computed from the daily discharge and suspended-sediment concentration records provided by the Dutch Ministry of Public Works (Rijkswaterstaat) for the gauging station Lobith, near the German–Dutch border. River discharge is estimated from water-level measurements using a stage–discharge relationship (Rijkswaterstaat, 1994). Suspended-sediment concentrations are obtained from point samples taken near the water surface, which are assumed to represent the average concentration of fine-grained suspended sediment over the cross-section.

### Flood magnitude

At VP, the measurements were carried out during a series of four floods of different magnitude and duration. The results of two floods will be discussed in more detail. The discharge and suspended-sediment transport characteristics of the floods studied, measured by Rijkswaterstaat,

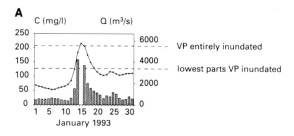

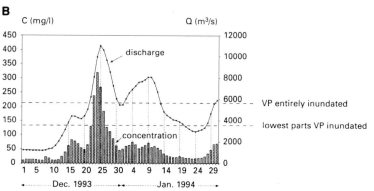

**Fig. 2.** Discharge ($Q$) and suspended sediment concentrations ($C$) measured by Rijkswaterstaat at Lobith: (A) during the flood of January 1993 and (B) during the flood of December 1993 to January 1994.

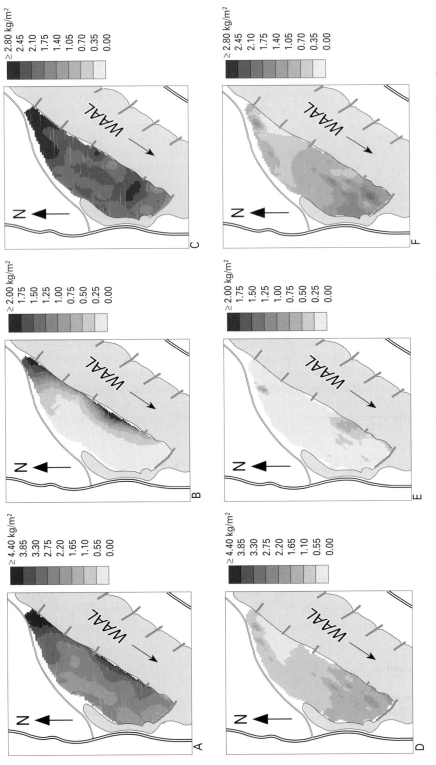

**Fig. 3.** Interpolated raster maps of sedimentation at the Variksche Plaat during floods of different magnitude. (A) December 1993, total; (B) December 1993, sand; (C) December 1993, silt and clay; (D) January 1993, total; (E) January 1993, sand; (F) January 1993, silt and clay.

**Table 1.** Sediment transport in the Rhine at Lobith and sedimentation at the Variksche Plaat during floods of different magnitude.

| | January 93 (low magnitude) | December 93–January 94 (high magnitude) | Ratio high/low |
|---|---|---|---|
| Sedimentation at the Variksche Plaat: | | | |
| total (t) | 303 | 781 | 2.6 |
| sand (t) | 28 | 131 | 4.7 |
| silt and clay (t) | 273 | 630 | 2.3 |
| Suspended sediment transport ($Q_s$) at Lobith: | | | |
| $Q_s$ during flood ($10^3$ t) | 300 | 1920 | 6.4 |

are given in Fig. 2. The flood of January 1993 was of low magnitude. Maximum discharge at Lobith was 5740 m$^3$ s$^{-1}$. Floods of this magnitude occur on average once a year. The flood of December 1993–January 1994 was of high magnitude (11 040 m$^3$ s$^{-1}$) and occurs about once every 40 yr.

## Results

To calculate the total amount of sediment deposited and to determine the spatial variability of sediment deposition, the sedimentation measurements were interpolated by means of block kriging (Burgess & Webster, 1980). To cope with minor trends in the data, a restricted search radius for interpolation was applied. Data sets, characterized by a considerable trend, were interpolated using universal block kriging (Burgess & Webster, 1980). Interpolation weights were derived from statistical analysis of the spatial dependence of the variable as indicated by the semivariograms. The variograms were fitted using a spherical model. To account for anisotropy of spatial variability, separate variograms were fitted for directions parallel and perpendicular to the main stream. The resulting raster maps are shown in Fig. 3.

After the high-magnitude flood, sediment accumulation varied between 1.5 and 5.5 kg m$^{-2}$ (Fig. 3A). Sedimentation was highest near the main channel and decreased in a downstream direction and in the direction of the river dyke. The pattern observed was the result mainly of variations in the deposition of sand, which decreased exponentially with distance from the channel (Fig. 3B). Deposition of silt and clay was distributed more evenly over the floodplain section (Fig. 3C).

During the low-magnitude flood, less sediment was deposited than during the high-magnitude flood. Maximum sedimentation equalled 2.4 kg m$^{-2}$ (Fig. 3D). Also, a different spatial distribution in sediment deposition was observed, with maximum sedimentation occurring in the lower downstream parts of the floodplain section, which were inundated for a longer period than the higher parts.

Similar patterns were found in the deposition of sand (Fig. 3E), and of clay- and silt-size material (Fig. 3F).

Differences in sediment transport and sediment accumulation between both floods are summarized in Table 1. Sedimentation at VP is computed using the raster maps shown in Fig. 3A–C. Hence, summation of the accumulation of sand, and of silt- and clay-size material do not automatically add up to the total sedimentation value given in the table. During the high-magnitude flood, 2.6 times as much sediment was deposited as during the low-magnitude flood. Deposition of sand was 4.7 times as much, whereas the deposition of silt and clay was only 2.3 times more. Thus it can be concluded that total sediment accumulation increases with flood magnitude, and that the increase is greater for sand than for silt and clay. The increase in sediment *deposition*, however, is less than would be expected from the increase in sediment *transport* (Table 1). During the high-magnitude flood of December 1993, about 1.9 10$^6$ t of suspended sediment were transported through the River Rhine at Lobith. During the low-magnitude flood of January 1993, the total sediment load was only one-sixth of this value. In other words, during the high-magnitude flood, 6.4 times as much sediment was transported as during the low-magnitude flood, whereas sediment deposition was only 2.6 times as much. This suggests that the efficiency with which the floodplain traps suspended sediment decreases at high discharge, probably because current velocities at high discharge exceed critical velocities for the deposition of fine suspended sediment.

## EFFECTIVE DISCHARGE FOR FLOODPLAIN SEDIMENTATION

### Effective discharge: a conceptual model

Long-term floodplain sedimentation rates are determined by sediment transport rates, flood frequency and by the efficiency with which the floodplain can trap suspended

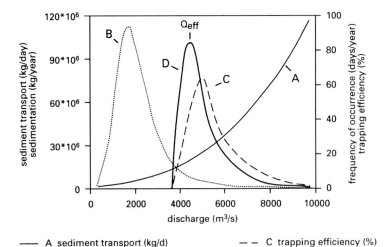

**Fig. 4.** Effective discharge for overbank sedimentation: a conceptual model (after Wolman & Miller, 1960). (A) Sediment rating curve that provides average sediment transport rates (kg day$^{-1}$) as a function of discharge. (B) Daily discharge frequency distribution (days yr$^{-1}$). Both curves are based on measurements carried out by Rijkswaterstaat at Lobith. (C) Percentage of the transported sediment that is deposited on the floodplain. The exact shape of this curve is unknown. Multiplication of curves A, B and C produces a floodplain-sedimentation curve (D) indicating the annual amount of sediment that is deposited on the floodplain at different discharge stages (kg yr$^{-1}$).

sediment (Fig. 4). Sediment transport rates (curve A) and the discharge frequency distributions (curve B) can be obtained from daily discharges and suspended-sediment concentrations measured by Rijkswaterstaat at Lobith. No information is available, however, on the shape of the sediment-trapping efficiency curve (curve C). It is expected that the ability of the floodplain to trap suspended sediment will increase rapidly with discharge at the onset of inundation. The sediment deposition measurements suggest that the trapping efficiency will decrease again at very high discharge stages. The exact shape of curve C, however, is unknown. When these three curves are combined (i.e. suspended-sediment transport × trapping efficiency × frequency of occurrence), a curve is obtained that shows the annual amount of sediment that is deposited on the floodplain sections studied. This relationship is given by curve D. The discharge at which this curve has its maximum value is called the 'effective discharge for floodplain sedimentation', representing the discharge at which most suspended sediment is deposited over a longer period of time. As the trapping efficiency of the floodplain depends on floodplain characteristics, the effective discharge will be different for floodplain sections that are enclosed by a minor river dyke, or that have a different topography.

Data on sediment transport rates (curve A) and the discharge frequency distribution (curve B) are available for the Rhine at Lobith. As the results of the sediment accumulation measurements carried out at VP show the overall effect of sedimentation during a range of discharge stages, they cannot be used to derive a detailed trapping efficiency curve for the VP and SU floodplain

sections. Therefore, the SEDIFLUX model developed by Middelkoop (1997) was applied to determine the sediment-trapping efficiency curves and to compute the effective discharge for sedimentation at the VP area and at two locations at SU.

**The SEDIFLUX model**

The SEDIFLUX model is a GIS based, two-dimensional sedimentation model that predicts the amount and pattern of floodplain sedimentation during a series of stationary discharge stages (Middelkoop, 1997). The model is based on the sediment balance within a raster cell, which is determined by the sediment deposition within a cell and horizontal sediment fluxes among adjacent cells. In this study, a rectangular raster was applied with grid cells of 50 m × 50 m. The model does not account for erosion of sediment deposited previously. Observed sedimentation patterns in combination with grain-size analyses, however, have shown that during high floods local erosion may occur (Asselman, 1997; Middelkoop & Asselman, 1998). In grass-covered areas no erosion of older deposits takes place. Sediment transport by turbulent diffusion is neglected as well. Previous modelling experiments with the WAQUA-DELWAQ model have shown that for the floodplain sections investigated, sediment transport by diffusion is much less important than sediment transport by convection (Middelkoop, 1997).

Input data for the SEDIFLUX model include raster maps of water level, flow velocities in the $x$ and $y$ directions, calculated with the hydrodynamic WAQUA model (Ubels, 1986), and floodplain elevation. Other parameters are the suspended-sediment concentration at the upstream model boundary, the effective settling velocity of the suspended sediment, $w_s$, and the critical shear stress for sediment deposition, $\tau_c$. In this study, the model is used to estimate long-term average sedimentation rates at given discharges. For this purpose, suspended-sediment concentrations are estimated using a sediment rating curve based on a 15–yr record of daily discharges and suspended-sediment concentrations measured near the German–Dutch border. Block effective values of $w_s$ and $\tau_c$ were obtained by model calibration (Middelkoop, 1997). In this study, $w_s$ was set equal to $7 \times 10^{-5}$ m s$^{-1}$ and $\tau_c$ to 2 N m$^{-2}$. As these values are block-effective values, they may differ significantly from point values determined from laboratory studies.

**Effective discharge: results**

The effective discharge for sedimentation can be determined from Fig. 5. Sediment accumulation was calculated with the SEDIFLUX model for seven stationary discharge stages ranging between 4400 and 9700 m$^3$ s$^{-1}$. Each stage is assumed to be representative for a discharge interval. Computed sedimentation rates were multiplied by the frequency of occurrence of each discharge interval, and summed to obtain the annual sedimentation. The resulting sedimentation curves are shown in Fig. 5. The discharge at which the curves peak is the effective

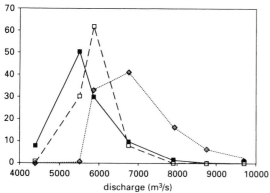

% of total annual sedimentation

- ■— VP - not bordered by minor dyke
- —□— ST - behind minor dyke
- ⋯◇⋯ ST - near major dyke

**Fig. 5.** Effective discharge for overbank sedimentation at the Variksche Plaat and the Stiftsche Uiterwaard computed with the SEDIFLUX model.

discharge for sedimentation for the specific area. The effective discharge for sedimentation at VP, which is not bordered by a minor dyke, is about 5500 m$^3$ s$^{-1}$. This discharge is of moderate magnitude and occurs relatively often (on average about 1.3 times per year). At higher discharge stages, less sediment is deposited because of a lower frequency of occurrence of these discharge intervals and because current velocities at high discharge may exceed critical velocities for the deposition of fine-grained suspended sediment. Behind the minor river dyke at SU a somewhat higher effective discharge is predicted (6000 m$^3$ s$^{-1}$). The highest effective discharge is predicted for locations close to the major river dyke. Here, discharges up to 7000 m$^3$ s$^{-1}$, which occur about once every 2 yr, are most effective in depositing suspended sediment. Apparently at these locations current velocities are sufficiently reduced even at high discharge for fine suspended sediment to settle.

**Discussion**

The results of this study indicate that low floodplain sections that are not protected from low floods by a minor river dyke have a lower effective discharge than floodplain sections that are enclosed by a minor river dyke. Also, areas located directly behind the lowest parts of the minor river dyke have a lower effective discharge than locations at greater distance from the main channel. The computed effective discharge for sedimentation varies between 5500 m$^3$ s$^{-1}$ at VP and 7000 m$^3$ s$^{-1}$ at some locations within SU. The effective discharge for sedimentation at the lowest parts of VP, however, may very well be overestimated. This partly may be the result of the small *number* of modelled stationary discharge stages. In this study, the model was run for seven discharge stages only. This yielded an effective discharge at VP of 5500 m$^3$ s$^{-1}$. If more discharge stages between 4400 and 5500 m$^3$ s$^{-1}$ had been modelled with the SEDIFLUX model, however, a lower effective discharge might have been found. Overestimation of the effective discharge for sedimentation at VP also can be related to the fact that the SEDI-FLUX model only allows for simulation of *stationary* discharges. According to the model computations, little sedimentation occurs at stationary discharges of about 4400 m$^3$ s$^{-1}$, when the lowest parts of the VP area are inundated by water flowing through the downstream part of the abandoned channel. Under stationary conditions, little sediment will enter these reaches. Spatial concentration gradients measured in the residual channel during the rising stages of a subsequent flood, however, indicate that the efficiency of the floodplain to trap suspended sediment at low, but *increasing* discharge, i.e. when water starts flowing into the downstream end of the abandoned channel, is high (Asselman, 1997). Sedimentation rates therefore may be much higher under non-stationary, rising discharge than under stationary discharge as computed with the SEDIFLUX model. This would result in an even lower effective discharge for overbank sedimentation at the VP area. Erosion during very high discharges, which is not accounted for by the model, might also lead to a lower effective discharge.

## CONCLUSIONS

Sedimentation measurements indicate that suspended-sediment accumulation increases with flood magnitude, and that the increase is greater for sand than for silt and clay. On low floodplain sections, however, the increase in sedimentation is not proportional to the increase in sediment transport in the main channel. Apparently, the efficiency of the floodplain to trap suspended sediment decreases at very high discharge. Model results confirm this hypothesis. The effective discharge for sedimentation at the floodplain section studied is of moderate magnitude. The effective discharge for sedimentation at the Variksche Plaat is about 5000 $m^3 s^{-1}$ or less. This discharge just exceeds bankfull discharge and occurs on average about 1.3 times a year. At the Stiftsche Uiterwaard, a higher effective discharge (up to 7000 $m^3 s^{-1}$) was found. This discharge occurs on average once every 2 yr. It therefore is concluded that the effective discharge for overbank sedimentation on floodplain sections along the River Waal is a discharge of moderate magnitude that occurs relatively often.

## ACKNOWLEDGEMENTS

This study is part of the National Research Programme (NRP-II) on the impact of climate change on lowland river systems, funded by the Dutch Ministry of the Environment (VROM). I would like to thank Hans Middelkoop, with whom I have carried out the field measurements, Gerard Ouwerkerk, Eric Faessen, Marjan Deurloo, Bastian van Dijck, Yolanda Onneweer and Gertjan van Manen for their assistance in the field and in the laboratory. Cees Klawer is thanked for his cooperation in the laboratory. Ir.N.G.M. Brink (Rijkswaterstaat/RIZA) provided the data on discharge and suspended-sediment transport near Lobith. Torbjörn Törnqvist commented on an earlier version of the manuscript. Two reviewers, Phil Ashworth and Sue Marriott, made useful suggestions at a later stage.

## REFERENCES

ALEXANDER, C.S. & PRIOR, J.C. (1971) Holocene sedimentation rates in overbank deposits in the black bottom of the Lower Ohio River, Southern Illinois. *Am. J. Sci.*, **270**, 361–372.

ANDREWS, E.D. (1980) Effective and bankfull discharges of streams in the Yampa River Basin Colorado and Wyoming. *J. Hydrol.*, **46**, 311–330.

ASSELMAN, N.E.M. (1997) *Suspended sediment in the River Rhine: the impact of climate change on erosion, transport, and deposition*. Thesis, Koninklijk Nederlands Aardrijkskundig Genootschap (Netherlands Geographical Studies 234), Utrecht, 257 pp.

ASSELMAN, N.E.M. & MIDDELKOOP, H. (1995) Floodplain sedimentation: quantities, patterns and processes. *Earth Surf. Process. Landf.*, **20**, 481–499.

BRAKENRIDGE, G.R. (1984) Alluvial stratigraphy and radio-carbon dating along the Duck River, Tennessee: implications regarding flood-plain origin. *Geol. Soc. Am. Bull.*, **95**, 9–25.

BRUNET, R.C., PINAY, G., GAZELLE, F. & ROQUES, L. (1994) Role of the floodplain and riparian zone in suspended matter and nitrogen retention in the Adour River, South-West France. *Regulated Rivers: Res. Manage.*, **9**, 55–63.

BURGESS, T. & WEBSTER, R. (1980) Optimal interpolation and isarithmic mapping II: block kriging. *J. Soil Sci.*, **31**, 505–524.

COSTA, J.E. (1975) Effects of agriculture on erosion and sedimentation in the Piedmont Province, Maryland. *Geol. Soc. Am. Bull.*, **86**, 1281–1286.

GOMEZ, B., MERTES, L.A.K., PHILLIPS, J.D., MAGILLIGAN, F.J. & JAMES, L.A. (1995) Sediment characteristics of an extreme flood: 1993 upper Mississippi River valley. *Geology*, **23**, 963–966.

GRETENER, B. & STRÖMQUIST, L. (1987) Overbank sedimentation rates of fine grained sediments: a study of the recent deposition in the lower river Fyrisan. *Geogr. Annaler Ser. A*, **69**, 139–146.

HE, Q. & WALLING, D.E. (1996) Use of fallout Pb-210 measurements to investigate longer-term rates and patterns of overbank sediment deposition on the floodplains of lowland rivers. *Earth Surf. Process. Landf.*, **21**, 141–154.

HEUSCH, K., BOTSCHEK, J. & SKOWRONEK, A. (1993) Fluviale Erosion und Sedimentation auf landwirtschaftlich genutzten Auenböden der Unteren Sieg. *Berl. Geogr. Arb.*, **78**, 175–192.

KESEL, R.H., DUNNE, K.C., McDONALD, R.C., ALLISON, K.R. & SPICER, B.E. (1974) Lateral erosion of overbank deposition on the Mississippi River in Louisiana, caused by 1973 flooding. *Geology*, **2**, 461–464.

KNOX, J.C. (1989) Long- and short-term episodic storage and removal of sediment in watersheds of southwestern Wisconsin and northwestern Illinois. In: *Sediment and the Environment* (*Proceedings of the Baltimore Symposium, May 1989*), (Eds Hadley, R.F. & Ongley, E.D.), pp. 157–164. IAHS Publication 184, International Association of Hydrological Sciences, Wallingford.

MANSIKKANIEMI, H. (1985) Sedimentation and water quality in the flood basin of the River Kyronjoki in Finland. *Fennia*, **163**, 155–194.

MIDDELKOOP, H. (1997) *Embanked floodplains in The Netherlands; geomorphological evolution over various time scales*. Thesis, Koninklijk Nederlands Aardrijkskundig Genootschap (Netherlands Geographical Studies 224), Utrecht, 341 pp.

MIDDELKOOP, H. & ASSELMAN, N.E.M. (1998) Spatial variability of floodplain sedimentation at the event scale in the Rhine–Meuse delta, The Netherlands. *Earth Surf. Process. Landf.*, **23**, 561–573.

RIJKSWATERSTAAT (1992) *Jaarboek Monitoring Rijkswateren 1991*. Rijkswaterstaat, Den Haag, 156 pp.

TRIMBLE, S.W. (1983) A sediment budget for Coon Creek basin in the driftless area, Wisconsin, 1853–1977. *Am. J. Sci.*, **283**, 454–474.

UBELS, J.W. (1986) *Verantwoording van het hoogwateronderzoek op de Boven-Rijn, de Waal, het Pannerdensch Kanaal, de Neder-Rijn, de Lek en de IJssel*. Nota 86.35, DBW/RIZA, Rijkswaterstaat, Arnhem.

WALLING, D.E. & BRADLEY, S.B. (1989) Rates and patterns of contemporary floodplain sedimentation: a case study of the River Culm, Devon, UK. *Geojournal*, **19**, 53–62.

WALLING, D.E., QUINE, T.A. & HE, Q. (1992) Investigating contemporary rates of floodplain sedimentation. In: *Lowland Floodplain Rivers: Geomorphological Perspectives* (Eds CARLING, P.A. & PETTS, G.E.), pp. 165–184. John Wiley & Sons, Chichester.

WEBB, B.W. & WALLING, D.E. (1982) The magnitude and frequency characteristics of fluvial transport in a Devon drainage basin and some geomorphological implications. *Catena*, **9**, 9–23.

WEBB, B.W. & WALLING, D.E. (1984) Magnitude and frequency characteristics of suspended sediment transport in Devon rivers. In: *Catchment Experiments in Fluvial Geomorphology* (Eds Burt, T.P. & Walling, D.E.), pp. 399–415. Geo Books, Norwich.

WOLMAN, M.G. & MILLER, J.P. (1960) Magnitude and frequency of forces in geomorphic processes. *J. Geol.*, **68**, 54–74.

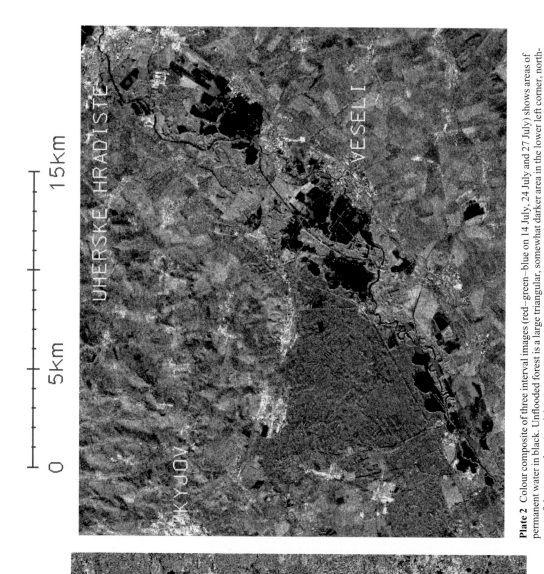

**Plate 2** Colour composite of three interval images (red—green—blue on 14 July, 24 July and 27 July) shows areas of permanent water in black. Unflooded forest is a large triangular, somewhat darker area in the lower left corner, northwest of river meanders (north is towards the top). Flooded forest occurs in the southeast on the opposite side of the river, as a brighter narrow strip continuing northeastward between two permanent water bodies.

**Plate 1** Colour composite of two time intervals with flooded areas shown in blue and permanent surface waters in black.

*Spec. Publs int. Ass. Sediment.* (1999) **28**, 71–75

# RADARSAT imaging of the 1997 Czech Republic flood

L. HALOUNOVÁ\*, R. DIXON†, H. POKRANT†, D. STRNAD‡, R. VAN WYNGAARDEN§,
V. KOLÁØ¶ *and* I. CÍCHA‖

\**Faculty of Civil Engineering, CTU Prague, Thákurova 7, Prague 6, Czech Republic (Email: Halounov@fsv.cvut.cz);*
†*Natural Resources, Remote Sensing Centre, Manitoba, Canada;*
‡*MGE DATA, Vrchlického 60, Prague, Czech Republic;*
§*Geomatics International, Burlington, Ontario, Canada;*
¶*Water Research Institute of T.G. Masaryk, Prague, Czech Republic; and*
‖*Czech Geological Survey, Prague, Czech Republic*

## ABSTRACT

Very intensive rainfall at the beginning of July 1997 caused extensive flooding in the eastern part of the Czech Republic. On Friday, 4 July, a low-pressure system, which originated in northern Italy, tracked northeast to the Moravia region of the Czech Republic. This system brought extreme rainfall as it moved to the southeast of Poland and Silesia on the morning of 6 July. Over a 5-day period the total rainfall for one-third of Moravia (10 000 km$^2$) was 500 mm. The average rainfall for the Czech Republic is 600 mm yr. This event caused 100–400-yr discharges in regional streams. On Tuesday, 8 July, the first RADARSAT image was ordered and the satellite imaged the northern part of Moravia on 10 July. An additional four images were ordered to archive intermediate stages of the flood. Areas of flooding were determined on all images. Delineation of flooded areas in post-flood images was possible owing to high radar sensitivity to the high soil-moisture contents. Changes in soil moisture, detectable over thousands of square kilometres, can also serve in studies of soil permeability or hydrogeological conditions, if imaged at successive time intervals.

## INTRODUCTION

Sedimentation processes are the result of thousands of years of mutual interaction between water flow and the geological and pedological conditions within river catchments, including the inundation of floodplains and relief of the catchment area. Changes in water discharge are reflected in water velocities, either mean cross-section velocities or real measured velocities within a cross-section. Water velocity is one of the most important variables affecting sedimentological processes. Floods represent extreme hydraulic states. River channels and their adjacent floodplains have very different depositional and erosional processes, as a consequence of varying flow velocities. Modelling of these processes requires calibration data. Satellite radar images, which can be obtained under any atmospheric situation at 1 : 25 000, 1 : 50 000 and smaller scales, can be a valuable data source for generating this information.

## AREA OF FLOOD

Moravia includes mountains in its northern and eastern areas. The southwestern part is hilly, and the central part is a lowland. The summits of the mountains exceed heights of 1000–1400 m. There are many deep narrow valleys with long steep slopes, which form large collection basins. Moravian lowlands consist of wide and long valleys situated within river systems.

Northwestern, central and southeastern regions of Moravia drain into the Moravia River, which flows into the Danube River and subsequently into the Black Sea. The northeastern part of this area drains into the Odra River and then into the Baltic Sea. The northwestern region belongs geologically to the Czech Massif, whereas the southeastern region is part of the Carpathians.

**Table 1.** Applied RADARSAT images.

| Data type | Date (1997) | Area | Image size (km$^2$) | Orbit | Incidence angle |
|---|---|---|---|---|---|
| Standard 2 | 10 July | North Moravia | $100 \times 100$ | Ascending | $24–31°$ |
| Standard 7 | 14 July | South Moravia | $100 \times 100$ | Descending | $45–49°$ |
| Standard 5 | 24 July | South Moravia | $100 \times 100$ | Descending | $36–42°$ |
| Wide 1 | 27 July | South Moravia | $150 \times 150$ | Descending | $20–31°$ |
| Wide 1 | 27 July | North Moravia | $150 \times 150$ | Descending | $20–31°$ |

## RELATED DATA

The area affected by the flood was about 25 000 km$^2$. Satellite images were a source of flood documentation, captured in near-real time. Satellite data bring an overview of the area at one moment in time. Table 1 summarizes the characteristics of the radar images obtained. Their spatial pre-launch nominal resolution is 30 m. This radar sensor operates at a single microwave frequency, known as the C-band (5.3 GHz frequency and 5.6 cm wavelength). RADARSAT transmits and receives its microwave energy in HH polarization (horizontal transmission and reception of waves). The spectral resolution of processed images is 16 bits per pixel. RADARSAT digital products can be delivered as six different data types. Path Image product (used in this case) is aligned parallel to the satellite orbit path. The product was calibrated, which refers primarily to the electrical stability of the radar sensor and its ability to provide repeatable measurements over time. The system was designed to achieve radiometric accuracy within one scene < 1.0 db, over 3 days < 2.0 db, with global dynamic range 30.0 db. Absolute radiometric calibration was required so that the magnitude of the digital data processed could be related to the radar back-scatter coefficients. To achieve this accuracy, detailed measurements of the radar and processing system performance were made on a regular basis (RADARSAT Illuminated, 1995).

The study area was also imaged by aerial photographs. There are hundreds of these photographs, and they will serve for future detailed evaluation.

## IMAGE PROCESSING AND INTERPRETATION

Radar images can be viewed as single-channel black and white images with a characteristic 'salt and pepper' appearance. The pixel values represent the strength of the returned radar signal from Earth's surface. For each surface feature there is a statistical distribution of the probable strength of that returned signal. Each pixel representing that surface is assigned a value selected randomly from the statistical distribution. Therefore, a seemingly homogeneous surface area has an irregular distribution of light and dark pixels, producing a granular effect. This effect is termed 'speckle' and is an inherent property of radar images. Original image data were compressed from 16 bits to 8 bits data. Results of data calibration control showed that the 24 July image values had 6–9% lower average values of back-scatter in selected targets. The same control, performed for 27 July, showed that average back-scatter values for the same targets were 10–15% higher, both compared with 14 July. To modify speckles, the images listed in Table 1 were smoothed by a $5 \times 5$ pixel spatial filter. All images were transformed into a Czech cartographic projection.

Pixel classification techniques, a method often used in image processing, was performed for surface water. To delineate flooded forest and areas that were flooded between image pairs, visual interpretation was used. In the lowlands, automated classification of the water surface present could be reliable for more than 90% of the area, excluding urban and forest regions. Whenever these two features occur, an important principle of radar back-scatter appears, i.e., a corner reflector. Two- or three-dimensional corner reflection is caused by the existence of buildings (two- and three-dimensional). Scattering from a forest canopy can present a complex case of volume scattering. Double-bounce scattering between tree trunks and the ground is one important effect of the volume scattering. This can give a very strong return if the ground is covered with water (Ahern, 1995). Double-bounce scattering is a geometrically similar situation to a two-dimensional corner reflection. Buildings and trees can redirect a radar beam, which was back-scattered from a smooth water surface, back to the radar sensor. This is why flooded towns and forests can look even brighter than unflooded areas.

## RESULTS

Northern Moravia was imaged twice with RADARSAT. The first image revealed the flood peak in the lowlands, whereas the second image featured the post-flood situation. The mountainous area of northern Moravia, which also was the main region of precipitation, suffered the most destruction. Destruction resulting from the floodwater affected roads, railways, bridges, houses and many other types of infrastructure features. The communities affected, within the narrow valleys of the high mountain ranges, could not be analysed by RADARSAT imagery for two reasons: (i) radar shadow was a factor and (ii) the duration of the flood in this region was short and had ended before the first RADARSAT image was obtained.

Central Moravia and southern Poland lowland regions were then studied on the image of 10 July. Flooded areas and permanent water bodies are shown as black areas (Fig. 1). Water surfaces without waves act as a smooth surface. When the radar sensor transmits a beam of radar energy towards this smooth surface, no back-scatter is returned to the radar sensor, but rather scattering of the radar energy away from the sensor occurs. Pixels for these areas have zero values and water areas are black, solid phenomena on images (Leconte & Pultz, 1991).

Pixel-classification techniques, often used in image processing, were performed for surface water. They detected not only areas with surface water excluding

forest and urban regions, but also shadows in high-relief areas. These shadows have the same values of reflection as water bodies and their measured values are the same: zero or very low values in both cases.

It was necessary to use two images from two different time intervals in order to distinguish flooded areas from permanent surface water. A colour composite (RGB = red–green–blue) of the two images (one of them must be used twice) can distinguish permanent versus floodwater immediately. Permanent surface water was black (Plate 1, facing p. 70), whereas flooded areas were lighter (blue in the colour version).

Figure 2 represents the same area on the image for 27 July 1997. Brighter features within the imagery coincided with previously flooded areas. Brighter features may be related to such factors as (i) terrain with greater surface roughness as a result of ploughing, (ii) sedimentation of coarse materials, or (iii) higher soil moisture content (RADARSAT Illuminated, 1995; Engman & Chauhan, 1995; Brown *et al.*, 1993). Sedimentation of coarse materials did not occur owing to rather low water velocities in this area. Nor was ploughing the reason for higher back-scatter values. The region was divided into small long private fields, which were not damaged significantly by the flood because crops continued in their growth after flood levels declined. It therefore was concluded that excessive soil moisture was the reason for the brighter back-scatter values.

Pixel values on the image of 24 July are lower by 6% compared with 14 July. Pixel values on the image of 27 July are higher by about 21% in comparison with 14 July in forest targets. Higher pixel values indicate higher reflectance of the flooded forest region. Comparison of unflooded and flooded forest is shown in Plate 2, facing p. 70.

An area around Olomouc (a town in central Moravia) did not show the same effect on the same RADARSAT images. This area is quite flat, similar to southern Poland, but probably with different soil permeability. Hydrogeological, geological and pedological conditions for the area around Olomouc are different as well. As a result, the previously flooded fields could not be detected on the post-flood image.

Southern Moravia was imaged by RADARSAT at three time intervals. On 14 July 1997, the flood peak was captured on the first image (Fig. 3A–C). The same area was brighter on the images of 24 July and 27 July. Comparison of these two images suggests that steep incidence angles (27 July) provide the greatest amount of information regarding soil moisture and also minimize roughness effects (Ulaby, 1974). The image of 24 July had a shallower incidence angle (36°–42°), which is why

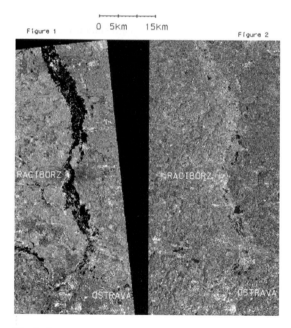

Figure 1                    0  5Km    15Km                    Figure 2

**Fig. 1.** Standard 2 image of southern Poland showing maximum flooded area, with surface water represented by solid black.
**Fig. 2.** Post-flood Wide 1 image of same area of southern Poland showing recently flooded areas in brighter hues, which coincide with flooded areas of Fig. 1.

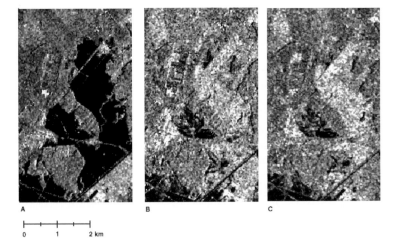

0    1    2 km

**Fig. 3.** Standard 7 image (14 July) shows flooded area in southern Moravia near Uhersky Ostroh (A). The same area on Standard 5 image (24 July) is in bright hue (B), and on Wide 1 (27 July) is in even brighter hue (C). Brighter colour is caused by higher soil moisture as a result of the previous flood. Soil moisture is a more decisive element than roughness at the incidence angle of Wide 1, which was steeper than that of Standard 5.

the area was not as bright as the same area on the image of 27 July (with an incidence angle = 20°–31°) if compared with its surroundings. This is an example of another way to delineate flooded areas on a post-flood radar image. To determine these areas reliably required images from a given area at the moment of existing higher soil moisture. This moment differs for various soil types, hydrogeological conditions, terrain slopes and canopy. To determine the time when soil-moisture levels reflected the previous flood must be a subject of more detailed studies in the areas of interest. More frequent post-flood images could offer this information.

A colour composite (RGB) of the three images can distinguish permanent and flood water immediately (Plate 2, facing p. 70). Permanent surface water shown as black versus flooded areas at different time intervals can be displayed in different colours. The RGB image was able to show the flood progress. RADARSAT images were the only images available for the flooded areas with such a short time lag after the flood onset. It was the only sensor able to repeat images in very short time intervals.

Radar sensors are the only instruments that can penetrate clouds, fog and smog. Moravia was covered by clouds nearly the whole time and thus no optical data were available.

## DISCUSSION

Any automated classification performed without visual interpretation can result in erroneous information. For example, radar shadows found in mountainous areas can have the same pixel values as a smooth water surface. A post-classification modification therefore must be applied.

Visual interpretation was necessary for forest areas and urban regions on single images. Single images display only surface-water bodies existing at the time of image capture and it is not possible to distinguish permanent water bodies from flooded areas. In contrast, image pairs at the time of flooding and before or after flooding enable discrimination between permanent and temporarily flooded areas. This task can be performed easily by creating colour composites.

Images from after a flood can be useful in cases where no images from the flood itself are available. Higher soil moisture as a consequence of flooding causes a higher back-scatter and thus can be interpreted as brighter regions on the post-flood image. To decide whether a brighter back-scatter value is the result of a high moisture content requires information about the locality in order to be able to exclude surface roughness, which might be related to field activities, such as ploughing, or to the previous flood in the case of coarse sediments. The incidence angle is another variable that must be taken into account. Steeper incidence angles emphasize soil-moisture influence on radar reflection. In contrast, shallower incidence angles are more influenced by surface roughness. Delineation of flooded areas can be one source of sedimentologically useful data. Detailed study of soil-moisture changes of flooded areas in short-time intervals after flooding can be a data source for models of determination of sediment permeability or their thickness.

## ACKNOWLEDGEMENTS

This project was co-sponsored by the Canadian International Development Agency, Ministry of Environment of the Czech Republic and MGE DATA, s.r.o. from Prague. The conference presentation was financed by grant 103/97/0669 of the Czech Grant Agency and by MGE DATA, s.r.o.

## REFERENCES

AHERN, F.J. (1995) *Fundamental Concepts of Imaging Radar: Basic Level*. Unpublished manual, Canada Centre for Remote Sensing, Ottawa.

BROWN, R.J., BRISCO, B., LECONTE, R., *et al.* (1993) Potential applications of RADARSAT data to agriculture and hydrology. *Can. J. Remote Sens.*, **4**, 317–329.

ENGMAN, E.T. & CHAUHAN, N. (1995) Status of microwave soil moisture measurements with remote sensing. *Remote Sens. Environ.*, **51**, 189–198.

LECONTE, R. & PULTZ, T.J. (1991) Evaluation of the potential of RADARSAT for flood mapping using simulated satellite imagery. *Can. J. Remote Sens.*, **3**, 241–249.

RADARSAT ILLUMINATED (1995) *User Guide, RADARSAT International Inc.*, Richmond, British Columbia.

ULABY, F.T. (1974) Radar measurement of soil moisture content. *IEEE Trans. Antennas Propag.*, **2**, 257–265.

*Spec. Publs int. Ass. Sediment.* (1999) **28**, 77–91

# The role of overbank flow in governing the form of an anabranching river: the Fitzroy River, northwestern Australia

C. F. H. TAYLOR

*Department of Geography, University of Western Australia, Nedlands, Western Australia 6907*
*(Email: clare@gis.uwa.edu.au)*

## ABSTRACT

The 85 000 km$^2$ Fitzroy River catchment lies in the semi-arid tropics of north-western Australia. Monsoon-associated rainfall events inundate the Fitzroy's largely unconfined, 300-km-long floodplain, with flows of 30 000 m$^3$ s$^{-1}$ ranking as 1 in *c.* 35 yr events. Adjustment of the main Fitzroy channel to convey these large flows efficiently is restricted by a low-gradient valley, an erosion-resistant floodplain, and both gradual and episodic constriction of the channel by sediment and vegetation. To compensate for the restricted capacity of the main channel, auxiliary channels form in response to water excesses, with high-magnitude overbank flows reactivating atrophying channels and/or scouring elongate depressions on distal floodplain, which can become preferred flow paths and eventually new channels. The planform of channels is governed by the interaction between the frequency and magnitude of overbank events and vegetation, and by the texture of channel banks and floors. An anabranching river results, which can exhibit adjacent braided, sinuous and straight channel reaches.

## INTRODUCTION

A better understanding of modern fluvial processes and their sedimentary styles is needed to predict the impact of future modification to river systems and to improve palaeoenvironmental interpretation. Channel and floodplain forms and processes in anabranching river systems are especially poorly understood and recent consideration in the literature has called for the investigation of more contemporary examples (e.g. Lewin, 1992; Hickin, 1993; Richards *et al.*, 1993; Nanson & Knighton, 1996).

The Fitzroy River of north-western Australia (Fig. 1) conforms to Nanson & Knighton's (1996) definition of anabranching in that it is a system of multiple channels characterized by stable alluvial islands that divide flow until discharges reach almost bankfull. The Fitzroy River, however, does not fit comfortably into any of the six types of anabranching rivers proposed by Nanson & Knighton (1996), and may require separate classification. The following study describes a new example of a multi-channel river system that incorporates various planform types, and attempts to explain this morphology. The modern Fitzroy River has not been examined in detail previously, and investigation is especially urgent because current proposals to dam river headwaters must address the potential impacts of flow regulation on the channel–floodplain system.

Overbank flow plays a significant role in shaping the contemporary floodplain and channel form of the Fitzroy River. Occasional high-magnitude rainfall events that extend over large areas of the 45 000 km$^2$ upper catchment can quickly produce flows exceeding channel capacities. The low-gradient valley, cohesive clay floodplain and an erosion-resistant palaeosurface underlying large areas of the floodplain restrict the channel's ability to adjust to these high flows. Furthermore, overbank flow is promoted by channel constriction via seasonal accretion on oblique clay banks and deposition of in-channel sediments, which are stabilized and encouraged by vegetation growth. Overbank flow is hydraulically inefficient (Leopold & Maddock, 1953), and the river responds according to principles of minimum energy (Leopold & Langbein, 1962; Yang, 1971; Chang, 1979), attempting to increase flow conveyance by increasing the number of channels draining the floodplain and enlarging its existing channels where possible. This leads to the development of a multiple-channel system displaying a variety of reach planforms.

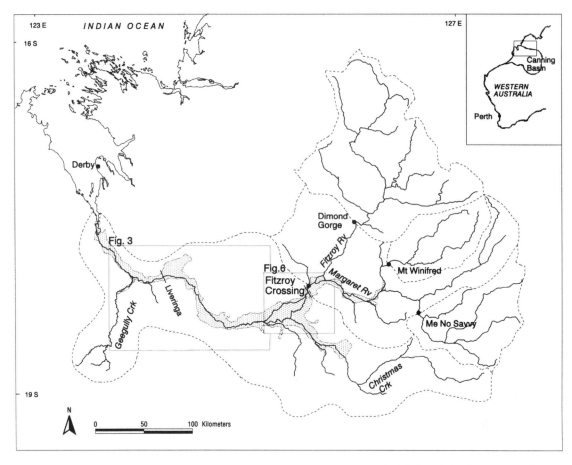

**Fig. 1.** Fitzroy River catchment showing major tributaries, subcatchment, the floodplain (shaded), boundaries, flow gauges and velocity recording sites.

## METHODS

Data have been collated from 4 months of field study, which included detailed stratigraphical logging of bank sections, as well as analysis of aerial photography and satellite imagery. Owing to the difficulties of recording processes active during high-magnitude floods, some ideas discussed in this paper must remain speculative. Previous investigation of the Fitzroy River is limited: flood hydrology has been described by Main Roads Western Australia (e.g. Goh, 1993), and floodplain soils and geomorphological units mapped for agricultural purposes (Speck *et al.*, 1964; Wright, 1964). The sedimentology of Geegully Creek (Wyrwoll *et al.*, 1992), sections of the main channel (Clark, 1991), and slackwater deposits in bedrock reaches (Wohl *et al.*, 1994a,b), have been interpreted for palaeohydrological information.

## STUDY AREA

The 85 000 km$^2$ Fitzroy catchment lies in the West Kimberley region of north-western Australia (Fig. 1). The upper Fitzroy River (Dimond Gorge subcatchment) drains quartzites, sandstones and dolerites of the Proterozoic Kimberley Block. To the south and east (Me No Savvy and Mount Winifred subcatchments) the Margaret River, the major tributary of the Fitzroy, drains Proterozoic ranges and laterite tablelands. Abundant outcrop and shallow soils contribute to high runoff, with extensive weathering of the generally low-relief catchment (average elevation 300 m above sea-level) providing a dominantly sand to clay sediment load. The single-channel Fitzroy and Margaret Rivers leave this Proterozoic terrain a few kilometres upstream of Fitzroy Crossing, where they unite to anabranch westwards over the wide, low-gradient plain of the Fitzroy Trough.

Subsidence in this sub-basin of the Canning Basin ceased during the Early Jurassic (Yeates *et al.*, 1984), and the region has been tectonically stable since the late Tertiary (Wright, 1964). Two ephemeral drainages, Christmas Creek and Geegully Creek, join the river before it discharges into the Indian Ocean.

## CLIMATE AND FLOOD HYDROLOGY

The Fitzroy catchment lies at the southernmost extent of the north-west Australian summer monsoon regime. Its annual rainfall averages 900 mm in the north and 350 mm in the south, with 90% of this rain falling between November and April (Bureau of Meteorology, 1996). The 35% coefficient of variation in annual rainfall is reflected in the discharge regime of the Fitzroy River, which includes droughts and extreme floods (Fig. 2). From May to October flow is minimal or absent. During the wet season, runoff from low-intensity monsoonal rainfall and thunderstorm activity generally remains within channels, but tropical cyclones or intense monsoonal depressions can cause discharges of greater than ~ 10 000 $m^3$ $s^{-1}$ at

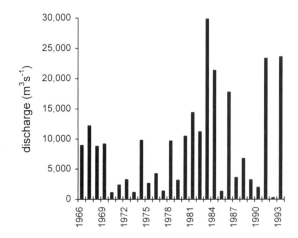

**Fig. 2.** Peak discharges recorded at Fitzroy Crossing (gauge 802055) 1966–1993 (data source: Water Authority of Western Australia; Goh, 1993).

Fitzroy Crossing, inundating the unconfined flood-plain downstream to depths averaging 3 m (KWRDO, 1993) (Fig. 3). A cluster of severe floods occurred between

**Fig. 3.** Satellite image showing large-scale inundation of the Fitzroy floodplain (1993 flood, peak discharge 25 000 $m^3$ $s^{-1}$). The valley floor acts as a broad flow zone (light grey area) within which the main channel is distinguished by riparian vegetation (black line). Flow from right to left. Scale bar is 15 km. (DOLA copy licence 540/98 TM 109–072 12/3/93.)

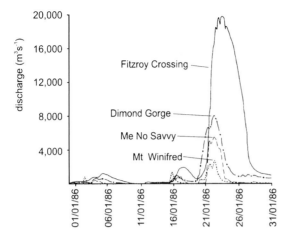

**Fig. 4.** Typical flood hydrographs at Fitzroy Crossing, Dimond Gorge, Me No Savvy and Mount Winifred (1 January 1986 to 31 January 1986) (data source: Water Authority of Western Australia).

1983 and 1993, with the highest estimated flow at Fitzroy Crossing of 30 000 $m^3$ $s^{-1}$ approaching global discharge maxima for this catchment area (Finlayson & McMahon, 1988). A flood of 30 000 $m^3$ $s^{-1}$ has been ranked as a 1 in *c*. 35 yr event (KWRDO, 1993).

The 1986 flood hydrograph (Fig. 4) is typical of recent events, and shows flow from Dimond Gorge, Me No Savvy and Mount Winifred subcatchments contributing to a peak discharge of 20 000 $m^3$ $s^{-1}$ at Fitzroy Crossing. The peaked shape of the hydrograph, principally caused by rainfall characteristics and the geography of the channel network, shows that an enormous amount of water is suddenly present at the eastern end of the floodplain. This sudden deluge easily exceeds the bankfull capacity of channels at Fitzroy Crossing and causes high down-valley overbank flow velocities despite floodplain widths averaging ~ 10 km (Fig. 5). Overbank flow velocities of up to 1.1 m $s^{-1}$ have been measured ~ 4 km from the main channel.

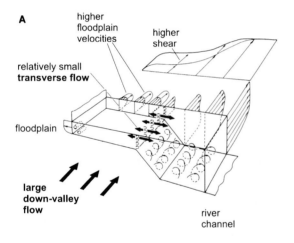

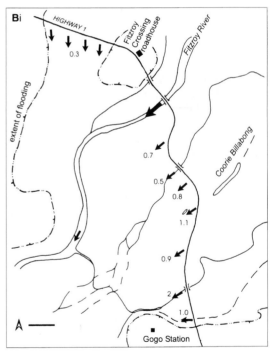

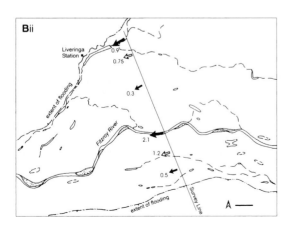

**Fig. 5.** (A) Conceptual diagram of hydraulic aspects of overbank flow, highlighting the large down-valley flow component of the Fitzroy River in flood despite an average valley width of ~ 10 km (adapted from Knight, 1989). (B) Flow velocities over the floodplain in $m^3$ $s^{-1}$ at (i) Fitzroy Crossing (1983 flood); (ii) Liveringa (full arrow = average of 1991 and 1993 floods, clear arrow = modelled flow velocity); scale bars are 1 km (data source: Water Authority of Western Australia (unpublished); Goh, 1993).

The sediment load transported by the Fitzroy River has not been measured during overbank events, but observed high turbidity and widespread deposition of in-channel and floodplain sediment suggest a large sand-size bedload and clay-size suspended load.

## CONTEMPORARY PLANFORM CHARACTERISTICS

### Floodplain setting

The Fitzroy floodplain considered here is the topographic floodplain (Graf, 1988) because the whole valley floor acts as a broad flow zone during large floods (Fig. 3). The floodplain is 300 km long, 3–20 km wide and has a low gradient of 0.0003 m m$^{-1}$. It is composed dominantly of swelling and cracking montmorillonite and illite clays, with a weak soil structure and varying degrees of self-mulching (Rutherford, 1964). A satellite image of part of the floodplain reveals the complexity of the anabranching channel system (Fig. 6A). The main Fitzroy River channel is distinguished by its relatively wide and continuous borders of dense phreatophytes. Enlargement of an area where the main channel bifurcates (Fig. 6B), along with a schematic cross-section (Fig. 6C), illustrate how channel planform and marginal stratigraphy vary between braided, sinuous and straight forms, both between channels and along reaches of the same channel (Fig. 6B, no. 1, 2, 3). Shallow scours and billabongs punctuate the plain, usually elongate subparallel to flow direction and often aligned in chains (Fig. 6B & C, no. 4). The modern floodplain overlies a 'base' of indurated orange alluvium (clay to coarse sand), which displays strong pedogenic features, such as dense root mats and tree trunks now replaced by carbonate. This erosion-retarding palaeosol (or palaeosols) undulates from the present floodplain height to at least 8 m below it, and forms the floor and/or partial walls of some channel reaches (Fig. 6C, no. 5). Floodplain vegetation is typically savanna or open eucalypt woodland. Riparian and in-channel vegetation consists mainly of mixed *Eucalyptus*, *Acacia* and *Casuarina*, which, in contrast to many humid climate species, are deep-rooted and have adapted to colonize channel sands and withstand high flow velocities.

### Main channel and Cunningham anabranch

The main Fitzroy River channel has a sinuosity of 1.2. It alternates between clay-lined reaches of approximately constant width (~ 70 m) (e.g. Fig. 6B, no. 2), and wider, sand-dominated reaches (e.g. Fig. 6B, top centre). Clay-lined reaches typically hold deep pools throughout the dry season, with mature vegetation growing on sloping banks. Bankfull channel depth in pooled reaches is estimated at ~ 12 m. Wider channel reaches are generally 500–1000 m long and 3–7 m deep. A continuum exists from reaches ~ 100 m wide, where low flow braids over transverse sand bars between pools, to reaches up to 700 m wide, where sand bedload forms large-scale bedforms and lateral bars up to 3 m high, sometimes stabilized by vegetation. The 53-km-long, low-sinuosity (1.08) Cunningham anabranch, where channel width can exceed 1000 m, represents the braiding extreme of this continuum (Fig. 6B, no. 3). Banks of wide channel reaches commonly contain cobbles and sands of palaeochannels and can be unstable, lacking mature riparian vegetation. Levees bordering wide channel reaches are disjunct or absent owing to both non-deposition and erosion on reaches where riparian vegetation roughness is low. Levees bordering clay-lined reaches of the main channel are topographically subtle. They are most common at the upstream end of the floodplain, where they are ~ 800 m wide and less than 1 m high. A single levee deposit is typically a subhorizontal couplet of silt and clay (Fig. 6C, no. 2), reflecting a change in flow hydraulics from rapid deposition during fast overbank flow (silt layer) to waning flood conditions (clay layer).

### Auxiliary channels

Today up to eight auxiliary channels drain a section across the floodplain. The dimensions and planform of auxiliary floodplain channels vary greatly, as do the scale and nature of any proximal deposits. Four main channel forms can be identified:

1 Most common are channels aligned with valley flow direction which generally have low sinuosity and minimal bedload (e.g. Melonhole Creek, Nuderone Greek, Figs 6A & B). Channel dimensions, and hence width/depth ratios, change from reach to reach. Deeper reaches (3–10 m deep, 5–15 m wide) have distinct banks, more riparian vegetation compared to in-channel vegetation, and can contain permanent billabongs. Shallower reaches (1–2 m deep, 5–25 m wide) have poorly defined or even discontinuous banks, and are often vegetated throughout.

2 Actively scouring floodplain channels have low sinuosity except where pre-existing channels are exploited, and are devoid of riparian vegetation. The indurated palaeosol retards incision, resulting in high width/depth ratios (~ 20) and irregular banks. Two such channels on the contemporary floodplain are Duckhole Creek (Fig. 6A, B, no. 5, C) and Quanbun Channel (Fig. 6A).

3 Channels draining upland areas adjacent to the floodplain can cross-cut valley flow direction and are more sinuous than form 1 and 2 channels. They are typically

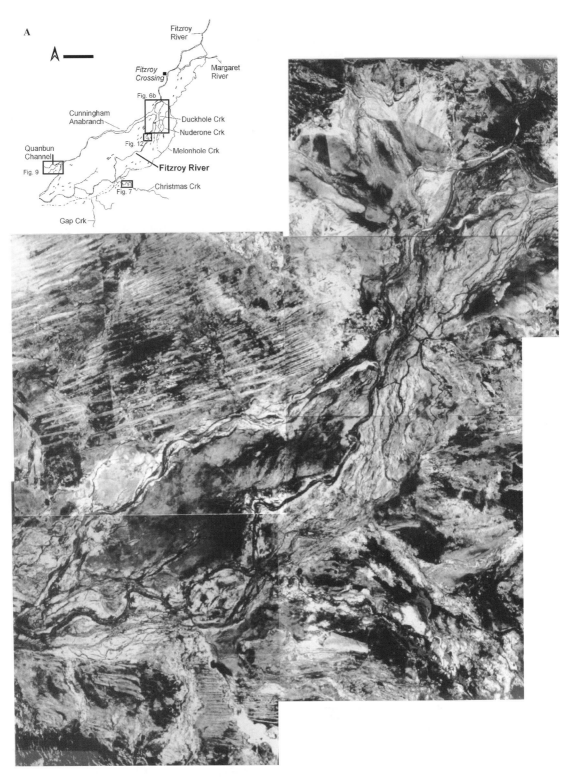

**Fig. 6.** (A) Satellite image of the upper Fitzroy floodplain showing the Fitzroy and Margaret Rivers, the Cunningham anabranch and numerous auxiliary channels, both active and atrophying. Flow from right to left. Scale bar is 5 km (adapted from S. Clark, unpublished data). (*continued*)

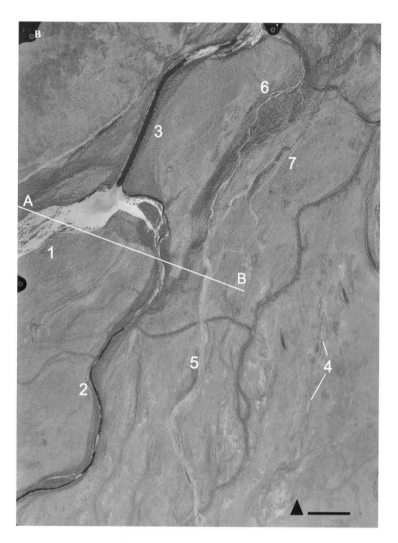

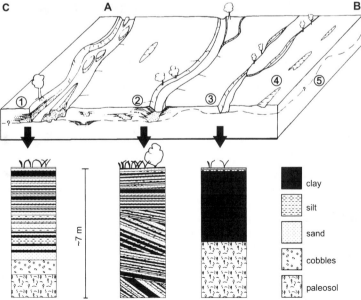

**Fig. 6.** (*continued*) (B) Aerial photograph showing the braided form of the Cunningham anabranch (1); sinuous (2) and straight (3) reaches of the main Fitzroy channel; chains of elongate billabongs (4); scour of Duckhole Creek (5), with the original channel joining the main river (6) and adjacent new floodplain scour (7). Flow from top to bottom. Scale bar is 500 m (DOLA copy licence 540/98: WA 3571, Run 5, Frame 5062, 950003, 15/07/95). (C) Cartoon of floodplain cross-section located on Fig. 6B showing: braided (1), sinuous (2) and straight (3) channel reaches; elongate depressions or billabongs aligned along valley-flow direction (4), which link and extend as illustrated in (3); undulating palaeosurface of indurated alluvium (thickness unknown) (5). Representative stratigraphy (from measured bank sections) is shown below each planform type.

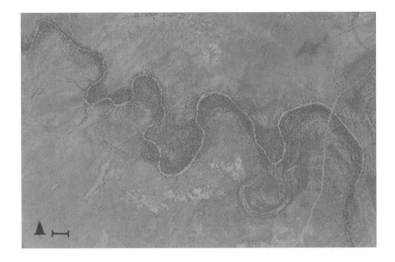

**Fig. 7.** Sinuous planform of the downstream reach of Christmas Creek. Flow right to left. Scale bar is 200 m. Note cut-off meander loop. (DOLA copy licence 540/98: WA 3571, Run 7, Frame 5206, 950003, 15/07/95).

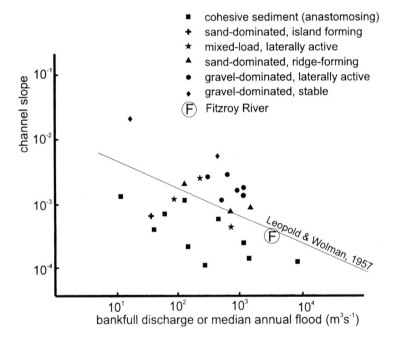

**Fig. 8.** Graph of slope versus discharge comparing the Fitzroy River with other anabranching rivers from Nanson & Knighton's (1996) classification, and Leopold & Wolman's (1957) threshold between meandering (below) and braiding (above) (adapted from Nanson & Knighton, 1996).

~ 2 m wide, ~ 1 m deep and contain sand or gravel bedload. Banks are fairly well vegetated (e.g. Gap Creek, Fig. 6A).

**4** Floodplain channels ~ 3–6 m wide exhibit steep clay banks up to ~ 5 m high supporting very mature riparian vegetation, and cross-cut valley flow direction. This low width/depth ratio channel form contrasts with most auxiliary channels on today's floodplain and is relatively uncommon. Reaches can be highly sinuous (e.g. the terminal reach of Christmas Creek, Figs 6A & 7).

## Classification

For comparison with other anabranching rivers, the Fitzroy River is plotted on a graph of slope versus discharge from Nanson & Knighton's (1996) classification of anabranching rivers (Fig. 8). The main channel plots near 'cohesive sediment' rivers of the classification, and indeed the Fitzroy has many features in common with the 'organo-clastic' subdivision of these 'cohesive sediment' rivers, including low gradient, cohesive banks, minimal lateral migration, abundant overbank deposition and sand bedload. The Fitzroy's specific stream power (a primary basis of the classification), however, is much greater ($\sim$ 75 W m$^{-2}$ versus 10 W m$^{-2}$), and the system lacks swampy, organic-rich islands. The Fitzroy River system also displays some features of 'sand-dominated, island-forming' and 'mixed-load laterally active' rivers, but until more data are available, classification is premature. Auxiliary channels are not included on this graph because insufficient data exist on their slope, dimensions and bankfull discharge.

## PROCESSES OF PLANFORM DEVELOPMENT

### Multiple channels

Today's actively scouring auxiliary channels (form 2) reveal mechanisms of anabranch formation. Duckhole Creek has scoured a fairly straight path $\sim$ 100 m wide and $\sim$ 6 m deep over 9 km of floodplain (Fig. 6A–C). The channel is devoid of riparian vegetation and contains islands of remnant floodplain. The narrow connection between the original Duckhole Creek and the main channel has filled with sand (Fig. 6B, no. 6), and scour is creating a new path extending into the floodplain directly up-valley (Fig. 6B, no. 7). In Duckhole's lower reaches scour is discontinuous but of similar dimensions. These features imply that floodwaters, rather than flow from the main channel, are scouring Duckhole Creek. Quanbun Channel has shown a marked acceleration of erosion coincident with the increased frequency of high-magnitude overbank events from 1983 to 1993 compared with 1964 to 1980 (Fig. 9). Channel width has increased from $\sim$ 10 m (1964), to 30 m (1982), to 100 m (1995); channel length has increased from discontinuous billabongs ($<$ 1 km long) (1964), to 3 km (1982), to 7 km (1995), with scour now extending into the floodplain both up- and down-valley. Quanbun's relatively narrow upstream end is marked by a 2-m-high headcut, 300 m away from the main channel. These characteristics again indicate that overbank flow, rather than flow from the main channel, plays an important role in initiating scour. In both newly scouring channels, incision and hence gradient change is retarded by the indurated palaeosol, resulting in a high width/depth ratio and uncertainty as to whether these channels will capture flow from the main river. Minimal bedload is present in the new channels, probably owing to lack of supply and the armouring effect of the palaeosol. Reworked sediment is forming broad levees over floodplain clays (Fig. 6C, section 3).

The development of new channels via levee breach, a commonly cited formation mechanism of multi-channelled rivers (e.g. Leeder, 1978; Smith & Smith, 1980; Richards *et al.*, 1993; Schumm *et al.*, 1996), is rare on the modern Fitzroy River. This implies that crevasse splay facies may be rare in the Fitzroy's recent alluvial record. The topographic subtlety of Fitzroy levee banks essentially eliminates the gradient differences normally inducing levee crevassing. Instead, the wide extent of floodwaters averaging over 3 m deep, combined with down-valley overbank velocities of up to 1.1 m s$^{-1}$ (Fig. 5Bii) and unit 'stream' powers of $>$ 30 W m$^{-2}$ (e.g. 1983 flood, Coorie Billabong, Fig. 5B (i)) cause new anabranches to form on distal floodplain by flood flow scouring shallow depressions (Fig. 10). Where vegetation is scarce, or where floodplain flow is constricted hydraulically, geologically or geomorphologically, scour is aggravated. The depressions become preferred flow paths, with chains of depressions eventually linking to form new channels (e.g. Fig. 9), which may or may not later join the main channel. Extension and (re)-excavation of existing or abandoned channels (the sandy deposits of which are less cohesive than floodplain clay) by flood flow also contribute to the development of anabranches. The formation of anabranches by distal overbank processes is not commonly reported, but Popov (1962) describes floodwaters scouring secondary floodplain channels that can enlarge to capture flow, and Nanson & Knighton (1996) suggest that in low-gradient valleys, the rising stage of low-velocity, long-duration flows can slowly scour new channels into the floodplain surface. The Fitzroy River demonstrates that avulsion of flow from, or a physical connection with, the main channel is not important for the formation of new channels in a flow regime that includes high-magnitude overbank events capable of floodplain scour.

### Planform variety

The sinuous planform exhibited by most of the main Fitzroy channel is characteristic of low-gradient rivers carrying a suspended load (Leopold & Wolman, 1957). Sinuosity remains low as a result of cohesive floodplain clays and well-vegetated banks limiting lateral meander migration: channel migration in clay-lined vegetated reaches is undetectable in 1 : 50 000 scale aerial photography taken 30 yr apart. The sloping banks accrete obliquely as flow recedes after each wet season, with occasional bank collapses revealing stacks of dipping clay or clay-silt laminae (Fig. 6C, section 2), similar to those described by Taylor & Woodyer (1978).

Braiding develops in wider channel reaches where high

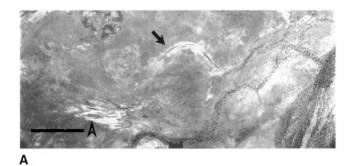

**A**

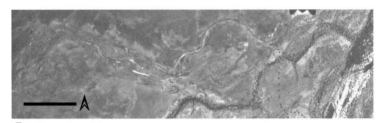

**B**

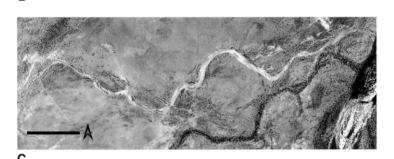

**C**

**Fig. 9.** Progressive scour of Quanbun Channel (arrow): (A) 1964, (B) 1982, (C) 1995. Erosion has accelerated between 1982 and 1995, a period of high flood frequency compared with 1964 to 1982. Flow from right to left. Scale bar is 1 km (DOLA copy licence 540/98 950003: WA 3571, Run 6, Frame 5096, 15/07/95; WA 2074, Run 7, Frame 5120, 26/07/82; WA 873, Run 12, Frame 5170, 10/09/64).

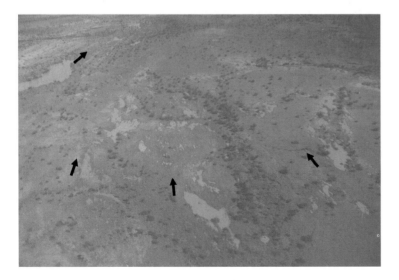

**Fig. 10.** Shallow elongate scours (centre) and lineations (top left) formed on distal floodplain during the 1993 flood (foreground ~ 150 m across) (photograph J. Henwood).

**Fig. 11.** Recent channel widening on the Cunningham anabranch owing to erosion of palaeochannel deposits (note hanging fence, and cobbles in left foreground). The indurated palaeosurface forms an armoured channel floor at this site. Flow from fore- to background. Vehicle is 1.8 m high.

flows have exploited weaknesses resulting from non-cohesive sands of abandoned or chute channels, and/or from loss of riparian vegetation by arid conditions, bank collapse, up-rooting, prolonged inundation, or recently, overgrazing (Fig. 11). Erosion can be aggravated by tributary in-flow, the indurated palaeosol and geomorphological features deviating flow, such as large vegetated bars. Abundant sand bedload is provided both by erosion of palaeodeposits, and by the expansion of channel width, reducing transport efficiency. The extent of braiding is partially determined by the colonization pattern and maturity of vegetation, which can either generate in-channel roughness, increasing channel width and braiding, or stabilize marginal sandy deposits to narrow the channel, which can result in downstream migration of the channel (Fig. 12).

Straight channel reaches (e.g. Fig. 6B, no. 3) occur where cohesive banks prevent meander migration and where overbank flow causes bend cut-offs. Also, floodplain scour elongate to flow direction creates initially straight channel reaches (e.g., Duckhole Creek, Fig. 6B, no. 5).

## CONTROLS ON PLANFORM DEVELOPMENT

Variables influencing the general form of any natural alluvial channel system include flow regime, tectonic setting, valley dimensions, sediment supply, inherited geomorphology and vegetation (Schumm & Lichty, 1965). For the Fitzroy River system, these variables relate to three factors identified by Nanson & Knighton (1996) as

common to anabranching rivers in a wide range of environments. They propose that a multi-channel planform is caused by frequent or high-magnitude flooding (factor 1), in channels that cannot readily alter their capacity or increase gradient. More specifically, they suggest that resistant banks (factor 2), and a mechanism to displace flow periodically from channel to floodplain (factor 3), are important to anabranching. All three factors promote overbank flow, the first by exceeding channel capacity, the second by limiting adjustment of channel dimensions to convey excess flow, and the third by reducing channel capacity. On the Fitzroy, high-magnitude rainfall events produce discharges that easily exceed channel capacities (i.e., factor 1), as demonstrated by floods between 1983 and 1993. The influence of this upper-catchment-generated discharge on the character of a channel–floodplain system depends on the division of flow between channel and floodplain (Brigza & Finlayson, 1990). This division is governed by the ability of a channel to enlarge its capacity, or straighten its course to increase its gradient, in order to improve hydraulic efficiency (Leopold & Maddock, 1953; Langbein, 1964; Chang, 1979). The Fitzroy's cohesive clays and deep-rooted riparian vegetation severely limit such channel adjustments on the low-gradient floodplain (i.e. factor 2), resulting in relatively high ratios of floodplain to channel flow. Thus, in an attempt to drain the floodplain more effectively, rivers scour new channels where gradient advantages exist or re-excavate less cohesive deposits of abandoned channels (Schumm *et al.*, 1996; Nanson & Huang, 1999). This generates multiple channels across the floodplain in an anabranching planform. The formation of new channels on the Fitzroy

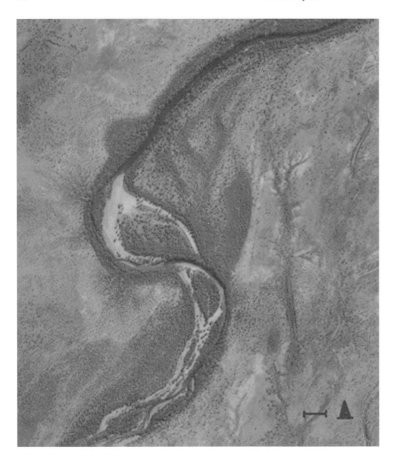

**Fig. 12.** Aerial photograph illustrating migration of meanders downstream, and the potential for braiding via vegetation loss and reworking of sand bedload. Flow from top to bottom. Scale bar is 200 m (DOLA copy licence 540/98; WA 3571, Run 6, Frame 5092, 950003, 15/07/95).

floodplain does not require that flow be displaced from one channel to another by avulsion (i.e. factor 3), because actively scouring drainages not connected to the main channel demonstrate that overbank flow is capable of significant erosion. Over the longer term, however, mechanisms displacing flow, such as accretion of channel sediment and growth of in-channel vegetation, may gradually reduce the capacity of a channel, leading to the scour of new channels. Such processes, which mostly act between floods, also influence individual reach planforms (e.g. Schumm & Lichty, 1963; Osterkamp & Costa, 1987).

The morphology of channels also may be related to attempts by the river to increase its capacity in order to drain the floodplain effectively during periods of increased flooding. Channel widening, and a resultant change in morphology from meandering to braiding, can occur where reach slope increases, or from the increased local availability of bedload sediment or decrease in bank silt–clay content (Ferguson, 1987). For channels close to

the meandering and braiding threshold (Fig. 8) only a small change in controlling variable is needed for a pattern transition (Schumm, 1988). Although the slopes of individual reaches have not been surveyed, the overall very low channel gradient and coarse palaeodeposits exposed in banks of some wide reaches (Fig. 6C, no. 1) point to the latter as a likely cause. Alternatively, or additionally, wide reaches may correspond with topographic highs in the indurated palaeosol, where vertical incision is retarded and enough energy is generated to erode channel walls. The palaeosol is present as the armoured floor of many sites of increased channel width, and, where evident in narrower reaches, it often forms a terrace 5–10 m wide, backed by eroded banks. More evidence is required to substantiate this hypothesis. Over long time-scales ($\gg 10^3$ yr) the palaeosol may only retard incision and cause temporary widening, because at two sites the main channel cuts through the palaeosol revealing it to be ~ 1 m thick. At other sites, however, the palaeosol is a minimum of 2.5 m thick.

The morphology of channels also seems closely related to the frequency and magnitude of floods occurring during their formative stage, prior to the establishment of vegetation. A channel forming during periods of frequent, high-magnitude floods can scour a path through the floodplain wider than the main channel (e.g. form 2 channels, and perhaps the Cunningham anabranch). If a source of sand is intersected (e.g. a palaeochannel or the main channel), and high-magnitude flows continue, it may adopt a braided planform. After a period with no high-magnitude overbank events, however, the channel is likely to stabilize and start to accrete, establishing riparian vegetation and a sinuous planform type. Any lateral migration will be retarded by the boundaries of the original channel. Anabranches forming during a low frequency of flooding may never reach the dimensions of channels such as Duckhole or Quanbun channels. Rather, irregular processes of flood erosion across the floodplain form deep billabongs, while other areas become shallow channels, stabilized by vegetation (form 1 channels). In addition, uneven filling and vegetation colonization of atrophying channels contribute to irregular channel dimensions. Flood frequency and magnitude have less influence on the form of floodplain channels that receive runoff from valley sides (form 3), because local rainfall, steeper gradients and higher sediment loads play a more dominant role. Low width/depth ratio channels cutting across the dominant down-valley drainage direction (form 4) are incongruous with floodplain features forming under the present flow regime and are interpreted to be inherited from a former flow regime. Excavation in Christmas Creek (Fig. 7) reveals that the indurated palaeosol here contains the template of a highly sinuous planform, on which 1–2 m of cohesive clay have accreted, stabilized by mature vegetation, preserving the planform despite transverse flood flows.

## CONCLUSION

Given the geological setting of the Fitzroy River (in terms of valley geometry, tectonic stability and sediment supply), it is the frequency and magnitude of overbank flow, combined with factors resisting flow, such as vegetation and the cohesive floodplain, that govern the river's anabranching pattern, and the number, longevity and planform of channels. The Fitzroy channel–floodplain system responds to an episodic generation of overbank events by scouring new channels in the floodplain or enlarging existing ones to increase drainage efficiency. High down-valley overbank flow means anabranches can develop on distal parts of the floodplain, independent of the main channel. Concurrent with these scouring processes, sediment accretes vertically over the floodplain, obliquely on channel banks, and as in-channel forms. This erosion and accretion, along with vegetation processes, contribute to a geomorphically effective feedback system that accounts for the river's contemporary form.

The Fitzroy River is in a constant state of adjustment and will remain so while its current flow regime persists. As these adjustments tend toward an average condition (anabranching), the Fitzroy River can be regarded as being in steady-state equilibrium. For the Fitzroy, the suggestion of Smith *et al.* (1989) that a multi-channel planform is a temporary stage in river evolution, and that eventually a single channel will be achieved probably will be appropriate only in the very long term. Not enough data are yet available for an assessment of a long-term equilibrium status (trends A, B or C, Fig. 13), although it seems closely linked to the activity of the northwest Australian monsoon regime, which governs the regime of high-magnitude overbank flows. Bank stratigraphy, combined with radiocarbon dating (unpublished), indicate

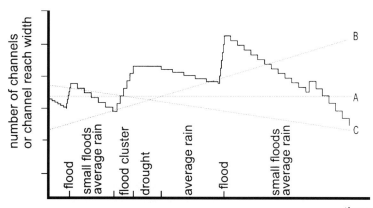

**Fig. 13.** Conceptual equilibrium status of the Fitzroy River, with possible long-term trends A, B, or C.

that the overbank-flow-dominated regime of the Fitzroy River has existed for over 3000 yr.

## ACKNOWLEDGEMENTS

This work forms part of a current doctorate project funded by a University Research Scholarship. Thanks go to A. George, M. Sivapalan, and two anonymous reviewers for advice that improved the manuscript. Comments from D. Blakeway and J. Goh were also most useful. K.-H. Wyrwoll is thanked sincerely for his help with the project. Agriculture Western Australia, Main Roads Western Australia, the Department of Land Administration and the Water and Rivers Commission very kindly provided data and/or logistic support, and field assistance from J. Carr, D. Goddard, E. Pickett, J. Sinclair and H. Zheng was invaluable. The hospitality, etc., of the wonderful West Kimberley locals is always greatly appreciated.

## REFERENCES

BRIGZA, S.O. & FINLAYSON, B.L. (1990) Channel avulsion and river metamorphosis: the case of the Thomson River, Victoria, Australia. *Earth Surf. Process. Landf.*, **14**, 391–404.

BUREAU OF METEOROLOGY (1996) *Kimberley Climatic Survey.* Australian Government Publishing Service, Canberra, 100 pp.

CHANG, H.H. (1979) Minimum stream power and river channel patterns. *J. Hydrol.*, **41**, 303–327.

CLARK, S.V. (1991) *The Holocene alluvial stratigraphy and paleohydrology of the Fitzroy River, Northwestern Australia.* Unpublished Honours Thesis, University of Western Australia.

FERGUSON, R. (1987) Hydraulic and sedimentary controls of channel pattern. In: *River Channels, Environment and Process* (Ed. Richards, K.), pp. 129–158. Blackwell, Oxford.

FINLAYSON, B.L. & McMAHON, T.A. (1988) Australia and the world: a comparative analysis of streamflow characteristics. In: *The Fluvial Geomorphology of Australia* (Ed. Warner, R.F.), pp. 17–40. Academic Press, Sydney.

GOH, J. (1993) *Fitzroy River Flooding 1983–1993.* Internal Report, Main Roads Western Australia, Perth.

GRAF, W.L. (1988) Definition of floodplains along arid-region rivers. In: *Flood Geomorphology* (Eds Baker, V.R., Kochel, C. & Patton, P.C.), pp. 231–242. John Wiley & Sons, Toronto.

HICKIN, E.J. (1993) Fluvial facies models: a review of Canadian research. *Progr. Phys. Geogr.*, **17**, 205–222.

KNIGHT, D.W. (1989) Hydraulics of flood channels. In: *Floods: Hydrological, Sedimentological and Geomorphological Implications* (Eds Beven, K. & Carling, P.), pp. 83–106. John Wiley & Sons, Chichester.

KWRDO (1993) *Fitzroy Valley Irrigation – a Conceptual Study.* Kimberley Water Resources Development Office, Perth. ISBN 0730955338.

LANGBEIN, W.B. (1964) The geometry of river channels. *Am. Soc. civ. Eng. Hydrol. Div. Pap.*, **3846**, 297–313.

LEEDER, M.R. (1978) A quantitative stratigraphic model for alluvium, with special reference to channel deposit density and interconnectedness. In: *Fluvial Sedimentology* (Ed. Miall, A.D.), Mem. Can. Soc. petrol. Geol., Calgary, **5**, 587–596.

LEOPOLD, L.B. & LANGBEIN, W.B. (1962) The concept of entropy in landscape evolution. *U.S. geol. Surv. Prof. Pap.*, **500A**, A1–A20.

LEOPOLD, L.B. & MADDOCK, T.J. (1953) The hydraulic geometry of stream channels and some physiographic implications. *U.S. geol. Surv. Prof. Pap.*, **252B**, 1–56.

LEOPOLD, L. B. & WOLMAN, M.G. (1957) River channel patterns – braided, meandering and straight. *U.S. geol. Surv. Prof. Pap.*, **282B**, 39–85.

LEWIN, J. (1992) Floodplain construction and erosion. In: *The Rivers Handbook; Hydrological and Ecological Principles* (Eds Calow, P. & Petts, G.E.), pp. 144–161. Blackwell Scientific Publications, Oxford.

NANSON, G.C. & HUANG, H.Q. (1999) Anabranching rivers: divided efficiency leading to fluvial diversity. In: *Varieties of Fluvial Form* (Eds Miller, A. & Gupta, A.). John Wiley & Sons, New York.

NANSON, G.C. & KNIGHTON, A.D. (1999) Anabranching rivers: their cause, character and classification. *Earth Surf. Process. Landf.*, **21**, 217–239.

OSTERKAMP, W.R. and COSTA, J.E. (1987) Changes accompanying an extraordinary flood on a sand-bed stream. In: *Catastrophic Flooding* (Eds Mayer, L. & Nash, D.), pp. 201–224. Allen & Unwin, Boston.

POPOV, I.V. (1962) Application of morphological analysis to the evaluation of the general channel deformations of the River Ob. *Soviet Hydrol.* (*Am. geophys. Union*), **3**, 267–324.

RICHARDS, K., CHANDRA, S. & FRIEND, P. (1993) Avulsive channel systems: characteristics and examples. In: *Braided Rivers* (Eds Best, J.L. & Bristow, C.S.), Spec Publ. Geol. Soc. London, No. 75, pp. 195–204. Geological Society of London, Bath.

RUTHERFORD, G.K. (1964) *Soils of the West Kimberley Area.* Land Research Series No. 9, pp. 119–133. Commonwealth Scientific and Industrial Research Organization, Melbourne, Australia.

SCHUMM, S.A. (1988) Variability of the fluvial system in time and space. In: *Scales and Global Change* (Eds Rosswall, T., Woodmansee, R.G. & Risser, P.G.), pp. 225–249. John Wiley & Sons, Chichester.

SCHUMM, S.A. & LICHTY, R.W. (1963) Channel widening and flood-plain construction along the Cimarron River in Southwestern Kansas. *U.S. geol. Surv. Prof. Pap.*, **352-D**, 72–88.

SCHUMM, S.A. & LICHTY, R.W. (1965) Time, space and causality in geomorphology. *Am. J. Sci.*, **263**, 110–119.

SCHUMM, S.A., ERSKINE, W.D. & TILLEARD, J.W. (1996) Morphology, hydrology, and evolution of the anastomosing Ovens and King Rivers, Victoria, Australia. *Geol. Soc. Am. Bull.*, **108**(10), 1212–1224.

SMITH, D.G. & SMITH, N.D. (1980) Sedimentation in anastomosed river systems: examples from alluvial valleys near Banff, Alberta. *J. Sediment. Petrol.*, **50**(1), 157–164.

SMITH, N.D., CROSS, T.A., DUFFICY, J.P. & CLOUGH, S.R. (1989) Anatomy of an avulsion. *Sedimentology*, **36**, 1–23.

SPECK, N.H., WRIGHT, R.L. & RUTHERFORD, G.K. (1964) *Land Systems of the West Kimberley Area.* Land Research Series No. 9, pp. 24–73. Commonwealth Scientific and Industrial Research Organization, Melbourne, Australia.

TAYLOR, G. & WOODYER, K.G. (1978) Bank deposition in suspended-load streams. In: *Fluvial Sedimentology* (Ed. Miall, A.D.), Mem. Can. Soc. petrol. Geol., Calgary, **5**, 257–275.

WAWA (1997) Recordings of discharge and flow velocity on the Fitzroy River and floodplain. Unpublished Reports, Water Authority of Western Australia, Perth.

WOHL, E.E., FUERTSCH, S.J. & BAKER, V.R. (1994a) Sedimentary records of late Holocene floods along the Fitzroy and Margaret Rivers, Western Australia. *Aust. J. Earth Sci.*, **41**, 273–280.

WOHL, E.E., WEBB, R.H., BAKER, V.R. & PICKUP, G. (1994b) *Sedimentary Flood Records in Bedrock Canyons of Rivers in Monsoonal Regions of Australia.* Water Resources Research Paper 107, pp. 1–102, Engineering Research Centre, Colorado State University, Fort Collins.

WRIGHT, R.L. (1964) *Geomorphology of the West Kimberley Area.* Land Research Series No. 9, pp. 103–118. Commonwealth Scientific and Industrial Research Organization, Melbourne.

WYRWOLL, K.-H., HOPWOOD, J. & McKENZIE, N.L. (1992) The Holocene paleohydrology and climatic history of the northern Great Sandy Desert–Fitzroy Trough: with special reference to the history of the northwest Australian monsoon. *Climatic Change*, **22**, 47–65.

YANG, C.T. (1971) On river meanders. *J. Hydrol.*, **13**, 231–253.

YEATES, A.N., GIBSON, D.L., TOWNER, R.R. & CROWE, R.W.A. (1984) Regional geology of the onshore Canning Basin, Western Australia. In: *The Canning Basin, Western Australia* (Ed. Purcell, P.G.), pp. 23–56. Geological Society of Australia and Petroleum Exploration Society of Australia Limited, Perth.

*Spec. Publs int. Ass. Sediment.* (1999) **28**, 93–112

# Downstream changes in floodplain character on the Northern Plains of arid central Australia

S. TOOTH*

*School of Geosciences, University of Wollongong, NSW 2522, Australia*

## ABSTRACT

Along the Sandover, Sandover–Bundey and Woodforde Rivers on the Northern Plains in arid central Australia, floodplain landforms, processes and sediments differ between confined upper and middle reaches (where channels and Holocene floodplains are flanked by indurated, Pleistocene alluvial terraces) and unconfined lower reaches (where channels are flanked by extensive Holocene floodplains). In confined reaches, terraces up to 5 m high restrict lateral channel migration and floodplains are typically less than 50 m wide. Channel avulsions, splays and distributary channels are rare and overbank vertical accretion is the main process of floodplain formation. Down-valley, terrace heights decline and they are eventually buried by younger, relatively erodible, floodplain silt and sand. In these unconfined reaches, termed *floodout zones*, channels are more laterally active, floodplains are up to 6 km wide and numerous splays, distributary channels and palaeochannels are present. Channels decrease in size downstream and end in *flood-outs*, a term used to describe sites where channelized flows terminate and floodwaters spill across adjacent alluvial surfaces. In flood-out zones, floodplains form mainly by vertical accretion, lateral point-bar accretion and abandoned-channel infilling. Stratigraphical and sedimentological features characteristic of flood-out zones include:

1 thin veneers of Holocene sediments over Pleistocene alluvium or bedrock;
2 a paucity of sedimentary structures preserved in channel or floodplain deposits;
3 the surficial nature of many floodplain features;
4 incorporation of aeolian sediments in the predominantly fluvial deposits;
5 channel sand and gravel bodies encased in fine-grained overbank deposits;
6 a general down valley decrease in the ratio of channel sands and gravels to overbank fines.

Although an absence of channels makes it difficult to reconcile floodouts with conventional definitions of 'floodplain' or with existing floodplain classifications, fluvial landforms and deposits in floodout zones are best regarded as part of a continuum of floodplain types.

## INTRODUCTION

Although a large number of studies have described the channel deposits resulting from floods in dryland rivers (e.g. McKee *et al.*, 1967; Williams, 1971; Karcz, 1972; Picard & High, 1973; Frostick & Reid, 1977, 1979; Langford & Bracken, 1987), the landforms, processes and sedimentology of dryland river floodplains have received relatively little attention in the literature. In part, this may result from the difficulty of generalizing about

dryland fluvial processes (Olsen, 1987), as floods may be largely channelled (e.g. Karcz, 1972; Picard & High, 1973), largely unchannelled (e.g. Jutson, 1919; Graf, 1988a) or show a combination of both channelled and unchannelled flow (e.g. Sneh, 1983; Pickup, 1991; Tooth, 1999). Better characterization of dryland floodplains is needed, however, because they may harbour proxy records of Quaternary hydrological and climatic change (Nanson & Tooth, in press) and also provide modern analogues for improved interpretation of the geological record (Miall, 1996).

Such diversity of fluvial process is characteristic of the Alice Springs region of arid central Australia, and has

*Present address: Department of Geology, University of the Witwatersrand, Johannesburg, Wits 2050, South Africa. (Email: 065tooth@cosmos.wits.ac.za)

resulted in a variety of floodplain types both along and between different drainage systems. This paper has four main aims:

**1** to describe downstream changes in the channels and floodplains of rivers draining the Northern Plains in the Alice Springs region;

**2** to illustrate differences in floodplain landforms, processes and sediments between confined middle reaches and unconfined lower reaches;

**3** to examine in detail the sedimentology in unconfined lower reaches;

**4** to draw comparisons with previous descriptions of floodplains in other dryland environments.

## REGIONAL SETTING

The 'Alice Springs area' of central Australia (Stewart & Perry, 1962) is an extensive, tectonically stable, arid region (Fig. 1). It is dominated by the central ranges, an east–west trending belt of Proterozoic and Palaeozoic crystalline and sedimentary rocks that cuts across the middle of the region (Fig. 1). The central ranges are broken by intermontane lowlands and are flanked by the extensive, low-relief Northern Plains and Southern Desert Basins (Mabbutt, 1962), where uplands are restricted in extent (Fig. 1).

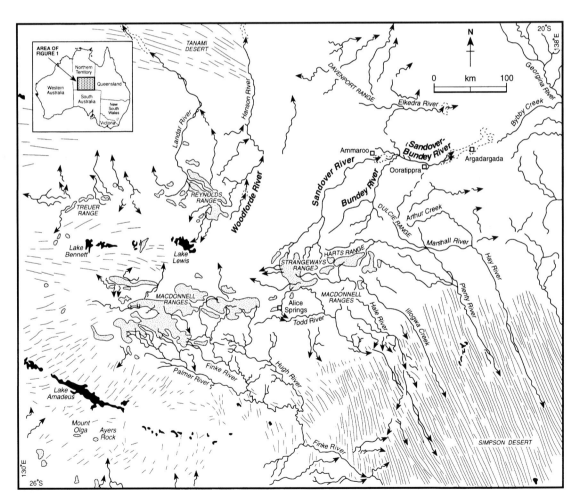

**Fig. 1.** Ephemeral drainage systems in the Alice Springs region of central Australia, highlighting the main rivers referred to in this paper. Stippled areas represent land over 750 m in elevation. The Northern Plains is the extensive area to the north of the central belt of ranges.

The findings presented here are drawn from study of the Sandover, Sandover–Bundey and Woodforde Rivers, three rivers on the Northern Plains with catchment areas of 10 600, 11 000 and 550 km$^2$, respectively (Fig. 1). Rainfall across the Northern Plains averages ~ 300 mm yr$^{-1}$ and pan evaporation is high (~ 3000 mm yr$^{-1}$). As a result, channels remain dry for much of the year and rarely flow throughout their length. Occasional heavy rainfalls (> 100 mm over 24 h), however, result in regionally widespread flooding. Although there are no rated gauging stations on the rivers of the Northern Plains, the last few decades have included some of the largest flood events in the 100–110 yr of European settlement in the region (Williams, 1970; Baker *et al.*, 1983; Pickup *et al.*, 1988; Pickup, 1991).

River channels on the Northern Plains are generally well-defined throughout their length but, in the lower reaches, channel capacities decline to a point at which an increasing proportion of large floods are diverted overbank. In this *floodout zone*, channels eventually disappear but the occasional large floods spread across broad, low-gradient, alluvial surfaces widely known in Australia as *floodouts* (Fig. 2). There are two types of floodout: *intermediate* floodouts, where unchannelled floodwaters spread out but channels reform further down-valley; and *terminal* floodouts, where unchannelled floodwaters spread out and ultimately dissipate (Tooth, 1999).

On the Sandover River, the channel initially disappears at Ammaroo station, some 250 km from the headwater ranges (Fig. 1). Large floods spread across an intermediate floodout that covers approximately 200 km$^2$ before several channels form again and carry floodwaters to the confluence with the Bundey River. Although the Bundey supplies the vast majority of flow and sediment downstream of the confluence, the channel continues as the Sandover River. It is referred to here as the Sandover–Bundey River to avoid confusion with the Sandover River upstream of Ammaroo (Fig. 1). The Sandover–Bundey continues some 55 km further to Ooratippra station, where it divides into several distributary channels which decrease in capacity and eventually disappear (Fig. 1). Large floods disperse across a terminal floodout approximately 800 km$^2$, with unchannelled flows only rarely continuing to Bybby Creek and thence to the Georgina River (Fig. 1), part of the centripetal drainage of Lake Eyre. On the far smaller Woodforde River, the channel initially disappears around 45 km from the ranges (Fig. 1) and large floods spread across an intermediate floodout approximately 40 km$^2$, before several channels form again. These channels carry flows a further 22 km before they disappear near the confluence with the larger Hanson River (Fig. 1).

Anecdotal accounts of local cattle pastoralists provide the only available evidence regarding flood frequency in the lower reaches. On the Sandover River, which receives few tributary contributions for some 135 km upstream of the channel terminus (Fig. 1), floods reach the floodout approximately every 3–4 yr. On the Sandover–Bundey River, which is joined by several small tributaries along its length (Fig. 1), flows reach the floodout at a more frequent interval of 1–2 yr, as is the case on the smaller Woodforde River where the floodout is located closer to the upland sources of runoff. Although there are no quantitative data, the infrequent flows and the large areas over which deposition occurs both result in low rates of sediment accumulation in floodout zones.

## METHODS

Sedimentary deposits along the Sandover, Bundey (Sandover–Bundey) and Woodforde Rivers were investigated by examining natural bank exposures and by hand augering and trenching where practicable. In the lower reaches of the Sandover at Ammaroo and the Sandover–Bundey at Ooratippra (Fig. 1), these methods were complemented by shallow drilling with an Edson 360 hydraulic, truck-mounted drill rig using both solid and hollow augers.

Limited preservation of sedimentary structures in channel or floodplain deposits, and problems of clean recovery of samples during augering and drilling, meant that standard fluvial facies coding schemes (e.g. Miall, 1996) were unsuitable for this study. Recovered or exposed sediments were described in the field by determining Munsell soil colours on air-dried samples and by using particle-size analysis cards. Samples were also collected for detailed grain-size analysis using standard techniques (Folk, 1974) and for thermoluminescence (TL) dating. The basic theory and methods of the TL dating technique are provided in Aitken (1985, 1990) and Nanson *et al.* (1991). In this study, difficulties in identifying suitable modern surface samples for the determination of residual TL in older sediments, means the reported ages are all uncorrected for residual TL and should be regarded as maximum ages. Full details of the TL method, TL ages and palaeoclimatic interpretations are provided in Tooth (1997).

## GEOMORPHOLOGICAL AND STRATIGRAPHICAL CONTEXT

The Sandover, Bundey (Sandover–Bundey) and Woodforde Rivers exhibit broadly similar patterns of

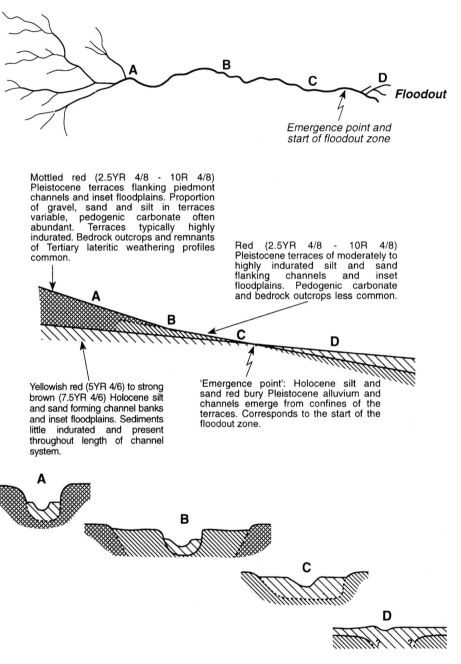

**Fig. 2.** Schematic illustrations of downstream changes in channel–floodplain morphology and alluvial sedimentary units along rivers on the Northern Plains. In the upper and middle reaches, Holocene channels and floodplains are confined by indurated Pleistocene alluvial terraces, bedrock, Tertiary weathering profiles, or aeolian sands (sites A, B, C). In the lower reaches, channels emerge into the unconfined settings of floodout zones, floodplains broaden markedly and Holocene silt and sand aggrades over Pleistocene alluvium (site D) (modified from Tooth, 1999).

downstream channel and floodplain change, as illustrated schematically in Fig. 2. These patterns are typical of many other rivers on the Northern Plains. In the upper reaches, upland networks of small, rocky channels transporting sand and gravel converge on the piedmont to give rise to well-defined channels that flow through low-lying depositional plains. In the uplands, alluvial deposits are restricted mainly to discontinuous pockets of gravel, sand and silt between bedrock spurs or at tributary mouths, but in the piedmont, Holocene floodplains of yellowish red to strong brown silt and sand (TL ages < *c*. 4 ka) are generally well-developed and inset within highly indurated, mottled red Pleistocene terraces of gravel, sand and silt (TL ages > 15 ka) (Fig. 2, site A). Although the terraces are inundated occasionally during extreme floods, in conjunction with bedrock outcrops and remnants of Tertiary

**Fig. 3.** Typical bank exposure in the vicinity of the emergence point of the Woodforde River. Up to 2 m of highly indurated, structureless, red Pleistocene alluvium is unconformably overlain by 1.5 m of little indurated, structureless, yellowish red Holocene alluvium. The contact between the two units is marked by the dotted line. Similar exposures can be identified on the larger Sandover River.

lateritic weathering profiles (Mabbutt, 1965, 1967), they confine channels and floodplains and contain most floods that exceed bankfull stage.

In the middle reaches, channels and Holocene floodplains are flanked by indurated, red Pleistocene terraces of silt and sand (TL ages > 15 ka) (Fig. 2, sites B and C). Throughout these reaches, channel gradients remain moderate to low (typically 0.0005–0.002) and river patterns vary from predominantly single-thread to anabranching. Beyond the limit of further tributary inflows, channel capacities generally decrease downstream owing to declining discharges resulting from transmission losses and floodwave attenuation. In many locations, the confining terraces are highly degraded and gullied or are obscured by aeolian sands, but they continue to provide a resistant boundary for channels and floodplains.

In the lower reaches, the height of Pleistocene terraces above the floodplains gradually declines until shallow burial by younger silt and sand occurs (Fig. 2, site D & Fig. 3). The term *emergence point* is used here to refer to the emergence of channels from the confines of previously enveloping terraces (Fig. 2). Downstream of the emergence point, in the floodout zone, overbank flows spread sediments for greater distances to either side of the channel margins and pockets of late Pleistocene and early Holocene alluvium (TL ages ranging from *c*. 13 to 9.5 ka) have been preserved in palaeochannels or overbank sequences alongside more widespread late Holocene deposits. Channel capacities rapidly decline downstream and channelized flow and bedload transport eventually terminate at floodouts.

## DOWNSTREAM CHANGES IN FLOODPLAIN CHARACTER

Along rivers on the Northern Plains, the character of floodplains in confined upper and middle reaches is different from that in unconfined lower reaches, as summarized in Table 1. Throughout confined reaches (Fig. 2, sites A–C), channels are mostly less than 300 m wide and floodplains rarely exceed 50 m in width, even on the large Sandover and Bundey Rivers. Channels are typically rectangular in cross-section, with flat, sandy floors and steep banks (bank angles > 40°) that range in height from ~ 0.5–3.0 m, laterally stable and low to moderately sinuous (< 1.2). Point bars are rarely developed, although in-channel benches are sometimes a localized but prominent feature of floodplain morphology. Benches abut the base of the channel banks and typically are < 1 m high, with widths ranging from 1 to 5 m (Fig. 4). Most benches are relatively short features but they sometimes persist for up

**Table 1.** Summary of the main differences between floodplains in confined and unconfined reaches of rivers on the Northern Plains.

|  | Confined reaches | Unconfined reaches |
|---|---|---|
| Channel-floodplain morphology | Single-thread or anabranching channels with floodplains typically less than 50 m wide | Single-thread or distributary channels with floodplains up to 6 km wide |
| Channel-floodplain features | In-channel benches, mid-channel ridges and islands, floodplain swales and scour channels | Point bars, in-channel benches, levees, splays, palaeochannels, transverse bedforms, waterholes, pans |
| Floodplain facies | Basal coarse sand and gravel with finer overburden of sand and silt | Mainly overbank sand and silt, occasional sandy to gravelly channel fills, local aeolian sands |
| Processes of floodplain formation | Dominantly overbank vertical accretion | Overbank vertical accretion, lateral point-bar accretion, abandoned-channel infilling |

**Fig. 4.** Vegetated, in-channel bench in a confined middle reach of the Sandover River (view downstream).

**Fig. 5.** Vegetated, mid-channel ridge separating anabranching channels in a confined middle reach of the Bundey River (view upstream).

to 200 m along the channels to form important sediment storage sites. In addition, along the extensive anabranching reaches of the Bundey (Sandover–Bundey) and Woodforde Rivers, elongate, vegetated, mid-channel ridges and broader islands also store sediment (Fig. 5).

Channel banks, benches, ridges, islands and floodplains are typically vegetated with a gallery of trees (principally *Eucalyptus* spp.), shrubs and grasses. Density of grass growth varies considerably along the rivers, usually in response to cattle grazing pressure but, in some reaches, dense growth helps to trap and bind floodplain sediments (Fig. 4). Floodplains tend to have little surface relief, being either flat-lying or gently sloping up to the higher confining surfaces, but in some locations bedrock outcrops introduce surface irregularities. Levees are rarely developed, and although occasional swales or larger scour channels result from overbank flows, there is little evidence of the uneven zone of alternate longitudinal banks and flood furrows described as typical of the floodplains of sand-bed channels in central Australia (Perry *et al.*, 1962; Mabbutt, 1977, 1986). With the exception of one short (17–18 km) reach of the Bundey River, splays and distributary channels are rarely present owing to confinement of channels and floodplains.

Typical floodplain facies in confined reaches include a basal layer of coarse sand and granule/pebble gravel with a finer overburden of sand and silt (Fig. 6). The percentage of silt–clay in channel banks is highly variable but averages around 30–40%. Internal sedimentary structures are rare owing to bioturbation from the colonizing vegetation, although planar and trough cross-bedding are sometimes preserved in benches and ridges.

The above morphological and sedimentological characteristics suggest that floodplain formation in confined reaches is dominated by overbank vertical accretion (Table 1). The high degree of lateral channel stability means lateral point-bar accretion is relatively unimportant, although the layer of coarse sand and gravel at the base of floodplains (Fig. 6) attests to its probable importance in the past. Lateral accretion resulting from the growth of in-channel benches plays only a minor role in floodplain formation, as does oblique accretion of channel banks resulting from muddy drapes of suspended material. Mid-channel ridges and islands can be seen as incipient floodplains in the sense that, if they continue to grow, or if intervening anabranching channels are abandoned, they eventually may become bank-attached and incorporated into floodplains at channel margins.

Floodplains in confined reaches can be modified or destroyed as a result of channel widening, floodplain scour or dissection by gullies and back-channels. Despite large floods in the last few decades, however, aerial photographs available since the 1950s show no evidence

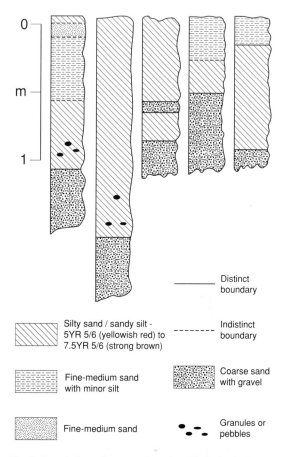

**Fig. 6.** Descriptions of excavated trenches, illustrating typical floodplain facies in confined middle reaches of the Woodforde River.

of substantial change to channels and floodplains (Nanson & Tooth, in press). Although localized instances of bank erosion can be identified in the field, the absence of substantial recent change may have resulted from the protection offered by bank-line vegetation, as well as the influence of the confining terraces in restricting channel migration, avulsion, and splay and distributary channel formation.

In unconfined reaches (floodout zones) (Fig. 2, site D), floodplain widths increase markedly (up to 6 km wide). Burial of Pleistocene terraces by relatively erodible, Holocene silt and sand means that channels are less stable laterally, with greater potential for bank erosion and channel change. Indeed, in unconfined reaches of the Sandover, Sandover–Bundey and Woodforde Rivers there is clear evidence of channel widening, avulsion and extension into floodouts during large floods in the last few decades. In the floodout zones of the Sandover and Woodforde Rivers, channels are single-thread but the

**Fig. 7.** Typical splay in the floodout zone of the Sandover–Bundey River, showing a well-defined feeder channel emanating from a breach in the parent channel bankline (flow direction in parent channel from top right to bottom left).

**Fig. 8.** Abandoned and largely infilled channel of the Sandover–Bundey River, near the terminus of a distributary channel.

Sandover–Bundey divides into three distributary channels. Bank angles and heights are similar to those in confined reaches, but increased lateral instability means that point bars up to 300 m wide are developed occasionally, most notably on the large Sandover and Sandover–Bundey Rivers. In-channel benches up to 1 m high and 5 m wide also are present in some reaches.

Throughout unconfined reaches, channel banks are typically well vegetated with trees, shrubs and grasses, except in several short reaches (< 500 m long) where there has been recent bank erosion. Floodplain vegetation varies considerably depending on grazing pressure and moisture availability, but following floods these areas often support rich carpets of ephemeral grasses and forbs (Lendon & Ross, 1978; Winstanley *et al.*, 1996). On the Woodforde River, low (< 0.5 m) levees are sometimes present but are discontinuous or rare along the Sandover and Sandover–Bundey. Along all three rivers, distributary channels, splays (Fig. 7) and abandoned channels (Fig. 8) introduce local relief to the extensive floodplains. Beyond channel termini, unchannelled floodwaters deposit fine sand and mud on the low-relief floodouts, with areas of aeolian sand up to 6 m high and occasional gullies and palaeochannels also present. Other surficial landforms characteristic of the floodplains and floodouts include transverse bedforms, waterholes and pans (Tooth, 1999).

The percentage of silt–clay in channel banks in unconfined reaches is similar to that in confined reaches. In addition to overbank vertical accretion, however, increased lateral instability in unconfined reaches means lateral point-bar accretion and abandoned-channel infilling are also processes of floodplain formation (Table 1). As a consequence, floodplain facies are more complex. Owing to the aggradational nature of these unconfined reaches, and the wider implications for improved understanding of the nature of low-gradient, dryland floodplains, the remainder of this paper focuses on the sedimentology of floodout zones.

## SEDIMENTOLOGY OF UNCONFINED REACHES

Fluvial depositional environments in unconfined reaches (floodout zones) can be subdivided into channels (active or abandoned), floodplains (flanking the channels) and floodouts (beyond the channel termini). Field investigations concentrated mainly on the Sandover and Sandover–Bundey Rivers but were supported by more limited field-work on the Woodforde. Figure 9 illustrates the location of drill holes, auger holes, excavation pits

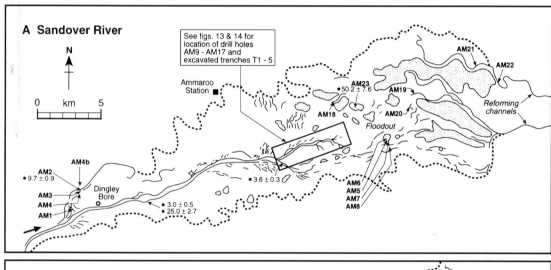

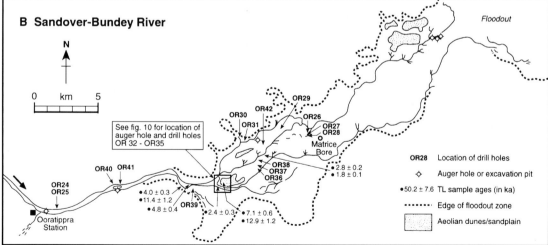

**Fig. 9.** Sketch maps of the floodout zones of the Sandover and Sandover–Bundey Rivers, showing the location of thermoluminescence (TL) samples, drill holes, auger holes and excavation pits. The floodout zones are surrounded by aeolian sands, red-earth plains or bedrock.

and TL sample ages in the floodout zones of the Sandover River at Ammaroo and the Sandover–Bundey River at Ooratippra. At Ooratippra, Aboriginal ownership of part of the extensive floodout restricted drilling to channels and floodplains leading into the floodout. At both locations, access with heavy drilling equipment was also often restricted by steep-sided channels or dense vegetation growth.

On the Sandover River, yellowish red to strong brown, Holocene silt and sand buries red Pleistocene terraces in the transition to the floodout zone (Fig. 2, site D & Fig. 3). Indeed, the vast majority of drill holes encountered indurated red alluvium at depths of 6 m or less. Bed and bank exposures in short sections of channel also indicated that this same alluvial unit was present at relatively shallow depths and showed a typically sharp contact with overlying channel and overbank deposits.

On the Sandover–Bundey River, Holocene deposits bury Pleistocene alluvium and bedrock in the floodout zone (Fig. 2, site D). Most drill holes reached calcareous bedrock (Arrinthrunga Formation) at depths between 3 and 10 m. Bedrock or weathered regolith exposed in channels further confirm the general shallowness of alluvium. Pleistocene and Holocene alluvium overlying bedrock and regolith typically consists of a 0.5–1.0 m thick basal layer of coarse sand and granule/pebble gravel, with a finer overburden of silty sand, which typically changes from yellowish red near the surface to red at depth. Unlike at Ammaroo, however, this colour change is usually very subtle with little evidence of a distinct stratigraphical break. This suggests that Pleistocene alluvium at Ooratippra may have been partially or completely eroded from many locations.

## Surficial floodplain features

The findings from the Sandover and Sandover–Bundey Rivers suggest that Holocene alluvial deposits in the floodout zone are little more than thin veneers over Pleistocene alluvium or bedrock, findings strongly supported by bank exposures in the floodout zone of the Woodforde. On all three rivers, the paucity of preserved sedimentary structures in alluvium probably results from deep weathering of older alluvium, bioturbation of freshly deposited sediments and low rates of sediment accumulation.

Many of the characteristic floodplain features are also only surficial. For instance, trenching and drilling of features such as swales and small depressions that feed water and sediment to pans located in the surrounding aeolian sands, reveal that many are incised only to shallow depths (< 0.5 m) into the underlying sediments. Hence, these features represent little more than a local reworking of older material or a local basin for more recent silt and sand. Similar findings emerge from a study of splays. Although splays are prominent landforms in floodout zones, most are small features, rarely exceeding 0.5 km$^2$ in area. In proximal and medial parts they are generally incised in floodplains in well-defined channels (Fig. 7) but the majority merge with the floodplain surface within 1 km of parent channel margins, sometimes ending in small, sandy lobes that prograde over the fine-grained floodplain sediments. Splay deposits are relatively thin, ranging from 0.5 m or less in proximal parts to 0.1 m or less in distal parts and are subject to aeolian reworking or disturbance by cattle. Hence, although splays are locally important accumulations of relatively coarse-grained sediments, their small size, limited thickness of transported sediments and aeolian reworking limit their volumetric significance within the overall fine-grained floodplain sediments.

## Fluvial–aeolian interactions

An example of a small-scale fluvial–aeolian interaction in floodout zones is the deposition of overbank fines along the margins of the surrounding aeolian sands (Tooth, 1999) and larger scale examples also can be identified. For example, on the Sandover–Bundey at Ooratippra, the initially single-thread channel abruptly bifurcates around a large (~ 1 km$^2$) aeolian dune complex and subsequently follows a distributary pattern towards the floodout. Field inspection of the 4.0–4.5-m-high channel bank-lines at the bifurcation point, together with auger and drill holes and TL ages for fluvial and aeolian sediments (Fig. 10; see Fig. 9B, OR32-35 for location), reveal a 2.5–3.0-m-thick basal layer of late Pleistocene sand and granules overlain by 6.0–7.0 m of slightly silty, fine to medium, aeolian sand, which is encircled by late Holocene silt and sand. The moderate sorting and coarse texture of the basal sediments indicate a fluvial origin. Stratigraphical relationships and TL ages (Fig. 10) suggest that the present-day distributary channel pattern developed as a result of the emplacement of aeolian sand across the Sandover–Bundey channel, with subsequent channel bifurcation and overbank vertical accretion resulting in the incorporation of aeolian sand within the floodplain sediments of the distributary channels.

At Ammaroo, kidney-shaped or circular aeolian dunes up to 1.5 km$^2$ and 5–6 m in height have been surrounded by fluvial deposits in the floodout zone (Fig. 11). A transect drilled across one of these dunes and on to the surrounding floodout (Fig. 12; see Fig. 9A, AM5-8 for location) indicates that around 3.5 m of fine to medium

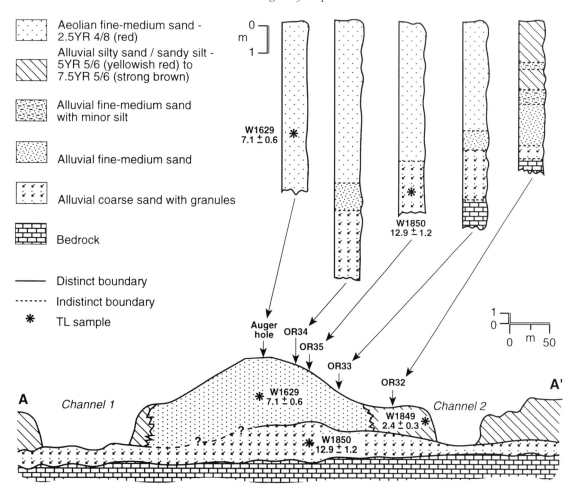

**Fig. 10.** Sedimentological logs and surveyed section showing the relation between aeolian and fluvial sediments in the floodout zone of the Sandover–Bundey River. See Fig. 9B for location.

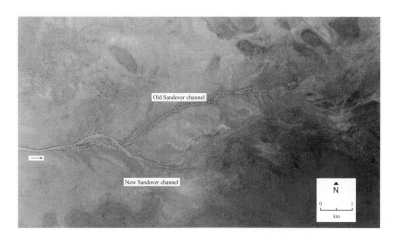

**Fig. 11.** Aerial photograph of the floodout zone of the Sandover River, showing areas of aeolian sand surrounded by fluvial deposits and the old and new line of the channel. (Part of colour aerial photograph: Elkedra–Derry Downs–Sandover, SF53-7, 1982, Ntc 790, Run 11, 010. Reproduced by permission of NT Department of Lands, Planning and Environment, Northern Territory Government.)

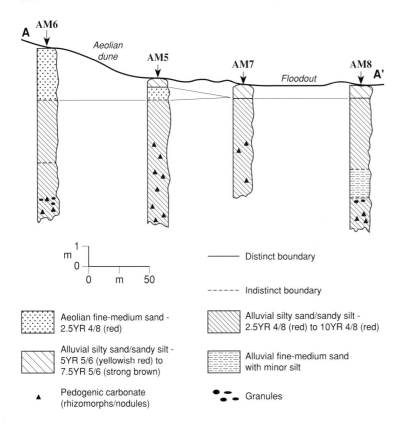

**Fig. 12.** Sedimentological logs showing the relationship between aeolian and fluvial sediments on the floodout of the Sandover River. See Fig. 9A for location.

aeolian sand overlies indurated red alluvial silt and sand. The stratigraphy suggests that the dune developed on the layer of Pleistocene alluvium that underlies the floodout zone. Near the dune margins, the aeolian sand decreases in thickness and is overlain by thin accumulations of recent alluvial silt and sand deposited in the floodout (Fig. 12). The transition is a subtle one, occurring over tens of metres, and suggests gentle onlap of alluvium. Further from the dune, on the surface of the floodout, the aeolian sand gradually pinches out and the alluvial silt and sand directly overlies older Pleistocene alluvium (Fig. 12). Similar examples of intercalated aeolian and fluvial sediments are common in the floodout zone at Ammaroo. For instance, avulsion of the lower section of the Sandover during the large 1974 flood and the cutting of a new channel has resulted in enhanced fluvial deposition at the southern margins of the floodout zone, isolating small areas of aeolian sand (Fig. 11).

**Channel fills**

Examples of fluvial–aeolian interactions in the floodout zones of the Sandover and Sandover–Bundey Rivers demonstrate how aeolian sands can be incorporated locally in alluvial deposits of the floodplains and floodouts. In addition, examples of channel diversion at Ooratippra (Fig. 10) and avulsion at Ammaroo (Fig. 11) show how channel fills can be incorporated in predominantly fine-grained floodplain deposits.

On the ground, older abandoned channels typically appear as lightly vegetated swales and are commonly flanked by recent floodplain deposits or aeolian sands. At Ammaroo, one prominent palaeochannel ~ 8.5 km in length is located just upstream of Dingley Bore (Fig. 9A, AM1-4b). Where it branches from the left-bank of the Sandover, the swale marking the line of the palaeochannel is about 1 km wide but abruptly narrows down to around 150 m. Two holes drilled in the centre of the swale some 2 km apart both showed 1.5–2.0 m of silty sand grading downwards into 2.5–4.0 m of poorly sorted sand and granule/pebble gravel. The poor sorting and coarse texture of the sand and gravel are interpreted as indicating a channel fill with a muted fining upwards succession. In both holes there was an abrupt junction between the channel fill and underlying Pleistocene alluvium. Three holes drilled through the floodplain and aeolian sands adjacent to the swale failed to intersect channel sand or gravel, even at depths of 7–8 m. This suggests that the present-day surface swale broadly follows the subsurface distribution of channel sand and gravel. Similar findings were made at Ooratippra, where drill holes in a small swale (2.5–3.0 km long, 50–80 m wide) near to Matrice Bore (Fig. 9B, OR26-28) intersected channel sand and gravel beneath a 2.0–2.5 m surface cover of silty sand. At both Ammaroo and Ooratippra, swales marking the location of palaeochannels indicate that channel sands and gravels are also buried beneath the fine-grained alluvium of the floodouts.

In addition to these older palaeochannels, channel fills resulting from more recent avulsions can be identified. At Ooratippra, for instance, aerial photographs taken since 1950 show that the distributary channel termini have extended short distances (700–900 m) into the floodout during recent large floods. On one of the channels, extension has resulted in avulsion and infilling of the former course. Today, the 700–750-m-long infilled section of channel is a narrow (3 m wide) depression lined with coolibah trees (*Eucalyptus microtheca*) and shrubs, and marked by low levees of structureless, silty sand (Fig. 8). The channel fill consists of 0.3–0.7 m of structureless, fine–medium sand overlain by 0.1–0.25 m of silty sand, illustrating the slow rates of deposition near and on the floodout.

At Ammaroo, an example of a recent, large-scale avulsion is the abandonment of the lowermost 5–6 km of the channel during the large 1974 flood. A new channel has been excavated in the floodplain sediments (Fig. 11) and the former channel is infilling. For much of its length, the low-sinuosity, single-thread palaeochannel is 160–180 m wide and marked by dead trees. Near its terminus, the palaeochannel breaks into a number of small (10 m wide) distributary channels (Fig. 11). Holes drilled along the palaeochannel (Fig. 13) show poorly sorted sand and occasional granules underlain by yellowish red silty sand, which in turn is underlain by a sharp contact with the red Pleistocene alluvium found at shallow depths throughout the floodout zone. At the point of avulsion, the palaeochannel is almost completely infilled by 3.0–3.5 m of poorly sorted sand and granules, with the uppermost 2.0 m exposed as a result of downcutting by the new Sandover channel. Sections excavated in the fill (Fig. 14, T2, T4 & T5) show structureless sand or silty sand passing downwards into horizontally laminated and planar cross-bedded, fine to medium sand with occasional granules. The fill rapidly decreases in thickness along the length of the palaeochannel to around 1 m within 3 km and to 0.5 m or less in the distributary channels (Fig. 13). Sediment samples from the surface of the fill typically consist of moderately sorted, very fine to medium sand. Although this sand is slightly finer than the bed material of the present Sandover River, to date there is little evidence of infilling by silt and clay.

The nature of the contact between the channel fill and adjacent floodplain deposits is partly obscured by bank slumping and fallen trees. Nevertheless, drill holes (Fig. 13) and excavated sections (Fig. 14, T1 & T3) on the palaeochannel margins show the lateral contact between channel and overbank facies to vary from sharp to gradational, with units of fine to medium sand sometimes interbedded with silty sand. Good exposure of the floodplain alluvium is provided by a large hole, approximately 8 m in diameter and more than 30 m deep, located just a few metres from the palaeochannel margin (Fig. 15). This hole is probably a collapsed hand-dug well, possibly dating to early European settlement in the region. The unstable edges of the hole limited detailed inspection but general observation showed only structureless fine-grained sediments with no evidence of coarse sand or granule units, despite close proximity to the channel fill.

Together, the data from Ammaroo and Ooratippra indicate the general nature of channel fills in floodout zones. More drill holes are needed before definitive statements can be made as to typical channel-fill geometry and size, but from the examples described, the thickness and lateral extent of channel-fill deposits appear to be relatively limited. Hence, channel fills represent locally important accumulations of coarse-grained sediment but form only a minor component of the predominantly fine-grained alluvial deposits in floodout zones.

## DISCUSSION

The term 'floodplain' is usually defined by reference to humid river floodplains, which occur adjacent to river channels, possess predominantly horizontal surfaces that are activated by the present flood regime, and typically occupy clearly delimited zones with morphological boundaries (e.g. Bates & Jackson, 1987; Nanson & Croke, 1992). As floods in drylands can be largely channelled, largely unchannelled, or show varying combinations of both (e.g. Jutson, 1919; Karcz, 1972; Picard & High, 1973; Sneh, 1983; Graf, 1988a; Pickup, 1991; Tooth, 1999), however, the depositional landforms and flood deposits of dryland rivers are more variable. Indeed, some authors even consider that many dryland rivers do not have conventional floodplains, being characterized either by braided channels that occupy the entire space between low terraces (Graf, 1988b, p. 217) or by 'less clearly defined' zones adjacent to the channel that consist

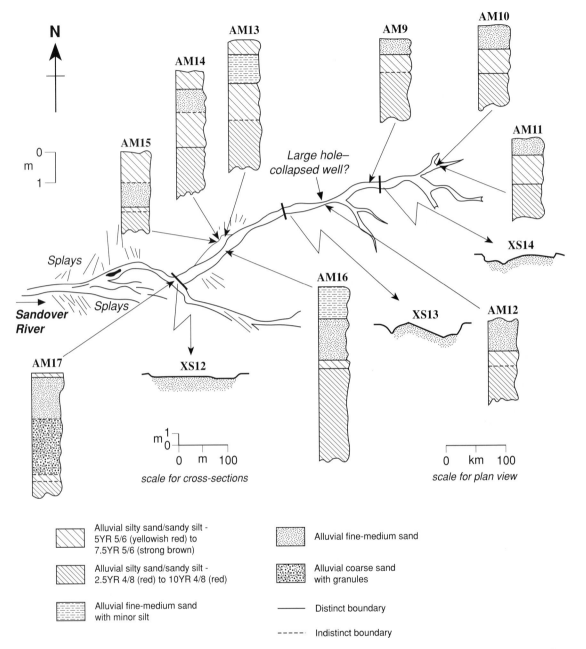

**Fig. 13.** Sketch map of the abandoned channel of the Sandover River showing sedimentological logs and surveyed cross-sections. See Fig. 9A for location.

of a 'composite mixture of alluvial features' (Cooke *et al.*, 1993, p. 155).

Some dryland rivers, however, do possess conventional floodplains. For example, in the semi-arid American south-west, Schumm & Lichty (1963), Burkham (1972) and Hereford (1984, 1986) have described flood-plain destruction resulting from channel widening during large floods, with later rebuilding during periods of lower discharge. In arid Australia, floodplain landforms, processes and deposits have been described in the

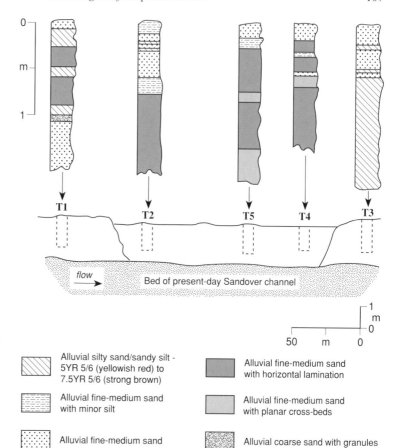

**Fig. 14.** Descriptions of excavated trenches in the abandoned channel of the Sandover River at the point of avulsion, illustrating channel fill and adjoining floodplain facies. See Fig. 9A for location. Note change in vertical scale of the sedimentological logs as compared with Figs 10, 12 & 13.

**Fig. 15.** Exposure of floodplain alluvium in a large hole near the margins of a recently abandoned channel of the Sandover River. The abandoned channel is marked by the line of dead trees visible in the background.

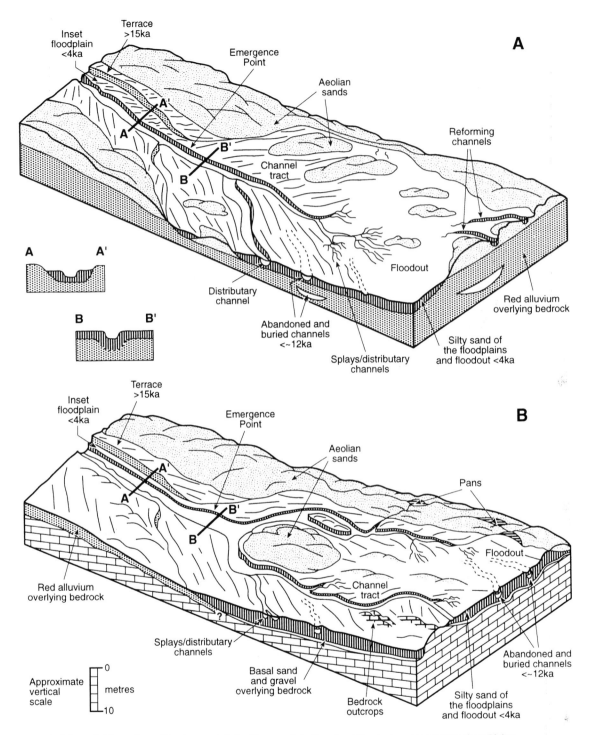

**Fig. 16.** Schematic illustrations of the characteristic landforms and stratigraphy of floodout zones on the Northern Plains: (A) intermediate floodouts of the Sandover and Woodforde Rivers; (B) terminal floodout of the Sandover–Bundey River. The diagrams illustrate how upstream confining terraces (and/or bedrock) are buried downvalley by younger silt and sand of the floodplains and floodouts. On large rivers such as the Sandover and Sandover–Bundey, channels traverse distances up to 40 km into the floodout zone, before eventually disappearing.

Channel Country, western Queensland (e.g. Rust, 1981; Nanson *et al.*, 1986, 1988; Rust & Nanson, 1986, 1989; Maroulis & Nanson, 1996) as well as in the Alice Springs region (e.g. Pickup, 1991; Patton *et al.*, 1993; Bourke, 1994).

Most previous descriptions of dryland river floodplains have been from settings where floods are largely channelled and there have been few descriptions where floods are only partially channelled or largely unchannelled (e.g. Jutson, 1919; Sneh, 1983; Graf, 1988a; Pickup, 1991; Patton *et al.*, 1993). In particular, little is known about floodplain landforms, processes and deposits in low-gradient, unconfined settings typical of lower reaches of rivers on the Northern Plains.

Floodout zones of rivers on the Northern Plains encompass channels, floodplains and floodouts and thus contain a mix of channel and overbank facies. Figure 16A illustrates features typical of intermediate floodouts, as found on the Sandover and Woodforde Rivers, where recent alluvial deposits are underlain by red Pleistocene alluvium and where channels reform further downstream. Figure 16B illustrates the terminal floodout of the Sandover–Bundey River, where alluvial deposits of the floodout zone are underlain by bedrock at shallow depths and where there are no channels further downstream. Despite differences in detail, the floodout zones of the three rivers show a number of similarities in stratigraphy and sedimentology. These include:

**1** thin veneers of Holocene sediments over Pleistocene alluvium or bedrock;

**2** a paucity of sedimentary structures preserved in channel or floodplain deposits;

**3** the surficial nature of many floodplain features;

**4** incorporation of aeolian sediments in the predominantly fluvial deposits;

**5** channel sand and gravel bodies encased in fine-grained overbank deposits;

**6** a general downvalley decrease in the ratio of channel sands and gravels to overbank fines.

Many stratigraphical and sedimentological features typical of floodout zones on the Northern Plains are also characteristic of other low-gradient, dryland fluvial systems, particularly those rivers transporting abundant bedloads and which are subject to declining downstream flows. In the Alice Springs region, thin veneers of younger (typically Holocene) alluvial sediments over heavily weathered deposits have been noted from a number of piedmont settings close to the central ranges (Litchfield, 1969; Pickup, 1991; Patton *et al.*, 1993). In many places, these younger deposits lack internal stratification and are associated with surface morphological features such as sand threads, sand sheets and

transverse bedforms (Pickup, 1991; Patton *et al.*, 1993). Similarly, alluvial sediments intercalating or interdigitating with aeolian sands have been described in many drylands, both from modern depositional settings (e.g. Glennie, 1970; Lancaster & Teller, 1988; Nanson *et al.*, 1988; Langford, 1989; Callen & Bradford, 1992) and the rock record (e.g. Ward, 1988; Langford & Chan, 1989; Olsen, 1989; Trewin, 1993; Smith *et al.*, 1993; Jones & Blakey, 1997). Predominantly fine-grained dryland floodplain sediments containing channel sand and gravel bodies also have been described widely, both from modern (e.g. Rust, 1981; Parkash *et al.*, 1983; Nanson *et al.*, 1988; McCarthy *et al.*, 1991, 1992) and from ancient examples (e.g. Friend, 1978; Friend *et al.*, 1979; Graham, 1983; Rust & Legun, 1983; Nichols, 1987; Olsen, 1987; Kelly & Olsen, 1993; Sadler & Kelly, 1993). Sand and gravel bodies are particularly common in the floodplains of anastomosing, anabranching and distributary dryland rivers, where the principal mode of floodplain formation is overbank vertical accretion and channel avulsion is common.

In terms of geomorphological setting, as well as in aspects of morphology and sedimentology, floodout zones have some similarities with terminal fans, which are characterized by extensive distributary channels that grade distally into sheet flood deposits with poorly defined channels (e.g. Mukerji, 1975, 1976; Friend, 1978; Graham, 1983; Abdullatif, 1989; Kelly & Olsen, 1993). A general absence of extensive distributary networks in floodout zones and a common location on middle or lower reaches of rivers where channels reform downstream of the flood-out, however, makes application of the term 'terminal fan' inappropriate (Tooth, 1999). In floodout zones, downstream disappearance of channelized flow means that floodplains (in the conventional sense of the term) grade into floodouts, which vertically accrete dominantly fine-grained alluvial sediments. Although an absence of channels makes it difficult to reconcile floodouts with conventional definitions of 'floodplain' or with existing floodplain classifications (e.g. Bates & Jackson, 1987; Nanson & Croke, 1992), fluvial landforms and deposits in floodout zones are best regarded as part of a continuum of floodplain types.

## CONCLUSION

Many previous studies of rivers in more humid environments have made a distinction between 'confined' and 'partially confined' or 'unconfined' reaches and have shown the degree of confinement to be significant for processes of floodplain formation or erosion (e.g. Lewin &

*S. Tooth*

Brindle, 1977; Milne, 1983; Sneh, 1983; Nanson, 1986; Brown, 1990; Warner, 1992; Miller & Parkinson, 1993; Rutherfurd, 1994). Similarly, on the Sandover, Sandover–Bundey and Woodforde Rivers, the distinction between confined upper and middle reaches and unconfined lower reaches has a number of implications for floodplain land-forms, processes and the resultant sedimentary deposits.

In confined reaches, narrow floodplains are inset within Pleistocene alluvial terraces, bedrock or aeolian sands, and consist of basal coarse sand and gravel with a finer overburden of largely structureless sand and silt. Floodplains typically have little surface relief, with levees, splays and distributary channels rarely developed, and overbank vertical accretion is the main process of floodplain formation. By contrast, in unconfined reaches, broad floodplains typically consist of thin veneers of Holocene sediments over Pleistocene alluvium or bed-rock. Holocene sediments largely consist of structureless, fine-grained sand and silt, but also incorporate sandy to gravelly channel fills as well as aeolian sands. Features such as splays and abandoned channels are common, and floodplains form by a number of processes, including ver-tical accretion, lateral point-bar accretion and abandoned-channel infilling.

The floodplains of rivers on the Northern Plains form by a number of different processes and are characterized by a variety of active and abandoned fluvial features. Their description provides a valuable addition to the rel-atively small body of knowledge on dryland river flood-plains. Growing interest in using the deposits of dryland rivers for reconstruction of Quaternary hydroclimatic change and as modern analogues for ancient hydrocarbon reservoirs means that characterization of floodplains in aggradational areas, such as unconfined lower reaches of rivers in central Australia, will become increasingly important. Study of the Sandover, Sandover–Bundey and Woodforde Rivers illustrates floodplains typical of rivers on the Northern Plains, but investigations of rivers in other geomorphological settings, such as those termin-ating among longitudinal dunefields in the western Simpson Desert (Fig. 1), will provide further examples of floodplain landforms, processes and deposits character-istic of low-gradient, dryland rivers.

## ACKNOWLEDGEMENTS

The findings presented form part of a PhD thesis super-vised by Gerald Nanson.

Research was supported by a postgraduate scholarship from the University of Wollongong, by the Quaternary Environments Research Centre at the University of Wollongong and by the Australian Research Council via grants to Gerald Nanson and David Price. Thermo-luminescence ages were provided by David Price and José Abrantes (School of Geosciences, University of Wollongong). The Weirs (Ammaroo station) and the Fulchers (Ooratippra station) generously allowed access to their properties with drilling equipment, and Damien Kelleher (Department of Archaeology and Natural His-tory, Australian National University) operated a drilling rig under difficult conditions. Richard Walsh, Steve Beaman, Amanda Hajnal, He Qing Huang and Alice Turkington helped during field-work. Cartographic assistance was provided by Chris Lewis and Elaine Watts (Department of Geography, University of Nottingham). Constructive criticism by reviewers Michael Blum and Lawrence Jones helped to improve an earlier version of the manuscript. The manuscript was prepared while the author was employed as a temporary lecturer by the Department of Geography, University of Nottingham.

## REFERENCES

ABDULLATIF, O.M. (1989) Channel-fill and sheet-flood facies sequences in the ephemeral River Gash, Kassala, Sudan. *Sediment. Geol.*, **63**, 171–184.

AITKEN, M.J. (1985) *Thermoluminescence Dating*. Academic Press, London.

AITKEN, M.J. (1990) *Science-based Dating in Archaeology*. Longman, London.

BAKER, V.R., PICKUP, G. & POLACH, H.A. (1983) Desert palaeofloods in central Australia. *Nature*, **301**, 502–504.

BATES, R.L. & JACKSON, J.A. (1987) *Glossary of Geology* (3rd edn.). American Geological Institute, Alexandria, Virginia.

BOURKE, M.C. (1994) Cyclical construction and destruction of flood dominated flood plains in semiarid Australia. In: *Variability in Stream Erosion and Sediment Transport*, pp. 113–123. Publication No. 224, International Association of Hydrological Sciences, Wallingford.

BROWN, A.G. (1990) Holocene floodplain diachronism and inherited downstream variations in fluvial processes: a study of the River Perry, Shropshire, England. *J. Quat. Sci.*, **5**, 39–51.

BURKHAM, D.E. (1972) Channel changes of the Gila River in Safford Valley, Arizona 1846–1970. *U.S. geol. Surv. Prof. Pap.*, **655-G**.

CALLEN, R.A. & BRADFORD, J. (1992) The Cooper Creek fan and Strzlecki Creek — hypsometric data, Holocene sedimentation, and implications for human activity. In: *Mines and Energy Review 158*, pp. 52–57. South Australia Department of Mines and Energy, Adelaide.

COOKE, R.U., WARREN, A. & GOUDIE, A.S. (1993) *Desert Geomorphology*. University College London Press, London.

FOLK, R.F. (1974) *Petrology of Sedimentary Rocks*. Hemphill Publishing Company, Austin, TX.

FRIEND, P.F. (1978) Distinctive features of some ancient river systems. In: *Fluvial Sedimentology* (Ed. Miall, A.D.), Mem. Can. Soc. petrol. Geol., Calgary, **5**, 531–542.

FRIEND, P.F., SLATER, M.J. & WILLIAMS, R.C. (1979) Vertical and lateral building of river sandstone bodies, Ebro Basin, Spain. *J. Geol. Soc. London*, **12**, 36–46.

FROSTICK, L.E. & REID, I. (1977) The origin of horizontal laminae in ephemeral stream channel-fill. *Sedimentology*, **24**, 1–9.

FROSTICK, L.E. & REID, I. (1979) Drainage-net control of sedimentary parameters in sand-bed ephemeral streams. In: *Geographical Approaches to Fluvial Processes* (Ed. Pitty, A.F.), pp. 173–201. GeoAbstracts, Norwich.

GLENNIE, K.W. (1970) *Desert Sedimentary Environments.* Elsevier Publishing Company, Amsterdam.

GRAF, W.L. (1988a) Definition of flood plains along arid-region rivers. In: *Flood Geomorphology* (Eds Baker, V.R., Kochel, R.C. & Patton, P.C.), pp. 231–242. John Wiley & Sons, New York.

GRAF, W.L. (1988b) *Fluvial Processes in Dryland Rivers.* Springer-Verlag, Berlin.

GRAHAM, J.R. (1983) Analysis of the Upper Devonian Munster Basin, an example of a fluvial distributary system. In: *Modern and Ancient Fluvial Systems* (Eds J.D. Collinson, J.D. & Lewin, J.), Spec. Publs Int. Ass. Sediment., No. 6, pp. 473–483. Blackwell Scientific Publications, Oxford.

HEREFORD, R. (1984) Climate and ephemeral-stream processes: twentieth-century geomorphology and alluvial stratigraphy of the Little Colorado River, Arizona. *Geol. Soc. Am. Bull.*, **95**, 654–668.

HEREFORD, R. (1986) Modern alluvial history of the Paria River drainage basin, southern Utah. *Quat. Res.*, **25**, 293–311.

JONES, L.S. & BLAKEY, R.C. (1997) Eolian–fluvial interaction in the Page Sandstone (Middle Jurassic) in south-central Utah, USA — a case study of erg-margin processes. *Sediment. Geol.*, **109**, 181–198.

JUTSON, J.T. (1919) Sheet-flows, or sheet-floods, and their associated phenomena in the Niagara District of sub-arid south-central Western Australia. *Am. J. Sci.*, **48**, 435–439.

KARCZ, I. (1972) Sedimentary structures formed by flash floods in southern Israel. *Sediment. Geol.*, **7**, 161–182.

KELLY, S.B. & OLSEN, H. (1993) Terminal fans — a review with reference to Devonian examples. In: *Current Research in Fluvial Sedimentology* (Ed. Fielding, C.R.). *Sediment. Geol.*, **85**, 339–374.

LANCASTER, N. & TELLER, J.T. (1988) Interdune deposits of the Namib Sand Sea. *Sediment. Geol.*, **55**, 91–107.

LANGFORD, R.P. (1989) Fluvial–aeolian interactions: Part 1, modern systems. *Sedimentology*, **36**, 1023–1035.

LANGFORD, R.P. & BRACKEN, B. (1987) Medano Creek, Colorado, a model for upper-flow-regime fluvial deposition. *J. sediment. Petrol.*, **57**, 863–870.

LANGFORD, R.P. & CHAN, M.A. (1989) Fluvial–aeolian interactions: Part II, ancient systems. *Sedimentology*, **36**, 1037–1051.

LENDON, C. & ROSS, M.A. (1978) Vegetation. In: *The Physical and Biological Features of Kunoth Paddock in Central Australia* (Ed. Low, W.A.), pp. 66–80. Division of Land Resources Management, Technical Paper No. 4, CSIRO, Melbourne.

LEWIN, J. & BRINDLE, B.J. (1977) Confined meanders. In: *River Channel Changes* (Ed. Gregory, K.J.), pp. 221–233. John Wiley & Sons, Chichester.

LITCHFIELD, W.H. (1969) *Soil Surfaces and Sedimentary History near the Macdonnell Ranges, N.T.* Soil Publication no. 25, CSIRO, Melbourne.

MABBUTT, J.A. (1962) Part VII. Geomorphology of the Alice Springs area. In: *General Report on Lands of the Alice Springs Area, Northern Territory, 1956–57* (Ed. Perry, R.A.), pp. 163–184. Land Research Series, No. 6. CSIRO, Melbourne.

MABBUTT, J.A. (1965) The weathered land surface in central Australia. *Z. Geomorphol. NF*, **9**, 82–114.

MABBUTT, J.A. (1967) Denudation chronology in central Australia: structure, climate and landform inheritance in the Alice Springs area. In: *Landform Studies from Australia and New Guinea* (Eds Jennings, J.N. & Mabbutt, J.A.), pp. 144–181. Australian National University Press, Canberra.

MABBUTT, J.A. (1977) *Desert Landforms.* Australian National University Press, Canberra.

MABBUTT, J.A. (1986) Desert lands. In: *Australia — A Geography*, Vol. 1, *The Natural Environment* (Ed. Jeans, D.N.), pp. 180–202. Sydney University Press, Sydney.

MAROULIS, J.C. & NANSON, G.C. (1996) Bedload transport of aggregated muddy alluvium from Cooper Creek, central Australia: a flume study. *Sedimentology*, **43**, 771–790.

MCCARTHY, T.S., STANISTREET, I.G. & CAIRNCROSS, B. (1991) The sedimentary dynamics of active fluvial channels on the Okavango fan, Botswana. *Sedimentology*, **38**, 471–487.

MCCARTHY, T.S., ELLERY, W.N. & STANISTREET, I.G. (1992) Avulsion mechanisms on the Okavango fan, Botswana: the control of a fluvial system by vegetation. *Sedimentology*, **39**, 779–795.

MCKEE, E.D., CROSBY, E.J. & BERRYHILL, H.L., JR. (1967) Flood deposits, Bijou Creek, Colorado, June 1965. *J. sediment. Petrol.*, **37**, 829–851.

MIALL, A.D. (1996) *The Geology of Fluvial Deposits: Sedimentary Facies, Basin Analysis, and Petroleum Geology.* Springer-Verlag, Berlin.

MILLER, A.J. & PARKINSON, D.J. (1993) Flood hydrology and geomorphic effects on river channels and flood plains: the flood of November 4–5, 1985, in the South Branch Potomac River basin of West Virginia. In: *Geomorphic Studies of the Storm and Flood of November 3–5, 1985, in the Upper Potomac and Cheat River Basins in West Virginia and Virginia* (Ed. Jacobson, R.B.). *U.S. geol. Surv. Bull.*, **1981**, E1–E96.

MILNE, J.A. (1983) Patterns of confinement in some stream channels of upland Britain. *Geogr. Annaler Ser. A*, **65**, 67–83.

MUKERJI, A.B. (1975) Geomorphic patterns and processes in the terminal triangular tract of inland streams in Sutlej-Yamuna Plain. *J. geol. Soc. India*, **16**, 450–459.

MUKERJI, A.B. (1976) Terminal fans of inland streams in Sutlej-Yamuna Plain, India. *Z. Geomorphol. NF*, **20**, 190–204.

NANSON, G.C. (1986) Episodes of vertical accretion and catastrophic stripping: a model of disequilibrium flood-plain development. *Geol. Soc. Am. Bull.*, **97**, 1467–1475.

NANSON, G.C. & CROKE, J.C. (1992) A genetic classification of floodplains. In: *Floodplain Evolution* (Eds Brackenbridge, G.R. & Hagedorn, J.). *Geomorphology*, **4**, 459–486.

NANSON, G.C. & TOOTH, S. (in press) Arid-zone rivers as indicators of climate change. In: *Palaeoenvironmental Reconstruction in the Arid Zone* (Eds Singhvi, A.K. & Derbyshire, E.). Oxford and IBH Publishing Co., New Delhi.

NANSON, G.C., RUST, B.R. & TAYLOR, G. (1986) Coexistent mud braids and anastomosing channels in an arid-zone river: Cooper Creek, central Australia. *Geology*, **14**, 175–178.

NANSON, G.C., YOUNG, R.W., PRICE, D.M. & RUST, B.R. (1988) Stratigraphy, sedimentology and late-Quaternary chronology of the Channel Country of western Queensland. In: *Fluvial*

*Geomorphology of Australia* (Ed. Warner, R.F.), pp. 151–175. Academic Press, Sydney.

NANSON, G.C., PRICE, D.M., SHORT, S.A., YOUNG, R.W. & JONES, B.G. (1991) Comparative uranium–thorium and thermoluminescence dating of weathered Quaternary alluvium in the tropics of northern Australia. *Quat. Res.*, **35**, 347–366.

NICHOLS, G.J. (1987) Structural controls on fluvial distributary systems — the Luna system, Northern Spain. In: *Recent Developments in Fluvial Sedimentology* (Eds Ethridge, F.G., Flores, R.M. & Harvey, M.), Spec. Publ. Soc. econ. Paleont. Mineral., Tulsa, **39**, 269–277.

OLSEN, H. (1987) Ancient ephemeral stream deposits: a local terminal fan model from the Bunter Sandstone Formation (L. Triassic) in the Tønder-3, -4 and -5 wells, Denmark. In: *Desert Sediments: Ancient and Modern* (Eds Frostick, L.E. & Reid, I.), Spec. Publ. geol. Soc. London, No. 35, pp. 69–86. Blackwell Scientific Publications, Oxford.

OLSEN, H. (1989) Sandstone-body structures and ephemeral stream processes in the Dinosaur Canyon Member, Moenave Formation (Lower Jurassic), Utah, USA. *Sediment. Geol.*, **61**, 207–221.

PARKASH, B., AWASTHI, A.K. & GOHAIN, K. (1983) Lithofacies of the Markanda terminal fan, Kurukshetra district, Haryana, India. In: *Modern and Ancient Fluvial Systems* (Eds Collinson, J.D. & Lewin, J.), Spec. Publs int. Ass. Sediment., No. 6, pp. 337–344. Blackwell Scientific Publications, Oxford.

PATTON, P.C., PICKUP, G. & PRICE, D.M. (1993) Holocene paleofloods of the Ross River, central Australia. *Quat. Res.*, **40**, 201–212.

PERRY, R.A., MABBUTT, J.A., LITCHFIELD, W.H. & QUINLAN, T. (1962) Part II. Land systems of the Alice Springs area. In: *General Report on Lands of the Alice Springs Area, Northern Territory, 1956–57* (Ed. Perry, R.A.), pp. 20–108. Land Research Series, No. 6. CSIRO, Melbourne.

PICARD, M.D. & HIGH, L.R., JR. (1973) *Sedimentary Structures of Ephemeral Streams.* Elsevier, Amsterdam.

PICKUP, G. (1991) Event frequency and landscape stability on the floodplain systems of arid central Australia. *Quat. Sci. Rev.*, **10**, 463–473.

PICKUP, G., ALLAN, G. & BAKER, V.R. (1988) History, palaeochannels and palaeofloods of the Finke River, central Australia. In: *Fluvial Geomorphology of Australia* (Ed. Warner, R.F.), pp. 105–127. Academic Press, Sydney.

RUST, B.R. (1981) Sedimentation in an arid-zone anastomosing fluvial system: Cooper Creek, central Australia. *J. Sediment. Petrol.*, **51**, 745–755.

RUST, B.R. & LEGUN, A.S. (1983) Modern anastomosing-fluvial deposits in arid Central Australia, and a Carboniferous analogue in New Brunswick, Canada. In: *Modern and Ancient Fluvial Systems* (Eds Collinson, J.D. & Lewin, J.), Spec. Publs int. Ass. Sediment., No. 6, pp. 385–392. Blackwell Scientific Publications, Oxford.

RUST, B.R. & NANSON, G.C. (1986) Contemporary and palaeochannel patterns and the late Quaternary stratigraphy of Cooper Creek, southwest Queensland, Australia. *Earth Surf. Process. Landf.*, **11**, 581–590.

RUST, B.R. & NANSON, G.C. (1989) Bedload transport of mud as pedogenic aggregates in modern and ancient rivers. *Sedimentology*, **36**, 291–306.

RUTHERFURD, I.D. (1994) Inherited controls on the form of a large, low energy river: the Murray River, Australia. In: *The Variability of Large Alluvial Rivers* (Eds Schumm, S.A. & Winkley, B.R.), pp. 177–197. American Society of Civil Engineers, New York.

SADLER, S.P. & KELLY, S.B. (1993) Fluvial processes and cyclicity in terminal fan deposits: an example from the Late Devonian of southwest Ireland. In: *Current Research in Fluvial Sedimentology* (Ed. Fielding, C.R.). *Sediment. Geol.*, **85**, 375–386.

SCHUMM, S.A. & LICHTY, R.W. (1963) Channel widening and flood-plain construction along Cimarron River in southwestern Kansas. *U.S. geol. Surv. Prof. Pap.*, **352-D**, 71–88.

SMITH, R.M.H., MASON, T.R. & WARD, J.D. (1993) Flash-flood sediments and ichnofacies of the Late Pleistocene Homeb silts, Kuiseb River, Namibia. In: *Current Research in Fluvial Sedimentology* (Ed. Fielding, C.R.). *Sediment. Geol.*, **85**, 579–599.

SNEH, A. (1983) Desert stream sequences in the Sinai Peninsula. *J. sediment. Petrol.*, **53**, 1271–1279.

STEWART, G.A. & PERRY, R.A. (1962) Part I. Introduction and summary description of the Alice Springs area. In: *General Report on Lands of the Alice Springs Area, Northern Territory, 1956–57* (Ed. Perry, R.A.), pp. 9–19. Land Research Series, No. 6. CSIRO, Melbourne.

TOOTH, S. (1997) *The morphology, dynamics and Late Quaternary sedimentary history of ephemeral drainage systems on the Northern Plains of central Australia.* Unpublished PhD thesis, University of Wollongong.

TOOTH, S. (1999) Floodouts in central Australia. In: *Varieties of Fluvial Form* (Eds Miller, A. and Gupta, A.), pp. 219–247. John Wiley & Sons, Chichester.

TREWIN, N.H. (1993) Controls on fluvial deposition in mixed fluvial and aeolian facies within the Tumblagooda Sandstone (Late Silurian) of Western Australia. In: *Current Research in Fluvial Sedimentology* (Ed. Fielding, C.R.). *Sediment. Geol.*, **85**, 387–400.

WARD, J.D. (1988) Eolian, fluvial and pan (playa) facies of the Tertiary Tsondab Sandstone Formation in the central Namib Desert, Namibia. *Sediment. Geol.*, **55**, 143–162.

WARNER, R.F. (1992) Floodplain evolution in a New South Wales coastal valley, Australia: spatial process variations. In: *Floodplain Evolution* (Eds Brackenbridge, G.R. & Hagedorn, J.). *Geomorphology*, **4**, 447–458.

WILLIAMS, G.E. (1970) The central Australian stream floods of February–March 1967. *J. Hydrol.*, **11**, 185–200.

WILLIAMS, G.E. (1971) Flood deposits of the sand-bed ephemeral streams of central Australia. *Sedimentology*, **17**, 1–40.

WINSTANLEY, D., MAHNEY, T. & IRRMARN COMMUNITY (1996) *Irrmarn: Apmer Anwernekakerrenh, Land Resource Assessment of Irrmarn Aboriginal Land Trust.* Land Resource Assessment Unit, Technical Report No. 1, Central Land Council, Alice Springs.

*Spec. Publs int. Ass. Sediment.* (1999) **28**, 113–130

# Confined meandering river eddy accretions: sedimentology, channel geometry and depositional processes

L. M. BURGE *and* D. G. SMITH

*Department of Geography, University of Calgary, Calgary, Alberta T2N 1N4, Canada*
*(Email: dgsmit@acs.ucalgary.ca)*

## ABSTRACT

Eddy accretions occur along margins of confined channel-meander belts where unusually deep scours, eroded by river flow impacting valley sides at right angles, infill primarily with sand. In unconfined meanders (e.g. Mississippi River), eddy accretions can form where channels impinge against resistant valley sides (bedrock, diamicton, silt–clay) or silt–clay oxbow fills within the valley. In confined settings, eddy-accretion deposits commonly occupy 25% of the floodplain and form where the ratio of floodplain-width to channel-width varies between 5 and 10.

From vibracores in confined meandering river valleys, stratigraphical profiles indicate that eddy-accretion deposits are considerably thicker than adjacent point bars, e.g., 17 m versus 6 m and 9 m versus 5 m for the Kootenay and Beaver rivers, respectively. In confined valleys, cross-valley geometries of eddy-accretion and point-bar deposits resemble dumb-bells, thick on the ends but thin across the middle. In one case (Kootenay River), a single eddy-accretion fill consists of two upward-fining sequences: a pebbly channel lag that crudely fines upward to a mid-sequence thick silt, above which rests another sandy sequence on an erosive base, similar to the lower, but capped by a rooted overbank silt. This apparent double upward-fining succession could be misinterpreted as the product of two superimposed point-bar upward-fining sequences. Acoustic bottom profiles of unusually deep scour holes adjacent to eddy accretions confirm the great thickness of fills as compared with nearby point bars.

## INTRODUCTION

Eddy accretions are one of the few remaining unstudied sedimentary deposits within meandering river systems (Fig. 1). In this paper the term eddy accretion is used to define morphology, a group of depositional processes and sedimentary deposits that form up-valley from abrupt channel bends along valley sides. Eddy accretions commonly form as unusually thick deposits (1.5–3 times thicker than point bars on the same river) along flanks of confined meandering river valleys where parallel resistant valley walls are close together. They also occur in unconfined meander belts, where rivers flow against resistant valley banks, as observed in the Mississippi and Red rivers, USA (Carey, 1969) and Peace and Liard rivers, Canada (Burge, 1997). In confined reaches of the Kootenay, Fort Nelson, Muskwa, Clearwater and Beaver rivers in British Columbia and Alberta, Canada, we have observed that eddy-accretion morphology constitutes 25–40% of the floodplain area (not including channel

area). Deep scour holes form where channel flow impacts the valley wall at high angles, causing flow separation and a reverse-flow eddy on the up-valley channel side (Carey, 1969). This immediate change in flow direction absorbs as much energy as several meander wave lengths (Carey, 1969). In confined meanders, eddy accretions migrate down-valley simultaneously with down-valley migration of scour holes and adjacent mid-valley point bars. Eddy accretion scroll-bar patterns are reverse to point-bar scroll patterns (Fig. 1), that is, eddy-accretion scrolls are concave whereas point-bar scrolls are convex when looking down-valley.

Despite their widespread occurrence, eddy-accretion sedimentology and details of their depositional processes remain largely unstudied. Accretion eddies were first identified and studied in the Mississippi River (Carey, 1969) and later in other meandering rivers (Lewin, 1978; Hickin, 1979, 1986; Page & Nanson, 1982; Page, 1983).

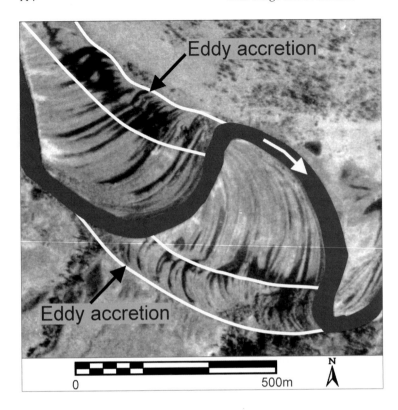

Eddy accretion

Eddy accretion

0                                    500m          N

**Fig. 1.** Typical morphology of eddy accretions on the Beaver River floodplain in Saskatchewan, located 48 km east of the Alberta border. (Photograph courtesy of Energy, Mines and Resources of Canada, no. A13163-44.)

Several researchers have studied the floodplain ridge-and-swale pattern associated with reverse eddies, terming the ridges within the channel 'concave benches' (Woodyer, 1975; Hickin, 1979; Farrell, 1987). Although Woodyer (1975) claimed that eddy accretions were similar to his concave-bank benches on the Barwon River, Australia, we believe they are different features, because concave-bank benches lack six of the nine features that all Mississippi eddy accretions have in common. More recently, some workers have termed eddy-accretion-like features as counterpoint deposits (counter or opposite to point bars; S.A. Smith, 1987; Nanson & Croke, 1992; Hesselink & Berendsen, 1997; Hickin, 1997). In spite of these later terms, the original term, eddy accretion, will be used throughout the remainder of this paper.

Eddy-accretion deposits are difficult to access, because they are not easily drilled and suitable geophysical methods are still unavailable. In earlier studies, sedimentological data were limited to the upper few metres by trenching and augering (Woodyer, 1975; Taylor & Woodyer, 1978; Woodyer *et al.*, 1979; Page & Nanson, 1982; Page, 1983; Hickin, 1986). Recent innovations in vibracoring technology, however, have allowed cores to be retrieved from considerably deeper sediment (up to 25 m; Smith, 1992, 1998).

Our objectives are fourfold:
**1** to investigate the sedimentology and geometry of eddy-accretion deposits in confined valley settings;
**2** to determine the depths of scour and associated flow velocities immediately ahead (down-valley) of eddy accretions;
**3** to understand the depositional processes of eddy accretions;
**4** to compare and contrast the nature of eddy accretions with adjacent point-bar deposits.

## PREVIOUS RESEARCH

Carey (1963) suggested that meandering river channels change direction in two distinct ways:
**1** in its own alluvial sediments, a meandering channel makes sweeping gentle curves as it wanders (termed a regular bend);
**2** when a meandering river channel encounters resistant bank material, such as a resistant valley wall or cohesive sediment (oxbow-fill of clay), it can change direction as an abrupt bend (90°).
Carey (1969) compared the processes involved in an abrupt bend on the Mississippi River to a water jet

impinging directly on a wall, dividing the flow equally in two directions. The part of the flow that moves up-valley is the reverse flow eddy that forms the concave morphology as compared to common convex-shaped point-bar scrolls. Carey (1969) estimated that the total energy consumed by an abrupt bend probably equalled the energy required to overcome the flow resistance of several kilometres of a normal meandering river channel. Carey (1969) indicated that as the entire eddy accretion migrated down-valley, the eddy may form either a concave accretion or a concave island, which eventually is left behind and incorporated into the floodplain, with depressions filled with arcuate-shaped residual lakes.

Carey (1963) noted that the unconfined Mississippi and Red rivers impinge on erosion-resistant valley sides. The Mississippi River, for example, impinges on valley walls at 17 locations between Cairo, Illinois and Baton Rouge, Louisiana, a river distance of 460 km (Carey, 1963). The Red River impinges on its valley walls at 11 sites between Shreveport and the Mississippi confluence, a 173-km channel distance (Carey, 1963). Some accretions migrate rapidly; for example at Port Hudson on the Mississippi River, one migrated 2.4 km down-valley in 80 yr, an average of 30 m yr$^{-1}$ (Carey, 1969). Carey stated that all eddy accretions have the following features:

**1** abrupt-angle change of channel direction, as a result of impingement on an erosion-resistant valley wall, with the angle generally greater than 90°;

**2** a powerful pressure eddy (reverse flow) upstream of the impingement;

**3** a suction eddy (normal flow) just upstream of the impingement;

**4** contours and other topographic features are concave with respect to the flow direction of the active river channel within the valley (down-valley);

**5** high silt and clay content and organic material at and near the surface of accretions (higher than any other alluvial deposit except 'back-swamp' clays);

**6** greatest river-scour depths occurring at abrupt bends (60 m on the Mississippi, below Old River);

**7** at slightly lower elevations than point-bars with overbank deposits or natural levees and often remain low and swampy;

**8** small streams commonly flow from the valley side, then flow along the depression and the contact between the point-bar and eddy-accretion scrolls.

Eddy accretions in confined meandering river valleys occur along both floodplain margins, as demonstrated by a large number of Pleistocene meltwater spillway valleys in western Canada (Page & Nanson, 1982), where the downstream limb of each point bar grades into an eddy accretion. Nanson & Page (1983) were the first to suggest

a geometry of juxtaposed eddy-accretion and point-bar deposits. Hickin (1986) estimated the geometry of eddy accretions and adjacent point-bar deposits in the subsurface, but no subsurface data were presented. Hickin (1979), however, does show an unusually deep bottom profile from an active channel forming an eddy accretion. This profile shows flow separation, a deep scour in the main channel and reverse flow in the accretion zone. Page & Nanson (1982) examined the formation of eddy accretions (there termed concave-bank benches) on a reach of the Murrumbidgee River in New South Wales, Australia, where aerial photographs show apparent sequential development. From shallow cores, they produced a general model for eddy accretion (separation zone islands) at different stages.

## RESEARCH SITES

In this paper, eddy accretions in the Kootenay River, British Columbia, and Beaver River, Alberta, Canada, were studied. In the south-eastern Canadian Cordillera, the Kootenay is a tributary of the Columbia River, which discharges into the Pacific Ocean, whereas the Beaver, in the plains of east-central Alberta, is a tributary of the Churchill River, which discharges into Hudson Bay (Fig. 2).

The Kootenay River enters the Rocky Mountain Trench at Canal Flats, where it turns south and flows between the Rocky Mountains to the east and Columbia Mountains to the west, before crossing the Canada–USA border into Montana. The Kootenay River study area is located 3 km south of the town of Wasa, British Columbia and immediately west of Highway 93. Here,

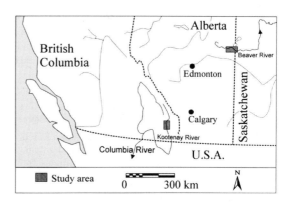

**Fig. 2.** Locations of study areas on the Kootenay River, southeast British Columbia, and Beaver River, east-central Alberta.

the Kootenay floodplain is 1 km wide and confined on both sides by bedrock and Pleistocene till, with eddy accretions on both sides of the floodplain. The Kootenay River drainage basin is 7120 km$^2$, measured from the Skookumchuck gauging station 12 km upstream of the study area. Monthly discharge hydrographs for the Kootenay show maximum flow in June (average 400 m$^3$ s$^{-1}$, maximum 1150 m$^3$ s$^{-1}$).

The Beaver River study area is 5 km south of the hamlet of Cherry Grove, Alberta, which straddles the Alberta–Saskatchewan border. The Beaver occupies a deep glacial meltwater channel cut into bedrock, gravel and till. Aerial photographs show eddy accretions on nearly every meander throughout the valley length. The Beaver drainage basin is 14 500 km$^2$, measured from the Cold Lake gauging station 20 km upstream. Monthly discharge hydrographs for the Beaver River show the maximum flow in June (average 50 m$^3$ s$^{-1}$, maximum 615 m$^3$ s$^{-1}$).

## METHODS

### Vibracoring and lithostratigraphical logging

Although vibracoring technology has been available for two decades, recent equipment and procedural modifications by Smith (1992, 1998) has allowed for deeper penetration (up to 25 m) in favourable conditions (muddy sand), while maintaining portability and low cost. After vibrating the first 6 m length of core-tubing into saturated sand and silt, a coupler is used to attach two tubes together, then coring continues. If the core-pipe penetration stops, but a greater depth is desired, the tube is extracted, the core is vibrated out and the pipe is reinserted into the same hole. This technique continues until no additional penetration occurs or the desired depth is achieved. If the depth objective has not been achieved with the 7.5 cm diameter core pipe using couplers and the reinserting technique, an additional method is used. A 15-cm diameter length of core-casing tube is vibrated 6, 8 or 10 m into the hole left by the extracted 7.5 cm core pipe. After the tube is vibrated into place, it is flushed of sediment with water from a pump and a plastic-pipe flushing system, so that the tube now becomes a casing. Next, a 7.5-cm diameter pipe is inserted into the casing and coring with couplers and reinserting continues until no additional depth is possible. Using this process, a maximum depth of 17 m was reached in the Kootenay River deposits.

Graphical and descriptive lithostratigraphical logs were constructed in the field from every vibracore. Sand-grain sizes were determined by visually comparing sediment with an Amstrat (American/Canadian Stratigraphic) grain-size chart using a 30-power field microscope. The silt–clay fraction was classified using qualitative tests such as consistency, feel and grittiness between teeth. The integrity of this method of estimating grain size has been tested successfully against laboratory sieving (Folk, 1974) and hydrometer methods, with field results yielding 90% accuracy (Calverley, 1984; Piet, 1992). The occurrence of shells, woody debris and fibrous organics was also recorded.

### Acoustic bottom profiling and flow velocity measurements

Channel cross-sectional geometry and depths were determined using a Raytheon DE-719E depth sounder from a Zodiac-type inflatable boat. On the Kootenay River, two eddy-accretion channels and one point bar were profiled at bankfull discharge. Six eddy-accretion and corresponding point-bar channels were profiled on the Beaver River below bankfull discharge (−1.7 m). Profiles were taken perpendicular to the channel edge in order to obtain true cross-sections. One cross-sectional flow-velocity profile was measured on a Kootenay River eddy accretion at bankfull discharge. Velocity readings were taken with a Price current meter at every 10% of channel depth for most of 20 cross-channel sites for a total of 177 point measurements, then processed following standard hydrometric procedures (Burge, 1997).

## LITHOFACIES DESCRIPTION AND INTERPRETATIONS FROM VIBRACORES

A total of 19 lithostratigraphical logs were recorded from vibracores, 9 taken from the Kootenay River and 10 from the Beaver River. The Kootenay logs were taken along the channel margin of the eddy accretion and point bar (Figs 3 & 4), whereas the Beaver logs were obtained along a straight transect across the floodplain from the north valley side (Figs 5 & 6). Seven facies, their number, description and interpretation, were identified from vibracores (Table 1).

Facies 1, a massive silt, varies from 0.1 to 1.7 m thick. The colour of facies 1 is shiny grey in the Kootenay and black in the Beaver and commonly contains minor amounts of shell and organic material. When facies 1 is present at the base of cores it frequently is overlain by facies 3 (coarse sand with pebbles). Facies 1 can be found at any stratigraphical level, but commonly occurs

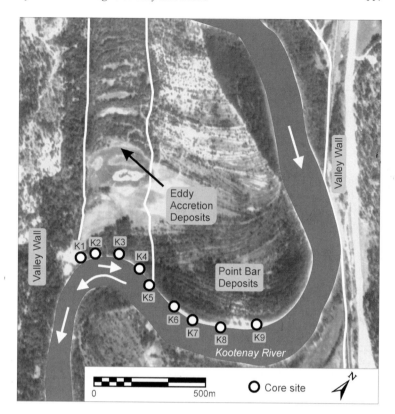

**Fig. 3.** Locations of vibracores (K1 to K9) taken along the Kootenay River. Eddy accretion is located on the west side of the valley and the point bar is located mid-valley. (Photograph courtesy of British Columbia Surveys and Mapping Branch, roll BC77036, no. 006, 1977.) All lithostratigraphical logs used floodplain surfaces as 0 datum.

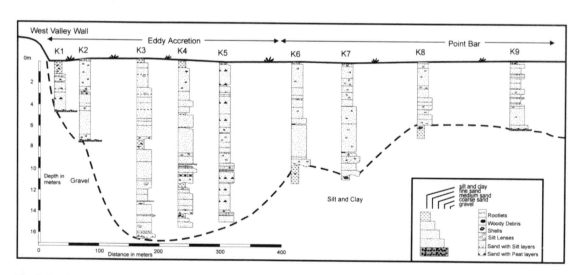

**Fig. 4.** Stratigraphical cross-sectional profile, from west to east along the Kootenay River, shown in Fig. 3. Note the greater thickness of eddy-accretion sediment as compared with point-bar sediment. Vertical exaggeration is ×17.

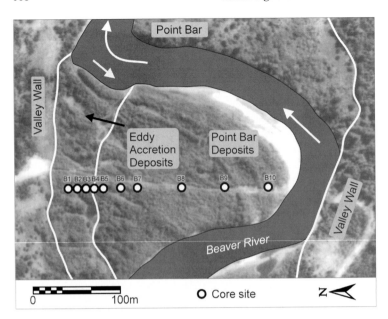

**Fig. 5.** Locations of vibracores (B1 to B10) taken across the Beaver River floodplain. Eddy-accretion deposits are located on both the north and south sides of the valley and point-bar at mid-valley. (Photograph courtesy of Alberta Energy Mines and Resources, roll 1633, no. 316, 1977.)

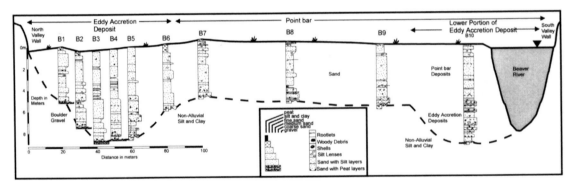

**Fig. 6.** Stratigraphical cross-sectional profile taken from north to south across the Beaver River floodplain. Core locations are shown in Fig. 5. Eddy-accretion deposits occur on the north and south sides of the valley and point-bar deposits at mid-valley. Vertical exaggeration is ×6.

**Table 1.** Summary of the seven facies, descriptions and interpretation.

| Facies | Description | Interpretation |
|---|---|---|
| 1 | Massive silt | Glacio-lacustrine or in-channel slack deposition |
| 2 | Gravel | Glacio-fluvial or alluvial fan |
| 3 | Coarse sand with pebbles | Channel bed lag |
| 4 | Organic litter and wood | Lateral accretion |
| 5 | Clean sand | Lateral accretion |
| 6 | Sand with silt and/ or organic litter | Lateral accretion |
| 7 | Silt and sand with rootlets | Overbank deposition |

mid-sequence, where it overlies clean sand (facies 5) or sand with silt and/or organic litter layers (facies 6). Facies 1 occurs up to seven times in Kootenay cores and three times in Beaver cores. Depending on its location within a core, facies 1 is interpreted in one of two ways. If located stratigraphically beneath a basal lag (facies 3), it is interpreted as pre-fluvial glacio-lacustrine in origin. As many glacial lakes formed during deglaciation (Clague, 1975; Sawicki & Smith, 1991), the contemporary Kootenay River sediments may rest on lacustrine silt. When facies 1 is present above the basal lag, it is interpreted as slack-water fallout of suspended sediment deposited at low stage.

Facies 2, a gravel, varies from 0.1 to 0.2 m thick and contains pebble clasts, usually within a sandy matrix. When present, it occurs at the base of cores and is overlain by sand with pebbles (facies 3), clean sand (facies 5), or sand with silt and/or organic litter layers (facies 6). This facies is interpreted as glacio-fluvial or alluvial fan in origin, depending on its location in the valley. Gravel is found at two locations on the Kootenay River, core K9, located on the east side of the valley, and cores K1 and K2 on the west side of the valley. Gravel on the east side of the valley is interpreted as glacio-fluvial, whereas on the west side of the valley it is from a small tributary creek, which is actively depositing an alluvial fan. Gravel on the north side of the Beaver River valley is interpreted as glacio-fluvial and is thought to be a basal lag deposit from larger post-glacial flows that eroded the valley.

Facies 3, sand with pebbles, varies from 0.6 to 1.1 m in thickness and occurs at the base of fluvial successions and near mid-sequence in eddy-accretion deposits. The sand usually is either coarse or medium, contains minor to numerous pebbles with occasional silt–clay rip-up clasts and may contain variable amounts of wood fragments, organic litter or shell fragments. Clean sand (facies 5) usually overlies facies 3, but less frequently below sand with silt and/or organic litter layers (facies 6). When occurring in mid-sequence of eddy-accretion deposits, facies 3 may be underlain or overlain by massive silt (facies 1), but also may occur overlain or underlain by facies 5 and 6. Facies 3 is interpreted as a channel-bed lag, deposited mostly in thalwegs. When present at mid-sequence in eddy accretions, it is deposited by reverse flows in perched eddy-accretion channels.

Facies 4, organic litter and wood fragments, is a minor facies, varying in thickness from 0.05 to 0.2 m. It is black in colour, generally fibrous with wood fragments and may contain minor medium sand. Facies 4 is overlain and underlain by clean sand (facies 5) or sand with silt and/or organic litter layers (facies 6). Facies 4 is interpreted as waterlogged organic litter and wood fragments, deposited within the channel on the downstream side of dunes or sand waves during low flows and subsequently buried by downchannel readvance of the bedform in the next high flow.

Facies 5, clean sand, occurs in all cores in both rivers. It is the most abundant facies, varies in thickness from 0.1 to 4.2 m and is dominated by medium sand, but sometimes fine sand. Facies 5 may contain minor silt layers or minor organics and commonly is overlain and underlain by facies 6 (sand with silt and/or organic litter layers), but occasionally it is found in contact with any other facies. Facies 5 is interpreted as lateral-accretion point-bar or eddy-accretion sand deposited at or near bankfull discharge.

Facies 6, sand with silt and/or organic litter layers, occurs in all cores in both rivers. It is the second most abundant facies and varies in thickness from 0.1 to 5.8 m. Grain size ranges from silty fine sand to medium sand. The facies contains minor to numerous silt layers up to 5 cm thick, variable organic litter layers, and occasional shell fragments. It occurs most commonly in the upper half of cores and usually is overlain and underlain by facies 5 (clean sand). It commonly is overlain by silt and sand with rootlets (facies 7) when at the top of a core. Facies 6 is interpreted as a lateral-accretion deposit and occurs on accretion surfaces of point bars or eddy accretions high upon the accretion slope.

Facies 7, silt and sand with rootlets, occurs in all cores in both rivers with thicknesses varying from 0.15 to 1.7 m. Grain size ranges from clayey silt to silty fine sand. This facies contains numerous rootlets and minor to numerous organics. It occurs at the top of all cores, at the floodplain surface, and generally is underlain by sand with silt and/or organic litter layers (facies 6). Facies 7 is interpreted as overbank sediment deposited during floods. Roots found in this facies are from active growing plants on the floodplain or from plants that were buried by vertical accretion during flooding.

## LITHOFACIES SUCCESSIONS AND CROSS-SECTIONAL PROFILES

Facies successions are vertically stacked facies deposited by a distinct sequence of depositional processes. Two facies successions, (i) point bar and (ii) eddy accretion, are recognized from the floodplain logs of the Kootenay and Beaver rivers.

### Point-bar succession

Representative point-bar successions from the Kootenay and Beaver rivers are shown in logs K8 and B8 (Fig. 7), which are similar to published facies successions of sandy point bars (Allen, 1965; Bernard *et al.*, 1970; Smith, 1987a; Jordon & Pryor, 1992). At the base of the succession, sand with pebbles (facies 3) is in erosional contact with underlying lacustrine silt (facies 1) or fluvial gravel (facies 2). This facies passes upward to a thick sequence of clean sand and sand with silt and/or organic litter layers (facies 5 or 6). Organic litter and woody material (facies 4) may occur at any interval in the sequence. The sequence is always capped by silt and sand with rootlets (facies 7). The succession generally fines upward, but may contain thin interbeds of coarser grained sand. Sedimentary structures were not preserved in vibracores.

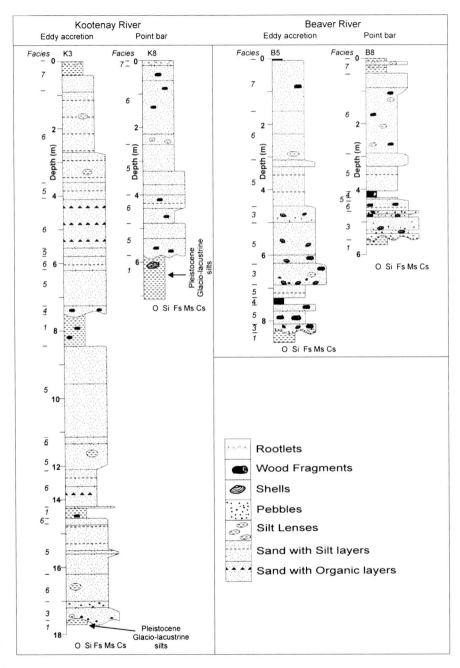

**Fig. 7.** Comparison of stratigraphical logs taken from eddy accretions and point bars on the Kootenay and Beaver rivers. Eddy-accretion deposits are significantly thicker than adjacent point-bar deposits.

## Eddy-accretion succession

Eddy-accretion successions from the Kootenay and Beaver rivers, shown in logs K3 and B5 (Fig. 7), are similar sedimentologically but are thicker than the nearby point bars. At the base, a channel lag of sand with pebbles and clay rip-up clasts erosively overlies lacustrine silt or fluvial gravel. Up-profile is a thick sequence of clean sand and/or sand with silt and organic litter. At mid-sequence in the Kootenay cores, a massive silt layer (facies 1) nearly 1 m thick usually overlies facies 5 or 6. This silt may occur more than once within a sequence. Commonly overlying the massive silt bed is another sand with pebbles (a lag), clean sand, or sand with silt and/or organic litter layers. The sequence continues with facies 5 and 6 and is capped by silt and sand with rootlets (facies 7), which always occurs at the top.

Some logs show two upward-fining sequences, one stacked above the other. If seen in outcrop or core from ancient rocks, this apparent double upward-fining sequence may be misinterpreted as two superimposed point-bar deposits.

A sedimentological comparison of eddy accretions and point bars is shown in logs K3 and K8 for the Kootenay and B5 and B8 for the Beaver (Fig. 7). Eddy-accretion and point-bar sequences have four important differences. First, eddy-accretion deposits are 1.5–3 times thicker than adjacent point-bar deposits. Second, eddy-accretion sedimentology may include massive silt layers at any interval, but most commonly they occur mid-sequence; similar thick layers are absent in point-bar deposits. Third, eddy-accretion deposits contain more silt and organic litter in the upper sequence than do point-bar deposits. Fourth, eddy-accretion deposits exhibit a higher degree of grain-size heterogeneity than do point bars. This heterogeneity probably results from the complex nature of flow conditions in reverse-flow eddy channels, where high and low velocities occur in close proximity.

## Cross-sectional profiles of fluvial fills

A Kootenay River stratigraphical cross-section was compiled from nine vibracores taken along the channel margin across the 1-km-wide floodplain (Fig. 3). The profile is deepest (17 m) on the west side, corresponding to the eddy-accretion deposit and shallows to 6 m at the mid-floodplain position, corresponding to the point bar (Fig. 4). This eddy accretion extends over 40% of the floodplain, whereas point-bar deposits cover the remaining 60% (percentages do not include channel area).

The Beaver River stratigraphical cross-section was compiled from 10 vibracores taken across the 0.3-km-wide floodplain (Fig. 5). The profile is deepest (9 m) on both floodplain margins and shallowest (5 m) at mid-valley (Fig. 6). Here eddy-accretion deposits form 30% of the total alluvial surface in cross-valley profile, whereas point-bar deposits form the remaining 70%. The Beaver River profile shows two thick eddy accretions, one on each margin of the floodplain.

# CHANNEL BED SCOURS AND FLOW VELOCITY

## Channel bottom profiles

In the Kootenay River, two eddy-accretion scours and one point bar were extensively bottom-profiled at bankfull stage. Six eddy-accretion channels and adjacent point-bar channels were bottom-profiled on the Beaver River at 1.7 m below bankfull discharge. Profiles were taken perpendicular to channel banks in order to obtain true cross-sections.

The channel-bottom profiles (echograms) are displayed in Fig. 8. All point-bar channel-bottom profiles clearly show a gently sloping lateral-accretion surface on the up-valley side and a steep cutbank on the down-valley bank (K2, B2, B4, B6; Fig. 8). Eddy-accretion profiles usually show one perched-channel segment on the up-valley bank, a ridge near mid-channel, and a deep scour hole with a steep cutbank on the down-valley bank (Fig. 9). An island and/or mid-channel ridge, depending on stage, was present in profile B7 (Fig. 8). During bankfull discharge, this island is slightly submerged (1 m).

All profiles at eddy accretions show unusually deep scours, a mid-channel ridge and a perched-channel segment. On the Kootenay River, eddy-accretion scour depths vary between 14 and 16 m, whereas adjacent to point bars, channel depths are 8 m. On the Beaver River, eddy-accretion scour depths vary between 5.5 and 8 m, whereas channels adjacent to point-bars range from 4 to 4.5 m below bankfull. Depth to the eddy-accretion ridges varied between 6 and 8.5 m on the Kootenay and 2–3 m on the Beaver. Depth of the eddy-accretion perched channel averaged 10–11 m on the Kootenay and 3–4 m on the Beaver.

## Flow-velocity profile

A cross-channel velocity profile was measured in an eddy-accretion scour of the Kootenay River at location K1 (Fig. 8). The velocity profile in Fig. 9 shows normal

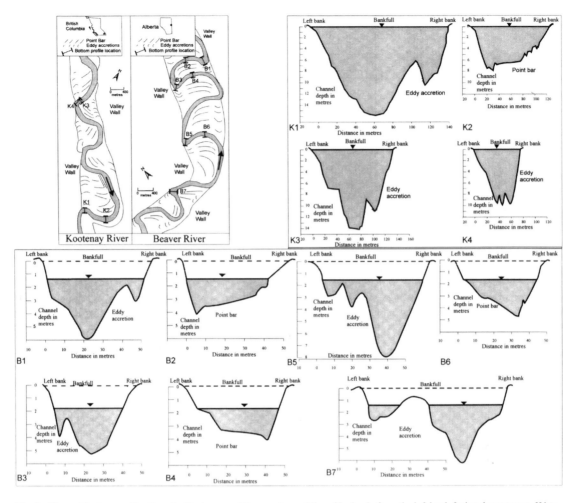

**Fig. 8.** Channel-bottom profiles from the Kootenay and Beaver rivers. All profiles begin from the left bank facing downstream. K1 to K4 bottom profiles taken from the Kootenay River at bankfull stage. B1 to B7 bottom profiles are from the Beaver River taken at 1.7 m below bankfull stage.

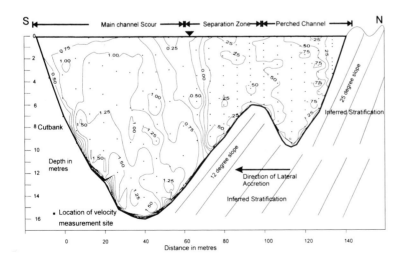

**Fig. 9.** Flow velocities (m s$^{-1}$) of the Kootenay River eddy-accretion channel (location K1 on Fig. 8). Highest flow velocities occur near the bed and left (south) bank. Slowest velocities are not directly above the channel ridge, but farther south into the main channel. This out-of-phase association may be the result of a delayed response of the channel to a rapid rise in discharge and velocity.

flow in the main channel scour and reverse flow over the perched channel (negative velocities). Severe turbulence in the separation zone separates the opposing flows. Discharge in the main channel (normal flow) was $810 \text{ m}^3 \text{ s}^{-1}$, whereas the perched-channel (reverse flow) discharge was $190 \text{ m}^3 \text{ s}^{-1}$, 23% of the main channel discharge. At bankfull discharge, highest point velocities ($1.6 \text{ m s}^{-1}$) occurred along the bed of the main channel (Fig. 9). These abnormally high bed velocities create the greatest shear stress and scour in the main channel.

Flow behaviour in an eddy-accretion channel may be separated into three main parts:
1 normal (downstream) flow in the main channel;
2 reverse flow (eddy) in the perched channel;
3 severe turbulent separation zone characterized by large eddies and upwelling.

Flow reversal occurs when flow impacts a resistant bank surface at 90°, usually the valley wall (Carey, 1969). Within the channel, water 'piles up' against the resistant valley side, forcing upstream (reverse) flow on the up-valley portion of the channel. The impingement-induced flow separation in the main channel creates a large reverse eddy. The eddy flows within the perched channel and erodes a semi-circular cutbank into the floodplain. Energy in the reverse flow eventually dissipates and rejoins the main-channel flow. Hence, the eddy recirculates flow into the main channel, thereby increasing discharge and velocity of the flow and causing the greatest depths of scour in the system.

## DISCUSSION

### Eddy-accretion depositional processes

Depositional processes of eddy accretions and their migration down-valley is interpreted from aerial photographs, lithostratigraphical logs, channel-bottom profiles, one flow-velocity profile and results of previous work. When combined, the data exhibit four depositional zones:
1 the main-channel zone and accretion slope;
2 the separation-zone ridge;
3 the perched-channel and second accretion zone;
4 the overbank zone (Fig. 10).

### *Main-channel zone*

The main-channel zone is located on the down-valley or south side of the channel profile (Fig. 10), bounded by a cutbank (left) and separation-zone ridge (or island). Two types of deposits occur within the main channel: (i) a coarse channel lag and (ii) an upward-fining laterally accreted sand. Deposition occurs on the lateral-accretion surface, probably during waning flows, as the channel migrates in accordance with cutbank erosion. Lateral-accretion slopes in the main channel are similar to slope angles of point bars, averaging 12°. The similar slope angles suggest similar depositional processes. Owing to the complex nature of the channel, we suggest that sand is not only deposited on the lateral-accretion slope of the main channel as migrating bedforms, but also sand is lifted into suspension by turbulent eddies and deposited

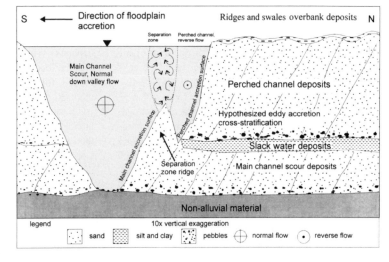

**Fig. 10.** A schematic cross-sectional profile of the Kootenay River scour channel and eddy accretion. Normal flow occupies the left main-channel and reverse flow (eddy) occupies the right perched channel. Deposition progrades down-valley (left or south) with active accretion surfaces sloping between 12 and 25° down-valley. Mud is deposited in the perched channel during lower stage flows. Two upward-fining sequences result from successive deposition of main-channel and perched-channel accretion surfaces.

**Fig. 11.** Main-channel flow in the Beaver River is towards the lower left in the photograph. Reverse channel flow occurs on the right side of the island. For scale note the inflatable boat in the lower right.

on to the accretion slope. As sand falls out of suspension when velocities decrease near the separation zone, avalanching occurs down the separation zone ridge because of its oversteep upper slope. This suspension fall-out helps form the ridge between normal and reverse flows.

### Separation-zone ridge

The separation-zone ridge is the shallowest part of the eddy-accretion channel deposits, has the lowest velocities and is characterized by significant upwelling and turbulence at the water surface (Fig. 10). This turbulence and upwelling is accompanied by low horizontal velocities and a decrease in flow competence, allowing deposition. Sand carried in suspension into the separation zone is deposited as a temporary ridge (in the case of the Kootenay River) between the main and perched channels. In the case of the Kootenay River study site, we believe the position of the ridge also migrates down-valley along with the main and perched channels. In the case of the Beaver River, this ridge may be preserved when it breaks the water surface and becomes vegetated. Many active Mississippi River and Red River eddy accretions (Carey, 1969), as well as the studied active eddy accretion down-valley of the Beaver River core sites contain islands (Fig. 11). Islands form when the main channel, causing the eddy-accretion channel, widens significantly enough to accommodate an island within the separation zone. The perched channel eventually becomes abandoned and infills, usually with mud, leaving the island as part of the floodplain. Former islands are visible on aerial photography as scroll patterns, and former reverse-flow channels appear as swales, arcuate lakes or wetlands.

### Perched-channel zone

Perched channels are located on the up-valley portion of the eddy-accretion channel, bounded by the separation-zone ridge or island on one side and a second lateral-accretion surface on the other bank (Fig. 10). Three types of deposits are associated with the perched channel: (i) slackwater massive silt (facies 1), (ii) coarse channel lag (facies 3), and (iii) lateral-accretion deposits (facies 5 and 6). Channel-lag and lateral-accretion deposits are similar to those deposited in the main channel. Slackwater deposits (facies 1), associated with perched channels, are formed during lower flow stages when the perched channel is partially cut off from the main channel by the separation-zone ridge (Fig. 12). At this time, flow velocities fall to almost zero, allowing deposition of silts from suspension on to the base of the perched channel (Fig. 12B). These thick massive silt layers (up to 70 cm preserved in the Kootenay) appear similar to lacustrine deposits. When river stage increases and the perched channel becomes reactivated, some massive silts are eroded, followed by deposition of coarse channel sediment (Fig. 12C).

As suspended sand is transported into the perched lateral-accretion slope by reverse flows, sediment settles, then avalanches down the accretion surface, depositing slope angles of up to 25°. Much of the suspended sand is probably not deposited on the accretion surface, but instead is transported through the perched channel to re-enter the main-channel flow.

Thick silt layers (facies 1) in eddy accretions would be thin to absent in rivers low in suspended silt. The eddy-accretion deposit studied in the Kootenay River contains

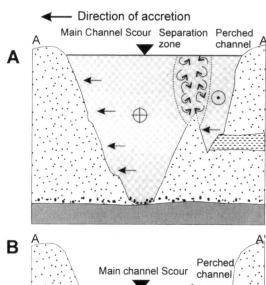

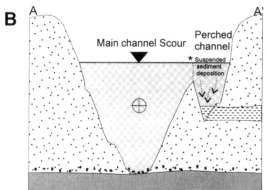

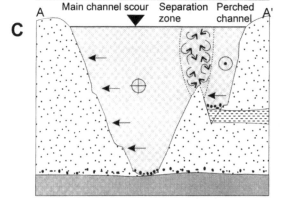

A. Bankfull discharge.
Eddy accretion is active and shifting laterally;
channel separation zone is large and significant mixing
of the normal and reverse flow directions occurs.
Velocities in the perched channel are high and
the bed is actively scoured.

B. Low stage.
The eddy is inactive because flow separation
is reduced and velocities fall to almost zero
in the perched channel. (*) In rivers with high
suspended sediment loads, silt falls from
suspension to form a thick silt layer on the
base of the perched channel.

C. Subsequent bankfull discharge.
Flow separation occurs and the eddy
accretion activates. The separation zone
and perched channel zone are active. The
perched channel scours into the silt layer and
deposits a channel lag overtop as it progrades
downvalley (to the left).

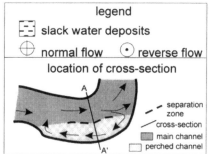

**Fig. 12.** Schematic cross-sections of a hypothesized three-step process for deposition of perched-channel silt layers in eddy accretions. (A) At bankfull discharge, eddy is actively accreting and scouring laterally. (B) Somewhat cut off from the main channel at low stage, low velocity in the perched channel permits deposition of suspended silts. (C) Subsequent high discharge resumes scour conditions; flow separation resumes and the perched channel reactivates, scouring into the silt layer. Coarse channel sediment deposits on top of the silt.

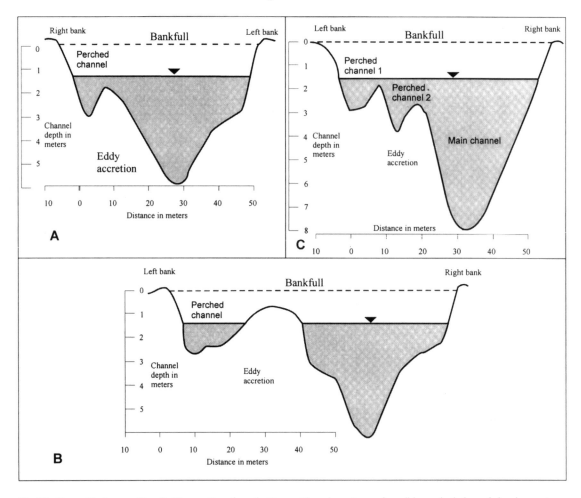

**Fig. 13.** Channel bottom profiles of eddy accretions from the Beaver River show stages of possible perched-channel abandonment. Profile locations are shown in Fig. 8. (A) Typical eddy-accretion channel with a reverse-flow perched channel. (B) Large separation-zone ridge, which becomes an island at low stage. It also shows incipient formation of a second perched channel within the main channel. (C) Channel with two perched channels and separation-zone ridges.

thick silt layers at approximately mid-sequence, whereas the Beaver River contains thinner silt layers. The Kootenay River is glacially fed, whereas the Beaver River water is relatively clear, receiving little silt from a muskeg-dominated watershed.

Over time, perched channels in the Beaver River do not move down-valley, but rather new perched channels and separation zones are created episodically, probably during extreme flows. The formation of new perched channels leaves behind former islands and perched channels within the floodplain, creating ridge-and-swale patterns. This sequence occurs when the main channel has

migrated down-valley, leaving the perched channel far enough behind in the floodplain so that the separation zone and reverse flow breaks down. The original perched channel becomes abandoned, and a new perched channel develops on the up-valley portion of the main channel. Normally, eddy-accretion channels contain one perched channel (Fig. 13A), but, secondary perched channels were observed forming and fully developed on the Beaver River within the main channel (Fig. 13B). During episodic movements, the first perched channel may be abandoned by the main channel as it migrates down-valley and a second perched channel may develop

(Fig. 13C). Abandoned perched channels are commonly observed around islands in the scroll pattern of the Beaver River floodplain.

## Overbank zone

Overbank vertical-accretion deposits consist of silt, clay and organic material (driftwood, leaf litter, needles and twigs). These deposits can be up to several metres thick in both rivers and form a concave arcuate scroll-and-swale morphology, when looking down-valley. Overbank eddy-accretion deposits are often topographically the lowest floodplain areas and contain the finest grained sediment of the floodplain.

Depositional processes in all four zones are probably more complex than we have described. We believe, however, that the ridge-and-swale scroll morphology associated with eddy accretions results from some unspecified magnitude, frequency and duration of high- (near bankfull) and low-water discharges, although we do not have data to support this view. High discharges may widen and scour the main channel and deposit islands, whereas low-discharge periods may infill perched channels with silt–clay and allow time for vegetation to stabilize islands. Does each eddy-accretion ridge represent flows of near bankfull or greater? This hypothesis could be investigated with long records of river-gauging data, coupled with aerial photography and [14]C-dated material from vibracores on rivers such as the Beaver.

## Confined valley-fill architectural geometry

A schematic three-dimensional model of eddy-accretion alluvial geometry for confined meanders is developed from the Kootenay and Beaver rivers on the basis of previously published research, stratigraphical cross-sections from vibracores and aerial photographs (Fig. 14). The data indicate that, in confined valleys, 25–40% of the total floodplain surface and up to 50% of the valley cross-sectional area may contain eddy-accretion deposits. In such sites, eddy-accretion deposits are approximately twice as thick as adjacent mid-valley point bars, with alluvial sand bodies having a 'dumb-bell shaped' sediment geometry in the subsurface. The bulging weights on each side of the 'dumb-bell' represent thicker eddy-accretion deposits along valley margins, whereas the dumb-bell handle represents thinner mid-valley point-bar deposits.

There are four features within the eddy-accretion facies succession that may indicate the presence of eddy accretions in ancient rocks.

**1** Eddy-accretion deposits may contain a massive, mid-sequence silt layer without rootlets or evidence of exposure (unlike overbank deposits). A channel lag may be preserved above this silt layer.

**2** Two different stratigraphical palaeocurrent zones may be preserved within an eddy accretion and may show a 180° difference, above and below a mid-sequence silt layer and/or lag deposits, if preserved.

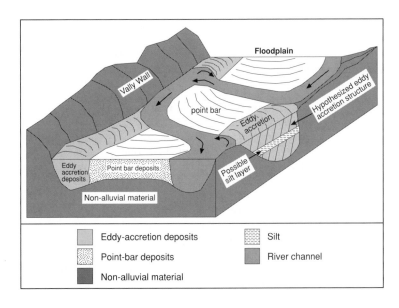

**Fig. 14.** Three-dimensional model of a hypothetical confined meandering river in which eddy-accretion deposits flank the sides of the valley, whereas point-bar deposits occur mid-valley.

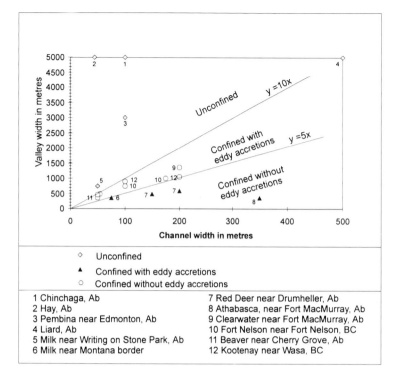

**Fig. 15.** Confined meandering river floodplain-width plotted against channel-width from a sample of Alberta (Ab) and British Columbia (BC) rivers. River reaches were recorded as unconfined, confined with eddy accretions, and confined without eddy-accretion patterns. Valley-width to channel-width ratio for confined-valley eddy accretions tend to occur between 5 : 1 and 10 : 1. In unconfined rivers, eddy accretions may form where channel meanders impinge at 90° against erosion-resistant valley walls. Valley widths and corresponding channel widths were measured from 1 : 50 000 topographic maps in 17 river reaches of 11 rivers.

**3** The eddy-accretion sand, above the mid-sequence silt layer and/or lag deposit, is muddier and finer grained than sand below the silt layer and/or lag.

**4** Although not observed in vibracores as a result of disturbance, steeply dipping inclined stratification (up to 25°) may be present within the upper sequence in eddy accretions and such stratification may be present in outcrops or rock cores.

### Prediction of eddy accretion occurrence in confined meander settings

Hickin (1986) was the first to use the ratio of floodplain-width to channel-width (between 3 : 1 and 6 : 1) to predict where eddy accretions may form in confined valleys, slightly less than our results of 5 : 1 to 10 : 1. To expand the data base, eddy accretions, valley widths and corresponding channel widths from 17 reaches of 11 rivers in Alberta and British Columbia were measured from 1 : 50 000 topographic maps. The floodplain and valley morphology were noted as (i) unconfined, (ii) confined with eddy accretions or (iii) confined without eddy accretions. Channel width was used as a proxy of bankfull discharge. All measurements were taken on a reach-by-reach basis as the same river may contain more than one valley morphology. The Chinchaga, Hay, Pembina,

Liard and Milk rivers contained unconfined reaches. The Red Deer, Milk and Athabasca rivers contained confined reaches without eddy accretions. The Kootenay, Fort Nelson, Beaver and Clearwater rivers contained confined reaches with eddy accretions. Based on these few rivers, some relationships between eddy-accretion formation and channel-width to valley-width ratio may be suggested. In unconfined rivers with floodplain-width to channel-width ratios greater than 10 : 1, eddy accretions may form where meander belts impinge upon resistant valley walls, but are unrelated to valley width. In confined rivers with floodplain-width to channel-width ratios less than 5 : 1, eddy accretions do not form, but instead river channels 'ricochet' from valley wall to valley wall. Confined-valley eddy accretions seem to form in a narrow range of floodplain-width to channel-width ratio between 5 : 1 and 10 : 1 (Fig. 15). In valleys with such floodplain-width to channel-width ratios, eddy accretions are predicted to occur.

### SUMMARY COMMENTS

The geometric nature and cause of eddy accretions within confined valleys are now somewhat predictable. Eddy accretions form in two types of meandering river valleys:

1 unconfined floodplains where meander-belt channels impinge on resistant valley walls at 90°;

2 confined meanders, where meander bends regularly impinge on resistant valley walls at 90°.

Eddy accretions are not formed in confined valleys that are too narrow relative to flow discharge because meandering channels ricochet off each valley wall at approximately 30–45° and reverse flow does not occur (e.g. Red Deer River near Dinosaur Provincial Park; Smith, 1987b).

This is the first study to investigate the sedimentology in the deeper portions of eddy-accretion deposits within meandering rivers. Owing to the greater depth of eddy-accretion scour holes, compared with depths of point-bar channels, eddy-accretion scours may be an important fish overwintering habitat in northern rivers. Fish require deep pools to survive harsh northern winters and eddy-accretion scour holes may be suitable sites. In addition, they may form an important summer (low-flow) resting habitat for fish. Understanding of eddy-accretion depositional processes and channel environments provides a good argument against partial river stabilization (Carey, 1969). Also, knowledge of the probable depth of scour may aid in better location of buried pipelines and cables at river crossings.

Finally, this research is important because eddy accretions are potentially thick aquifers (up to 60 m in Mississippi River eddy accretions) and major fairways for fluid movement in modern and deeply buried ancient rocks. In ancient rock successions, equivalent thick fluvial sandstones may be considered as preferred targets for hydrocarbon exploration. Also, thick massive silt layers located mid-sequence in eddy-accretion deposits may separate upper and lower upward-fining trends. This apparent double upward-fining sequence may be misinterpreted as the product of two superimposed point bars. Understanding the sedimentology and depositional processes involved within eddy accretions provides insight into their identification and location in subsurface rocks.

## ACKNOWLEDGEMENTS

Funding for this project and equipment was provided by a NSERC operating grant to D.G. Smith. The Department of Geography, University of Calgary, is thanked for the use of field equipment, office space and T.A. support. Barbara Ramovs is thanked for field assistance, encouragement and editing. Duane Froese and Rick Meyers are thanked for editing the paper. Aaron Clack greatly improved the diagrams.

## REFERENCES

ALLEN, J.R.L. (1965) Fining upward cycles in alluvial successions. *Geol. J.*, **4**, 229–246.

BERNARD, H.R., MAJOR, C.F., PARROTT, B.S. & LE BLANC, R.J. (1970) *Recent Sediments of Southeast Texas: a Field Guide to the Brazos Alluvial and Deltaic Plains and the Galveston Barrier Island Complex.* Bureau of Economic Geology, Guidebook 11, University of Texas at Austin, 16 pp.

BURGE, L.M. (1997) *Meandering river eddy accretions: sedimentology, morphology, architectural geometry, and depositional processes.* MSc thesis, University of Calgary, 99 pp.

CAREY, W.C. (1963) Turn mechanisms of alluvial rivers. *Military Eng.*, **Jan–Feb.**, 14–16.

CAREY, W.C. (1969) Formation of floodplain lands. *Proc. Am. Soc. civ. Eng. J. Hydraul. Div.*, **95**, 981–994.

CALVERLEY, E.A. (1984) *Sedimentology and geomorphology of the modern epsilon cross-stratified point bar deposits in the Athabasca upper delta plain.* MSc thesis, University of Calgary, 116 pp.

CLAGUE, J.J. (1975) Late Quaternary sediments and geomorphic history of the southern Rocky Mountain Trench, British Columbia. *Can. J. Earth Sci.*, **12**, 595–605.

FARRELL, K.M. (1987) Stratigraphy of a Mississippi River concave-bench deposit, lower Mississippi valley, Louisiana. In: *Abstracts, Society of Economic Paleontologists and Mineralogists Annual Midyear Meeting*, Tulsa, pp. 26.

FOLK, R.L. (1974) *Petrology of Sedimentary Rocks.* Hemphill, Austin, TX, 182 pp.

HESSELINK, A.W. & BERENDSEN, H.J.A. (1997) Morphological evolution of enbanked floodplains in the Rhine–Meuse delta, The Netherlands. In: *Abstracts, 6th International Conference on Fluvial Sedimentology* (Ed. Rogers, J.), 22–26 September, University of Cape Town, Cape Town, South Africa, p. 85.

HICKIN, E.J. (1979) Concave-bank benches on the Squamish River, British Columbia. *Can. J. Earth Sci.*, **16**, 200–203.

HICKIN, E.J. (1986) Concave-bank benches in the floodplains of Muskwa and Fort Nelson Rivers, British Columbia. *Can. Geogr.*, **30**, 111–122.

HICKIN, E.J. (1997) Counterpoint deposition in confined meandering river floodplains, northeastern British Columbia, Canada. In: *Abstracts, 6th International Conference on Fluvial Sedimentology* (Ed. Rogers, J.), 22–26 September, University of Cape Town, Cape Town, South Africa, p. 86.

JORDAN, D.W. & PRYOR, W.A. (1992) Hierarchical levels of heterogeneity in a Mississippi River meander belt and application to reservoir systems. *Bull. Am. Assoc. petrol. Geol.*, **76**, 1601–1624.

LEWIN, J. (1978) Meander development and floodplain sedimentation: a case study from mid-Wales. *J. Geol.*, **13**, 25–36.

NANSON, G.C. & CROKE, L.C. (1992) A genetic classification of floodplains. *Geomorphology*, **4**, 459–486.

NANSON, G. & PAGE, K.J. (1983) Lateral accretion of fine-grained concave-benches on meandering rivers. In: *Modern and Ancient Fluvial Systems* (Eds Collinson, J.D. & Lewin, J.), Spec. Publs Int. Ass. Sediment., No. 6, pp. 133–143. Blackwell Scientific Publications, Oxford.

PAGE, K.J. (1983) Concave-bench evolution and sedimentation on the Manawatu River, New Zealand. *N. Z. Geogr.*, **39**, 59–63.

PAGE, K.J. & NANSON, G. (1982) Concave-benches and associated floodplain formation. *Earth Surf. Process*, **7**, 529–543.

PIET, L.J.M. (1992) *Sedimentology of point bars and oxbow-fills.* MSc thesis, University of Calgary, 107 pp.

SAWICKI, O. & SMITH, D.G. (1991) Glacial Lake Invermere, upper Columbia River valley, British Columbia: a paleogeographic reconstruction. *Can. J. Earth Sci.*, **29**, 687–692.

SMITH, D.G. (1987a) Meandering river point bar lithofacies models: modern and ancient examples compared. In: *Recent Developments in Fluvial Sedimentology* (Eds Ethridge, F.R., Flores, R.M. & Harvey, M.D.), Spec. Publ. Soc. econ. Paleont. Miner., Tulsa, **39**, 83–91.

SMITH, D.G. (1987b) *Landforms of Alberta: interpreted from airphotos and satellite imagery.* Publication 87-1, Alberta Remote Sensing Centre, Edmonton, Alberta, Canada, pp. 1–105.

SMITH, D.G. (1992) Vibracoring: recent innovations. *J. Paleolimnol*, **7**, 137–143.

SMITH, D.G. (1998) Vibracoring: a new method for coring deep lakes. *Palaeogeogr., Palaeoclimatol., Palaeoecol.*, **140**, 433–440.

SMITH, S.A. (1987) Gravel counterpoint bars: examples from the River Tywi, South Wales. In: *Recent Developments in Fluvial Sedimentology* (Eds Ethridge, F.R., Flores, R.M. & Harvey, M.D.), Spec. Publ. Soc. econ. Paleontol. Miner., Tulsa, **39**, 75–81.

TAYLOR, G. & WOODYER, K.D. (1978) Bank deposition and suspended load in streams. In: *Fluvial Sedimentology* (Ed Miall, A.D.), Mem. Can. Soc. petrol. Geol., Calgary, **5**, 257–275.

WOODYER, K.D. (1975) Concave-bank benches on the Barwon River, New South Wales. *Austral. Geogr.*, **13**, 36–40.

WOODYER, K.D., TAYLOR, G. & GROOK, K.A.W. (1979) Sedimentation and benches in a very low gradient suspended load stream: the Barwon River, New South Wales. *Sediment. Geol.*, **22**, 97–120.

*Spec. Publs int. Ass. Sediment.* (1999) **28**, 131–145

# The influence of flooding on the erodibility of cohesive sediments along the Sabie River, South Africa

G. L. HERITAGE*, A. L. BIRKHEAD†, L. J. BROADHURST† *and* B. R. HARNETT‡

*\*Department of Geography, Peel Building, Salford University, Manchester M5 4WT, England*
*(Email: George.Heritage@Geography.Salford.ac.uk)*
*†Centre for Water in the Environment, Department of Civil Engineering, Hillman Building, University of the*
*Witwatersrand, Private Bag 3, WITS 2050, Johannesburg, South Africa*
*(Email: Birkhead@civen.civil.wits.ac.za); and*
*‡Kings College, London University, London, England*

## ABSTRACT

The dominance of extreme flood events in controlling geomorphological change in incised semi-arid river systems has been demonstrated by many authors. Catastrophic stripping of cohesive sediment may occur, particularly where vegetation cover is sparse. This paper investigates the existing potential for large-scale stripping of the cohesive fluvial sediment deposited in the incised macrochannel of the Sabie River, South Africa. The frequency distribution of critical resistance of the cohesive bed material was estimated using *in situ* shear-vane strength measurements, combined with laboratory testing of plasticity index and grain-size distributions. Topographical and hydraulic data were used to compute the distribution of applied hydraulic shear stress, corresponding to a large flood event of between 1705 and 2259 $m^3 s^{-1}$ (1-in-60 yr return period on the annual maximum flow series at the downstream reaches), for cross-sections along representative reaches of bedrock-anastomosing, mixed-anastomosing, pool–rapid, braided and alluvial single-thread channel types. The probability of eroding the cohesive sediments was determined by statistical analysis of all possible combinations, where the applied hydraulic shear stress was greater than the critical resisting shear strength of the bed material. The results show that 68% of the area of cohesive bed material along the Sabie River potentially may be eroded by such an extreme flood. Field observations indicated that this did not occur during a recent flood of this magnitude. The important role of in-channel vegetation in dissipating energy by increased resistance, thereby reducing boundary shear, was identified as a major factor preventing widespread cohesive sediment stripping. This was clearly evident in the densely vegetated bedrock-anastomosing reaches, where flow resistance was found to be highest and geomorphological change as a result of erosion of the cohesive sediment was minimal. The inability of the flood to erode sediment also may be attributed partly to the river transporting material at capacity. This is likely to be a function of antecedent conditions, where a recent prolonged drought resulted in sediment accumulating over the catchment. Furthermore, the short duration of the flood may have contributed, by inundating the large-scale consolidated deposits for a relatively short period, insufficient to instigate widespread erosion. This study demonstrates that there is considerable potential for large-scale stripping of the macrochannel deposits along the Sabie River and emphasizes the need to manage the riparian vegetation, which acts to protect the cohesive bed from entrainment, if the physical and ecological diversity of the river is to be maintained.

## INTRODUCTION

Fluvial geomorphological work is defined in terms of the volume of sediment transported through a reach (Wolman & Miller, 1960), whereas geomorphological effectiveness is defined as the degree of modification of a landform (Wolman & Gerson, 1978). Major floods can both transport large quantities of sediment and significantly modify the fluvial geomorphology if the threshold for erosion is exceeded. River response to large floods has been shown to be highly variable, however. Schumm & Lichty (1963) report large-scale sediment loss in response to flooding in the semi-arid Cimarron River, south-west Kansas, USA. Nanson (1986) describes the 'catastrophic stripping' of reaches of Charity Creek on the humid coastal region of New South Wales, Australia. Catastrophic floods in areas of high flow variability have been shown to have a major geomorphological role, for example, in desert channels (Schick, 1974), in the Eldorado Canyon, Nevada, USA (Glancy & Harmsen, 1975), in the semi-arid West Nueces River, Texas, USA (Baker, 1975), and in Coffee Creek, north California, USA (Stewart & LaMarche, 1967). Dramatic landscape modification was caused by Hurricane Camille in Virginia, USA (Williams & Guy, 1973). In contrast, Moss & Kochel (1978) and Costa (1974) noted only minor morphological changes to the lower reaches of the Conestoga River, Pennsylvania and Appalachian Piedmont, USA, respectively, following the Hurricane Agnes flood. Kochel & Baker (1982) report only minor changes to the distribution of in-channel bars following a large flood on the Pecos River, Texas, USA.

A large number of factors have been suggested as influential in determining the geomorphological effectiveness of large floods (Kochel, 1988). Channels are readily prone to morphological change if they are characterized by a flashy flow regime, have a high channel gradient, coarse unconsolidated bed material, low bank cohesion and a deep confined channel cross-section generating high flood-flow velocities and shear stresses. Consolidated silts and clays also may undergo erosion if the applied hydraulic shear stress exceeds the shear strength of the material imparted by the cohesive properties arising from structural and physiochemical forces. Highly cohesive channel banks can play a dominant role in determining channel form. This is demonstrated by the Carl Beck, a tributary of the River Tees in northern England with cohesive till banks, which constrained channel erosion over a range of discharges (Carling, 1988). Pickup & Warner's (1976) study of rivers in the Cumberland Basin, New South Wales, Australia, also identifies highly cohesive banks, in addition to a lack of fine sediment and a highly variable flow regime, causing the capacity of the channel to be related to two flow populations. They found that a group of large, infrequent floods controlled bank erosion and floodplain deposition, whereas lower flows subsequently reworked in-channel sediments. More recently, the 1993 Mississippi River flood had remarkably little geomorphological impact on the floodplain, principally because the cohesive soils resisted erosion (Gomez et al., 1995).

This paper describes an investigation of the potential for eroding the cohesive fluvial sediment deposited as large-scale consolidated bar forms within the incised 'floodplain' of the semi-arid Sabie River, South Africa, which has been termed the macrochannel by van Niekerk et al. (1995). The critical resistance to erosion of the cohesive sediments was determined from in situ shear-vane strength measurements, combined with laboratory testing of disturbed samples. The spatial distribution of maximum shear stresses applied during a recent flood event is determined using measured data on channel geometry, flow resistance, flood stages and water-surface slopes. A measure of the potential for erosion is provided by a statistical analysis of the frequency distribution of resisting and applied shear stresses, and is discussed with reference to field observations following a large flood in February 1996.

## CHARACTERISTICS OF THE SABIE RIVER AND LOCATION OF STUDY SITES

The Sabie River drains approximately 6000 km$^2$ of the Mpumalanga Province in the north-east of South Africa (Fig. 1). It is a semi-arid catchment with a mean annual precipitation varying from 1800 mm at the Drakensberg escarpment in the west to 400 mm in the east. This is in sharp contrast to the mean annual evapotranspiration losses of 1400 mm in the west, to 1700 mm in the east. Rainfall and consequently discharge, are highly variable, with discharge displaying seasonal minima (0.5 m$^3$ s$^{-1}$ to 1 m$^3$ s$^{-1}$) in the dry winter months (April–September) and mean summer flows of 15 m$^3$ s$^{-1}$ to 20 m$^3$ s$^{-1}$. Extreme flood events have been gauged in excess of 2000 m$^3$ s$^{-1}$ (February 1996).

The river has incised over the last 10 000 to 100 000 yr, to create a bedrock macrochannel (van Niekerk et al., 1995), which extends across the width of the incised 'valley' and contains the full extent of sedimentary deposits, channels and riparian vegetation. Within the macrochannel, one or more active channels carry water throughout the year and seasonal channels become active during the elevated summer flows. Macrochannel deposits, however, will be inundated only by rare, large-magnitude flood events. The degree of bedrock influence varies along the river as a function of the geology (Cheshire, 1994). This has resulted in a series of distinct channel types within the macrochannel that display characteristic morphological assemblages and hydraulic characteristics (van Niekerk et al., 1995; Broadhurst et al., 1997; Heritage et al., 1997). The five major channel types that have been identified within the Kruger National Park study area on the Sabie River (Fig. 1) are bedrock-anastomosing, pool–rapid, mixed-anastomosing, single-thread and braided. The diversity of channel types along the river is reflected in the wide range of geomorphological units identified. Each is described briefly in Table 1, adapted from a

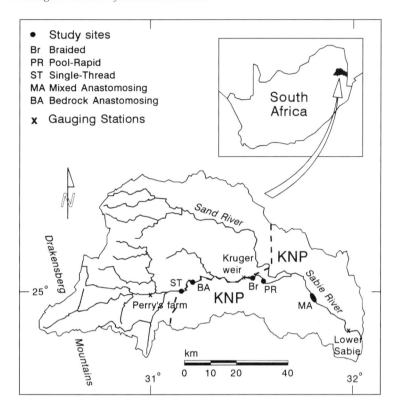

**Fig. 1.** The Sabie River catchment, showing the major geomorphological zones, location of representative study reaches and permanent gauging stations along the river.

**Table 1.** A brief description of the principal geomorphological units observed on the Sabie River in the Kruger National Park (modified from van Niekerk *et al.*, 1995).

| Morphological unit | Description |
| --- | --- |
| Rapid | Steep bedrock sections, high velocity, concentrated flow |
| Bedrock pavement | Horizontally extensive area of exposed rock |
| Isolated rock | Discrete small outcrop of bedrock |
| Pool (bedrock, mixed and alluvial) | Topographic low point in the river channel associated with a downstream bedrock or alluvial control |
| Braid bar | Accumulation of sediment in mid-channel causing the flow to diverge over a scale that approximates to the channel width |
| Lateral bar | Accumulation of sediment attached to the side of the channel, may occur sequentially downstream as alternate bars |
| Point bar | Accumulation of sediment on the inside of a meander bend |
| Bedrock core bar | Accumulation of finer sediment on top of bedrock in bedrock-anastomosing areas |
| Lee bar | Accumulation of sediment in the lee of flow obstructions |
| Anabranch (bedrock, mixed and alluvial) | Individual active channel in an anastomosing system |
| Island | Large mid-channel sediment accumulation that is rarely inundated |
| Anastomosing bar | Accumulation of coarser sediment on top of bedrock in bedrock-anastomosing areas |
| Macrochannel lateral bar | Large accumulation of fine sediment on the sides of the incised macrochannel |

comprehensive review of the geomorphology of the
Sabie River in van Niekerk *et al.* (1995). The active-
channel bed material and that forming in-channel bar
deposits along the Sabie River consists predominantly of
non-cohesive medium to coarse sands and gravels.

The bedrock-anastomosing, mixed-anastomosing and
pool–rapid channel types are all heavily influenced by
bedrock, cropping out within the macrochannel as areas
of bedrock pavement, rapids, isolated rocks and bedrock anabranch channels. Extensive cohesive deposits are
found in the anastomosed sections, forming large-scale
bedrock core bars, where accumulations of fine sediment
cover the underlying core of bedrock, between the anabranching channels (van Niekerk *et al.*, 1995). These
features are heavily vegetated with a community structure
dominated by the closed evergreen canopy of Mingerhout
trees (*Breonadia salicina*), and an understory shrub layer
dominated by the 'potato bush' (*Phyllanthus reticulatus*)
(van Coller *et al.*, 1997). The bedrock anabranches are
characteristically free of sediment or contain minor unconsolidated bar deposits. The pool–rapid channel type
is characterized by cohesive macrochannel lateral bars,
colonized by 'jackal berry' (*Diospyros mespiliformis*), a
tree species that is not exclusive to riparian zones. Active
channels display bedrock rapids and some unconsolidated bar features colonized by reeds (*Phragmites mauri-
tianus*), together with occasional consolidated bedrock
core bars, colonised by *P. reticulatus* and 'river bush-
willow' (*Combretum erythrophyllum*), an open-canopy
deciduous tree. The single-thread alluvial channel type is
uncommon on the Sabie River within the Kruger National
Park. Where single-thread channel types are found
they are characterized by extensive alluvial infill of the
macrochannel as macrochannel lateral bars and terraces
adjacent to the main active channel. These bar features
are commonly colonized by *C. erythrophyllum* and *D.
mespiliformis*. The braided channel type is characterized
by cohesive macrochannel lateral bar deposits and unconsolidated braid bars, dominated by *P. reticulatus* and
*C. erythrophyllum*. The unconsolidated in-channel sandy
deposits are often covered by reeds.

## HYDRAULIC SHEAR RESISTANCE OF COHESIVE SEDIMENTS

The erodibility of channels with cohesive sediments has
received much attention, owing to its importance for the
design of stable engineered channels. A literature survey
by the American Society of Civil Engineers (ASCE) Task
Committee on Erosion of Cohesive Sediments (1968)
reveals the interdisciplinary nature of the problem, with

much of the work having been carried out by hydraulic
and agricultural engineers, as well as by soil scientists.
Owing to the complexity of the subject, most studies have
been directed towards field observations and laboratory
investigations to determine empirical relationships for
critical velocity or shear stress required to initiate erosion.
Partheniades & Paasewell (1970) summarize and critic-
ally evaluate work ranging from early empirical studies
to more recent investigations, directed at predicting critical resistance criteria, which must be exceeded for erosion to occur. The majority of the studies relate critical
shear stress required to initiate erosion to a variety of
gross soil properties, including the Atterburg limits (particularly the plasticity index), composition and particle-
size distribution, compressive strength, bulk density,
hydraulic conductivity, moisture content and void ratio.

Investigations of the resistance of channels with cohesive sediments to erosion consider different combinations
of soil parameters in their analyses, making comparisons between studies difficult. Furthermore, the laboratory investigations use differing experimental apparatus
to model erosion, including hydraulic-jet (Dunn, 1959),
rotating cylinder (Moore & Masch, 1962), rotating
impeller (Thomas & Enger, 1961; Carlson & Enger, 1963)
and flume tests (Smerdon & Beasley, 1959; Kamphuis
& Hall, 1983). Additionally, in many of these studies,
the selection of the erosion threshold at which scour is
considered to occur is relatively subjective. The results,
however, generally indicate an increase in the critical
shear stress required to initiate erosion, with an increase
in plasticity index, shear-vane strength, compressive
strength, clay content and consolidation pressure; whereas,
increasing moisture content and void ratio result in a
reduction in the shear stress required to initiate erosion.

The critical shear stress required to initiate erosion
is correlated most often with the shear-vane strength
resistance (Dunn, 1959; Espey, 1963; Flaxman, 1963;
Rectoric & Smerdon, 1964; Kamphuis & Hall, 1983)
and plasticity index (Bureau of Reclamation, 1953;
Sundborg, 1956; Dunn, 1959; Smerdon & Beasley,
1959; Carlson & Enger, 1963; Lyle & Smerdon, 1965;
Kamphuis & Hall, 1983). An empirical relationship
derived by Smerdon & Beasley (1959) is often used to
quantify the critical shear stress required to initiate erosion (e.g. ASCE Task Committee on Erosion of Cohesive
Sediments, 1968; Partheniades & Paasewell, 1970; Graf,
1972). This relationship relates the critical shear stress
($\tau_c$) to the plasticity index ($I_w$):

$$\tau_c = 0.0034\, I_w^{0.84} \tag{1}$$

The plasticity index is an empirical measure of the
consistency of cohesive soils and no unique value can

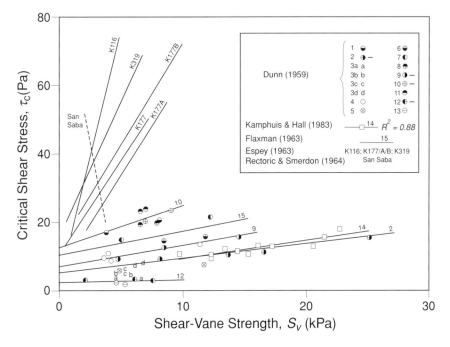

**Fig. 2.** Critical hydraulic shear stress necessary to initiate erosion as a function of shear-vane strength. (Individual data points and strength relationships are given where these were quoted by the various authors.)

characterize a particular cohesive material (Graf, 1972). Equation (1) predicts critical shear stresses required to initiate erosion that are considerably lower than those of other investigations, suggesting that other factors, including soil shear strength, influence a material's resistance to erosion (Partheniades & Paasewell, 1970). Other investigations, outlined below, have sought to incorporate these factors, utilizing the shear-vane apparatus to quantify soil shear strength. This shear strength may be attributed to three basic components: the frictional resistance to sliding between particles, cohesion and adhesion between particles, and interlocking and bridging between particles to resist deformation (Cernica, 1982). The shear-vane apparatus is appropriate for *in situ* testing of shear strength and incorporates some measure of the moisture content (Espey, 1963; Rectoric & Smerdon, 1964; Otsubo & Muraoka, 1988), unconfined compressive strength (Flaxman, 1963; Kamphuis & Hall, 1983; Kamphuis, 1990) and void ratio (Chow, 1959). The shear strength obtained by the shear-vane method depends on the rate at which torsion is applied to the soil and this is seldom controlled or standardized under field conditions. Notwithstanding this shortcoming and the need to exercise control during the testing procedure, *in situ* testing is likely to provide information that is more indicative

of the resistance of the material to entrainment or actual sediment-transport potential under field conditions.

The relationship between the critical shear stress necessary to initiate erosion and shear-vane strength was investigated by Dunn (1959), who conducted a laboratory study on consolidated-clay samples subjected to erosion by a submerged water-jet. The relationship between critical hydraulic shear stress necessary to initiate erosion and shear-vane strength was assumed to be linear and in accordance with the form of a generalized theoretical relationship derived for the specific testing procedure given in Dunn (1959). Partheniades & Paasewell (1970) point out that the linear relationship of Dunn (1959) was based on an inadequate number of data points (two to four per soil type), and therefore is questionable. The data points obtained by Dunn (1959) for the 16 soil types considered, containing fitted relationships for four selected soil types (2, 9, 10 and 12), where more than two data points were collected, are plotted in Fig. 2. The general relationship between critical shear stress and shear-vane strength is given by:

$$\tau_c = (S_v + 8.62) \tan \theta \qquad (2)$$

where $S_v$ is the shear-vane strength (kPa) and $\theta$ is the slope of the linear relationship between the critical shear

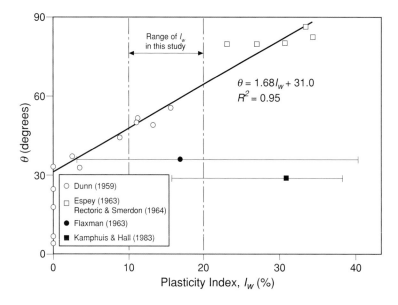

$$\theta = 1.68 I_w + 31.0$$
$$R^2 = 0.95$$

**Fig. 3.** The slope of the linear relationship, θ, between the critical shear stress required to initiate erosion and sample shear-vane strength, expressed as a function of plasticity index (data from various sources).

stress required to initiate erosion and sample shear-vane strength.

Although the linear relationship derived experimentally by Dunn (1959) (eqn 2) was shown to be consistent with the expected theoretical function, it requires further justification, as more than two data points were obtained for only four of the soils tested (soil types 2, 4, 9, and 12). The validity of Dunn's (1959) formulation was assessed in this study by analysing data from various sources and testing procedures, including, rotating cylinder (Espey, 1963; Rectoric & Smerdon, 1964), flume (Kamphuis & Hall, 1983), and field observations (Flaxman, 1963; as reported by Kamphuis & Hall, 1983) (Fig. 2). With the exception of the San Saba silty clay, which displayed a large amount of scatter (ASCE Task Committee on Erosion of Cohesive Sediments, 1968) the additional seven soil types (denoted 14, 15, K116, K117/A/B and K319) generally confirm the linear relationship developed by Dunn (1959). Unfortunately, the data points used to develop the relationships (with the exception of the data of Kamphuis & Hall (1983)) are not available in the literature to allow an assessment of the degree of scatter. Nevertheless, the additional relationships extend the range of applicability for the relationship between the critical shear stress required to initiate erosion and sample shear-vane strength beyond the limits given by Dunn (1959).

It is clear from Fig. 2 that the slope of the linear relationship (θ) between the critical shear stress required to initiate erosion and sample shear-vane strength is variable. Dunn (1959) related the slope of the linear

relationship (θ) to three additional indices, namely the plasticity index, percentage of sediment finer than 60 μm and statistical parameters describing the grain-size distribution. Of these identifying indices, the plasticity index provided a useful determinant for θ > 30° (Fig. 3), with the gradient unrelated to plasticity index below this value. Furthermore, the value of the plasticity index is not reproduced accurately by different investigators below 5%, owing to the subjective nature of the test in the field. The plasticity index data of Espey (1963) and Rectoric & Smerdon (1964) corroborates and extends the range of validity of the linear relationship suggested by Dunn (1959). This is not the case for the data of Flaxman (1963) and Kamphuis & Hall (1983), who found that the plasticity index varied across a large range for a given θ value (3–40% and 16–38%, respectively) making direct comparison using these data difficult. Dunn (1959) also developed a linear relationship between θ and percentage sediment finer than 60 μm over a plasticity index range of 5–95%, which is applicable over a wider range of gradients (4° ≤ θ ≤ 50.5°).

A single robust empirical relationship to determine the slope of the linear relationship (θ) between the critical shear stress required to initiate erosion and sample shear-vane strength, was derived using multiple regression on the sample plasticity index and percentage of sediment finer than 60 μm (uf). Data from the literature (Dunn, 1959; Rectoric & Smerdon, 1964; Kamphuis & Hall, 1983) in the range $0\% \leq I_w \leq 34\%$ and $5\% \leq U_f \leq 95\%$ were used in the analysis, and provided an acceptable linear relationship ($R^2 = 0.97$) given by eqn (3). The value of θ

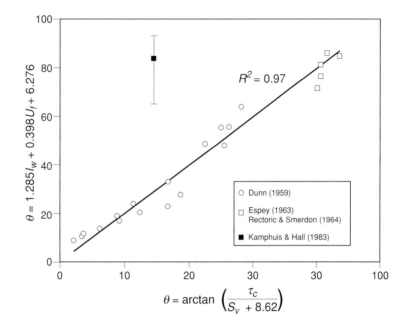

**Fig. 4.** Value of θ determined from eqn (2) and empirically as a function of plasticity index and fraction finer than 60 μm (eqn 3).

$$\theta = \arctan \left( \frac{\tau_c}{S_v + 8.62} \right)$$

determined from eqn (2) is plotted against the empirical relationship (eqn 3) in Fig. 4.

$$\theta = 1.285\, I_w + 0.398\, U_f + 6.276 \qquad (3)$$

Using eqns (2) and (3) as the basis for estimating critical shear stress required to initiate erosion, *in situ* shear strengths of the cohesive sediments along the Sabie River in the Kruger National Park were determined by performing 300 shear-vane tests. Data collection covered representative examples of the five major channel types (Heritage *et al.*, 1997) and also included a range of macrochannel and active-channel geomorphological features. As no spatial relationship between shear strength and channel type or geomorphological unit was evident (Fig. 5), the data were analysed statistically and are presented as a frequency distribution (Fig. 6). Disturbed samples, corresponding to a range of *in situ* shear strengths, were extracted from the field for laboratory analysis of plasticity index and fraction finer than 60 μm. These data were used to compute the distribution of θ (eqn 3) and are plotted in Fig. 7.

The distribution of critical shear for the initiation of erosion for the cohesive sediments was determined by computing all possible occurrences of $\tau_c$ according to eqn (2), based on the discrete distributions for shear-vane strength (Fig. 6) and θ (Fig. 7). For example, for $S_v = 5$ kPa (Fig. 6) and θ = 47.5° (Fig. 7), eqn (2) gives $\tau_c = 14.9$ kPa, this has a frequency of occurrence given by

the product of the individual probabilities of occurrence (i.e. 0.74% and 14.3%, respectively = 10.6%). The resulting frequency distribution for the critical shear stress necessary to initiate erosion is plotted as Fig. 8A by channel type and Fig. 8B for the whole river.

## APPLIED SHEAR STRESSES DURING EXTREME FLOOD EVENTS

The average applied shear stress ($\tau_a$) acting on the surface of the consolidated sediment may be calculated as follows:

$$\tau_a = \gamma R S_f \qquad (4)$$

where γ is the unit weight of water (N m$^{-3}$), R is the hydraulic radius (m) and $S_f$ is the energy slope.

Equation (4) assumes uniform flow, and many authors contend that this assumption is reasonable under medium- and high-flow conditions in alluvial rivers (Bathurst, 1982; Bhowmik, 1982). The large-scale non-uniformity of bedrock-influenced reaches and the effect of dense vegetation along the Sabie River, however, may result in the relationship underestimating the applied shear stresses locally, where the vegetation cover is less dense. The approximate distribution of total boundary shear stress across a channel cross-section may be determined by assuming that the total shear stress is distributed linearly, according to the variation of flow depth (i.e. flow

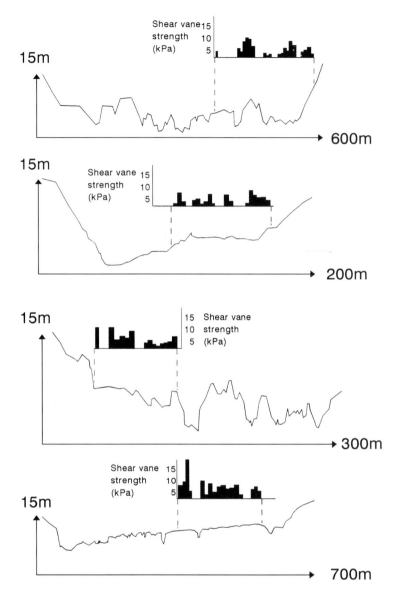

**Fig. 5.** Distribution of cohesive strength across the macrochannel at several locations along the Sabie River, expressed as a function of the *in situ* shear-vane strength.

depth is substituted for hydraulic radius in eqn (4), to obtain the local boundary shear stress).

The change in flow resistance, concomitant with discharge for five representative reaches, one for each of the channel types, was determined by Broadhurst *et al.* (1997). Channel geometry and hydraulic data for several cross-sections per reach were collected, enabling quantification of energy slope through the reach and reach-average flow resistance. Manning's *n* flow resistance is presented (Fig. 9) for these reaches, up to the discharge magnitude of between 1705 and 2259 m³ s⁻¹. Topographic and

hydraulic data from the five representative reaches and 17 additional single cross-sections surveyed throughout the Sabie River in the Kruger National Park were used to compute the distribution of boundary shear stress, corresponding to the peak discharge. Where single, isolated cross-sections prohibited the direct measurement of energy slope (owing to the lack of information on local velocity acceleration or deceleration through the cross-section), it was back-calculated from the Manning's *n* flow resistance equation (Fig. 9). This method substituted a representative flow resistance value for a given channel type, quantified

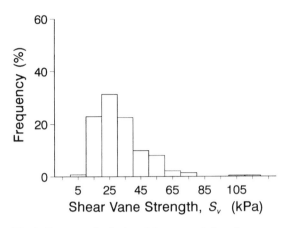

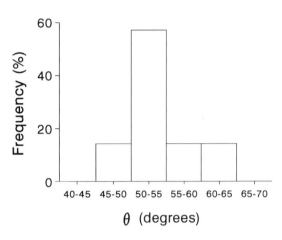

**Fig. 6.** Frequency distribution of shear strength from shear-vane tests.

**Fig. 7.** Frequency distribution of θ, determined empirically (eqn 3).

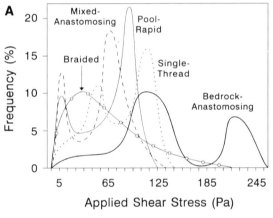

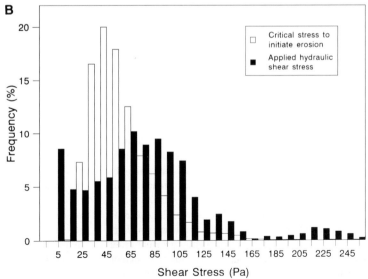

**Fig. 8.** (A) Frequency distribution of critical applied shear stress (discharge of between 1705 and 2259 m³ s⁻¹, depending on position along river) acting to erode the cohesive sediments, as a function of channel type (braided, mixed-anastomosing, pool–rapid, bedrock-anastomosing and single-thread. (B) Combined frequency distribution of critical resisting and applied hydraulic shear stress (discharge of between 1705 and 2259 m³ s⁻¹, depending on position along river) for all cohesive sediments along the Sabie River in the Kruger National Park.

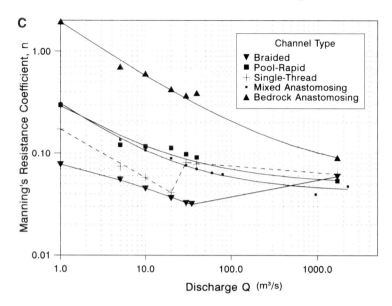

**Fig. 9.** Variation in Manning's resistance coefficient with discharge for the major channel types along the Sabie River (extended from Broadhurst *et al.*, 1997).

**Table 2.** Probability of eroding cohesive sediments (discharge in the range 1705 m³ s⁻¹ to 2259 m³ s⁻¹, according to position along river). The overall erosion figure is a summation of the probability of erosion multiplied by the spatial composition of each channel type.

| Channel type | Spatial composition (%) | Probability of erosion (%) | Average applied shear stress (Pa) |
|---|---|---|---|
| Braided | 14 | 63 | 70 |
| Pool–rapid | 28 | 66 | 63 |
| Single-thread | 3 | 75 | 80 |
| Mixed-anastomosing | 35 | 63 | 59 |
| Bedrock-anastomosing | 20 | 91 | 145 |
| Sabie River | 100 | 68 | 70 |

at the representative reaches, and used known channel geometry variables for the cross-section in question.

The frequency distribution of applied shear stress within the cross-sections was determined by dividing the macrochannel into equal increments and including areas along the profile where cohesive sediments occur as surface material. In this way, a frequency distribution of applied shear over non-cohesive sediments may be computed for each channel type (Fig. 8a) and also for the whole Sabie River in the Kruger National Park (Fig. 8b), taking account of the proportional occurrence of the channel types (Table 2).

The distribution of applied shear stress (Fig. 8A & B) is bimodal, reflecting the shape influence of large macrochannel deposits and consequently irregular cross-sectional geometry of the profiles. The lower mode corresponds to the macrochannel bank and elevated features

along the macrochannel floor, whereas the higher mode corresponds to the more incised areas. Similar distributions occur for each of the channel types, with the exception of braided (Fig. 8A), which is characterized by a more prismatic cross-sectional geometry and hence unimodal distribution.

Given the apparent lack of a direct spatial relationship between the applied shear stress and the critical shear stress required to initiate erosion (Fig. 5), the probability of eroding the cohesive sediments along the river may be determined statistically. This is achieved by summing the individual probabilities of all possible combinations, where the applied shear stress is greater than the critical resisting shear stress. The results of this analysis are given in Table 2 for each of the representative channel types, giving a composite picture of the whole river in the Kruger National Park.

**Table 3.** Summary of the flow characteristics of the Sand River in the Kruger National Park (based on calibrated ACRU simulated data).

| Years | Mean flow ($m^3 s^{-1}$) | Maximum flow ($m^3 s^{-1}$) | Standard deviation ($m^3 s^{-1}$) | Years with no-flow events | Number of no-flow events | Total no-flow days |
|---|---|---|---|---|---|---|
| 1934–1943 | 12.5 | 850.0 | 35.4 | 2 | 5 | 66 |
| 1944–1953 | 7.4 | 20.3 | 20.3 | 3 | 14 | 150 |
| 1954–1963 | 10.8 | 844.0 | 31.2 | 2 | 7 | 176 |
| 1964–1973 | 8.0 | 290.0 | 19.0 | 5 | 7 | 194 |
| 1974–1983 | 10.7 | 475.0 | 23.8 | 3 | 17 | 385 |
| 1984–1993 | 5.9 | 263.0 | 15.0 | 2 | 6 | 580 |

## DISCUSSION AND CONCLUSIONS

The analysis shows that 68% of the area of cohesive sediments along the Sabie River may potentially be eroded by a discharge of between 1705 and 2259 $m^3 s^{-1}$ (Table 2). This value remains relatively consistent (63–75%) for all the channel types, with the exception of bedrock-anastomosing, where the deposits have a 91% chance of being eroded at this discharge. The higher probability in the bedrock-anastomosing channel types arise from the significantly elevated shear stresses (Table 2).

Field observations following a flood experienced in February 1996 (peak was gauged at between 1705 $m^3 s^{-1}$ and 2259 $m^3 s^{-1}$, upstream and downstream of the Sand River tributary, respectively, Fig. 1) revealed that the removal of cohesive material was patchy for the sites investigated along the length of the river. The influence of vegetation colonizing the macrochannel floor was apparent, with vegetational cover providing increased protection from erosive forces. The aerial proportion of the cohesive bed material that was subjected to actual erosion during this flood appeared to be less than 68%, although it was difficult to assess owing to its patchy nature and evidence of overlying sandy fluvial deposits, deposited by the receding flood, that covered many areas of the macrochannel.

This study's analysis is based on the assumption that the total resisting force is dissipated along the channel perimeter, as boundary shear acting on the surface of the cohesive sediment. The Sabie River is characterized by well-developed riparian vegetation, particularly in bedrock-anastomosing reaches. In-channel vegetation increases the total flow resistance and acts to reduce boundary shear at the bed. In addition, the root masses bind the sediments, further reducing entrainment. The proportion of the total resisting force that may be dissipated by vegetational resistance was apparent in the bedrock-anastomosing reaches, where extensive riparian tree mortalities were recorded, with trees commonly uprooted by the flow.

Also, the assumption of cross-sectional flow uniformity in the estimation of shear stress is questionable, given the irregular cross-sectional bed profile. Under these conditions, the local shear stresses may be higher or lower than the value calculated, further increasing the likelihood of erosion in certain places.

Catchment antecedent conditions also may have influenced the potential for extensive erosion of in-channel cohesive sediments. Bevan (1981) demonstrated the importance of temporal ordering of events when considering likely geomorphological response on the River Exe, England. Additionally, Newson (1980) noted the importance of event queuing or sequencing in influencing the geomorphological effectiveness of flood flows in studies on the rivers Wye and Severn in Wales. This was a result of the degree of channel recovery from previous floods and the lack of available sediment supply from the catchment, as the slopes had been denuded only a few years earlier. Sediment production across the Sabie River drainage basin is concentrated in the centre and east of the catchment, particularly in the area drained by the Sand River (van Niekerk & Heritage, 1994; Donald *et al.*, 1995; Donald, 1997), and a recent extended drought between 1986 and 1996 (Table 3) resulted in the build-up of material across the catchment. Much of this material would have entered the Sabie River during the February 1996 flood, resulting in the river transporting at capacity, leaving little excess energy to erode the previously deposited consolidated sediment. Support for this argument also comes from evidence of extensive new sandy deposits, which were preserved as 'stalled' dune fields across many of the macrochannel features (typical dune amplitudes were of the order of 1 m). The effect of a series of three major floods between 1937 and 1939 (Fig. 10) resulted in the significant loss of cohesive sediment from the Sabie River, as proportionally less material was being delivered from the catchment with each subsequent event, this is evident from historic aerial photographic records (Fig. 11). This contrasts with the February 1996 event, where abundant sediment was supplied from the catchment.

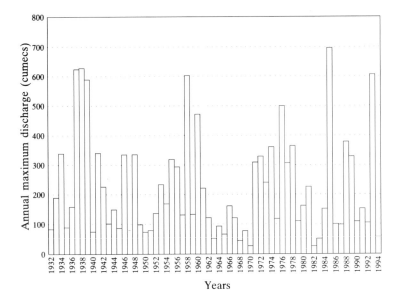

**Fig. 10.** Sequence of annual maximum flows determined from mean daily flow data, illustrating the three large-magnitude events that occurred between 1937 and 1939.

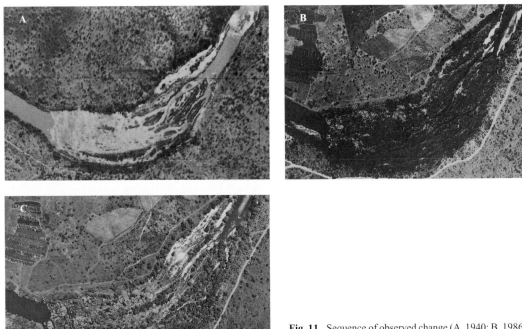

**Fig. 11.** Sequence of observed change (A, 1940; B, 1986; C, 1996) within the representative bedrock-anastomosing channel type, showing the large-scale stripping of cohesive sediment following floods in the late 1930s.

The short period during which the flood inundated the large-scale consolidated macrochannel deposits may have further contributed to the noted low occurrence of erosion. Longer duration floods, or a series of high-magnitude events, have the capacity to transport more sediment over time, and continued inundation may also act to alter the cohesive strength of the consolidated deposits, enhancing the probability of erosion. The floods of the late 1930s inundated the macrochannel cohesive deposits for an average of 5–7 days, longer than the 2–3 day inundation period associated with the February 1996 event, which did not result in the same degree of sediment removal.

Although the majority of studies concerning the effect of extreme floods indicate some erosion, Ritter (1988) observed the deposition of in-channel and overbank gravels in the Gasconade River, Missouri, USA. Similarly, Ritter (1975) noted in-channel and floodplain deposition, following a 1-in-100 yr flood in Sexton Creek, Illinois, USA in 1973. Kochel *et al.* (1982) describe extreme erosion, but subsequent widespread deposition following an extreme flood on the Devils River, Texas, USA, which they estimated had not experienced a significant flood in the last 1200 yr. Gupta (1975, 1983) reports similar responses on the Yallahs River, Jamaica. These studies, along with some of those detailed in the introduction, demonstrate that widespread erosion does not necessarily result from large-magnitude flood events. In particular, the presence of cohesive bed and banks can play a dominant role in determining channel form (Pickup & Warner, 1976; Carling, 1988; Gomez *et al.*, 1995) and inhibiting erosion.

The erosion of cohesive materials in a mixed alluvial–bedrock river containing cohesive and mobile non-cohesive granular sediments and significant in-channel riparian vegetation is complex. This study quantifies the aerial probability of eroding the cohesive sediments along the Sabie River, based on analyses of critical resistance of the local bed material and applied shear stresses exerted by the flow. Although the estimated probability of eroding cohesive material from the Sabie River bed for a large flood event appears to be higher than that observed following an extreme recent flood, owing to other complicating factors, the analysis nevertheless shows that there is considerable potential for large-scale stripping of the macrochannel deposits. The reduction in boundary shear stress resulting from in-channel vegetation, the river transporting at capacity, antecedent catchment conditions and the short duration of the flood wave are possible reasons for the erosion potential being higher than erosion actually observed. This emphasizes the need to manage the riparian vegetation adequately to enable it to continue affording protection to the cohesive bed during extreme flood events, as this will maintain the physical and ecological diversity of the Sabie River system.

## ACKNOWLEDGEMENTS

Funding by the Water Research Commission is gratefully acknowledged. The project is logistically supported by the National Parks Board. Much appreciated support has been provided by G. Strydom, P. Mdlovo, G. Mauleke and J. Maboso of the Kruger National Park and P. Frost and K. Kapur of King's College (London University). We also thank N. Alexander for his everwilling assistance and advice in the Civil Engineering soils laboratory and J. Galpin for her guidance in the statistical minefield. B. Moon assisted with *in situ* shear-vane testing. C. James took part in fruitful discussions and provided useful comments. Reviewer Leal Mertes is acknowledged for her contributions and comments.

## REFERENCES

AMERICAN SOCIETY OF CIVIL ENGINEERS TASK COMMITTEE ON EROSION OF COHESIVE SEDIMENTS (1968) Erosion of cohesive sediments. *Proc. Am. Soc. civ. Eng. J. Hydraul. Div.*, **94**(HY4), 1017–1049.

BAKER, V.R. (1975) Flood hazards along the Balcones Escarpment in Central Texas: alternative approaches to their recognition, mapping, and management. *Tex. Univ. Austin, Bur. econ. Geol. Circ.*, **75–5**, 22 pp.

BATHURST, J.C. (1982) Discussion on: Shear stress distribution and secondary currents in straight open channels. In: *Gravel-bed Rivers* (Eds Hey, R.D., Bathurst, J.C. & Thorne, C.R.), pp. 57–58. John Wiley & Sons, Chichester.

BEVAN, K. (1981) The effect of ordering on the geomorphic effectiveness of hydrologic events. In: *Erosion and Sediment Transport in Pacific Rim Steeplands* (Eds Davies, T.R.H. & Pearce, A.J.), pp. 510–526. Publication 132, International Association of Hydrological Sciences, Wallingford.

BHOWMIK, N.G. (1982) Shear stress distribution and secondary currents in straight open channels. In: *Gravel-bed Rivers* (Eds Hey, R.D., Bathurst, J.C. & Thorne, C.R.), pp. 31–55. John Wiley & Sons, Chichester.

BROADHURST, L.J., HERITAGE, G.L., VAN NIEKERK, A.W., JAMES, C.S. & ROGERS, K.H. (1997) *Translating Local Discharge into Hydraulic Conditions on the Sabie River: An Assessment of Channel Flow Resistance.* WRC Report No. 474/2/96, Water Research Commission, Pretoria.

BUREAU OF RECLAMATION (1953) *Interim Report on Channel Stability of Natural and Artificial Drainageways in Republican, Loup and Little Sioux River Areas, Nebraska and Iowa.* U.S. Department of Interior, Washington, D.C.

CARLING, P. (1988) The concept of dominant discharge applied to two gravel bed streams in relation to channel stability thresholds. *Earth Surf. Process. Landf.*, **13**, 355–367.

CARLSON, E.J. & ENGER, P.F. (1963) *Studies of Tractive Forces of Cohesive Soils in Earth Canals.* Hydraulic Branch Report No. Hyd-504, Bureau of Reclamation, Denver, CO.

CERNICA, J.N. (1982) *Geotechnical Engineering*. Holt, Rhinehart and Winston, New York, 488 pp.

CHESHIRE, P. (1994) *Geology and Geomorphology of the Sabie River, Kruger National Park and its Catchment Area*. Centre for Water in the Environment Report 1/1994, Johannesburg.

CHOW, V.T. (1959) *Open Channel Hydraulics*. McGraw-Hill, New York.

COSTA, J.E. (1974) Response and recovery of a piedmont watershed from tropical storm Agnes, June 1972. *Water Resour. Res.*, **10**, 106–112.

DONALD, P.D. (1997) *GIS modelling of erosion and sediment yield in a semi-arid environment*. MSc thesis, University of the Witwatersrand, Johannesburg.

DONALD, P.D., VAN NIEKERK, A.W. & JAMES, C.S. (1995) GIS modelling of sediment yields in semi-arid environments. *Seventh South African National Hydrology Symposium*, Grahamstown, 4–6, September.

DUNN, I.S. (1959) Tractive resistance of cohesive channels. *Proc. Am. Soc. civ. Eng. J. Soil Mech. Found. Div.*, **85** (SM3, Paper 2062), 1–24.

ESPEY, W.H. JR. (1963) *A New Test to Measure the Scour of Cohesive Sediment*. Hydraulics Engineering Laboratory, Department of Civil Engineering, Technical Report HYD01-6301, The University of Texas, Austin, TX.

FLAXMAN, E.M. (1963) Channel stability in undisturbed cohesive soils. *Proc. Am. Soc. civ. Eng. J. Hydraul. Div.*, **HY2**, 87–96.

GLANCY, P.A. & HARMSEN, L. (1975) A hydrologic assessment of the September 14, 1974 flood in Eldorado Canyon, Nevada. *U.S. geol. Surv. Prof. Pap.*, **930**, 28 pp.

GOMEZ, B., MERTES, L.A.K., PHILLIPS, J.D., MAGILLIGAN, F.J. & JAMES, L.A. (1995) Sediment characteristics of an extreme flood: 1993 upper Mississippi River valley. *Geology*, **23**(11), 963–966.

GRAF, W.H. (1972) *Hydraulics of Sediment Transport*. Series in Water Resources and Environmental Engineering, McGraw-Hill, New York, 513 pp.

GUPTA, A. (1975) Stream characteristics in eastern Jamaica, an environment of seasonal flow and large floods. *Am. J. Sci.*, **275**, 825–847.

GUPTA, A. (1983) High magnitude floods and stream channel response. In: *Modern and Ancient Fluvial Systems* (Eds Collinson, J.D. & Lewin, J.), Spec. Publs int. Ass. Sediment, No. 6, pp. 219–227. Blackwell Scientific Publications, Oxford.

HERITAGE, G.L., VAN NIEKERK, A.W., MOON, B.P., BROADHURST, L.J., ROGERS, K.H. & JAMES, C.S. (1997) The geomorphological response to changing flow regime of the Sabie and Letaba river systems. WRC Report No. 376/1/96, Water Research Commission, Pretoria.

KAMPHUIS, J.W. (1990) Influence of sand or gravel on the erosion of cohesive sediment. *J. Hydraul. Res.*, **28**, 43–53.

KAMPHUIS, J.W. & HALL, K.R. (1983) Cohesive material erosion by unidirectional current. *J. hydraul. Eng.*, **109**(1), 49–61.

KOCHEL, R.C. (1988) Geomorphic impact of large floods: review and new perspectives on magnitude and frequency. In: *Flood Geomorphology* (Eds Baker, V.R., Kochel, R.C. & Patton, P.C.), pp. 169–188.

KOCHEL, R.C. & BAKER, V.R. (1982) Paleoflood hydrology. *Science*, **215**, 353–361.

KOCHEL, R.C., BAKER, V.R. & PATTON, P.C. (1982) Palaeohydrology of Southwest Texas. *Water Resour. Res.*, **18**, 1165–1183.

LYLE, W.M. & SMERDON, E.T. (1965) Relation of compaction and other soil properties to the erosion resistance of soils. *Trans. Am. Soc. agric. Eng.*, **8**, 419–422.

MOORE, W.L. & MASCH, F.D. (1962) Experiments on the scour resistance of cohesive sediments. *J. geophys. Res.*, **67**(4), 1437–1449.

MOSS, J.H. & KOCHEL, R.C. (1978) Unexpected geomorphic effects of the Hurricane Agnes storm and flood, Conestoga drainage basin, south-eastern Pennsylvania. *J. Geol.*, **86**, 1–11.

NANSON, G.C. (1986) Episodes of vertical accretion and catastrophic stripping: a model of disequilibrium floodplain development. *Geol. Soc. Am. Bull.*, **97**, 1467–1475.

NEWSON, M. (1980) The geomorphological effectiveness of floods — a contribution simulated by two recent events in mid-Wales. *Earth Surf. Process.*, **5**, 1–16.

OTSUBO, K. & MURAOKA, K. (1988) Critical shear stress of cohesive bottom sediments. *J. hydraul. Eng.*, **114**(10), 1241–1256.

PARTHENIADES, E. & PAASEWELL, R.E. (1970) Erodibility of channels with cohesive boundary. *Proc. Am. Soc. civ. Eng. J. Hydraul. Div.*, **96**(HY3), 755–771.

PICKUP, G. & WARNER, R.F. (1976) Effects of hydrologic regime on magnitude and frequency of dominant discharge. *J. Hydrol.*, **29**, 51–75.

RECTORIC, R.J. & SMERDON, E.T. (1964) Critical shear stress in cohesive soils from a rotating shear apparatus. *Am. Soc. civ. Eng. Pap.*, **64–216**(June 21–24).

RITTER, D.F. (1975) Stratigraphic implications of coarse-grained gravel deposited as overbank sediment, Southern Illinois. *J. Geol.*, **83**, 645–650.

RITTER, D.F. (1988) Floodplain erosion and deposition during the December 1982 floods in Southeast Missouri. In: *Flood Geomorphology* (Eds Baker, V.R., Kochel, R.C. & Patton, P.C.), pp. 243–260.

SCHICK, A. (1974) Formation and obliteration of desert storm terraces — a conceptual analysis. *Z. Geomorphol. N.F.*, **21**, 88–105.

SCHUMM, S.A. & LICHTY, R.W. (1963) Channel widening and floodplain construction along the Cimarron River in southwestern Kansas. *U.S. geol. Surv. Prof. Pap.*, **352-D**, 71–88.

SMERDON, E.T. & BEASLEY, R.P. (1959) *Tractive force theory applied to stability of open channels in cohesive soils*. Research Bulletin No. 715, Agricultural Experimental Station, University of Missouri, Colombia, MO.

STEWART, J.H. & LAMARCHE, V.C. (1967) Erosion and deposition produced by the flood of December 1964, on Coffee Creek, Trinity County, California. *U.S. geol. Surv. Prof. Pap.*, **422-K**, 22 pp.

SUNDBORG, Å. (1956) The River Klarälven; a study of fluvial processes. *Geogr. Annaler*, **38**, 125–316.

THOMAS, C.W. & ENGER, P.F. (1961) Use of an electric computer to analyse data from studies of critical tractive forces from cohesive soils. *International Association for Hydraulic Research, 9th Congress*, Dubrovnik, pp. 760–771.

VAN COLLER, A.L., ROGERS, K.H. & HERITAGE, G.L. (1997) Linking riparian vegetation types and fluvial geomorphology along the Sabie River within the Kruger National Park, South Africa. *Afr. J. Ecol.*, **35**, 194–212.

VAN NIEKERK, A.W. & HERITAGE, G.L. (1994) The use of GIS techniques to evaluate channel sedimentation patterns for a bedrock controlled channel in a semi-arid region. In: *Proceedings, International Conference on Basin Development* (Eds Kirby, C. & White, W.R.), pp. 257–271. John Wiley & Sons, Chichester.

VAN NIEKERK, A.W., HERITAGE, G.L. & MOON, B.P. (1995) River classification for management: the geomorphology of the Sabie River. *S. Afr. geogr. J.*, **77**(2), 68–76.

WILLIAMS, G.P. & GUY, H.P. (1973) Erosional and depositional aspects of Hurricane Camille, in Virginia, 1969. *U.S. geol. Surv. Prof. Pap.*, **804**, 1–80.

WOLMAN, M.G. & GERSON, R. (1978) Relative scales of time and effectiveness of climate in watershed geomorphology. *Earth Surf. Proc.*, **3**, 189–203.

WOLMAN, M.G. & MILLER, J.C. (1960) Magnitude and frequency of forces in geomorphic processes. *J. Geol.*, **68**, 54–74.

*Spec. Publs int. Ass. Sediment.* (1999) **28**, 147–160

# Erosion of sediments between groynes in the River Waal as a result of navigation traffic

W. B. M. TEN BRINKE*, N. M. KRUYT†, A. KROON† *and* J. H. VAN DEN BERG†

*\*National Institute for Inland Water Management and Waste Water Treatment (RIZA), P.O. Box 9072, 6800 ED Arnhem, The Netherlands; and*
*†Department of Physical Geography, Utrecht University, P.O. Box 80115, 3508 TC Utrecht, The Netherlands*

## ABSTRACT

The River Waal is the largest river in The Netherlands, and its shipping density is among the highest of all the inland waterways of the world. The river bed sediments of the Waal are mainly sandy. To protect the river banks from erosion, groynes have been built all along the river. The groynes are submerged at high discharges only, and it is hypothesized that the sandy beaches between the groynes are the result of a balance between sand deposition at high discharges and sand erosion by currents induced by navigation traffic at moderate and low discharges.

In the summer of 1996, currents and sediment resuspension resulting from navigation traffic were measured between the groynes. The largest vessels had the strongest impact, typically creating a water-level depression of 15–20 cm and currents of 30–40 cm s$^{-1}$ at 10 cm above the bed. Scaling sand transport from groyne fields up to the entire river and a time-scale of 1 yr results in overestimates of sand losses from the groyne fields, suggesting that more measurements in groyne fields with different orientations and at different discharge conditions have to be carried out. Models in the literature that relate vessel characteristics to currents induced near the bank do not seem to serve for the conditions of the navigation traffic in the River Waal.

## INTRODUCTION

The Netherlands is made up largely of sediments deposited by rivers and the sea. Although at present, civil engineering works along the rivers and the coast prevent the lower areas from being flooded, the rivers and the sea still determine the Dutch landscape. In addition, they contribute significantly to the Dutch economy. In a densely populated country such as The Netherlands, water systems serve many purposes, which have to be combined such that conflicts are avoided. In particular, the combination of shipping interests with other functions, such as nature and flood protection, has resulted in a typical Dutch river landscape where rivers are embanked and groynes are built perpendicular to the river banks at regular distances of generally 200 m.

The groynes are an example of human impact on the river system, resulting in a fixed river planform, a navigation channel that is relatively deep over a large part of its cross-section, and sandy beaches between the groynes. Thus several functions seem to be combined without mutual conflict. The success of the groynes in fulfilling

these functions depends on the balance of the hydrodynamic forces acting on the sandy deposits between these groynes, resulting in net erosion or deposition of sand at the beaches. Erosion of sand is thought to take place as a result of currents and waves induced by navigation traffic. Deposition of sand probably takes place mainly at times of high discharge, when the groynes are completely submerged. During these flood events, sand is transported from the navigation channel to the groyne field beaches and further landward on to the natural levees (Ten Brinke *et al.*, 1998). As long as erosion and deposition are in equilibrium on a time-scale of a couple of years, the beaches between the groynes are in dynamic equilibrium. This has been the case over the last several decades. The situation may change, however, if the balance of the hydrodynamic forces changes. For the preservation of the sandy beaches, the change of the composition of the inland navigation fleet towards ever-increasing carrying capacity (Fig. 1) is of major concern. In addition, groyne fields in a dynamic equilibrium may imply a significant

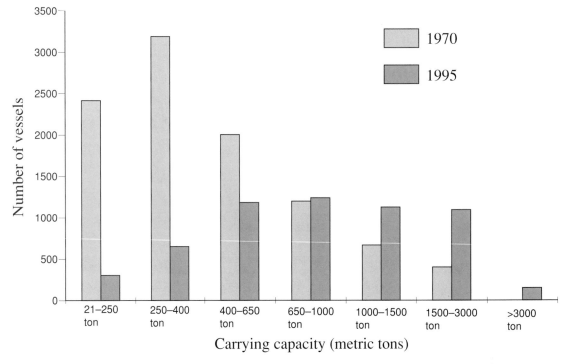

**Fig. 1.** The composition of the Dutch inland navigation fleet, expressed in carrying capacity of the vessels, in 1970 and 1995. (Source: Centraal Bureau voor de Statistiek, 1970–1995.)

source and sink of sand at specific times. At present, these terms are unknown and therefore not incorporated into two-dimensional morphological models of rivers. The success of these models depends partly on the knowledge of the sediment fluxes in and out of the groyne fields.

Publications in the literature deal mainly with modelling hydrodynamics behind groynes (Tingsanchali & Maheswaran, 1990). In The Netherlands, studies were carried out to model the currents in the area between the groynes by means of laboratory experiments (Termes *et al.*, 1991). These studies have greatly contributed to the knowledge of circulation cells and their hydrodynamic forcing. Although these studies are essential for understanding the processes that take place, they cannot be used at present to estimate the role of sediment transport induced by navigation traffic for the sediment budget for the entire River Waal on a long time-scale. There is a strong need for field data that comprises all the controlling variables on transport volumes and changes in bed level in a set of groyne fields.

A study therefore was carried out in the groyne fields to quantify the impact of different types of vessels on currents and the resuspension of sediments. The measurements were conducted in the River Waal during 4 weeks

of the summer of 1996 in a relatively straight reach at a river discharge slightly less than the yearly average. The measurements are part of a large programme in which measurements in inner and outer bends and measurements at high discharges are also carried out. This total programme aims at:

1 quantifying the sediment exchange between the channel and the groyne fields in time and space;

2 analysing the processes that govern the sediment transport to and from the groyne fields;

3 modelling the groyne field sediment transport.

The present paper aims at quantifying the impact of different types of vessels on currents and the erosion of sediments in groyne fields. It focuses on a relatively straight section of the river and conditions where the groynes are not submerged. The analyses should result in a sediment budget for these groyne fields.

## AREA OF RESEARCH

The River Rhine originates in the Alps and flows through Switzerland and Germany to The Netherlands (Fig. 2A). The average discharge of the Rhine near the

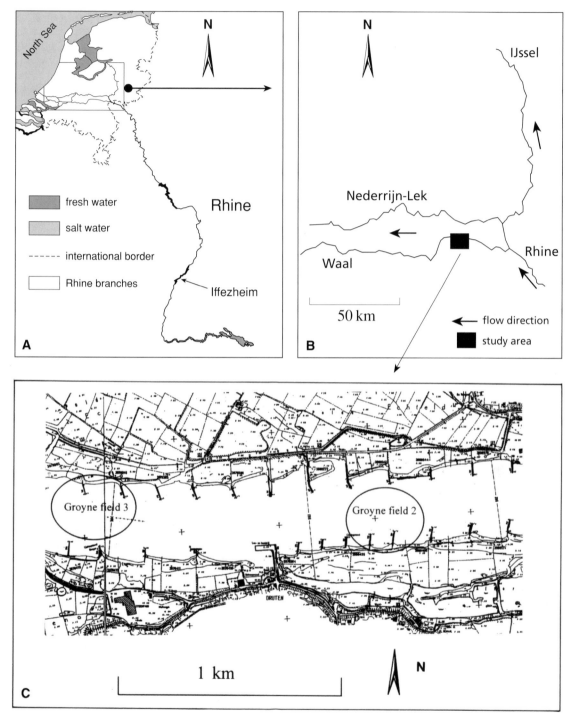

**Fig. 2.** The Rhine River system and its tributaries in The Netherlands. The study area is indicated.

Dutch–German border is 2300 m$^3$ s$^{-1}$, derived from both rain and snowmelt. Maximum discharges are up to 12 000 m$^3$ s$^{-1}$. In The Netherlands, the Rhine divides into three distributaries: the Waal, Nederrijn-Lek and IJssel (Fig. 2B). These branches have embankments constructed along their entire length. The discharge ratio between these distributaries is approximately 6 : 2 : 1.

The shipping density of the Rhine and the Waal, connecting the world's largest harbour of Rotterdam with the major industrial area of Germany, is among the highest of all the inland waterways of the world. The fresh water of the Rhine branches serves industrial, domestic and agricultural purposes, and prevents saline groundwater seepage into the Dutch polders. Human activity has affected and changed the river system over centuries, including dredging and excavation works, artificial meander cutoffs, groynes, rip-rap on the river banks, weirs and dams at some of the (former) river outlets. The present Dutch Rhine riparian landscape is characterized mainly by pastures, separated from the main channel by groynes. Parts of these pastures and the sandy river banks and beaches are increasingly turned into natural riparian zones. Doing this without compromising the river's other functions means carrying out research on natural and human-induced (shipping) morphological processes.

Measurements were carried out in four groyne fields in a relatively straight section of the Waal. In this section the groynes are generally 50 m long. In this paper, the results of the measurements in two of these groyne fields are presented (Fig. 2C). The average water depth of the Waal is of the order of 5 m. The length and width of this reach are 84 km and 260–340 m, respectively, and bed gradient is $1.1 \times 10^{-4}$. The Waal is a sand–gravel-bed river with a median particle size of bed material of about 0.5–4 mm. The sand of the beaches between the groynes in the area of research is finer grained than the river bed itself, with $D_{50} = 0.35$–0.40 mm.

## FLOW PATTERN IN GROYNE FIELDS

Under conditions where the groynes are not submerged, the groyne fields are not part of the wetted cross-section, so the currents in the groyne fields are not directly the result of the discharge in the main channel. Indirectly there is an effect of this discharge, however, because some water is diverted by the groynes from the main channel into the groyne fields. This water flows into the groyne fields with low current velocities through the downstream half of the wetted cross-section between the groynes. This water flows back into the river over a small width just downstream of the upstream groyne of the

groyne fields (Fig. 3B; Termes *et al.*, 1991). As a result of the small width of this so-called outflow point, current velocities just downstream of the groynes are high, resulting in scour holes 1–2 m deep (Fig. 4). Thus the discharge in the main channel induces a circulation flow in the groyne field, the strength of which is influenced by vessels passing the groyne field. The circulation flow is in the form of a circulation cell directed towards the outflow point and varies in strength mainly as a result of navigation traffic. Earlier investigations (Termes *et al.*, 1991) have shown that when the distance between the groynes is 200 m or more, one circulation cell no longer covers the entire groyne field, and a second smaller circulation cell is present near the downstream groyne (Fig. 3A). The currents of this smaller cell are generally not strong enough to create a second scour hole.

The major variables that are assumed to control sediment exchange between the groyne fields and the main channel are the characteristics of the navigation traffic and the distance between the groynes. The characteristics of the navigation traffic vary between the north and south side of the river because the vessels going upstream are generally loaded and follow the south bank, whereas the empty vessels returning downstream follow the north bank.

## METHODS AND ANALYSES

Four groyne fields were selected which varied in size and in characteristics of vessels passing by. Two of these fields were on the north side and two on the south side of the Waal. The distance between the groynes varied from 100 m to 250 m for the south side to 200 m for the north side.

The measurements were carried out with optical backscatter turbidity sensors (OBS), electromagnetic flowmeters (EMF) and pressure sensors. These sensors were attached to three tripods. One tripod was equipped with a computer and a set of batteries to allow for 4 weeks of semi-continuous measurements. The tripod measured independently and was always placed in the outflow point of the groyne field at the beginning of the week. This tripod had two sets of flowmeters and turbidity sensors: one flowmeter and one turbidity sensor were positioned at 8.5 cm above the bed, the other set was positioned at 34 cm above the bed. The pressure sensor was placed at 40 cm above the bed. The tripod had a compass for orientation and two devices for measuring the inclination of the bed in two directions. The tripod dimensions were $4 \times 4 \times 2.5$ m and its weight was 600 kg (Fig. 5). The other two tripods were much smaller, with dimensions of

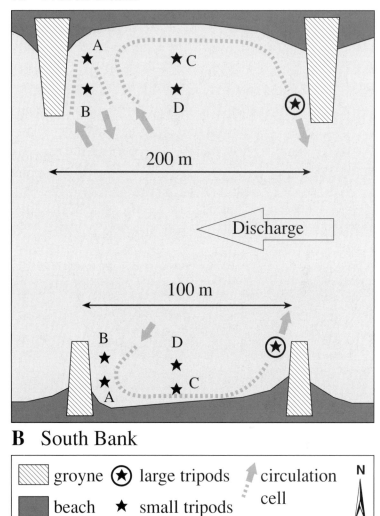

**Fig. 3.** Flow circulation cells in groyne fields with a large (A) and small (B) distance between the groynes.

$1.5 \times 1.5 \times 0.5$ m and a weight of 70 kg. Each of these small tripods had only one flowmeter and one turbidity sensor, both positioned 10 cm above the bed, and one pressure sensor 50 cm above the bed. These tripods had no computers and batteries; the sensors instead were connected to computers on a research vessel. The connecting cables were 100 m so that the research vessel could be positioned far enough from the tripods so as not to influence the measurements. The tripods had no compass; the orientation of the tripods and hence the sensors was determined by placing a 3.5-m-long stake with a horizontal piece on top of the tripods, and determining

the orientation of this stake with a tachymeter positioned on the river bank. The tripods were positioned on the beaches of the groyne fields such that the top piece of the stake was always exposed. The co-ordinates of these tripods were determined by a differential global positioning system (DGPS) by positioning a DGPS receiver on top of the crane aboard the vessel just above the tripods. Current velocity was measured in two horizontal directions perpendicular to each other.

The entire campaign was carried out from 17 June to 12 July 1996. In each groyne field, measurements were carried out for four successive days (Monday through to

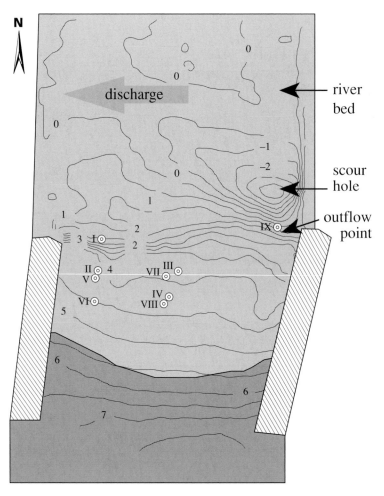

I–IX: tripods positions     ☐ water     ▨ groyne

—2— depth-contour     ▨ beach

**Fig. 4.** Typical bed topography of a groyne field, including the beaches and a part of the river bed. Depths are in metres with respect to Dutch Ordnance Datum.

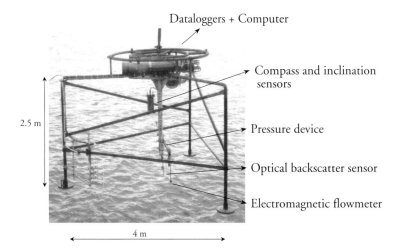

**Fig. 5.** The large tripod with sensors, used for measuring sand transport and hydrodynamics in the outflow point of the groyne fields.

Thursday). The large tripod was positioned in the outflow point of the groyne field near the upstream groyne and the two small frames were positioned at locations A and B (Fig. 3) on Mondays and Wednesdays and at locations C and D on Tuesdays and Thursdays. Locations A and C were shallow and were chosen such that the resuspension of sediment by surging waves could be measured. The locations B and D were at water depths of some 2 m. Measurements were carried out during periods (sampling intervals) of 35 min, starting on the hour. Care was taken that the measurements of the three tripods were done synchronously by adjusting the clock aboard to the clock in the computer of the large tripod at the beginning of the week. The sensors of the small tripods sampled with a frequency of 4 Hz. The sampling frequency of the sensors of the large tripod was only 2 Hz because this tripod was not intended to be used for measuring waves. With the small frames, four to five sampling intervals were measured each day, whereas 24 sampling intervals each day were measured with the large frame. The success of a series of measurements in a groyne field depends strongly on the position of the large tripod because the outflow is concentrated over a small width. Whether this tripod was positioned successfully at exactly the right spot can be evaluated only from the measurements at the end of the week. Unfortunately, in two of the four groyne fields, the position of the large tripod was not exactly in the middle of the outflow current. In this paper, therefore, only the results for two groyne fields will be presented: a 100 m groyne field at the south side and a 200 m groyne field at the north side.

For all turbidity sensors, calibration curves were determined in the laboratory using sediment from the beaches of the study area. This sediment was sand without an admixture of mud. The suspended sediment in the field, however, was a mixture of sand and mud. The suspended mud concentration was wash load and relatively constant throughout the study interval. A correction for this mud concentration had to be made because this mud concentration partly determined the OBS results, whereas this effect was not taken into account in the calibration curves. The turbidity values of the small tripod sensors were corrected for the mud concentration by taking water samples *in situ* with a peristaltic pump attached to a tube located next to the turbidity sensor. The water samples of the small tripods were then filtered over 50 µm plankton gauze and 0.4 µm filters to determine the suspended sand and mud concentration. For the large tripod, no samples were taken. From the recordings of the turbidity sensors at depths of 8.5 cm and 34 cm above the bed, however, it was clear that at 34 cm no response to vessel passages was measured, whereas turbidity peaks were

clearly present in the data of 8.5 cm above the bed. Thus, the turbidity at 34 cm was assumed to be wash load only, and the turbidity at 8.5 cm was corrected for the influence of this wash load by subtracting the results at the 34 cm level from the results at the 8.5 cm level. The data of the pressure sensors were corrected for air pressure to obtain data on water depth.

During each sampling interval with the small frames, the characteristics of all vessels passing the area of research were noted. These characteristics included direction of navigation, time of passage of bow and stern, length, width and depth (sometimes noted on the vessels, sometimes estimated), cargo (loaded, half-loaded, empty: estimated), and distance from the bank (estimated). From these characteristics velocity and volume of displaced water were calculated.

## CALCULATION PROCEDURES

### Sediment fluxes

Tripod data showed that turbidity peaks generally coincided with peaks in current velocity and strong depression of water level. Mostly, at these times a vessel (just) passed the groyne field. Sometimes, however, no vessel was observed for up to several minutes before these peaks. As the weather was calm, river discharge was constantly low, and navigation traffic density was always high. With ships leaving their signatures on hydrodynamics for quite some time after they passed, it may be safely assumed that these peaks also result from navigation traffic. Probably, interference of waves and reflection at groynes and river banks play a role. Therefore, sand transport fluxes during these peaks were quantified and summed for all the sampling intervals at a specific location in order to obtain the sand transport induced by navigation traffic. Peaks were defined as concentrations higher than the average plus one standard deviation.

Sand transport at time *t* follows from

$$q_t = \int_0^h U_t C_t \, dz \tag{1}$$

where $q_t$ is sand transport (kg m$^{-1}$ s$^{-1}$), U is current velocity (m s$^{-1}$), $C$ is suspended sand concentration (kg m$^{-3}$), $h$ is water depth (m) and $z$ is height above the bed (m).

Vertical profiles of current velocity and concentration could not be measured with the facilities available. Sand transport therefore had to be estimated from measurements at one or two depths only. Assumptions therefore had to be made about the representativeness of these depths in describing depth-averaged values. For the current velocity at a depth of 10 cm above the bed, a logarithmic current velocity profile was assumed

$$U_z = \frac{U_*}{\kappa} \ln\left(\frac{z}{z_0}\right) \tag{2}$$

where $U_*$ is shear velocity (m s$^{-1}$), $\kappa$ is 0.4, $z$ is height above the bed (m) and $z_0$ is $z$ with $U = 0$. From this relationship it can be shown that depth-averaged current velocity is

$$\bar{U} = U_{(z=0.37h)} \tag{3}$$

and the ratio between the current velocity at a depth of 10 cm above the bed and depth-averaged current velocity is

$$\frac{U_{(z=0.1)}}{U_{(z=0.37h)}} = \frac{\ln\left(\dfrac{0.1}{z_0}\right)}{\ln\left(\dfrac{0.37h}{z_0}\right)} \tag{4}$$

Under the assumption that $z_0 = 1/30 \times D_{65}$ and $D_{65} = 1.3 \times 10^{-3}$ m ($D_{65}$ is the grain size for which 65% of the grains by weight are smaller), eqn (4) becomes

$$U_{(z=0.1m)} = \frac{\ln\left(\dfrac{0.1}{4.34\times10^{-5}}\right)}{\ln\left(\dfrac{0.37h}{4.34\times10^{-5}}\right)} \times \bar{U} = \frac{7.74 \times \bar{U}}{\ln\left(\dfrac{0.37h}{4.34\times10^{-5}}\right)} \tag{5}$$

At water depths of 1–2 m, which was the case in this study, this implies that the current velocity at a depth of 10 cm above the bed is 80–85% of the depth-averaged current velocity.

Sand concentration decreases exponentially with height above the bed. Thus, sand concentration at a depth of 10 cm above the bed will be a lower estimate of sand concentration near the bed. The results of the large tripod showed that sand concentration at 34 cm above the bed was negligible at all times. It will be assumed, therefore, that sand transport takes place over a depth of 34 cm above the bed and that the product of current velocity ($U$) and concentration ($C$) at a height of 10 cm is a good approximation of depth-averaged sediment flux. From the tripod data, sand transport was calculated as

$$\bar{q}_t = \frac{\sum\limits_{t=\text{beginpeak}}^{t=\text{endpeak}} (U_{t,10\text{cm}} \times C_{t,10\text{cm}} \times 0.34)}{\sum\limits_{t=\text{beginpeak}}^{t=\text{endpeak}} t} \tag{6}$$

where '$t$ = begin peak' and '$t$ = end peak' define the beginning and end of a turbidity peak, and $U_{t,10\text{cm}}$ and $C_{t,10\text{cm}}$ are current velocity and sediment concentration at time $t$ at 10 cm above the bed.

Equation (6) results in average sand transport fluxes, $q_t$ (kg s$^{-1}$ m$^{-1}$), induced by navigation traffic. These fluxes were calculated for all five positions in the groyne field and were decomposed into fluxes perpendicular to and parallel to the river bank. These flux components are transport volumes per running metre. In order to arrive at a sediment budget for the groyne field, these components were multiplied by a width over which the sand flux was assumed constant. This was done by dividing the groyne field into five parts such that each of the five tripod positions was in the middle of one of these parts. The flux components at the tripod positions in the directions parallel to and perpendicular to the river bank were then multiplied by the width of the matching parts to obtain transport volumes (kg s$^{-1}$) from one part to another.

## Return flow induced by vessels

The transport flux induced by a vessel is related to the characteristics of this vessel through the return flow near the vessel and the bankward reduction of this return flow. The physics behind these processes are complicated, and predictive equations are not derived easily. The variability of the characteristics of the Dutch inland navigation fleet is large, and for the Waal no information is available on how to combine all of these vessel characteristics such that a meaningful regression with sediment fluxes may be derived. As a first approach, therefore, relationships from the literature were used. From a comparison by Bhowmik *et al.* (1995) of several models in the literature, the model of Hochstein & Adams (1989) was thought to be the most useful for the River Waal. Hochstein & Adams (1989) developed a method to compute return flow directly from the vessel characteristics, based on a combination of theory and observations:

$$U_{ar} = V_s\{[(a-1)B + 1]^{0.5} - 1\} \tag{7}$$

where $U_{ar}$ is the average return flow velocity in the wetted cross-section between the vessel and the bank (m s$^{-1}$), with

$$a = \left[\frac{n}{(n-1)}\right]^{2.5} \tag{8}$$

$$n = \frac{A_C}{A_M} \tag{9}$$

$B = 0.3e^{1.8(V_s/V_{cr})}$ for $V_s/V_{cr} \leq 0.65$ or $B = 1$ for $0.65 < V_s/V_{cr} \leq 1$ \hfill (10)

$$V_{cr} = K \left( \frac{gA_C}{B_C} \right)^{0.5} \tag{11}$$

where $V_s$ is vessel speed (m s$^{-1}$), $n$ is blocking ratio, $A_c$ is wetted cross-section of the river (m$^2$), $A_M$ is underwater cross-section of the vessel (m$^2$), $V_{cr}$ is critical velocity (m s$^{-1}$) (defined as such by Hochstein (1967); in Hochstein & Adams (1989)), $K$ is constrainment factor defined as a function of the blocking ratio, $g$ is gravitational acceleration (m$^2$ s$^{-1}$), and $B_c$ is surface width of the wetted cross-section (m). For the Waal, $K \approx 0.7$ (Hochstein & Adams, 1989) and $0.65 < V_s/V_{cr} \leq 1$, so that $B = 1$. Thus

$$U_{ar} = V_s \left\{ \left( \frac{A_C}{A_C - A_M} \right)^{1.25} - 1 \right\} \tag{12}$$

Hochstein & Adams (1989) assumed an exponential shape of the lateral return flow distribution towards the river bank:

$$U_{ar}(y) = \alpha U_{ar} e^{-y/k_2} \tag{13}$$

$$k_2 = \frac{W_S}{\alpha[1 - e^{-\alpha F(\alpha)}]} \tag{14}$$

$$F(\alpha) = 0.42 + 0.5\ln\alpha \tag{15}$$

$$\alpha = \frac{0.114W_T}{b} + 0.715 \tag{16}$$

where $y$ is distance measured from the vessel (m), $W_s$ is distance from the vessel to the river bank (m), $W_T$ is river width (m) and $b$ is vessel width (m).

Equations (12)–(16) were used to calculate the theoret-ical vessel-induced current velocity near the groyne fields for all vessels that passed the area of research during the measurements.

## RESULTS

An example of a record of current velocity, suspended sand concentration and water depth measurements is shown for one of the small tripods (Fig. 6) and the large tripod (Fig. 7). Figures 6 & 7 refer to the same sampling interval, measured in the groyne field on the south side of the Waal. The times a vessel passed the study area are also indicated. The long-lasting passage at 11.15 refers to a barge-tow combination and the other passages are smaller vessels. The barge-tow combination induced a drawdown current of 30–40 cm s$^{-1}$, resulting in a depression of water level of 15–20 cm (Fig. 6). This current was strong enough to resuspend the sediment of the beach. The impact of the other vessels was generally smaller. Figure 7 shows the same sampling interval for the large tripod for the sensors at the depth of 8.5 cm above the bed. The strong impact of the barge-tow combination clearly results in a transport of resuspended sediment upstream towards the channel. Current velocities were comparable to the ones measured by the small tripod, but the sand concentration was higher. At this position, the current velocity before and after the passage of the barge-tow combination varied less than at the position of the small tripod.

The results in Figs 6 & 7 are characteristic examples of the measured influence of navigation traffic. The

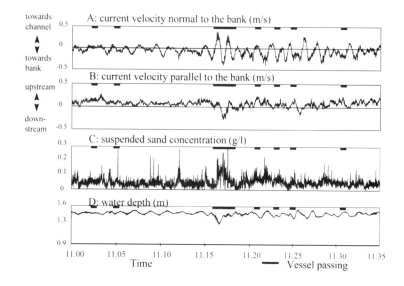

**Fig. 6.** An example of the record of current velocity perpendicular to (A) and parallel to (B) the river bank, the sand concentration calculated from turbidity (C), and water depth (D) during one sampling interval, measured by the small tripod located at D on Fig. 3. The current velocities perpendicular to and parallel to the river bank are defined positive for the directions towards the river and upstream, respectively.

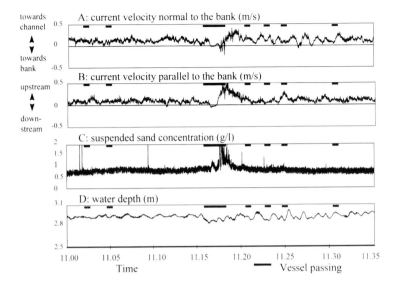

**Fig. 7.** An example of the record of current velocity perpendicular to (A) and parallel to (B) the river bank, the sand concentration calculated from turbidity (C), and water depth (D) during one sampling interval, measured by the large tripod in the outflow point at the south bank at the same time as the record in Fig. 6. The current velocities perpendicular to and parallel to the river bank are defined positive for the directions towards the river and upstream, respectively.

measured effect on current velocity at the positions of the small tripods was always a current perpendicular to the bank and moving to and from the groyne field beach a couple of times, whereas the effect in the outflow point was always a current upstream and towards the channel.

For all the passages that induced currents and resuspension, the resulting transport fluxes were calculated for the two components perpendicular to and parallel to the river bank, and for each of the five positions in the groyne field these transport fluxes were summed. For each of the five positions, the size of the transport fluxes perpendicular to and parallel to the river bank is expressed relative to the sum of these four components, for the results on the south side (week 2) in Fig. 8, and for the results on the north side (week 3) in Fig. 9.

Figure 8 shows that sediment transport at positions A and B of the small tripods (near the downstream groyne) on the south side of the river was dominated by fluxes perpendicular to the bank, the fluxes to (southward) and from (northward) the beach being about equal. At positions C and D of the small tripods, the ratio of fluxes to and from the beach, perpendicular to the bank, is comparable to positions A and B, but these positions also experience a significant component of fluxes parallel to the bank and upstream. At the outflow point, sand transport is generally upstream and towards the river.

Figure 9 shows that sediment transport on the north side of the river at the outflow point is similar to the south side, sediment transport being directed upstream and towards the river. At the small tripod positions A and B, however, results are different from the south side, the transport component towards the river being larger than the component towards the bank.

The net transport volumes at the five positions, averaged over all sampling intervals, and calculated for components perpendicular to and parallel to the river bank, are shown in Fig. 10. The results for the groyne field on the south side confirm the presence of a single current circulation cell, indicated by a dashed line. Net transport volumes at the positions of the small tripods were small compared with the volume at the outflow point. The results for the groyne field on the north side point to the presence of two circulation cells. The transport volume at the outflow point equals this volume in the groyne field on the south side.

Figure 10 shows the averaged effect of all the vessels that passed the groyne field during the measurements. Generally, 4 days of measurements (one groyne field) included 250–300 vessel passages. For the entire campaign of 4 weeks, the characteristics of 1044 vessel passages and their corresponding hydrodynamics and sediment concentration in the groyne field are known. For these passages, the theoretical return flow near the groyne field was calculated using eqns (12)–(16). In Fig. 11 this return flow is plotted versus the maximum current velocity measured in the groyne field during the passage of the vessels. This is done for the average return flow (Fig. 11A) in the river according to eqn (12) and the return flow near the bank (Fig. 11B) according to eqns (13)–(16). The measured current velocities in this figure refer to one of the small tripods. Actual maximum current velocities are much higher than the near-bank current velocities according to the model of Hochstein & Adams (1989). In fact they are even higher than the average return flow in the river. The results for the other two tripods were similar.

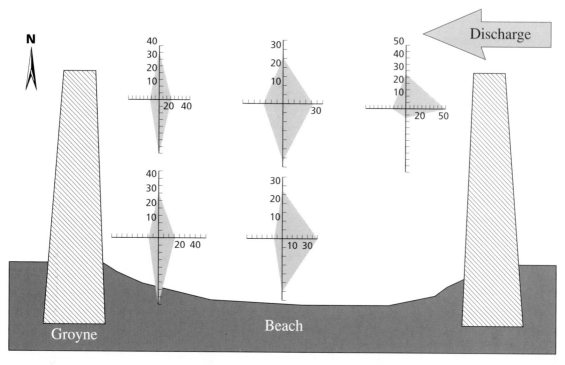

**Fig. 8.** Radar plots showing the variation of the direction of sand transport at the five positions in the groyne field on the south side of the river for 4 days of measurements.

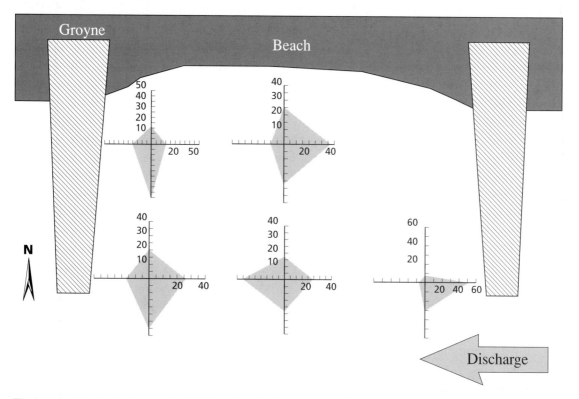

**Fig. 9.** Radar plots showing the variation of the direction of sand transport at the five positions in the groyne field on the north side of the river for 4 days of measurements.

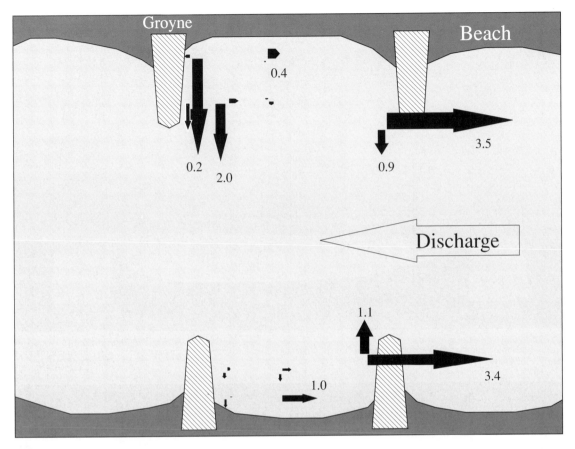

**Fig. 10.** The transport volumes at the five positions in the groyne fields on the south and north side of the river, integrated over all sampling intervals, and calculated for components perpendicular to and parallel to the river bank. The values are kg s⁻¹.

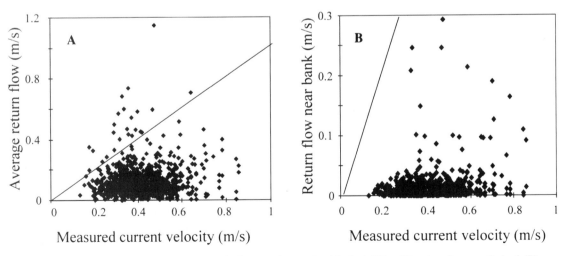

**Fig. 11.** Average return flow in the wetted cross-section between the vessel and the bank (A) and the return flow near the bank (B), according to the model of Hochstein & Adams (1989), versus measured maximum current velocities between the groynes along the Waal during the passage of 1044 vessels in 4 weeks of 1996. The straight line represents $Y = X$.

## DISCUSSION

Understanding the impact of navigation traffic on the sediment budget of the sandy groyne fields requires an understanding of the impact of navigation traffic on the pattern and strength of the circulation cells. Although the pattern and strength of these cells vary as a result of the passage of vessels, they are also controlled by characteristics of the river (discharge, morphology) and the groynes (distance, length, orientation). Thus, modelling the sediment transport in groyne fields is complicated. On the other hand, these groyne fields may play a significant role in the sediment budget of the river. Yearly sand transport volumes in the main channel at the upstream end of the Waal are in the order of 500 000 m$^3$ (Kleinhans, 1996), which corresponds with a bed elevation change of only 5–10 cm between the groynes along the Waal. Clearly, the sediment budget of these groyne fields cannot be neglected.

These first measurements show that *in situ* measurements successfully contribute to understanding the sediment budget of groyne fields. With the chosen combination of sensors, the impact of vessels on hydrodynamics and sediment transport was clear. The circulation pattern was also reflected in the data at the five positions of the tripods. There are, however, shortcomings to this approach, most of which cannot be dealt with properly. These limitations refer to the way current velocity and sand concentration are measured and sediment transport is calculated. It was difficult to position the large tripod in the outflow point precisely. Sediment transport should be measured at several depths at the same time, but this was not feasible. Therefore, assumptions were made on the variation of current velocity and concentration with depth. The assumption that the vertical current velocity profile has a logarithmic form, however, is unclear in wave-induced currents. The impact of navigation traffic was mainly through drawdown currents, which lasted several tens of seconds. The transport fluxes calculated with eqn (6) may be conservative estimates because both the current velocity and the suspended sand concentration at 10 cm above the bed underestimate depth-averaged values. It is not clear, however, whether the depth of 34 cm for sand transport overestimates field conditions. The success of using optical sensors also depends on the variability of the mud concentration. When mud concentration varies considerably, this variability dominates the variations in sand concentration, and optical turbidity sensors cannot be used. The peaks that were measured with the large tripod at 8.5 cm above the bed but did not occur 34 cm above the bed, were attributed entirely to sand, but these peaks may have

resulted partly from resuspended mud occurring as a thin layer near the bed. These layers cannot be sampled with common sampling techniques, and therefore it could not be ascertained whether or not they were present in the groyne fields. Transport volumes were calculated by multiplying the fluxes with the dimensions of the five parts into which the groyne field was divided. This schematization probably oversimplifies reality, especially near the outflow point, where sand transport is overestimated.

Vessels that are relatively large, fast, or pass the groyne field at a close distance may be expected to have a marked influence on the hydrodynamics of the water between the groynes. Indeed, the impact of these vessels on water level, current velocity and turbidity was generally most pronounced. The smaller vessels generally did not measurably influence hydrodynamics. On the other hand, peaks in current velocity and turbidity were observed that did not coincide with the passage of a vessel. In fact, more concentration peaks were identified than vessel passages. This seems to be the result of interference of waves of different vessels and reflection of waves on the river banks and the groynes. From the time series of water level and current velocity, it is clear that there are always fluctuations of water level and currents in the periods between the passages of vessels. These fluctuations are similar during both days with calm and days with windy weather, and thus must be the result of shipping. During the 4 weeks of sampling, a vessel passed the measurement stations an average of every 2 min. A more-or-less continuous impact of navigation traffic on hydrodynamics, therefore, seems to be the case.

The comparison of predictions from the Hochstein & Adams model (1989) with measured current velocities shows that this model cannot be applied to navigation traffic in the River Waal. These formulations were derived for large barge-tow combinations on the Mississippi. Both the dimensions of the navigation traffic and the dimensions and characteristics of the Mississippi differ from those in the River Waal. The formulations of Hochstein & Adams (1989) were not derived for the hydrodynamics between groynes, and their model should be used only to calculate the currents in the wetted cross-section between the vessel and the head of the groynes. In addition, the currents between the groynes were so much stronger than the ones calculated by the model that it may be safely assumed that the model does not hold for the wetted cross-section either. In the model, an exponential reduction of lateral return flow towards the bank was assumed. For navigation traffic on the Waal, this results in current velocities near the bank that are negligible for all vessels except the barge-tow combinations. Bhowmik *et al.* (1995) tested four models on the return flow distribution from barge-tows to the shoreline, including Hochstein & Adams (1989), on the River Illinois and found that all of these models underestimated the actual current velocities. As far as we know, there are no other relationships available in the literature that might be useful for the Dutch rivers, so a new formulation has to be derived, based on the conditions in the River Waal.

Sand was resuspended 14% of the time for the southern and 15% of the time for the northern groyne field. Along the banks of the Waal, there are roughly 500 sandy groyne fields with lengths of 100–250 m. An average sand export out of these groyne fields of 4.5 kg s$^{-1}$ for 15% of the time with a dry density of 1500 kg m$^{-3}$, taking place throughout the year at the same rate, would result in a sediment input into the river of 7 million m$^3$ yr$^{-1}$. Compared with a yearly sand transport of 0.5 million m$^3$ yr$^{-1}$, this is clearly far too much. The overestimation is possibly the result of a number of factors that include:

**1** the sediment budget varies with river discharge, and the sediment loss resulting from navigation traffic is probably relatively high at low discharges;

**2** the measurements were carried out in groyne fields with relatively wide sandy beaches, whereas a large part of the groyne fields along the Waal are filled with rip-rap;

**3** the measurements were carried out in a straight section of the Waal and therefore may not be representative for inner and outer bends;

**4** the width of the sediment transport in the outflow current in the schematization is incorrect.

Clearly, more data from groyne fields in other parts of the river (inner and outer bends) and at higher discharges are needed for the yearly sediment budget to be estimated with reasonable accuracy. These data should be compared with data on temporal bed level variability for groyne fields in different sections of the river.

## CONCLUSIONS

The hydrodynamics in the area between the groynes along the banks of the River Waal are influenced indirectly by the discharge in the main channel. This results from some of the discharge leaking into the groyne field and leaving it as a concentrated outflow near the upstream groyne. When the distance between the groynes is sufficiently large, there is also an outflow near the downstream groyne. Thus, one or two circulation cells exist, depending on the size of the groyne field. The currents in these cells are strengthened when a vessel passes and some of these vessels, especially push-towing, generate strong enough currents to resuspend some of the sediment of the sandy beaches, part of which is transported to the main channel through the outflow. The largest vessels

typically cause a water-level depression of 15–20 cm and currents of 30–40 cm s$^{-1}$ 10 cm above the bed.

Although it is clear that large and fast vessels have a stronger impact than small and slow ones, the characteristics of the vessels cannot be related to sand fluxes at the moment. Models in the literature do not serve the conditions of the navigation traffic in the River Waal. Many characteristics of the vessels and the river play a role, which have to be combined in an as yet unknown way.

Sand transport volumes for the groyne fields appear to be overestimated, in particular because of a lack of knowledge of sediment transport in groyne fields with orientations (inner and outer bends) different from the ones studied in this paper, and a lack of knowledge of vessel-induced sediment transport at higher discharges.

## ACKNOWLEDGEMENTS

This research was supported financially by the Directorate Eastern Netherlands (DON) and the Head Office of Rijkswaterstaat. The authors wish to thank the crew of the M.V. *Conrad* as well as their supervisors for their assistance in the organization and implementation of the field campaign, B. van Maren, J.W. Mol and R.J.M. Lenders for carrying out most of the field measurements, M.C.G. van Maarseveen of the technical staff of the Department of Physical Geography of Utrecht University for his assistance in operating the tripods, and E. van Velzen, R.A. Kostaschuk, L.E. Frostick and N.D. Smith for their comments on an earlier draft of the manuscript.

## REFERENCES

BHOWMIK, N.G., XIA, R., MAZUMDER, B.S. & SOONG, T.W. (1995) Return flow in rivers due to navigation traffic. *J. hydraul. Eng.*, **121**, 914–918.

CENTRAAL BUREAU VOOR DE STATISTIEK (1970–1995) *Statistical Annals*, Voorburg.

HOCHSTEIN, A. (1967) *Navigation use of Industrial Canals*. Water Transportation, Moscow Publishing House, U.S.S.R.

HOCHSTEIN, A.B. & ADAMS, C.E. (1989) Influence of vessel movements on stability of restricted channels. *J. Waterw. Port, Coast. Ocean Eng.*, **115**, 444–465.

KLEINHANS, M.G. (1996). *Sediment Transport in the Dutch Rhine Branches*. Report ICG 96/9, Utrecht University.

TEN, BRINKE, W.B.M., SCHOOR, M.M., SORBER, A.M. & BERENDSEN, H.J.A. (1998) Overbank sand deposition in relation to transport volumes during large-magnitude floods in the Dutch sand-bed Rhine river system. *Earth Surf. Process. Landf.*

TERMES, A.A.P., VAN DER WAL, M. & VERHEIJ, H.J. (1991) *Hydrodynamics by Navigation Traffic on Rivers and in Groyne Fields*. Report Q1046, Delft Hydraulics. (Text in Dutch.)

TINGSANCHALI, T. & MAHESWARAN, S. (1990) 2-D depth-averaged flow computation near groyne. *J. Hydraul. Eng.*, **116**, 71–86.

*Spec. Publs int. Ass. Sediment.* (1999) **28**, 161–168

# The geochemical and mineralogical record of the impact of historical mining within estuarine sediments from the upper reaches of the Fal Estuary, Cornwall, UK

S. H. HUGHES

*Camborne School of Mines, University of Exeter, Pool, Redruth, Cornwall TR15 3SE, UK*
*(Email: suehughes@btinternet.com)*

## ABSTRACT

The Fal Estuary, in southwest England, drains areas of extensive metal mining on the eastern side and china-clay mining to the west. As a consequence, the estuarine sediments provide a record of the historical pollution from mining. Sediment cores from the intertidal mudflats of tributaries at the northern end of the Fal Estuary have been examined mineralogically and geochemically. Initial analyses reveal a significant pulse of mine-waste contamination at approximately 30–50 cm below the present-day sediment surface. High metal levels of 2300 p.p.m. Sn, 430 p.p.m. As, 1590 p.p.m. Pb, 1970 p.p.m. Zn and 1070 p.p.m. Cu occur within this interval, together with a heavy mineral suite comprising chalcopyrite, arsenopyrite, pyrite, cassiterite, Fe–Ti oxides (ilmenite and rutile), wolframite, sphalerite, barite, zircon, monazite and xenotime. In addition, human-made smelt products also occur. This contamination probably correlates with the discharge of mine waste either during, or immediately after, the peak of mining activity, which occurred between 1853 and 1893.

## INTRODUCTION

Estuaries receive waters and sediments from both inland via river systems and offshore via tidal currents, and are therefore efficient sediment traps for both terrestrial and marine-derived products. Deposition of approximately 95% of fine-grained suspended sediment occurs at high and low tides, when the water velocity is almost zero (Andrews *et al.*, 1996). Estuarine sediments can, therefore, act as a chronicle of past environmental conditions (Buckley, 1994) and examination of sediment cores recovered from intertidal mudflats may reveal a record of anthropogenic activity. Contamination of estuarine sediments has been examined by few workers (e.g. Buckley & Winters, 1992; Buckley, 1994; Buckley *et al.*, 1995; Dickinson *et al.*, 1996) and by combining geochemical, mineralogical and dating techniques with historical records, the identification of potential sources is possible. Little of this type of research has been undertaken in the UK.

South-west England has had a long association with mining. The extraction of tin dates back to the Bronze Age (2000 BC) and the mining of copper occurred in the 18th to 19th centuries, together with other metals such as arsenic, tungsten, iron and uranium. These activities, however, produced a large amount of waste through inefficient techniques in mineral processing, up to 33% in

the case of tin in 1905 (Thomas, 1913). This unwanted material was often washed into the rivers (Healy, 1996). The impact of these activities on stream sediments, soils and water quality within the region has been studied by several workers (e.g. Davies, 1971; Aston *et al.*, 1975; Yim, 1981; Abrahams & Thornton, 1987). Little work, however, has been carried out on estuarine sediments.

The impact of Cornwall's industrial heritage on the Fal Estuary may be 'retained' as a record of pollution within estuarine sediments. It is therefore the aim of this paper to present initial results from geochemical and mineralogical analyses on recent sediments from the upper reaches of the Fal Estuary. This will allow the spatial and temporal distribution of mine-waste contaminants, as well as their source, to be determined.

## STUDY AREA

The Fal Estuary is one of the largest estuarine systems in southwest England and is located on the south coast of Cornwall (Fig. 1). It has been described as a ria, or a drowned dendritic river valley system (Stapleton & Pethick, 1995). The estuary can be divided into two parts:

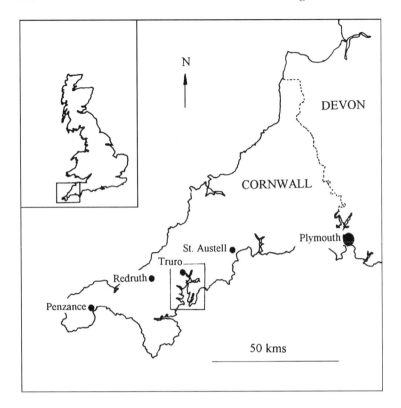

**Fig. 1.** Location of the Fal Estuary in southwest England.

the outer estuary, known as Carrick Roads, which is a deep tidal basin and the inner tidal tributaries (Fig. 2). Carrick Roads forms the main body of water, with approximately 80% of the water volume (Sherwin, 1993; in Stapleton & Pethick, 1995), and because of a lack of a sediment source and low tidal currents has experienced only limited sedimentation (Stapleton & Pethick, 1995). The six major tributaries and their constituent creeks and rivers, which have experienced significant siltation during the Holocene, all eventually flow into Carrick Roads.

The Fal Estuary has a shoreline length of 127 km and a total area of 2500 km$^2$, of which 70% is subtidal, 26% intertidal mudflats and 4% supratidal saltmarsh, which is situated mainly at the heads of the major creeks. The maximum tidal limit extends from the mouth of the estuary, 18.1 km inland, to the village of Tresillian in the north. The spring tidal range of 5.3 m at Falmouth, decreases to 3.5 m at Truro (Fig. 2).

### Regional geological setting

The region is dominated by Devonian sedimentary rocks, which consist mainly of metamorphosed sandstones and siltstones. These were intruded subsequently by granite

towards the end of the Carboniferous and Early Permian (Evans, 1990). Two such granite masses, within the area of the Fal Estuary, are the Carnmenellis Granite to the west and the St Austell Granite to the east. Associated with the granite intrusion, polymetallic mineralization occurred between 240 and 190 Ma (Evans, 1990). Main-stage tin and copper mineralization occurred in the Early Permian and developed in steeply dipping lodes in zones or belts within the metasediments or at the margins of the granite. Lead–zinc mineralization, in NNW–SSE trending lodes, developed in the Triassic at lower temperatures, and frequently cross-cut the tin–copper lodes (Camm & Hosking, 1984). The main minerals found within the mineralized districts are cassiterite ($SnO_2$), wolframite (($Fe,Mn)WO_4$), chalcopyrite ($CuFeS_2$), arsenopyrite ($AsFeS$) and galena ($PbS$), together with gangue minerals such as quartz, feldspar, mica, tourmaline, chlorite, haematite and fluorspar (Leveridge *et al.*, 1990). Alteration of the St Austell Granite by kaolinization has led to the formation of world-class china-clay deposits.

Within the region, Quaternary periglacial sediments (known locally as 'head') unconformably overlie the Devonian metasediments. These head deposits (produced by solifluction and mass wasting) served to transport cassiterite that was eroded from primary tin lodes during the

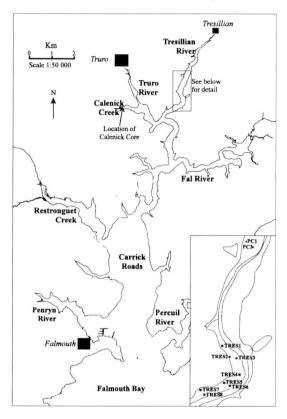

**Fig. 2.** Components of the Fal Estuary showing the outer estuary, Carrick Roads — a deep tidal basin — and the inner tidal tributaries, which all eventually flow into Carrick Roads. The location of the core from the tidal limit of Calenick Creek is shown, together with the core locations from the mid-reaches of the Tresillian River (inset not to scale).

Tertiary to the Quaternary, and subsequently accumulated as fluviatile placers (Camm & Hosking, 1984). During the Holocene, sea-level rose following the last (Devensian) glaciation and drowned many river valleys on the south coast of Cornwall. Holocene sediments are known to underlie recent estuarine sediments within the area.

### Human activity

Tin has been extracted from south-west England for 4000 yr, initially through the recovery of placer cassiterite from streams, by a simple mining method known as streaming. Up to the 17th century, the majority of Cornwall's tin came from this practice. Subsequently, underground mining of tin took over from streaming, although streaming continued and also included the reworking of mine tailings. In 1872, Cornwall became the world's premier tin field (Barton, 1961). Foreign com-

petition, however, reduced this industry to just one mine in the Camborne–Redruth area, which closed in March 1998. Submarine mining was undertaken in Restronguet Creek (Fig. 2) in the mid-19th century, where tin was extracted from tin-bearing gravels beneath 18 m of mud. There are records of four such mines having existed (Simpson, 1993) and the sites of these can still be seen. The tin was smelted within the county, with the establishment of 27 smelting houses (Barton, 1961), 12 of which were located at the tidal limits of Truro River, Calenick Creek, Restronguet Creek and Penryn River, on the western side of the Fal Estuary.

Copper mining was far more important to Cornwall's economy, surpassing tin in value and tonnage, but lasted only two centuries between 1700 and 1900, with production peaking in 1855 (Barton, 1967). In many cases, once the copper industry collapsed, mines were able to continue by extracting tin from below the copper zone. Mainly as by-products and in many cases providing additional revenue, other metals such as arsenic and tungsten were also mined when demand deemed it economical. Iron and uranium were mined mainly on the eastern side of the estuary, within the vicinity of the St Austell Granite. China clay has been extracted from the western side of the St Austell Granite since the mid-18th century and still remains a world-class industry today, with the UK being the second largest producer of kaolin after the USA.

### METHODS

Sediment cores have been collected from the intertidal mudflats, using a 2-m-long plastic tube with a diameter of 6.5 cm. The resultant cores were extracted, described, photographed and then split into separate 5-cm-long stratigraphical intervals. Each subsample was split further for geochemical and mineralogical analyses. Bulk geochemistry was analysed by X-ray fluorescence (XRF) using pressed powder pellets for Sn, As, Pb, Zn and Cu. Analytical precision is of the order 10 p.p.m. For mineralogical examination, the subsample was wet-sieved to obtain three grain-size fractions (> 63 μm, 63–20 μm, < 20 μm). The > 63 μm and 63–20 μm grain-size fractions were set into resin blocks and polished for examination by scanning electron microscope (SEM) using a JEOL SEM with a LINK Energy Dispersive X-ray Spectrometry and back-scattered electron detector. The < 20 μm grain-size fraction was examined by X-ray diffraction (XRD) using a SIEMENS diffractometer. In addition, subsamples from two cores (PC1 and PC3) from the Tresillian River (Fig. 2) were subdivided into

+ 63 µm, 63–4 µm and < 4 µm by beaker decantation using Stokes' law of settling.

## RESULTS

### Sediment geochemistry

Initially, 10 cores were recovered from the mid-reaches of the Tresillian River, the northernmost tidal tributary in the Fal Estuary (Fig. 2). The range of total metal concentrations, together with the mean, is given in Table 1. Typically, all cores display relatively low metal levels at the base (Fig. 3), then all metal values increase rapidly to

**Table 1.** Metal concentrations (in p.p.m.) of Sn, As, Pb, Zn and Cu for the 10 cores recovered from the mid-reaches of the Tresillian River, the northernmost tributary of the Fal Estuary.

| Metal | Range (p.p.m.) | Mean (p.p.m.) |
|-------|----------------|---------------|
| Sn    | 35–2300        | 830           |
| As    | 4–430          | 170           |
| Pb    | 20–1590        | 220           |
| Zn    | 100–1970       | 685           |
| Cu    | 15–1070        | 430           |

pronounced peaks, with the maximum values occurring at depths of 30–50 cm within the sediments. Metal concentrations then decrease gradually towards the present-day sediment surface, although the surficial sediments do not attain the initial low metal levels found in older sediment at the base of each core. From a north (PC1 and PC3) to south (TRES8) aspect, there are no apparent changes in the geochemical profiles, although slightly higher metal levels are found within the southernmost cores. The peaks of metal contamination display a lateral variation when compared with core locations. Cores recovered from the bank display a sharp geochemical spike, whereas those taken farther out on the mudflat, nearer the active subtidal channel, show a much broader diffuse signature.

By way of contrast, a core was recovered from near the head of Calenick Creek (Fig. 2), which drains the mineralized district on the eastern side of the Fal Estuary. The geochemical profile is shown in Fig. 4, together with the core sedimentology. All metals show highly elevated concentrations and, with the exception of Sn, follow each other throughout the core. The Sn concentrations remain very high throughout, with a range of 3190–7850 p.p.m., and follows a completely different pattern to the other metals. The two peaks in the Sn profile at depths of 10 cm and 25 cm correspond with the coarse sand beds.

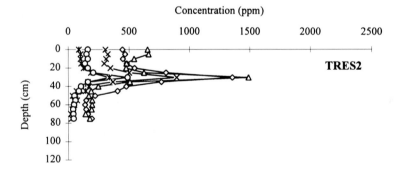

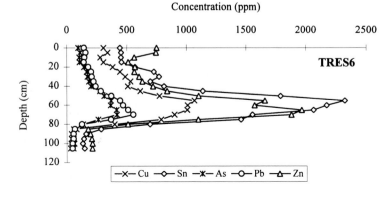

**Fig. 3.** Two typical geochemical profiles of cores TRES2 and TRES6 from the mid-reaches of the Tresillian River, showing the variations of Sn, Cu, Zn, Pb and As (p.p.m.) versus depth. Core TRES2 was taken close to the estuary margin, whereas TRES6 was from farther out on the mudflats. See Fig. 2 for core locations.

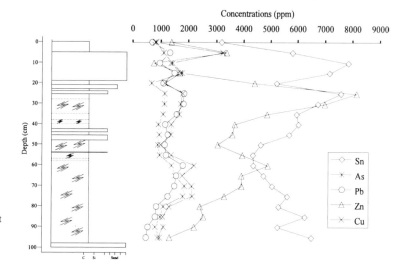

**Fig. 4.** Core sedimentology and geochemical profile showing the variations of metals Sn, Cu, Zn, Pb and As (p.p.m.) in a core from the tidal limit of Calenick Creek, which drains the mineralized district on the west side of the Fal Estuary.

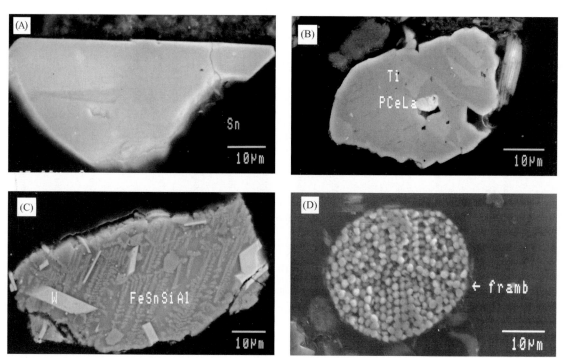

**Fig. 5.** Scanning electron microscopy photomicrographs of detrital and diagenetic minerals within the estuarine sediments: (A) liberated cassiterite; (B) rutile and monazite locked within a silicate host; (C) polymetallic smelt product with tungsten locked within — this shows the typical texture of the human-made products examined; (D) diagenetic framboidal pyrite. Scale bars = 10 μm.

## Sediment mineralogy

Selected samples of 63 μm and 63–20 (and 63–4) μm grain-size fractions from the Tresillian River cores were examined for their heavy mineral content using the back-scattered electron detector of the SEM. The dominant minerals present included cassiterite, pyrite, Fe oxides, zircon, sphalerite, Fe–Ti oxides, monazite, chalcopyrite and xenotime together with lesser amounts of arsenopy-rite, barite and wolframite. Cassiterite dominates in the 63–20 μm size fraction (Fig. 5A), with an average size of 10 μm (Fig. 6). The other dominant minerals examined

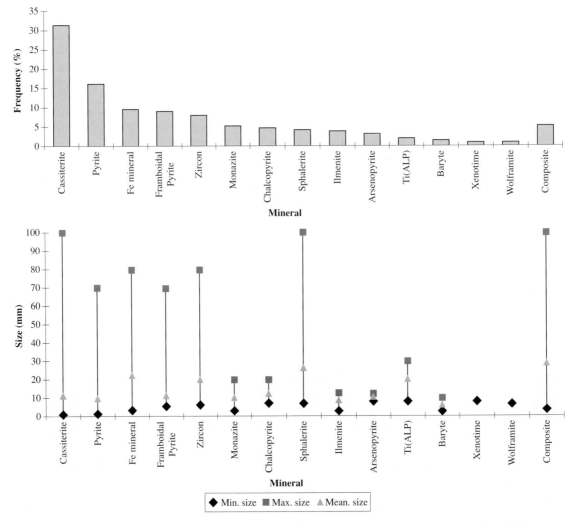

**Fig. 6.** Scanning electron microscopy data from one sample: (A) graph showing the frequency (%) of each mineral identified; (B) graph displaying the range of sizes for each mineral together with their mean size. The term 'composite' has been given to polymetallic minerals (usually human-made products). Note that although cassiterite dominates, its mean size is only approximately 10 µm.

occur mainly within the 63–20 µm grain-size fraction as liberated grains, but in both grain-size fractions are associated more commonly with silicate hosts (Fig. 5B). Rare polymetallic and Fe oxide slag products also have been found and are generally > 63 µm in diameter (Fig. 5C). These typically occur below a depth of 30 cm from the present-day sediment surface. Pyrite occurs as both detrital grains and as early diagenetic framboidal pyrite (Fig. 5D). Analyses of the < 20 µm size fraction by XRD has revealed a typical mineral assemblage comprising quartz, kaolinite, illite, chlorite and mica, which does not vary stratigraphically throughout the core.

## DISCUSSION

The geochemical and mineralogical characteristics of the cores examined from the intertidal mudflats of the Tresillian River suggest the presence of considerable mine-waste contamination. The distinctive geochemical pulse, revealed at a depth of 30–50 cm below the present-day sediment surface, is related to the sudden release of mine tailings that occurred during the peak of mining activity (1850s to 1890s) or immediately post-mining (Pirrie *et al.*, 1997). The nature of the mineralogy reflects the local geology, including the main ore minerals such as cassiterite, sphalerite, chalcopyrite, pyrite and arsenopyrite. The minerals zircon, monazite and xenotime derive from granite and are known to occur within the Carnmenellis Granite to the west of the Fal Estuary (Pirrie *et al.*, 1997). The presence of smelt/human-made products, found within the cores below a depth of 30 cm, confirms that the contamination is the result of historical mining activity within the region.

The supply of sediment to the Tresillian River is either from its catchment area, or it may have been transported upstream on tidal currents from rivers to the south. It is known that tin streaming was practiced within the catchment area, together with minor Pb, Cu, Ag and Sn mining and, as a result of this activity, Whitley (1883) calculated that in 1880 alone, 10 413 m³ of sediment was supplied to the estuary. Historical records suggest that there were no tin-smelting houses within the catchment area, therefore the source of smelt products within the estuarine sediments, together with the total concentration of metals, must come from rivers to the south, such as either Truro River (Calenick Creek) or Restronguet Creek.

It is suspected that the peaks of elevated metal values from Calenick Creek will only be found at the head of the creek, as they coincide with the presence of coarse-sand beds within the core. These coarse-sand beds are not present farther downstream and they may therefore relate to smelting activity that was undertaken in the valley between 1711 and 1791. These sediments, however, have not been examined mineralogically.

## CONCLUSION

Within an estuarine environment, sediment is transported either by fluvial or tidal currents and, within the catchment area of the Fal Estuary, possible sediment sources are:

1 natural weathering and erosion of the granite and metasediments;

2 reworking of Quaternary sediments, principally the head deposits;

3 the discharge of mine waste.

It can be suggested that the low metal values at the base of the cores from the Tresillian River may correspond with products of natural weathering (Pirrie *et al.*, 1996) or early mining. The distinctive geochemical pulse, comprising the highest levels of the metals Sn, Cu, As, Pb and Zn, occurring at 30–50 cm below the present-day sediment surface, however, cannot be a natural phenomenon and therefore is the result of mining contamination. The exact source and timing of this contamination is poorly constrained at present. The mineral assemblage, however, including the presence of human-made smelt products, does indicate a provenance other than the catchment of the Tresillian River, which suggests sediment input from the south, either from Truro River or Restronguet Creek. The geochemical profile of the core from the tidal limit of Calenick Creek has revealed highly elevated metal levels and is attributed to the direct release of mine waste that came from within the creek's catchment area.

A study of cores from the intertidal mudflats of the upper reaches of the Fal Estuary revealed a historical record of anthropogenic impact, resulting primarily from mining activities within the region. At present, the mudflats provide a stable sink for this contamination (Pirrie *et al.*, 1996). Any future disturbance of these sediments, however, will have a significant affect on the quality of waters and biota (Healy, 1996), owing to the remobilization of heavy metals (Dickinson *et al.*, 1996). Changes in the coastal zone could result from anthropogenic activity, for example, land reclamation and dredging or natural responses to changing conditions, such as sea-level rise, which will cause the erosion of mudflats and saltmarshes (Stapleton & Pethick, 1995) within the estuarine environment.

## ACKNOWLEDGEMENTS

This PhD research project is sponsored by the Camborne School of Mines' Trust Fund. Dr Duncan Pirrie and Professor Peter Scott are gratefully acknowledged for their supervision and guidance throughout this project. Tony Ball, Fiona Thomas and Tony Clarke provided laboratory assistance.

## REFERENCES

ABRAHAMS, P.W. & THORNTON, I. (1987) Distribution and extent of land contaminated by arsenic and associated metals in mining regions of southwest England. *Trans. Inst. Min. Metal. Sect. B*, **96**, B1–B8.

ANDREWS, J.E., BRIMBLECOMBE, P., JICKELLS, T.D. & LISS, P.S. (1996) *An Introduction to Environmental Chemistry.* Blackwell Science, Oxford, 209 pp.

ASTON, S.R., THORNTON, I., WEBB, J.S., MILFORD, B.L. & PURVES, J.B. (1975) Arsenic in stream sediments and waters of south west England. *Sci. Total Environ.*, **4**, 347–358.

BARTON, D.B. (1961) *A History of Copper Mining in Cornwall and Devon.* D. Bradford Barton, Truro.

BARTON, D.B. (1967) *A History of Tin Mining and Smelting in Cornwall.* D. Bradford Barton, Truro.

BUCKLEY, D.E. (1994) 25 years of environmental assessment of coastal estuarine systems: lessons for environmental quality management. In: *Coastal Zone Canada 94. Co-operation in the Coastal Zone, Conference Proceedings*, Vol. 3 (Eds Wells, P.G. & Ricketts, P.J.), pp. 1304–1340.

BUCKLEY, D.E. & WINTERS, G.V. (1992) Geochemical characteristics of contaminated surficial sediments in Halifax Harbour: impact of waste discharge. *Can. J. Earth Sci.*, **29**, 2617–2639.

BUCKLEY, D.E., SMITH, J.N. & WINTERS, G.V. (1995) Accumulation of contaminant metals in marine sediments of Halifax Harbour, Nova Scotia: environmental factors and historical trends. *Appl. Geochem.*, **10**, 175–195.

CAMM, G.S. & HOSKING, K.F.G. (1984) The stanniferous placers of Cornwall, south west England. *Bull. geol. Soc. Malaysia*, **17**, 323–356.

DAVIES, B.E. (1971) Trace metal content of soils affected by base metal mining in West of England. *Oikos*, **22**, 366–372.

DICKINSON, W.W., DUNBAR, G.B. & McLEOD, H. (1996) Heavy metal history from cores in Wellington Harbour, New Zealand. *Environ. Geol.*, **27**, 59–69.

EVANS, C.D.R. (1990) *The Geology of the Western English Channel and its Western Approaches*. British Geological Society, HMSO, London, 94 pp.

HEALY, M.G. (1996) Metalliferous mine waste in West Cornwall: the implications for coastal management. In: *Studies in European Coastal Management* (Eds Jones, P.S., Healy, M.G. & Williams, A.T.), pp. 147–155. Samara Publishing, Cardigan.

LEVERIDGE, B.E., HOLDER, M.T. & GOODE, A.J.J. (1990) *Geology of the Country around Falmouth, Memoir of the*

British Geological Survey, Sheet 352. Natural Environment Research Council, HMSO, London.

PIRRIE, D., HUGHES, S.H. & PULLIN, H. (1996) The effect of mining on sedimentation over the last 500 years in the Fal Estuary, Cornwall, UK. In: *Late Quaternary Coastal Change in West Cornwall, U.K. Field Guide* (Ed. Healy, M.G.), 75–82. Environmental Research Centre, Research Publication 3, University of Durham.

PIRRIE, D., CAMM, G.S., SEAR, L.G. & HUGHES, S.H. (1997) Mineralogical and geochemical signature of mine waste contamination, Tresillian River, Fal Estuary, Cornwall, UK. *Environ. Geol.*, **29**, 58–65.

SHERWIN, T.J. (1993) *The Oceanography of Falmouth Harbour*. Report (U93–4) to the National Rivers Authority (South West Region). Unit for Coastal and Estuarine Studies, Menai Bridge.

SIMPSON, B. (1993) *Mining History in Restronguet Creek*. Restronguet Creek Society, Devoran, 38 pp.

STAPLETON, C. & PETHICK, J. (1995) *The Fal Estuary: Coastal Processes and Conservation*. Report to English Nature, Institute of Estuarine and Coastal Studies, Hull.

THOMAS, W. (1913) Losses in the treatment of Cornish tin ores. *Trans. Cornish Inst. Min. Mech. Metal. Eng.*, **1**, 56–74.

WHITLEY, H.M. (1883) The silting up of the creeks of Falmouth Haven. *J. R. Inst. Cornwall*, **VII**, 12–17.

YIM, W.W.-S. (1981) Geochemical investigations on fluvial sediments contaminated by tin mine tailings, Cornwall, England. *Environ. Geol.*, **3**, 245–256.

Avulsion: Modern and Ancient

*Spec. Publs int. Ass. Sediment.* (1999) **28**, 171–178

# Causes of avulsion: an overview

L. S. JONES\*,† *and* S. A. SCHUMM‡

\* *Department of Geology and Geophysics, University of Wyoming, Laramie, WY 82071, USA; and*
‡*Colorado State University, Department of Earth Resources, Fort Collins, CO 80523, USA*

## ABSTRACT

Avulsion, i.e. the relatively sudden displacement of a river channel, has an important effect on sediment distribution and on architecture of fluvial deposits because avulsion is a primary control on channel location on a floodplain. Most avulsions occur when a triggering event, commonly a flood, forces a river across a stability threshold. The closer the river is to the threshold, the smaller is the flood discharge needed to initiate an avulsion.

Avulsions can be categorized by the processes or events that decrease stability and move the river toward the avulsion threshold, and/or serve as avulsion triggers. These processes or events produce one or more of the following:
**1** an increase in the ratio of avulsion course slope to existing channel slope caused by a decrease in gradient of the existing channel;
**2** an increase in the ratio of avulsion course slope to existing channel slope caused by an increase in gradient away from the existing channel;
**3** a non-slope-related reduction in the capacity of the existing channel to carry all the water and sediment delivered to it.
In most cases several of these causes will combine to bring a river close to a threshold, after which the next triggering event of sufficient magnitude will push the system across the threshold and avulsion will occur.

Avulsion frequency is controlled by the interaction between the rate at which various processes combine to move a river toward the avulsion threshold (instability) and the frequency of triggering events. If the combined processes that lead to instability proceed rapidly relative to triggering events, then the frequency of the triggering events will control avulsion frequency. In such settings, avulsion frequency may be a predictable function of flood frequency. If triggering events occur frequently relative to the rate at which the river becomes unstable, however, then the rate at which a combination of processes leads to instability will control avulsion frequency.

## INTRODUCTION

Avulsion, the relatively rapid shift of a river to a new channel on a lower part of a floodplain, alluvial plain, delta or alluvial fan (Allen, 1965), is a primary process that determines channel location over the long term, and therefore has an important effect on the large-scale distribution of river sediment. Under conditions of long-term aggradation, avulsion has 'a profound effect on the resulting stratigraphic architecture' (Miall, 1996, pp. 317–318) because the interaction of avulsion frequency, channel migration rate and net sedimentation rate across the floodplain appears to determine the relative abundance and geometry of abandoned channels, channel deposits

and deposits derived from overbank flows. Increasing interest in avulsion has resulted in considerable recent work that allows some generalizations to be made regarding the nature and causes of avulsion in various river settings. Our purpose in this brief overview is to use documented examples and geometric arguments to classify and discuss the various causes of avulsions, which may be controlled either by natural cycles within the fluvial system or by external events (Miall, 1996, pp. 317–327), and to suggest how these causes may interact to affect avulsion frequency.

In braided streams, the term avulsion is sometimes used to describe the shift of the main thread of current to the other side of a mid-channel bar (Leedy *et al.*, 1993; Miall, 1996, p. 317), but here the term is restricted to the

† Current address: 1967 South Van Gordon Street, Lakewood, CO 80228, USA. (Email: larjones@earthlink.net)

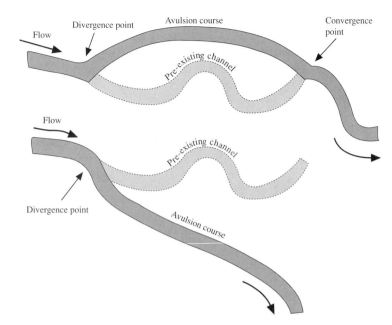

**Fig. 1.** Map view of hypothetical rivers showing how the terms avulsion, divergence point, and convergence point are used in the text. In the upper example, avulsion course rejoins pre-existing channel at convergence point. In lower example, avulsion channel does not rejoin, so no convergence point exists.

complete shift of the entire channel (Fig. 1). The point upstream, where the new (avulsion) channel leaves the old channel, is the divergence point. The point downstream where the new channel rejoins an old channel is the convergence point. The convergence point is not present where the avulsed channel enters a sea or lake or occupies an entirely new area of floodplain and does not rejoin the original channel. Avulsion is distinguished from the other primary method of channel relocation, lateral migration, which involves gradual removal of material from one channel bank concurrent with sediment deposition on the opposite bank.

We attribute no particular size to avulsions, so the length of channel reach that is affected by an avulsion can vary from a few tens of metres (Jones & Harper, 1998), to hundreds of kilometres (Fisk, 1944). The rapidity with which discharge shifts completely from an old to a new channel varies from as little as 1 day (Ning, 1990) to a decade (Jones & Harper, 1998) to perhaps millennia (Schumm *et al.*, 1996). Avulsion may result in the rapid shift of flow to a single new channel (Brizga & Finlayson, 1990) or in a gradual process of crevasse splay formation, wetland infilling and channel coalescence (Smith *et al.*, 1989).

Avulsion is of direct concern to the study of ancient fluvial sediments because it locates the channel on the floodplain and thereby controls the large-scale geometry of channel and channel-belt deposits. Modelling results confirm that the nature of the relationship between

avulsion frequency and sedimentation rate has a strong effect on hypothetical fluvial depositional geometries (Allen, 1965, 1974, 1978; Leeder, 1978; Bridge & Leeder, 1979; Bridge & Mackey, 1993a,b; Mackey & Bridge, 1995; also see discussion in Heller & Paola, 1996). For example, in one set of experiments in which avulsion frequency remained constant while sedimentation rate varied, channel belt interconnectedness was shown to be inversely proportional to sedimentation rate (Bridge & Leeder, 1979, p. 632). Until the response of avulsion frequency to changes in sedimentation rate in aggradational avulsive natural rivers is understood, however, the nature of the relationship, if any, between channel-belt interconnectedness and sedimentation rates in ancient rocks is uncertain.

## CAUSES OF AVULSIONS

Avulsion commonly results when an event (usually a flood) of sufficient magnitude occurs along a reach of a river that is at or near an avulsion threshold. As demonstrated for geomorphological systems in general, the closer a river is to the threshold, the smaller is the event needed to trigger the avulsion (Fig. 2). This is why avulsions are not always triggered by the largest floods on a given river (Brizga & Finlayson, 1990; Ethridge *et al.*, this volume, pp. 179–191). Nevertheless, some avulsions appear to occur without the evolution of channel

**Fig. 2.** Schematic graph showing increase in channel instability (approach to avulsion threshold) over time. Line 1 shows increasing propensity of channel to avulse as it approaches the avulsion threshold, which indicates certainty of avulsion at time B. Superimposed on line 1 are vertical lines representing flood magnitude. The large flood at time A causes avulsion, but if it had not occurred, smaller floods would lead to avulsion between times A and B. Note that even large floods do not cause avulsion until the line representing channel instability approaches the avulsion threshold (from Schumm, 1977). The shape, curvature and slope of line 1 will vary for different rivers.

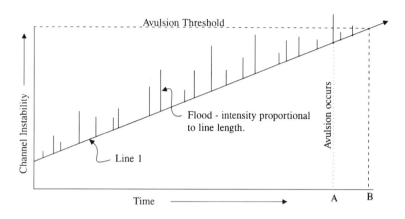

**Table 1.** Causes of avulsion.

| Processes and events that create instability and lead toward an avulsion threshold, and/or act as avulsion triggers | | Can act as trigger? | Ability of channel to carry sediment and discharge |
|---|---|---|---|
| Group 1. Avulsion from increase in ratio, $S_a/S_e$*, owing to decrease in $S_e$ | a. Sinuosity increase (meandering) | No | Decrease |
| | b. Delta growth (lengthening of channel) | No | Decrease |
| | c. Base-level fall (decreased slope†) | No | Decrease |
| | d. Tectonic uplift (resulting in decreased slope) | Yes | Decrease |
| Group 2. Avulsion from increase in ratio, $S_a/S_e$, owing to increase in $S_a$ | a. Natural levee/alluvial ridge growth | No | No change |
| | b. Alluvial fan and delta growth (convexity) | No | No change |
| | c. Tectonism (resulting in lateral tilting) | Yes | No change |
| Group 3. Avulsion with no change in ratio, $S_a/S_e$ | a. Hydrological change in flood peak discharge | Yes | Decrease |
| | b. Sediment influx from tributaries, increased sediment load, mass failure, aeolian processes | Yes | Decrease |
| | c. Vegetative blockage | No | Decrease |
| | d. Log jams | Yes | Decrease |
| | e. Ice jams | Yes | Decrease |
| Group 4. Other avulsions | a. Animal trails | No | No change |
| | b. Capture (diversion into adjacent drainage) | – | No change |

* $S_a$ is the slope of the potential avulsion course, $S_e$ is the slope of the existing channel.
† In settings where the up-river gradient is greater than the gradient of the lake floor or shelf slope, base-level fall may result in river flow across an area of lower gradient.

characteristics to a sensitive condition, and many avulsions result from a combination of causes. Avulsion cannot easily occur where a river has incised a narrow valley or canyon. The river must have another path to follow in a wide valley, or on an alluvial fan, alluvial plain or delta.

Any event that causes flooding can probably serve as an avulsion trigger. 'Normal' hydrological floods are the most common triggers, but flooding caused by ice jams, sudden tectonic movement, landslides or similar events can also locate and trigger avulsions.

The underlying causes of avulsions (i.e. those processes or events that move a river toward an avulsion threshold) can be organized into four groups (Table 1). Some of these processes and events can also act as avulsion triggers. The first two groups involve an increase in the ratio, $S_a/S_e$, of the slope ($S_a$) of the potential avulsion course to the slope ($S_e$) of the existing channel. In group 1, the increase in $S_a/S_e$ results from a decrease in $S_e$ (Fig. 3A–C). The decrease in $S_e$ may also reduce the ability of the channel to carry water and sediment. In group 2,

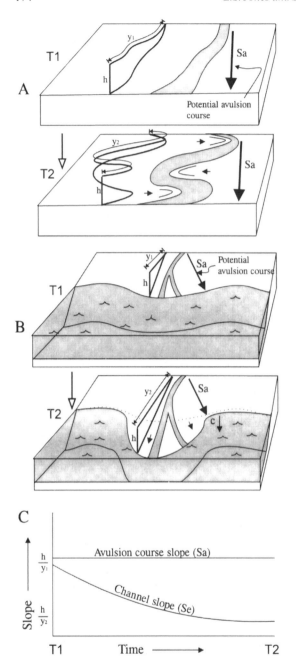

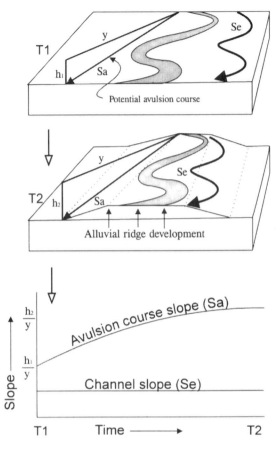

**Fig. 4.** Schematic block diagram and graph demonstrating slope changes with time for alluvial ridge avulsions (Table 1, 2a). As the natural levees and channel aggrade, creating an alluvial ridge, the slope of a potential avulsion course ($S_a$) increases from $h_1/y$ to $h_2/y$. At the same time, channel slope ($S_e$) remains constant (or nearly so) and the ratio $S_a/S_e$ therefore increases. The curve of $S_a$ is shown to flatten with time because it appears likely that the net sedimentation rate on most natural levees decreases as levee relief increases owing to reduced sedimentation rate on the levee top and levee subsidence into floodplain muds (cf. Fisk, 1944). Compare graph with Fig. 3C.

**Fig. 3.** Schematic block diagrams and graph demonstrating slope changes with time for sinuosity and deltaic avulsions (Table 1, 1a, 1b). (A) As sinuosity increases from time 1 (T1) to time 2 (T2), length $y_1$ increases to $y_2$ causing a decrease in channel slope, $S_e$, from $h/y_1$ to $h/y_2$. (B) As the delta lobe progrades from T1 to T2 at constant relative sea-level, length $y_1$ increases to $y_2$, causing a decrease in $S_e$ from $h/y_1$ to $h/y_2$. (C) In both cases, the slope of a potential avulsion course, $S_a$, down the floodplain remains constant, and the ratio $S_a/S_e$ increases. If, in (B) subsidence occurs at point c, the avulsion course slope, $S_a$, may simultaneously increase.

the increase in $S_a/S_e$ is caused by an increase in $S_a$ and does not necessarily result in channel clogging (Fig. 4). In the third group the ability of the channel to carry water and sediment decreases for reasons unrelated to slope changes. The fourth group is composed of causes that do not fall into the first three categories.

In group 1 (Table 1), $S_e$ can decrease from four processes, leading to in-channel deposition and avulsion. A pattern change (1a) can increase sinuosity (Fig. 3A). On a coarse-grained reach of the Rio Grande that does not have natural levees, increased sinuosity led to slope reduction, resulting in bedload deposition where the slope decreased, clogging of the channel, and finally avulsion. In two avulsions of that river, the ratio $S_a/S_e$ was approximately 2. The local flux of coarse sediment into the channel appears to have controlled the rate at which meander growth occurred and how quickly the avulsion threshold was reached (Jones & Harper, 1998). In south-eastern Australia, the Ovens and King rivers avulse from a sinuous channel into a steeper straight channel. The greater velocity and power in the straight channel causes incision and bank erosion that eventually forms a sinuous channel, which is again susceptible to avulsion (Schumm *et al.*, 1996).

The extension of deltas (1b) at constant relative sea-level also decreases gradient of the existing channel (Fig. 3B). Where distributaries are relatively straight and laterally stable (e.g. Mississippi River delta), delta lobe growth lengthens the channel and reduces slope, creating a threshold condition. Flooding may then trigger avulsion down a shorter and steeper slope to the sea. This type of avulsion appears to be common on fluvial-dominated deltas, where delta lobes extend seaward. Examples include the Mississippi (Fisk, 1944), the Po (Nelson, 1970) and Yellow rivers (Ning, 1990; Van Gelder *et al.*, 1994). In many deltas, net subsidence away from the active delta front may simultaneously increase the slope $S_a$.

A base-level fall (1c) that exposes a lake floor or sea shelf of lower gradient than that of the existing river channel may create a reach of decreased river gradient where the channel meets the exposed lake floor or sea shelf and may result in channel clogging. In a similar manner, uplift or faulting (1d) across a river course can decrease slope and cause a river to avulse around the obstruction (see examples of Indus and Murray rivers and Amazon River tributaries in Schumm & Winkley, 1994, pp. 103–184). Classification of tectonism as either a cause or a trigger depends on rate, e.g. uplift that occurs relatively slowly can move the river toward instability by decreasing slope, whereas a rapid tectonic event that results in an immediate diversion would be a trigger.

Avulsions in the second group (Table 1) result from increased lateral slope away from the existing channel (Fig. 4). A major cause is the development of an alluvial ridge (2a) on rivers where the channel and levees aggrade more rapidly than the surrounding floodplain (Fisk, 1952). Locally, the ratio $S_a/S_e$ (here, $S_a$ is defined as the steepest local slope down the levee) can be as high as 30 (Elliott, 1932). Mackey & Bridge (1995) use the ratio, $S_{cv}/S_{dv}$, of the cross-valley slope ($S_{cv}$, controlled at least in part by alluvial ridge relief) to the down-valley channel belt slope ($S_{dv}$) to help determine avulsion frequency and location in some model simulations. The channel base (thalweg) of a river on an alluvial ridge is commonly

shown above the level of the floodplain (Schumann, 1989, fig. 6; Brizga & Finlayson, 1990, fig. 6; Bryant *et al.*, 1995, fig. 1; Heller & Paola, 1996, fig. 2A), although elevation of the thalweg above the adjacent floodplain is not necessary in order to reach the avulsive threshold (Schumm *et al.*, 1996, fig. 4). Alluvial ridge development (2a) and sinuosity increase (1a) commonly combine to create a perched channel of lower gradient, as described for the Ovens and King rivers in Australia (Schumm *et al.*, 1996) and Red Creek in Wyoming (Schumann, 1989). In this case, decrease of $S_e$ and increase of $S_a$ combine to increase the slope ratio $S_a/S_e$.

An upward convex shape of a delta or alluvial fan (2b) creates a setting (similar to natural levee and alluvial ridge development) in which the channel becomes perched on the highest point of the deposit, then avulses down a steeper course. This has been demonstrated experimentally by Schumm *et al.* (1987). Some of the avulsions documented by Blair & McPherson (1994) on Walker River deltas during very rapid base-level fall of Walker Lake, Nevada, may have been caused by this process.

Increase of $S_a$ as a result of lateral tilting (2c) of a valley floor from tectonism or other causes can produce repeated avulsions in one direction, as has been described for the South Fork Madison River (Leeder & Alexander, 1987), St Francis River (Boyd & Schumm, 1995) and Owens River (Reid, 1992). As discussed for group 1d, a sudden tectonic tilt that resulted in an immediate avulsion could be classified as an avulsion trigger.

The third group (Table 1) involves a reduction in the capacity of a channel to convey all of the water and sediment delivered to it. In this group, sudden channel blockage, e.g. a channel-blocking mass movement, can actually trigger an avulsion, whereas more gradual channel clogging moves the river toward a threshold condition in which a periodic flood may actually serve as the trigger. Although slope changes are not required, the presence of a more favourable, steeper slope down a potential avulsion course is probably a necessary precondition for avulsion to occur in these cases.

A hydrological change to an increase of flood peaks (3a) can cause increased overbank flooding, incision and avulsion if the existing channel cannot change to accommodate the increased flow.

In-channel deposition (3b) from a variety of causes other than slope change can produce channel blockage and result in avulsion. Schumann (1989) describes how bank failure along Red Creek, an ephemeral stream in Wyoming, leads to the formation of within-channel benches of fine sediment that reduce channel width and cause overtopping and avulsion during floods. Along Red Creek, the divergence point is at the downstream end of a meander bend where super elevation of the water surface promotes overbank flooding. Sediment delivered by tributaries, mass failure or a general increase in sediment load from upstream can cause in-channel deposition, clogging and avulsion. Aolian dune migration into a channel may result in avulsion (Jones & Blakey, 1997), particularly where ephemeral streams are present.

Vegetation encroachment (3c) may decrease channel capacity. McCarthy *et al*. (1992) showed how growth of channel-lining vegetation restricted channel flow causing in-channel sedimentation and eventual avulsion on the Okavango fan. The colonization of mid-channel sand bars by vegetation to form islands can have the same result. This type of biological activity may be important in some settings.

Log jams (3d) can cause avulsion by blocking the channel and forcing overbank flow. One massive jam, the great Red River Raft in Louisiana, was nearly 160 km long and was not cleared to permit navigation of the river until 1876. For a fascinating discussion of the problems involved in the removal of this impressive feature see McCall (1984).

Ice-jam induced flooding (3e) is a very common occurrence and may be an important contributing factor to avulsion in some climates. Ethridge *et al*. (this volume, pp. 179–191) document an ice-jam induced avulsion on the Niobrara River in Nebraska. A major ice jam in 1928 may have triggered an avulsion of the Yellow River. Human activity returned the river to the original channel, but renewed flooding in 1937 led to a permanent avulsion in the same area (Todd & Elliassen, 1940).

Other causes of avulsion are relegated to group 4 (Table 1). Animal trails (4a) can lead to overbank flow and avulsion (McCarthy *et al*., 1992). On the Okavango fan, flood flow follows hippopotamus trails, resulting in scour, and the eventual relocation of the channel (in combination with vegetative blocking described above). In this particular setting, hippopotamus highways appear to have a significant control on the location of avulsion courses.

Lateral shift of a channel until it intersects an adjacent, steeper abandoned or functioning channel may lead to capture (4b) and avulsion of the main channel. In a similar manner, secondary channels formed on alluvial fans, deltas and alluvial valleys by surface runoff and/or groundwater sapping may eventually capture the main channel and divert it into the steeper captor stream course (Denny, 1965, p. 58). This process, which differs from the others genetically because it requires the interaction of two separate drainage systems, may be considered a separate type of avulsion.

## DISCUSSION AND CONCLUSIONS

Many different natural processes and causes can produce the same result (Schumm, 1991), and this review emphasizes that there are a variety of causes of avulsion. As a consequence, avulsions are common in many different settings (see examples in Richards *et al*., 1993). Most avulsions can be grouped according to the causal process or event that leads to instability, and/or serves as an avulsion trigger:

**1** increased ratio of avulsion course slope to existing channel slope resulting from a decrease in gradient of the existing channel;

**2** increased ratio of avulsion course slope to existing channel slope resulting from an increase in gradient away from the existing channel;

**3** non-slope related reduction in the capacity of a channel to carry all the water and sediment delivered to it.

Generally, these processes will combine to produce an avulsion. For example, increased sinuosity or delta extension may combine with increased alluvial ridge relief to simultaneously decrease existing channel slope and increase lateral slope, both of which cause an increase in the ratio of the avulsion course slope to the existing channel slope. Avulsions from channel clogging alone are probably rare, and may require that other processes provide a slope advantage for the avulsion course. At this time, the lack of an extensive data base makes any ranking of the relative importance of the various causes of avulsion irresponsible. We speculate, however, that no single process appears to be dominant in causing avulsions because no single process that leads toward instability is present or dominant in all fluvial settings in which avulsions occur.

This review suggests that avulsion frequency is controlled by the interaction between the rate at which various combinations of group 1, 2 and 3 processes and events lead toward an avulsion threshold, and the frequency of triggering events (cf. Heller & Paola, 1996). If the combined processes that lead to the threshold condition proceed rapidly relative to triggering events, then the frequency of the triggering events will control avulsion frequency. In such settings avulsion frequency may be a predictable function of flood frequency. If, however, triggering events occur frequently relative to the rate at which the river moves toward an avulsion threshold (Fig. 2), then the rate at which a combination of processes leads to the threshold condition will control avulsion frequency. The factors that control the complex interaction of the various processes and events to produce the rate at which a river moves toward the avulsion threshold are at present not well known or understood.

Some of the processes that lead to the avulsion threshold (sinuosity increase, delta growth, natural levee growth, alluvial fan growth and increase in delta convexity) are integral (intrinsic) to the river and appear to be related to sedimentation. If the rate at which those processes proceed is a predictable function of sedimentation rate, then a relationship between avulsion frequency and sedimentation rate may exist in some settings. Other processes (tectonism, sea- or lake-level change, mass failure, aeolian dune migration, log jams, vegetative blocking and presence of animal trails) are the result of external (extrinsic), non-fluvial influences. In settings where these types of processes occur, avulsion frequency may be less predictable and may be unrelated to sedimentation rate.

## ACKNOWLEDGEMENTS

We are grateful for numerous helpful and constructive comments in reviews by Frank Ethridge, Scudder Mackey, IAS Volume Editor Norm Smith, and Ronald Steel.

## REFERENCES

ALLEN, J.R.L. (1965) A review of the origin and characteristics of recent alluvial sediments. *Sedimentology*, 5, 89–191.

ALLEN, J.R.L. (1974) Studies in fluviatile sedimentation: implications of pedogenic carbonate units, Lower Old Red Sandstone, Anglo-Welsh outcrop. *Geol. J.*, 9(2), 181–208.

ALLEN, J.R.L. (1978) Studies in fluviatile sedimentation: an exploratory quantitative model for the architecture of avulsion-controlled alluvial suites. *Sediment. Geol.*, 21, 129–147.

BLAIR, T.C. & McPHERSON, J.G. (1994) Historical adjustments by Walker River to lake level fall over a tectonically tilted half-graben floor, Walker Lake basin, Nevada. *Sediment. Geol.*, 192, 7–19.

BOYD, K.F. & SCHUMM, S.A. (1995) Geomorphic evidence of deformation in the northern part of the New Madrid seismic zone. *U.S. geol. Surv. Prof. Pap.*, 1538-R, 35.

BRIDGE, J.S. & LEEDER, M.R. (1979) A simulation model of alluvial stratigraphy. *Sedimentology*, 26, 617–644.

BRIDGE, J.S. & MACKEY, S.D. (1993a) A revised alluvial stratigraphy model. In: *Alluvial Sedimentation* (Eds Marzo, M. & Puigdefabregas, C.), Spec. Publs int. Ass. Sediment., No. 17, pp. 319–336. Blackwell Scientific Publications, Oxford.

BRIDGE, J.S. & MACKEY, S.D. (1993b) A theoretical study of fluvial sandstone body dimensions. In: *The Geological Modelling of Hydrocarbon Reservoirs and Outcrop Analogues* (Eds Flint, S.S. & Bryant, I.D.), Spec. Publs int. Ass. Sediment., No. 15, pp. 21–56.

BRIZGA, S.O. & FINLAYSON, B.L. (1990) Channel avulsion and river metamorphosis: the case of the Thomson River, Victoria, Australia. *Earth Surf. Process. Landf.*, 15, 391–404.

BRYANT, M., FALK, P. & PAOLA, C. (1995) Experimental study

of avulsion frequency and rate of deposition. *Geology*, 23, 365–368.

DENNY, C.S. (1965) Alluvial fans in the Death Valley region California and Nevada. *U.S. geol. Survey Prof. Pap.*, 466, 1–62.

ELLIOTT, D.O. (1932) *The Improvement of the Lower Mississippi River for Flood Control and Navigation*, Vol. 1. U.S. Army Corps of Engineers, Waterways Experimentation Station, Mississippi River Commission, Vicksburg, MS.

ETHRIDGE, F.G., SKELLY, R.L. & BRISTOW, C.S. (1999) Avulsion and crevassing in the sandy, braided Niobrara River: complex response to base-level rise and aggradation. In: *Fluvial Sedimentology VI* (Eds Smith, N.D. & Rogers, J.), Spec. Publs int. Ass. Sediment., No. 28, pp. 179–191, Blackwell Science, Oxford.

FISK, H.N. (1944) *Geological Investigations of the Alluvial Valley of the Lower Mississippi River*. U.S. Army Corps of Engineers, Waterways Experimentation Station, Mississippi River Commission, Vicksburg, MS.

FISK, H.N. (1952) *Geological Investigation of the Atchafalaya Basin and the Problem of Mississippi River Diversion*. U.S. Army Corps of Engineers, Waterways Experimentation Station, Mississippi River Commission, Vicksburg, MS.

HELLER, P.L. & PAOLA, C. (1996) Downstream changes in alluvial architecture: an exploration of controls on channel-stacking patterns. *J. sediment. Res.*, 66, 197–306.

JONES, L.S. & BLAKEY, R.C. (1997) Eolian–fluvial interaction in the Page Sandstone (Middle Jurassic) south-central Utah, USA — a case study of erg-margin processes. *Sediment. Geol.*, 109, 181–198.

JONES, L.S. & HARPER, J. (1998) Channel avulsions and related processes, and large-scale sedimentation patterns since 1875, Rio Grande, San Luis Valley, Colorado. *Geol. Soc. Am. Bull.*, 110, 411–421.

LEEDER, M.R. (1978) A quantitative stratigraphic model for alluvium, with special reference to channel deposit density and interconnectedness. In: *Fluvial Sedimentology* (Ed. Miall, A.D.), Mem. Can. Soc. petrol. Geol., Calgary, 5, 587–596.

LEEDER, M.R. & ALEXANDER, J. (1987) The origin and tectonic significance of asymmetric meander belts. *Sedimentology*, 34, 217–226.

LEEDY, J.O., ASHWORTH, P.J. & BEST, J.L. (1993) Mechanisms of anabranch avulsion within gravel-bed braided rivers: observations from a scaled physical model. In: *Braided Rivers* (Eds Best, J.L. & Bristow, C.S.), Spec. Publ. geol. Soc. London, No. 75, pp. 119–127. Geological Society of London, Bath.

MACKEY, S.D. & BRIDGE, J.S. (1995) Three-dimensional model of alluvial stratigraphy: theory and application. *J. Sediment. Res.*, B65, 7–31.

McCALL, E. (1984) *Conquering the Rivers: Henry Miller Shreve and the navigation of America's inland waterways*. Louisiana State University Press, Baton Rouge, LA.

McCARTHY, T.S., ELLERY, W.N. & STANISTREET, I.G. (1992) Avulsion mechanisms on the Okavango fan, Botswana: the control of a fluvial system by vegetation. *Sedimentology*, 39, 779–796.

MIALL, A.D. (1996) *The Geology of Fluvial Deposits*. Springer-Verlag, Berlin.

NELSON, B.W. (1970) Hydrography, sediment dispersal and recent historical development of the Po River delta, Italy. In: *Deltaic Sedimentation, Modern and Ancient* (Eds Morgan, J.P. & Shaver, R.H.), Spec. Publ. Soc. econ. Paleont. Miner., Tulsa, 15, 52–184.

NING, Q. (1990) Fluvial processes in the lower Yellow River after levee breaching at Tongwaxiang in 1855. *Int. J. Sediment. Res.*, **5**, 1–13.

REID, J.B., JR. (1992) The Owens River as a tiltmeter for Long Valley Caldera, California. *J. Geol.*, **100**, 353–364.

RICHARDS, K., CHANDRA, S. & FRIEND, P. (1993) Avulsive channel systems: characteristics and examples. In: *Braided Rivers* (Eds Best, J.L. & Bristow, C.S.), Spec. Publ. geol. Soc. London, No. 75, 195–203. Geological Society of London, Bath.

SCHUMANN, R.R. (1989) Morphology of Red Creek, Wyoming, an arid-region anastomosing channel system. *Earth Surf. Process. Landf.*, **14**, 277–288.

SCHUMM, S.A. (1977) *The Fluvial System*. John Wiley & Sons, New York.

SCHUMM, S.A. (1991) *To Interpret the Earth*. Cambridge University Press, Cambridge.

SCHUMM, S.A. & WINKLEY, B.R. (Eds) (1994) *The Variability of Large Alluvial Rivers*. American Society of Civil Engineers Press, New York.

SCHUMM, S.A., MOSLEY, M.P. & WEAVER, W.E. (1987) *Experimental Fluvial Geomorphology*. Wiley, New York.

SCHUMM, S.A., ERSKINE, W.D. & TILLEARD, J.W. (1996) Morphology, hydrology, and evolution of the anastomosing Ovens and King Rivers, Victoria, Australia. *Bull. Geol. Soc. Am.*, **108**, 1212–1224.

SMITH, N.D., CROSS, T.A., DUFFICY, J.P. & CLOUGH, S.R. (1989) Anatomy of an avulsion. *Sedimentology*, **36**, 1–23.

TODD, O.J. & ELIASSEN, S. (1940) The Yellow River problem. *Trans. Am. Soc. civ. Eng.*, **105**, 346–416.

VAN GELDER, A., VAN DEN BERG, J.H., CHENG, G. & XUE, C. (1994) Overbank and channelfill deposits of the modern Yellow River delta. *Sediment. Geol.*, **90**, 293–305.

*Spec. Publs int. Ass. Sediment.* (1999) **28**, 179–191

# Avulsion and crevassing in the sandy, braided Niobrara River: complex response to base-level rise and aggradation

F. G. ETHRIDGE*, R. L. SKELLY*† *and* C. S. BRISTOW,‡

*\*Department of Earth Resources, Colorado State University, Fort Collins, Colorado 80523-1482, USA*
*(Email: fredpet@cnr.colostate.edu)*
*†Department of Geology, Birkbeck College, University of London, Malet Street, London WC1E 7HX, UK*
*(Email: c.bristow@ucl.ac.uk)*

## ABSTRACT

During a study of the alluvial architecture of the lower portion of the sandy, braided Niobrara River, north-eastern Nebraska, a series of crevasse splays formed and several avulsions occurred. The majority of these events happened between 1995 and 1997, and they can be related to a maximum 2.9 m base-level rise and aggradation of the main Niobrara channel belt, which began in the 1950s following damming of the Missouri River. Crevasses and avulsions, along with a rising groundwater table, have turned the lower 3.3 km of the Niobrara, above its confluence with the Missouri, into an extensive wetland with characteristics and processes similar to those found in some anabranching rivers.

Following 43 yr of aggradation, the Niobrara channel has become elevated above its floodplain, which has led to a series of avulsions, often initiated by the development of crevasse splays. Floodplain aggradation is occurring rapidly with up to 1.5 m of crevasse-splay deposition in a year. Crevasses have formed at low points in the river banks and levees and are locally constrained by floodplain topography and human-made structures. The timing of crevasse initiation may be linked to localized bank erosion or the presence of ice dams rather than increased discharge. The river appears to be evolving into an anabranching or distributary system, reactivating old channels and flowing across former islands and floodplains. The major changes have occurred very rapidly, over a 2-yr period (1995–1997), following more gradual systematic changes in channel-belt width over the previous 41 yr (1954–1995). This pattern of gradual change, followed by dramatic short-term change, is interpreted to indicate the crossing of a geomorphological threshold, beyond which the river behaviour changed from aggradational to avulsive. Our data support models for avulsion occurring when there is a decrease in channel-belt slope and/or an increase in cross-valley slope and indicate that aggradation and superelevation of the channel belt is a major factor in allowing avulsion to occur and persist. In the Niobrara, the major external factor forcing channel change has been a significant rise in base level.

## INTRODUCTION AND OBJECTIVES

Models of alluvial stratigraphy have changed dramatically over the past two decades. Early models decoupled avulsion frequencies from sedimentation rates (Allen, 1978; Leeder, 1978; Bridge & Leeder, 1979) and floodplain aggradation was often considered to be related directly to overbank flooding (Nanson & Croke, 1992). Recent research and modelling suggest that a relationship exists between avulsion frequency and aggradation rates (Mackey & Bridge, 1995; Bryant *et al.*, 1995; Heller & Paola, 1996) and that topography has a strong influence on channel avulsion (Leddy *et al.*, 1993). In a general overview paper, Jones & Schumm (this volume, pp. 171–178) attributed avulsions to decreased channel slope leading to channel blockage, increased lateral slope, and to channel blockage resulting from non-slope-related causes (jams, vegetation, etc.). Recent research on anastomosing systems suggests that a great deal of floodplain deposition occurs during short periods of crevassing and

† Present address: Exxon Exploration Company, P.O. Box 4778, Houston, TX 77210-4778
(Email: Raymond.L.Skelly@Exxon.sprint.com).

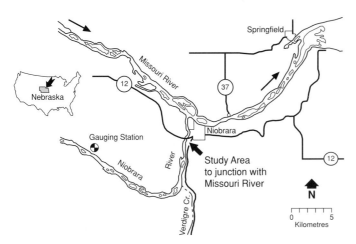

**Fig. 1.** Location of the Niobrara River and study area near the confluence with the Missouri River. Arrows along the Missouri River indicate flow direction. Arrow in insert map of USA shows location of study area in north-eastern portion of Nebraska. Gauging station is located at latitude 42°44′23″N, longitude 98°13′26″W.

avulsion, rather than during periods of major overbank flooding (Smith, 1986; Schumann, 1989; Smith *et al.*, 1989, 1997; Smith & Pérez-Arlucea, 1994). This concept has been applied to ancient meandering river deposits by Kraus & Aslan (1993) and Kraus (1996). Nanson & Croke (1992) and Reinfelds & Nanson (1993) review various hypotheses for floodplain aggradation in braided rivers and Coleman (1969) and Bristow *et al.* (in press) discuss crevasse splays in braided rivers. These hypotheses regarding crevassing, avulsion and floodplain aggradation are tested with data from a rapidly aggrading, sandy, braided river; the Lower Niobrara River, north-east Nebraska.

The headwaters of the Niobrara River are located near Lusk, in southeastern Wyoming. Along most of its course, the Niobrara flows along the northern margin of the Sand Hills of Nebraska, before emptying northwards into the Missouri River near the town of Niobrara (Fig. 1). Throughout much of its length, the Niobrara cuts deeply into sedimentary rocks beneath the Sand Hills (Bentall, 1989). The river is one of the last relatively unregulated rivers in the northern Great Plains. The lower reaches of the river are characterized by perennial flow and exhibit a variety of morphologies over a range of discharges (Buchanan & Schumm, 1990). In the winter the river freezes over. At high discharges, during the spring ice break-up, most flow is concentrated in a single, sinuous channel that forms around alternate bars. During lower discharges the river takes on a more characteristic braided pattern with flow divided among multiple channels. The lower 4.8 km of the Niobrara River occupies a 1.25-km- to 1.6-km-wide valley between bedrock bluffs. The active, braided channel belt ranges from 88 to 329 m wide and consists of a network of channels, active bars and bedforms, and vegetated islands. The channel belt is flanked by a floodplain and/or terraces that extend to the valley walls.

At the confluence with the Missouri, the Niobrara has formed a delta following construction of dams on the Missouri (Livesey, 1976). Upstream of the confluence, Fort Randall Dam has resulted in reduced mean and peak discharge, and Gavins Point Dam, downstream, has created a lake that extended to within 4.8 km of the mouth of the Niobrara by 1992 (Resource Consultants and Engineers, Inc., 1993). The competence of the Missouri River to transport sediment has been greatly reduced (Livesey, 1976). As a result of these conditions, base level (the level of the Missouri River) has risen and aggradation of up to 2.9 m has occurred at the mouth of the Niobrara River (Resource Consultants and Engineers, Inc., 1992). Aggradation of the Niobrara River channel, relative to the adjacent floodplains, has created an alluvial ridge since initiation of the base-level rise. Surveys of channel cross-sections between 1956 and 1983 upstream of the Highway 12 bridge (Fig. 1), indicate that channel bed elevations have increased over 2 m, from a level 1.5 m below to a level 0.6 m above the adjacent floodplain (Resource Consultants and Engineers, Inc., 1993). Channel surveys also suggest that aggradation has extended 21 km up the river during the past 43 years (Resource Consultants and Engineers, Inc., 1993). This setting provides an opportunity to study the morphology, dynamics and stratigraphy of an aggrading braided river.

During a vibracoring and ground-penetrating radar (GPR) study in the reach of the Niobrara River, in and near Niobrara State Park, a series of crevasse splays and a major avulsion occurred, which appears to be the result of base-level rise and channel-belt aggradation. The objectives of this contribution are:

**1** to describe the recent history of the lower Niobrara River, emphasizing the crevassing and avulsion that occurred on the floodplain in the 1990s;

**2** to compare models of crevassing, avulsion and flood-plain aggradation and the concept of thresholds and complex response to actual events on the Niobrara River.

The characteristics and development of two of the major crevasse splays and the internal architecture of the channel-belt deposits are described elsewhere (Bristow *et al.*, 1997; Skelly *et al.*, 1997; Bristow *et al.*, in press).

## RECENT HISTORY OF THE LOWER NIOBRARA RIVER

### Lewis and Clark, 1804

The Niobrara was first described by the pioneering explorers, Lewis and Clark, who visited the study area on 4 September 1804. An excerpt from William Clark's diary gives a graphic account of the river, which he called the Que Courre (i.e. river which runs or flows). 'Continuing one and a half miles came at the mouth of Que Courre. This river comes roleing its sand which is corse into the Missoures from the SW by W. This river is 152 yards across the water and not exceeding 4 feet Deep. It does not rise high. When it does it spreds over a large Surface, and is not navagable. It has a Great many Small Islands and Sand Bars' (Moulton, 1987, p. 47). Despite his poor grammar, chaotic use of capitals and phonetic spelling, it is clear from his detailed notes that the Niobrara was a shallow, sand-bed, braided river in 1804.

### General observations pre-dam to late 1990s

The lower portion of the Niobrara River valley has undergone dramatic changes in the past 43 yr, as a direct result of a major base-level rise caused by damming of the Missouri River. General conditions prior to the initiation of this base-level rise are illustrated by a 1938 aerial photograph (Fig. 2). At that time, the lower Niobrara River flowed within well-defined terraces (used for agricultural purposes) or against a bedrock valley wall. At the mouth of the Niobrara River, there is no evidence of any storage of sediment. All sediment exiting the Niobrara River was carried downstream by the Missouri River. Niobrara State Park was located on Niobrara Island, a large island between the Niobrara River and a natural, sinuous anabranch (i.e. the so-called Mormon Canal). This sinuous anabranch exited the main channel 4.8 km upstream from the railroad bridge (south of the area shown in Fig. 2) and re-entered the river just upstream (south) from

this bridge (Fig. 2). The town of Niobrara was located on a major terrace of the Missouri River (Fig. 2).

This situation remained relatively stable until the early to middle 1950s when Fort Randall Dam, located upstream on the Missouri River and Gavins Point Dam, located downstream on the Missouri River, were completed. Reduced peak and mean discharges, rising water levels and aggradation, resulting from these dams have produced significant changes within the lower Niobrara River. By the 1990s, the lower Niobrara River had risen above the low terrace (Fig. 3) and a major delta had built almost completely across the main channel of the Missouri River (Fig. 4). Niobrara Island has been partially submerged, as the groundwater level within the valley has risen. Because of this rise in ground water, Niobrara State Park was relocated to the bedrock uplands west of the Niobrara River valley in 1987 (Fig. 4). The Mormon Canal had undergone 1.5 km of shortening at the upper end as a result of avulsion. Rising groundwater level and increased occurrence of flooding on the Missouri River terrace required the relocation of the town of Niobrara to the bedrock uplands south of the former town site between 1975 and 1980.

### Changes between 1938–1995

A clear picture of changes in the lower Niobrara River valley emerges from a series of aerial photographs taken between 1938 and 1995. Sketches of important features and quantification of data such as topographic surveys, channel-belt width, terrace versus floodplain area, flow partition between the Niobrara River and the Mormon Canal and development of crevasse splays and avulsions provide the evidence for these changes. As base level rose, aggradation exceeded 2.9 m at the Niobrara River mouth and 2.3 m at the Highway 12 bridge (Resource Consultants and Engineers, Inc., 1992). Average channel width decreased by 45% in a steady and systematic manner within the study area between 1938 and 1996 (Fig. 5). Pinning points with no change in width are the lower end of the Niobrara River that abuts the valley wall, the Highway 12 bridge and the area upstream of the study area that is characterized as island-braided (Fig. 5A). The decrease in channel width is reflected by the linear regression line (Fig. 5B). Possible explanations for this systematic decrease in width include reduced peak or mean discharge or sediment bedload during the time period in question. Historic decreases in channel width have been reported for other braided rivers of the Great Plains (Williams, 1978; Nadler & Schumm, 1981). In both cases, decreases in channel width were related to decreases in water discharge. Niobrara River peak and

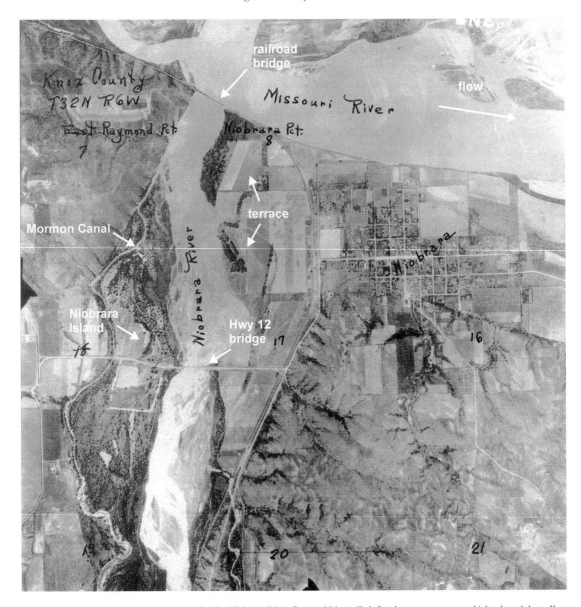

**Fig. 2.** Aerial photograph from 1938. Note that the Niobrara River flows within well-defined terraces, vegetated islands and the valley wall. The town of Niobrara is located on a major terrace of the Missouri River. Niobrara Island lies between the Niobrara River and the narrow, sinuous Mormon Canal, which exits the river upstream (south) of the area covered by the photograph. Niobrara Island is dry and the east floodplain upstream from the railway and road bridges appears to be cultivated agricultural land. There is no obvious delta at the mouth of the Niobrara River. The width of the main channel belt at Highway 12 bridge is approximately 200 m.

mean discharge, during the period 1938–1996, are shown on Fig. 6. These data are from the Verdel gauging station, 21 km upstream of the study area (Fig. 1). Annual peak-discharge data show an overall decrease, but most of the change occurs between 1959 and 1967 (Fig. 6). Data for the period 1967 to 1996 show no trend. Available mean

annual discharge data show no systematic changes (Fig. 6). Bedload data are not available for the lower Niobrara River. The systematic decrease in average channel width in the study area during the period 1938–1996 is probably related to diversion of flow into the Mormon Canal and on to the floodplain. During this period, floodplain area,

**Fig. 3.** Aerial view in August 1994 of Niobrara upstream of Highway 12 bridge. The majority of flow is still confined to the main channel belt, where bars appear to alternate along channel banks and are dissected by braided-channel networks. The single, well-defined thalweg and alternate bars are characteristic of higher discharges. Niobrara Island is now a low-lying wetland area. The photograph was taken prior to the avulsion of the main channel belt at a narrow point in the channel bank (arrow). The active channel is approximately 250 m wide at the Highway 12 bridge. River flow is from right to left (south to north). The Missouri River can be seen in the upper left.

**Fig. 4.** Aerial view in August 1994 of the Niobrara River delta, looking south, up the Niobrara River. The Missouri River is in the foreground and flows from right to left (west to east). The main channel belt of the Niobrara River is on the eastern side of the valley and abuts against older terrace deposits to the east. The Mormon Canal flows along a bedrock valley wall to the west and re-enters the Niobrara River just upstream of the railroad bridge at the mouth of the Niobrara River. The railroad bridge at the mouth of the river is approximately 170 m wide.

**Fig. 5.** (A) Graph of width of the main channel belt of the Niobrara River versus distance upstream from the river mouth at various times between 1938 and 1996. Measurements were taken from topographic maps and aerial photographs. The plot starts at the railroad bridge at the confluence of the Niobrara and Missouri rivers (0 m upstream). Convergence of points just below 500 m is the location where the Mormon Canal rejoins the Niobrara River and the river abuts against the west valley wall. Convergence of points between 2000 and 2500 m upstream is the location of Highway 12 bridge. Convergence of points at 4000 m upstream is upstream of the main study reach in an island braided portion of the Niobrara River.

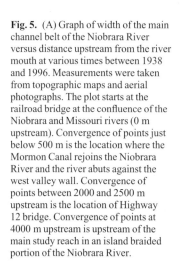

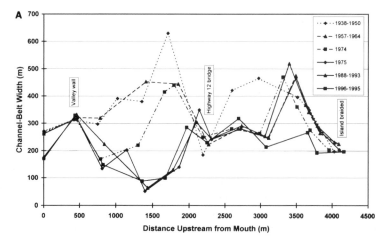

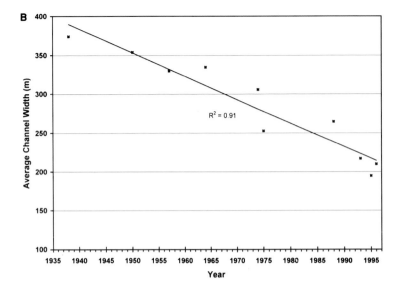

**Fig. 5.** (B) Graph of average channel width of the lower Niobrara River within the study reach between 1938 and 1996. Note that the significant linear regression has an $R^2$ value of 0.91.

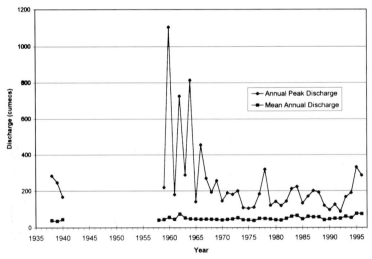

**Fig. 6.** Annual peak discharge and mean annual discharge between 1938 and 1940 and between 1959 and 1996 at the Verdel gauge on the lower Niobrara River above the study area. Note the significantly higher values for annual peak discharge between 1960 and 1966 and lower and more consistent values between 1967 and 1996. Note the lack of any significant trend between 1959 and 1996 for mean annual discharge.

adjacent to the channel belt, increased from 8% to 57% of the total alluvial valley surface, just upstream of the Highway 12 bridge. This significant increase came at the expense of terrace area. The total combined area of the Niobrara River and Mormon Canal remained unchanged for the 1938–1996 period. Discharge into the Mormon Canal from the Niobrara River has obviously increased since 1988, but no data for partitioning of flow are available for the entire 48-yr period.

The Mormon Canal was affected by a series of avulsions that took place over relatively short periods of time in 1988 and 1995. During the 1988 avulsion, flow continued in both the upstream and downstream divergent points for a period of time before the upstream divergent

point was abandoned completely (Fig. 7A). During the 1995 avulsion, flow also continued through both the upstream and the downstream divergent points (Fig. 7B). In 1995, only a small amount of flow from the main channel belt was diverted into the Mormon Canal on the west (left) bank (Fig. 8). A third avulsion also occurred, sometime between 1994 and 1995, downstream of the Highway 12 bridge and on the east (right) bank, with water flowing in a new anabranch that follows the old course of a contact between a low and a high terrace (see Figs 2 & 8). The second and third avulsions were preceded and initiated by crevasse splays. It is unknown whether this is true for the 1988 avulsion. Each avulsion was directed toward the lowest area of the floodplain

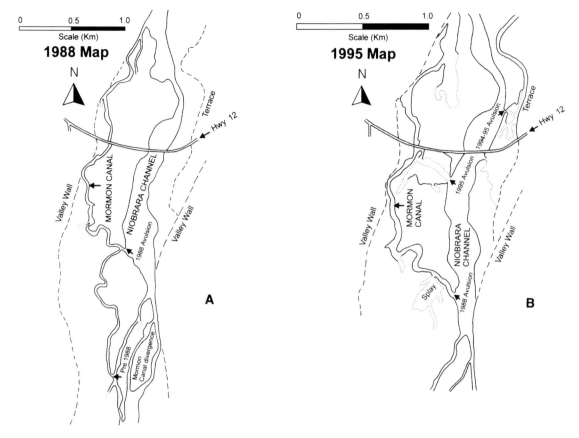

**Fig. 7.** (A) Sketch map showing location of 1988 avulsion of Niobrara River into the Mormon Canal. At this time flow from the Niobrara River to the Mormon Canal was through both the 1988 avulsion site and the former upstream divergent point. The upstream divergent point was abandoned sometime later. (B) Sketch map showing 1995 avulsion of the Niobrara River into the Mormon Canal just above the Highway 12 bridge. Note, flow from the Niobrara River into the Mormon Canal continues through the 1988 avulsion site and also downstream through the 1995 avulsion site. A third avulsion site, which formed sometime during 1994 or 1995, can be seen on the east side of the Niobrara River belt, downstream (north) of the Highway 12 bridge.

**Fig. 8.** Vertical aerial photograph of the Niobrara River and the town of Niobrara taken in the summer of 1995. Flow in the main channel belt is to the north (top of photograph). The photograph shows the location and extent of two of the major crevasse splays. The splay south of the Highway 12 bridge, west of the channel belt (1), was initiated in the spring of 1995 and has already covered a sizeable area. Water from the Niobrara River passes through this crevasse to the Mormon Canal (seen on the left side of the photograph). The second crevasse splay downstream (north) of the Highway 12 bridge and east of the Niobrara River, is also allowing the movement of some flow from the Niobrara River (2) into a new anabranch formed along an old channel at the edge of a terrace. The main channel belt of the Niobrara River is approximately 250 m wide at the Highway 12 bridge.

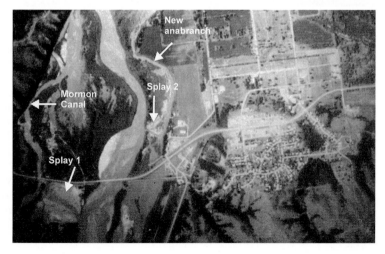

**Fig. 9.** Aerial view of the Niobrara River upstream of the Highway 12 bridge taken in August 1996. View is looking east and flow in the Niobrara is generally from right to left (south to north). The majority of flow in the Niobrara River is diverted through the 1995 avulsion into the Mormon Canal (seen in the lower portion of the photograph). The former channel belt of the Niobrara River is mostly exposed above the level of the groundwater table, downstream of the avulsion. Small crevasse splays have developed along the Mormon Canal both upstream and downstream of the Highway 12 bridge. Figure 3 shows a similar view prior to avulsion. The bridge across the Niobrara River is approximately 250 m wide.

**Fig. 10.** Aerial view of the Niobrara River in August 1996 looking south across Highway 12. East is to the left and flow is from top to bottom (south to north). View shows major crevasse splays upstream and downstream of Highway 12 and flow into the old channel along the terrace on the east side of the river valley. The distance from the east bank of the main channel to edge of the eastern terrace along the highway is approximately 375 m.

in that particular area of the alluvial valley (former or active channel or anabranch). The number and frequency of crevasse splays has increased with continued aggradation, especially in that portion of the river directly above and below the Highway 12 bridge.

### Changes between 1995 and 1997

By 1995, the lower end of the Niobrara River, Mormon Canal and the entire alluvial valley from just above the Highway 12 bridge to the mouth had become a very dynamic depositional system. This portion of the river appears to be undergoing a major change in behaviour as a result of the base-level rise event that began in 1954. A minimum of eight crevasse splays were active along both sides of the main channel belt and the Mormon Canal. The major crevasse just upstream of the Highway 12 bridge was beginning to convey ever-increasing quantities of water from the main channel belt to the Mormon Canal on the western side of the valley. By 1996, enough of the flow from the main channel belt had been diverted into the Mormon Canal, that the majority of the main channel belt was higher than the water level (Fig. 9). The canal re-entered the main Niobrara River channel 0.4 km upstream of the railroad bridge at the mouth of the Niobrara River valley. Crevasse splays upstream and downstream of the Highway 12 bridge and on the east side of the channel belt continued to grow and water from the main channel belt continued to flow into the new anabranch along the margin of a high terrace (Fig. 10).

**Fig. 11.** Aerial view in August 1997, looking west (top of photograph) across the mouth of the Niobrara River toward the new site of Niobrara State Park on the upland. Flow is from left to right (south to north) into the Missouri River. View shows wetlands that have developed upstream of the Niobrara delta and the railroad bridge, shown as cultivated fields in the 1938 aerial photograph (Fig. 2). Distance across the wetlands along the railroad bridge and the embankment is approximately 800 m.

Flow in this channel continued downstream to the Missouri River. Niobrara Island was largely inundated, primarily from a rising groundwater table on the north end of the island. By 1996, the majority of flow in the main channel belt of the Niobrara River had been diverted through the major splay above the Highway 12 bridge and into the Mormon Canal. A large number of crevasse splays originated from the Mormon Canal downstream and upstream of the Highway 12 bridge (Fig. 9). These splays accelerated the inundation of Niobrara Island and the area to the west of the canal. Wooden buildings within the old park area were filled with sediment up to 0.6–0.8 m deep. By the summer of 1997, the entire lower end of the Niobrara River below the Highway 12 bridge had been turned into a wetlands with water extending almost continuously from the bedrock valley wall on the western side of the valley to an old high terrace on the east (Fig. 11). River level was only 0.5 m below the top of the support piers for the abandoned railroad bridge. In the summer of 1997, the level of Highway 12 was being raised west of the main bridge across the channel belt of the Niobrara River and a new bridge was being built across the Mormon Canal.

## CREVASSING AND AVULSION IN THE LOWER NIOBRARA RIVER VALLEY: COMPLEX RESPONSE TO AN AGGRADATIONAL EVENT

Abrupt landform changes can result with or without a

**Table 1.** Changes in downstream gradient of the channel belt versus cross-valley gradient in lower Niobrara River, north-eastern Nebraska (1950 data from U.S. Geological Survey; Niobrara and Vertigre topographic quadrangle sheets).

| Gradients | 1950 | 1995–1996 |
|---|---|---|
| Channel belt | 0.14% | 0.04% |
| Cross-valley | 0.00%* | 0.18% |

\* Estimate of 0% gradient across Niobrara Island in vicinity of 1995 avulsion is based on data from a contour map with a contour interval of 10 ft (3 m). At that time, terraces flanked the active channel belt. Aerial photographs from 1938 and 1957 suggest that the terraces were relatively flat.

change in external control. Abrupt landform change resulting from progressive change in an external control occurs as an extrinsic threshold is crossed (Schumm, 1979). Major morphological changes are taking place in the lower 5 km of the Niobrara River. These changes are related directly to 43 yr of aggradation, caused by a significant rise in base level. The main channel belt has aggraded to a point where the gradient along this belt was significantly less than the across-valley gradient in 1996 (Table 1). In the Mackey & Bridge (1995) model of alluvial architecture, avulsion is favoured when there is a decrease in channel-belt slope and/or an increase in cross-valley slope. Bryant *et al.* (1995) and Heller & Paola (1996) suggest that avulsion is related to aggradation of a channel and superelevation of the channel bed above the surrounding floodplain. Our abservations on the Niobrara

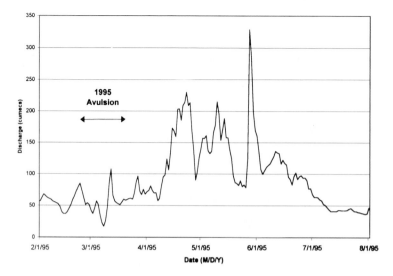

**Fig. 12.** A graph of daily discharge records for the Niobrara River at the Verdel Gauge from 1 February 1995 to 1 August 1995. Note that maximum discharges occurred in May and not in late February to early March when the avulsion of the Niobrara River just above the Highway 12 bridge was initiated. Data downloaded from U.S. Geological Survey, Water Resources Division.

River suggest that the relationship between the channel-belt and the cross-valley gradients caused by aggradation and superelevation of the channel belt is the underlying cause that permits avulsions to occur and to persist. Triggers that determine the exact location and timing of a particular avulsion include discharge and ice blocks or ice sheets, and are discussed here in terms of the 1995 avulsion, just upstream of the Highway 12 bridge.

Continuous U.S. Geological Survey discharge records are available for the Verdel gauging station (Fig. 1) from 1959. Mean daily discharge for the Niobrara at Verdel, during the period 1938 to 1996, was 46.7 cumecs (1650 cfs). The highest flow on record of 710.8 cumecs (25 100 cfs) occurred on 27 March 1960. Higher than normal flows occurred in 1995. The highest flows, however, did not coincide with the avulsion. The avulsion occurred between late February and early March 1995, but the highest flow of 327.1 cumecs (11 600 cfs) did not occur until 28 May (Fig. 12). In March, however, and coinciding with the avulsion, there were ice blocks and sheets in the river (Tom Moteck and a state wildlife official, pers. comm., 1995). Ice and water impacted a low area (Fig. 3) on the west bank of the Niobrara River upstream from an old access road that reinforces the river bank. The high cross-valley gradient and the higher than normal discharge probably enhanced flows from the point of avulsion to the lowest part of the floodplain (i.e. the Mormon Canal). Within 4 months of the initial break-through, splay deposits, up to 1.5 m thick covered an area of 0.125 km$^2$. By the summer of 1996, most of the flow from the channel belt was being diverted into the Mormon Canal (Fig. 9). In the summer of 1997, flow in the Niobrara was divided among the former channel

belt and various avulsion pathways (Fig. 13). Whether or not the 1988 avulsion into the Mormon Canal or the 1994–1995 avulsion (Figs 7 & 8) toward a terrace on the east floodplain were initiated by ice blocks and sheets is unknown. It is clear that the 1994–1995 avulsion did originate as a crevasse and that the anabranch, which now carries water to the Missouri River along the east side of the valley, owes its origin primarily to this crevasse and possibly to others along the eastern side of the Niobrara (Fig. 10). By 1996, crevasse splays were also occurring along both banks of the Mormon Canal upstream and downstream of the Highway 12 bridge.

Human-made structures have had an effect on observed morphological changes. The embankment of Highway 12 has led to ponding on the floodplain on the east side of the river upstream of the road (Figs 10 & 13) and diversion of natural flowpaths. For example, there is a small crevasse in the river bank, where flow from a large splay (East Splay in Bristow *et al.*, in press) returns to the main channel. If the embankment of Highway 12 had not been present, flow from the East Splay may well have continued downstream along the east side of the river. The splays downstream from the bridge (Figs 7B & 10) would then form part of a larger splay complex and the river might have made a more complete avulsion into the abandoned channel on the eastern floodplain.

The 1995 avulsion site occurs on the outside of a channel bend immediately upstream of a section of the river bank, which had been protected by concrete blocks to prevent erosion of a roadway. The avulsion was caused by bank erosion, but its location was partly determined by the end of the bank defences. The embankment of Highway 12 across Niobrara Island prevented flow from

**Fig. 13.** Aerial view of the Niobrara River in the vicinity of Highway 12 bridge, looking east-northeast. Flow in the river is to the north (right to left in the photograph). The photograph was taken in August 1997. The view shows flow divided among the main channel, the 1995 avulsion upstream of Highway 12 bridge into Mormon Canal and the 1994–1995 avulsion downstream of Highway 12 bridge into a former channel. Bridge across the Niobrara channel belt is approximately 250 m wide.

the 1995 avulsion continuing downstream along the abandoned channel, which extends through the old abandoned state park, and forced flow to join the Mormon Canal immediately upstream of the canal bridge. As discharge in the Mormon Canal increased, the bridge acted as a flow constriction and crevasse splays formed on both sides of the canal immediately downstream, with flow diverging into a pond area (West Splay in Bristow *et al.*, in press) and into the old state park area of Niobrara Island.

In the model studies of Mackey & Bridge (1995), avulsion sites shift up-valley and decrease in abundance with time. No such relationship has been observed in the lower Niobrara River. The divergent points where flow leaves the Niobrara to join the Mormon Canal have shifted downstream between 1985 and 1995. Crevasse and avulsion sites have increased between 1995 and 1997. The timeframe examined here is three to five orders of magnitude less than that modelled by Mackey & Bridge (1995), and their model does not deal with specific triggers.

Processes and deposits in the lower Niobrara River are beginning to take on the character of a distributary or an anabranching system, which appears to have some similarities with the Cumberland Marshes of east-central Saskatchewan, Canada (Smith *et al.*, 1989; Smith & Pérez-Arlucea, 1994). The area from the railroad bridge to just above the Highway 12 bridge is rapidly becoming a wetland (Figs 11 & 13). This portion of the river is dominated by a rising groundwater table and by avulsions initiated at crevasse splays. These crevasses funnel water to low areas of the floodplain, such as current subordinate channels (i.e. the Mormon Canal) or former channels. Floodplain aggradation is occurring, primarily as a result of crevasse splay deposition, rather than from major overbank flooding of a master channel. In this sense the lower Niobrara River is also similar to the Saskatchewan River (Smith & Pérez-Arlucea, 1994). We concur with Kraus & Aslan (1993), Kraus (1996), and Smith & Pérez-Arlucea (1994) that avulsion-related deposits may be more common in the rock record than realized previously.

It appears that Richards *et al.* (1993) were correct in suggesting that a transition exists between anastomosing systems and other types of avulsive systems, such as the sandy, braided, lower Niobrara River. Changes in the lower Niobrara River have been very dramatic since 1994, and the system appears to be undergoing complex response after having exceeded a threshold (Schumm, 1977). The exact trigger for most crevasse and avulsion events has not been observed; however, it is certainly possible that the higher than normal peak discharges of 1995–1996 could be a general cause. Forty-three years of aggradation have created a system that is sensitive to change (Schumm, 1991). Higher than normal peak discharges may have forced this system over a threshold, creating the complex response and evolution we observe. The ultimate course of the lower Niobrara River remains uncertain. It may evolve into a fully anastomosing river, or form a distributive channel system, or after a phase of instability settle into a single channel adjusted to the new conditions.

## SUMMARY AND CONCLUSIONS

Aggradation in the lower Niobrara River, as a direct result of the construction of Gavins Point Dam and flow regulation on the Missouri River, has extended approximately

20.9 km upstream from the confluence of the two rivers. The lower 3.3 km have experienced the most change, with major crevassing, avulsions and a rising groundwater table, turning this portion of the river into an extensive wetland. All of these changes have forced the movement of Niobrara State Park and the town of Niobrara on to the uplands adjacent to the Niobrara River Valley. In addition, a railroad bridge was abandoned and a highway bridge has been relocated and raised twice. Beginning in 1938 and continuing to 1996, there has been a steady and systematic decrease in average channel width of the Niobrara River in the same lower 3.3 km. There is no direct correlation between these changes and systematic changes in mean or peak annual discharge recorded from an upstream gauging station. Increased water discharge from the Niobrara River into the Mormon Canal has occurred and probably explains the decrease in Niobrara channel width, but no records exist for determining partitioning of flow between the two channels.

Avulsions of the lower Niobrara River are initiated and develop as crevasse splays in a manner similar to that documented in some anastomosing systems. Avulsions include anabranches that re-enter the Niobrara and distributaries that continue to the Missouri River. All the avulsions have emptied into former or current minor channels occupying lower topographic areas of adjacent floodplains. In the best-documented avulsion, just upstream of the highway bridge, up to 1.5 m of sediment has been added to the floodplain over an area of 0.125 km². Other splays have also added sediment to the floodplain during a period when little or no widespread, conventional overbank flooding has taken place. In general, these avulsions are related to aggradation of the main channel belt above the adjacent floodplain. Initiation of one well-documented avulsion suggests no correlation between avulsions and high discharge. The trigger for this avulsion was an ice jam that forced water against a low area in the levee. There is no evidence of upstream or downstream progression of avulsions during the period of study. The increase in crevasse and avulsion frequency during the 1995–1997 period may be the result of a very sensitive system that has exceeded an extrinsic threshold and is undergoing complex response to a progressive base-level rise. The lower Niobrara appears to be changing from an aggradational to an avulsion system. The ultimate fate of this portion of the river is unknown at this time.

## ACKNOWLEDGEMENTS

Acknowledgement is made to the Donors of the Petroleum Research Fund, administered by the American Chemical Society, for support of this research. Ray Skelly acknowledges partial support from the Geological Society of America and Chevron USA. Initial work by Jonathan Cole provided data and insight into river processes and changes. Matt Wheaton, Jarett Zuboy, Jay Cederberg, Mike Beshore and Brendan Kelly provided valuable field assistance during several field seasons. Robert F. Diffendal, Jr. and the Nebraska Conservation and Survey Division supplied a number of critical aerial photographs. Tom Moteck, superintendent of Niobrara State Park and his staff provided encouragement, access and support during the conduct of the research. Numerous discussions with Stan Schumm have clarified our thinking on river processes. We thank Larry Jones and Norm Smith for constructive criticism of an earlier version of this paper.

## REFERENCES

ALLEN, J.R.L. (1978) Studies in fluviatile sedimentation; an exploratory quantitative model for the architecture of avulsion-controlled alluvial suites. *Sediment. Geol.*, **21**(2), 129–147.

BENTALL, R. (1989) Streams. In: *An Atlas of the Sand Hills, Vol. Resource Atlas No. 5a* (Eds Bleed, A. & Flowerday, C.), pp. 93–114. Conservation and Survey Division, Institute of Agricultural and Natural Resources Division, University of Nebraska, Lincoln, NE.

BRIDGE, J.S. & LEEDER, M.R. (1979) A simulation model of alluvial stratigraphy. *Sedimentology*, **26**(5), 617–644.

BRISTOW, C.S., SKELLY, R.L. & ETHRIDGE, F.G. (1997) Crevasse splays from a rapidly aggrading sand-bed braided river, Niobrara River, Nebraska. In: *Abstracts Sixth International Conference on Fluvial Sedimentology* (Ed. Rogers, J.), University of Cape Town, Cape Town, p. 30.

BRISTOW, C.S., SKELLY, R.L. & ETHRIDGE, F.G. (in press) Crevasse splays from a rapidly aggrading sand-bed braided river, Niobrara River, Nebraska. *Sedimentology*.

BRYANT, M., FALK, P. & PAOLA, C. (1995) Experimental study of avulsion frequency and rate of deposition. *Geology*, **23**(4), 365–368.

BUCHANAN, J. & SCHUMM, S.A. (1990) The riverscape — Niobrara River. In: *Surface Water Hydrology*, Vol. O-1 (Eds Wolman, M.G. & Riggs, H.C.), pp. 314–321. Geological Society of North America, The Geology of North America, Boulder, CO.

COLEMAN, J.M. (1969) Brahmaputra river: channel processes and sedimentation. *Sediment. Geol.*, **3**, 129–239.

HELLER, P.L. & PAOLA, C. (1996) Downstream changes in alluvial architecture: an exploration of controls on channel-stacking patterns. *J. sediment. Res. B: Stratigr. Global Stud.*, **66**(2), 297–306.

JONES, L.S. & SCHUMM, S.A. (1999) Causes of avulsion: an overview. In: *Fluvial Sedimentology VI* (Eds Smith, N.D. & Rogers, J.), Spec. Publs int. Ass. Sediment., No. 28, pp. 171–178. Blackwell Science, Oxford.

KRAUS, M.J. (1996) Avulsion deposits in Lower Eocene alluvial rocks, Bighorn Basin, Wyoming. *J. sediment. Res. B: Stratigr. Global Stud.*, **66**(2), 354–363.

KRAUS, M.J. & ASLAN, A. (1993) Eocene hydromorphic paleosols: significance for interpreting ancient floodplain sequences. *J. sediment. Petrol.*, **63**, 453–463.

LEDDY, J.O., ASHWORTH, P.J. & BEST, J. (1993) Mechanisms of anabranch avulsion within gravel-bed braided rivers: observations from a scaled physical model. In: *Braided Rivers* (Eds Best, J.L. & Bristow, C.S.), Spec. Publ. geol. Soc. London, No. 75, pp. 119–127. Geological Society of London, Bath.

LEEDER, M.R. (1978) A quantitative stratigraphic model for alluvium, with special reference to channel deposits density and interconnectedness. In: *Fluvial Sedimentology* (Ed. Miall, A.D.), Mem. Can. Soc. petrol. Geol., Calgary, **5**, 587–596.

LIVESEY, (1976) The sedimentary influence of a tributary stream, growth of the Niobrara delta. In: *Proceedings of the Third Federal Inter-Agency Sedimentation Conference*, March 22–25, Denver, CO, pp. 4-127–4-137.

MACKEY, S.D. & BRIDGE, J.S. (1995) Three-dimensional model of alluvial stratigraphy: theory and application. *J. sediment. Res.*, **B65**(1), 7–31.

MOULTON, G.E. (Ed.) (1987) *The Journals of the Lewis & Clark Expedition*, Vol. 3, *August 25, 1804 – April 6, 1805*. University of Nebraska Press, Lincoln, NE, 544 pp.

NADLER, C.T. & SCHUMM, S.A. (1981) Metamorphosis of South Platte and Arkansas rivers, eastern Colorado. *Phys. Geogr.*, **2**, 95–115.

NANSON, G.C. & CROKE, J.C. (1992) A genetic classification of floodplains. *Geomorphology*, **4**, 459–486.

REINFELDS, I. & NANSON, G. (1993) Formation of braided river floodplains, Waimakariri River, New Zealand. *Sedimentology*, **40**, 1113–1127.

RESOURCE CONSULTANTS AND ENGINEERS, INC. (1992) *Sedimentation near the Confluence of the Missouri and Niobrara Rivers 1954 to 1990*. M.R.D. Sediment Memorandum No. 12, U.S. Army Corps of Engineers, Omaha District, Omaha, NE (paged by section).

RESOURCE CONSULTANTS AND ENGINEERS, INC. (1993) *Niobrara River Sedimentation Impacts Study*, Phase II. Unpublished report prepared for the U.S. Army Corps of Engineers, Omaha District, Omaha, NE (paged by section).

RICHARDS, K., CHANDRA, S. & FRIEND, P. (1993) Avulsive channel systems: characteristics and examples. In: *Braided Rivers* (Eds Best, J.L. & Bristow, C.S.), Spec. Publ. geol. Soc. London, No. 75, pp. 195–203. Geological Society of London, Bath.

SCHUMANN, R.R. (1989) Morphology of Red Creek, an arid region anastomosing channel system. *Earth Surf. Process. Landf.*, **14**, 277–288.

SCHUMM, S.A. (1977) *The Fluvial System*. John Wiley & Sons, New York, 338 pp.

SCHUMM, S.A., (1979) Geomorphic thresholds: the concept and its applications. *Trans. Inst. Br. Geogr., New Ser.*, **4**, 485–515.

SCHUMM, S.A. (1991) *To Interpret the Earth: Ten Ways to be Wrong*. Cambridge University Press, Cambridge, 133 pp.

SKELLY, R.L., BRISTOW, C.S. & ETHRIDGE, F.G. (1997) Architecture and evolution of channel-belt deposits, lower Niobrara River, Nebraska, USA. In: *Abstracts, Sixth International Conference on Fluvial Sedimentology* (Ed. Rogers, J.), University of Cape Town, Cape Town, p. 186.

SMITH, D.G. (1986) Anastomosing river deposits, sedimentation rates, and basin subsidence, Magdalena River, northwestern Columbia, South America. *Sediment. Geol.*, **46**, 177–196.

SMITH, N.D. & PÉREZ-ARLUCEA, M. (1994) Fine-grained splay deposition in the avulsion belt of the Lower Saskatchewan River, Canada. *J. Sediment. Res.*, **B64**, 159–168.

SMITH, N.D., CROSS, T.A., DUFFICY, J.P. & CLOUGH, S.R. (1989) Anatomy of an avulsion. *Sedimentology*, **36**(1), 1–23.

SMITH, N.D., MCCARTHY, T.S., ELLERY, W.N., MERRY, C.L. & RUTHER, H. (1997) Avulsion and anastomosis in the panhandle region of the Okavango Fan, Botswana. *Geomorphology*, **20**, 49–65.

WILLIAMS, G.P. (1978) The case of the shrinking channels — the North Platte and Platte Rivers in Nebraska. *U.S. geol. Surv. Bull.*, **781**, 1–48.

*Spec. Publs int. Ass. Sediment.* (1999) **28**, 193–209

# Contrasting styles of Holocene avulsion, Texas Gulf Coastal Plain, USA

A. ASLAN* *and* M. D. BLUM†

*\*Department of Physical and Environmental Sciences, Mesa State College, Grand Junction,
CO 81502, USA (Email: aaslan@mesa5.mesa.colorado. edu)*
*\*Department of Geosciences, University of Nebraska-Lincoln, Lincoln, NE 68588, USA (Email: mblum@unl.edu)*

## ABSTRACT

Examination of outcrops, satellite imagery and shallow (< 25 m long) floodplain cores shows that rivers of the Texas Gulf Coastal Plain undergo two distinct avulsion styles: (i) avulsion by channel reoccupation and (ii) avulsion by diversion into flood basins. Holocene avulsion histories of rivers with large sediment supplies, such as the Colorado, and rivers with small sediment supplies, such as the Trinity and Nueces rivers, further suggest that different styles of avulsion occur during different stages of incised-valley filling.

The Nueces and Trinity river valleys represent early stages of filling, in response to the Holocene transgression, and these rivers avulse by reoccupying segments of Late Pleistocene falling-stage and lowstand channel courses that are buried by thin veneers of Holocene sediment. Because these rivers have small sediment supplies, floodplain aggradation is slow and this factor, in addition to the abundance and large size of Late Pleistocene palaeochannels, favours channel reoccupation. This style of avulsion is accomplished primarily by erosion and reworking of channel sediments and avulsion deposits are rare, especially in the vicinity of the avulsion node.

In contrast to the Trinity and Nueces rivers, the Colorado River has almost completely filled the accommodation space produced during the Holocene transgression and avulsion deposits comprise a significant portion (*c.* 50%) of the fill. Repeated episodes of channel diversion into flood-basin depressions accompanied rapid floodplain aggradation and valley filling and produced metres-thick successions of massive or laminated floodbasin muds that encase crevasse-splay sands. Deposits of individual avulsions are separated by slickensided muds or buried A horizons of soils, which represent periods of floodplain stability between episodes of avulsive deposition.

The modern Colorado River represents a late stage of valley filling and avulses by channel reoccupation. During the most recent event, the Colorado River abandoned its alluvial valley and reoccupied a Pleistocene channel belt of the previous interglacial highstand. Similarly, during the preceding avulsion, the Colorado River incised and reoccupied a buried Pleistocene palaeochannel, that was active during the falling stage of the last glacial cycle. In a river with a large sediment supply, such as the Colorado, late stages of valley filling coincide with sea-level highstand, which limits accommodation space and slow floodplain aggradation leading to avulsions by channel reoccupation.

The significance of avulsion during valley filling by these coastal-plain rivers is threefold. First, avulsion deposits represent a large portion of the valley fills, especially of rivers with large sediment supplies such as the Colorado. Second, avulsion by channel reoccupation during early and late stages of valley filling produces multilateral and multistorey sheet sands. Third, the most recent avulsion of the Colorado River demonstrates that avulsion during late stages of valley filling can determine the location of future incised alluvial valleys and controls the preservation of older valley fills.

## INTRODUCTION

Avulsion has long been recognized as an important process in alluvial rivers (e.g. Fisk, 1944, 1947; Bernard *et al.*, 1970), but only over the past decade have field and experimental studies begun to highlight the complexity and far-ranging significance of this process in the fields of fluvial sedimentology and geomorphology (e.g. Wells & Dorr, 1987; Smith *et al.*, 1989; Brizga & Finlayson, 1990; Smith & Pérez-Arlucea, 1994; Törnqvist, 1994; Bryant *et al.*, 1995; Kraus, 1996). The growing body of information on avulsion origins, processes, and stratigraphy is especially important to studies of alluvial architecture because these studies indicate that avulsion is responsible for the accumulation of much larger proportions of floodplain deposits in modern and ancient settings than generally recognized (e.g. Smith *et al.*, 1989; Kraus & Aslan, 1993; Kraus, 1996).

This paper examines avulsion processes and deposits of Late Quaternary fluvial systems of the Texas Gulf Coastal Plain. This area has long been used as a natural laboratory for the study of fluvial processes and facies (e.g. Doering, 1956; Bernard & LeBlanc, 1965; Bernard *et al.*, 1970; McGowen & Garner, 1970; Winker, 1979; Blum & Valastro, 1994; Blum & Price, 1998), and we build upon the sedimentological and stratigraphical frameworks established by previous workers to examine avulsion styles during incised-valley filling. The term 'avulsion style' simply refers to the conditions and processes by which a river avulses. Our discussion focuses on the Holocene avulsion history of the Colorado, Trinity and Nueces rivers, which have different sediment supplies and represent different stages of valley filling. We use these rivers to describe two styles of avulsion and show that these styles relate to specific stages of valley filling. Because alluvial deposits of the Texas Gulf Coastal Plain represent the up-dip portion of the subsiding passive margin Gulf of Mexico Basin, the avulsion deposits described here have a high preservation potential

and should represent good analogues for interpreting the transgressive and highstand components of ancient incised-valley fills from similar passive-margin settings (Blum, 1994).

## GEOLOGICAL SETTING

The Texas Gulf Coastal Plain consists of a series of low-gradient, fan-shaped, alluvial–deltaic plains that originate within each of the major river valleys (Fig. 1). The characteristics of each of the major rivers are strongly influenced by drainage-basin size and sediment load. Extrabasinal rivers (*sensu* Galloway, 1981) such as the Colorado, Brazos and Rio Grande drain tectonic hinterlands, have large sediment supplies and construct laterally extensive alluvial–deltaic plains. Because these rivers have large sediment supplies, floodplain aggradation has kept pace with Holocene sea-level rise and valleys that were incised during the Last Glacial Maximum have completely or almost completely filled. In contrast to the extrabasinal streams, basin-fringe rivers (*sensu* Galloway, 1981), such as the Trinity, Guadalupe and Nueces, drain the coastal plain, have smaller sediment loads, commonly discharge into estuaries (i.e. the wave-dominated estuaries of Dalrymple *et al.*, 1992) and are presently constructing small bay-head deltas (Fig. 1). Accommodation space is abundant in these river valleys and they represent early stages of valley filling.

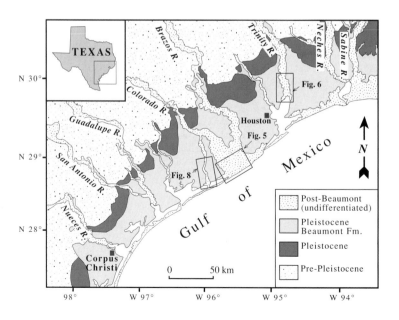

**Fig. 1.** Geological map of the Texas Gulf Coastal Plain between the Sabine and Nueces rivers, showing the distribution of Late Quaternary deposits and major river valleys (simplified from DuBar *et al.*, 1991).

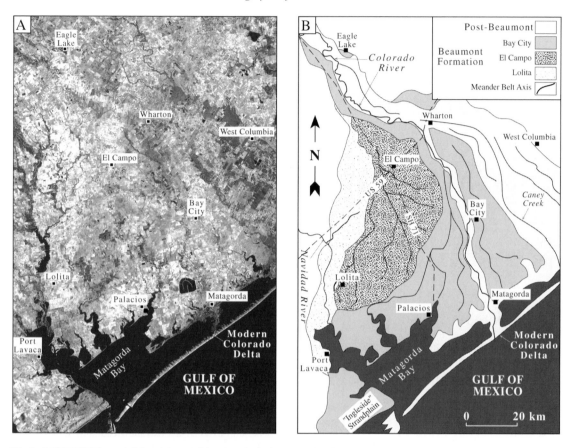

**Fig. 2.** LANDSAT Thematic Mapper image (A) and geological map (B) showing Beaumont and post-Beaumont alluvial deposits of the Colorado River (from Blum & Price, 1998). Note that the Colorado River's Beaumont and post-Beaumont alluvial plains are subdivided into four valley fills that span the last 400 000 yr. Also note that the present-day Colorado River diverges from the post-Beaumont alluvial valley near Wharton, Texas and traverses south across Late Pleistocene Beaumont deposits to the Gulf of Mexico.

Alluvial plains of the Texas Gulf Coast were subdivided initially into three 'morphostratigraphic units' of presumed Pleistocene age and designated as the Willis (oldest), Lissie and Beaumont (youngest) formations (see Morton & Price, 1987; DuBar *et al.*, 1991; Blum & Price, 1998, for reviews) (Fig. 1). Detailed mapping during the 1960s and 1970s (e.g. Bernard *et al.*, 1970; Aronow, 1971; Fisher *et al.*, 1972; Brown *et al.*, 1976; McGowen *et al.*, 1975, 1976; Winker, 1979) showed that Beaumont alluvial plains consist of numerous cross-cutting meander belts and intervening flood basins and suggested that alluvial plains were constructed by a series of autocyclic meander-belt avulsions. Beaumont strata were correlated with the 'Sangamon' interglacial period (now marine Oxygen Isotope Stage 5). Blum & Price (1998) later showed that the Beaumont alluvial plain of the Colorado

River represents three distinct valley-fill complexes (Fig. 2), which span a much longer time interval than recognized previously (Oxygen Isotope Stages 9–5). Only the youngest valley-fill complex of the Beaumont Formation (Bay City, Fig. 2) represents the Oxygen Isotope Stage 6 to 5 glacial–interglacial cycle.

Post-Beaumont valley fills of rivers such as the Colorado, Trinity and Nueces accumulated during the last glacial cycle to the present (Oxygen Isotope Stages 4 through to 1) (Figs 3 & 4). The valleys formed as sea-level fell below Oxygen Isotope Stage 5 interglacial positions, which caused the rivers to incise and abandon Beaumont alluvial plains and extend across the subaerially exposed shelf to mid-shelf or shelf-edge positions (see Suter & Berryhill, 1985; Suter, 1987; Morton & Price, 1987; Anderson *et al.*, 1992, 1996; Thomas, 1990;

SERIES   LITHOLOGY   STRATIGRAPHY

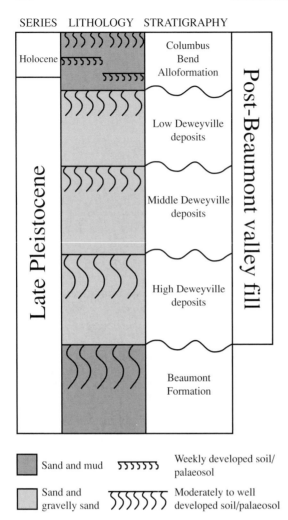

**Fig. 3.** Late Quaternary alluvial stratigraphy of the Colorado River in south Texas.

Thomas & Anderson, 1994; Morton & Suter, 1996). Basal sediments within the incised valleys were described initially by Bernard (1950) as the 'Deweyville' terraces (see Fig. 4), and these deposits are present within all the major river valleys of the Texas Gulf Coastal Plain. The terraces are at elevations intermediate between older Beaumont surfaces and Holocene floodplains. In rivers with large sediment supplies such as the Colorado, Holocene deposits bury Deweyville strata (Fig. 4A), whereas in small-sediment-load rivers, such as the Trinity and Nueces, only low Deweyville and in some instances, middle Deweyville surfaces are buried (Fig. 4B & C).

Although Bernard (1950) suggested that 'Deweyville' deposits represent valley filling during the initial stages of the post-glacial transgression, recent study by Blum *et al.* (1995) and Durbin *et al.* (1997) shows that 'Deweyville' strata accumulated during the falling stage and lowstand of Oxygen Isotope Stages 4, 3 and 2 (approximately 70 000–20 000 yr BP). Deweyville sediments consist of gravelly channel sands, 10 m or more in thickness and muddy lenticular channel fills. The upper boundary of each Deweyville unit is represented by a moderately to well-developed palaeosol. The gravelly sands and palaeosols represent periods of continuous lateral channel migration, sediment reworking and minor vertical aggradation, followed by renewed valley incision and terrace formation.

Post-glacial sea-level rise led to shoreline transgression, valley aggradation and, in some rivers (e.g., the Colorado River), burial of Deweyville deposits and palaeosols by Late Pleistocene and Holocene sediments. The maximum thickness of post-Deweyville strata in the Colorado River valley is at least 25 m, and stratigraphically these sediments are the Columbus Bend Alloformation (Blum & Valastro, 1994) (Fig. 3). Radiocarbon ages indicate that Columbus Bend sediments accumulated between 13 000 yr BP and the present (Blum & Valastro, 1994), and these deposits extend upward to the modern floodplain surface. The Holocene Colorado River alluvial plain is 20–30 km wide along strike and extends west to the Beaumont alluvial plain and east to the Brazos River (Fig. 5). The alluvial plain consists of multiple channel belts, separated by broad flood basins, with 2–3 m of relief between the tops of natural levees and flood-basin depressions (see McGowen *et al.*, 1976; Fig. 2). Channel belts range between 1 and 3 km in width and contain sandy point bars, flanked by levees and crevasse splays, as well as sinuous active and abandoned channels. Flood basins contain small floodplain streams and the deposits consist primarily of mud. The Colorado River avulsed from the main portion of the post-Beaumont alluvial valley near Wharton, Texas and presently flows south across the Beaumont Formation to the Gulf of Mexico (Blum & Valastro, 1994) (Fig. 2). Radiocarbon ages of $390 \pm 50$ yr BP (Tx-8566), $230 \pm 30$ yr BP (Tx-8565) and $200 \pm 50$ yr BP (Tx-8567) from tree stumps in growth position, which are buried by post-avulsion sediments, indicate this event occurred during the very late prehistoric to early historic time period (M.D. Blum, unpublished data).

In contrast to the Colorado River, Holocene floodplains of the Trinity and Nueces rivers do not have prominent alluvial ridges and the small, highly sinuous

## A  Colorado River

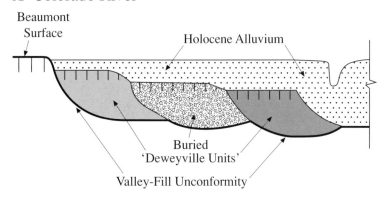

## B  Nueces River

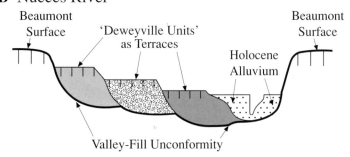

**Fig. 4.** Schematic valley cross-sections showing stratigraphical relationships among post-Beaumont deposits of the Colorado, Nueces and Trinity rivers (from Blum *et al.*, 1995). (A) All 'Deweyville' surfaces are onlapped and buried by Holocene strata in the Colorado River post-Beaumont valley fill, and modern Colorado River deposits represent the latest stage of valley filling. (B) The steep-gradient basin-fringe Nueces River represents an early stage of valley filling. Owing to steeper floodplain gradients, the low Deweyville surface is a terrace upstream of the modern bay-head delta. (C) The Trinity River is a low-gradient basin-fringe stream that also represents an early stage of valley filling. High and middle Deweyville surfaces are terraces, and low Deweyville surfaces are onlapped and buried by Holocene floodplain strata. 'Deweyville' allostratigraphical units are shown occurring on one side of the valley for illustration purposes only. Vertical lines represent soils and palaeosols. Relative scale of valley-fill sequences as indicated.

## C  Trinity River

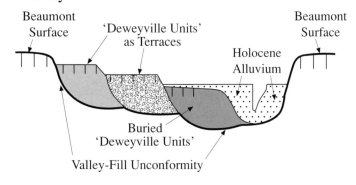

channels flow within abandoned Deweyville channel courses (Fig. 6). Post-Deweyville strata of the Trinity and Nueces rivers are presently unnamed, but probably correlate with the Columbus Bend Alloformation of the Colorado River. In the lowermost reaches of the Trinity River, these deposits form a thin blanket of sediment that rests unconformably on Deweyville deposits (Aten, 1983) (Figs 4C & 6).

## METHODS

A combination of aerial photographs, satellite imagery, floodplain cores and stratigraphical sections from river cutbanks were used to study Holocene avulsion deposits. Fourteen floodplain borings were made in the Colorado River Valley using a truck-mounted rotary drill rig, and

0            **10 km**

**Fig. 5.** LANDSAT Thematic Mapper image showing the Holocene Colorado River alluvial plain and the location of the transect line and core-sampling sites for the cross-section shown in Fig. 10. The image is a grey-scale print of Thematic Mapper bands 2, 3 and 4.

most cores were located along a cross-valley transect that extends from the western valley wall of the Holocene alluvial plain, east to the San Bernard River, an older Colorado River channel belt (Fig. 5). Individual cores ranged from 12 to 27 m in length, with a 2 km spacing between cores. An additional 10 cores, up to 5 m in length, were acquired using a Giddings hydraulic soil probe, and the total length of all the cores is *c.* 300 m. Data from the Trinity and Nueces rivers are derived primarily from satellite imagery and channel-cutbank exposures. Cores and outcrops were described in the field or in the laboratory using standard nomenclature (e.g. Miall, 1996). Description of soils and soil features in cores and cutbank exposures followed the terminology of Birkeland (1984).

## AVULSION STYLES

Holocene deposits of the Colorado, Trinity and Nueces rivers show evidence for two distinct styles of avulsion. In some instances, the rivers avulse by simply reoccupying Holocene or Late Pleistocene abandoned channel courses, whereas in other cases, the rivers avulse by diversion from an elevated alluvial ridge into a flood-basin depression. Each avulsion style involves different processes and produces different types of deposits.

### Avulsion by channel reoccupation

The Trinity and Nueces rivers provide numerous examples of avulsion by reoccupation of Late Pleistocene Deweyville channel courses that were active during the falling stage and lowstand of the last glacial cycle. Deweyville palaeochannel widths and meander wavelengths are substantially greater than those of the modern Trinity and Nueces rivers and abandoned channels represent large depressions on the modern floodplains (Blum *et al.*, 1995) (Fig. 6). In the low-gradient Trinity valley, Holocene channels reoccupy both middle and low

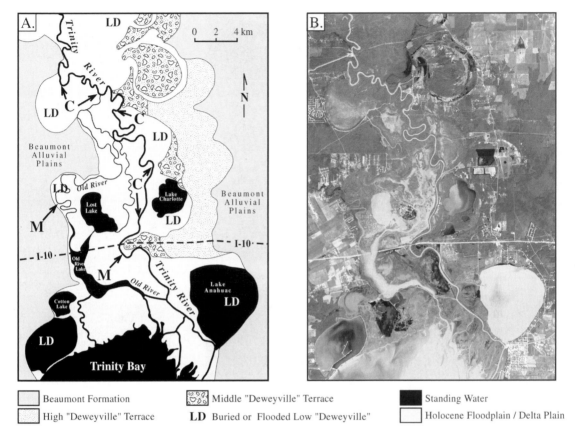

| | | | |
|---|---|---|---|
| Beaumont Formation | Middle "Deweyville" Terrace | Standing Water | |
| High "Deweyville" Terrace | **LD** Buried or Flooded Low "Deweyville" | Holocene Floodplain / Delta Plain | |

**Fig. 6.** (A) Geological map of the Lower Trinity River valley, showing the distribution of high and middle Deweyville terraces. Low Deweyville (LD) surfaces are buried locally and onlapped by Holocene deposits or flooded in the lowermost reaches. Holocene Trinity River channels locally reoccupy Deweyville meanders (M) and cutoff channels (C). See text for further discussion. (B) LANDSAT Thematic Mapper image of Trinity River valley, illustrating general valley morphology, the large meander scars characteristic of Deweyville surfaces, and the sinuous floodplain channels that developed during the present sea-level highstand.

Deweyville abandoned channel courses (Figs 6 & 7). By contrast, the steeper gradient Nueces River reoccupies low Deweyville channel courses only. Middle Deweyville palaeochannels are present on terrace surfaces that are several metres above the modern floodplain.

Holocene Trinity and Nueces River channels produce narrow (< 1 km wide) channel belts, lack broad levees, and do not form alluvial ridges. These observations suggest that the Trinity and Nueces channels are short-lived, probably because the abundance of Deweyville palaeochannels and erodible channel deposits lead to frequent avulsion. Furthermore, the reoccupation of abandoned Deweyville channel courses by Trinity and Nueces channels is discontinuous in the downstream direction. Holocene channels repeatedly avulse into and exit from Deweyville abandoned meanders and cutoff channels

and typically reoccupy the palaeochannels for distances of only 5–10 km before departing from the Deweyville courses (Fig. 6). Field relationships suggest that Holocene channels avulse by migrating laterally and intersecting a palaeochannel, or by overflowing into Deweyville channel courses during flood events. The abundance of unconsolidated coarse sand and gravel in the Deweyville deposits probably leads to rapid incision by Holocene channels and reoccupation.

Post-avulsion channel deposits are inset within and cross-cut Deweyville channel-fill muds and point-bar sands and gravels (Fig. 7) and consist of steeply dipping, interbedded fine sands and muds that are 8–12 m thick. The inclined interbedded sands and muds are lateral accretion deposits associated with point-bar migration and are overlain by up to 3 m of vertical accretion muds

**Fig. 7.** Photographs of Trinity River cutbanks located upstream from the area shown in Fig. 6, illustrating reoccupation of Deweyville channels during Holocene avulsions. The modern floodplain surface is marked by triangles. (A) Holocene reoccupation of a middle Deweyville abandoned channel is shown by muddy and sandy Late Holocene (LH) channel deposits that cross-cut middle Deweyville channel-fill (MD/C) and point-bar (MD/PB) facies. Note person (*c.* 2 m tall) for scale. (B) Muddy and sandy Late Holocene (LH) channel deposits fill a low Deweyville (LD) abandoned channel, and the entire sequence is overlain by modern (M) overbank deposits. A well-developed palaeosol in low Deweyville deposits dips to the left along the unconformity shown by the dashed line, and late Holocene deposits onlap this unconformity. The boundary between late Holocene and modern sediments is represented by a weakly developed palaeosol (dashed line above LH). The top of the cutbank is approximately 5 m above the water surface.

and fine sands. In many examples, the vertical accretion deposits form thin (< 2 m thick) veneers that bury palaeosols developed in uppermost Deweyville strata.

Satellite imagery and field relationships indicate that the most recent avulsion of the Colorado River was also accomplished by channel reoccupation. In this example, the Colorado River reoccupied a Pleistocene Beaumont channel belt that was active during the Oxygen Isotope Stage 5 highstand (Blum & Price, 1998). During this event, the Colorado River abandoned its alluvial valley near Wharton, Texas and followed a slightly shorter route to the Gulf of Mexico (Fig. 8). Palaeomeanders on the Beaumont surface are part of the Pleistocene channel belt that extends *c.* 1 km east and west of the modern channel.

The north–south distribution of the palaeomeanders demonstrates that the modern channel generally coincides with the location of the Late Pleistocene Beaumont channel belt. Cutbank exposures along the Colorado River show that the palaeomeanders are underlain by black muddy channel-fill deposits with a well-developed vertisol and channel-bar sands that dip towards the present-day channel axis (Fig. 9). Lateral tracing of strata in the field demonstrates that the highest elevation of channel-fill mud is typically 5 m lower than the top of correlative channel-bar sand. These observations indicate that prior to avulsion, this area was a trough-shaped depression on the Beaumont surface and the depression, along with the presence of erodible sands, led to the avulsion.

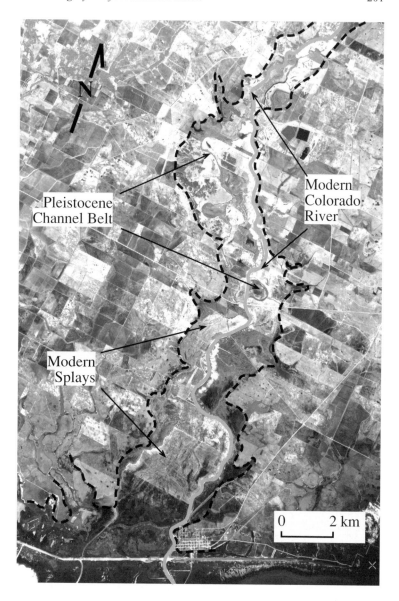

**Fig. 8.** Black-and-white print of NASA high-altitude colour-infrared aerial photograph showing the modern Colorado River channel reoccupying a Pleistocene Beaumont channel belt. The black dashed lines define the western and eastern boundaries of the Pleistocene channel belt, and the Colorado River avulsed by reoccupying this channel belt. Note the modern crevasse splays along the lower reaches of the Colorado River, which locally bury the Pleistocene channel belt. Flow is towards the base of the photograph. See text for further discussion.

The scarcity of Holocene sediments along the modern channel indicates that the avulsion was accomplished largely by channel enlargement and sediment bypass. Post-avulsion deposits and landforms consist of small sandy point bars and several metres of overbank fine sands and muds inset against Beaumont strata. Although point-bar sediments are locally 10–12 m thick, these deposits are volumetrically insignificant compared with older Colorado River deposits elsewhere. Erosion and sediment bypassing near the avulsion node is also demonstrated by the development downstream of crevasse splays and a delta at the mouth of the Colorado River near Matagorda, Texas (Figs 2 & 8). The delta prograded south across East Matagorda Bay during the early part of this century and is now attached to the Holocene transgressive barrier island complex (Wadsworth, 1966; Kanes, 1970). This avulsion and episode of delta progradation is similar to the historic development of the Atchafalaya River delta in Louisiana (Fisk, 1952; Tye & Coleman, 1989; Roberts & Coleman, 1996).

**Fig. 9.** Photographs showing Pleistocene Beaumont deposits exposed along the banks of the Colorado River north of Bay City, Texas. The photographs show that the Colorado River avulsed by reoccupying a Pleistocene Beaumont palaeochannel. For scale, each cutbank is a maximum of 12–13 m high. (A) Pleistocene Beaumont channel fills are locally incised by the modern Colorado River and buried by overbank deposits. (B) View showing Pleistocene channel-fill deposits along both banks of the modern-day channel with inset modern point bars. Note the Beaumont channel fills dip gently towards the axis of the modern channel.

**Avulsion by diversion into flood basins**

Core data show that the Colorado River has also avulsed repeatedly by diversion into flood basins, and suggest that avulsion plays a major role in valley filling. Sediments that accumulate during the diversion (i.e. avulsion deposits) are intercalated with pedogenically modified overbank deposits and channel-belt sands, which indicate that individual episodes of avulsive deposition are followed by the establishment of a new channel belt and soil formation (Fig. 4A). Furthermore, vertical changes in sediment types, sand-body geometries and the abundance of soil features are used to subdivide the valley fill

into three stratigraphical packages. Collectively, these deposits record changes in avulsion frequencies and processes during Late Pleistocene through to Holocene valley filling (Fig. 10).

The oldest stratigraphical package is at least 5 m thick, occupies the base of the valley fill (see Fig. 10, lower unit, cores 1 and 9), and consists of reddish-brown flood-basin muds with common slickensides, grey root mottles and calcite nodules (Fig. 11A). The overlying stratigraphical package is 10–15 m thick (see Fig. 10, middle unit) and consists of reddish-brown flood-basin muds that encase crevasse-splay sands (Fig. 12). Muds are laminated or massive and contain few calcite nodules and grey root

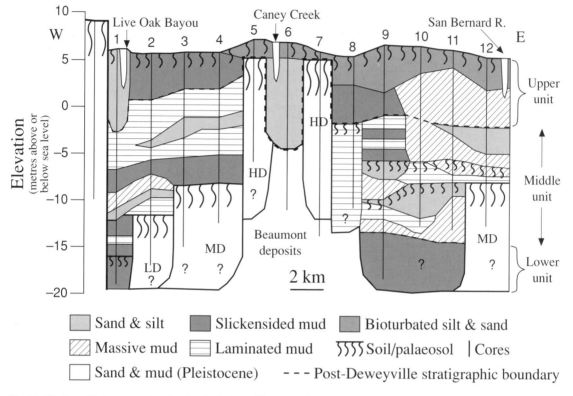

**Fig. 10.** Stratigraphical cross-section showing the facies architecture of the post-Beaumont Colorado River valley fill southeast of Bay City, Texas. The transect line for this cross-section is shown in Fig. 5. Note that the post-Deweyville valley-fill sediments are subdivided into lower, middle and upper units. The three post-Deweyville units are not, however, equivalent stratigraphically to the allostratigraphical units of the Columbus Bend Alloformation (Blum & Valastro, 1994). HD, high Deweyville deposits; MD, middle Deweyville deposits; LD, low Deweyville deposits.

**Fig. 11.** Core photographs showing representative sedimentological and pedological features of the post-Deweyville Colorado River valley fill. All elevation values are depths below the alluvial plain surface. (A) Reddish-brown flood-basin mud with slickensides. Three large slickensided fractures that dip from right to left are shown (white arrows). Core 4, 12.15–12.50 m. (B) Reddish-brown massive flood-basin mud (lower arrow) overlain by reddish silt and sand (upper arrow). Core 11, 7.25–7.60 m. (C) Ripple cross-stratified crevasse-splay sand and silt with sharp lower boundary (arrow) overlying reddish-brown massive flood-basin mud. Core 11, 8.85–9.20 m. (D) Black organic mud (deposits above the arrow) interpreted as a buried-soil A horizon developed in marsh deposits. The organic mud overlies laminated reddish-brown clay and silt. Core 14, 8.5–8.8 m. (E) Blocky soil (ped) structure and white calcite nodules (black arrows) in the Btk horizon of a weakly developed palaeosol in natural-levee sands and silts. Core 9, 1.1–1.4 m.

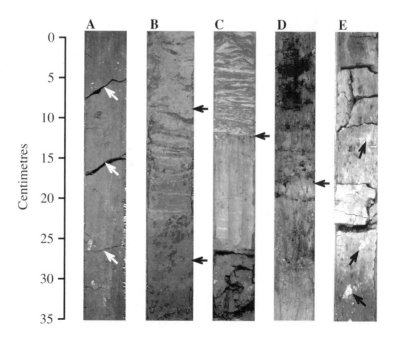

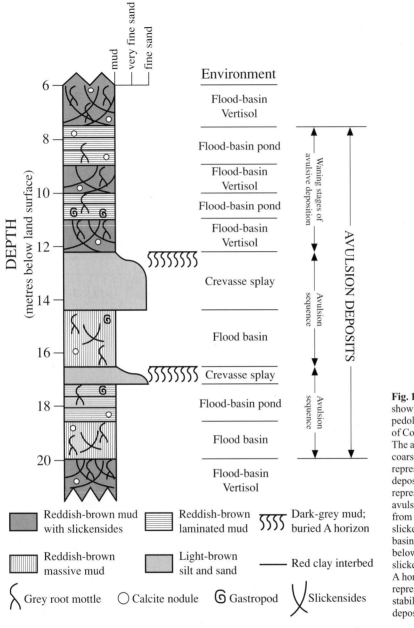

**Fig. 12.** Stratigraphical section showing representative lithofacies, pedologic features and environments of Colorado River avulsion deposits. The avulsion deposits consist of two coarsening-upward sequences that represent discrete episodes of avulsive deposition, and a third sequence representing the waning stages of avulsive sedimentation. Data are from Core 9 in Fig. 10. Note that slickensided muds representing flood-basin vertisols are present above and below the avulsion deposits. The slickensided muds and the buried-soil A horizons within the avulsion deposits represent short periods of floodplain stability between episodes of avulsive deposition.

mottles, and the root mottles are distributed uniformly (Fig. 11B). Laminated muds also contain a few 5–15-cm-thick zones with common terrestrial gastropod shells. Crevasse-splay sands typically consist of sheets 1–2 m thick and lenticular channel shapes 3–4 m thick with sharp lower boundaries. The sands are very fine to fine grained and massive, weakly laminated, or ripple cross-stratified (Fig. 11C). Individual crevasse splays can be

traced laterally in the subsurface for up to 4 km (Fig. 10). The middle stratigraphical unit also contains pedogenically modified deposits including several thin (< 2 m thick) zones of slickensided flood-basin muds that separate packages of massive or laminated flood-basin muds and crevasse-splay sands (Figs 10 & 12). The eastern half of the valley fill contains two laterally persistent units of black, organic-rich mud, which represent buried organic

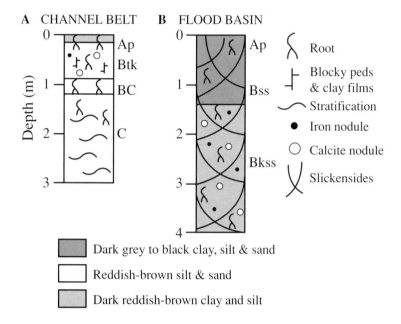

**Fig. 13.** Schematic profiles of typical Holocene Colorado River alluvial soils sampled using a hydraulic soil probe. The soils are associated with the Live Oak Bayou channel belt (see Fig. 10). (A) Soil developed in channel-belt silt and sand. (B) Soil developed in flood-basin clay and silt. Ap, ploughed A horizon; Btk, B horizon with pedogenic clay and carbonate accumulations; Bss, B horizon with pedogenic slickensides; Bkss, B horizon with pedogenic slickensides and carbonate accumulations; BC, BC horizon; C, C horizon.

horizons (A horizons) of weakly developed soils that formed in flood-basin marshes (Figs 10 & 11D). The middle unit also contains a 10-m-thick accumulation of laterally restricted laminated and interbedded muds and sands that are interpreted as a channel fill (see Fig. 10, core 8).

The youngest stratigraphical package (see Fig. 10, upper unit) is 5–10 m thick and extends upward to the surface of the recently abandoned floodplain. Deposits consist of channel-belt sands, natural-levee silts and sands, and flood-basin muds with many slickensides, roots, calcite nodules and blocky ped structures (Fig. 11E). Soil features are more abundant in these deposits than in the underlying sediments, and soil profiles represent a combination of inceptisols, entisols and vertic mollisols (USDA, 1974; Fig. 13). The presence of (i) clay films along ped faces of sandy and silty B horizons of channel-belt and natural-levee soils, and (ii) slickensides in the flood-basin soils provide evidence of pedogenic processes such as clay illuviation and seasonal soil shrinking and swelling.

The abundance of mud, the presence of thin isolated sand bodies, and the scarcity of soil features suggest that the middle package of the post-Deweyville valley fill are avulsion deposits (Figs 10 & 12). These deposits represent a time of rapid floodplain aggradation, probably associated with prograding crevasse-splay complexes. Slickensided flood-basin muds and buried-soil A horizons that separate the 1–2-m-thick packages of massive or laminated flood-basin muds and crevasse-splay sands represent short periods of floodplain stability that occurred between avulsions. In contrast to the avulsion deposits, the greater abundance of soil features in the lower and upper units of the post-Deweyville valley fill, plus the channel-belt sands in the upper valley-fill unit, indicate that these deposits represent periods of slower floodplain aggradation accompanied by lateral channel migration and overbank deposition.

Sedimentological and stratigraphical similarities between Colorado River sediments and avulsion deposits of the Saskatchewan River in Canada (Smith *et al.*, 1989; Smith & Perez-Arlucea, 1994) further support the idea that a significant proportion of the post-Deweyville valley fill consists of avulsion deposits. Saskatchewan River avulsion deposits overlie peat and consist commonly of 1–3 m of mud that encases thin sand bodies. Channels of new alluvial ridges incise avulsion deposits, and overbank muds veneer avulsion sediments. A significant aspect of the Saskatchewan River avulsion study is the recognition that the width of the avulsion belt (*sensu* Smith *et al.*, 1989) is greater than the width of the younger channel belt. Thus the channel belt incises avulsion deposits locally, but a large percentage of the avulsion deposits are preserved despite the development of laterally mobile channels. These observations suggest that avulsion deposits may or may not be incised by channel-belt sands and explain why, in some instances, Colorado River avulsion deposits are incised by channel-belt sands

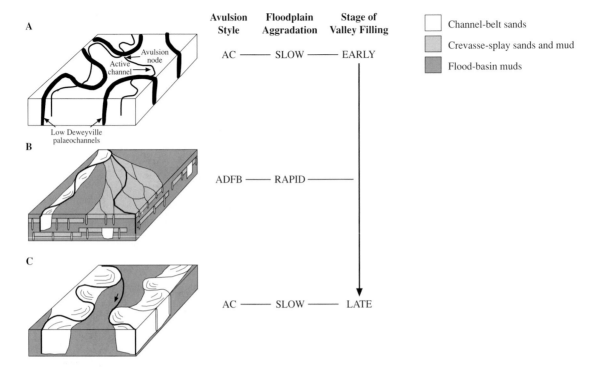

**Fig. 14.** Schematic block diagrams showing different styles of avulsion associated with different stages of valley filling and rates of floodplain aggradation. (A) Avulsion by channel reoccupation (AC) of Deweyville palaeochannels occurs during periods of slow aggradation and early stages of valley filling. (B) Avulsion by diversion into flood basins (ADFB) occurs during rapid aggradation and valley filling. (C) Avulsion by channel reoccupation (AC) of highstand channel belts occurs during periods of slow aggradation and late stages of valley filling.

whereas in others, avulsion deposits are simply bounded by slickensided flood-basin muds or buried-soil A horizons. These pedogenically modified deposits probably represent slow post-avulsion overbank sedimentation and soil formation.

## CHANGES IN AVULSION STYLE DURING VALLEY FILLING

Late Quaternary alluvial deposits of the Texas Gulf Coastal Plain show that differences in avulsion style are linked closely to rates of floodplain aggradation and different stages of valley filling (Fig. 14). In the examples described here, rates of sea-level rise and production of accommodation space essentially have been the same for all rivers. The rivers, however, represent different stages of valley filling owing to differences in sediment supply and rates of aggradation. These differences provide a basis for examining how avulsion style varies during early, intermediate and late stages of valley filling.

### Avulsion during early stages of valley filling

The Trinity and Nueces rivers show that during early stages of valley filling and periods of slow aggradation, rivers avulse by reoccupying abandoned channel courses (Fig. 14A). Because these rivers have smaller sediment loads than the Colorado River and unfilled accommodation space, the Trinity and Nueces rivers represent a stage of valley filling that the Colorado River experienced some time after the Last Glacial Maximum (*c.* 20 ka) and prior to the Holocene. As these rivers continue to avulse by channel reoccupation they will produce multilateral and multistorey sheet sands that will extend across the entire valley width (5–10 km). This channel activity will continue until overbank muds bury Deweyville palaeochannels sufficiently so that they no longer cause channel reoccupation.

Differences among the floodplain and terrace gradients between the Trinity and Nueces rivers suggest that the Trinity River may be near a threshold of floodplain aggradation that will lead to a new style of avulsion. The gradient of the low Deweyville surface and the modern Trinity River floodplain are approximately the same (Blum *et al.*, 1995) and the river is slowly burying low Deweyville palaeochannels as it fills accommodation space produced during the Holocene transgression. Once low Deweyville channels are buried and no longer lead to frequent Trinity avulsions, the river will begin to divert into flood-basin depressions. In contrast to the Trinity River, the gradient of the low Deweyville terrace in the Nueces River valley is greater than that of the modern floodplain (Blum *et al.*, 1995). The steeper terrace gradient will continue to favour avulsion by channel reoccupation because it will take longer to bury Deweyville palaeochannels and eliminate their influence on Nueces channels.

**Avulsion during rapid valley filling**

Colorado River avulsion by diversion into flood-basin depressions accompanied rapid floodplain aggradation and valley filling during the Holocene transgression (Fig. 14B). The scarcity of soil features and the absence of Holocene estuarine facies in the post-Deweyville fill (see Fig. 10) suggests that accommodation space was filled as rapidly as it was generated. The avulsion deposits consist of mud and isolated sand bodies and represent a significant proportion of the entire post-Deweyville fill (Fig. 10). The thickness and lateral continuity of the avulsion deposits further suggests that rapid aggradation and valley filling was accomplished by continuous avulsive deposition. This phrase does not mean that discrete avulsions (i.e. individual breakout events) were continuous. Instead, we suggest that avulsive deposition occurred on some portion of the Colorado River alluvial plain for the majority of time that the valley was filling. Rapid aggradation caused channel instability and frequent avulsions, which resulted in short-lived channel belts and inhibited the development of well-defined alluvial ridges. Natural-levee deposits in the valley fill also provide evidence for continuous avulsion. Natural levee silts and sands form a 3–4-m-thick sheet adjacent to the Colorado River channel belts near the alluvial plain surface (Fig. 10). Widespread natural levee deposits, however, are not recognized at greater depths as would be expected if channel belt locations had remained constant as the alluvial plain aggraded.

Studies of avulsion elsewhere support the suggestion that entire floodplains aggrade in instances where avulsive deposition occurs continuously. For instance, the Saskatchewan River avulsion (Smith *et al.*, 1989) shows that individual avulsions deposit sediment only over a portion of an entire floodplain. For an *entire* floodplain to aggrade for thousands of years, as in the case of the Colorado River, avulsive deposition must occur more or less continuously. This interpretation is supported by field and experimental studies by Törnqvist (1994) and Bryant *et al.* (1995) who show rapid floodplain aggradation causes high avulsion frequencies. Frequent avulsion by diversion into flood basins would, in turn, favour continuous avulsive deposition.

**Avulsion during late stages of valley filling**

During the late stages of valley filling, the Colorado River has avulsed by channel reoccupation (Fig. 14C). The change from avulsion by diversion into flood basins, to avulsion by channel reoccupation accompanied the establishment of sea-level highstand conditions 5–6 ka ago' and decreasing rates of floodplain aggradation. In addition to the most recent Colorado River avulsion, the previous avulsion also occurred by channel reoccupation. This avulsion produced the Caney Creek channel belt of the Colorado River (see Fig. 5), which was abandoned during the late prehistoric period. Stratigraphical relationships show that the Caney Creek channel belt incised high Deweyville channel sands and a well-developed palaeosol that are present at shallow depths beneath the modern alluvial plain surface (Fig. 10, cores 5–7). The presence of Deweyville palaeochannels and erodible channel sands probably facilitated this avulsion and led to development of the Caney Creek channel belt. Similar to early stages of valley filling, avulsion by channel reoccupation during late stages of valley filling can also produce multilateral and multistorey sheet sands.

The close correlation between the location of the Caney Creek channel belt and the high Deweyville channel sands further suggests that the presence of erodible sands is an important factor that contributes to the success of avulsions by channel reoccupation. Whether or not this factor is more important during the late stages of valley filling than at other times is unclear. If, however, low floodplain gradients and slow aggradation during sea-level highstands inhibit avulsion, as suggested by field and experimental studies (Törnqvist, 1994; Bryant *et al.*, 1995), then perhaps the presence or absence of erodible strata does play a more important role in successful avulsions during late stages of valley filling compared with other times.

**SUMMARY**

Differences in the Holocene avulsion histories of the Colorado, Trinity and Nueces rivers illustrate both the complexity and significance of avulsion during valley filling and alluvial plain construction on the Texas Gulf Coastal Plain. Avulsion by channel reoccupation is accomplished primarily by the enlargement of older channels, and avulsion deposits are rare, at least in the vicinity of the avulsion node. Sediment reworking produces multilateral and multistorey sheet sands and sediment bypassing leads to the development of crevasse splays, bay-head deltas and coastal deltas. The Texas rivers suggest that this style of avulsion is most prevalent during early and late stages of valley filling when rates of aggradation are low, owing to low sediment supply (e.g. the Trinity and Nueces Rivers) or limited accommodation space (e.g. the Colorado River). By contrast, avulsion by diversion into flood basins deposits large volumes of floodplain muds and lesser amounts of sand, and occurs during periods of rapid accommodation-space production and valley filling. Based on an estimated thickness of 10–15 m for the avulsion deposits in the Colorado River post-Beaumont valley fill, avulsion sediments represent a significant proportion (*c.* 50%) of the valley fill.

Lastly, avulsion also controls the location of future incised valleys and the preservation of valley-fill deposits. For instance, the Colorado River avulsed from the main portion of the post-Beaumont alluvial valley as valley filling neared completion (see Fig. 2). If sea-level were to fall and valley incision begin, it is likely the post-Beaumont valley fill would be mostly preserved. Furthermore, the presence of several older Pleistocene Colorado River valley fills suggests that this process has occurred previously, probably during late stages of valley filling and sea-level highstands (see Blum & Price, 1998).

## ACKNOWLEDGEMENTS

We thank Norm Smith and Mary Kraus for organizing the Avulsion Symposium at the 6th International Conference on Fluvial Sedimentology held in Cape Town. We also thank Marta Pérez-Arlucea, Galina Morozova and Norm Smith for reviews and comments that significantly improved this manuscript. This research was supported by the Petroleum Research Fund of the American Chemical Society, the National Science Foundation, and a consortium of industry associates, with participants including Amoco Production Co., Conoco Inc., Exxon Production Research, Mobil Research and Development, and Union Pacific Resources.

## REFERENCES

ANDERSON, J.B., THOMAS, M.A., SIRINGIN, F.P. & SMYTH, W.C. (1992) Quaternary evolution of the Texas Coast and Shelf. In: *Quaternary Coasts of the United States: Marine and Lacustrine Systems* (Eds Fletcher, III, C.H. & Wehmiller, J.F.), pp. 253–265. Spec. Publ. Soc. econ. Palaent. Miner., Tulsa, **48**, 253–256.

ANDERSON, J.B., ABDULAH, K., SARZALEJO, S., SIRINGIN, F. & THOMAS, M.A. (1996) Late Quaternary sedimentation and high-resolution sequence stratigraphy of the East Texas Shelf. In: *Geology of Siliciclastic Shelf Seas* (Eds De Batist, M. & Jacobs, P.), Spec. Publ. geol. Soc. London, No. 117, pp. 95–124, Geological Society of London, Bath.

ARONOW, S. (1971) Nueces River delta plain of the Pleistocene Beaumont Formation. *Bull. Am. Assoc. petrol. Geol.,* **55**, 1231–1248.

ATEN, L.E. (1983) *Indians of the Upper Texas Coast.* Academic Press, New York.

BERNARD, H.A. (1950) *Quaternary geology of southeast Texas.* Unpublished PhD dissertation, Louisiana State University, Baton Rouge, LA, 240 pp.

BERNARD, H.A. & LEBLANC, R.J. (1965) Resume of the Quaternary geology of the northwestern Gulf of Mexico Province. In: *The Quaternary of the United States* (Eds Wright, H.E. & Frey, D.G.), pp. 137–185. Princeton University Press, Princeton, NJ.

BERNARD, H.A., MAJOR, C.F., PARROTT, B.S. & LEBLANC, R.J. SR. (1970) *Recent Sediments of Southeast Texas: Field Guide to the Brazos Alluvial Deltaic Plain and Galveston Barrier Island Complex,* Bureau of Economic Geology, The University of Texas at Austin, TX.

BIRKELAND, P.W. (1984) *Soils and Geomorphology.* Oxford University Press, Oxford.

BLUM, M.D. (1994) Genesis and architecture of incised valley fill sequences: a Late Quaternary example from the Colorado River, Gulf Coastal Plain of Texas. In: *Siliciclastic Sequence Stratigraphy: Recent Developments and Applications* (Eds Weimer, P. & Posamentier, H.W.), Mem. Am. Assoc. petrol. Geol., Tulsa, **58**, 259–283.

BLUM, M.D. & PRICE, D.M. (1998) Quaternary alluvial plain construction in response to interacting glacio-eustatic and climatic controls, Texas Gulf Coastal Plain. In: *Relative Role of Eustasy, Climate, and Tectonism in Continental Rocks* (Eds Shanley, K. & McCabe, P.), Spec. Publ. Soc. econ. Paleont. Miner., Tulsa, **59**, 31–48.

BLUM, M.D. & VALASTRO, S. JR. (1994) Late Quaternary sedimentation, Lower Colorado River, Gulf Coastal Plain of Texas. *Geol. Soc. Am. Bull.,* **106**, 1002–1016.

BLUM, M.D., MORTON, R.A. & DURBIN, J.M. (1995) 'Deweyville' terraces and deposits of the Texas Gulf Coastal Plain. *Trans. Gulf Coast Assoc. geol. Soc.,* **45**, 53–60.

BRIZGA, S.O. & FINLAYSON, B.L. (1990) Channel avulsion and river metamorphosis: the case of the Thomson River, Victoria, Australia. *Earth Surf. Process. Landf.,* **15**, 391–404.

BROWN, L.F., BREWTON, J.L., MCGOWEN, J.H., EVANS, T.J., FISHER, W.L. & GROAT, C.G. (1976) *Environmental Geologic Atlas of the Texas Coastal Zone: Corpus Christi Area.* Bureau of Economic Geology, The University of Texas at Austin, TX.

BRYANT, M., FALK, P. & PAOLA, C. (1995) Experimental study of avulsion frequency and rate of deposition. *Geology*, **23**, 365–368.

DALRYMPLE, R.W., ZAITLIN, B.A. & BOYD, R. (1992) Estuarine facies models: conceptual basin and stratigraphic implications. *J. sediment. Petrol.*, **62**, 1130–1146.

DOERING, J.A. (1956) Review of Quaternary surface formations of the Gulf Coast Region. *Bull. Am. Assoc. petrol. Geol.*, **40**, 1816–1862.

DUBAR, J.R., EWING, T.E., LUNDELIUS, E.L., OTVOS, E.G. & WINDER, C.D. (1991) Quaternary geology of the Gulf of Mexico Coastal Plain. In: *Quaternary Non-Glacial Geology of the Conterminous United States* (Ed. Morrison, R.B.), pp. 583–610. Geology of North America Volume K-2, Geological Society of America, Boulder, CO.

DURBIN, J.M., BLUM, M.D. & PRICE, D.M. (1997) Late Pleistocene stratigraphy of the Nueces River, Corpus Christi, Texas: climatic and glacio-eustatic control on valley fill architecture. *Trans. Gulf Coast Assoc. geol. Soc.*, **47**, 119–130.

FISHER, W.L., MCGOWEN, J.H., BROWN, L.F. & GROAT, C.G. (1972) *Environmental Geologic Atlas of the Texas Coastal Zone: Galveston–Houston Area*. Bureau of Economic Geology, The University of Texas at Austin, TX.

FISK, H.N. (1944) *Geological Investigations of the Alluvial Valley of the Lower Mississippi River*. U.S. Army Corps of Engineers, Mississippi River Commission, Vicksburg, MS.

FISK, H.N. (1947) *Fine-grained Alluvial Deposits and their Effect on Mississippi River Activity*. U.S. Army Corps of Engineers, Mississippi River Commission, Vicksburg, MS, 78 pp.

FISK, H.N. (1952) *Geological Investigation of the Atchafalaya Basin and the Problem of the Mississippi River Diversion*. U.S. Army Corps of Engineers, Waterways Experiment Station, Vicksburg, MS, 82 pp.

GALLOWAY, W.E. (1981) Depositional architecture of Cenozoic Gulf Coastal Plain fluvial systems In: *Recent and Ancient Non-Marine Depositional Environments* (Eds Ethridge, F.G. & Flores, R.M.), Spec. Publ. Soc. econ. Paleont. Miner., Tulsa, **31**, 127–156.

KANES, W.H. (1970) Facies and development of the Colorado River delta in Texas. In: *Deltaic Sedimentation: Modern and Ancient* (Eds Morgan, J.P. & Shaver, R.H.), Spec. Publ. Soc. econ. Paleont, Miner., Tulsa, **15**, 78–106.

KRAUS, M.J. (1996) Avulsion deposits in Lower Eocene alluvial rocks, Bighorn Basin, Wyoming. *J. sediment. Res.*, **66**, 354–363.

KRAUS, M.J. & ASLAN, A (1993) Eocene hydromorphic paleosols: significance for interpreting ancient floodplain processes. *J. sediment. Petrol.*, **63**, 453–463.

MCGOWEN, J.H. & GARNER, L.E. (1970) Physiographic features and stratification types of coarse-grained point bars, modern and ancient examples. *Sedimentology*, **14**, 86–93.

MCGOWEN, J.H., BROWN, L.F., EVANS, T.J., FISHER, W.L. & GROAT, C.G. (1975) *Environmental Geologic Atlas of the Texas Coastal Zone: Bay City–Freeport Area*. Bureau of Economic Geology, The University of Texas at Austin, TX.

MCGOWEN, J.H., PROCTOR, C.V., BROWN, L.F., EVANS, T.J.,

FISHER, W.L. & GROAT, C.G. (1976) *Environmental Geologic Atlas of the Texas Coastal Zone: Port Lavaca Area*. Bureau of Economic Geology, The University of Texas at Austin, TX.

MIALL, A.D. (1996) *The Geology of Fluvial Deposits: Sedimentary Facies, Basin Analysis, and Petroleum Geology*. Springer-Verlag, New York.

MORTON, R.A. & PRICE, W.A. (1987) Late Quaternary sea-level fluctuations and sedimentary phases of the Texas Coastal Plain and Shelf. In: *Sea-level Fluctuations and Coastal Evolution* (Eds Nummedal, D. & Pilkey, O.H.), Spec. Publ. Soc. econ. Paleont. Miner., Tulsa, **41**, 181–198.

MORTON, R.A. & SUTER, J.R. (1996) Sequence stratigraphy and composition of Late Quaternary shelf margin deltas, northern Gulf of Mexico. *Bull. Am. Assoc. petrol. Geol.*, **80**, 505–530.

ROBERTS, H.H. & COLEMAN, J.M. (1996) Holocene evolution of the deltaic plain: a perspective — from Fisk to present. *Eng. Geol.*, **45**, 113–138.

SMITH, N.D. & PEREZ-ARLUCEA, M. (1994) Fine-grained splay deposition in the avulsion belt of the Lower Saskatchewan River, Canada. *J. sediment. Res.*, **B64**, 159–168.

SMITH, N.D., CROSS, T.A., DUFFICY, J.P. & CLOUGH, S.R. (1989) Anatomy of an avulsion. *Sedimentology*, **36**, 1–23.

SUTER, J.R. (1987) Fluvial systems. In: *Late Quaternary Facies and Structure, Northern Gulf of Mexico — Interpretations from Seismic Data* (Ed. Berryhill, H.L.), Am. Assoc. petrol. Geol. Stud. Geol., Tulsa, **23**, 81–129.

SUTER, J.R. & BERRYHILL, H.L. (1985) Late Quaternary shelf-margin deltas, Northwest Gulf of Mexico. *Bull. Am. Assoc. petrol. Geol.*, **69**, 77–91.

THOMAS, M.A. (1990) *The impact of long-term and short-term sea level change on the evolution of the Wisconsinan–Holocene Trinity/Sabine incised valley system, Texas Continental Shelf*. Unpublished PhD Dissertation, Rice University, Houston, TX.

THOMAS, M.A. & ANDERSON, J.B. (1994) Sea level controls on the facies architecture of the Trinity/Sabine incised valley system, Texas Continental Shelf. In: *Incised Valley Systems: Origins and Sedimentary Sequences* (Eds Dalrymple, R.M., Boyd, R. & Zaitlin, B.A.), Spec. Publ. Soc. econ. Paleont. Miner., Tulsa, **51**, 63–82.

TÖRNQVIST, T.T. (1994) Middle and late Holocene avulsion history of the River Rhine (Rhine–Meuse delta, Netherlands). *Geology*, **22**, 711–714.

TYE, R.S. & COLEMAN, J.M. (1989) Depositional processes and stratigraphy of fluvially dominated lacustrine deltas: Mississippi delta plain. *J. sediment. Petrol.*, **59**, 973–996.

USDA (1974) *Soil Survey of Wharton County, Texas*. U.S. Department of Agriculture, Soil Conservation Service, 43 pp.

WADSWORTH, A.H. (1966) Historical deltation of the Colorado River, Texas. In: *Deltas in Their Geologic Framework* (Eds Shirley, M.L. & Ragsdale, J.A.), pp. 99–105. Houston Geological Society, Houston, TX.

WELLS, N.A. & DORR, J.A. (1987) Shifting of the Kosi River, northern India. *Geology*, **15**, 204–207.

WINKER, C.D. (1979) *Late Pleistocene fluvial–deltaic deposition on the Texas coastal plain and shelf*. Unpublished MA thesis, University of Texas at Austin, TX, 187 pp.

*Spec. Publs int. Ass. Sediment.* (1999) **28**, 211–220

# Pemiscot Bayou, a large distributary of the Mississippi River and a possible failed avulsion

M. J. GUCCIONE*, M. F. BURFORD* and J. D. KENDALL†

*Geology Department, University of Arkansas, Fayetteville, Arkansas, USA (Email: guccione@comp.uark.edu); and
†Marathon Oil Company, Tyler, Texas, USA*

## ABSTRACT

Pemiscot Bayou, the largest of four major distributary systems in the northern Lower Mississippi Valley (LMV) captured approximately 25% of the Mississippi discharge 5000 yr ago. While active, the bayou developed a 1.5–5-km-wide meander belt and associated point bar, overbank and abandoned meander channel-fill deposits. Typical of distributaries, its size decreased downstream. At 27 km down-valley from the avulsion junction, the channel was 600 m wide and 12 m deep. At 58 km down-valley, the channel width was reduced to 200–300 m and 7–9 m depth. Overbank silt, deposited by sheetflow as a unit up to 3 m thick and 10 km wide, buried adjacent point-bars of fine-to-medium sand and backswamp clay. Thin-bedded fine-to-very-fine sand and silt, laminated silt and massive clay infilled the abandoned channel.

This meandering distributary system was active for approximately 2000–3000 yr, but never fully avulsed for four possible reasons:
1 the slope ratio (cross-valley gradient : down-valley gradient = 4.2) was too low;
2 the discharge ratio (relative magnitude of annual flood discharge) was too low;
3 the suspended-sediment load was too large to expand the distributary channel;
4 tectonic influence in the New Madrid seismic zone controlled development of the distributary.
These conditions were adequate to capture only 25% of the Mississippi discharge. The fate of Pemiscot Bayou as a failed avulsion was certain when the Mississippi River meander at the distributary node was cut off, but it is possible that the distributary ceased to function earlier. By 2200 yr ago the distributary channel was largely infilled. A small amount of additional channel-fill has accumulated in the past 2000 yr, the most recent of which is fluvial reworking of vented sand during seismic events. Today Pemiscot Bayou is inactive and is < 1% of its former size.

## INTRODUCTION

The present meandering channel pattern of the Mississippi River reflects the cessation of glacial meltwater discharge and delivery of large sediment loads into the alluvial valley by 10–11 ka (Guccione *et al.*, 1988; Royall *et al.*, 1991; Saucier, 1994). Since the river has meandered, the Mississippi River has changed its channel position by avulsion, producing five to six different meander belts in the alluvial valley (Fig. 1) (Saucier, 1994). These meander belts are found only in the central and southern portions of the Lower Mississippi Valley (LMV), whereas in the northern portion of the valley a single meander belt has been active for the past 10 000–11 000 yr. In contrast, numerous distributary channels are present throughout the length of the LMV, including the northern portion of the valley. These distributaries have diverted discharge into and through the backswamp. Some distributaries have developed into new meander belts, but others have become inactive.

This study focuses on the characterization, genesis and chronology of the largest Holocene distributary in the northern LMV in an effort to address possible reasons for its failure to avulse and form a new meander belt. This study also provides quantitative data to refine avulsion models (Mackey & Bridge, 1995; Paola & Mohrig, 1997; Slingerland & Smith, 1998).

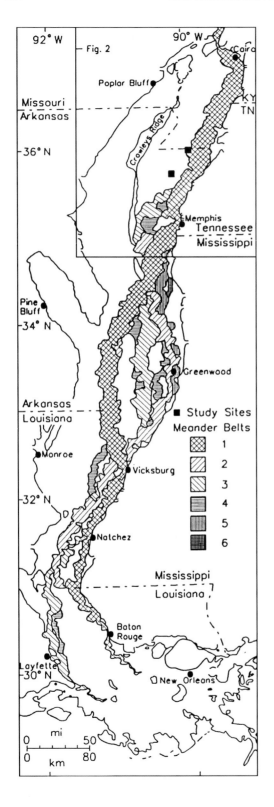

**Fig. 1.** Mississippi River meander belts. Note the single meander belt in the northern portion of the valley. Black rectangles denote the two study sites. (After Saucier, 1994.)

## AVULSIONS

An avulsion is a stage in the evolution of an aggrading meander belt. Typically, aggradation of natural levees along the channel is produced by both sheet flow and by crevassing (Farrell, 1987; Smith & Pérez-Arlucea, 1994). When a crevasse channel breaches the natural levee and extends into the adjacent lowland, it may capture enough flow from the main river to become a distributary channel. Although most of these distributaries subdivide and disperse flow into the backswamp, a few may capture a significant portion of the flow. If this captured flow exceeds 40% of the main river discharge (Fisk, 1952), an avulsion may occur. To capture this much flow, several conditions must be present. First, the meander belt must have aggraded as an alluvial ridge to form a steeper slope between the river surface and adjacent backswamp that results in a gradient advantage relative to the main channel (Smith & Smith, 1980; Mackey & Bridge, 1995; Slingerland & Smith, 1998). Second, the relative magnitude of flood discharge must be high enough that adequate discharge with a relatively high shear velocity is periodically diverted into the crevasse channel (Mackey & Bridge, 1995; Slingerland & Smith, 1998). Third, the concentration of suspended load must be less than the carrying capacity of the crevasse channel so that erosion will occur (Slingerland & Smith, 1998). Such a crevasse channel can enlarge to become a distributary channel and ultimately fully avulse into the beginning of another meander belt.

Where the crevasse channel gradient is insufficient for avulsion, the crevasse channel will continue to exist in a state of equilibrium or eventually heal, depending on the shear velocity of the major stream and the grain size of the suspended load. If the suspended load is coarse grained, the carrying capacity in both channels will be approximately equal and the channels can exist in a state of equilibrium (Slingerland & Smith, 1998). In contrast, where the suspended load is fine grained, the sediment-carrying capacity of the main channel will not be attained and alluviation will not occur along the main channel, but will occur in the crevasse channel allowing it to heal.

Aggradation along the Mississippi River and development of the maximum relief between the meander belt and backswamp has been greatest in the southern portion of the LMV (Aslan, 1994), where the river is responding

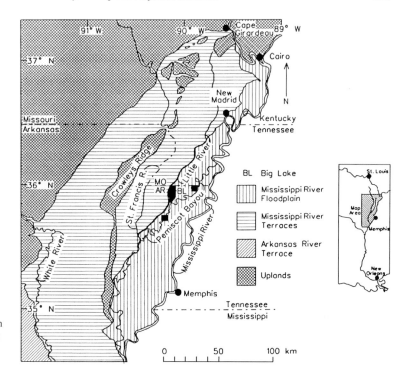

**Fig. 2.** Geomorphology of the northern Lower Mississippi River Valley. Black rectangles denote the two study sites. (After Saucier, 1994.)

to sea-level rise (Fisk, 1944; Saucier, 1994). This aggradation may be responsible for the corresponding frequency of avulsion and abundance of meander belts in the southern and central portions of the LMV. Conversely, the lack of both avulsions and multiple meander belts in the northern part of the LMV may be caused by inadequate gradient, caused by relatively low channel-aggradation rates (Bridge, 1997).

## STUDY SITE

The fluvial architecture of a meander belt is relatively well known, but the architecture of a distributary within the Mississippi River backswamp has not been documented in detail (Saucier, 1994) and is one focus of this study. The study site is Pemiscot Bayou, the youngest and largest of the four distributaries recognized in the northern LMV (Saucier, 1994). Enlargement of a crevasse splay within the natural levee of the Mississippi River in southeast Missouri is the probable origin of Pemiscot Bayou. From its nodal point in Missouri, the bayou flows south-west into Arkansas across the Mississippi River floodplain toward the tectonically subsided Big Lake basin within the backswamp (Fig. 2) (Guccione, 1993).

Here the valley trend changes to south-south-west, and the bayou flows along a distal backswamp to its convergence with the St Francis River and ultimately rejoins the Mississippi River. This downstream reach of the bayou, termed the Left Hand Chute of the Little River, is parallel and only a few kilometres east of an older meandering distributary, the Right Hand Chute of the Little River. For simplification the entire river is termed the Pemiscot Bayou in this paper. The Mississippi River never avulsed into the Pemiscot Bayou channel to create a new meander belt. Eventually the bayou ceased to function as an active distributary and today the channel is largely infilled.

## METHODS

Topographic maps, aerial photographs, undisturbed continuous sediment cores collected by the hydraulically powered Giddings Soil Sampling Machine and trenches were used to examine the geomorphology and sedimentology of Pemiscot Bayou at two locations (Fig. 2). At 27 km down-valley, sediment from 29 cores, three backhoe trenches and one borrow pit along two transects perpendicular to the channel (Kendall, 1997) was used to characterize the fluvial architecture and size of the

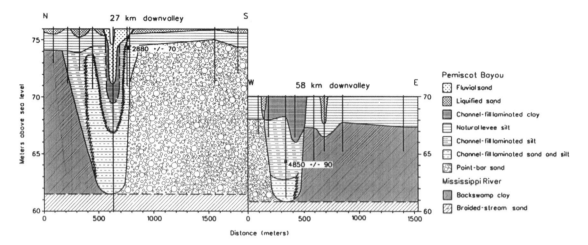

**Fig. 3.** Cross-section of Pemiscot Bayou channel at the two study sites shown in Figs 1 & 2. The upstream site (at 27 km) is the present channel. The downstream site (at 58 km) is an abandoned channel. Although the land surface is shown as horizontal, the transects may have up to 0.5 m of relief. Core locations and depths are shown as vertical lines. Locations of conventional radiocarbon dates that date Pemiscot Bayou are shown.

upstream portion of the channel. A second location, 58 km down-valley, also was used to characterize sediment, the fluvial architecture and to define the downstream channel size. Here, 10 trenches and 25 cores were taken along three transects. A transect perpendicular to an abandoned channel (Fig. 3) illustrates the channel size during an early phase of a distributary. A second transect crosses the present channel of Pemiscot Bayou and documents the channel infill at the time it ceased to be an active distributary. The third transect is parallel to the course of Pemiscot Bayou and was used to confirm the scale and fluvial architecture of the depositional units along the edge of the meander belt.

The exposures and cores were described using Munsell Soil Colour Charts in accordance with the U.S. Department of Agriculture Soil Survey Manual (Soil Survey Staff, 1981). Two hundred and sixty-five sediment samples were taken from selected cores and exposures. One sample was taken from each stratum that was thick enough to sample and multiple samples were taken from any stratum > 30 cm thick. Textural analysis was accomplished by dry sieving for gravel (> 2.0 mm) and five sand-size fractions (2.0–0.0625 mm). Three silt fractions (0.0625–0.002 mm) and one clay fraction (< 0.002 mm) were characterized by the pipette method (Day, 1965). Two textural designations, one based on geological criteria (Folk, 1968) and the other based on the U.S. Department of Agriculture Soil Survey Manual (Soil Survey Staff, 1981), were applied to these samples to

allow both pedogenic and geological determinations. Folk's (1968) textural classification is used in this report.

Gradients (Mackey & Bridge, 1995) of the LMV at the study area were measured using topographic maps with 5 ft (1.5 m) contour intervals. Pemiscot Bayou is no longer a continuous flowing channel as a result of infilling and channel diversion into drainage ditches, so its channel gradient cannot be determined along numerous sections of the bayou. As an estimate of water-surface altitude, the altitude of the infilled channels was used to calculate the bayou gradient. The cross-valley gradient was measured from the nodal point along an abandoned Mississippi River meander to the most distal position in the backswamp at Big Lake, where the bayou changes from a south-westerly orientation to a southern orientation (Fig. 2). To be consistent in measuring the gradient, the down-valley gradient was measured from the node to an infilled meander channel on the floodplain at the same latitude as the cross-valley measurement. The slope ratio is the cross-valley gradient : down-valley gradient.

Chronology of the deposits and landforms was established by radiocarbon dating of organic matter and by analysis of diagnostic human artifacts from archaeological sites occurring along Pemiscot Bayou.

## RESULTS

Pemiscot Bayou is a distributary with a meandering

channel pattern. In the study area, the cross-valley gradient along the bayou is 0.25 m km$^{-1}$, 4.2 times the 0.06 m km$^{-1}$ down-valley gradient of the Mississippi River. Because the bayou has a sinuosity of 2.54 (channel length : valley length), the river gradient (0.09 m km$^{-1}$) is much lower than the valley gradient and the gradient of the channel base is quite low (Fig. 3). Pemiscot Bayou exhibits all the landforms commonly associated with a meandering river system, including point bars, natural levees and abandoned meanders. There are no major distributaries between the node and the study site 27 km down-valley (Saucier, 1964), where the channel was 600 m wide and 12 m deep. Based on this measurement and interpretation of topographic maps, the distributary channel is estimated to have had a similar cross-sectional area at its node. The channel width at the node is 300 m, suggesting that the depth may be greater than that at the downstream location. Typical of distributaries, however, the channel size does decrease downstream. At 58 km down-valley and downstream of a channel bifurcation (Saucier, 1964), an abandoned meander of the bayou is only 200–300 m wide and 7–9 m deep.

Deposits associated with these meander-belt landforms have been characterized at both study locations (Fig. 3) (Table 1). Point-bar deposits of bedded fine- to medium-sand occur along the concave side of the channel. The natural levee of Pemiscot Bayou and its associated deposit changes down-valley from where the bayou crosses the Mississippi River meander belt to where it crosses the backswamp. Within the Mississippi River meander belt the distributary channel was deeply incised so that flooding of the distributary was relatively uncommon. As a result, Pemiscot Bayou natural levees are difficult to distinguish as a landform and the overbank deposit is probably derived from flooding of both the Mississippi River and the Pemiscot Bayou. Two metres of massive silt to clayey silt buries point-bar sand. In contrast, within the Mississippi River backswamp the distributary channel is less deeply incised and flooding of the backswamp was more common than upstream. As a result up to 3 m of interbedded to massive silt, mud and very fine sandy silt were deposited by sheetflow to form a natural levee of Pemiscot Bayou. The deposit is at least 10 km wide, burying point-bar sand of the bayou meander belt and adjacent Mississippi River backswamp clay.

The present channel is mostly infilled with interbeds of silt and sandy silt, with a few beds of fine sand (Fig. 3). This interbedded sequence grades upward to laminated silt. Today the channel is < 1% of its original cross-section. The abandoned channel, examined at 58 km down-valley is infilled with thinner bedded and finer grained sediment than that which fills most of the present

channel. It is comprised of uniformly dark grey silty clay to clay beds, some of which are laminated. The clay content varies, but generally increases upward. There is no soil development or bioturbation within the sediment, indicating that the sedimentation rate was more rapid than any mixing processes and that the channel continued to receive some discharge and sediment after the initial cut off. As the channel infilled and the water became shallow, sediment influx into the channel decreased and water-tolerant vegetation could be established, resulting in 2 m of homogeneous dark grey (4N) lacustrine silty clay to clay that fills the channel to within 1 m of the channel bank. The high shrink–swell activity of this poorly drained, gleyed sediment and the growth of roots has destroyed any bedding that may have been present within this deposit. Today this abandoned channel is still a topographically low, poorly drained area that holds water periodically, but it is no longer a permanent lake.

Within the Mississippi River backswamp, clay is present on both sides of the Pemiscot Bayou meander belt (Fig. 3). This massive, homogeneous dark grey clay becomes slightly more silty in the upper 0.5–2.0 m of the deposit adjacent to the Pemiscot Bayou meander belt. This upward coarsening probably reflects the transition from typical dispersed-sediment accumulation within a low-relief backswamp to funnelling of increased sediment into the lowest portion of the backswamp as the floodplain relief became greater. A poorly drained, vertic buried soil has root pores with iron concentrations along the pores and slickensides at depths > 1 m below the palaeosol surface. The presence of this soil in the backswamp clay indicates a very slow aggradation rate in the backswamp, prior to the development of Pemiscot Bayou. The absence of soil features within the lower portion of the overbank sediment reflects more rapid deposition after the distributary developed.

Relatively small-scale liquefaction features are common throughout the area, but only those present at the land surface are shown in the cross-section (Fig. 3). These are more common at the upstream study site because the epicentre of the December 1811 New Madrid earthquake was approximately 10 km distant (Johnston & Schweig, 1996). Lignite and inclusions of clay are common within the generally massive, medium-grained, yellowish brown (10YR5/4) sand. These features intrude all the fluvial and lacustrine deposits in the area and the boundary of the liquefied sand and the intruded deposit is abrupt. Saturated sand was liquefied by earthquakes within the New Madrid seismic zone. Many liquefaction features are probably the result of the 1811–1812 great earthquakes (Fuller, 1912), but some have formed during earlier seismic events (Tuttle *et al.*, 1996).

**Table 1.** Characteristics of Pemiscot Bayou and surrounding sediment.

| Environment | Texture | Colour | Sedimentary structures | Pedologic structures | Width (km) | Thickness (m) |
|---|---|---|---|---|---|---|
| Upper point bar*, † | Silty clay, muddy fine sand and sand (> 90% sand) | Brown to dark grey to greenish grey hue = 2.5Y, 10YR value = 4–5 chroma ≤ 3 | Upper 2 m massive to laminated beds 4–10 cm thick, lower — massive to laminated beds 0.2–0.6 cm thick | Sporadic root pores, iron concentrations | ≤ 0.6 | ≤ 10 |
| Abandoned-channel fill† | Silty clay to clay (46–70% clay) | Dark grey to greenish grey hue = 5Y, 4–5N value = 4 chroma ≤ 1 | Laminated beds ≤ 2.8 m thick, few thin (0.1 m thick) sandy silt beds | Disseminated organic matter | 0.2–0.3 | 7.1+ |
| Active-channel fill*, † | Silt and sandy silt (42–90% silt) in lower and intermediate portions of deposit, coarse silt (80–87% silt) at top of unit | Brown hue = 10YR, 2.5Y, 2.5/5GY value = 5 chroma = 3 | Lower beds 0.05–0.2 m thick at base, intermediate beds 0.2–0.6 m thick, upper 0.8 m laminated | None | ≤ 0.6 | 11 |
| Lacustrine† | Clay and silty clay (56–74% clay) | Dark grey hue = 2.5–5Y, 10YR value = 3–4 chroma = 1 | Massive | Slickensides, root pores, strong angular blocky structure | 0.2–0.3 | 1.8 |
| Overbank — Mississippi River and Pemiscot Bayou* | Silt to clayey silt (50–76% silt) | Very dark grey to dark brown hue = 10YR value = 3–5 chroma = 1–3 | Massive | Strong subangular blocky structure, root pores, iron concentrations | 20 | ≤ 2 |
| Overbank — Pemiscot Bayou† | Silt to mud and sandy silt (45–83% silt) in proximal locations, sandy silt to clayey silt (50–76% silt) in distal locations | Dark grey to greyish brown hue = 10YR, 2.5Y value = 3–6 chroma = 1–2 | Massive to laminated and thin (< 6 cm) bedded | Moderate subangular blocky structure, root pores, iron concentrations | > 10 | 2.8 |
| Proximal backswamp* | Silty clay (50–67% clay) | Very dark grey to dark grey hue = 10YR value = 4–5 chroma ≤ 2 | Massive | Slickensides, strong subangular blocky structure, root pores, iron concentrations | 15 | 7 |
| Distal backswamp† | Clay (72–78% clay) | Olive grey to dark grey hue = 4–5N, 10G, 2.5–5Y, 10YR value = 4–5 chroma ≤ 2 | Massive | Slickensides, strong subangular blocky structure, root pores, iron concentrations | 15 | 8 |

* Upstream location.
† Downstream location.

**Table 2.** Radiocarbon dates from Pemiscot Bayou.

| Laboratory number | Material | Location | Environment | Conventional radiocarbon age (BP) | Calibrated age 2σ (BC/AD) |
|---|---|---|---|---|---|
| Minimum age of Pemiscot Bayou distributary | | | | | |
| B-96374 | Charred wood† | 58 km down-valley | Hearth intrusive into overbank sediment | 680 ± 70 | AD 1240–1420 |
| B-96375 | Wood* | 58 km down-valley | Root intrusive into backswamp sediment | 1240 ± 60 | AD 670–96 |
| Age of Pemiscot Bayou distributary | | | | | |
| B-89857 | Wood | 27 km down-valley | Chute channel | 2880 ± 70 | 1265–885 BC |
| B-17026 | Whole soil | Big Lake, 49 km down-valley | Abandoned meander | 3500 ± 150‡ | 2200–1440 BC§ |
| B-107365 | Organic sediment* | 58 km down-valley | Infilled abandoned meander | 4850 ± 90 | 3380–3490 BC 3460–3380 BC |

* Accelerator mass spectrometer (AMS) analysis.
† Small sample given extended counting time.
‡ Guccione *et al.* (1988).
§ Calibrated ages calculated using CALIB rev 3.0 (Stuiver & Reimer, 1993) and rounded to the nearest decade.

## CHRONOLOGY

Relative age dating and diagnostic cultural artifacts along Pemiscot Bayou allow a first approximation of the distributary age and indicate that the bayou was an active distributary between 4000 and 2500 yr BP (Saucier, 1994). To test this estimate and to provide an absolute time-frame for the development and demise of this distributary, five radiocarbon dates were obtained from deposits associated with Pemiscot Bayou (Table 2). Two dates from within distributary channels are the most direct dates of distributary activity. The earlier date of 4850 ± 90 yr BP from fill within the abandoned channel at the downstream locality (Fig. 3) indicates that a distributary pre-dates this date. The second date of 2880 ± 70 yr BP is from detrital wood in the fill of a chute channel, associated with the final position of the active channel at the upstream study locality, and provides a maximum age for the active Pemiscot Bayou, immediately prior to its infilling. An additional date of 3500 ± 150 yr BP is from a deposit adjacent to an abandoned meander of the bayou near Big Lake (Fig. 2). This deposit contains pollen of open-water plants (low-spine Compositae, including *Ambrosia, Franseria, Iva* and *Xanthium* and *Isoetes* microspores), interpreted to be from a meander of Pemiscot Bayou at the time it was an oxbow lake. The date is intermediate between the two dates from the channel and thus is consistent with that interpretation.

Finally, minimum dates for the activity of Pemiscot Bayou are 1240 ± 60 yr BP and 680 ± 70 yr BP on materials that intrude deposits associated with the bayou and are also consistent with the dates discussed previously.

There are two interpretations of these dates. First, the abandoned meander is a cutoff meander bend of Pemiscot Bayou and therefore the bayou was an active distributary for more than 2000 yr. This interpretation would confirm the relative age estimates of Saucier (1994) and extend the initiation of the distributary more than 900 yr earlier than estimated originally. An alternative hypothesis is that abandoned meanders south of Big Lake, including the one examined in this study are remnants of an older distributary that the bayou occupied. The younger Pemiscot Bayou distributary channel followed some sections of the older channel and in other localities the bayou crossed meander bends, leaving abandoned meanders. In support of the second interpretation, there are only a few cutoff meanders of Pemiscot Bayou upstream of Big Lake and they become more numerous south of Big Lake. In addition, the cutoff meander bends are smaller than the bends of the present channel, indicating that the discharge was different (Saucier, 1994). Regardless of which interpretation is correct, the Pemiscot Bayou was an active distributary for 2000 yr, if meander abandonment was caused by the formation of Pemiscot Bayou. In contrast, if the older distributary channel was already infilling prior to the formation of the Pemiscot Bayou, the bayou was active for < 2000 yr.

The bayou became an inactive distributary between 2900 and 2200 yr BP based on both the radiocarbon date and archaeology of the region. Ninety-four per cent of the 79 surface archaeological sites along the bayou are from the Woodland and younger cultural periods (Davidson, 1996). These cultural periods are younger than 500 yr BC, based on more than 200 radiocarbon dates of associated organic material from sites between Memphis, Tennessee and Cairo, Illinois (Tuttle *et al.*, 1996). At the upstream study site (Fig. 2), an intensive archaeological survey identified only Middle Woodland and younger sites that are younger than 200 yr BC (Tuttle *et al.*, 1996). This archaeological site distribution indicates that the land surface along the bayou, including most of the channel fill, was stable by 2200 yr BP. Using these bracketing radiocarbon and archaeological dates, the Pemiscot Bayou distributary was active for 2000–3000 yr.

## DISCUSSION

Based on sediment patterns and radiocarbon dates, the early stage of the meandering Mississippi River did not include distributary channels. Distributaries could not develop in the region until the Mississippi River became a meandering river, aggraded its channel and developed natural levees with a significantly greater slope than that of the main channel. Tectonic subsidence within the backswamp at Big Lake (Guccione *et al.*, 1993) also may have contributed to a relatively steep cross-valley gradient. Clay began to accumulate in the low-relief backswamp more than 9500 yr ago as the river shifted from a braided pattern, associated with Pleistocene glacial outwash, to the Holocene meandering pattern (Guccione *et al.*, 1988). The gradient advantage that caused flow to be diverted into the backswamp may have taken 4000 yr to develop in the northern portion of the LMV, based on age estimates of the oldest distributaries (Saucier, 1994). The first increase in grain size and an increase in sedimentation rate, indicating that flow and sediment were being directed into the backswamp at Big Lake, is estimated from radiocarbon dates to have begun approximately 5400 yr ago. Although the Pemiscot Bayou may not have formed quite that early, it probably had become a major distributary by 4900 yr ago.

Pemiscot Bayou was a large distributary in the northern LMV, yet it never captured the entire flow of the Mississippi River. During a relatively long period of stability (2000 yr), the distributary did not become a new Mississippi River meander belt, but it did develop a meandering pattern and at least a few cut offs of meander bends formed. If these cutoff meander bends are part of Pemiscot Bayou, rather than an older distributary, the early phase of the bayou probably had less discharge than at some later time, because the channel cross-section, wavelength and wave amplitude of the meanders are smaller. This interval was not long enough for more than one period of cut offs to develop and no cross-cutting abandoned meanders are present.

The ultimate demise of Pemiscot Bayou could have resulted from healing of the nodal channel, but was probably caused by the cut off of the Mississippi River meander bend, along which the distributary node was located. With either interpretation, discharge into the Mississippi channel, and therefore into the distributary channel, was reduced and the bayou infilled rapidly. Most of the fill accumulated in 700 yr and only a small amount has accumulated during the last 2200 yr.

Although its fate was sealed with the cut off of the Mississippi meander bend, Pemiscot Bayou probably would not have captured the entire flow of the Mississippi River unless some boundary condition(s) such as slope ratio, relative magnitude of the annual flood discharge, or suspended-sediment load changed. First, and probably of prime importance, was the relatively low cross-valley slope. Assuming that the palaeoslope ratio was similar to the modern ratio of 4.2, which is less than 5, the ratio estimated by Slingerland & Smith (1998) to be necessary for a channel to avulse. Although the slope ratio could have changed during the lifespan of the distributary, if aggradation or degradation along the Mississippi River were different from that along the bayou, aggradation of overbank sediment along the bayou and the stability of the channel suggests that the slope ratio probably decreased, if any change at all occurred. This change would make channel avulsion even less likely.

Second, the relative magnitude of annual flood discharges may not have been adequate to cause an avulsion (Mackey & Bridge, 1995). Regional studies show that the warmest and driest interval in the mid-western USA (drainage basin of the Mississippi River) was at approximately 6000 yr ago (Webb *et al.*, 1993) during the Hypsithermal Period. In the northern LMV, drought conditions were present between 8000 and 5000 yr ago (King & Allen, 1977). This would suggest that discharge of the Mississippi River was minimal at about the time the distributary was initiated, but the relative magnitude of the annual flood discharge is unknown. Mississippi River discharge probably increased during the several thousand years that the bayou was an active distributary because regional moisture increased during this interval. Apparently, any increase in flood discharge or relative magnitude of annual flood discharge was not adequate to cause avulsion.

The stability of Pemiscot Bayou as an active distributary channel for several thousand years suggests that the grain size of the suspended-sediment load may have been coarser than that at present. For a grain size of 0.4 mm (medium sand), a distributary is modelled numerically to remain in equilibrium at slope ratios < 5 (Slingerland & Smith, 1998). For smaller suspended-load grain sizes, however, such as 0.1 mm (very fine sand), a distributary will not remain in equilibrium, but will heal as a result of sedimentation within the crevasse channel. Presently 58–98% of the Mississippi River suspended load at Memphis, Tennessee (125 km down-valley from the Pemiscot Bayou node) is silt and clay (Moore *et al.*, 1988; Westerfield *et al.*, 1993; Flohr *et al.*, 1994). If Slingerland & Smith (1998) are correct, crevasse channels would heal with this size of suspended sediment and a change in suspended sediment size during the late Holocene may be responsible for the demise of Pemiscot Bayou. Indeed, no large distributaries have formed in the northern LMV during the past four to five millennia. This hypothesis of channel healing provides an alternative to our interpretation that a Mississippi River meander cut off at the distributary node was responsible for the demise of Pemiscot Bayou as an active distributary and infilling of its channel.

## CONCLUSIONS

Although the northern quadrant of the LMV contains only a single Mississippi River meander belt, the back-swamp region contains additional landforms and deposits generated by the activity of distributaries during the middle Holocene. The largest of these, Pemiscot Bayou, captured approximately 25% of the Mississippi River discharge and probably was active for 2000–3000 yr. Apparently, either the slope ratio (4.2), the relative magnitude of annual flood discharge (unknown), or the suspended-sediment load (unknown), was incapable of eroding a channel deep enough to capture all of the Mississippi River flow and to cause an avulsion. The final demise of Pemiscot Bayou probably occurred after the cut off of the Mississippi River meander where the node was located, but could have occurred in response to a decrease in suspended-sediment grain size, or in response to tectonic influence (Mackey & Bridge, 1995) along the New Madrid seismic zone. This resulted in reduced flow and infilling of most of the channel in less than 1000 yr. For approximately the past 2000 yr the channel has continued to infill slowly.

## ACKNOWLEDGEMENTS

This research was funded by the U.S. Geological Survey under Contract USGS-1434-93- G2352, by the U.S. Army Corps of Engineers, Memphis District Contract DACW66-93-D-0119 D.O. 14, to Mid-Continental Research Associates, a Geological Society of America student research grant to James Kendall Jr. and a W. Joe Edwards Scholarship from the University of Arkansas to Minnie Burford. The views and conclusions contained in this document are those of the authors and should not be interpreted as necessarily representing the official policies of the U.S. Government. We would like to thank Annaick Chauvet, Phillip Owens, Jay Fairbanks, Marion Haynes, E. Moye Rutledge and Martitita Tuttle for assistance in the field. Access to critical field areas was provided by Michael Besharse, Blytheville-Gosnell Regional Airport Authority, Perry Dixon, the Haynes family, Charles Perry and Matt A. Stallings. Whitney J. Autin, John J. Rogers, Derald Smith and Norman Smith provided insightful reviews of the original manuscript.

## REFERENCES

ASLAN, A. (1994) *Holocene sedimentation, soil formation, and floodplain evolution of the Mississippi River floodplain, Ferriday, Louisiana*, PhD dissertation, Department of Geosciences, University of Colorado, 260 pp.

BRIDGE, J.S. (1997) Avulsion behavior of the Mississippi River: simulations using a 3D-model of alluvial architecture. *Abstracts, Sixth International Conference on Fluvial Sedimentology* (Ed. Rogers, J.), University of Cape Town, Cape Town, p. 26.

DAVIDSON, J.M. (1996) *Preferential settlement patterns from the Archaic to Protohistoric times, as observed along the Left Hand Chute of the Little River, Mississippi County, Arkansas.* Unpublished report, University of Arkansas, Fayetteville, AR, 22 pp.

DAY, R.A. (1965) Partial fractionation and particle size analysis. In: *Methods of Soil Analysis*, Part I (Ed. Black, C.A.). *Agronomy*, **9**, 552–562.

FARRELL, K.M. (1987) Sedimentology and facies architecture of overbank deposits of the Mississippi, False River Region, Louisiana, In: *Recent Developments in Fluvial Sedimentology* (Eds Ethridge, F.G., Flores, R.M. & Harvey, M.D.), Spec. Publ. Soc. econ. Paleont. Miner., Tulsa, **39**, 111–120.

FISK, H.N. (1944) *Geological Investigation of the Alluvial Valley of the Lower Mississippi River.* Mississippi River Commission, Vicksburg, MS.

FISK, H.N. (1952) *Geological Investigation of the Atchafalaya Basin and the Problem of Mississippi River Diversion.* U.S. Army Corps of Engineers, Mississippi River Commission, Vicksburg, MS.

FLOHR, D.F., HAMILTON, J.T., LEWIS, J.G. & Thomas, L.B. (1994) *Water Resources Data: Tennessee Water Year 1994.*

U.S. Geological Survey Water-data Report TN-94-1, Nashville, TN, 399 pp.

FOLK, R.L. (1968) *Petrology of Sedimentary Rocks*. Hemphills, Austin, TX.

FULLER, M.L. (1912) The New Madrid earthquake. *U.S. Geol. Surv. Bull.*, **494**, 119 pp.

GUCCIONE, M.J. (1993) Grain-size distribution of overbank sediment and its use to locate channel positions. In: *Alluvial Sedimentation* (Eds Marzo, M. & Puigdefabregas, C.), Spec. Publs int. Ass. Sediment., No. 17, pp. 185–194. Blackwell Scientific Publications, Oxford.

GUCCIONE, M.J., LAFFERTY, III, R.H. & CUMMINGS, L.S. (1988) Environmental constraints of human settlement in an evolving Holocene alluvial system, the Lower Mississippi Valley. *Geoarchaeology*, **3**, 65–84.

GUCCIONE, M.J., VANARSDALE, R.B., HEHR, L.H., *et al.* (1993) *Origin and Age of the 'sunklands' using Drainage Patterns, Sedimentolotgy, Dendrochronology, Archeology, and History*. Unpublished report submitted to the U.S. Geological Survey, Award 14-08-0001-G1997, Washington, DC, 61 pp.

JOHNSTON, A.C. & SCHWEIG, III, E.S. (1996) The enigma of the New Madrid earthquakes of 1811–1812. *Ann. Rev. Earth planet. Sci.*, **24**, 339–384.

KENDALL, J.D. 1997. *Sedimentation and seismic modification of Pemiscot Bayou, Yarbro, Arkansas*. Unpublished MS thesis, University of Arkansas, Fayetteville, AR, 117 pp.

KING, J.E. & ALLEN, W.H., JR. (1977) A Holocene vegetation record from the Mississippi River valley, southern Missouri. *Quat. Res.*, **8**, 307–323.

MACKEY, S.D. & BRIDGE, J.S. (1995) Three-dimensional model of alluvial stratigraphy: theory and application. *J. sediment. Res.*, **B65**, 7–31.

MOORE, M.A., PORTER, J.E., WESTERFIELD, P.W. & YOUNG, K. (1988) *Water Resources Data: Arkansas Water Year 1988*. U.S. Geological Survey Water-data Report AR-88-1, Little Rock, AR. 622 pp.

PAOLA, C. & MOHRIG, D.C. (1997) Jump-length scaling and channel reoccupation in river avulsion. *Abstracts, Sixth International Conference on Fluvial Sedimentology* (Ed. Rogers, J.), University of Cape Town, Cape Town, p. 161.

ROYALL, P.D., DELCOURT, P.A. & DELCOURT, H.R. (1991) Late

Quaternary paleoecology and paleoenvironments of the central Mississippi Alluvial Valley. *Geol. Soc. Am. Bull.*, **103**, 157–170.

SAUCIER, R.T. (1964) *Geologic Investigation of the St. Francis Basin*. Mississippi River Commission and the U.S. Army Corps of Engineers Waterways Experiment Station Technical Report 3–659, Vicksburg, MS.

SAUCIER, R.T. (1994) *Geomorphology and Quaternary Geologic History of the Lower Mississippi Valley*. U.S. Army Corps of Engineers, Mississippi River Commission, Vicksburg, MS.

SLINGERLAND, R. & SMITH, N.D. (1998) Necessary conditions for a meandering-river avulsion. *Geology*, **26**, 435–438.

SMITH, D.G. & SMITH, N.D. (1980) Sedimentation in anastomosed river systems: examples from alluvial valleys near Banff, Alberta. *J. sediment. Petrol.*, **50**, 157–164.

SMITH, N.D. & PÉREZ-ARLUCEA, M. (1994) Fine-grained splay deposition in the avulsion belt of the lower Saskatchewan River, Canada. *J. sediment. Res.*, **B64**, 159–168.

SOIL SURVEY STAFF (1981) *Examination and Description of Soils in the Field, Draft Revisions of Chapter 4*. Soil Survey Manual, Agricultural Handbook No. 18, U.S. Department of Agriculture, U.S. Government Printing Office, Washington, DC.

STUIVER, M. & REIMER, J.R. (1993) Extended C14 data base and revised CALIB 3.0 C14 age correlation program. *Radiocarbon*, **35**, 215–230.

TUTTLE, M.P., LAFFERTY, III, R.H., GUCCIONE, M.J. *et al.* (1996) Use of archaeology to date liquefaction features and seismic events in the New Madrid seismic zone, central United States. *Geoarchaeology*, **11**, 451–480.

WESTERFIELD, P.W., EVANS, D.A. & PORTER, J.E. (1993) *Water Resources Data: Arkansas Water Year 1993*. U.S. Geological Survey Water-data Report AR-93-1, Little Rock, AR, 528 pp.

WEBB, III, T., BARTLEIN, P.J., HARRISON, S.P. & ANDERSON, K.H. (1993) Vegetation, lake levels, and climate in eastern North America for the past 18 000 years. In: *Global Climates since the Last Glacial Maximum* (Eds Wright, H.E., Kutzbach, J.E., Webb, III, T., Ruddiman, W.F., Street-Perrott, F.A. & Bartlein, P.J.), pp. 415–467. University of Minnesota Press, Minneapolis, MN.

*Spec. Publs int. Ass. Sediment.* (1999) **28**, 221–230

# Gradual avulsion, river metamorphosis and reworking by underfit streams: a modern example from the Brahmaputra River in Bangladesh and a possible ancient example in the Spanish Pyrenees

C. S. BRISTOW

*Research School of Geological and Geophysical Sciences, Birkbeck College and UCL,*
*Gower Street, London WC1E 6BT, UK (Email: c.bristow@ucl.ac.uk)*

## ABSTRACT

Avulsion of rivers from one course to another may result in metamorphosis of channel pattern. The Brahmaputra River, a large sand-bed braided river, avulsed into its present course along the Jamuna channel over 100 yr ago. Cartographic evidence indicates that the avulsion was gradual rather than instantaneous and that the new course of the river has changed from sinuous to braided. The former course is occupied by the Old Brahmaputra, a meandering river that is reworking the top of the deposits of the abandoned braided channel belt. This situation, where an underfit stream reworks a partially abandoned channel belt, is probably quite common, but not typically recognized in the rock record. River abandonment is unlikely to be instantaneous and reworking of channel-belt sediments by an underfit stream following avulsion should be expected. A multistorey sandstone capped by an exhumed meander bend in the Oligocene Campodarbe Group, in the Spanish Pyrenees, may be an example. The superposition of a fine-grained meander belt on top of a coarse-grained multistorey sandstone, which probably was deposited by a braided river, can be explained by avulsion. The implications are that braided-river sand bodies may be capped by meandering-river deposits, as a consequence of avulsion, and that river metamorphosis can be autocyclic, rather than allocyclic, in avulsive river systems.

## INTRODUCTION

The term avulsion is widely used to describe a change in channel location that is achieved by a jump rather than a progressive shift. There is some confusion over whether the term applies to changes in location of a whole river or of channels within a river. Definitions include: 'Avulsions: aggrading streams may break out of levees or former channel zones completely and adopt an entirely new course' (Kellerhals *et al.*, 1976, p. 827); 'Avulsion is the term used to describe a river's relatively sudden abandonment of a channel belt in favour of a new course' (Bridge & Leeder, 1979, p. 627); '. . . the diversion of a river channel to a new course at a lower elevation on its floodplain' (Smith *et al.*, 1989, p. 1), all of which clearly indicate movement of the whole river or channel belt. On the other hand, the term avulsion has also been applied to channel switching within a braided river, which has been termed anabranch avulsion (Leddy *et al.*, 1993). In this paper, I use the term avulsion to describe switching of a whole river or channel belt to a different course.

River metamorphosis has been defined as a major change of river morphology; for example, from meandering to braided and vice versa (Schumm, 1969). Recent examples include the South Platte River and Arkansas River, which have had changes in discharge and sediment load that induced dramatic changes in channel pattern (Nadler & Schumm, 1981). In the case of the South Platte and the Arkansas rivers, Nadler & Schumm (1981) argue that the rivers were close to a threshold value at which channel metamorphosis would occur. The change in the channel pattern of the Brahmaputra River, described in this paper, is the result of dramatic changes in discharge following river avulsion. Reworking of channel-belt sediments has occurred after the avulsion, because flow continued at a lower discharge. Consequently, underfit meander belts may overlie or overprint earlier channel geometries. In an avulsive river system, river metamorphosis may be autocyclic rather than allocyclic.

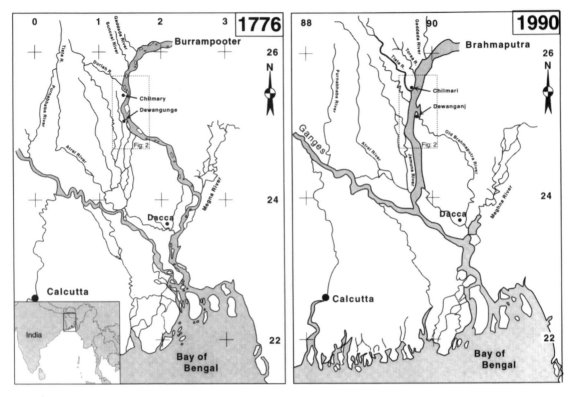

**Fig. 1.** The Brahmaputra river drainage system (A) redrawn from Rennell's map of 1776, (B) redrawn from *The Times Atlas of the World* (1990), showing the major changes in the Brahmaputra and its tributaries over the past 200 yr.

The purpose of this paper is to show that when an avulsion occurs, the former river course may continue to carry a reduced flow, or form an alternative flow path along a topographic low that can be used during floods, and to suggest how these observations could be applied to palaeochannels in the rock record. When an avulsion is not complete and instantaneous, some residual flow along the former river course should be expected. This flow typically will be less than the original channel-forming discharge, which can lead to the development of an underfit stream within the abandoned channel belt. Examples include the partially abandoned river courses of the Kosi Fan (Wells & Dorr, 1987; Gole & Chitale, 1996), the Rapti River (Richards *et al.*, 1993) and the Brahmaputra. The avulsion of the Brahmaputra River (Fig. 1), described in this paper, has led to major changes in channel pattern. The new course of the river, the Jamuna, has changed from sinuous to braided, whereas the former braided channel belt is occupied by a meandering river, the Old Brahmaputra. The meandering Old

Brahmaputra is reworking the braided-river deposits and has overprinted the geomorphology of the abandoned braided-river channel belt.

An outcrop in the Spanish Pyrenees is described, where a meandering-river sandstone, on top of a multistorey sandstone, indicates a major change in palaeochannel pattern. This can be interpreted as channel metamorphosis, which might have resulted from an avulsion. The meander-bend sandstone is less than 1 m thick and composed of fine- to very fine-grained sand, whereas the underlying 7-m-thick multistorey sand body is coarse grained with larger channels, up to 3 m preserved thickness and larger bedforms. These differences suggest that the meander-belt sandstone was deposited by a narrower, shallower and lower energy river than the underlying sand body. The differences may be explained if a braided-river sand body was reworked by a smaller, underfit, meandering stream, following avulsion or partial abandonment. The ancient avulsion event itself, however, has not been proven.

## AVULSION OF THE BRAHMAPUTRA RIVER

The earliest study of the Brahmaputra was made by Rennell (1781) who mapped the Brahmaputra (Rennell, 1776) as a braided river flowing into the Meghna River and following a course now occupied by a much smaller river known as the Old Brahmaputra. Since then the river has changed course and avulsed into its present course known as the Jamuna River. The exact timing and cause of the avulsion remains uncertain. Theories include tectonic tilting, an earthquake trigger, river capture, flooding or increased discharge resulting from an upstream tributary avulsion. The main theories are reviewed here and discussed with respect to the cartographic evidence (Fig. 2), although it is possible that the maps themselves may not be completely accurate.

### Tectonic influence

Morgan & McIntire (1959) assert some tectonic influence; 'Probably the diversion of the Brahmaputra was gradual, as most river diversions are, and was caused in part by gradual tilting of the Madhupur block. This tilting caused the Old Brahmaputra River to become antecedent in places, necessitating river scour into slowly or periodically rising, comparatively resistant Pleistocene sediments.' (Morgan & McIntire, 1959, p. 331). Coleman (1969, p. 336) says that '. . . faulting was probably the major cause of the recent shift of the Brahmaputra River from its course east of the Madhupur Jungle to its present position.' Winkley *et al.* (1994) state that tectonic activity in 1772 altered the course of the Brahmaputra, which seems unlikely because this pre-dates Rennell's map (1776). Bangladesh is tectonically active and the rivers appear to follow structural trends (Morgan & McIntire, 1959); tectonic subsidence has almost certainly affected the river. There is no clear link, however, between the avulsion and a specific earthquake.

### Tributary switching

It has been suggested (Morgan & McIntire, 1959; Monsur, 1995) that the avulsion of the Brahmaputra River followed avulsion of the Tista River, a major tributary of the Brahmaputra, which has its confluence just upstream from the avulsion node of the Brahmaputra. 'The sudden change of course by the Tista River with resulting addition of waters to the Brahmaputra River may well have been a contributing factor towards diversion' (Morgan & McIntire, 1959; p. 331). Using evidence from Rennell's map of 1776, Morgan & McIntire (1959) suggest that the Tista flowed into the Purnabhaba and Atrai rivers, which are tributaries of the River Ganges (Fig. 1). They suggest that during the 1770s, when Rennell made his map, the Atrai was the major channel. Nowadays, the Tista flows directly into the Brahmaputra. Neither Wilcox's map of 1828 nor Allen's map of 1843 show the Tista as a significant tributary of the Brahmaputra, however, which raises a question over the assertion that avulsion of the Tista caused the avulsion of the Brahmaputra. Allen (1843) shows that the Tista River was still largely flowing into the Purnabhaba and Atrai Rivers, whereas the Brahmaputra was starting to change course (Fig. 2C). The Tista appears in its present position on a map of 1872, based on a survey of 1868 (Thuillier, 1872). The cartographic evidence indicates that the Tista avulsion occurred after the avulsion of the Brahmaputra.

### Flood event

According to LaTouche (1910), in Morgan & McIntire (1959), the Tista River changed course suddenly in 1787 during a single flood. This report is not supported by the maps of Wilcox (1828) or Allen (1843), neither of which show the Tista as a significant tributary of the Brahmaputra (Fig. 2B & C). This does not rule out a later flood event on the Brahmaputra causing the avulsion.

### River Capture

On Rennell's map there is a distributary of the Brahmaputra called the Jenni River, which exits the Brahmaputra downstream from the avulsion node. Other authors (Morgan & McIntire, 1959; Monsur, 1995) refer to the 'Jenai' as a precursor of the Jamuna, which opens up the possibility of river capture as a cause of the avulsion. Rennell's and Wilcox's maps, however, show the towns of Chilmari (Chilmary, Chilmaree) and Diwanganj (Dewangunge) on the west (right) bank of the Brahmaputra, upstream from the Jenni (Fig. 2A & B). At the present day the town of Diwanganj is on the east (left) bank of the Jamuna and downstream from the Jamuna/Old Brahmaputra junction. Thus the avulsion node is upstream from the Jenni. This interpretation is supported by Allen's map of 1843, which shows the Jenni River to the east of the Jamuna, indicating that the Jenni was not the site of the Brahmaputra avulsion. This observation raises questions over Coleman's (1969) interpretation of westward migration of the Jamuna River. The large meander loops (B-6 of Coleman, 1969) may not have been formed by the Jamuna; they may relate to an earlier river, or an earlier phase of the Jenni with a higher discharge. The

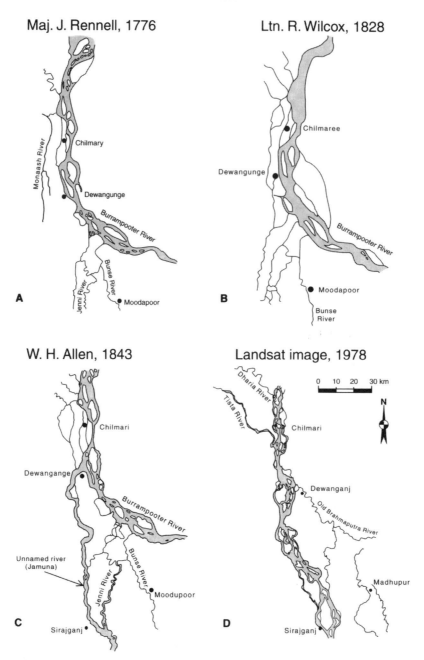

**Fig. 2.** Details of the Brahmaputra avulsion site redrawn from maps by (A) Rennell (1776), (B) Wilcox (1828), (C) Allen (1843) and (D) LANDSAT satellite image (1978). The maps indicate that the avulsion started between 1828 and 1843. Since then the Jamuna River has become braided whereas the Old Brahmaputra has changed from braided to meandering.

Jenni appears to have been decreasing in importance following the Brahmaputra avulsion (Fig. 2D) and might have been a more significant distributary channel with larger meander loops before the avulsion.

The towns of Chilmari and Diwanganj appear on Rennell's map as Chilmary and Dewangunge respectively; both are shown on the west bank of the Burampooter (Fig. 2A). Today, Chilmari is still on the west (right) bank, but Diwanganj is on the east (left) bank. As such, it seems most likely that the avulsion node was between these two towns, with the Brahmaputra switching into an unnamed tributary of the Monaash and Joobnee Rivers, which are shown on Rennell's map (Rennell, 1776), but do not appear on modern maps. These rivers probably have been eroded by the Jamuna.

## A Gradual Avulsion

The lack of agreement over a flood or tectonic trigger mechanism for the avulsion of the Brahmaputra may indicate that there was no significant trigger. An alternative explanation for the avulsion may be inferred from the maps of Allen (1843), Wilcox (1828) and Rennell (1776) (Fig. 2), all of which show large mid-channel islands within the 'Burampooter' River downstream from Chilmari, close to the avulsion site. It is suggested here that flow divergence around the mid-channel islands, with one channel directed towards the west (right) bank, caused bank erosion, which led to flow diversion into an existing floodplain channel, and which the Brahmaputra exploited and enlarged to form the Jamuna.

Allen's map (Allen, 1843) appears to show the early stages of the avulsion, with Chilmary on the west (right) bank of the river, upstream from the junction and Dewangange on the east (left) bank, downstream from the junction, as they are today (Fig. 2C & D). Allen's map also shows the original Brahmaputra River as the larger river with a braided pattern, whereas the new, as yet unnamed, Jamuna channel has a broadly sinuous course flowing past Sirajganj. The Jamuna, which initially was sinuous, still flows along this course, but has now become braided (Fig. 2D). The change in channel pattern is interpreted to indicate a gradual transfer of flow from the Old Brahmaputra to the Jamuna, with the Jamuna changing from sinuous to braided as discharge increased. This supports the report of Hirst (1916; in Coleman, 1969) that the changes took place gradually between 1720 and 1830. As the Jamuna has expanded, it has removed the evidence of the earlier channel system. Borehole evidence from the Jamuna floodplain, however, indicates that this is not the first time that the Brahmaputra has flowed down this valley (Umitsu, 1993).

## Old Brahmaputra

Although the Jamuna has been studied by Coleman (1969), Bristow (1987, 1993), Klaassen & Vermeer (1988), Klaassen & Masselink (1992) and Thorne *et al.* (1993), the Old Brahmaputra has received relatively little attention. There are now two rivers running along the old course of the Brahmaputra, the Old Brahmaputra River and the Dhashani River, both of which are meandering and partially anastomosed. The Old Brahmaputra is active during the flood season, although it has a discharge approximately 100 times less than that of the Jamuna. It is almost dry at low-flow stage (Fig. 3), with water ponded in pools along the river, and parts of the bed are cultivated for rice. The river is approximately 200 m wide and is reworking the deposits of the Brahmaputra, superimposing the record of a meandering river over those of a much larger braided river, which was up to 10 km wide. LANDSAT and SPOT images of the Old Brahmaputra do not show braid-bar morphology preserved on the floodplain. Coleman (1969) describes five abandoned courses (B-1 to B-5), each of which is described as a meander belt with some evidence for a few braided reaches. The latest, B-5,

**Fig. 3.** Photograph of the Old Brahmaputra River during the dry season, the river is reduced to a series of pools and the river bed is being cultivated, note the rice planting on mud drapes in the troughs of dune bedforms and vegetables within the river bank. Over 100 yr after the avulsion, the old channel still offers a conduit for flood waters during the wet season.

exhibits very small, tightly curved meander loops from the remnant channel, showing total overprinting of the earlier braided channel pattern.

## A POINT BAR IN THE PYRENEES

The Campodarbe Group is an Oligocene succession of fluvial sediments, which were deposited in the foreland basin of the Spanish Pyrenees. The group is at least 2500 m thick in the study area (Puigdefabregas, 1975), increasing in thickness to 3600 m along strike near Javierrelatre. Palaeocurrents recorded by Puigdefabregas (1975) are generally towards the north-west, around 300°.

The exposure described here is just one of many fluvial

sand bodies within the Campodarbe Group, but it stands out because of the exhumed depositional topography preserved on the top of the sand body, which is described by Puigdefabregas & Van Vliet (1978) as beautifully preserved scroll-bar topography. The exposure, which is close to the abandoned village of Escusaguas (42° 22′ 29.9″ N 0° 23′ 14.2″ W), is cut by a N–S trending road-cut on a disused section of the N330 between Huesca and Sabinanigo (Fig. 4). The road-cut provides an excellent section through the sand body, which is 8 m thick and has a sharp erosive base that cuts into an underlying heterolithic channel fill (Figs 4 & 5).

Sets of trough cross-stratification are common and generally indicate palaeoflow towards the north, although there are marked local changes, with one set of flute casts

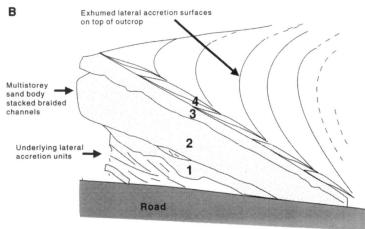

**Fig. 4.** (A) Photograph of the Escusaguas road-cut exposure looking towards the south shows concentric ridges of fine grained sandstone on top of the outcrop, which are interpreted as part of a meander bend. (B) Synoptic sketch of the west side of the Escusaguas exposure illustrating the units of the multistorey sand body and the exhumed topography on top.

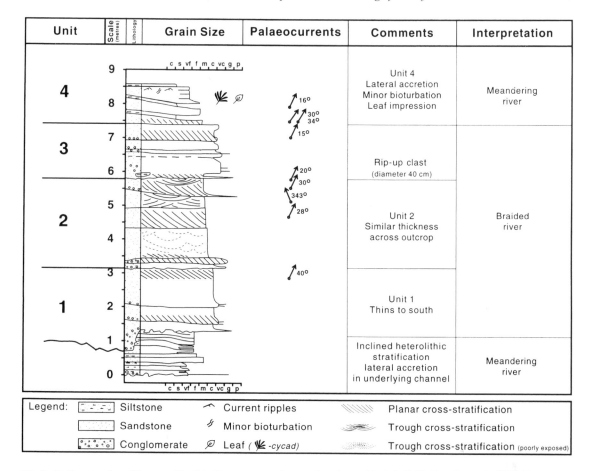

**Fig. 5.** Sedimentary log of the west side of the Escusaguas road-cut section, the sand body is divided into four units divided by bounding surfaces that can be traced along the exposure.

indicating flow towards the west. Erosion surfaces within the sand body can be traced along the exposure on both sides of the road-cut. Puigdefabregas & Van Vliet (1978) use the presence of two internal scour surfaces, overlain by gravel, to divide the sand body into three units. They note that palaeocurrents vary between these units with a trend towards 330° in the lower sand body, a trend towards 250° in the middle sand body and 20° in the upper sand body. Palaeocurrent measurements in this study agree with the broad trend towards the north, but do not show such a clear difference between the units (Fig. 5). In this study, the same erosion surfaces are recognized and the division of the sand body into three coarse-grained units is maintained. The upper unit 3 of Puigdefabregas & Van Vliet (1978), however, is divided into two, with the top half now identified as a separate unit 4 (Figs 4 & 5).

The three lower units are each about 2–3 m thick and have sharp erosive bases with gravel lags, including extra formational clasts and intraformational mudstone clasts up to 40 cm in diameter. The basal unit 1 is coarse grained, with local concentrations of granules and small pebbles. Unit 2, fines up from very coarse to coarse- and medium-grained sandstone, and unit 3 is composed of coarse- to very coarse-grained sandstone with local concentrations of granule to pebble conglomerate (Fig. 5). In contrast, unit 4 is much finer grained, passing up from medium to fine and very fine sandstone with mudstone laminae. The mudstones show minor bioturbation and preserve the impressions of leaves, one of which resembles a cycad. Differential erosion of the muds has produced an undulating topography, with the sandstones forming arcuate ridges on the top of the exposure (Fig. 4). This topography has been described previously as scroll bars (Puigdefabregas & Van Vliet, 1978), but is a partial exhumation of heterolithic lateral accretion surfaces. In cross-section, the sands form sigmoid-shaped lenses, around 3 m wide, that dip towards the outside of the curve, which swings around from E–W to N–S. The radiating ridges change orientation slightly owing to expansion and translation of a meander bend, and palaeocurrent measurements, taken from sets of trough cross-stratification at the base of unit 4 show a broad swing from east to north around the exposure, following the trend of the lateral accretion surfaces. The thickness of unit 4 indicates a bankfull depth of only 1 m, which is less than the preserved thickness of units 1, 2 and 3. The reduction in thickness indicates that the river that deposited unit 4 was shallower than the river that deposited units 1, 2 and 3.

The lack of preserved channel margins in units 1, 2 and 3 indicates channel widths greater than the width of the outcrop. In comparison, unit 4 contains lateral accretion units only 3 m wide, which suggest that the meandering river was much narrower. The finer grain size and smaller bedforms in unit 4 indicate a lower energy river, the alternating sands and muds suggest deposition over several flood events, rather than low-stage reworking. Puigdefabregas & Van Vliet (1978) suggest that the units within the sand body represent individual active meandering-channel deposits, where only the uppermost sand body was preserved completely. They attribute the multistorey aspect of the sand body to downstream meander migration, with partial preservation of the lower parts of meandering-channel deposits. They suggest that the partial preservation results from a relatively slow aggradation rate. This opinion is repeated in Jolley & Hogan (1989).

An alternative explanation is that the preserved lateral-accretion surfaces are part of a fourth channel sand body, which represents a different channel pattern to that of the underlying very coarse-grained sandstones. The lower units 1, 2 and 3 do not contain any lateral-accretion surfaces, which might provide evidence of meandering, although their absence does not necessarily mean that the river was not meandering (Jackson, 1978; Bridge, 1985). There is no conclusive evidence of braiding either, such as the presence of preserved mid-channel bars (Bridge, 1993). The coarse-grained multistorey sand body that includes units 1, 2 and 3, however, is very different from unit 4, so even if there were not a transition from braided to meandering, a significant reduction in grain size, channel depth, flow velocity and discharge is required to account for the change in sedimentary structures and sandbody character.

It is suggested that the whole sand body represents the

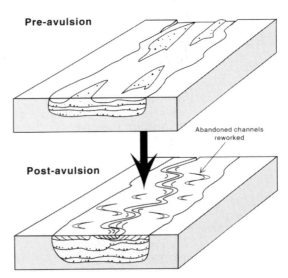

**Fig. 6.** Cartoon illustrating the potential change in channel pattern that might result from the avulsion of a large braided river. After the avulsion, discharge is reduced and an underfit meandering river reworks the top of a braided-river sand body (e.g. Old Brahmaputra). This results in the preservation of a meander belt from a small meandering river on top of a multistorey sand body from a larger braided river (e.g. Escusaguas).

deposits of a coarse-grained multichannel braided river, overlain by the deposits of a much smaller, finer grained, meandering river. It is possible that the meandering channel is an underfit stream, which occupied the channel belt after avulsion of the coarser grained braided river (Fig. 6). Owing to the limitations of the exposure it is not possible to prove that there was a change from braided to meandering as a result of avulsion. Some significant change in river morphology, a river metamorphosis, however, did occur. Alternative explanations, such as a climatically driven reduction in discharge, or a decrease in slope resulting from a rise in base level could be invoked to explain the change in channel pattern (Shanley & McCabe, 1993). These explanations, however, would require a basin-wide event, which has not been recorded in the field.

## CONCLUSIONS

Changes in channel pattern, as a consequence of avulsion, are predicted to be quite common and may result in underfit meandering rivers reworking the top of braided-river sand bodies. The Brahmaputra River avulsed over 100 yr ago, probably in the early 1800s. The cause of the avulsion is uncertain, but historic maps indicate that it was gradual, with increasing flow in the Jamuna and reduced flow in the Old Brahmaputra. As a consequence, the Jamuna River changed from meandering to braided, whereas the Old Brahmaputra River changed from braided to meandering. The Old Brahmaputra is an underfit meandering river, which has reworked the top of the abandoned braided-river channel belt. The occurrence of a meandering-river sand body on top of a multistorey, possibly braided-river sand body may occur as a consequence of avulsion.

One possible example has been recognized in the Oligocene, Campodarbe Group of the Southern Pyrenees foreland basin. At the Escusaguas road-cut a coarse-grained multistorey sandstone is overlain by a fine-grained sandstone interpreted as an exhumed meander bend. The meandering-channel sandstone is thinner than the underlying channel sandstones and is fine- to very fine-grained, with mud drapes containing leaf impressions and minor bioturbation. This contrasts strongly with the underlying 7-m-thick multistorey sand body, which contains cross-stratified coarse-grained sandstones with pebble lags and mudstone clasts, and individual storeys up to 3 m thick. The change is interpreted as an example of autocyclic river metamorphosis resulting from avulsion, with a small meandering river reworking the top of the deposits of an earlier, larger, higher energy river, which may have been braided. Although no shift in base level or climate change has been identified in this case, the effects of tectonic activity influencing palaeochannel pattern within this foreland basin setting cannot be ruled out.

## ACKNOWLEDGEMENTS

This paper has benefited from critical and constructive review by Peggy Guccione, Andres Aslan and Norman Smith. David Upton is thanked for his help in drafting figures.

## REFERENCES

ALLEN, W.H. (1843) *A Newly Constructed and Improved Map of India Compiled Chiefly from Surveys Executed by Order of the East India Company.* W.H. Allen & Co., London.

BRIDGE, J.S. (1985) Palaeochannels inferred from alluvial deposits: a critical evaluation. *J. sediment. Petrol.*, **55**, 579–589.

BRIDGE, J.S. (1993) The interaction between channel geometry, water flow, sediment transport and deposition in braided rivers. In: *Braided Rivers* (Eds Best, J.L. & Bristow, C.S.), Spec. Publ. geol. Soc. London, No. 75, pp. 13–71. Geological Society of London, Bath.

BRIDGE, J.S. & LEEDER, M.R. (1979) A simulation model of alluvial stratigraphy. *Sedimentology*, **26**, 617–644.

BRISTOW, C.S. (1987) Brahmaputra River: channel migration and deposition. In: *Recent Developments in Fluvial Sedimentology* (Eds Ethridge, F.G., Flores, R.M. & Harvey, M.D.), Spec. Publ. Soc. econ. Paleont. Miner., Tulsa, **39**, 63–74.

BRISTOW, C.S. (1993) Sedimentary structures exposed in bar tops in the Brahmaputra River, Bangladesh. In: *Braided Rivers* (Eds Best, J.L. & Bristow, C.S.), Spec. Publ. geol. Soc. London, No. 75, pp. 277–289. Geological Society of London, Bath.

COLEMAN, J.M. (1969) Brahmaputra River channel processes and sedimentation. *Sediment. Geol.*, **3**, 129–239.

GOLE, C.V. & CHITALE, S.V. (1966) Inland delta-building activity of Kosi River. *Proc. Am. Soc. civ. Eng., J. Hydraul. Div.*, **HV2**, 111–126.

HIRST, F.C. (1916) *Report on the Nadia Rivers, 1915.* Bengal Secretariat Book Depot, Calcutta, 39 pp.

JACKSON, R.G. (1978) Preliminary evaluation of lithofacies models for meandering alluvial streams. In: *Fluvial Sedimentology* (Ed. Miall, A.D.), Mem. Can. Soc. petrol. Geol., Calgary, **5**, 543–576.

JOLLEY, E.J. & HOGAN, P.J. (1989) The Campodarbe Group of the Jaca Basin. In: *Pyrenean Tectonic Control of Oligo-Miocene River Systems, Huesca, Aragon, Spain* (Ed. Friend, P.F.), pp. 93–120. 4th International Conference on Fluvial Sedimentology Guidebook, 4, Publications del Servei Geologic de Catalunya Barcelona.

KELLERHALS, R., CHURCH, M. & BRAY, D.I. (1976) Classification and analysis of river processes. *Am. Soc. civ. Eng., J. Hydraul. Div.*, **102**, 813–829.

KLAASSEN, G.J. & MASSELINK, G. (1988) Planform changes of a braided river with fine sand as bed and bank material. *5th International Symposium on River Sedimentation*, April 1992 Karlsruhe, Germany.

KLAASSEN, G.J. & VERMEER, K. (1988) Channel characteristics of the braiding Jamuna River, Bangladesh. In: *International Conference on River Regime* (Ed. White, W.R.), pp. 173–189. John Wiley & Sons, Chichester.

LATOUCHE, T.H.D. (1910) *Relics of the Great Ice Age in the Plains of Northern India.* Reprinted (1919) in *Report on the Hooghly River and its Headwaters.* The Bengal Secretariat Book Depot, Calcutta V1, pp. 21–22.

LEDDY, J.O., ASHWORTH, P.J. & BEST, J.L. (1993) Mechanisms of anabranch avulsion within gravel-bed braided rivers: observations from a scaled physical model. In: *Braided Rivers* (Eds Best, J.L. & Bristow, C.S.), Spec. Publ. geol. Soc. London, No. 75, pp. 119–127. Geological Society of London, Bath.

MONSUR, M.H. (1995) *An Introduction to the Quaternary Geology of Bangladesh.* Rehana Akhter, Dhaka, Bangladesh, pp. 70.

MORGAN, J.P. & MCKINTIRE, W.G. (1959) Quaternary geology of the Bengal Basin, East Pakistan and India. *Geol. Soc. Am. Bull.*, **70**, 319–342.

NADLER, C.T. & SCHUMM, S.A. (1981) Metamorphosis of South Platte and Arkansas Rivers, Eastern Colorado. *Phys. Geogr.*, **2**, 95–115.

PUIGDEFABREGAS, C. (1975) *La Sedimentacion molasica en la Cuenca de Jaca.* Monografias del Instituto de Estudios Pirenaicos 104, Jaca, 188 pp.

PUIGDEFABREGAS, C. & VAN VLIET, A. (1978) Meandering stream deposits from the Tertiary of the southern Pyrenees. In: *Fluvial Sedimentology* (Ed. Miall, A.D.), Mem. Can. Soc. petrol. Geol., Calgary, **5**, 469–485.

RENNELL, J. (1776) *An actual survey of the provinces of Bengal, Bahar etc. by Major James Rennell, Engineer, Surveyor General to the Honourable East India Company*. Published by the permission of the committee of directors by Andrew Dury.

RENNELL, J.J. (1781) The Ganges and Brahmaputra Rivers. *Philos. Trans. R. Soc. London*, **81**, 91–103.

RICHARDS, K., CHANDRA, S. & FRIEND, P. (1993) Avulsive channel systems: characteristics and examples. In: *Braided Rivers* (Eds Best, J.L. & Bristow, C.S.), Spec. Publ. geol. Soc. London, No. 75, pp. 195–203. Geological Society of London, Bath.

SCHUMM, S.A. (1969) River metamorphosis. *Proc. Am. Soc. civ. Eng., J. Hyraul. Div.*, **95**, 255–273.

SHANLEY, K.W. & MCCABE, P.J. (1993) Alluvial architecture in a sequence stratigraphic framework: a case history from the Upper Cretaceous of southern Utah, USA. In: *The Geological Modelling of Hydrocarbon Reservoirs and Outcrop Analogues* (Eds Flint, S.S. & Bryant, I.D.), Spec. Publs int. Ass. Sediment., No. 15, pp. 21–55. Blackwell Scientific Publications, Oxford.

SMITH, N.D., CROSS, T.A., DUFFICY, J.P. & CLOUGH, S.R. (1989) Anatomy of an avulsion. *Sedimentology*, **36**, 1–23.

THUILLIER, H.L. (1872) *Sketch map of the provinces comprising the Lieutt Governorship of Bengal shewing provincial and district divisions*. Taken from map of Bengal Behar and Orissa compiled and published in 1868. Published under the direction of H.L. Thuillier, Surveyor General of India.

*Times Atlas of the World* (1990) Times Books, London.

THORNE, C.R., RUSSELL, A.P.G. & ALAM, M.K. (1993) Planform pattern and channel evolution of the Brahmaputra River, Bangladesh. In: *Braided Rivers* (Eds Best, J.L. & Bristow, C.S.), Spec. Publ. geol. Soc. London, No. 75, pp. 257–276. Geological Society of London, Bath.

UMITSU, M. (1993) Late Quaternary sedimentary environments and landforms in the Ganges Delta. *Sediment. Geol.*, **83**, 177–186.

WELLS, N.A. & DORR, J.A., JR. (1987) A reconnaissance of sedimentation on the Kosi alluvial fan of India. In: *Recent Developments in Fluvial Sedimentology* (Eds Ethridge, F.G., Flores, R.M. & Harvey, M.D.), Spec. Publ. Soc. econ. Paleont. Miner., Tulsa, **39**, 51–61.

WILCOX, R. (1828) *Map of the countries lying between $21\frac{1}{2}°$ and $29\frac{1}{2}°$ North Latitude and $90°$ and $98°$ East Longitude, showing the sources of the Irawady River and the eastern branches of the Brahmaputra, comprising Assam, Muneepoor, and the hilly districts of the Singphos, part of the Sham, and of the Chinese provinces of Yunan and Thibet*. By Lieut. R. Wilcox, B.N.I. Scale 16 inches to 1 mile. Office of the Surveyor General of India (July 1830).

WINKLEY, B.R., LESLEIGHTER, E.J. & COONEY, J.R. (1994) Instability problems of the Arail Khan River, Bangladesh. In: *The Variability of Large Alluvial Rivers* (Eds Schumm, S.A. & Winkley, B.R.), pp. 269–284. American Society of Civil Engineers Press, New York.

*Spec. Publs int. Ass. Sediment.* (1999) **28**, 231–249

# Holocene avulsion history of the lower Saskatchewan fluvial system, Cumberland Marshes, Saskatchewan–Manitoba, Canada

G. S. MOROZOVA* *and* N. D. SMITH†

*University of Illinois at Chicago, Department of Earth & Environmental Sciences (m/c 186), 845 West Taylor Street, Chicago, IL 60607-7059, USA*
†*Department of Geosciences, 214 Bessey Hall, University of Nebraska, Lincoln, NE 68588-0340, USA*
*(Email: gmorozl@uic.edu and nsmith1@unl.edu)*

## ABSTRACT

The Holocene avulsion history and alluvial stratigraphy of the lower Saskatchewan fluvial system in the Cumberland Marshes, central Canada, were studied from nearly 200 boreholes. These data, together with 48 radiocarbon dates of organic-rich deposits, allowed access to fluvial history back to *c.* 5400 yr BP. Throughout the middle and late Holocene, repeated avulsions, the most recent initiated in the 1870s, have formed prominent channel belts, most of which now stand as abandoned alluvial ridges separated by floodbasin depressions. Nine principal avulsions have occurred in the past 5400 yr. Most avulsions originated near the western edge of the Cumberland Marshes, where the Saskatchewan River undergoes an abrupt reduction in gradient. Major channel belts were active for up to 2400 yr, indicating that channel belts at times coexisted. Holocene sediments show evidence of two basic avulsion styles, channel reoccupation and splay-complex progradation. Avulsion styles and alluvial stratigraphy were affected by climate change, which influenced levels and areal extents of floodplain lakes and wetlands. Climate changes also influenced the discharge, flood regime and sediment load of the Saskatchewan River. Holocene alluvium that formed before *c.* 5400 yr BP is composed of mainly silty and sandy sediments and represents drier and warmer climates of the Hypsithermal Interval. Following a change toward cooler and wetter conditions between 5400 and *c.* 2600 yr BP, floodplain lakes became higher and more extensive, resulting in widespread deposition of a middle layer of lacustrine and wetland deposits. Beginning approximately 2600 yr BP, the lakes became increasingly filled by fluvial deposits, resulting in the formation of an upper complex of silty and sandy deposits. Avulsion by splay-complex progradation characterized floodplain evolution during the period of lake-level rise between 5400 and 2600 yr BP, whereas avulsions since then involved both reoccupation and splay-complex progradation.

## INTRODUCTION

Although avulsions often have been described from modern and Quaternary settings (Smith, 1983, 1986; Wells & Dorr, 1987; Schumann, 1989; Smith *et al.*, 1989; Brizga & Finlayson, 1990; Autin, 1992; McCarthy *et al.*, 1992; Richards *et al.*, 1993; Törnqvist, 1993, 1994; Blum & Valastro, 1994; Saucier, 1994), our understanding of the controls of avulsions as well as their sedimentological and stratigraphical effects on floodplain development is still incomplete. Models of alluvial stratigraphy are also hampered by our poor understanding of the underlying mechanisms and origins of avulsions (Heller & Paola, 1996).

Avulsion leads to the development of a new channel belt and is thought to result mainly from channel-belt superelevation above the flood-basin surface, which creates a local gradient advantage. During avulsion, relatively coarse sandy and silty sediments may be delivered into a new area of the flood-basin and bury older fine-grained floodplain deposits. New channels may form within a few years of the initial breakout if the new diversion is rapid and pre-existing channels are reoccupied (e.g. Brizga & Finlayson, 1990; Smith *et al.*, 1998). Avulsion of rivers such as the Mississippi appears to involve more gradual flow diversion and slower development of a new channel (Autin *et al.*, 1991; Saucier, 1994). External controls of most avulsions are not well understood, although variable subsidence rates, base-level changes, tectonism and changing sediment supply have been suggested as factors influencing avulsion frequency

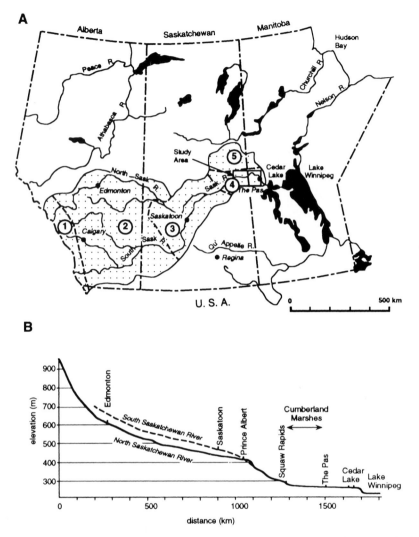

**Fig. 1.** (A) Map showing the drainage basin of the Saskatchewan River (stippled) and the location of the Cumberland Marshes study area. Letters in circles indicate different physiographic provinces of the Saskatchewan River basin: 1, eastern slopes of the Rocky Mountains; 2, Alberta Plains; 3, Saskatchewan Plains; 4, Manitoba Lowlands; 5, Canadian Shield (after Ashmore, 1986). (B) Longitudinal profile of the Saskatchewan River (after PFRA, 1954).

(Smith, 1983, 1986; Alexander & Leeder, 1987; Törnqvist, 1993, 1994; Bryant *et al.*, 1995). Limited evidence from modern and Holocene alluvial systems suggest avulsion frequencies of the order $10^2$–$10^3$ yr (Bridge & Leeder, 1979; Törnqvist, 1994), confirming that avulsions are rare events.

The Saskatchewan River at the Cumberland Marshes underwent a major avulsion in the 1870s that resulted in alluviation of about 500 km$^2$ of the adjoining floodplain (Smith *et al.*, 1989, 1998). This avulsion not only deposited large quantities of sediment in a relatively short

period of time (*c.* 120 yr), but also profoundly affected floodplain morphology and the subsequent evolution of channels. As a result of this avulsion, the Saskatchewan River abandoned its former course, the present Old Channel, in favour of a new course, now the New Channel. The Cumberland Marshes contain several large abandoned Holocene channel belts representing former courses of the Saskatchewan River (Kuiper, 1960), which indicates that the 1870s event is the youngest of a series of avulsions. This paper describes the deposits and ages of the older channel belts, discusses avulsion history and

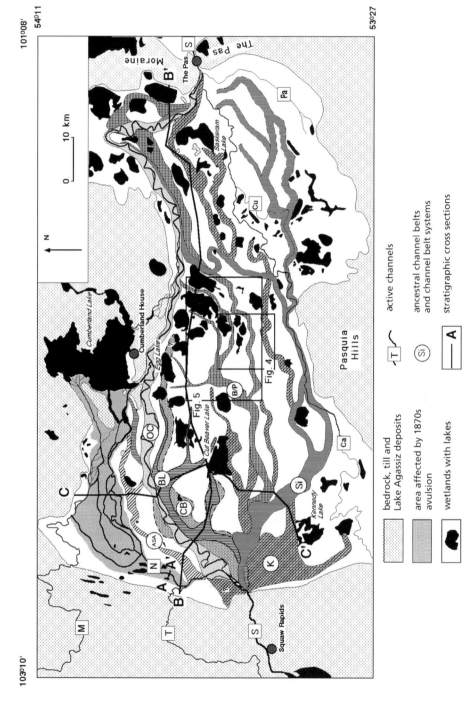

**Fig. 2.** Major geomorphological elements of the Cumberland Marshes area. Letters in squares indicate active channels: S, Saskatchewan River; T, Torch River; Pa, Pasquia River; M, Mossy River; Ca, Carrot River; Cu, Culdesac River; NC, New Channel. Letters in circles indicate abandoned channel belts: K, Kennedy; Si, Sipanok; B/P, Birch–Petabec; ASA, Ancestral South Angling; BL, Bloodsucker; CB, Cut Beaver; OC, Old Channel.

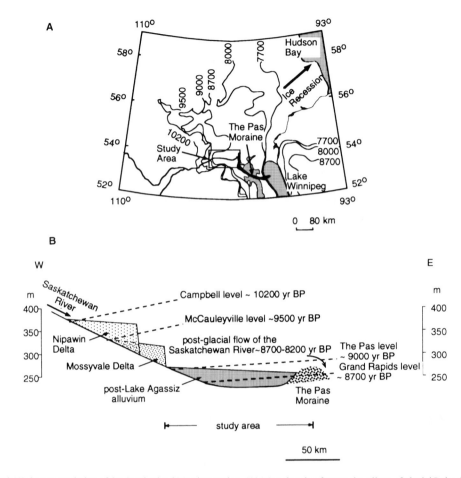

**Fig. 3.** Early Holocene evolution of the Cumberland Marshes region. (A) Map showing former shorelines of glacial Lake Agassiz. Ages (yr BP) of shorelines decrease to the north-east (after Klassen, 1983; Schreiner, 1983). (B) Schematic cross-section showing Lake Agassiz shorelines, now tilted by rebound, and Saskatchewan River deltas (after Grice, 1970; Klassen, 1983; Schreiner, 1983; Christiansen *et al.*, 1995).

considers this history in the context of Holocene climate and lake-level changes.

## LOCATION AND
## GEOLOGICAL SETTING

The Cumberland Marshes, sometimes referred to as the 'Saskatchewan Delta' (Kuiper, 1960; Dirschl, 1972), represents a large (about 8000 km²) floodplain region of the Saskatchewan River in central Canada (Fig. 1A). The Marshes are situated just downstream of a sharp gradient reduction of the Saskatchewan River (Fig. 1B) and consist of extensive wetlands and shallow lakes separated by several partly and completely abandoned channel belts

of the Saskatchewan River (Fig. 2). The elevation of the Marshes decreases in an eastward direction from 275 to 260 m over about 120 km. The Marshes are surrounded by uplands composed of bedrock and till, in some places mantled by deposits of former glacial Lake Agassiz (Fig. 2). Upstream and, locally, downstream, the Saskatchewan River flows in incised valleys (Kuiper, 1960; Klohn, 1966).

The area was covered by ice until approximately 11 000 yr BP. Retreating glaciers deposited the The Pas moraine (Christiansen, 1979; Klassen, 1983; Schreiner, 1983), which defines the eastern border of the study area (Figs 2 & 3). As the glaciers retreated, large proglacial lakes were formed at and near the margins of the melting ice. Glacial Lake Saskatchewan (Christiansen, 1979,

1992) existed west of the study area about 11 500–10 200 yr BP. Glacial Lake Agassiz began to form in this area at approximately 10 200 yr BP (Elson, 1965, 1967; Fenton *et al.*, 1983; Klassen; 1983; Schreiner, 1983). The lake retreated to the north-east, as shown by relict shorelines and deltas at different elevations (Fig. 3), until it finally drained into the Tyrell Sea about 7500 yr BP (Klassen, 1983). Four regressive deltas of the lower Saskatchewan River exist west of the study area (Christiansen *et al.*, 1995). Two of these are related to Lake Saskatchewan (Fort a la Corne and Prince Albert deltas) and two (Nipawin, Mossyvale deltas) to Lake Agassiz (Fig. 3B). The Nipawin delta correlates with the highest Campbell level of glacial Lake Agassiz, with the shoreline at 375 m, and the Mossyvale delta corresponds to the McCauleyville level at about 335 m (Christiansen *et al.*, 1995). Between 8700 and 8000 yr BP, glacial Lake Agassiz retreated east of the The Pas moraine (Fig. 3A) (Klassen, 1983; Schreiner, 1983). The lower course of the Saskatchewan River, east of the The Pas moraine, was established between 8200 and 7750 yr BP (Grice, 1970). Therefore, post-glacial flow of the Saskatchewan River through the Cumberland Marshes area, following the withdrawal of the glacial Lake Agassiz, probably began some time between 8700 and 8200 yr BP (Fig. 3B).

Withdrawal of continental glaciers caused significant isostatic adjustments. As a result of rebound, former shorelines of the glacial Lake Agassiz show significant deformation (Fig. 3B). The possible effects of isostatic rebound on the alluvial history of the Saskatchewan River is considered later in this paper.

## CLIMATIC AND HYDROLOGICAL SETTING

The Saskatchewan River basin extends from the continental divide in the Rocky Mountains eastward to Cedar Lake and Lake Winnipeg in western Manitoba (Fig. 1). The largest portion of the basin is occupied by the interior plains of Alberta and Saskatchewan. About 70% of the mean annual flow in the Saskatchewan River is derived from the slopes and foothills of the Rocky Mountains. The plains contribute only 7% of the flow and 23% originates from the Cumberland Marshes and Manitoba Lowlands downstream (Ashmore, 1986). In contrast to the mean annual flow, most of the annual sediment yield is derived from the interior plains. The largest part of the sediment load is transported during spring–early summer peak discharges, originating from the snowmelt in the mountains and on the plains (Ashmore, 1986; Ashmore & Day, 1988). Prior to the construction of upstream dams,

more than half of the annual sediment load of the lower Saskatchewan River was deposited in the Cumberland Marshes (Kuiper, 1960; Ashmore, 1986).

During the Holocene, the Saskatchewan River basin underwent significant climatic and corresponding environmental changes (Ritchie, 1976, 1983; Martini & Glooschenko, 1985; Barnosky, 1989; Zoltai & Vitt, 1990; Ritchie & Harrison, 1993; Webb *et al.*, 1993; Vance *et al.*, 1995). Most workers believe that the mid-Holocene climates were warmer and generally drier than late Holocene climates and refer to this period as the Hypsithermal Interval (Ritchie, 1976, 1983; Martini & Glooschenko, 1985; Barnosky, 1989; Zoltai & Vitt, 1990). In the interior of Canada, this period was characterized by 17–29% greater aridity compared with the present (Zoltai & Vitt, 1990). Peak aridity occurred about 6000 yr BP (Ritchie & Harrison, 1993; Webb *et al.*, 1993; Vance *et al.*, 1995). The present climatic conditions were established between 5500 yr BP and 3000 yr BP, although the onset of these conditions varied locally (Ritchie & Hadden, 1975; Ritchie, 1976, 1983; Teller & Last, 1981; Zoltai & Vitt, 1990; Vance *et al.*, 1995). Data from the eastern slopes of the Rocky Mountains and foothills demonstrate that mid-Holocene aridity also extended there. This region experienced both early Holocene and mid-Holocene aridity prior to 5500–5000 yr BP, separated by a 1000-yr period of moist climate between approximately 9400 yr BP and 8400 yr BP (MacDonald, 1989).

Vegetational adjustments in central North America lagged about 50–200 yr behind the climatic changes (Bryson *et al.*, 1970). Following the withdrawal of glaciers and subsequent changes towards a warmer and drier climate, most of the drainage basin within the interior plains became occupied by prairie grassland (Ritchie, 1976, 1983; Ritchie & Harrison, 1993). After a shift to wetter conditions, vegetation over the northern part of the basin was replaced by forest and parkland, and the transition to modern climate and vegetation in the interior plains was accomplished by 3000–3500 yr BP (Ritchie & Hadden, 1975; Ritchie, 1976, 1983). In the Canadian Rocky Mountains, modern climate and vegetation patterns have been established since 5500–5000 yr BP (MacDonald, 1989).

Climate change greatly influenced the evolution of lakes in the interior of Canada (Heinselman, 1970; Teller & Last, 1981, 1982; Last & Schweyen, 1985; Hickman, 1987; Schweger & Hickman, 1989; Zoltai & Vitt, 1990; Vance *et al.*, 1992, 1993; Ritchie & Harrison, 1993; Vance & Last, 1994; Hickman & Schweger, 1996). During the Hypsithermal Interval, these lakes commonly were low, saline and subject to drying as a result of both low precipitation/evaporation ratios and, for lakes fed by streams, decreased stream discharges and flood magnitudes (Teller & Last, 1981). The end of the mid-Holocene period of drier and warmer climate was characterized by a transition to relatively deeper freshwater lakes.

The evolution of wetlands and peatlands was also affected by climate change. There is an almost complete absence of peatlands prior to 6000 yr BP in the interior plains south of latitude 54°30′, which may be explained by the general aridity, when evapotranspiration exceeded precipitation and water tables fluctuated seasonally (Zoltai & Vitt, 1990). Large fluctuations in groundwater tables favour increased decomposition, resulting in little organic accumulation (Zoltai & Vitt, 1990). Peatlands appeared south of latitude 54°30′, including the study area, only after 6000 yr BP, and rapid development of peatlands took place from 5000 to 4500 yr BP. Bogs dominated by *Sphagnum* moss appeared only after the transition to present-day climatic conditions (Zoltai & Vitt, 1990).

Climatic and related environmental changes in the Saskatchewan River basin are thought to have affected the Holocene fluvial history and stratigraphy of the Cumberland Marshes.

## METHODS

Interpretations of fluvial history were based on auger boreholes, aerial photographs, radiocarbon dates, crosscutting relationships of channel belts and the degree of channel-belt burial by wetland sediments. Nearly 200 manual boreholes, with depths up to 13 m, were obtained by gouge auger. This instrument recovered relatively undisturbed 1 m by 2-cm-diameter samples, which were logged in the field for sediment texture, sedimentary structures and contact relationships between different lithofacies. A total of 149 sediment samples were taken from these borings to the laboratory for grain-size analyses by pipette and sieve methods (Lewis, 1984), and 268 samples were examined for total organic content (TOC) by loss-on-ignition methods (Gross, 1971; Dean, 1974; Andrejko *et al.*, 1982; Gale & Hoare, 1991).

Forty-one samples of peat and organic-rich mud were collected for radiocarbon dating, and seven additional radiocarbon dates were provided by other sources (E.A. Christiansen, personal communication; Ducks Unlimited Canada, personal communication; Smith, 1983). Radiocarbon dates were considered consistent with each other if they overlapped within two standard deviations (Taylor, 1987). The dates for the beginning and end of fluvial activity for individual channel belts were taken as averages of the principal consistent radiocarbon dates. Sampling strategies for radiocarbon dating were similar to those applied by Törnqvist & Van Dijk (1993) and Törnqvist (1994). To date the beginning of fluvial activity for a particular channel belt, organic-rich deposits were sampled from directly beneath floodplain deposits associated with that channel belt. The end of fluvial activity was assumed to be represented by similar samples taken directly above siliciclastic floodplain deposits or, where available, from the base of channel-fill deposits of abandoned channel belts. In most cases this strategy provides consistent results for the beginning of the fluvial activity (Törnqvist & Van Dijk, 1993). Inferring the end of fluvial activity for a particular channel belt, however, is commonly more difficult because:

**1** abandonment may be very slow and gradual, with no clear-cut time of final cessation of alluvial deposition;

**2** organic sedimentation may not begin until many years after abandonment (Törnqvist & Van Dijk, 1993).

To minimize the possible diachroneity of clastic–organic transitions with increasing distance from the channel belt, organic-rich deposits were sampled less than 0.5 km from the channel of the associated channel belt. To check for the consistency of radiocarbon dates, the samples were taken, when possible, from two or more widely spaced boreholes on the same channel belt. In order to obtain a necessary amount of organic-rich sediment for conventional dating, about 10-cm thicknesses of samples were obtained from auger borings in most cases. This corresponds to about 100 yr of deposition for peat and about 60–70 yr of deposition for organic-rich mud (Table 1). These values are within typical standard deviation values (40–100 yr) for samples up to 5000 yr old using conventional decay-counting methods. For calculation of sedimentation rates (Table 1), radiocarbon dates (yr BP) were converted to calendar ages using the calibration program of Stuiver & Reimer (1993). All other dates in this paper are uncalibrated radiocarbon dates.

## STRATIGRAPHY AND FACIES OF HOLOCENE ALLUVIUM

Eight former courses of the Saskatchewan River are identified in the Cumberland Marshes. Seven of these are exposed and shown in Fig. 2. The oldest belt, the Saskeram (see later) has been identified solely from drilling results. On aerial photographs, channel belts are apparent as forested alluvial ridges separated by floodplain wetlands (flood basins) (Fig. 4). Some of these abandoned channel belts are solitary (e.g. Old Channel). Others, at least partly, represent channel-belt systems that

**Table 1.** Sedimentation rates of organic-rich deposits, Cumberland Marshes.

| Sample | Lithology | Thickness (m) | Radiocarbon date top (yr BP) | Radiocarbon date bottom (yr BP) | Principal calibrated dates top (yr BP) | Principal calibrated dates bottom (yr BP) | Accumulation rates (mm yr⁻¹) |
|---|---|---|---|---|---|---|---|
| 1 | Peat | 1.05 | 630 ± 40 | 1760 ± 80 | 563–627 | 1630–1690 | 0.93–1.05 |
| 2 | Organic mud | 0.70 | 2320 ± 70 | 2665 ± 120 | 2340 | 2760 | 1.67 |
| 3 | Organic mud | 0.70 | 2665 ± 70 | 3060 ± 75 | 2760 | 3210–3220 | 1.52–1.56 |
| 4 | Peat | 0.50 | N/A | 620 ± 40 | 60* | 561–524 | 0.87–1.0 |

* Age of top was determined from old maps. N/A, not applicable.

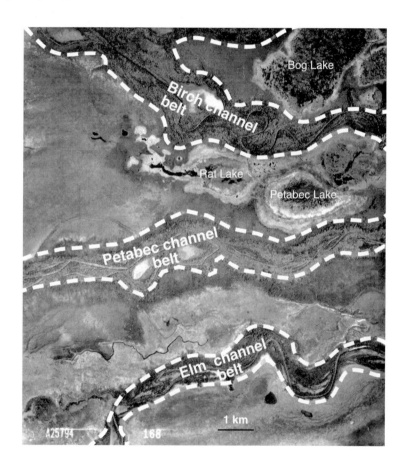

**Fig. 4.** Aerial photograph showing three former Holocene channel belts of the lower Saskatchewan River (Elm, Petabec, Birch) (white dashed outlines). Floodplain areas between channel belts contain wetlands and shallow lakes. Location is shown in Fig. 2.

consist of several channels diverging from the main channel (e.g. Birch–Petabec, Bloodsucker) (Fig. 5). Channel belts were delineated by present levee and channel positions. With increasing degree of burial, the apparent width of a channel belt decreases as forested levees are replaced by wetland environments.

Channel belts in the Cumberland Marshes commonly comprise well-sorted to moderately sorted fine and medium sand deposited in channels, moderately to very poorly sorted fine sand to sandy silt in splay complexes, poorly and very poorly sorted fine sand to medium silt, commonly rooted, in levees and silty proximal backswamp deposits (Fig. 6). In the area affected by the 1870s avulsion (Smith *et al.*, 1989, 1998), initial avulsive incursion into the floodbasin resulted in a belt of splay, channel and lacustrine delta deposition up to 10–12 km wide (Fig. 2) and averaging approximately 2 m thick. Flood-basin areas of some abandoned channel belt systems, e.g. the Birch–Petabec, are in places underlain by silty and sandy deposits in distal backswamp areas, which

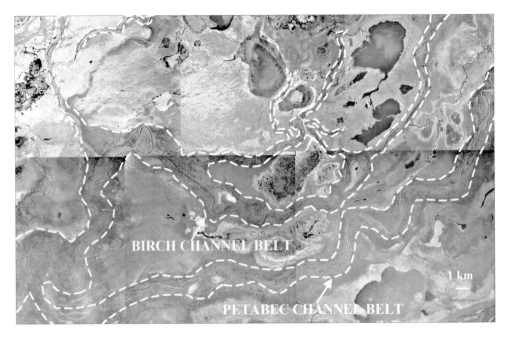

**Fig. 5.** Aerial photograph of the Birch–Petabec channel system (white dashed outlines). Floodplain areas between channel belts contain wetlands and shallow lakes. Location is shown in Fig. 2.

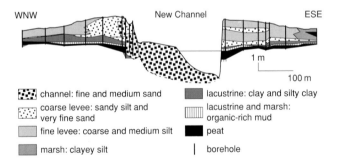

**Fig. 6.** Stratigraphical cross-section of the New Channel along line A–A' showing major facies of alluvial deposits. Section location is shown in Fig. 2.

suggests that the present width of an abandoned channel belt may represent only a part of the corresponding avulsion belt of relatively coarse deposition from which this channel belt evolved. Silty and sandy sediments of the avulsion belt, together with channel, levee, splay and proximal backswamp facies associated with large dominant channels, form in the areas of the floodplain directly affected by, or proximal to, avulsions. Channel belts in the Cumberland Marshes commonly are underlain by, and abandoned channel belts are overlain by, very fine silts and clays as well as organic-rich sediments, which

represent deposition in areas either distal to, or unaffected by, avulsion. Based on the modern sedimentary environments in the Cumberland Marshes, these deposits are inferred to have originated in wetland environments. Four principal wetland classes occur in the Cumberland Marshes: fens, bogs, marshes and shallow open lakes (Dirschl, 1972; Martini & Glooschenko, 1985; Zoltai & Polett, 1983). Fens and bogs commonly are peat-forming environments. Marshes typically are rich in fine-grained siliciclastic deposits, and shallow lakes may accumulate either siliciclastic sediments or organic-rich muds and

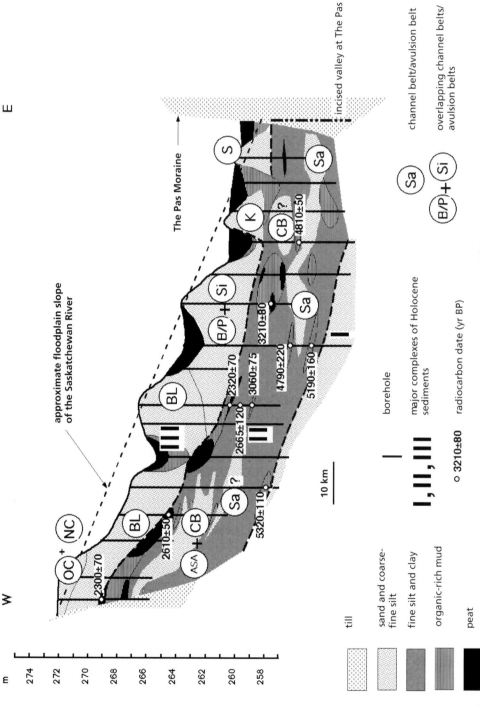

**Fig. 7.** Longitudinal cross-section of Holocene alluvium along line B–B' (section location is shown in Fig. 2). Letters in circles indicate avulsion/channel belts: Sa, Saskeram; CB, Cut Beaver; ASA, Ancestral South Angling; K, Kennedy; B/P, Birch–Petabec; BL, Bloodsucker; Si, Sipanok; OC, Old Channel; NC, New Channel; S, Saskatchewan River.

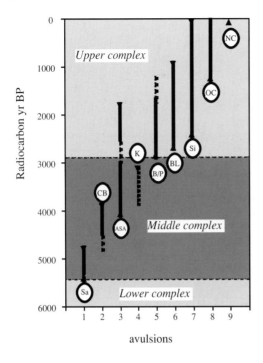

**Fig. 8.** Diagram showing periods of activity of main channels. Letters in circles indicate avulsion/channel belts: Sa, Saskeram; CB, Cut Beaver; ASA, Ancestral South Angling; K, Kennedy; B/P, Birch–Petabec; BL, Bloodsucker; Si, Sipanok; OC, Old Channel; NC, New Channel.

lake peats (Dirschl, 1972; Zoltai & Polett, 1983; Martini & Glooschenko, 1985), depending on their degree of isolation from floodwaters.

Examination of lithofacies in boreholes indicates a broad three-part stratigraphy of middle to late Holocene alluvium in the Cumberland Marshes: a lower complex (older than *c.* 5400 yr BP) dominated by silty and sandy deposits, a middle complex (5400–2600 yr BP) composed of mainly clay/very fine silt, and an upper complex (younger than *c.* 2600 yr BP) dominated by silty and sandy deposits. This stratigraphical ordering is especially apparent in the eastern two-thirds of the Marshes, away from the nodal region of dominant avulsions (Fig. 7).

## Lower complex

The lower complex was penetrated by a few boreholes only, and its character is thus not well known. The upper portion of this unit is composed mainly of coarse silt and very fine sand. In one borehole in the central part of the Marshes, this complex is composed of more than 8 m of medium to coarse sand, possibly a former channel deposit. The lower complex is older than *c.* 5400 yr BP, based on four radiocarbon dates from the base of organic-rich clays that directly overlie it (5320 ± 110, 5190 ± 160, 5750 ± 340 and 5360 ± 60 yr BP). Avulsion history prior to 5400 yr BP remains unknown.

## Middle complex

The middle complex of predominantly fine-grained deposits was formed between 5400 and 2800–2400 yr BP (Figs 7 & 8). Four avulsions occurred during the deposition of the middle complex: Saskeram, Cut Beaver, Ancestral South Angling and Kennedy (Fig. 8).

The earliest dated avulsion event was identified wholly from drilling results. This event, named the *Saskeram*, produced a belt of coarse deposits at least 45 km long and is related to a buried channel belt in the eastern part of the Cumberland Marshes west of Saskeram Lake (Figs 2 & 7). The westward extent of this channel belt has not been identified with certainty. Silty and sandy deposits in three sections in the western part of the Marshes, however, probably belong to the same belt (Fig. 7). Three samples of organic-rich clays, underlying silty and sandy floodplain deposits, provide dates of 5190 ± 160, 5360 ± 60 and 5750 ± 340 yr BP (average 5430 yr BP) to mark the beginning of fluvial activity. Two dates from organic-rich deposits above the Saskeram-equivalent floodplain sediments indicate the end of fluvial activity: 4790 ± 220 and 4810 ± 50 yr BP (average 4800 yr BP) (Figs 7 & 8).

The *Cut Beaver* channel belt is represented by several large and nearly buried meander loops, north-west of Cut Beaver Lake. This channel belt was traced further eastward, south of the Old Channel, from borehole data (Fig. 2). In the eastern and, possibly, western parts of the Marshes, the deposits of the Cut Beaver belt partially overlie the deposits of the older Saskeram belt. A small abandoned channel belt, west of Cut Beaver Lake, is probably a distributary that resulted from the same avulsion (Fig. 2). The end of fluvial activity is dated as 3930 ± 80 yr BP, based on organic-rich mud overlying Cut Beaver overbank sediments (Figs 8 & 9).

The *Ancestral South Angling* channel-belt system consists of three channel belts. Two exposed distributaries are located north of the Old Channel (Fig. 2), and one buried distributary was identified close to the present Mossy River (Fig. 9). This avulsion occurred approximately 4070 yr BP, based on two dates of 4190 ± 70 and

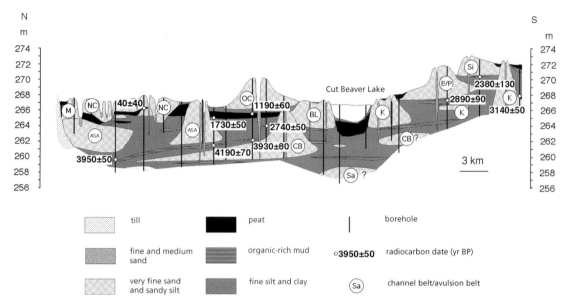

**Fig. 9.** Transverse cross-section of Holocene alluvium along line C–C′ (section location is shown in Fig. 2). Letters in circles indicate avulsion/channel belts: Sa, Saskeram; CB, Cut Beaver; ASA, Ancestral South Angling; K, Kennedy; B/P, Birch–Petabec; BL, Bloodsucker; Si, Sipanok; OC, Old Channel; NC, New Channel; M, Mossy River.

$3950 \pm 50$ yr BP (Figs 8 & 9). One sample was taken from directly beneath upward-coarsening proximal back-swamp sediments associated with this avulsion, the final single-thread channel of which is now partially reoccupied by the South Angling Channel (Smith *et al.*, 1998). The second sample was taken from beneath overbank sediments of a buried distributary of this channel system (Fig. 9). The sample marking the end of fluvial activity ($1730 \pm 50$ yr BP) came from organic-rich sediments directly above levee deposits of the Ancestral South Angling (Fig. 9).

The *Kennedy* avulsion is named after the prominent channel belt of Kennedy Creek in the western portion of the Marshes north of Kennedy Lake (Fig. 2). Based on cross-cutting relationships and the stratigraphical position of channel-belt deposits, smaller channel belts were probably distributaries of the Kennedy channel belt. The time of initial avulsion is uncertain. In the eastern part of the Marshes, however, the deposits of the Kennedy belt overlie those of the older Cut Beaver and Saskeram belts (Fig. 7). The base of the organic-rich clay above the levee of the main Kennedy channel belt yields an age of $3140 \pm 50$ yr BP for the end of major fluvial activity (Figs 8 & 9).

## Upper complex

Deposition of the upper complex began about 2600 yr BP, during which five avulsions occurred (Fig. 8). Its lower boundary is time-transgressive and corresponds, in different parts of the Marshes, to the onset of the Birch–Petabec, Sipanok and Bloodsucker avulsions, which interrupted fine-grained deposition between 2835 and 2390 yr BP. Two additional avulsions, the Old Channel and the New Channel, occurred after the initiation of the upper complex.

The *Birch–Petabec* is a complex channel-belt system that consists of two main alluvial ridges, which diverge from a single ridge 40 km east of the main avulsion node (Figs 2 & 5). The Birch belt consists of several distributaries, whereas the Petabec comprises a single belt (Fig. 5). The main Birch–Petabec avulsion invaded a flood-basin depression between two channel belts of the Kennedy system (Fig. 2). Two boreholes through levee deposits provide the dates for the beginning of fluvial activity: $2890 \pm 90$ and $2780 \pm 70$ yr BP (average 2835 yr BP). Although the Birch–Petabec system is now essentially abandoned, the age of channel abandonment is uncertain.

The *Bloodsucker* avulsion produced several distributary channels and involved partial reoccupation of the Cut Beaver channel belt. This avulsion took place between the channel belts of the Kennedy and Birch systems to the south and the Ancestral South Angling to the north (Fig. 2). Two samples date the beginning of channel belt activity: $2610 \pm 50$ (Fig. 7) and $2740 \pm 50$ yr BP (Fig. 9), averaging about 2675 yr BP. The end of fluvial activity of the Bloodsucker channel belt, dated by one organic-rich silt sample from above associated overbank deposits, is 940 ± 40 yr BP (Fig. 8).

The *Sipanok* avulsion took place between the Kennedy and main Birch–Petabec alluvial ridges (Fig. 2). Aerial photographs suggest that part of the Sipanok flow probably was diverted through the present Culdesac River channel belt, although we have no borehole data from the south-eastern region of the Marshes to confirm this. The proximal part of the Sipanok channel belt is solitary, whereas closer to the Culdesac River several distributaries can be identified (Fig. 2). The Sipanok avulsion involved appropriation of part of the Carrot River and the lower reaches of the Petabec channel belt. Two samples date the beginning of fluvial activity: 2400 ± 60 and 2380 ± 130 yr BP (Fig. 9), averaging about 2390 yr BP. The present Saskatchewan River channel belt, east of its junction with the Petabec belt, was established at about 1790 ± 70 yr BP, during a period of major activity of the Sipanok channel. The Sipanok channel carried significant flows at the beginning of the present century (Horsey, 1924), but today it is virtually abandoned.

The *Old Channel* was the main course of the Saskatchewan River prior to the 1870s avulsion, and it still remains partly active, carrying about 5–10% of the annual flow. Five samples from beneath associated levee and flood-basin deposits yield consistent dates for the beginning of activity of the Old Channel: 1390 ± 50, 1230 ± 40, 1190 ± 60, 1230 ± 40 and 1390 ± 40 yr BP, averaging about 1290 yr BP (Fig. 8). In addition, a wood fragment from near the base of sand in the Old Channel was dated 890 ± 140 yr BP (E.A. Christiansen, personal communication). The Old Channel is a solitary channel belt that probably reoccupied the Bloodsucker belt upstream and the Birch–Petabec and Sipanok channel belts downstream, east of Cumberland House (Fig. 2). Thus, the present course of the Saskatchewan River downstream of the Petabec junction is approximately 1790 yr old, whereas upstream of the Petabec belt to the 1870s avulsion node, it is younger than 1290 yr BP.

The date of the most recent avulsion is known from historical records (Smith *et al.*, 1998). Descriptions of the *New Channel* avulsion and the evolution of the breakout area following the avulsion are provided elsewhere (Smith *et al.*, 1989, 1998; Smith & Pérez-Arlucea, 1994; Pérez-Arlucea & Smith, 1999).

## AVULSION STYLES

The principal middle and late Holocene avulsions of the Saskatchewan River are mainly *nodal* avulsions (Leeder, 1978; Mackey & Bridge, 1995), which occurred near the western edge of the Cumberland Marshes (Fig. 2), where the slope decreases abruptly as the river enters the flat lacustrine plain of former Lake Agassiz (Fig. 1b). This imposed reduction in gradient has presumably resulted in greater alluvial aggradation throughout the Holocene, which, combined with the unconfined character of the lacustrine basin, resulted in most major avulsions occurring near this entry point. Nine avulsions have occurred during the past 5400 yr (Fig. 8, Table 2), indicating an average avulsive frequency of about 600 yr. For those channel belts for which both the beginning and the end of fluvial activity are dated, periods of activity lasted up to 2300–2400 yr (e.g. Ancestral South Angling, Sipanok). Significant overlap between periods of activity of different channel belts suggests that, in most cases, new avulsions were accompanied by gradual abandonment of older channels. This is well demonstrated by the present coexistence of the New and the Old Channels since the beginning of the 1870s avulsion.

Two basic avulsion styles can be identified from the Holocene record of the Saskatchewan River (Table 2):
**1** channel reoccupation;
**2** extensive deposition and progradation into adjacent floodbasins.
Cases in which avulsions resulted in reoccupation of older channels include:
**1** the Sipanok Channel reoccupied segments of both the Carrot River and the lower Birch–Petabec belts;
**2** the Old Channel reoccupied portions of the Birch–Petabec–Sipanok belts and probably the Bloodsucker belt;
**3** the Bloodsucker partly reoccupied the Cut Beaver belt;
**4** the New Channel in the present proximal part of the 'breakout area' presently occupies about 4 km of the proximal Ancestral South Angling belt and 13 km of the former Torch channel (Smith *et al.*, 1998).
Reoccupation was accompanied by modification of solitary channel belts and did not involve extensive avulsive deposition in adjoining floodplains. This style thus does not produce wide avulsion belts from which single dominant channels eventually develop.

In other cases, avulsions occurred by flow diversion on to adjacent flood basins, similar to the breakout area of the 1870s avulsion and with little or no obvious appropriation of pre-existing channels. These avulsions resulted in multiple channel systems (e.g. the Kennedy, Ancestral South Angling, Birch–Petabec and Bloodsucker systems). They were characterized by the formation of distributary channel systems and progradation of splay complexes and small deltas into flood basins, producing belts of avulsive deposits with overall widths that may greatly exceed the width of the subsequent dominant channel belts. For example, the Cut Beaver avulsion belt attains a width

**Table 2.** Summary of the Holocene alluvial history of the Saskatchewan River at the Cumberland Marshes.

| Avulsion event | Type | Beginning of activity (yr BP) | End of activity (yr BP) | Avulsion style | Lake levels | Floodplain stratigraphy |
|---|---|---|---|---|---|---|
| New Channel (1870s) | Solitary and distributary | 120 | Active | Reoccupation and progradation | | |
| Old Channel | Mainly solitary | 1290 | Still partly active | Mainly reoccupation | Stable | Upper Complex |
| Sipanok | Solitary and distributary | 2390 | Beginning of 20th century | Reoccupation and progradation | Stable | Upper Complex |
| Bloodsucker | Distributary | 2675 | 940 | Reoccupation and progradation | | |
| Birch–Petabec | Distributary | 2835 | ? | Mainly progradation | | |
| Kennedy | Distributary | ? | 3140 | Progradation | | |
| Ancestral South Angling | Distributary | 4070 | 1730 | Progradation | Rising | Middle Complex |
| Cut Beaver | Distributary | ? | 3930 | Progradation | Rising | Middle Complex |
| Saskeram | ? | 5430 | 4800 | ? | | |

of at least 18 km (Fig. 9), considerably greater than the exposed channel belt (Fig. 2). Avulsion style during the formation of the middle complex was dominated by splay-complex progradation and distributary-type channel systems (Table 2). The Saskeram avulsion lacks sufficient data, however, to confirm its dominant style. Avulsions during the deposition of the upper complex involved both progradation and reoccupation (Table 2).

Thus, the broad three-part alluvial stratigraphy, combined with the record of avulsions, indicate continuous change for the Saskatchewan River floodplain through the middle and late Holocene. These changes are discussed next in the context of climatic, vegetational and isostatic adjustments in the Saskatchewan River basin following the withdrawal of continental glaciers.

## MAJOR CONTROLS OF ALLUVIAL HISTORY

### Influence of climate on discharge and sediment load of the Saskatchewan River

Climate affects the mean discharge, magnitude and frequency of floods, and sediment load, both directly and indirectly, through vegetation that controls overland flow and sediment yields (Schumm, 1977; Knox, 1983). In the long term, the transition to cooler and wetter climates results not only in higher mean discharges, but also more frequent large floods (Knox, 1993). Mean annual flow

and the flood regime of the Saskatchewan River during the Holocene was determined largely by climatic changes in the Rocky Mountains. Since the establishment of modern climate and vegetation in the Rocky Mountains from 5000–5500 yr BP to present, the mean annual flow and flood regimes of the Saskatchewan River probably have not changed much. On the other hand, the sediment load of the Saskatchewan River was controlled largely by climatic and vegetational changes in the interior plains. Compared with forest and parkland vegetation, prairie vegetation provides higher amounts of annual sediment yield per unit area and increased stream sediment loads (Knox, 1983). For example, the two major branches of the Saskatchewan River, the North and South Saskatchewan rivers, account for 41% and 36% of the total discharge of the Saskatchewan River at The Pas, respectively (Ashmore, 1986). Prior to reservoir construction, however, the South Saskatchewan River, draining grasslands, provided two-thirds of the annual sediment load, compared with one-third from the North Saskatchewan River, flowing mainly through forest and parkland (Ashmore, 1986). Thus, the dry Hypsithermal Interval on the interior plains probably provided higher sediment yields and suspended-sediment concentrations in the Saskatchewan River.

The overall effect of climate and vegetation change on the hydraulic regime of the Saskatchewan River may be summarized as follows: a period of low discharges, smaller floods, and probably higher sediment concentrations occurred before 5500–5000 yr BP, during which the lower complex of Holocene alluvium was deposited. The limitations of our data do not permit us to infer the fluvial character of this lower coarse complex, but we speculate that the channels may have been braided. A period of present-day discharges and flood regimes, with possibly higher-than-present sediment loads, occurred between 5500–5000 and 3500–3000 yr BP, accompanied by formation of the middle complex, when the interior plains underwent transition to modern climatic and vegetational conditions. Present hydrological and sediment-supply conditions have been present since 3500–3000 yr BP, corresponding to the deposition of the upper complex. Variations in hydrology and sediment load alone, however, do not explain adequately the differences between the middle and the upper complexes or the differences in dominant avulsion styles.

## Climatic control on lake-level change

During the Holocene, varying proportions of the Cumberland Marshes have been occupied by shallow floodplain lakes. Elevations of Holocene lacustrine deposits plotted against their radiocarbon ages (Fig. 10) may thus be used to infer changes in lake-bottom elevations with time. In most cases, these deposits are represented by very fine grey silt and clay with a pelecypod shells and fine-grained dispersed organic matter. Black, fine-grained peat also has a lacustrine origin (Dirschl,

1972; Martini & Glooschenko, 1985). These pre-avulsive deposits generally are similar to lacustrine sediments presently forming in floodplain lakes of the Marshes. To correct for variation in elevation resulting solely from regional floodplain slope, each observed elevation was adjusted to a horizontal plane. This was done by adding the elevation of each dated lacustrine sample to the product of floodplain gradient (0.00013 based on PFRA, 1954) and distance from the avulsion node. Results indicate that the elevations of lake bottoms generally increased between 5700 and *c.* 2500 yr BP and either did not change or dropped slightly from *c.* 2500 yr BP until present (Fig. 10). Lake-level rise was accompanied by the deposition of the middle complex of Holocene alluvium. The upper complex was formed when lake levels remained stable or dropped slightly. The scatter of data points requires additional explanation. Modern lacustrine deposits were sampled from Cut Beaver, Egg, Cumberland and Saskeram lakes (Fig. 2). Present floodplain lakes in the Cumberland Marshes are situated at different elevations, but typically have small depth ranges of about 1–3 m. As a consequence, present-day lake bottoms are distributed over a narrow adjusted elevation range. The increase in the wet floodplain area during the formation of the middle complex probably was accompanied both by formation of new lakes at higher elevations and increased depths of existent lakes. From this point of view, it is reasonable that the maximum scatter is observed between

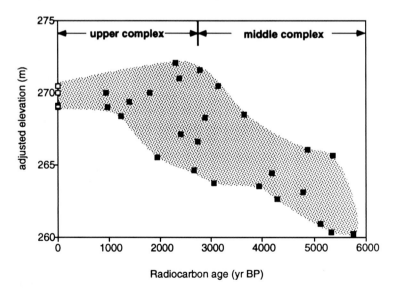

**Fig. 10.** Elevations of Holocene and present lake-bottom deposits in the Cumberland Marshes. Lake-bottom elevations are adjusted for regional gradient.

2500 and 3000 yr BP, when the lakes and other wetland areas reached their maximum extents (Fig. 10). The slight drop in lake-bottom elevation maxima during the formation of the upper complex probably is caused by progressive lake infilling by avulsions.

Thus, at the end of Hypsithermal Interval, the lake levels and the proportion of the Cumberland Marshes occupied by floodplain lakes and wetlands began to increase. This trend corresponds to the evolution of other lakes in central Canada (Fig. 11). The transition for most lakes began between 4000 and 5000 yr BP, with earlier changes reported for central Alberta lakes (Schweger & Hickman, 1989). Rising lake levels accompanied higher precipitation/evaporation ratios and groundwater levels related to the climatic transition toward wetter and cooler climate. Consequently, the relative proportion of fine-grained flood-basin sediments in the Cumberland Marshes increased also. Possible longitudinal gradient reduction and decreased competence of flow as a result of rising lake levels also contributed to the finer grained texture of alluvium. Generally higher-than-present sediment loads of the Saskatchewan River between 5500 and *c.* 3000 yr BP compensated for the increasing accommodation space created by lake-level rise. Beginning approximately 3500–3000 yr BP, after present-day climatic and vegetational conditions became established, lacustrine and wetland areas ceased to enlarge and began to fill with deposits of the Saskatchewan River, resulting in increasing proportions of relatively coarse-grained sediments, beginning with the Birch–Petabec avulsion (Fig. 8).

The formation of organic-rich sediments in the Cumberland Marshes was influenced both by regional climatic conditions (Zoltai & Vitt, 1990) and by local variations in the intensity of alluvial deposition. Organic-rich deposits older than 3000 yr BP are dominated by organic-rich clays and black, fine-grained lake peats. The later appearance of coarse sedge peats is believed to reflect in part the shallowing and infilling of floodplain lake basins by avulsive deposits. Subsequent abandonment and isolation from active channels was followed by formation of extensive peat layers, with ages younger than approximately 3000 yr BP (Figs 7 & 9).

Avulsions occurred throughout the middle and late Holocene, apparently independent of climatic changes. The styles of these avulsions and sedimentation patterns, however, are believed to have been controlled, in part, by changing levels and areal extents of floodplain lakes. During the period of lake expansion, avulsions were commonly progradational in style, producing laterally widespread deposits and multichannel distributary networks. During the subsequent period of generally stable lake levels, avulsions took place both by reoccupation of older channel belts and by diversion into flood basins bounded by elevated alluvial ridges. Locally, water levels and areal extents of floodplain lakes were undoubtedly increased by the avulsions themselves (cf. Smith *et al.*, 1998).

## Isostatic adjustments

Based on the deformation of glacial Lake Agassiz strandlines (Fig. 3B), the maximum gradient of isostatic rebound from south-west to north-east is approximately $0.5 \text{ m km}^{-1}$ (Elson, 1967; Teller & Thorleifson, 1983), with the gradient in the west–east direction equal to approximately $0.35 \text{ m km}^{-1}$. The width of the Cumberland Marshes floodplain west of the The Pas moraine is about 120 km. This implies that the eastern edge of the study area rebounded about 42 m more than the western edge. The bottom of the incised channel at The Pas, however, is only about 13 m below the top of the The Pas moraine (Fig. 7). Even if all of this incision were attributed solely to rebound, its amount appears to be only 13 m during the past 8200–8700 yr, suggesting that most rebound took place immediately after deglaciation. This observation is supported by the data from other formerly glaciated areas (Andrews & Peltier, 1989; Tushingham & Peltier, 1991; Clark *et al.*, 1994). In addition, the stratigraphy of Holocene alluvium since approximately 5400 yr BP shows no significant effects of isostatic rebound. The slopes of the boundaries between the three major complexes of Holocene sediments, the slopes of the lines connecting organic-rich layers of the same age, and the present regional gradient of the Cumberland Marshes area are all more or less parallel to each other (Fig. 7). If rebound effects were significant during floodplain deposition, the slopes could be expected to reduce with time. This implies that rebound did not affect lake-level changes during the deposition of the middle and upper complexes of Holocene alluvium. We therefore conclude that isostatic rebound has had little or no effect on middle and late Holocene deposition patterns in the Cumberland Marshes. The influence of rebound, however, was probably still significant immediately after the withdrawal of glacial Lake Agassiz and contributed to the uplift of the The Pas moraine, which contains the only outlet for the Cumberland Marshes.

## CONCLUSIONS

**1** Abandoned channel belts of the Saskatchewan River in the Cumberland Marshes indicate a history of Holocene avulsions that pre-date the 1870s avulsion. At least nine

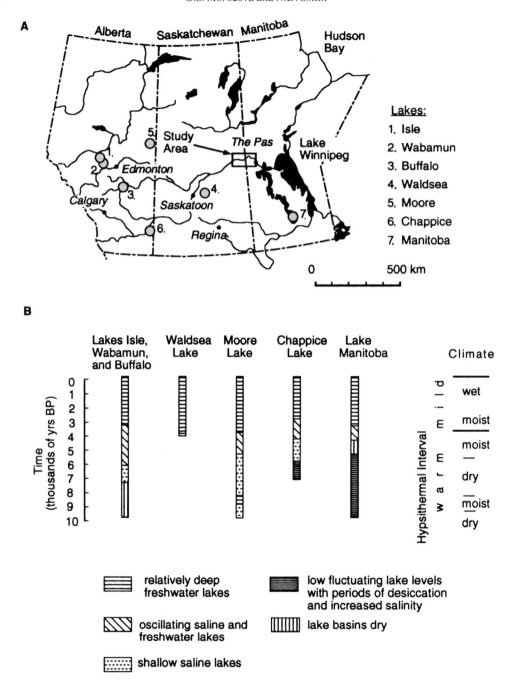

**Fig. 11.** Holocene climate change and lake evolution in south-central Canada. (A) Map showing location of lake basins. (B) Holocene climate and relative stages for selected lakes (after Teller & Last, 1981; Last & Schweyen, 1985; Schweger & Hickman, 1989; Vance *et al.*, 1992, 1993, 1995; Ritchie & Harrison, 1993; Hickman & Schweger, 1996).

major avulsions have taken place since about 5400 yr BP, indicating an average avulsive frequency of about 600 yr. The periods of major activity commonly overlapped, reflecting coexistence of more than one active channel belt at a time. Middle and late Holocene avulsions were mainly nodal, originating at the point where the Saskatchewan River enters the Marshes from the west.

**2** Holocene alluvium in the eastern two-thirds of the Marshes is composed of three major complexes: a lower complex older than 5400 yr BP, dominated by silty and sandy sediments, a middle complex (from 5400 to 2600 yr BP) of predominantly fine silt and clay deposits, and an upper complex (younger than 2600 yr BP) of mainly silty and sandy deposits. Fluvial character and avulsion history during the formation of the lower complex is unknown. Four main avulsions occurred during the deposition of the middle complex, and five avulsions occurred during the formation of the upper complex.

**3** During the past 5400 yr, the interval embraced by the data of this study, the alluvial history was affected by climatic changes. The lower complex of Holocene alluvium was formed during a period of warmer and drier climate, characterized by lower discharges, possibly higher sediment concentrations, lower precipitation/evaporation ratios, lower groundwater levels and grassland vegetation over most of the Saskatchewan River drainage basin. The middle complex was formed during the transition to the present cooler and wetter climate, accompanied by increasing discharges and flood magnitudes, higher precipitation/evaporation ratios, higher groundwater levels and transition to present vegetation patterns. These changes were accompanied by an increase in abundance and size of floodplain lakes and wetlands. The upper complex was formed under present climatic and vegetational conditions, when existing wetlands and lakes were filled in during avulsions.

**4** Avulsion styles were controlled largely by levels and areal extents of floodplain lakes. Two basic avulsion styles were identified from the Holocene record: channel reoccupation and splay-complex progradation. The first style led to the modification of existing channel belts, not preceded by widespread deposition and resulting in relatively little floodplain aggradation. The second style is characterized by widespread deposition of silty and sandy sediments in flood-basin depressions, followed by multichannel distributary patterns and, eventually, a single dominant channel. The period of lake-level rise between 5400 and about 2600 yr BP was dominated by progradational avulsions, whereas both reoccupation and progradational styles were typical for avulsions that occurred after 2600 yr BP, when lake levels remained stable.

## ACKNOWLEDGEMENTS

This study was supported by NSF grant EAR-9304104 and Geological Society of America grant 5891–96. $^{14}$C dates were determined by Salvatore Valastro, Radiocarbon Laboratory, University of Texas at Austin. Seven additional radiocarbon dates and core descriptions were provided by E.A. Christiansen (personal communication), Ducks Unlimited Canada (personal communication), and Saskatchewan Highways and Transportation. We acknowledge the interest and assistance of the people of Cumberland House. Special thanks to Gary Carriere and Morley Hoksbergen for help with field work; to Donald Fiddler, Gilbert Gibb and Ducks Unlimited Canada for providing technical support; and to Stella Hoksbergen for her hospitality. We thank Andres Aslan, Earl Christiansen and Bryan Schreiner for constructive criticism of an earlier version of the manuscript.

## REFERENCES

ALEXANDER, J. & LEEDER, M.R. (1987) Active tectonic control on alluvial architecture. In: *Recent Developments in Fluvial Sedimentology* (Eds Ethridge, F.G., Flores, R.M. & Harvey, M.D.), Spec. Publ. Soc. econ. Paleont. Miner., Tulsa, **39**, 243–252.

ANDREJKO, M.J., FIENE, F. & COHEN, A.D. (1982) Comparison of ashing techniques for determination of the inorganic content of peats. In: *Testing of Peats and Organic Soils* (Ed. Jarett, P.M.). *Am. Soc. Test. Mater. Spec. Tech. Publ.*, **820**, 5–20.

Andrews, J.T. & PELTIER, W.R. (1989) Quaternary geodynamics in Canada. In: *Quaternary Geology of Canada and Greenland*, Vol. 1, *Geology of Canada* (Ed. Fulton, R.J.), pp. 543–572. Geological Survey of Canada, Ottawa.

ASHMORE, P.E. (1986) *Suspended Sediment Transport in the Saskatchewan River Basin*. Sediment Survey Station, Water Survey of Canada, Water Resources Branch, Inland Waters Directorate, Environment Canada, Winnipeg 204 pp.

ASHMORE, P.E. & DAY. T.J. (1988) Spatial and temporal patterns of suspended-sediment yield in the Saskatchewan River basin. *Can. J. Earth Sci.*, **25**, 1450–1463.

AUTIN, W.J. (1992) Use of alloformations for definition of Holocene meander belts in the middle Amite River, southeastern Louisiana. *Geol. Soc. Am. Bull.*, **104**, 233–241.

AUTIN, W.J., BURNS, S.F., MILLER, B.J., SAUCIER, R.T. & SNEAD, J.I. (1991) Quaternary geology of the lower Mississippi valley. In: *Quaternary Nonglacial Geology: Conterminous U.S.*, Vol. K-2, *The Geology of North America* (Ed. Morrison, R.B.), pp. 547–582. The Geological Society of America, Boulder, CO.

BARNOSKY, C.W. (1989) Postglacial vegetation and climate in the northwestern Great Plains of Montana. *Quat. Res.*, **31**, 57–73.

BLUM, M.D. & VALASTRO, S., JR. (1994) Late Quaternary sedimentation, lower Colorado River, Gulf Coastal Plain of Texas. *Geol. Soc. Am. Bull.*, **106**, 1002–1016.

BRIDGE, J.S. & LEEDER, M.R. (1979) A simulation model of alluvial stratigraphy. *Sedimentology*, **26**, 617–644.

BRIZGA, S.O. & FINLAYSON, B.L. (1990) Channel avulsion and river metamorphosis: the case of the Thompson River, Victoria, Australia. *Earth Surf. Process. Landf.*, **15**, 391–404.

BRYANT, M., FALK, P. & PAOLA, C. (1995) Experimental study of avulsion frequency and rate of deposition. *Geology*, 23, 365–368.

BRYSON, R.A., BAERRIS, D.A. & WENDLAND, W.M. (1970) Tentative climatic patterns for some late glacial and postglacial episodes in central North America. In: *Life, Land and Water* (Ed. Mayers-Oakes, W.J.), pp. 271–298. University of Manitoba, Winnipeg.

CHRISTIANSEN, E.A. (1979) The Wisconsinan deglaciation of southern Saskatchewan and adjacent areas. *Can. J. Earth Sci.*, **16**, 913–938.

Christiansen, E.A. (1992) Pleistocene stratigraphy of the Saskatoon area, Saskatchewan, Canada: an update. *Can. J. Earth Sci.*, **29**, 1767–1778.

CHRISTIANSEN, E.A., SAUER, E.K. & SCHREINER, B.T. (1995) Glacial Lake Saskatchewan and Lake Agassiz deltas in east-central Saskatchewan with special emphasis on the Nipawin delta. *Can. J. Earth Sci.*, **32**, 334–348.

CLARK, J.A., HENDRIKS, M., TIMMERMANS, T.J., STRUCK, C. & HILVERDA, K.J. (1994) Glacial isostatic deformation of the Great Lakes region. *Geol. Soc. Am. Bull.*, **106**, 19–31.

DEAN, W.E.J. (1974) Determination of carbonate and organic matter in calcareous sediments and sedimentary rocks by loss on ignition: comparison with other methods. *J. sediment. Petrol.*, **44**, 242–248.

DIRSCHL, H.J. (1972) Geobotanical processes in the Saskatchewan River Delta. *Can. J. Earth Sci.*, **9**, 1529–1549.

ELSON, J.A. (1965) Soils of the Lake Agassiz region. In: *Soils in Canada* (Ed. Legget, R.F.), *R. Soc. Can. Spec. Publ.*, **3**, 51–79.

ELSON, J.A. (1967) Geology of glacial Lake Agassiz. In: *Life, Land and Water* (Ed. Mayers-Oakes, W.J.), pp. 37–96. University of Manitoba Press, Winnipeg.

Fenton, M.M., MORAN, S.R., TELLER, J.T. & CLAYTON, L. (1983) Quaternary stratigraphy and history in the southern part of the lake Agassiz basin. In: *Glacial Lake Agassiz* (Ed. Teller, J.T. & Clayton, L.). *Geol. Assoc. Can. Spec. Pap.*, **26**, 49–74.

GALE, S.J. & HOARE, P.G. (1991) *Quaternary Sediments*. Belhaven Press, New York–Toronto, 323 pp.

GRICE, R.H. (1970) Quaternary geology of the Grand Rapids area, Manitoba. *Can. J. Earth Sci.*, **7**, 853–857.

GROSS, M.G. (1971) Carbon determination. In: *Procedures in Sedimentary Petrology* (Ed. Carver, R.E.), pp. 573–596. John Wiley & Sons, New York.

HEINSELMAN, M.L. (1970) Landscape evolution, peatland types, and the environment in the Lake Agassiz peatland natural area, Minnesota. *Ecol. Monogr.*, **40**, 235–261.

HELLER, P.L. & PAOLA, C. (1996) Downstream changes in alluvial architecture: an exploration of controls on channel-stacking patterns. *J. sediment. Res.*, **66**, 297–306.

HICKMAN, M. (1987) Paleolimnology of a large shallow lake: Cooking Lake, Alberta, Canada. *Arch. Hydrobiol.*, **111**, 121–136.

HICKMAN, M. & SCHWEGER, C.E. (1996) The late Quaternary palaeoenvironmental history of a presently deep freshwater lake in east-central Alberta, Canada and palaeoclimate implications. *Palaeogeogr. Palaeoclimatol. Palaeoecol.*, **123**, 161–178.

HORSEY, G.F. (1924) *Carrot River Reclamation Project*. Dominion Water Power and Reclamation Service, Department of Interior, Canada, 45 pp.

KLASSEN, R.W. (1983) Assiniboine delta and Assiniboine–Qu'Appelle valley system—implications concerning the history of Lake Agassiz in southern Manitoba. In: *Glacial Lake Agassiz* (Eds Teller, J.T. & Clayton, L.). *Geol. Assoc. Can. Spec. Pap.*, **26**, 211–229.

KLOHN, J.K. (1966) Design and performance of earthwork and foundations for the Squaw Rapids development. *Can. Geotech. J.*, **IV**, 244–300.

KNOX, J.C. (1983) Responses of river systems to Holocene climates. In: *Late-Quaternary Environments of the United States* (Ed. Wright, H.E., Jr.), pp. 26–41. University of Minnesota Press, Minneapolis.

KNOX, J.C. (1993) Large increases in flood magnitude in response to modest changes in climate. *Nature*, **36**, 430–432.

KUIPER, E. (1960) Sediment transport and delta formation. *Proc. Am. Soc. civ. Eng. J. Hydraul. Div.*, **86**, 244–300.

LAST, W.M. & SCHWEYEN, T.H. (1985) Holocene history of Waldsea Lake, Saskatchewan, Canada. *Quat. Res.*, **24**, 219–234.

LEEDER, M.R. (1978) A quantitative stratigraphic model for alluvium with special reference to channel deposit density and interconnectedness. In: *Fluvial Sedimentology* (Ed. Miall, A.D.), Mem. Can. Soc. petrol. Geol., Calgary, **5**, 587–596.

LEWIS, D.W. (1984) *Practical Sedimentology*. Hutchinson Ross, New York, 229 pp.

MacDONALD, G.M. (1989) Postglacial palaeoecology of the subalpine forest–grassland ecotone of southwestern Alberta: new insights on vegetation and climate change in the Canadian Rocky Mountains and adjacent foothills. *Palaeogeogr. Palaeoclimatol. Palaeoecol.*, **73**, 155–173.

MACKEY, S.D. & BRIDGE, J.S. (1995) Three-dimensional model of alluvial stratigraphy: theory and application. *J. sediment. Res.*, **B65**, 7–31.

MARTINI, I.P. & GLOOSCHENKO, W.A. (1985) Cold climate peat formation and its relevance to Lower Permian coal measures of Australia. *Earth-Sci. Rev.*, **22**, 107–140.

McCARTHY, T.S., ELLERY, W.N. & STANISTREET, I.G. (1992) Avulsion mechanisms on the Okavango fan, Botswana: the control of a fluvial system by vegetation. *Sedimentology*, **39**, 779–795.

PÉREZ-ARLUCEA, M. & SMITH, N.D. (1999) Depositional patterns following the 1870s avulsion of the Saskatchewan River (Cumberland Marshes). *J. sediment. Res*, **69**, 62–73.

PFRA (1954) *Interim Report No. 8. Saskatchewan River Reclamation Project, Hydrometric Surveys 1954*. Department of Agriculture, Prairie Farm Rehabilitation Administration, Engineering Branch, Winnipeg, Manitoba, 36 pp.

RICHARDS, K., CHANDRA, S. & FRIEND, P. (1993) Avulsive channel systems: characteristics and examples. In: *Braided Rivers* (Eds Best, J.L. & BRISTOW, C.S.), Spec. Publ. geol. Soc. London, No. 75, 195–203. Geological Society of London, Bath.

RITCHIE, J.C. (1976) The late Quaternary vegetational history of the Western Interior of Canada. *Can. J. Bot.*, **54**, 1793–1818.

RITCHIE, J.C. (1983) The paleoecology of the central and northern parts of the glacial lake Agassiz basin. In: *Glacial Lake Agassiz* (Ed. Teller, J.T. & Clayton, L.), *Geol. Assoc. Can. Spec. Pap.*, **26**, 157–170.

RITCHIE, J.C. & HADDEN, K.A. (1975) Pollen stratigraphy of Holocene sediments from the Grand Rapids area, Manitoba, Canada. *Rev. Palaeobot. Palynol.*, **19**, 193–202.

RITCHIE, J.C. & HARRISON, S.P. (1993) Vegetation, lake levels, and climate change in western Canada during Holocene. In: *Global Climates Since the Last Glacial Maximum* (Eds Wright, H.E., Kutzbach, J.E., Webb, III, T., Ruddiman, W.F., Street-Perrot, F.A. & Bartlein, P.J.), pp. 401–414. University of Manitoba Press, Minneapolis–London.

SAUCIER, R.T. (1994) *Geomorphology and Quaternary Geologic History of the Lower Mississippi Valley.* U.S. Army Corps of Engineers, Vicksburg, MS, 364 pp.

SCHREINER, B.T. (1983) Lake Agassiz in Saskatchewan. In: *Glacial Lake Agassiz* (Eds Teller, J.T. & Clayton, L.). *Geol. Assoc. Can. Spec. Pap.*, **26**, 75–96.

SCHUMANN, R.R. (1989) Morphology of Red Creek, Wyoming, an arid-region anastomosing channel system. *Earth Surf. Proc. Landf.*, **14**, 277–288.

SCHUMM, S.A. (1977) *The Fluvial System.* John Wiley & Sons, New York, 338 pp.

SCHWEGER, C.E. & HICKMAN, M. (1989) Holocene paleohydrology of central Alberta: testing the general-circulation-model climate simulations. *Can. J. Earth Sci.*, **26**, 1826–1833.

SMITH, D.G. (1983) Anastomosed fluvial deposits: modern examples from western Canada. In: *Modern and Ancient Fluvial Systems* (Eds Collinson, J.D. & Lewin, J.), Spec. Publs. int. Ass. Sediment., No. 6, pp. 155–168. Blackwell Scientific Publications, Oxford.

SMITH, D.G. (1986) Anastomosing river deposits, sedimentation rates and basin subsidence, Magdalena River, northwestern Columbia, South America. *Sediment. Geol.*, **46**, 177–196.

SMITH, N.D. & PÉREZ-ARLUCEA, M. (1994) Fine-grained splay deposition in the avulsion belt of the lower Saskatchewan River. *J. sediment. Res.*, **B64**, 159–168.

SMITH, N.D., CROSS, T.A., DUFFICY, J.P. & CLOUGH, S.R. (1989) Anatomy of an avulsion. *Sedimentology*, **36**, 1–23.

SMITH, N.D., SLINGERLAND, R.L., PÉREZ-ARLUCEA, M. & MOROZOVA, G.S. (1998) The 1870s avulsion of the Saskatchewan River. *Can. J. Earth Sci.* **35**, 453–466

STUIVER, M. & REIMER, P.J. (1993) University of Washington, Quaternary isotope Lab, Radiocarbon calibration program Rev.3.0. *Radiocarbon*, **35**, 215–230.

TAYLOR, R.E. (1987) *Radiocarbon Dating. An Archeological Perspective.* Academic Press, Orlando, FL, 212 pp.

TELLER, J.T. & LAST, W.M. (1981) Late Quaternary history of Lake Manitoba, Canada. *Quat. Res.*, **16**, 97–116.

TELLER, J.T. & LAST, W.M. (1982) Pedogenic zones in post-glacial sediment of Lake Manitoba, Canada. *Earth Surf. Process. Landf.*, **7**, 367–379.

TELLER, J.T. & THORLEIFSON, L.H. (1983) The Lake Agassiz–Lake Superior connection. In: *Glacial Lake Agassiz* (Eds Teller, J.T. & Clayton, L.). *Geol. Assoc. Can. Spec. Pap.*, **26**, 261–290.

TÖRNQVIST, T.E. (1993) Holocene alternation of meandering and anastomosing fluvial systems in the Rhine–Meuse delta (Central Netherlands) controlled by sea-level rise and subsoil erodibility. *J. sediment. Petrol.*, **63**, 683–693.

TÖRNQVIST, T.E. (1994) Middle and late Holocene avulsion history of the River Rhine (Rhine–Meuse delta, Netherlands). *Geology*, **22**, 711–714.

TÖRNQVIST, T.E. & VAN DIJK, C.J. (1993) Optimizing sampling strategy for radiocarbon dating of Holocene fluvial system in a vertically aggrading setting. *Boreas*, **22**(2), 129–145.

TUSHINGHAM, A.M. & PELTIER, W.R. (1991) Ice-3G: A new global model of late Pleistocene deglaciation based upon geophysical predictions of post-glacial relative sea level change. *J. geophys. Res.*, **96**, 4497–4523.

VANCE, R.E. & LAST, W.M. (1994) Paleolimnology and global change on the southern Canadian prairies. In: *Current Research 1994-B*, pp. 49–58. Geological Survey of Canada, Ottawa.

VANCE, R.E., MATHEWES, R.W. & CLAGUE, J.J. (1992) 7000 year record of lake-level change on the northern Great Plains: a high-proxy of past climate. *Geology*, **20**, 879–882.

VANCE, R.E., CLAGUE, J.J. & MATHEWES, R.W. (1993) Holocene paleohydrology of a hypersaline lake in southeastern Alberta. *J. Paleolimnol.*, **8**, 103–120.

VANCE, R.E., BEAUDOIN, A.B. & LUCKMAN, B.H. (1995) The paleoecological record of 6 ka BP climate in the Canadian Prairie Provinces. *Geogr. Phys. Quat.*, **49**, 81–98.

WEBB, T.I., BARTLEIN, P.J., HARRISON, S.P. & ANDERSON, K.H. (1993) Vegetation, lake levels, and climate in eastern North America for the past 18,000 years. In: *Global Climates Since the Last Glacial Maximum* (Eds Wright, H.E., Kutzbach, J.E., Webb, III, T., Ruddiman, W.F., Street-Perrot, F.A. & Bartlein, P.J.), pp. 4415–467. University of Manitoba Press, Minneapolis–London.

WELLS, N.A. & DORR, J., JR. (1987) A reconnaissance of sedimentation on the Kosi alluvial fan of India. In: *Recent Developments in Fluvial Sedimentology* (Eds Ethridge, F.G., Flores, R.M. & Harvey, M.D.), Spec. Publ. Soc. econ. Paleont. Miner., Tulsa, **39**, 51–61.

ZOLTAI, S.C. & POLETT, F.C. (1983) Wetlands in Canada: their classification, distribution, and use. In: *Mires: A. Swamp, Bog, Fen, and Moor; B. Regional Studies* (Ed. Gore, A.J.P.), pp. 245–268. Elsevier, Amsterdam.

ZOLTAI, S.C. & VITT, D.H. (1990) Holocene climatic change and the distribution of peatlands in the western Interior of Canada. *Quat. Res.*, **33**, 231–240.

*Spec. Publs int. Ass. Sediment.* (1999) **28**, 251–268

# Recognizing avulsion deposits in the ancient stratigraphical record

M. J. KRAUS* *and* T. M. WELLS†

*\*Department of Geological Sciences, University of Colorado, Boulder, CO 80309-0399, USA
(Email: Mary.Kraus@colorado.edu); and
†BP Exploration (Alaska) Inc., P.O. Box 196612, Anchorage, AK 99519-6612, USA (Email: wellstm@bp.com)*

## ABSTRACT

Comparison of avulsion deposits in Palaeogene deposits from the Bighorn Basin, Wyoming and in the Saskatchewan River system of Canada shows common facies and facies architecture that has helped establish a more detailed model for avulsion deposits. Ancient avulsion deposits can be recognized in other alluvial basins using the following criteria:

1 moderately well-developed to well-developed palaeosols or coals underlie and overlie the avulsion deposits;
2 sandstones deposited by the trunk channels locally overlie the avulsion deposits;
3 the avulsion deposits are lithologically heterogeneous and consist of fine-grained deposits that surround ribbon and thin sheet sandstones;
4 the fine-grained deposits show only weak pedogenic modification;
5 many of the ribbon sandstones show palaeoflow subparallel to or in the same direction as palaeoflow in the trunk channels;
6 ribbons generally have width thickness (W/T) ratios of less than 10;
7 some of the ribbon sandstones cluster at particular stratigraphical levels and are laterally connected by thin sandstone sheets to form 'tiers' that suggest that networks of crevasse channels once occupied particular areas of the floodplain;
8 the heterolithic avulsion deposits are extensive laterally and form a significant part of the stratigraphical succession, suggesting that they were deposited over large areas of the floodplain.

The heterolithic avulsion deposits can be distinguished from ordinary crevasse-splay deposits developed off stable trunk channels on the basis of their spatial scale, the presence of truncating channel sandstones and attributes of the ribbon sandstones. The Palaeogene avulsion deposits are not only extensive laterally, they also dominate the stratigraphical sections, suggesting that they covered much larger areas than is typical for crevasse-splay deposits. Sandstone bodies deposited by the trunk channels are always underlain by the heterolithic deposits, which is consistent with an avulsion interpretation for the heterolithic intervals. The ribbon sandstones in the avulsion deposits have smaller W/T ratios than is characteristic for the feeder channels of crevasse splays, and their palaeoflow trends tend to parallel flow in the main channel rather than being oblique or perpendicular to it.

Various features also distinguish avulsion deposits from flood-basin deposits. The ribbon sandstones are an integral part of the avulsion deposits. In contrast, fine-grained flood-basin deposits may be cut by some of the ribbon sandstones; however, those sandstones scour down from the overlying avulsion interval. More obvious is the presence of well-developed, cumulative palaeosols on the flood-basin deposits and only weakly developed, generally compound, palaeosols on the avulsion deposits. The avulsion deposits show only weak pedogenic development because accumulation rates were so rapid. In the field, the sharp contrast between the well-developed palaeosols and the weakly developed palaeosols highlights the presence of avulsion deposits.

Differences in climatic conditions and proximity to sediment sources produced lithological and pedologic differences in the avulsion deposits and flood-basin deposits among the study areas. The overall thickness of avulsion intervals also varies among the modern and ancient examples. Thickness appears to be influenced by both levee height and local palaeotopography. The time between successive avulsion deposits was estimated to be of the order of $10^4$ yr. This is considerably longer than interavulsion periods that have been determined from Holocene systems, showing that a vertical section records only a fraction of the avulsions that actually took place.

## INTRODUCTION

Although avulsion has an important impact on the alluvial stratigraphical record, particularly on the stacking patterns of channel sand bodies, it remains a poorly understood process (e.g. Smith *et al.*, 1989; Mackey & Bridge, 1995; Heller & Paola, 1996). Smith *et al.* (1989) and Smith & Pérez-Arlucea (1994) described the process of avulsion and its depositional record for one modern system, the Saskatchewan River. These studies showed that fine-grained sediment accumulates on the Saskatchewan floodplain by two distinct processes. First, overbank flooding of a trunk channel inundates flood basins, and suspended sediment is deposited. Second, fine-grained deposits result from channel avulsion and are deposited with splay sands at all stages of flow, not just flood stage, through channel walls, rather than over channel banks. In this model, flow from the main channel is being diverted into splay systems, which are expanding into low-lying floodplain areas to create an avulsion complex (Smith *et al.*, 1989). As development continues, older splay systems are abandoned and flow in the avulsion complex is gradually concentrated in fewer but larger channels. Eventually, a new meander belt develops, which occupies only a portion of the old avulsion belt. The avulsion deposits are dominated by silty clays and silts that encase sandy splay-channel and thin sheet deposits.

Because the Saskatchewan avulsion belt is being deposited by crevasse-splay systems, the avulsion deposits are lithologically similar to the crevasse-splay deposits described by some other workers. For example, crevasse-splay deposits in the Rhine–Meuse system consist of fine sand, sandy clay and silty clay intervals (Törnqvist *et al.*, 1993; Weerts & Bierkens, 1993). Despite the similarities, avulsion differs from ordinary crevasse-splay deposition in spatial and temporal scale. Crevasse splays are channel marginal deposits, and although they can extend out into the floodplain, most of those described cover limited areas, generally less than several tens of km$^2$ (e.g. Farrell, 1987; Mjøs *et al.*, 1993; Weerts & Bierkens, 1993; Jorgensen & Fielding, 1996). In contrast, the Saskatchewan avulsion complex is no longer marginal to the trunk channel, and it covers a vast area, 500 km$^2$. Crevasse splays normally have short lifespans (days to years). Although individual areas of the Saskatchewan avulsion complex have lifespans comparable to those of ordinary crevasse splays, the entire avulsion complex has been building for over 100 yr.

As Bridge (1984) noted, lithological evidence for avulsion should be expected in the alluvial stratigraphical record. The Saskatchewan River studies are important because they provide clues as to what those avulsion deposits might look like and because they predict that fine-grained deposits produced by avulsion should be common in the stratigraphical record. The studies of Bridge (1984), Smith *et al.* (1989), and Smith & Pérez-Arlucea (1994) are excellent starting points for recognizing ancient avulsion deposits; however, they do not provide a complete model. Analysis of well-exposed ancient examples can help improve the model by providing additional information on the facies and their organization.

Ongoing studies have shown that avulsion deposits are widespread in the Palaeocene Fort Union Formation and latest Palaeocene to early Eocene Willwood Formation in the Bighorn Basin, Wyoming (Kraus & Aslan, 1993; Gwinn, 1995; Kraus, 1996; Wells, 1997; Kraus & Gwinn, 1997). The major goal of this paper is to begin to develop a more complete model of avulsion deposits by summarizing the features that are common to these Palaeogene avulsion deposits. Avulsion deposits in different parts of the Palaeogene sequence had different climatic and tectonic settings; however, they have certain facies and facies arrangements in common. We also summarize the evidence that supports the interpretation of avulsion deposits in the Palaeogene examples. The second goal of this paper is to contrast the modern and ancient examples to understand the factors that controlled differences in the avulsion-belt deposits. Although the floodplain deposits in the various cases have similar facies and three-dimensional facies arrangements, the avulsion deposits show differences including the degree of pedogenic development, thickness and recurrence frequency. To compare and contrast the Palaeogene examples and the Saskatchewan modern analogue, we begin by summarizing their depositional settings.

## DEPOSITIONAL SETTINGS OF THE MODERN AND ANCIENT EXAMPLES

### Saskatchewan River

The Saskatchewan River in Canada is situated in a tectonically quiescent area, however, the region has undergone isostatic rebound since recession of the late Wisconsonian ice sheets (Walcott, 1972; Vincent & Hardy, 1979; Smith *et al.*, 1989) (Table 1). The rate of uplift increases progressively to the northeast, and Smith & Putnam

**Table 1.** Characteristics of modern and ancient avulsion deposits.

| Study area | Saskatchewan (Smith *et al.*, 1989) | Fort Union lower | Fort Union upper | Willwood lower | Willwood upper |
|---|---|---|---|---|---|
| Climatic setting | Subhumid MAT 0–10°C* | Humid MAT 10°C† | Humid MAT 10°C† | Drier, seasonal MAT 13°C‡ | Drier, seasonal MAT 15°C‡ |
| Tectonic setting | Passive (isostatic rebound)§ | Active subsidence | Active subsidence | Active subsidence | Active subsidence |
| Areal extent | 500 km$^2$ | 1.5 km$^2$ | 1.5 km$^2$ | 3 km$^2$ | 7 km$^2$ |
| Thickness of avulsion deposits | 1–3 m | 1–5 m n = 8 | 1–4 m n = 11 | 4–7 m up to *c.* 10 m n = 15 | 3–4.5 m up to *c.* 7 m n = 19 |
| Grain size of fine-grained deposits¶ | | Average clay content = 66% 43% of samples > 67% clay n = 49 | Same as Fort Union lower | Average clay = 64% 51% > 67% clay n = 49 | Average clay = 77% 86% > 67% clay n = 41 |
| Colours of mature palaeosols | Peats, reddish-brown to black | Black, browns, yellows, greys | Black, browns, yellows, greys | Reds, oranges, greys | Purple, reds, orange, greys |
| Colour of avulsion fines | | Greys, greens | Greys, greens | Greys, yellow-brown | Greys, yellow-brown |
| Thickness of ribbons | *c.* 1–5 m | 0.2–8 m 86% < 3 m n = 71 | 0.4–6 m 90% < 3 m n = 66 | 1–9 m 70% < 3 m n = 50 | 0.3–5 m 97% < 3 m n = 66 |
| Sediment-accumulation rate | 0–33 mm yr$^{-1}$ (avulsion belt only) | *c.* 0.1 mm yr$^{-1}$ averaged over 3 Myr | *c.* 0.3 mm yr$^{-1}$ averaged over 1.2 Myr | *c.* 0.5 mm yr$^{-1}$ averaged over 0.44 Myr | *c.* 0.5 mm yr$^{-1}$ averaged over 0.9 Myr |
| Interavulsion frequency | *c.* 600–800 yr‖ | *c.* 41 000 yr n = 8 | *c.* 12 500 yr n = 11 | *c.* 13 000 yr** n = 7 *c.* 16 000 yr n = 8 | *c.* 14 000 yr n = 19 |

\* Dirschl & Coupland (1972), Tricart & Cailleux (1972) and Smith *et al.* (1989). † Hickey (1980). ‡ Wing (1981) and Greenwood & Wing (1995). § Walcott (1972) and Vincent & Hardy (1979). ¶ Includes fine-grained deposits in both heterolithic intervals and well-developed palaeosols. ‖ Morozova *et al.* (1996). ** Two stratigraphical sections, in areas separated by *c.* 1.5 km, were examined (Kraus, 1996).

(1980) suggested that greater uplift down-river may be a primary influence on the aggradation and large-scale anastomosed channel patterns seen in the Cumberland Marsh region, where Smith *et al.* (1989) studied the depositional response to a modern avulsion.

The Saskatchewan avulsion belt formed in a subhumid climate with long cold winters and short warm summers (Smith *et al.*, 1989). The mean annual precipitation is 43 cm (Dirschl & Coupland, 1972; Smith *et al.*, 1989), and the area falls into the continental climate zone of Tricart & Cailleux (1972), which has a mean annual temperature (MAT) between 0 and 10°C.

**Palaeogene study areas**

Two stratigraphical intervals in the Fort Union Formation (Fm) were studied, both along Polecat Bench, and two

Willwood stratigraphical intervals were studied, one in the northern part of the basin and one in the central part of the basin (Fig. 1). Both formations consist of sandstones and colour-banded mudrocks, although mudrock colours in the Fort Union Fm are sombre (black, browns, yellows, greys) compared with the bright orange, red and purple Willwood mudrocks. Deposition of the Palaeogene sections accompanied the tectonic development of the Bighorn Basin during the latest Cretaceous to early Tertiary Laramide orogeny. The Bighorn, Beartooth and Owl Creek mountains were formed by compression in a NE–SW horizontal thrust field (Brown, 1993). The Absaroka Mountains were created by volcanic activity that post-dated the deposits described here (Bown, 1980).

Tertiary continental rocks in North America can be subdivided into biostratigraphical zones on the basis of mammalian faunas. The two Fort Union intervals are

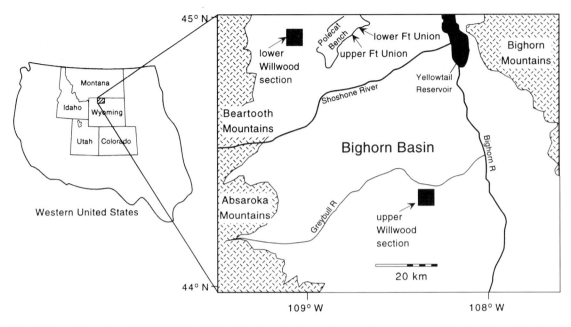

**Fig. 1.** Map of the northern half of the Bighorn Basin, Wyoming, showing major mountain ranges surrounding the basin and location of the four Palaeogene study areas. Two sections in the Fort Union Formation were studied along the south side of Polecat Bench. Two sections were examined in the Willwood Formation: one in the northern part of the basin and one in the central part of the basin, just south of the Greybull River. (Modified from Clyde, 1997.)

assigned to the Tiffanian 3 (Ti3) and Tiffanian 5 (Ti5) zones (e.g. Gingerich, 1983; P.D. Gingerich, personal communication, 1996), and the Willwood sections are assigned to the middle of the Clarkforkian land mammal age (e.g. Rose, 1981) and the middle of the Wasatchian land mammal age (Bown *et al.*, 1994) (Fig. 2). Palaeomagnetic analyses (Butler *et al.*, 1981; Tauxe *et al.*, 1994) allow the study sections to be correlated to the Global Polarity Time Scale (GPTS) of Cande & Kent (1995).

Palaeobotanical evidence indicates that the Tiffanian climate had a MAT of 10°C and an annual temperature range of 25°C (Hickey, 1980) (Table 1). Mean annual precipitation has not been estimated; however, coals in the Fort Union Fm in the northernmost part of the basin (Hickey, 1980) indicate that the Palaeocene climate was humid (e.g. Lottes & Ziegler, 1994). By Clarkforkian time, the vegetation included palms and cycads, which are intolerant of long frosts (Hickey, 1980), and plant physiognomy suggests that MAT had risen to 13°C. The vegetation indicates that, by Wasatchian time, the MAT was 13–18°C and rainfall was seasonal (Wing, 1981; Greenwood & Wing, 1995).

*Fluvial systems*

Both Palaeogene formations can be subdivided into three

major facies (Kraus & Aslan, 1993; Kraus, 1996; Kraus & Gwinn, 1997; Wells, 1997) (Figs 3 & 4). Thick, laterally extensive sheet sandstones represent the deposits of trunk channels. The major sheets locally truncate and are also surrounded by the other two facies:

1 fine-grained deposits, on which relatively well-developed palaeosols formed;

2 heterolithic deposits that consist of ribbon sandstones and thin sheet sandstones surrounded by mudrocks with weakly developed palaeosols.

The relatively well-developed palaeosols indicate that those fine-grained sediments accumulated relatively slowly and episodically; thus they have been interpreted as flood-basin deposits (Kraus & Aslan, 1993; Kraus, 1996). The heterolithic deposits are interpreted as avulsion deposits.

The major sheet sandstones make up about 20% of measured sections in the Fort Union and Willwood formations (e.g. Jepson, 1940; Neasham & Vondra, 1972). The lower Willwood section contains the largest; it has a maximum preserved thickness of 25 m and a preserved lateral extent perpendicular to palaeoflow of 1.5 km (Kraus & Gwinn, 1997). The largest sheets in the other sections are approximately 10 m thick and range from about 0.8 to 1.2 km wide (transverse to palaeoflow). Width/thickness (W/T) ratios calculated for these sands

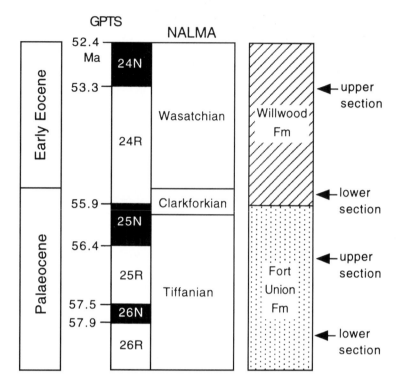

**Fig. 2.** Diagram showing Palaeogene magnetostratigraphic correlations of study sections in the Fort Union and Willwood formations with the Global Polarity Time Scale (GPTS) of Cande & Kent (1995). The Fort Union and lower Willwood correlations are from Butler *et al.* (1981) and the upper Willwood section correlation is from Tauxe *et al.* (1994). North American Land Mammal Age (NALMA) correlations also are shown (Rose, 1981; Gingerich, 1983; Bown *et al.*, 1994).

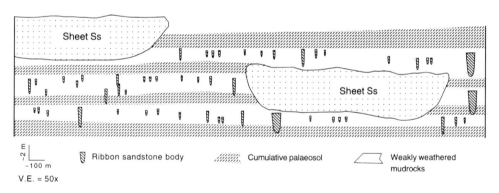

**Fig. 3.** Cross-section showing alluvial architecture in the upper Willwood section, which is representative of the Palaeogene sections. The thicknesses of the sheet sandstones, cumulative palaeosols and heterolithic intervals are depicted accurately; however, the locations of ribbon sandstones are schematic. Fine-grained deposits with cumulative palaeosols are interpreted as overbank deposits. Heterolithic intervals consisting of ribbon sandstones enveloped by incipient palaeosols developed on fine-grained facies are interpreted as avulsion deposits. Both the avulsion and overbank deposits are truncated locally by major sheet sandstones. Vertical exaggeration is ×50.

range from 75 to 200. The sheets are multistoried with two to three stories, which have a maximum preserved thickness range between 3 and 4.5 m.

Grain size varies from very fine to coarse sand, but most is very fine to medium sand. Internally, sheets are dominated by massive bedding and large-scale trough cross-stratification, with sets ranging between 10 and 100 cm thick. The massive bedding appears to be the result of intense soft-sediment deformation. Lateral accretion sets are not evident in all of the major sheets; however, the abundant massive and/or contorted bedding may obscure their presence.

**Fig. 4.** Field view showing the alluvial architecture in the lower Willwood section. A major sheet sandstone (arrows) caps the ridge. Cumulative palaeosols (P) form the darker bands and are interpreted as overbank deposits. The cumulative palaeosols alternate with lighter coloured heterolithic intervals (H). The heterolithic deposits have only weakly developed palaeosols surrounding ribbon and thin sheet sandstones, and they are interpreted as avulsion deposits. Sloping ridge in foreground of photograph is 75 m high.

The thick sheet sandstones were produced by meandering rivers, as indicated by the local presence of well-developed lateral accretion deposits and highly variable palaeoflow directions in vertically successive stories. The preserved thickness of individual stories suggests that the rivers were of the order of 3–5 m deep. Major sheet sandstones in all four sections show palaeoflow to the north-west, paralleling the structural axis of the basin.

## FEATURES COMMON TO AVULSION DEPOSITS

### Saskatchewan River

Smith *et al.* (1989) found that the Saskatchewan River avulsion complex, which began about 120 yr ago, grows by the progradation of splay systems, which were divided into three types. Stage I splays are simple lobes of sand with a wedge-shaped cross section in a transect perpendicular to flow in the main channel (Fig. 5). These are equivalent to what many other workers have described as crevasse splays (e.g. Allen, 1965; O'Brien & Wells, 1986; Farrell, 1987). Stage I splays have feeder and distributary channels that are broad and shallow (W/T > 40), and flow directions are highly variable, ranging from perpendicular to flow in the main channel to up- and down-valley. Stage I splays evolve into stage II or stage III splay complexes (Fig. 5). Stage II splays are more elongate than stage I splays and have numerous channels that produce sand bodies with variable geometries – isolated lenses or sheets that, in the Saskatchewan system, are up to 2–3 m thick. Stage III splays commonly develop from either

stage I or stage II splays and are characterized by stable anastomosed channels that produce ribbon sands with W/T ratios less than about 10. The splay channels have small levees, and flow in the splay channels is subparallel or in the same direction as flow in the trunk channel, i.e. down the floodplain gradient.

Despite the presence of numerous channel sands, the Saskatchewan avulsion deposits are actually dominated by fine-grained deposits that surround the sand bodies (Fig. 5). The avulsion deposits overlie peats that formed during prior wetland conditions. Locally, the avulsion deposits are truncated by the trunk channel that becomes established following avulsion.

### Palaeogene avulsion deposits

The Palaeogene heterolithic intervals have facies and facies architecture that resemble the Saskatchewan avulsion deposits. Furthermore, as described below, the heterolithic deposits share certain lithological features, despite being deposited in different parts of the Palaeogene succession with different climatic conditions and different accumulation rates.

### *Fine-grained deposits*

Because the fine-grained deposits (both flood-basin and avulsion deposits) are clay-rich, clay content (with the clay–silt break at 4 μm) provides the best measure of grain size. Of the 49 samples analysed in the Fort Union Fm, the average clay content was 66%, and 43% of the samples contain at least 67% clay (Table 1). The other samples were mudstones with a clay content between

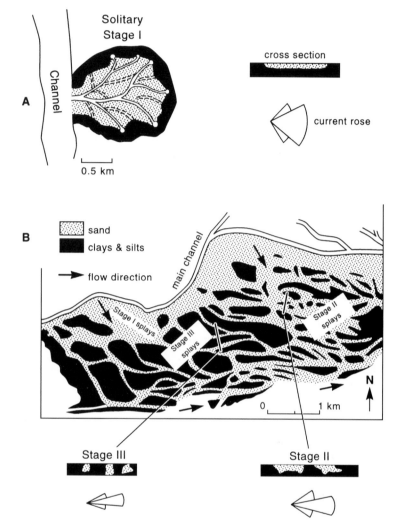

**Fig. 5.** Schematic diagram showing map views and cross-sections for the three stages of avulsion recognized by Smith *et al.* (1989). Avulsion begins with stage I splays, which are lobes of sand with wedge-shaped cross-sections. Stage II splays are more elongate than stage I splays and have channels that produce sand bodies of various geometries — sheets that are several metres thick or isolated lenses. Stage III splays have stable anastomosed channels that produce ribbon sands with W/T ratios less than about 10. Hypothetical current roses show a greater spread for stage I than stage II or III splays. (Modified from Smith *et al.*, 1989; current roses from Kraus & Gwinn, 1997.)

33% and 67%. The samples from the lower Willwood section are similar. Only the upper Willwood section is slightly finer, which Kraus & Gwinn (1997) attributed to being farther from mountain sediment sources.

### Stratigraphical relations—bounding palaeosols

In the field, the heterolithic intervals are conspicuous because of their marked colour contrast from the well-developed palaeosols (flood-basin deposits) that bound them (Figs 6 & 7). That colour difference reflects the better pedogenic development of the flood-basin deposits.

The well-developed palaeosols formed on clay-rich sediment. Individual profiles can be several metres thick and have well-differentiated horizons (Fig. 6). Gradual downward changes in grain size and geochemistry indicate that the parent material was homogenized by pedogenic processes, as is characteristic of cumulative soils that form when sedimentation is steady but slow enough that new sediment is incorporated into the developing profile (Kraus & Aslan, 1993; Kraus, 1997). Cumulative floodplain soils commonly form because overbank floods gradually add thin increments of alluvium to the soil profile over a relatively long time.

Despite the advantage that pedogenic overprinting provides for differentiating avulsion deposits from flood-basin deposits, establishing the contact between the two is difficult because the pedogenic properties can obscure the depositional origins of the sediment. In particular, the lower parts of many cumulative palaeosols are developed on older avulsion deposits, which are visually included with the cumulative palaeosols and which

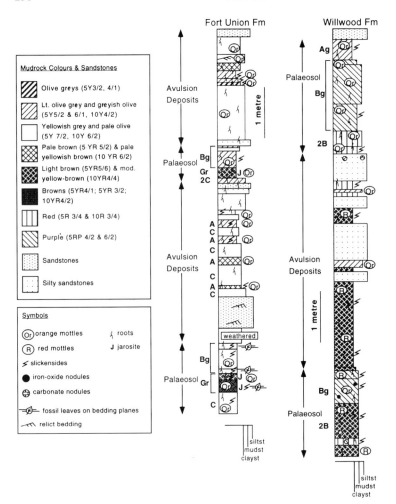

**Fig. 6.** Measured stratigraphical sections through representative deposits from the lower Fort Union section and upper Willwood section showing morphological and textural data. Cumulative palaeosols (indicated as 'palaeosol') are interpreted as flood-basin deposits. They alternate with lithologically heterogeneous intervals (sandstones and mudrocks) showing only weak pedogenic modification. The heterolithic intervals are interpreted as avulsion deposits. Although individual profiles are designated for the weakly developed palaeosols of the Fort Union Formation (A/C profiles), they are difficult to discriminate in the Willwood section and are thus not designated. Ag, gleyed A horizon; Bg, gleyed B horizon; Gr, permanently reduced subsurface horizon; 2B, subdivision of the B horizon, differs morphologically from the overlying Bg horizon; 2C, C horizon developed on different parent material than higher in the profile.

**Fig. 7.** Field view of lower Willwood section showing the darker coloured cumulative palaeosols (P) that bound the light-coloured heterolithic intervals. Thin sheet sandstones form narrow ledges in the heterolithic intervals, and a ribbon sandstone is indicated by an arrow. Note that the upper and lower cumulative palaeosols are developed directly above thin sheet sandstones, which belong to the underlying heterolithic interval. Entire section is approximately 20 m thick.

**Fig. 8.** Schematic diagram showing how pedogenesis can obscure the depositional origins of avulsion and flood-basin deposits. (A) Because pedogenesis resulted in a reddish colour, the 2C horizon is included with the cumulative palaeosol. Detailed section measuring shows that it has a coarser grain size and that, on the basis of its depositional origins, the 2C horizon should be included with the underlying avulsion deposits not with the flood-basin deposits on which the bulk of the cumulative soil developed. This situation is common in the Willwood Formation. (B) Detailed section measuring shows that the sediment recognized as a Bg horizon is coarser than the underlying Gr horizon and that, on the basis of its depositional origins, it should be included with the avulsion deposits above the cumulative palaeosol. Because of its palaeosol properties, this unit has been included with the Gr horizon as a very water-logged profile. This situation is common in the Fort Union Formation. Ag, gleyed A horizon; Bg, gleyed B horizon; Bw, weakly developed B horizon that shows only development of colour or structure, not illuviation; Gr, reduced gley horizon; 2B, subdivision of the B horizon, differs morphologically from the overlying Bg horizon; 2C, C horizon developed on different parent material than higher in the profile.

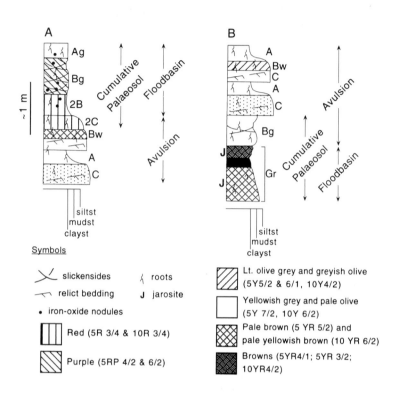

can be distinguished only by detailed section measuring (Fig. 8A). The Fort Union palaeosols are even more complicated than the Willwood palaeosols. These palaeosols are very poorly drained and have either a Bg (gleyed B) horizon or A/Bg sequence above a Gr (permanently reduced subsurface) horizon (Fig. 6). The clay-rich nature of the Gr horizons indicates that the parent material was deposited by overbank flooding and far from the active channel. The upper, lighter coloured parts of many profiles were probably deposited by avulsion, but their palaeosol properties lead to their inclusion with the cumulative palaeosol (Figs 6 & 8B). This pedogenic blurring of the boundary between the flood-basin deposits and avulsion deposits is important when trying to measure the thickness and thickness variations of avulsion deposits.

### Geometry

Despite the difficulties in measuring thickness variations, the heterolithic deposits from the Palaeogene examples

have been described as having a sheet-like geometry because they are laterally extensive relative to their thickness (e.g. Kraus & Aslan, 1993). In the upper Willwood section, where outcrops are most laterally continuous, individual avulsion deposits can be traced for more than 5 km both parallel and perpendicular to palaeoslope. Individual avulsion deposits in the Willwood Fm show lateral variations in thickness, and some show overall directional thinning suggesting a lateral termination of the avulsion belt (Kraus & Aslan, 1993). So far, however, exposures in the Bighorn Basin have not allowed us to trace an individual avulsion deposit to a lateral termination.

Except for where ribbon sandstones locally scour into older deposits, the bases of avulsion deposits do not show signs of erosion. In fact, in the Palaeogene examples, the contact between what is interpreted as an avulsion deposit and the underlying flood-basin deposit is difficult to pinpoint. Thicknesses of the heterolithic intervals range from 1 to 10 m in the examples described, but, because thickness is so variable among the case studies, it is discussed further in a later section.

## Truncating sheet sandstones

Wherever a major sheet sandstone is present, it overlies and locally truncates a heterolithic interval (Fig. 3). Although an overlying sheet sandstone cannot be found for each heterolithic interval, this probably reflects the lateral scale of the depositional system relative to the scale of the exposures. Because even the best-exposed heterolithic intervals cannot be traced more than about 5 km, the exposures probably sample a very small part of the original heterolithic deposit.

## Weakly developed palaeosols

Fine-grained deposits in the heterolithic intervals range from claystone to siltstone (Fig. 6). Fine-grained deposits in the Saskatchewan avulsion belt commonly coarsen upward (either in a subtle or more pronounced fashion) and consist of silty clay, silt and very fine sand (Smith *et al.*, 1989; Smith & Pérez-Arlucea, 1994). Similarly, the lower parts of the heterolithic intervals are dominated by claystone and mudstone and have lesser proportions of siltstone and sandstone, whereas siltstone and sandstone increase upwards (Kraus, 1996).

Each bed may or may not have preserved primary stratification, and many beds show evidence for weak pedogenic modification, including mottling, root traces and burrows, and slickensides (Fig. 6). Vertical sequences show erratic downward changes in grain size and geochemical properties (e.g. iron, manganese, and total organic carbon content) (Kraus & Aslan, 1993). Horizon development is weak; in fact, recognizing individual profiles can be difficult. Where they can be delimited, profiles are thin (< 1 m thick) and vertically stacked, indicating that these are compound palaeosols, which form when sedimentation is rapid and unsteady, as is characteristic of the Saskatchewan River avulsion deposition (e.g. Smith *et al.*, 1989; Smith & Pérez-Arlucea, 1994).

## Ribbon and thin sheet sandstones

Ribbon sandstones are well documented in the Fort Union and Willwood formations (Gwinn 1995; Kraus 1996; Kraus & Gwinn, 1997; Wells, 1997). The ribbons are lenticular in cross-section, generally have W/T ratios less than 10, and range from 0.3 to 9 m in thickness (Table 1). With the exception of the lower Willwood section, where 70% of the 50 ribbons measured are less than 3 m thick, at least 85% of ribbons in the other three

Palaeogene sections are less than 3 m thick and have one story. Ribbons thicker than about 3 m usually have two stories separated by an erosion surface or a thin, bioturbated siltstone. Lateral accretion deposits are absent, and, where good lateral exposures are preserved, individual ribbons are straight to gently curved in plan view (Fig. 9).

Most ribbon sandstones consist of very fine- and fine-grained sand. Although internal stratification in the ribbons is commonly poorly preserved, where bedding is present, most structures are generally small-scale and large-scale trough cross-bedding. Bioturbation and contorted bedding are locally abundant, especially in the upper parts of ribbon sandstones.

Palaeocurrents were determined from large-scale cross-bedding and, where exposures permitted, from the trends of laterally traceable ribbons. In the Fort Union Fm, most of the larger ribbons show palaeoflow to the north-west, parallel to flow in the major sheet sandstone. Smaller ribbons have variable palaeocurrent trends. An example from the lower section shows a major ribbon, with a maximum thickness of about 7 m, which forms a ridge that trends north-west for 0.6 km (Fig. 9). Smaller ribbons, between 1 and 2.5 m thick, are situated at the same stratigraphical level and form a channel network with the larger ribbon. Cross-bedding in the major ribbon shows that palaeoflow was to the north-west; some of the smaller ribbons were converging with the major ribbon, others were diverging. In the upper Willwood section, palaeoflow of most large and small ribbons is north-north-west. Fewer ribbons in the lower Willwood section are amenable to measuring palaeocurrents, and palaeoflow directions there are more variable.

Where ribbon margins are preserved, many have thin (0.3 to 1 m thick) sandstone or siltstone wings extending from the top. The sheets become thinner and finer grained away from their associated ribbon sandstone and eventually grade into mudstones and claystones. Many of the thin sheet sandstones or siltstones have been bioturbated; less commonly they show small-scale cross-stratification and/or parallel bedding. Although most can be traced only 10 to 15 m laterally, some examples are more extensive and connect stratigraphically equivalent ribbon sandstones in what have been termed ribbon 'tiers' (Gwinn, 1995; Kraus & Gwinn, 1997) (Fig. 10). In the example shown, palaeoflow indicators show that the small channels flowed subparallel to parallel to one another and to the trunk channel.

Most of the Fort Union and Willwood ribbons resemble stage III crevasse channels, which have W/T ratios less than 10 and flow in the same direction or subparallel to flow in the main channel (Smith *et al.*, 1989)

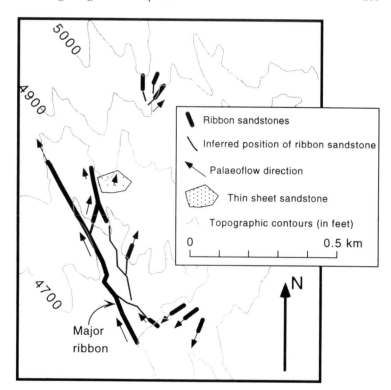

**Fig. 9.** Mapped positions of ribbon sandstones in one heterolithic interval from the lower Fort Union section. Thick lines show where the ribbons are continuously exposed. Narrow lines indicate where ribbons are mapped with less confidence because they are more poorly exposed or eroded. (Modified from Wells, 1997.)

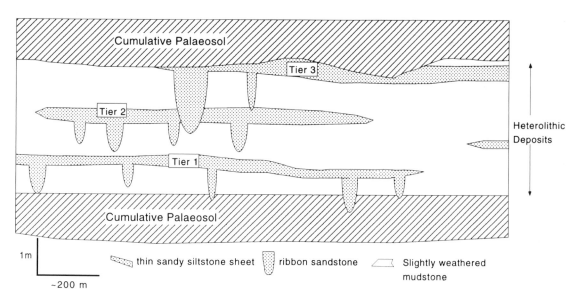

**Fig. 10.** Diagram showing three ribbon tiers in a heterolithic interval from the upper Willwood section. Thin sheet sandstones or siltstones laterally connect the ribbon sandstones in each tier. (Modified from Gwinn, 1995.)

(Fig. 5). The ribbons are interpreted as ancient splay channels that fed the developing avulsion belt; the thin sheets probably formed as sheet floods or overbank deposits from the splay channels. The Palaeogene ribbon tiers show that small channels coexisted on a particular area of the floodplain, and palaeocurrent data show that flow in the small channel networks was in the same direction or subparallel to flow in the larger trunk channels, as it is in stage III splays (Fig. 5). The tiers and the intense bioturbation of the thin sheet sandstones at the tops of the tiers also suggest that a particular splay lobe was active, then abandoned, and then another area of the avulsion belt became active.

The plan views of exhumed ribbons show that the channels were straight to moderately sinuous; however, the absence of any evidence for lateral migration indicates that the channels were fixed and that aggradation was primarily vertical. The reconstructed ribbon networks and the palaeoflow directions imply that the channel patterns were anastomosed, which is characteristic of the Saskatchewan avulsion belt. Direct evidence for anastomosis was not observed, however, except for rare exposures showing the confluence of two ribbon sandstones.

Not all ribbon sandstones in the Palaeogene sections are as narrow as typical stage III channel sands, and not all sheet sandstones are thin (< 1 m thick) or wings from ribbon sandstones. A few ribbons (*c.* 6%) have W/T ratios greater than 10, and many of those W/T values are only minima because the margins of the sandstone bodies have been eroded. A few sheet sandstones are up to 2 m thick, although thickness varies laterally because of irregular basal surfaces. The wider ribbons and the thicker sheet sandstones resemble the sands deposited by stage II splay channels in the Saskatchewan system (Smith *et al.*, 1989). Many of the wider ribbons and thicker sheet sandstones also show palaeoflow oblique to the north-north-west palaeoslope, which is more consistent with flow trends seen in stage II crevasse channels. Nonetheless, deposits interpreted as stage II splays are far less common than those interpreted as stage III splays.

The dominance of stage III splays in the Palaeogene sections should not be surprising. In the Saskatchewan system, they occupy greater areas than either stage I or stage II splays. In fact, Smith & Pérez-Arlucea (1994) estimated that between 50% and 70% of the Saskatchewan avulsion belt consists of stage III splay deposits. They also have a greater chance of preservation because they accumulate in the more distal and topographically lower parts of the floodplain. Consequently, we expect that stage III splays should dominate other ancient avulsion deposits as well.

## Discussion

On the basis of lithology, an alternative explanation for the heterolithic deposits is that they represent crevasse-splay deposits developed off stable trunk channels. The Saskatchewan avulsion deposits resemble crevasse-splay deposits because the avulsion belt is being created by splay systems as they prograde into low areas of the floodplain (Smith *et al.*, 1989); however, in this case, splay deposits cover a vast area (*c.* 500 km$^2$) and splay deposition proceeds to avulsion. Certain attributes of the Palaeogene heterolithic intervals, including lateral extent and thickness, the presence of truncating channel sandstone bodies and features of the ribbon sandstones, support an avulsion origin for these deposits.

Although crevasse-splay deposits can extend out into the floodplain, most of those studied cover limited areas. For example, Mjøs *et al.* (1993) found splays up to 2.2 km wide and 1.75 km long, Jorgensen & Fielding (1996) described splays 4 km long, and Weerts & Bierkens (1993) figured splays that extend about 1 km out from the channel. Even the splay described by Farrell (1987) from the Mississippi River, a much larger system than the rivers that deposited the Palaeogene sections, covers only approximately 25 km$^2$. Two features indicate that the Palaeogene heterolithic deposits were of a scale larger than typical crevasse-splay deposits. First, individual heterolithic intervals have been traced for distances up to 5 km, both transverse and parallel to the general palaeodrainage direction. Over those distances, they show no evidence for thinning, suggesting that originally they extended much farther, although they are now either eroded or covered. Second, heterolithic deposits are the dominant facies in all four Palaeogene sections. For example, randomly measured sections show that, if major sheet sandstones are ignored, heterolithic intervals make up 59% of the upper Fort Union section (135 m thick with 12 heterolithic intervals) and 65% of the upper Willwood section (130 m thick with 19 heterolithic intervals). The abundance of this kind of deposit also argues that individual intervals must have been areally widespread. Because they have a restricted areal extent, ordinary crevasse-splay deposits usually form only a comparatively small fraction of alluvial deposits, as exemplified by cross-sections of the Rhine–Meuse system (Weerts & Bierkens, 1993).

A second feature that is more consistent with an avulsion origin than a crevasse-splay origin is the fact that major sheet sandstones, which represent the trunk channels, are directly underlain by heterolithic deposits. The Saskatchewan River avulsion model predicts that, in a fluvial system undergoing avulsion, the major channel sandstones should overlie avulsion deposits, although they may locally scour lower, into pre-avulsion deposits. Without avulsion, there is no reason for channel-sandstone bodies to overlie crevasse-splay deposits instead of fine-grained flood-basin deposits.

Finally, features of the ribbon sandstones support an avulsion origin for the heterolithic deposits. The stage I splays of Smith *et al.* (1989), which are equivalent to what many other workers (e.g. Allen, 1965; O'Brien & Wells, 1986) regard as crevasse splays, have channels with W/T ratios (> 40) that would produce sandstones that are much wider than the Palaeogene ribbons (W/T < 10). Furthermore, flow in those feeder and distributary channels is typically perpendicular or oblique to flow in the main channel, not parallel to it as it is in many Palaeogene ribbons.

### Other examples

Other examples of heterolithic deposits that are stratigraphically bounded by palaeosols are now starting to be described. Thick sequences (10–15 m) of muds encasing small (1–3 m thick) ribbon and sheet sandstones were described from Holocene deposits of the Colorado River (Aslan & Blum, this volume, pp. 193–209). These heterolithic deposits also are bounded vertically by pedogenically modified sediment, and Aslan & Blum interpret them as avulsion deposits.

In the Chinji Fm of the Himalayan Basin, Willis & Behrensmeyer (1994) recognized 'palaeosol-bounded sequences' that appear to be similar to what we term heterolithic deposits. These palaeosol-bounded sequences are described as wedge- or lobe-shaped features with variable thickness, and they can be traced laterally for as much as 10 km, although some of these pinch out over distances of 4–10 km (Willis & Behrensmeyer, 1994; Behrensmeyer *et al.*, 1995). Each sequence contains mudstones showing evidence for incipient pedogenic development and ribbon and thin sheet sandstones, which were not described in detail. Although Willis & Behrensmeyer (1994) concluded that the sequences were the result of laterally extensive crevasse-splay lobes that filled in low areas of the floodplain, the crevasse-splay deposition was not believed to be necessarily associated with avulsion of a trunk channel. Another interpretation that might be considered for these palaeosol-bounded sequences, especially considering their lateral extent and stratigraphical abundance, is that they represent ancient avulsion deposits.

## PALAEOGENE DEPOSITS VERSUS THE MODERN ANALOGUE

Kraus & Gwinn (1997) summarized the differences between the Willwood avulsion deposits and the stage III splays of Smith *et al.* (1989). These differences extend to the Fort Union Fm as well. First, there are lithological differences. Whereas Smith *et al.* (1989, p. 8) found that their stage III channels 'are starved of coarse sediment (fine to medium sand)', the Palaeogene ribbon sandstones consist of very fine to fine sandstone. Stage III splays commonly form where splays build into floodplain lakes (Smith & Pérez-Arlucea, 1994), and the pre-avulsion surface in the Saskatchewan River case is characterized by a continuous peat layer that underlies most of the avulsion belt. In contrast, the Palaeogene avulsion deposits overlie palaeosols developed on claystones and mudstones.

These differences probably reflect both sediment source influences and climatic differences. The mountains surrounding the Bighorn Basin provided abundant clastic sediment, whereas the Saskatchewan River, where it was studied, is distant from mountain sources. The combination of a wetlands setting and limited clastic sediment may have led to peats and the paucity of sand in the stage III channels. Although the Fort Union climate was as wet as that of the Saskatchewan system, the soils were waterlogged and vegetation was abundant, the palaeosols are mineral soils rather than organic soils, indicating that clastic input was high. The Willwood Fm, on the other hand, had a drier climate that was not conducive to peat formation, in addition to abundant clastic input.

## DIFFERENCES AMONG THE ANCIENT EXAMPLES

### Weakly developed palaeosols

The weakly developed palaeosols in the Willwood Fm show different degrees of pedogenic development. Those in the younger section show features indicative of greater pedogenic development, including intense mottling, brunification or reddening, nodule development, rooting and burrowing. Such features usually are weakly developed or even absent from a significant proportion of the fine-grained avulsion deposits in the lower section. Assuming that the rates of pedogenesis were comparable, these differences suggest more time for soil development in the upper Willwood avulsion deposits.

This suggests that avulsions were completed (and a new channel established) more rapidly in the lower Willwood section than in the upper section. Few studies provide information on the speed at which avulsion proceeds. On the basis of [14]C dating, Törnqvist (1994) concluded that the Rhine–Meuse delta has been dominated by geologically instantaneous avulsions over the past 9000 yr, although he found several episodes of gradual avulsion. Using information from Autin *et al.* (1991), Törnqvist also suggested that gradual avulsion (of the order of 500 yr) characterized the Lower Mississippi Valley during the Holocene. The Saskatchewan River avulsion studied by Smith *et al.* (1989) has not yet reached completion and it began over 100 yr ago. Computer simulation models of alluvial architecture assume that avulsions are instantaneous (e.g. Bridge & Leeder, 1979; Mackey & Bridge, 1995). What may have caused different rates of avulsion in the two study sections is not known.

## Thickness of avulsion deposits

Despite the difficulties in recognizing the vertical boundaries of avulsion deposits and, thus, their true thicknesses, differences in thickness do appear to exist among the ancient examples (Table 1). One control on the thickness of avulsion deposits is the height of levees associated with the avulsing channel. A second control is topographic irregularities on the floodplain, which should influence lateral variations in the thickness of individual avulsion belts. These two features provide accommodation space for the avulsion deposits. For example, in the Saskatchewan system, avulsion deposits are up to 3.3 m thick (Smith *et al.*, 1989). Their thickness is roughly comparable to levee height, levees rise 2–3 m above the surrounding floodplain; however, avulsion deposits tend to be thicker in lower areas of the floodplain and thinner in higher areas (Smith *et al.*, 1989). Similarly, topographic irregularities have been invoked to explain lateral variations in the thickness of heterolithic deposits in the Chinji Fm (Behrensmeyer *et al.*, 1995) and in the lower Willwood section (Kraus, 1996).

Because no direct evidence of levee height was found in the Willwood Fm, Gwinn (1995) and Kraus (1996) used the average thickness of individual avulsion intervals to estimate ancient levee heights. This relationship seems appropriate for the avulsion deposits in the Fort Union sections and in the upper Willwood section, most of which have an average thickness between 1 and 5 m (Table 1). Recent compilations of modern and ancient levees cited in the literature show a remarkable range of levee height and indicate that levee heights of 4–5 m are

not uncommon (Brierley *et al.*, 1997). More problematic are the thicker heterolithic deposits in the lower Willwood section, some of which are of the order of 10 m. Similarly, the avulsion deposits described by Aslan & Blum (this volume, 193–209) from the Colorado River are 10 to 15 m thick. There are examples of unusually elevated levees. Along the Mississippi River, levees can be as high as 9 m (Farrell, 1987), and, in the ancient record, levee deposits as thick as 10 m or 12 m have been described by Fielding (1986) and Fielding *et al.* (1993), although whether the levees actually stood that high above the floodplain is not indicated. Nonetheless, the unusually thick examples in the Willwood Fm may not represent a single avulsion deposit but:

**1** two or more avulsion-belt deposits that are vertically stacked;

**2** two avulsion deposits that are separated by a cumulative palaeosol that has not been recognized, possibly because it is poorly developed;

**3** avulsion deposits overlain by levee deposits generated after avulsion and establishment of the new channel (Kraus & Gwinn, 1997).

## Times between successive avulsion intervals

The time intervals separating vertically successive avulsion deposits also differ among the ancient examples. To calculate the time intervals between avulsion deposits, sediment accumulation rates were first determined for each Palaeogene section on the basis of magnetostratigraphical studies and the revised geomagnetic time-scale of Cande & Kent (1995). Unfortunately, all rates are time-averaged over different lengths of time, with both of the Fort Union sections averaged over long periods (1 or 3 Myr) (Table 1).

The lower Fort Union falls into C26R, which totals about 265 m thick in the study area (Butler *et al.*, 1981) and spanned 3 Myr (Cande & Kent, 1995). The upper Fort Union section falls into C25R, which is 330 m thick (Butler *et al.*, 1981) and occupied *c.* 1.2 Myr (Cande & Kent, 1995). From these values, the time-averaged accumulation rate for the lower Fort Union section was slightly less than 0.1 mm yr and, for the upper Fort Union section, ~ 0.3 mm yr$^{-1}$. The upper Willwood section falls within the middle of C24N.3, which totals about 210 m in thickness (Tauxe *et al.*, 1994) (Fig. 2). This section spanned 0.44 Myr (Cande & Kent, 1995) and, thus, had a rock-accumulation rate of nearly 0.5 mm yr$^{-1}$. The lower section falls within C25R (Butler *et al.*, 1981) and below the Clarkforkian–Wasatchian (Cf–Wa) boundary. The section between the base of C25N and the Cf–Wa boundary is 440 m thick. The top of C25N is placed at 55.90 Ma

(Cande & Kent, 1995). The Palaeocene–Eocene boundary, as currently defined, falls in the early Wasatchian (Koch *et al.*, 1995) and is dated at 54.80 Ma (Cande & Kent, 1995). Therefore, the Cf–Wa boundary slightly precedes the Palaeocene–Eocene boundary and is estimated as 55 Ma. With those assumptions, the accumulation rate for the lower Willwood section was ~ 0.5 mm yr$^{-1}$ (averaged over 0.9 Myr).

Once accumulation rates had been calculated, the amount of time represented by each of the four Palaeogene sections was determined by dividing section thickness by the sediment-accumulation rate for each section (Table 1). That value was then divided by the number of avulsion intervals within a particular section. For example, the lower Fort Union section is 33 m thick and thus took 330 000 yr to accumulate at a rate of 0.1 mm yr$^{-1}$. In this section, we have recognized eight avulsion intervals with eight interbedded cumulative palaeosols. The average time between avulsion deposits is thus 330 000/8 or *c.* 41 000 yr (Table 1). This method yields an average time between avulsion deposits of 12 500 yr for the upper Fort Union section. The lower Willwood section was studied at two localities for which values of approximately 13 000 yr and 16 000 yr were determined. The average time between successive avulsion deposits for the upper Willwood section was calculated originally as 19 000 yr (Kraus & Aslan, 1993). On the basis of the recent magnetostratigraphical work (Tauxe *et al.*, 1994) and revised geomagnetic polarity time-scale of Cande & Kent (1995), this value has been revised to 14 000 yr. Behrensmeyer *et al.* (1995) estimated that the times between successive palaeosol-bounded intervals in the Chinji Fm were between 13 000 and 23 000 yr. The accumulation rate calculated for the Chinji Fm is 0.21 mm yr$^{-1}$; however, this value is time-averaged over 3.2 Myr.

The times between vertically successive avulsion *deposits* are considerably longer than the interavulsion periods that have been reported from Holocene river systems, which are of the order of $10^2$–$10^3$ yr (e.g. Autin *et al.*, 1991; Törnqvist, 1994; Morozova *et al.*, 1996). This discrepancy shows that only a fraction of the avulsions that actually took place are recorded by the avulsion deposits seen in a single vertical section. This is not surprising, because an avulsion belt covers only part of the floodplain.

Nonetheless, the time between avulsion deposits should be related to the interavulsion period of the river system. Except for the lower Fort Union section, the values fall within a fairly narrow range of 12 000 to 23 000 yr. Thus, the time between successive avulsion deposits is of the order of $10^4$ yr, which is probably an order of magnitude larger than the typical avulsion frequency of *c.* $10^3$ yr. Experimental work suggests that avulsion frequency increases with increasing sedimentation rate (Bryant *et al.*, 1995). Estimated interavulsion periods for the Palaeogene sections are roughly consistent with that hypothesis (Table 1). The lowest avulsion frequency (or largest interavulsion periods) characterizes the lower Fort Union section, which has the lowest accumulation rate. As accumulation rates increased into the lower Eocene, the frequency of avulsion also increased in the Willwood section, although the interavulsion period calculated for the upper Fort Union section is too rapid for its accumulation rate (Table 1).

## CONCLUSIONS

On the basis of similarities among avulsion deposits found in the Saskatchewan River, Fort Union and Willwood formations, supplemented with information from the Chinji Fm and Colorado River, a more complete model for recognizing avulsion deposits in the ancient record can be established. The fundamental criteria that can be used to recognize ancient avulsion deposits in other alluvial basins include:

1 cumulative palaeosols (or coals), which are moderately well developed to well developed, bound the avulsion deposits;

2 sandstones deposited by the trunk channels directly overlie the avulsion deposits;

3 the avulsion deposits are heterolithic and consist of fine-grained deposits that surround ribbon and thin sheet sandstones;

4 the fine-grained deposits contrast with the flood-basin deposits because they show only incipient pedogenic modification;

5 many of the ribbon sandstones show palaeoflow subparallel to or in the same direction as palaeoflow in the trunk channels;

6 ribbons generally have W/T ratios less than 10;

7 some of the ribbon sandstones, together with thin sandstone sheets, form tiers that suggest that networks of crevasse channels once occupied particular areas of the floodplain;

8 the heterolithic (avulsion) deposits are laterally extensive and form a significant part of the stratigraphical succession, indicating that they were deposited over large areas of the floodplain. The overall thickness of avulsion intervals appears to be controlled by levee height and floodplain depressions, which determine the short-term accommodation space in which sediment can accumulate.

Features including attributes of the ribbon sandstones, spatial scale and the presence of truncating channel sandstones help to distinguish avulsion deposits from ordinary crevasse-splay deposits developed off stable trunk channels. The stage I splays of Smith *et al.* (1989), which are equivalent to ordinary crevasse splays, have produced sands that are broader and thinner than most of the Palaeogene ribbons, and flow in those feeder and distributary channels is perpendicular or oblique to flow in the main channel, not parallel to it. Second, the Palaeogene avulsion deposits are laterally extensive and form a significant percentage of the measured stratigraphical sections, suggesting that they covered much larger areas than is typical for crevasse-splay deposits. Finally, the fact that sandstone bodies deposited by the trunk channels directly overlie the heterolithic deposits argues for an avulsion interpretation.

Avulsion deposits can be distinguished from flood-basin deposits on the basis of the ribbon sandstones and their pedogenic properties. The flood-basin deposits may be cut by ribbon sandstones; however, those sandstones scour downwards from the overlying heterolithic interval and thus are younger than the flood-basin deposits. Because of slow, sporadic deposition, flood-basin deposits are characterized by well-developed palaeosols. The avulsion deposits show only weak pedogenic development because accumulation rates were so rapid. In the field, the sharp contrast between the well-developed palaeosols and the weakly developed palaeosols enhances recognition of the avulsion intervals. At the same time, pedogenic overprinting obscures the depositional origin of some sediments, making it difficult to identify the contact between avulsion and flood-basin deposits.

Avulsion deposits make up more than half of the overbank deposits in all four Palaeogene sections studied. Avulsion deposits also comprise approximately 50% of the Colorado valley-fill deposits described by Aslan & Blum (this volume, 193–209), and the Chinji Fm sections of Willis & Behrensmeyer (1994) contain significant amounts of similar palaeosol-bounded deposits. Avulsion deposits also have been described from ancient braided river deposits (Bentham *et al.*, 1993). These examples support the view of Smith *et al.* (1989) that, in many meandering systems, a significant portion of the fine-grained sediment was deposited as the trunk channel avulsed. They also suggest that avulsion deposits should be commonly preserved in the rock record, although they have not yet been generally recognized.

## ACKNOWLEDGEMENTS

This research was supported by National Science Foundation Grant EAR-9303959. P.D. Gingerich and the Churchill family provided invaluable logistical support for field-work in Powell. We thank Paul Koch and Scott Wing for helpful discussions on the stratigraphy, particularly the magnetostratigraphy, of the Palaeogene rocks and for calculations of sediment-accumulation rates. MJK thanks Norm Smith for informative discussions throughout the course of this research. John Drexler provided geochemical analyses. Constructive reviews were provided by Greg Nadon and T.E. Törnqvist.

## REFERENCES

ALLEN, J.R.L. (1965) A review of the origin and characteristics of recent alluvial sediments. *Sedimentology*, **5**, 89–191.

ASLAN, A. & BLUM, M.D. Contrasting styles of Holocene avulsion, Texas Gulf Coastal Plain, USA. In: *Fluvial Sedimentology VI* (Eds Smith, N.D. & Rogers, J.), Spec. Publs Int. Ass. Sediment., No. 28, pp. 193–209. Blackwell Science, Oxford.

AUTIN, W.J., BURNS, S.F., MILLER, B.J., SAUCIER, R.T. & SNEAD, J.I. (1991) Quaternary geology of the Lower Mississippi Valley. In: *Quaternary Non-glacial Geology of the Conterminous United States*, Vol. K-2, *The Geology of North America* (Ed. Morrison, R.B.), pp. 547–582. The Geological Society of America, Boulder, CO.

BEHRENSMEYER, A.K., WILLIS, B.J. & QUADE, J. (1995) Floodplains and paleosols of Pakistan Neogene and Wyoming Paleogene deposits: a comparative study. *Palaeogeogr., Palaeoclimatol., Palaeoecol.*, **115**, 37–60.

BENTHAM, P.A., TALLING, P.J. & BURBANK, D.W. (1993) Braided stream and flood-plain deposition in a rapidly aggrading basin: the Escanilla Formation, Spanish Pyrenees. In: *Braided Rivers* (Eds Best, J.L. & Bristow, C.S.), Spec. Publ. geol. Soc. London, No. 75, 177–194. Geological Society of London, Bath.

BOWN, T.M. (1980) Summary of latest Cretaceous and Cenozoic sedimentary, tectonic, and erosional events, Bighorn Basin, Wyoming. In: *Early Cenozoic Paleontology and Stratigraphy of the Bighorn Basin, Wyoming* (Ed. Gingerich, P.D.). *Univ. Mich. Pap. Paleontol.*, **24**, 5–32.

BOWN, T.M., ROSE, K.D., SIMONS, E.L. & WING, S.L. (1994) Distribution and stratigraphic correlation of fossil mammal and plant localities of the Fort Union, Willwood, and Tatman Formations (upper Paleocene–lower Eocene), central and southern Bighorn Basin, Wyoming. *U.S. geol. Surv. Prof. Pap.*, **1540**, 1–103.

BRIDGE, J.S. (1984) Large-scale facies sequences in alluvial overbank environments. *J. sedim. Petrol.*, **54**, 583–588.

BRIDGE, J.S. & LEEDER, M.R. (1979) A simulation model of alluvial stratigraphy. *Sedimentology*, **26**, 599–623.

BRIERLEY, G.J., FERGUSON, R.J. & WOOLFE, K.J. (1997) What is a fluvial levee? *Sediment. Geol.*, **114**, 1–9.

BROWN, W.G. (1993) Structural style of Laramide basement-cored uplifts and associated folds. In: *Geology of Wyoming* (Eds Snoke, A.W., Steidtmann, J.R. & Roberts, S.M.). *Mem. geol. Surv. Wyo.*, **5**, 312–371.

BRYANT, M., FALK, P. & PAOLA, C. (1995) Experimental study of avulsion frequency and rate of deposition. *Geology*, **23**, 365–368.

BUTLER, R.F., GINGERICH, P.D. & LINDSAY, E.H. (1981) Magnetic polarity stratigraphy and biostratigraphy of

Paleocene and Lower Eocene continental deposits, Clark's Fork basin Wyoming. *J. Geol.*, **89**, 299–316.

CANDE, S.C. & KENT, D.V. (1995) Revised calibration of the geomagnetic polarity timescale for the Late Cretaceous and Cenozoic. *J. geophys. Res.*, **100**, 6093–6095.

CLYDE, W.C. (1997) *Stratigraphy and mammalian paleontology of the McCullough Peaks, northern Bighorn Basin, Wyoming: Implications for biochronology, basin development, and community reorganization across the Paleocene–Eocene boundary*. PhD thesis, University of Michigan, Ann Arbor, Michigan, 271 pp.

DIRSCHL, H.J. & COUPLAND, R.T. (1972) Vegetation patterns and site relationships in the Saskatchewan River Delta. *Can. J. Bot.*, **50**, 647–675.

FARRELL, K.M. (1987) Sedimentology and facies architecture of overbank deposits of the Mississippi River, False River Region, Louisiana. In: *Recent Developments in Fluvial Sedimentology* (Eds Ethridge, F.G., Flores, R.M. & Harvey, M.D.), Spec. Publ. Soc. econ. Paleont. Miner., Tulsa, **39**, 111–120.

FIELDING, C.R. (1986) Fluvial channel and overbank deposits from the Westphalian of the Durham coalfield, NE England. *Sedimentology*, **33**, 119–140.

FIELDING, C.R., FALKNER, A.J. & SCOTT, S.G. (1993) Fluvial response to foreland basin overfilling: the Late Permian Rangal Coal Measures in the Bowen Basin, Queensland, Australia. *Sediment. Geol.*, **85**, 475–497.

GINGERICH, P.D. (1983) Paleocene – Eocene faunal zones and preliminary analysis of Laramide structure deformation in the Clark's Fork Basin, Wyoming. In: *Guidebook of the Wyoming Geological Association, 34th Annual Field Conference*, 1983, Bighorn Basin, WY, pp. 185–195.

GREENWOOD, D.R. & WING, S.L. (1995) Eocene continental climates and latitudinal temperature gradients. *Geology*, **23**, 1044–1048.

GWINN, B.M. (1995) *Avulsion deposit architecture and controls on floodplain construction in the Lower Eocene Willwood Formation, Bighorn Basin, Wyoming*. MS thesis, University of Colorado, Boulder, CO, 125 pp.

HELLER, P.L. & PAOLA, C. (1996) Downstream changes in alluvial architecture: an exploration of controls on channel stacking patterns. *J. sediment. Res.*, **66**, 297–306.

HICKEY, L.J. (1980) Paleocene stratigraphy and flora of the Clark's Fork Basin. In: *Early Cenozoic Paleontology and Stratigraphy of the Bighorn Basin, Wyoming* (Ed. Gingerich, P.D.). *Univ. Mich. Pap. Paleont.*, **24**, 33–49.

JEPSON, G.L. (1940) Paleocene faunas of the Polecat Bench Formation, Park County, Wyoming. *Proc. Am. Philos. Soc.*, **83**, 217–340.

JORGENSEN, P.J. & FIELDING, C.R. (1996) Facies architecture of alluvial floodbasin deposits: three-dimensional data from the Upper Triassic Callide Coal Measures of east-central Queensland, Australia. *Sedimentology*, **43**, 479–495.

KOCH, P.L., ZACHOS, J.C. & DETTMAN, D.L. (1995) Stable isotope stratigraphy and paleoclimatology of the Paleogene Bighorn Basin (Wyoming, USA). *Palaeogeogr., Palaeoclimatol, Palaeoecol.*, **115**, 61–89.

KRAUS, M.J. (1996) Avulsion deposits in lower Eocene alluvial rocks, Bighorn Basin, Wyoming. *J. sediment. Res.*, **66B**, 354–363.

KRAUS, M.J. (1997) Early Eocene alluvial paleosols: pedogenic development, stratigraphic relationships, and paleosol/landscape associations. *Palaeogeogr., Palaeoclimatol., Palaeoecol*, **129**, 387–406.

KRAUS, M.J. & ASLAN, A. (1993) Eocene hydromorphic paleosols: significance for interpreting ancient floodplain processes. *J. sediment. Petrol.*, **63**, 453–463.

KRAUS, M.J. & GWINN, B.M. (1997) Controls on the development of early Eocene avulsion deposits and floodplain paleosols, Willwood Formation, Bighorn Basin. *Sediment. Geol.*, **114**, 33–54.

LOTTES, A.L. & ZIEGLER, A.M. (1994) World peat occurrence and the seasonality of climate and vegetation. *Palaeogeogr., Palaeoclimatol., Palaeoecol.*, **106**, 23–27.

MACKEY, S.D. & BRIDGE, J.S. (1995) Three-dimensional model of alluvial stratigraphy: theory and application. *J. sedim. Res.*, **B65**, 7–31.

MJØS, R., WALDERHAUG, O. & PRESTHOLM, E. (1993) Crevasse splay sandstone geometries in the Middle Jurassic Ravenscar Group of Yorkshire, UK. In: *Alluvial Sedimentation* (Eds Marzo, M. & Puigdefabregas, C.), Spec. Publs int. Ass. Sediment., No. 17, pp. 167–184. Blackwell Scientific Publications, Oxford.

MOROZOVA, G.S., SMITH, N.D. & SLINGERLAND, R.L. (1996) Holocene avulsion history and facies of the Lower Saskatchewan fluvial system in the Cumberland Marshes area, Saskatchewan, Canada. *Geol. Soc. Am. Abstr. Program*, **28**, A-471.

NEASHAM, J.W. & VONDRA, C.F. (1972) Stratigraphy and petrology of the lower Eocene Willwood Formation, Bighorn Basin, Wyoming. *Geol. Soc. Am. Bull.*, **83**, 2167–2180.

O'BRIEN, P.E. & WELLS, A.T. (1986) A small alluvial crevasse splay. *J. sediment. Petrol.*, **56**, 876–879.

ROSE, K.D. (1981) The Clarkforkian land-mammal age and mammalian faunal composition across the Paleocene–Eocene boundary. *Univ. Mich. Pap. Paleont.*, **26**, 1–196.

SMITH, D.G. & PUTNAM, P.E. (1980) Anastomosed river deposits: modern and ancient examples in Alberta, Canada. *Can. J. Earth Sci.*, **17**, 1396–1406.

SMITH, N.D. & PÉREZ-ARLUCEA, M. (1994) Fine-grained splay deposition in the avulsion belt of the lower Saskatchewan River, Canada. *J. sediment. Res.*, **B64**, 159–168.

SMITH, N.D., CROSS, T.A., DUFFICY, J.P. & CLOUGH, S.R. (1989) Anatomy of an avulsion. *Sedimentology*, **36**, 1–24.

TAUXE, L., GEE, J., GALLET, Y., PICK, T. & BOWN, T. (1994) Magnetostratigraphy of the Willwood Formation, Bighorn Basin, Wyoming: new constraints on the location of Paleocene/Eocene boundary. *Earth planet. Sci. Lett.*, **125**, 159–172.

TÖRNQVIST, T.E. (1994) Middle and late Holocene avulsion history of the River Rhine (Rhine–Meuse delta, Netherlands). *Geology*, **22**, 711–714.

TÖRNQVIST, T.E., VAN REE, M.H.M. & FAESSEN, E.L.J.H. (1993) Longitudinal facies architectural changes of a Middle Holocene anastomosing distributary system (Rhine–Meuse delta, central Netherlands). *Sediment. Geol.*, **85**, 203–219.

TRICART, J. & CAILLEUX, A. (1972) *Introduction to Climatic Geomorphology* (translated by C.J. Kiewiet de Jonge). Longman, London, 295 pp.

VINCENT, J.S. & HARDY, L. (1979) The evolution of glacial Lakes Barlow and Ojibway, Quebec and Ontario. *Geol. Surv. Can. Bull.*, **6**, 1–18.

WALCOTT, R.L. (1972) Late Quaternary vertical movements in eastern North America: quantitative evidence of glacio-isostatic rebound. *Rev. geophys. Space Phys.*, **10**, 849–884.

WEERTS, H.J.T. & BIERKENS, M.F.P. (1993) Geostatistical analysis of overbank deposits of anastomosing and meandering fluvial systems; Rhine-Meuse delta, The Netherlands. *Sediment. Geol.*, **85**, 221–232.

WELLS, T.M. (1997) *Alluvial architecture of avulsion deposits in the Paleocene Fort Union Formation, Clark's Fork Basin, Wyoming, USA*. MS thesis, University of Colorado, Boulder, CO, 150 pp.

WILLIS, B.J. & BEHRENSMEYER, A.K. (1994) Architecture of Miocene overbank deposits in northern Pakistan. *J. sediment. Res.*, **64B**, 60–67.

WING, S.L. (1981) *A study of paleoecology and paleobotany in the Willwood Formation (early Eocene, Wyoming)*. PhD thesis, Yale University, New Haven, CT, 391 pp.

Controls on River Systems and Alluvial Successions

*Spec. Publs int. Ass. Sediment.* (1999) **28**, 271–281

# The use of models in the interpretation of the effects of base-level change on alluvial architecture

S. B. MARRIOTT

*School of Geography and Environmental Management, University of the West of England, Bristol,
Coldharbour Lane, Bristol BS16 1QY, UK (Email: Susan.Marriot@uwe.ac.uk)*

## ABSTRACT

Base-level changes occur over a broad range of time-scales and may result from several interacting factors. As a result, relative sea-level has changed considerably over geological time, at times of intense glaciation and at times when the Earth was ice-free. Several models have been proposed that examine the effects of base-level change on river systems and the expected impact on alluvial architecture. These models tend to be qualitative and their aim was to give guidance in the interpretation of seismic stratigraphy. From Allen's pioneering work in the 1970s, a new generation of alluvial-architecture models has evolved. These are now solved by computer and provide an insight into a range of intrinsic and extrinsic influences on the development of fluvial sequences. None of these quantitative models, however, has been used to examine the effects of base-level change on alluvial architecture. This paper reviews literature on fluvial response to base-level change, then illustrates the use of a computer model developed by Crane in 1982 to examine this response in terms of a fluvial sequence. The resulting computer output of the alluvial architecture is then compared with a conceptual model and field data from southern France and Argentina. The general pattern of sandbody geometry and connectedness, simulated by computer, is in broad agreement with the conceptual model and with the outcrop patterns described. It is considered unreasonable to suggest that the models discussed would be applicable to all fluvial systems, because base-level change is only one of many extrinsic and intrinsic influences.

## INTRODUCTION

Base-level changes occur over a broad range of time-scales and may result from several interacting factors, for example, local or regional uplift or subsidence; changes in the mass of ocean water as a result of wastage or accumulation of glaciers and ice sheets (Geophysics Study Committee, 1990). As a result, sea-level has changed considerably over geological time, with repeated variations of over 100 m from its present level (Chappell & Shackleton, 1986). These large changes have occurred at times of intense glaciation and at times when the Earth was ice-free. The major factors dominating sea-level changes over the past 250 million yr were climate change, tectonic processes and sedimentation (Geophysics Study Committee, 1990). Haq *et al.* (1987), however, have attributed the events of coastal onlap and offlap (related to marine transgression and regression) to eustatic variation only.

Unconformity-bounded sequences form in response to changes in the depositional base level (effectively the same as sea-level; see below and discussion in Schumm, 1993. Seismic-stratigraphical methods, which have been developed over the past 20 yr and applied to the analysis of sedimentary successions, rely on the identification of unconformity-bounded sequences in seismicreflection profiles (Vail *et al.*, 1977). Most sequence boundaries record times at which the rate of sea-level fall increased (or reached a maximum) or the rate of tectonic subsidence decreased (Christie-Blick *et al.*, 1990; Christie-Blick, 1991). The recognition of depositional sequences, bounded by unconformities, forms the basis for sequence stratigraphy.

A qualitative sequence-stratigraphical model for fluvial successions was put forward by Posamentier & Vail (1988). This has been criticized by (among others) Miall (1991), Schumm (1993) and Leeder & Stewart (1996), particularly with regard to misconceptions of equilibrium profile changes in response to changes of base level. A series of quantitative models, commencing

with work by Allen (1978) and developing to sophisticated models of alluvial architecture solved by computer
(e.g. Mackey & Bridge, 1995; and see below) also have
been used to aid interpretation of fluvial successions. It is
apparent, however, that no attempt has been made to use a
quantitative model to investigate the behaviour of fluvial
systems in relation to base-level changes, nor to add
strength to conceptual models. It is the aim of this paper,
therefore, to present an investigation into the use of a
computer simulation model for alluvial stratigraphy to
interpret the effects of changing base level. The resulting alluvial architecture is compared with a published
record of field observations from France and Argentina.
Before discussing the possible sequences generated in
alluvial systems during base-level change, the major processes which result in floodplain accumulation and river
response to base-level change are reviewed.

## SEDIMENT ACCUMULATION ON FLOODPLAINS

A wide variety of modern floodplain types has been
defined by various workers (Wolman & Leopold, 1957;
Schumm, 1968; Lewin, 1978; Nanson & Croke, 1992),
and their formation can be regarded, in most cases, as the
product of the interaction of vertical- and lateral-accretion processes (Nanson, 1986). Lateral accretion on point
bars during meander migration occurs from deposition
of bedload material and results in reworking of existing channel and floodplain sediments. In addition, some
modern floodplains consist of finer sediments vertically
accreted during floods, with lateral accretion deposits of
minor importance (Marriott, 1996, 1998; see also Brown
(1996) for discussion of other sedimentation processes
on floodplains). In effect, floodplains are acting as stores
for fine-grained material (Walling & Bradley, 1989), with
the result that, although floodplains on average have
low accretion rates over time, many have relatively high
short-term rates (Bridge (1984) records an average value
of 2 mm yr$^{-1}$).

Accumulation and storage of sediment on floodplains
is limited by several factors. The rate of accretion usually
declines with time and elevation of the surface, as then
only high-magnitude (long return-period) discharges will
cause flooding. The storage of sediment is also associated
with the bed level of the related channel. If the channel is
confined by prominent levees, the bed of the channel may
build up above the level of the surrounding floodplain.
Major floods can then result in crevassing and severe
erosion (Nanson, 1986; Walling & Bradley, 1989). If,
however, the bed of the channel also aggrades vertically,

to keep pace with accretion on the floodplain, the whole
system will store sediment. If the amount of vertical
accretion possible in a system is constrained, lateral
accretion by rivers can result in more extensive bedload
deposits, with corresponding shifts in the fine-to-coarse
ratio of the resulting alluvial suite.

The concept of accommodation as used in sequence
stratigraphy also can be applied to deposition on floodplains. The accommodation available for sediment accumulation on a floodplain is controlled by the elevation of
the channel and its bankfull depth. Floodplain deposits
cannot build up above bankfull level and this ultimately is
controlled by base level (see below). If bankfull level is
static, floodplains can rapidly accrete vertically to this
stage, and then may become reworked and incorporated
into lateral accretion deposits if meandering progresses
(Wright & Marriott, 1993). The relationship between base-
level changes and the accumulation of vertically accreted
sediment and lateral reworking is critical in understanding how the architecture of alluvial suites can change
during base-level changes induced by sea-level fluctuations (see also Allen, 1974; Allen & Williams, 1982).

## FLUVIAL RESPONSE TO BASE-LEVEL CHANGE

The term 'base level' as used here, is interpreted as
sea-level. Davis (1902) reviewed previous definitions of
base level and proposed that, to avoid confusion, base
level should be defined as 'the level base with respect to
which normal subaerial erosion proceeds', i.e. sea-level
(although near their mouths, rivers erode their channel
below sea-level). Davis (1902) also used the term 'grade'
to refer to the balanced condition of a 'mature' river. The
development of this balanced condition is effected by
changes in the river's ability to do work, and in the quantity of work that the river has to do. These two quantities,
initially unequal, reach equality when the river is at grade
(Davis, 1902). A river grades its course by a process of
cutting and filling, until a slope is developed along which
the river can transport its load in the most efficient way. A
mature river system is one in which all parts of the river
have reached a graded condition. In explaining the concepts of river grade and maturity, Davis (1902) also used
base level as a temporary, local factor, controlling the
river's development at reach scale. This, and other con-
fusing uses of 'base level' in describing the development
of fluvial systems, were further reviewed by Schumm
(1993), who prefers to relate the term to sea-level.

Changes of extrinsic variables, such as climate, tectonic uplift and relative sea-level (Fig. 1), will alter the

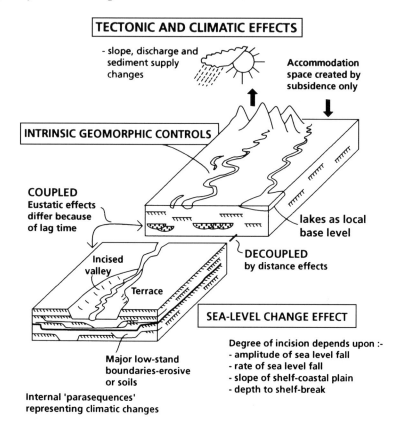

**TECTONIC AND CLIMATIC EFFECTS**

- slope, discharge and sediment supply changes

Accommodation space created by subsidence only

**INTRINSIC GEOMORPHIC CONTROLS**

**COUPLED** Eustatic effects differ because of lag time

lakes as local base level

**DECOUPLED** by distance effects

Incised valley

Terrace

**SEA-LEVEL CHANGE EFFECT**

Degree of incision depends upon :-
- amplitude of sea level fall
- rate of sea level fall
- slope of shelf-coastal plain
- depth to shelf-break

**Major low-stand boundaries-erosive or soils**

**Internal 'parasequences' representing climatic changes**

**Fig. 1.** Regional and local controls on base level. (Diagram supplied courtesy of V.P. Wright.)

progress of the geomorphological cycle and, while progressing towards grade (or equilibrium profile), rivers constantly have to make adjustments in order to compensate for these extrinsic changes (Davis, 1899; Schumm, 1975).

Base-level lowering will rejuvenate a river, causing an increase in gradient and the delivery of increased amounts of sediment downstream. This has been shown experimentally by Holland & Pickup (1976) and from observations of the effects of Pleistocene sea-level lowering on valley systems such as the Mississippi (Fisk, 1944), the Rhone (Tesson *et al.*, 1990) and the Gironde (G.P. Allen & Posamentier, 1993). These observations confirm Davis' (1899) assertion that a river will adjust by aggradation or erosion to restore its equilibrium profile. Sea-level, however, was lowered by 120 m during the Pleistocene (Fairbanks, 1989), and incision on the Texas continental shelf by the Sabine and Trinity rivers, was to a depth of only 40 m; yet the influence of sea-level fall on the Mississippi extended 370 km upstream (Schumm, 1993). The effects of base-level change may be minimal in some cases and significant in others, and Schumm (1993) suggested that other variables must be important in either

ameliorating or accentuating them. The significant controls were grouped into:

1 base-level controls – direction, magnitude, rate and duration;

2 geological controls – lithology, structure and nature of valley alluvium;

3 geomorphological controls – inclination of exposed surfaces, valley morphology, river morphology and adjustability.

The direction of base-level change affects whether a river aggrades or erodes. The magnitude and rate of change, particularly in the case of a base-level fall, can influence whether the river can make adjustments to accommodate the change without incision. If the fall is small, a channel can adjust by changing its pattern or shape; whereas, if the fall is large, incision is likely. Likewise, when a base-level fall occurs slowly, the channel can adjust its slope by lateral migration (Yoxall, 1969), although when the change is rapid, flow will be concentrated in a narrow deep valley and incision can migrate upstream.

The nature of the material forming the channel banks determines the upstream extent of the effects of base-level lowering. Experimental studies have been performed to examine the effects of base-level fall in low-cohesion sediments by Brush & Wolman (1960) and in cohesive sediments by Begin *et al.* (1981) and Begin (1982, 1988), who also incorporated field studies in the investigation. The experiments showed that, in non-cohesive sediment, knickpoints will not migrate indefinitely upstream. This is due to the slope of the knickpoint reach decreasing until it is nearly equal to that of the stream. The stream competence then declines, owing to reduction of slope and considerable bank caving, so that no bedload transport occurs (Brush & Wolman, 1960). In cohesive sediment, however, channel incision occurs and water becomes concentrated in the incised channel, increasing incision and promoting headward erosion (Begin *et al.*, 1981; Begin, 1988). Leeder & Stewart (1996) showed, by quantitative and conceptual modelling, that channel incision depends on sediment-transport capacity and the difference in slope between the river's profile and the continental shelf. They give a valuable check-list for recognition of incision in the stratigraphical record and discuss the important criteria required from field observations, as they propose that the assumption that fluvial incision has occurred, when sedimentary sequences are examined, is overestimated.

The sediment load a river is required to carry may be increased following incision (i.e. the river has more work to do in Davis' (1902) terms). The response of a channel to the change in gradient may, therefore, be a change in pattern (Schumm & Khan, 1972). Schumm (1993) gives an example from the Mississippi, where it is apparent that increases in the valley slope have caused the river pattern to change from low to high sinuosity. If the valley slope becomes too steep, a braided planform will result (Schumm *et al.*, 1987). Schumm (1993) also makes the point that, if the channel is confined, a river is not likely to adjust to an increased valley slope by becoming more sinuous. In this case, incision will occur. When base-level change results in a decrease in valley slope (e.g. when sea-level rises) sinuosity also may decrease.

The effects of base-level change on a fluvial system are therefore complex. Patton & Schumm (1981), Gardner (1983) and Schumm *et al.* (1987) have shown that erosion in one reach of a river may be contemporaneous with deposition in another reach, owing to the increased sediment loads generated by incision. Large rivers will be able to accommodate base-level changes by adjusting their planform (sinuosity) and channel dimensions and it is unlikely that a fall in base level will rejuvenate the whole system. This also is shown by Leeder & Stewart (1996) who propose that rivers may 'keep up' with falling sea-level and will construct a prograding alluvial plain over a gently sloping continental shelf. This is in general agreement with the results of a series of experiments conducted by Wood *et al.* (1993).

## THE USE OF MODELS TO EXAMINE THE EFFECTS OF BASE-LEVEL CHANGE

Models for alluvial architecture have been widely used to examine possible influences on the development of fluvial sequences. Since 1978, when Allen's quantitative model for the architecture of avulsion-controlled suites was published (see also Leeder, 1978; Bridge & Leeder, 1979), a new generation of sophisticated models refining the originals has evolved (Crane, 1982; Mackey & Bridge, 1992, 1995; Bridge & Mackey, 1993; Bryant *et al.*, 1995). The modelling of fluvial sequences over a long time-scale ($10^5$–$10^6$ yr) generally is based, necessarily, on simplification of (some of) the processes assumed to be operating, although the new generation of models, using extended computer capacities, are able to do away with some of these constraints (Howard, 1996). Further, despite the existence of these computational models for almost two decades, Leeder *et al.* (1996) were the first to use data from field observations to test an alluvial architecture model, owing primarily to the wide range of variables needed. Leeder & Stewart (1996) have also pointed out that these quantitative models for alluvial architecture do not examine the effects of base-level change on the modelled system. This neglect is rectified somewhat here.

The computer-simulation model for alluvial architecture used in this study was developed by Crane (1982) as an attempt to gain insight into the distribution and interconnectedness of sandstone bodies in fluvial suites, sand/fines ratios and to aid prediction of locality and extent of hydrocarbon reservoirs. It also provided a means of investigating alluvial architecture in relation to possible responses to changes in extrinsic variables, such as subsidence and climate and their interactions. Crane's model built on earlier quantitative models (Allen, 1978; Leeder, 1978; Bridge & Leeder, 1979) incorporating a wider range of processes (Fig. 2) and attempting to reduce arbitrary effects to a minimum. The main differences between Crane's (1982) simulation and the earlier ones are as follows:

**1** deposition on the floodplain varies with distance from the channel;
**2** new accretion on the floodplain is evaluated frequently (at 50-yr intervals of modelled time);
**3** lithology of the overburden is taken into account when calculating compaction (also on 50-yr cycles);

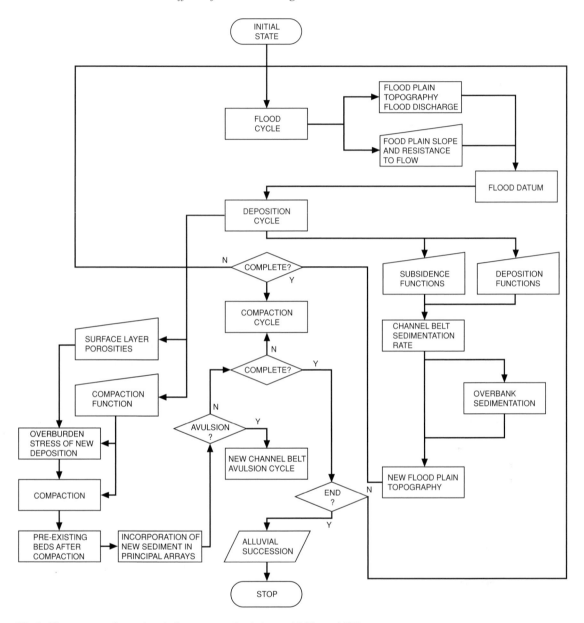

**Fig. 2.** The sequence of operations in the computer-simulation model (Crane, 1982).

4 occurrence of avulsion can be operator selected to occur either

(a) by reference to floodplain topography and the mean-flood flow-velocity required to erode the floodplain sufficiently to cause channel relocation,

(b) because the channel bed had built up to a level above that of the floodplain,

(c) because a pre-set period of modelled time (e.g. 1000 yr) had elapsed.

Crane (1982) compared model output with field observations of part of the Old Red Sandstone from South Wales, and examined the influence of environmental changes, such as mean annual flood discharge and subsidence rate. He considered that the sequences generated, particularly in relation to sandbody interconnectedness, are strongly influenced by the deposition rate. Within the program, deposition is calculated using a pseudologarithmic function that relates the deposition on the floodplain

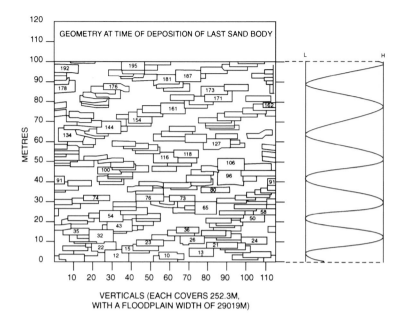

**Fig. 3.** Diagram of computer output from Crane's (1982) program using a variation in subsidence. The period covered is 474 650 yr. Some sandbody numbers have been excluded for clarity. H, highstand; L, lowstand. (Original printout in Marriott, 1993, p. 143.)

to that in the channel, which in turn is calculated according to the maximum flood discharge occurring in each 50-yr period. The peak discharge for such intervals is generated by a random-number-generating subprogram and is based on the operator-set mean annual flood variable. This is the only random element introduced into the program (for full details of program development and operation, see Crane, 1982).

With regard to the locality of the new channel following avulsion, Crane's program allows for the effects of floodplain topography, siting the new channel at the lowest point on the floodplain. If two localities are of equally low level, then the one nearest the old channel is chosen. This provision takes into account the effects of compaction in different lithologies and the influences of underlying lithologies, in particular old, abandoned channel belts. In this it develops Allen's (1978) model. In addition, in Crane's (1982) model, the development of the floodplain topography is directly proportional to the total quantity of subsidence occurring over the duration of each modelled channel system.

The effects of varying subsidence can be used to investigate the effects of variation in base level on the modelled alluvial suite. This is because the area covered is assumed to be on a coastal plain (Crane, 1982) and, in these areas depositional base level is the result of the relative interaction between variations in eustasy and subsidence (Posamentier & Vail, 1988). If eustatic sea-level is assumed to remain constant, then relative base level will be influenced entirely by variations in subsidence, which

in turn will affect the increase or decrease in available accommodation space. Crane (1982) uses a sine-curve function to represent the change in subsidence over time. This is similar to the curve used by Posamentier & Vail (1988) to represent variations in eustasy with time. Unlike their examples, however, the variables in Crane's program, which relate to the maximum, minimum and wavelength of the subsidence curve, cannot be varied during any one run of the program, although they can be changed by the operator to investigate the effects of different rates of subsidence and different time-scales.

The simulated alluvial suite is shown in Fig. 3. The period covered is 474 650 yr, relating to cycles of base-level variation of 10 m over 100 000 yr. This equates with fourth-order Milankovitch eccentricity megacycles (Kerr, 1987). The scale of sequences modelled (e.g. third- or fifth-order cycles) can be investigated by changing the variables relating to maximum and minimum subsidence rates and the wavelength of the sine curve (see above). The variation in base level, throughout the modelled time, is shown in Fig. 3 (the curve appears tighter in the lower portion, owing to the effects of compaction on the alluvial suite).

At lowstand, the channel belt is incised, sometimes several metres (e.g. sand bodies 54, 96 and 178 on Fig. 3) and thick sand bodies are deposited. During late lowstand or at the onset of the transgression, the incised channel has to be filled with sediment before deposition can become widespread on more distal parts of the floodplain.

The thin sandbodies that occur throughout the suite are

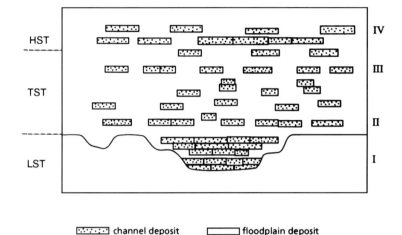

**Fig. 4.** Qualitative model of alluvial sequence stratigraphy (after Wright & Marriott, 1993). HST, highstand systems tract; TST, transgressive systems tract; LST, lowstand systems tract. See text for explanation of numerals I–IV.

related to transgressive depositional systems and highstand systems. The difference between the two systems appears to be that the thin units related to the transgression deposits have a widespread appearance (e.g. sand bodies 13–31), whereas following a highstand the sand bodies form stacked units (e.g. sand bodies 38–53 and 163–176).

During transgression or relative rise in base level, the rate of increase in accommodation is high, so avulsion is likely to be frequent. In these circumstances, channels are unlikely to occupy a stable position and little reworking of floodplain sediments occurs. This results in thin channel sand bodies and relatively low interconnectedness (e.g. 148–162 on Fig. 3). One of the features of Crane's (1982) model is that the sand bodies, numbered on the computer printout, actually represent the whole meander belt, rather than just the channel. The thin sand bodies that occur during the transgressive episodes, therefore, still represent some lateral migration before avulsion and a shift of the whole meander belt to a new locality. Additionally, the alluvial suite represented has no lateral boundaries, i.e. it is a cylindrical model, similar to a coastal plain of inundation, rather than a fluvial system contained within confining-valley sides. This accounts for sand bodies with the same number appearing on both sides of the diagram (e.g. 91 on Fig. 3).

During the highstand, there is little further increase in accommodation space and channel gradients will fall, so the river will rework its floodplain sediment as meanders migrate. Stacked sand bodies are likely to result. These are represented on Fig. 3, either by thick sand bodies (e.g. 32 and 116), which relate to episodes where reworking has taken place solely within the meander belt, or by stacked thin sand bodies (e.g. 163–176).

The output from the model is further complicated by the random element introduced by the scale of discharge during operation of the program. As detailed above, a random-number-generating subprogram is used to select the peak discharge, during each modelled 50-yr period. This does not allow for the influence of other parameters within the program and therefore may account for the thick sand bodies that occur throughout the suite (e.g. 65 and 127 on Fig. 3).

In order to assess the success of this simulation in indicating the likely development of an alluvial suite, under conditions of relative base-level change, it is necessary to compare it with other models and field data. A simple conceptual model for alluvial sequence stratigraphy was proposed by Wright & Marriott (1993), which relates changing alluvial architecture to increases and decreases in the availability of accommodation space, as base level varies. As such it provides a suitable model for comparison with the computer simulation. The illustration (Fig. 4) is necessarily more simple than that of the computer printout and illustrates a sequence of base-level change from incision (lowstand) up to late highstand. Wright & Marriott (1993) propose that the suite is likely to develop as follows:

**I** during lowstand and following incision, fluvial deposits will be restricted to the incised valley;

**II** during the early stages of the transgression, the rate of creation of accommodation space is low and deposition will continue to occur only in the incised valley until the valley is filled and then older surfaces will be flooded;

**III** during later stages of the transgression, the rate of provision of accommodation space is high, flooding and avulsion will be more frequent and isolated sand bodies are likely;

**IV** during highstand, floodplain-accretion rates drop, as flooding is less frequent and avulsion is less likely, and the result of reduced accommodation is the reworking of floodplain sediments by meander migration.

## COMPARISON WITH FIELD DATA

The concepts discussed by Wright & Marriott (1993) are in broad general agreement with the computer simulation. Further evidence, however, is available from field data. Marriott (1993) attempted a comparison with data from an outcrop of Late Cretaceous fluvial sediments in the Languedoc Basin. This comparison was restricted, however, because although both sheet sandstones, with evidence of lateral migration, and isolated or stacked sand bodies within thick floodplain sediments are present, there is no evidence of incision. The eustatic curve proposed for the Cretaceous (Haq *et al.*, 1987) shows that there was probably a major (75–100 m), relatively rapid sea-level rise followed by a slight fall (−20 m) before a further relatively minor rise (25–40 m) in the latest Maastrichtian. Between highstand and transgressive systems, the rate of change in base level reaches a minimum at the lowstand and Posamentier & Vail (1988) recognized two types of lowstand sequence boundary. Type 1 relates to a major, rapid fall in base level accompanied by incision and this is the type of sequence modelled conceptually by Wright & Marriott (1993) and in the computer simulation. Type 2 sequences have no incision, as the rate or magnitude of fall is not great enough. The sequence described by Marriott (1993) could be a type 2 sequence, as the base-level fall identified is relatively small in magnitude (Haq *et al.*, 1987), but equally, palaeogeographical reconstructions (Freytet & Plaziat, 1982, p. 5) show that the area was some considerable distance from the Gulf of Aquitaine. From discussions by Schumm (1993) and Leeder & Stewart (1996), therefore, channel incision may not be evident because of the distance from the coast.

More convincing field data are available from Argentina, as documented by Legarreta *et al.* (1993) and compared with their conceptual model for non-marine sequences on the same lines as that of Wright & Marriott (1993). Legarreta *et al.* (1993) propose an idealized sequence that is composed of three systems tracts, termed forestepping (FST), backstepping (BST) and aggradational systems tracts (AST), which relate in general terms to lowstand, transgressive and highstand systems tracts, respectively. The proposed FST has a prograding pattern of sandy to gravelly bedload deposits, organized as an upward-coarsening and -thickening channel-fill complex. These deposits are areally restricted and overlie a sharp

and locally incised bounding surface (sequence boundary). This pattern of deposition is deemed to arise as a consequence of limited accommodation space, during stages of falling base level.

The middle member of their proposed sequence is the BST, which is developed as a response to increasing accommodation space during base-level rise. The associated deposits are composed of mixed bedload and suspended-load sediments, organized as an upward-fining and -thinning complex that exhibits stepwise migration toward the basin margin. The AST overlies the BST and occurs as a widespread set of strata, dominated by suspended-load deposits. These deposits include soil profiles that show an upward increase in profile maturity (see also comments on soil development in Wright & Marriott, 1993) and are considered to develop under a depositional regime dominated by a decrease in the availability of accommodation space.

Legarreta *et al.* (1993) provide, as an example, the middle to Late Cretaceous fluvial sequence, the Chubut Group at Cerro Bayo in the San Jorge Basin, Argentina, which they interpret according to their model. The sequence is illustrated in Fig. 5 and can be seen to provide both extensive vertical (320 m) and lateral formation. Section A (85 m thick) consists of vertically isolated sandstone bodies, set within thick mudstones. The sandstones are thin and lensoid in shape and show a ribbon morphology. Thicker (1–1.5 m) lenses commonly have sharp erosional bases. Legarreta *et al.* (1993) have interpreted this section as a low-gradient alluvial setting dominated by suspended-load streams, with deposition on the floodplain occurring mainly by crevasse-splay processes.

Section B1 (100 m thick) contains sand bodies that show simple and multistorey geometry and which contain lateral-accretion surfaces with mud drapes. Many of the surrounding and interbedded mudstone units contain calcrete nodules, some in continuous horizons. This is interpreted as a low-gradient floodplain with sinuous mixed-bedload and suspended-load streams. Overlying is a thinner section B2 (50 m thick), which is dominated by vertically amalgamated sand bodies showing multistorey–multilateral geometry. Some occur as sheet sandstones and scoured surfaces are common. Each of the sandstones has an upward-fining texture, although the overall section has an upward-coarsening and -thickening trend. Pelletier & Turcotte (1997) have advised caution in interpreting bed-thickness variation or cyclicity as being the result of orbital forcing, which is often seen as a cause of base-level variation. At the top of section B2, mudrock interbeds are rarely preserved or present only as thin relicts. Legarreta *et al.* (1993) interpret the change to a dominance of bedload material and change in sandbody

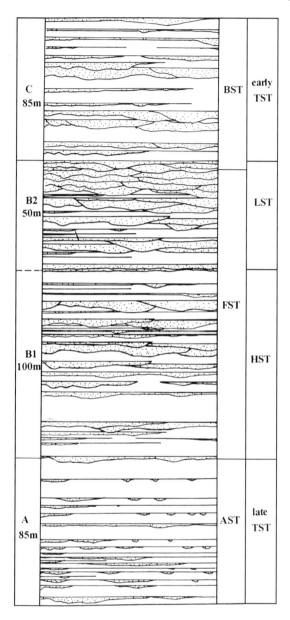

**Fig. 5.** Measured section from Cerro Bayo (Legarreta *et al.*, 1993) showing facies succession and interpretation in relation to base-level changes. For abbreviations and explanation see text and Fig. 4.

simple or multistorey geometry with lateral accretion surfaces nearer the top, where the sand bodies are set within thicker mudrock units. This suggests a return to highly sinuous, mixed-load streams on a low-gradient muddy floodplain. Legarreta *et al.* (1993) have related sections A, B and C of the section at Cerro Bayo to AST, FST and BST (Fig. 5), but there are some features of this interpretation that are at variance with the discussion on fluvial response given earlier. The FST is quite thick (150 m) and there is no incision, although this could then relate to a type-2 situation (see above). The lower part of section B has a low channel-to-floodplain-deposit ratio, suggesting at least moderate rates of aggradation during the phase, which Legarreta *et al.* (1993) suggest follows the lowstand erosive boundary. The upper part of Section B shows a decrease in accommodation space leading to a very high channel-to-floodplain-sediment ratio. The upward-coarsening trend appears to be one of their main criteria for identifying a lowering of base level, but it would seem more likely that in fact section B1 represents the late highstand, and section B2 represents a lowstand systems tract. This new interpretation is also shown on Fig. 5 for comparison.

Some of the differences in interpretation arise because Legarreta *et al.* (1993) consider that all of the FST is deposited during falling base level, whereas the corresponding LST of Wright & Marriott (1993) follows the unconformity (sequence boundary), which occurs after the late highstand. The upward-coarsening and -thickening sand bodies described by Legarreta *et al.* (1993) relate to the late highstand and would be the result of renewed erosion upstream, as base level starts to fall. Legarreta *et al.* (1993) describe upward-fining and -thinning deposits during the BST and relate this to transgression. At lowstand, however, when the rate of base-level change falls toward zero (stillstand), the fluvial system would work towards equilibrium and an upward-thinning and -fining sequence of deposits occurs, as sediment supply reduces and fines. If incision occurs, however, further deposition is less likely, as sediment will be carried out to form a clastic lowstand wedge (Posamentier & Vail, 1988). This has been identified on the French Mediterranean coast, following Pleistocene incision by the Rhône, and deposition of deltaic lowstand wedges on the continental shelf (Tesson *et al.*, 1990).

# SUMMARY

The general pattern of sandbody geometry and connectedness simulated by the run of Crane's (1982) program with varying subsidence does accord with the outcrop

style as indicative of braided rivers on a steeper gradient alluvial plain.

The thicker section, at the top of the exposure (C, 85 m thick), shows a further change to more vertically isolated sand bodies, with tabular geometries near the base and pattern from Cerro Bayo. Change in relative base level, however, is the only major control that is considered in the models discussed here and therefore it is unreasonable to suggest that the models would be applicable to all fluvial systems, in view of the range of extrinsic and intrinsic influences (see Fig. 1). Schumm (1993) has argued that base-level control is only one of at least 10 variables affecting a fluvial system and the extent of the effect on upstream areas depends on the magnitude, rate and duration of the change. If duration is long, the rate may be slow and the fluvial system can make adjustments to counteract the effects of long-term base-level change. Conversely, if base level falls or rises rapidly by a considerable amount, the effects will be significant, although sea-level changes, even large ones, might not affect the fluvial system several kilometres upstream. Channel incision could still be occurring upstream after base level has started to rise again (Schumm, 1993).

With the increasing sophistication of computers, simulation models will evolve to accommodate the advance in understanding of fluvial processes and response. These models may then enable better interpretation of the development of alluvial architecture. To use them to indicate future change, however, would be risky as a further factor, human impact, has had a dramatic influence on present-day river systems.

## ACKNOWLEDGEMENTS

The author wishes to thank Richard Crane for granting permission to use the program; Paul Wright for supplying Fig. 1; Paul Revell for drawing other diagrams; and Jim Pizzuto for a very constructive review of an earlier draft.

## REFERENCES

ALLEN, G.P. & POSAMENTIER, H.W. (1993) Sequence stratigraphy and facies model of an incised valley fill: the Gironde Estuary, France. *J. sediment. Petrol.*, **63**, 378–391.

ALLEN, J.R.L. (1974) Studies in fluviatile sedimentation: implications of pedogenic carbonate units, Lower Old Red Sandstone, Anglo-Welsh outcrop. *Geol. J.*, **9**, 181–208.

ALLEN, J.R.L. (1978) Studies in fluviatile sedimentation: an exploratory quantitative model for the architecture of avulsion-controlled alluvial suites. *Sediment. Geol.*, **21**, 129–147.

ALLEN, J.R.L. & WILLIAMS, B.P.J. (1982) The architecture of an avulsion-controlled suite: rocks between the Townsend Tuff and Pickard Bay Tuff Beds (Early Devonian), Southwest Wales. *Philos. Trans. R. Soc. London*, **B297**, 51–89.

BEGIN, Z.B. (1982) Application of 'diffusion' degradation to some aspects of drainage net development. In: *Badland Geomorphology and Piping* (Eds Bryan, R. & Yair, A.), pp. 196–179. Geobooks, Norwich.

BEGIN, Z.B. (1988) Application of a diffusion erosion model to alluvial channels which degrade due to base level lowering. *Earth Surf. Process. Landf.*, **6**, 49–68.

BEGIN, Z.B., MEYER, D.F. & SCHUMM, S.A. (1981) Development of longitudinal profiles in response to base level lowering. *Earth Surf. Process. Landf.*, **6**, 49–68.

BRIDGE, J.S. (1984) Large scale facies sequences in alluvial overbank environments. *J. sediment. Petrol.*, **54**, 583–588.

BRIDGE, J.S. & LEEDER, M.R. (1979) A simulation model of alluvial stratigraphy. *Sedimentology*, **26**, 617–644.

BRIDGE, J.S. & MACKEY, S.D. (1993) A revised alluvial stratigraphy model. In: *Alluvial Sedimentation* (Eds Marzo, M. & Puigdefabregas, C.), Spec. Publs int. Ass. Sediment., No. 17, 319–336. Blackwell Scientific Publications, Oxford.

BROWN, A.G. (1996) Floodplain palaeoenvironments. In: *Floodplain Processes* (Eds Anderson, M.G., Walling, D.E. & Bates, P.D.), pp. 95–138. John Wiley & Sons, Chichester.

BRUSH, L.M., JR. & WOLMAN, M.G. (1960) Knickpoint behaviour in non-cohesive material, a laboratory study. *Geol. Soc. Am. Bull.*, **71**, 59–74.

BRYANT, M., FALK, P. & PAOLA, C. (1995) Experimental study of avulsion frequency and rate of deposition. *Geology*, **23**, 365–369.

CHAPPELL, J. & SHACKLETON, N.J. (1986) Oxygen isotopes and sea level. *Nature*, **324**, 137–140.

CHRISTIE-BLICK, N. (1991) Onlap, offlap, and the origin of unconformity-bounded depositional sequences. *Mar. Geol.*, **97**, 35–56.

CHRISTIE-BLICK, N., MOUNTAIN, G.S. & MILLER, K.G. (1990) Seismic stratigraphic record of sea-level change. In: *Sea Level Change*, pp. 116–140. Geophysics Study Committee, National Academy of Sciences Press, Washington.

CRANE, R.C. (1982) *A computer model for the architecture of avulsion-controlled suites*. Unpublished PhD thesis, University of Reading, Reading.

DAVIS, W.M. (1899) The geographical cycle. *Geogr. J.*, **14**, 481–504.

DAVIS, W.M. (1902) Base-level, grade and peneplain. *J. Geol.*, **10**, 77–111.

FAIRBANKS, R.G. (1989) A 17,000 year glacio-eustatic sea level record: influence of glacial melting rates on the Younger Dryas event and deep ocean circulation. *Nature*, **342**, 637–642.

FISK, H.N. (1944) *Geological investigation of the alluvial valley of the Lower Mississippi River*. U.S. Army Corps of Engineers, Mississippi River Commission, Vicksburg, MS.

FREYTET, P. & PLAZIAT, J.C. (1982) *Continental Carbonate Sedimentation and Pedogenesis — Late Cretaceous and Early Tertiary of Southern France*. Contributions in Sedimentology, Vol. 12, E. Schwerzerbat'sch Verlagbuchandlung, Stuttgart, 212 pp.

GARDNER, T.W. (1983) Experimental study of knickpoint and longitudinal profile evolution in cohesive, homogeneous material. *Geol. Soc. Am. Bull.*, **94**, 664–672.

Geophysics Study Committee (1990) Overview and recommendations. In: *Sea Level Change*, pp. 3–34. National Academy of Sciences Press, Washington.

HAQ, B.U., HARDENBOL, J. & VAIL, P.R. (1987) Chronology of fluctuating sea levels since the Triassic. *Science*, **235**, 1156–1166.

HOLLAND, W.N. & PICKUP, G. (1976) Flume study of knickpoint development in stratified sediment. *Geol. Soc. Am. Bull.*, **87**, 76–82.

HOWARD, A.D. (1996) Modelling channel evolution and flood-plain morphology. In: *Floodplain Processes* (Eds Anderson, M.G., Walling, D.E. & Bates, P.D.), pp. 15–62. John Wiley & Sons, Chichester.

KERR, R.A. (1987) Milankovitch climate cycles through the ages. *Science*, **235**, 973–974.

LEEDER, M.R. (1978) A quantitative stratigraphic model for alluvium, with special reference to channel deposit density and interconnectedness. In: *Fluvial Sedimentology* (Ed. Miall, A.D.), Mem. Can. Soc. petrol. Geol., Calgary, **5**, 587–596.

LEEDER, M.R. & STEWART, M.D. (1996) Fluvial incision and sequence stratigraphy: alluvial responses to relative sea-level fall and their detection in the geological record. In: *Sequence Stratigraphy in British Geology* (Eds Hesselbo, S.P. & Parkinson, D.N.), Spec. Publ. geol. Soc. London, No. 103, 25–39. Geological Society of London, Bath.

LEEDER, M.R., MACK, G.H., PEAKALL, J. & SALYARDS, S.L. (1996) First quantitative test of alluvial stratigraphic models: Southern Rio Grande rift, New Mexico. *Geology*, **24**, 87–90.

LEGARRETA, L., ULIANA, M.A., LAROTONDA, C.A. & MECONI, G.R. (1993) Approaches to non-marine sequence stratigraphy – theoretical models and examples from Argentine basins. *Proceedings of 7th Institut Francais du Petrole, Conference on Exploration and Production: Subsurface Reservoir Characterisation for Surface Observations*, 12–17 April 1992, Scarborough, pp. 1–19.

LEWIN, J. (1978) Floodplain geomorphology. *Progr. phys. Geogr.*, **2**, 408–437.

MACKEY, S.D. & BRIDGE, J.S. (1992) A revised FORTRAN program to simulate alluvial stratigraphy. *Computers Geosci.*, **18**, 119–181.

MACKEY, S.D. & BRIDGE, J.S. (1995) Three dimensional model of alluvial stratigraphy: theory and application. *J. sediment. Res.*, **65**, 7–31.

MARRIOTT, S.B. (1993) *Floodplain processes, palaeosols and alluvial architecture: modelling and field studies.* Unpublished PhD thesis, University of Reading, Reading.

MARRIOTT, S.B. (1996) Analysis and modelling of overbank deposits. In: *Floodplain Processes* (Eds Anderson, M.G., Walling, D.E. & Bates, P.D.), pp. 63–93. John Wiley & Sons, Chichester.

MARRIOTT, S.B. (1998) Channel–floodplain interactions and sediment deposition on floodplains. In: *United Kingdom Floodplains* (Eds Bailey, R.G., José, P.V. & Sherwood, B.R.), pp. 43–61. Westbury Publishing, Otley.

MIALL, A.D. (1991) Stratigraphic sequences and their chronostratigraphic correlation. *J. sediment. Petrol.*, **61**, 497–505.

NANSON, G.C. (1986) Episodes of vertical accretion and catastrophic stripping: a model of disequilibrium flood-plain development. *Geol. Soc. Am. Bull.*, **97**, 1467–1475.

NANSON, G.C. & CROKE, J.C. (1992) A genetic classification of floodplains. *Geomorphology*, **4**, 459–486.

PATTON, P.C. & SCHUMM, S.A. (1981) Ephemeral stream processes: implications for studies of Quaternary valley fills. *Quaternary Research*, **15**, 24–43.

PELLETIER, J.D. & TURCOTTE, D.L. (1997) Synthetic stratigraphy with a stochastic diffusion model of fluvial sedimentation. *J. sediment. Res.*, **67**, 1060–1067.

POSAMENTIER, H.W. & VAIL, P.R. (1988) Eustatic controls on clastic deposition II – sequences and systems tracts models. In: *Sea Level Changes, an Integrated Approach* (Eds Wilgus, C.K., Hastings, B.S., Kendall, C.G.St C., Posamentier, H.W., Ross, C.A. & Van Wagoner, J.C.), Spec. Publ. Soc. econ. Paleont. Miner., Tulsa, **42**, 125–154.

SCHUMM, S.A. (1968) Speculations concerning paleohydraulic controls of terrestrial sedimentation. *Geol. Soc. Am. Bull.*, **79**, 1573–1588.

SCHUMM, S.A. (1975) Episodic erosion: a modification of the geomorphic cycle. In: *Theories of Landform Development* (Eds Melhorn, W.N. & Flemal, R.C.). Publications in Geomorphology, State University of New York, Binghampton, 306 pp.

SCHUMM, S.A. (1993) River response to baselevel change: implications for sequence stratigraphy. *J. Geol.*, **101**, 279–294.

SCHUMM, S.A. & KHAN, H.R. (1972) Experimental study of channel pattern. *Geol. Soc. Am. Bull.*, **83**, 1755–1770.

SCHUMM, S.A., MOSLEY, M.P. & WEAVER, W.E. (1987) *Experimental Fluvial Geomorphology*. John Wiley & Sons, New York, 413 pp.

TESSON, M., GENSOUS, B., ALLEN, G.P. & RAVENNE, C. (1990) Late Quaternary deltaic lowstand wedges on the Rhone continental shelf, France. *Mar. Geol.*, **91**, 325–332.

VAIL, P.R., MITCHUM, R.M., JR., TODD, R.G. *et al.* (1977) Seismic stratigraphy and global changes of sea-level. In: *Seismic Stratigraphy – Applications to Hydrocarbon Exploration* (Ed. Payton, C.E.), Mem. Am. Assoc. petrol. Geol., Tulsa, **26**, 49–204.

WALLING, D.E. & BRADLEY, S.B. (1989) Rates and patterns of contemporary floodplain sedimentation: a case study of the River Culm, Devon, UK. *Geojournal*, **19**, 53–62.

WOLMAN, M.G. & LEOPOLD, L.B. (1957) River floodplains: some observations on their formation. *U.S. geol. Surv. Prof. Pap.*, **282C**, 87–107.

WOOD, L.J., ETHERIDGE, F.G. & SCHUMM, S.A. (1993) The effects of rate of base-level fluctuation on coastal-plain, shelf, and slope depositional systems: an experimental approach. In: *Sequence Stratigraphy and Facies Associations* (Eds Posamentier, H.W., Summerhayes, C.P., Haq, B.U. & Allen, G.P.), Spec. Publs int. Ass. Sediment., No. 18, pp. 43–53. Blackwell Scientific Publications, Oxford.

WRIGHT, V.P. & MARRIOTT, S.B. (1993) The sequence stratigraphy of fluvial depositional systems: the role of floodplain sediment storage. *Sediment. Geol.*, **86**, 203–210.

YOXALL, W.H. (1969) The relationship between falling base level and lateral erosion in experimental streams. *Geol. Soc. Am. Bull.*, **80**, 1379–1384.

*Spec. Publs int. Ass. Sediment.* (1999) **28**, 283–304

# Subsidence rates and fluvial architecture of rift-related Permian and Triassic alluvial sediments of the southeast Iberian Range, eastern Spain

A. ARCHE *and* J. LÓPEZ-GÓMEZ

*Instituto de Geología Económica – Departamento de Estratigrafía, CSIC-UCM, Facultad de Geología, Universidad Complutense, 28040 Madrid, Spain*

## ABSTRACT

The Iberian Basin is an extensional rift basin in east central Spain that dates back to the Early Permian, but which underwent complex development during the Mesozoic and Cenozoic. The earliest sedimentary infill consists of alluvial deposits subdivided into three unconformity-bounded macrosequences of Early Permian, Late Permian and Early Triassic ages. The interplay of subsidence rate changes, sediment supply and regional gradient controlled fluvial styles, channel geometry and vertical stacking of sequences. Subsidence rates have changed over time by a factor of 35, depending on basin boundary-fault activity and multiphase extensional processes. Palaeocurrent changes reflect regional gradient changes, and sediment supply changed considerably from the early stages of infill, which were dominated by transverse transport from local sources, to the later stages, which were dominated by longitudinal transport from distal sources.

Relative sea-level changes did not affect the two early macrosequences deposited in interior drainage basins, but they may have affected the upper part of the third macrosequence deposited during a lowstand to transgressive period. Climatic changes are very difficult to discern, but a general trend from wet, seasonal climates to more arid conditions is suggested by palaeontological and sedimentological evidence. Subsidence curves are presented and related to the ratios of channel/overbank deposits, avulsion rates and the degree of amalgamation of channel deposits in each macrosequence. Periods of fast subsidence are characterized by isolated channels and predominance of overbank deposits, whereas periods of slow subsidence show lateral accretion channel deposits, scarce overbank deposits and amalgamation of channel sandstone bodies.

## INTRODUCTION

Permian and Triassic alluvial sediments of the Iberian Range, an intracratonic, linear alpine structure occupying central and eastern Spain (Fig. 1), were deposited in the fault-bounded Iberian Basin, during an extensional period. This basin is related to coeval basins such as the Pyrenean, Ebro, Catalan and Cuenca-Mancha (Arche & López-Gómez, 1996). Detailed sedimentological and stratigraphical descriptions of the sediments may be found in Virgili *et al.* (1983), Sopeña *et al.* (1988) and López-Gómez & Arche (1992).

The Iberian Basin developed over a Hercynian basement of Cambrian–Silurian slates and quartzites, deformed during the Hercynian orogeny. This basement was affected by very low-grade metamorphism (chlorite–pyrophyllite zone) (Capote & González-Lodeiro, 1983) along NW–SE trending late Hercynian or older structures inherited by the Iberian Basin master faults. Crustal

thickness is rather uniform in central Spain (Dañobeitia *et al.*, 1992) and extends to approximately 32 km. There is a lack of deep roots in the 1100 km long Hercynian belt of Iberia. The Permian–Triassic sedimentary record has been subdivided into six major sequences, bounded by unconformities or equivalent paraconformities. They may, thus, be considered as alloformations and are described and interpreted by López-Gómez & Arche (1993a).

River response to base-level change or vertical oscillations of the 'bay line' represents the basic mechanism controlling fluvial architecture (Ouchi, 1985; Schumm, 1993). Under purely continental conditions, distant from the sea, however, fluvial sedimentation depends on other mechanisms such as regional changes in slope, rates of subsidence and local base levels, each controlled by tectonics and avulsion frequency, and by sediment supply

A. Arche and J. López-Gómez

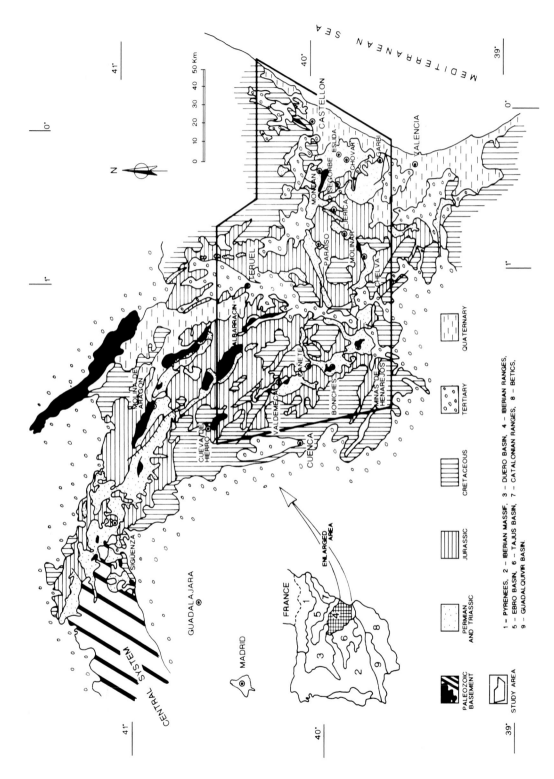

**Fig. 1.** The Iberian Range and the area studied.

controlled by both tectonics and climate. The characteristics of channel and overbank deposits, depocentre shifts and three-dimensional fluvial architecture depend ultimately on these tectonic and climatic factors (Allen, 1978; Leeder, 1978; Alexander & Leeder, 1987; Bridge & Mackey, 1993).

This paper attempts to illustrate tectonic controls on the fluvial style of well-dated Permian and Triassic sediments. These represent the early stages in development of the extensional Iberian Basin, mainly during the rift stage.

## THE SEDIMENTS

The stratigraphical succession under investigation corresponds to the first stage in the Permian–Triassic evolution of the Iberian Basin, which has been inverted structurally to form the present-day Iberian Range. These rocks are of alluvial origin and have been subdivided into four formations: Boniches Conglomerates, Alcotas

Siltstones and Sandstones, Cañizar Sandstones and Eslida Siltstones and Sandstones from bottom to top (Fig. 2). They have been dated by means of palynomorph assemblages (Boulouard & Viallard, 1982; Doubinger *et al.*, 1990). The ages used in Fig. 2 are based on the geological time-scale of Harland *et al.* (1990). Detailed sedimentological and stratigraphical studies of these individual units may be found in López-Gómez & Arche (1992, 1993a,b, 1997), López-Gómez *et al.* (1993) and Arche & López-Gómez (1996).

Figure 3 shows the distribution and correlation of the units in the study area. This figure also lists the main stratigraphical characteristics observed in 14 detailed field sections and one off-shore oil well. The top of the Cañizar Formation was chosen as the main correlation horizon (Fig. 3) and aids in the understanding of sediment distribution and the development of the basin.

Seven hundred and eighty-three palaeocurrent measurements were superimposed on the lithological synthesis of Fig. 3. These measurements were taken for each unit of the study (Fig. 4) and obtained from pebble imbrication

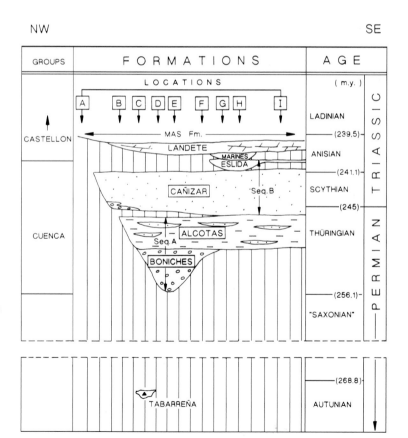

**Fig. 2.** Permian and Triassic sediments in the area studied. Areas with vertical hatching represent periods of non-deposition/erosion. Sequences A and B are described in detail in this paper. Localities: A, Cueva de Hierro; B, Valdemeca; C, Boniches; D, Minas de Henarejos; E, Landete; F, Chelva; G, Montán; H, Eslida; I, Benicasim. Absolute ages from Harland *et al.* (1990).

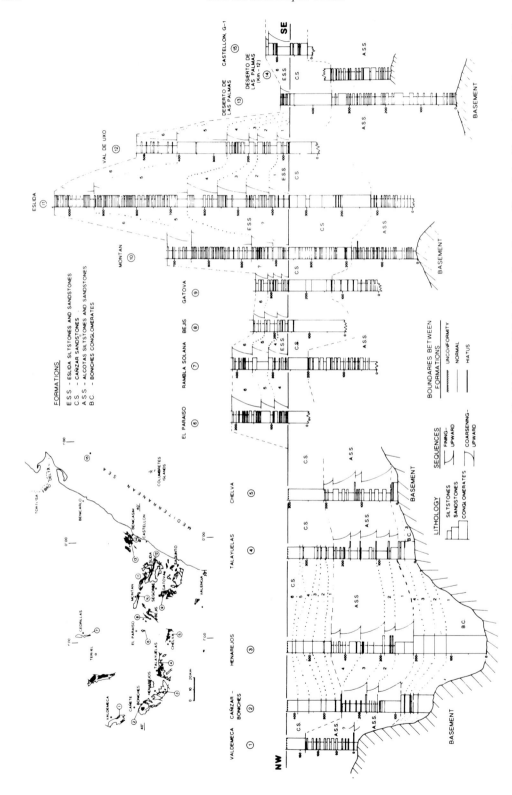

**Fig. 3.** Stratigraphical logs of the alluvial sediments corresponding to sequences A and B with details of lithologies, cycles and types of boundaries. The main correlation level is the top of the Cañizar Formation, well defined over all the area studied. Dotted lines separate members and dashed lines separate formations. Estimated absolute ages in Fig. 2.

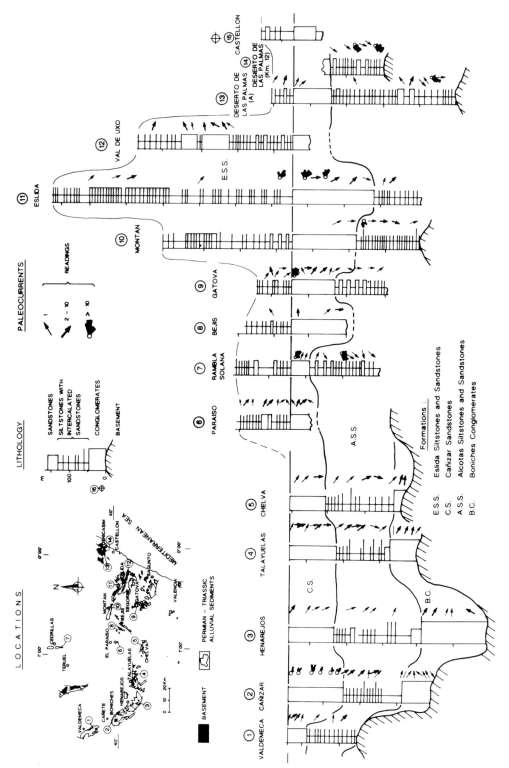

**Fig. 4.** Averaged palaeocurrents measured in the alluvial sediments of sequences A and B. For explanation see text. Number 15 is an oil-well log without available palaeocurrent data.

(4%), planar and trough cross-bedding (71%), channels and scours (18%), parting lineation (6%) and others (1%). The ambivalence in bipolar structures was solved comparing the two possible directions with those measured in unipolar structures (López-Gómez & Arche, 1993a, 1997). The present investigation involves the study of alluvial successions, A and B (Fig. 2), of the six main sequences of Permian and Triassic sediments recognized by López-Gómez & Arche (1993a).

## Sequence A

This sequence (Fig. 2) consists of the Boniches Conglomerates and the Alcotas Siltstones and Sandstones formations. The sediments are of Thüringian age (Upper Permian) and were deposited in about 8–10 Myr. Both units may well be correlated with the 'Saxonian' Facies described by Ramos (1979) for the north-west part of the Iberian Range. The Boniches Formation consists of 3–4-m-thick packets of conglomerates bounded by erosive surfaces, sometimes separated by thin sandstone layers up to 0.6 m thick, covering an area 72 km long and 9 km wide in the north-west of the study area (Fig. 5A). The

conglomerates consist of subangular to well-rounded quartzite pebbles (up to 40 cm) organized in upward-fining composite cycles. The total thickness is up to 85 m (Fig. 6A). The sandstone layers are composed of quartz, phyllosilicates, hematite and feldspars. The clay fraction is composed of illite, pyrophyllite and kaolinite, according to X-ray diffraction (XRD) analysis. Palaeocurrents trend towards the north-east in the lower part of the formation, with a radial dispersion of 40°, and towards the south-east in the upper part (Fig. 4).

The Alcotas Formation consists of red siltstones and clays and associated lenticular sandstones and conglomerate bodies and can reach 170 m in thickness (Figs 5B & 6B). Sandstones are mainly arkoses and the clay fraction consists of illite and quartz, based on XRD analysis. Sandstones and conglomerate bodies consist of upward-thinning and -fining sequences generally less than 1 m thick. They have, when well developed, concave-upward bases and do not generally exceed 550 m in lateral extension. The thinner bodies have a somewhat flatter base and exceed several tens of metres in lateral extent. Palaeocurrents are directed persistently towards the south-east (Fig. 4).

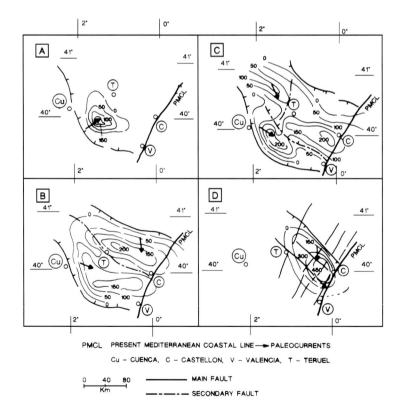

PMCL   PRESENT MEDITERRANEAN COASTAL LINE ──▶ PALEOCURRENTS

Cu – CUENCA,   C – CASTELLON,   V – VALENCIA,   T – TERUEL

0   40   80 ──────── MAIN FAULT
Km
─ ─ ─ ─ ─ ─ SECONDARY FAULT

**Fig. 5.** Isopach maps (in metres) of the Boniches Formation (A), Alcotas Formation (B), Cañizar Formation (C) and Eslida Formation (D) and the main synsedimentary basin boundary faults. See also Fig. 9 for the names of the main and secondary faults.

**Fig. 6.** Field aspects of the Permian and Triassic alluvial sediments: Boniches Formation (A), Alcotas Formation (B), Cañizar Formation (C) and Eslida Formation (D). Bar scales are: A, 50 m; B, 3 m; C, 12 m; D, 16 m.

## Sequence B

This sequence (Fig. 2) consists of the Cañizar Sandstones and the Eslida Siltstones and Sandstones formations. Sequence B is of late Thüringian (Upper Permian) to early Anisian (Lower Triassic) age and it was deposited in about 15–16 Myr.

The Cañizar Formation has been described in detail by López-Gómez & Arche (1993b). It consists of red to pink sandstones, which may be found over the entire study area and througout the wide Iberian Range (Figs 5C & 6C). It is an arkosic sandstone cemented by illite, K-feldspar and quartz, with a total thickness of 170 m with minor conglomerate and siltstone interbeds. The formation has a complex diagenetic history with early clay cements followed by younger quartz cements. Palaeocurrents are consistently to the south-east with 30° dispersion (Fig. 4). The top of the formation is of Anisian age, but the age of the base is less well constrained and could be Scythian (Lower Triassic) or even Thüringian (Upper Permian), indicating a long period of slow, intermittent deposition.

The Eslida Formation is the youngest alluvial unit preceding the first transgression of the Tethys (Palaeotethys) Sea in the Iberian Basin. It occupies a narrow zone of the south-east part of the basin where the Boniches Formation is absent (Fig. 5D). It consists of red siltstones and intercalated decimetre-scaled sandstone bodies of arkosic composition. It is up to 660 m thick, thinning to the south-east and north-west (Fig. 6D). Its age is Anisian (Lower Triassic) and palaeocurrents are dispersed generally to the east and south-east (Fig. 4).

## TECTONIC EVOLUTION OF THE IBERIAN BASIN DURING THE PERMIAN AND THE TRIASSIC PERIODS

The Iberian Basin commenced as a rift basin during Early Permian times and experienced several extensional periods (tens of millions years long) and compressive events (millions of years long) during the Mesozoic and Cenozoic Eras (Salas & Casas, 1993; Arche & López-Gómez,

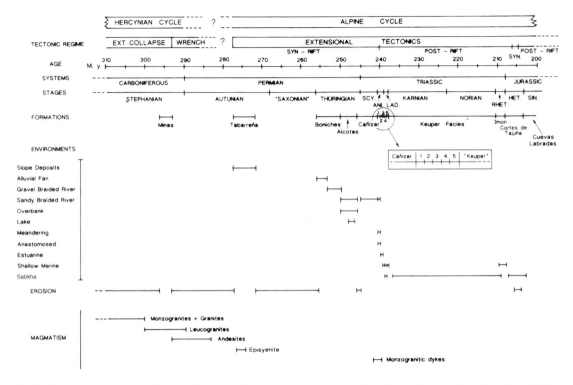

**Fig. 7.** Tectonic regime, associated magmatism and sedimentary environments in the Late Carboniferous to Early Jurassic period in the Iberian Range. Formations 1 to 5 are, respectively, the Eslida Formation, the Marines Formation, the Landete Formation, the Mas Formation and the Cañete Formation. Note the very rapid change of environments during the Anisian–Ladinian (Middle Triassic) period (ANI-LAD in the figure) as a result of tectonic activity and marine transgression. Magmatic data from González-Casado *et al.* (1996) and Lago *et al.* (1991, 1992). Absolute age scale from Harland *et al.* (1990).

1996). This led to its present-day configuration: a segment of the Alpine chain arising from structural inversion, folding and thrusting of moderate intensity. The Iberian Basin trends NW–SE and its boundary faults are generally interpreted as late Hercynian or older structures that were subsequently reactivated. It is bounded by the crystalline Iberian Central System to the south-west and by the Ebro Massif and the Catalan Range to the northeast (Fig. 1). The main stages in the development of this Iberian basin and related structures, magmatic episodes and ages of the sedimentary infilling of the basin have been relatively well documented (Sopeña *et al.*, 1988; Guimerà & Álvaro, 1990; Lago *et al.*, 1991, 1992; Aurell *et al.*, 1992; López-Gómez & Arche, 1992; Salas & Casas, 1993; Doblas *et al.*, 1994; Roca *et al.*, 1994; Arche & López-Gómez, 1996; González-Casado *et al.*, 1996). Controversy persists, however, over the timing of certain events and the stages of evolution from the late Hercynian episodes to the onset of the early Alpine extensional tectonics.

The Late Carboniferous to Late Triassic interval was a period of rapid tectonic and palaeogeographical change in the Iberian Plate: from continental collision to the extensional collapse of the Hercynian orogen, the readjustment of blocks by strike-slip faulting, and finally, to a period of widespread extension, rifting and marine transgression (López-Gómez & Arche, 1992; González-Casado *et al.*, 1996).

The late Carboniferous marked the final episode of the Hercynian orogeny: the extensional collapse of the Hercynian orogen probably caused uplift, widespread intrusion of monzonitic and granitic complexes in adjacent areas (including the Central System, see Fig. 7), and an extensional regime leading to the creation of small intermontane basins (Arche & López-Gómez, 1996; González-Casado *et al.*, 1996) (Fig. 7). Later, denudation following uplift eliminated most of the Carboniferous basins of the Iberian domain except for a relict basin in Minas de Henarejos (Fig. 1), which falls outside the scope of the present paper.

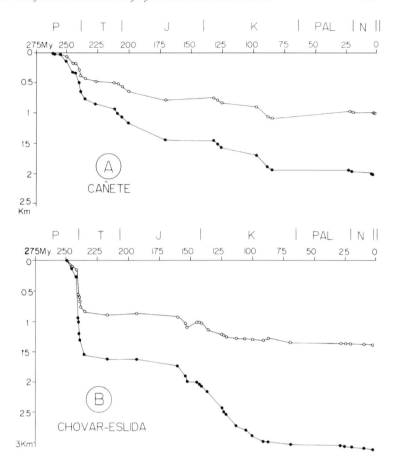

**Fig. 8.** Total (filled circles) and tectonic (open circles) subsidence curves for two representative sections in the south-east Iberian Range (see Fig. 1 for locations). Note the multiphase character of the synrift subsidence period (Early Permian to Middle Triassic) and the concave-upwards shape, characteristic of extensional periods. See discussion in text.

The Early Permian Epoch was a period of readjustment by local strike-slip faulting, creation of small transpressive basins, the intrusion of leucogranites and eruption of volcanic and volcaniclastic rocks of calc-alkaline affinities (andesites and subordinate rhyolites and basalts; González-Casado *et al.*, 1996). Thick alluvial and lacustrine sediments intermittently accumulated in some small basins. The transpressive basins developed mainly in the north-west part of the Iberian Basin, although some structures with a NW–SE orientation may be found in the Albarracín and Boniches areas (Fig. 1). These events were considered to be latest Hercynian events by Arthaud & Matte (1977), but the radical change in the tectonic regime, the unconformable nature of the basal contact, and the nature of the sediments probably indicate the onset of the Alpine cycle.

Episyenites were intruded along tensional structures in the nearby Central System at about 277 Ma (González-Casado *et al.*, 1996). They represent a magmatic episode contemporaneous with the infilling of the Tabarreña Formàtion (Autunian, Early Permian; Fig. 7). Volcanic

and volcaniclastic rocks of andesitic composition are found to the north-west of the study area, also of Autunian age (Lago *et al.*, 1991, 1992).

Widespread extension started in the Iberian Basin during Late Permian times and led to three phases of rifting during the Late Permian and Early Triassic Epochs (Fig. 8). The earliest phase led to the development of a set of normal faults along the Serrania de Cuenca (Fig. 9) and consisted of a series of NW–SE trending, normal, listric faults dipping to the north-east and linked by transfer faults trending NNE–SSW (i.e. at about 70°). Their plan shape was arcuate, and they acted as the basin boundary faults for a series of half-grabens that were filled by transverse alluvial fans and longitudinal, braided, proximal and distal river deposits. The subsidence rates during the deposition of the Boniches Formation was moderate and increased during the deposition of the Alcotas Formation (Fig. 8).

The second phase of rifting is separated from the first by a short period of uplift and denudation. This caused the Cañizar Sandstone Formation to lie unconformably on

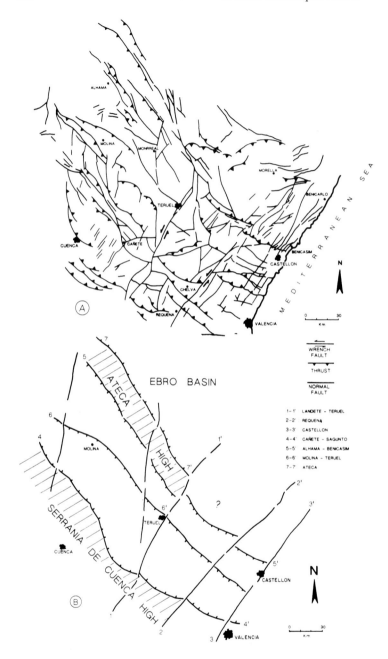

**Fig. 9.** (A) Present-day main alpine faults in the south-east Iberian Range. (B) Palinspastic reconstruction of the main basin boundary faults for the Late Permian to Early Triassic period and the associated transfer faults. Ruled areas are elevated Palaeozoic blocks. Names of the faults are indicated in (B).

different levels of the Alcotas Formation (Fig. 2). The basin had an almost symmetrical cross-section during latest Permian and earliest Triassic times, as extension was accommodated in the hangingwall block by the creation of an antithetic set of faults, the Maestrazgo system (Fig. 9). The subsidence rate was very slow, and infill consisted mainly of braided river deposits (Cañizar

Formation) organized into six depositional sequences bounded by prominent erosive surfaces, which represent minor regional tilting and changes in regional slope (López-Gómez & Arche, 1993b).

Magmatic activity took place at 245 Ma in the Central System, corresponding to the end of the Late Permian to Early Triassic period of sedimentation, and was related to the tectonic activity that led to the graben geometry of the basin and the change in location in the basin boundary fault from the south-west to the north-east margin of the basin.

The third phase of rifting during the Anisian (Early Triassic) affected only the north-east margin of the basin. The Ateca–Maestrazgo fault system was reactivated between Teruel and Sierra del Espadán as an antithetic fault system leading to rapid but local subsidence along this margin (Figs 5D & 8B); the basin was filled with alluvial sediments up to 600 m thick (Eslida Formation), but this phase was short-lived. During the Anisian, the synrift tectonic period ended and thermal, post-rift subsidence commenced (Fig. 8). The rate of subsidence rapidly decreased. The area of sediment accumulation expanded, drowning the basin margins as a marine transgression took place from the east. This latter phase is represented by the upper part of the Marines and Landete formations (Fig. 2), which are composed of shallow-marine to coastal sabkha evaporites and carbonates of Middle–Late Triassic age. During earliest Jurassic times, there was another short-lived rifting period, which led to extension forming synsedimentary half-grabens and an angular unconformity and breccia (Aurell *et al.*, 1992).

The late development of the Iberian Basin is beyond the scope of this study, but briefly consists of a post-rift, thermal subsidence period during Jurassic times, a synrift period during Early Cretaceous times, a post-rift period during middle Cretaceous to early Oligocene times, a compressive tectonic period during Oligocene to early Miocene times and an extensional tectonic period up to the present day.

## FLUVIAL ARCHITECTURE AND FLUVIAL STYLE

Stratigraphical and sedimentological investigations of fluvial style lead to an approximate reconstruction of the sedimentary architecture of fluvial deposits (e.g. Allen, 1983; Miall, 1985, 1988a, 1989, 1991, 1996; Soegaard, 1990; DeCelles *et al.*, 1991; Jordan & Pryor, 1992; Bridge, 1993; Bridge & Mackey, 1993). The Permian and Triassic alluvial sediments of the south-east Iberian Range, which exceed 1 km thick, were deposited over a period of 13–14 Myr. The sedimentology of these deposits has been studied in detail by López-Gómez & Arche (1992, 1993a,b, 1997) and will not be repeated in this paper. Only the relevant aspects of the fluvial architecture and palaeoclimatic interpretations will be discussed here.

The alluvial sediments have been subdivided by means of surfaces of different scales; the category and

characteristics of the bounding surfaces serve to estimate the interval of time they represent (i.e. McKee & Weir, 1953; Allen, 1980; Miall, 1988a,b). The most important are those that separate formations and are divided into normal, discordant and paraconformities or concordant contacts that represent a proven hiatus (Fig. 10).

Conformable contacts are found between the Boniches and Alcotas formations and the Eslida and Marines formations. The relationship between these formations is clearly gradational (Fig. 2), but they have a different significance; the first one represents a change from local to distal source areas and from transverse to longitudinal drainage. The second records a transition from alluvial to estuarine–mudflat sedimentation at the onset of a major marine transgression, with the rate of sediment supply keeping pace with the increased rate of creation of accommodation.

Discordant contacts or angular unconformities are found at the base of the Boniches and Alcotas formations, and at the base and the top of the Cañizar Formation. The base of the Boniches Formation represents the basal unconformity of the Permian–Triassic alluvial sediments on the lower Palaeozoic basement. Beyond its area of deposition to the north-west and the south-east, the Alcotas Formation lies unconformably on the basement. This is a type 8 discordant contact (Miall, 1985, 1988b) and represents the creation of the sedimentary basin in the different areas studied in this paper.

The base of the Cañizar Formation also forms a discordant contact with the Alcotas Formation in the Cañete–Boniches area (López-Gómez & Arche, 1993b) and the north-west Iberian Basin (Ramos, 1979). It is a type 7 surface and represents a fundamental reorganization of the sedimentary basin after a tectonic pulse of extensional character. The top of the Cañizar Formation is also a type 7 surface, from Chelva to Valdemeca (Fig. 1), as different marine transgressive units are deposited on it. Palaeosols, an iron-rich crust and bleaching of the uppermost 15 m of the fluviatile sandstones mark this surface, which represents another change in tectonic regime.

A paraconformity or concordant contact with a hiatus is found at the base of the Cañizar Formation from Cañete–Boniches to the south-east (Figs 2 & 3). The contact is sharp and across the surface there is a major change in fluvial environment of deposition. This also is a type 7 surface, in lateral continuity with the unconformity formed to the northeast.

Surfaces of type 6 also are found in every formation and have been mapped all over the study area (Fig. 10). There are three type 6 surfaces in the Boniches Formation, separating different alluvial fan and gravelly

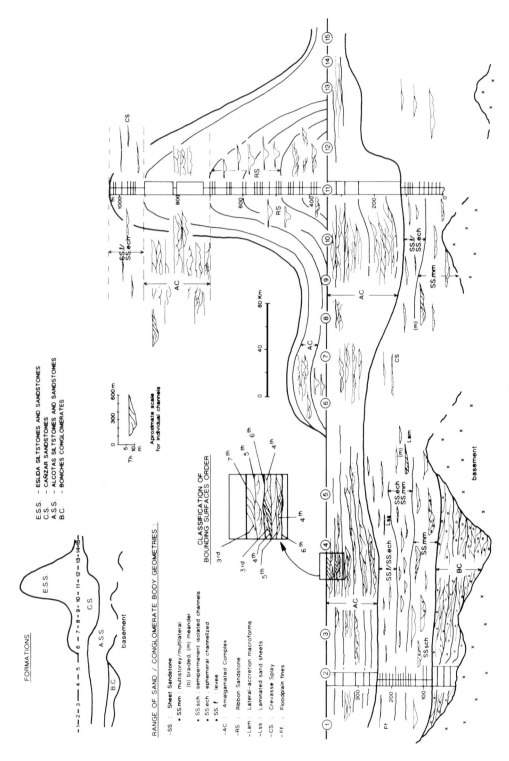

**Fig. 10.** Sedimentological characteristics and bounding surface hierarchy (Miall, 1988a,b) in the Permian to Triassic alluvial sediments of the south-east Iberian Basin. Localities 1 to 15 as in Fig. 3.

braided river composite sequences, two in the Alcotas Formation, separating fining and thinning distal braided–floodplain cycles, five in the Cañizar Formation, separating sandy braided composite sequences, and five in the Eslida Formation, separating aggradational–progradational sandy braided-river–meandering-river–floodplain composite sequences. Surfaces of lower order (fifth to first order) are not studied here and represent changes caused by autocyclic processes: alluvial fan lobe abandonment, avulsions, etc., and their period is estimated to be of the order of 10 000–100 000 yr (for the fifth- to third-order surfaces) (López-Gómez & Arche, 1993b, 1997).

Sixth-, seventh- and eighth-order surfaces represent various fundamental allogenic factors that controlled the fluvial architecture, with periods of different magnitude calculated according to the biostratigraphical data. The eighth- and seventh-order surfaces represent major tectonic events, such as the beginning of extension in the basin with the appearance of new basin boundary-fault systems, and cause major, fundamental changes in basin geometry and regional slope (changes in fluvial style, palaeocurrent patterns, etc.), with a period of 3–5 Myr. The sixth-order surfaces are also thought to be of tectonic origin (individual pulses of the basin boundary faults) leading to changes in the rate of creation of accommodation and of the relative importance of the proximal and distal source areas. Its period is about 1 Myr.

Fluvial styles have been inferred from the study of lithofacies and channel geometries. The different channel bodies observed in the formations fall into three main types: isolated bodies of lenticular geometry, interpreted as short-lived structures, ribbons of lenticular to tabular geometry with lateral accretion surfaces and amalgamated, multilateral, multistorey complexes or belts of tabular geometry, including different bedforms and channel forms (Fig. 10 & Table 1). Overall, meandering, braided, anastomosed and probably ephemeral systems have been recognized in the units. As the climatic evolution was gradual, the changes in channel pattern and its vertical stacking appear to result from tectonic movement. Aggradation and degradation occur as the river adjusts to valley slope changes (Ouchi, 1985), leading to meandering and anastomosed patterns during rapid subsidence periods, with ribbon geometries. Braided patterns develop during slow subsidence periods and generate sheet sandstone geometries.

The Boniches and Cañizar formations display the most consistent style, which may be described as braided–amalgamated (Fig. 6A & C). The rate of creation of accommodation was much greater during sedimentation of the Boniches Formation and the source area was

proximal, leading to the deposition of stream-flow conglomerates with rounded pebbles of quartzite, sometimes imbricated, which formed upward-fining metre-scale sequences, similar to those described by Rust (1984) for the Devonian, Malbaie Formation, Canada.

The rate of creation of accommodation was much less for the Cañizar Formation and the source area was distal, resulting in sandy braided river facies consisting of decimetre-scale accretionary bundles, separated by third-order erosion surfaces (Shepherd, 1987). Common, minor internal bounding surfaces (Miall, 1996) suggest seasonal discharge conditions under humid-temperate conditions.

In the Alcotas and Eslida formations (Fig. 6B & D), fluvial styles change vertically from braided to meandering and even to anastomosed patterns in the middle of the latter. The Alcotas Formation contains relatively well-preserved remains of vegetation, even large logs (López-Gómez & Arche, 1993b). The absence of any features indicating catastrophic runoff events indicates the absorption of moisture by soils and plants, which prevented flash floods. The Eslida Formation consists of aggradational–progradational fluvial cycles, accumulated during a period of rapid creation of accommodation. There is a clear variation in fluvial style. Amalgamated bodies were formed under humid, temperate conditions similar to those described for the Cañizar and Alcotas formations. The presence of calcic soils, salt-hopper crystals near the top, abrupt changes in style from meandering to braided, or even the presence of flash-flood deposits, however, indicate an evolution towards a more arid climate at the beginning of Middle Triassic times. These interpretations are based on the criteria used by Friend *et al.* (1979), Friend (1983), Galloway (1977), Hirst (1991) and Schumm (1963, 1981), and also because the Iberian Basin was distant from coastal regions, near the Equator (Scotese & Langford, 1995), and had a seasonal temperate climate (Crowley *et al.*, 1989; Barron & Fawcett, 1995). Table 2 summarizes the different fluvial styles observed in the units.

## TECTONIC CONTROL OF FLUVIAL ARCHITECTURE

The depositional record of rift basins is controlled by two major processes:
1 general base-level changes, or bay-line changes;
2 the rate and rate of change of subsidence.
The eustatic or general base-level changes were irrelevant to the fluvial architecture of the Permian–Triassic deposits in the Iberian Basin because, for the most part,

**Table 1.** Summary of sedimentological characteristics of the Permian–Triassic alluvial deposits, south-east Iberian Range.

| Body type: channels, bedforms and sheets | Lithology (after Miall, 1996) | Characteristics | Interpretation | Location (in the study area) and general references |
|---|---|---|---|---|
| Amalgamated complexes of gravel bedforms (AC) | Gm, Gp, Gt | Amalgamation of massive or crudely bedded bodies. Fining-up sequence < 1 m. Storeys: 2–5 m thick | Braided fluvial system with longitudinal and transverse bars. Channel fill. Presence of waning flows | Boniches Formation (Middleton & Trujillo, 1984; Nemec & Postma, 1993; Ore, 1964) |
| Amalgamated complexes of sand bedforms (AC) | St, Sp, Sh, Sl, Sr, Se, Ss | Amalgamation of different channels and elementary and composite bedforms | Braided fluvial systems with transverse and linguoid bedforms. Channel fill | Eslida and Cañizar Formations (Allen, 1983; Bluck, 1976; Marzo & Anadón, 1987) |
| Sheet sandstone, multilateral/multistorey (SS.mm) | St, Sp, Sh, Sl, Sr, Se, Ss | Lenticular bedded with mainly longitudinal progradation. Width < 600 m. Many storeys compose the sheets | Isolated lenticular bodies composed of transverse and linguoid bars and channels | Eslida and Alcotas Formations (Horne & Fern, 1976; Røe & Hermansen, 1993) |
| Semi-permanent isolated channels (SS.sch) | St, Sp, Sh, Sl, Se and less commonly Gp, Gt | Width < 200 m, 2–4 m thick, erosive base, lenticular shape. Three to five stages fill the channel | Development and evolution of channels crossing the floodplain, probably as a result of avulsion | Eslida and Alcotas Formations (Allen, 1984; Ashley, 1990; Collinson & Thompson, 1982) |
| Ephemeral isolated channels (SS.ech) | St, Sp, Sh, Sl, Se | Width < 20 m, 0.2–0.8 m thick. Very gentle erosive base and almost tabular shape. A single stage fills the channel | Development and evolution of channels produced during short periods of flow following local and intense rainfall | Eslida and Alcotas Formations (Picard & High, 1973; Dreyer, 1993) |
| Ribbon sandstone (RS) | Se, St, Sh, Sl | W/D < 15, with 'wings'. Storeys < 4.5 m thick. Isolated channels bounded by floodplain deposits | Isolated anastomosed fluvial channels | Eslida Formation (Eberth & Miall, 1991; Miall & Gibling, 1978; Long, 1978) |
| Lateral-accretion macroforms (Lam) | St, Sh, Sl, Sp | Isolated examples. Fine grained facies, associated wedge shape | Isolated meandering fluvial channels | Eslida and Alcotas Formations (Puigdefábregas & van Vliet, 1978) |
| Laminated sand sheets (Lss) | Sh, Sl and less commonly Sp | < 30 m long. Thin (< 0.3) tabular lamina | Flash-floods depositing sand under upper flow regime, plane bed conditions | Eslida, Cañizar and Alcotas Formations (Rust & Nanson, 1989) |
| Levee (SS.l) | Sh, Sl | Wedge shape, up to 3 m thick and 300 m wide | Overbank flooding | Alcotas, Eslida and Cañizar Formations (Rust *et al.*, 1984; Farrell, 1987; Kraus & Bown, 1988) |
| Crevasse splay (c.s) | St, Sr | Sheet-like bodies < 1 m thick and 10–100 m across. Laterally pass to fine deposits. Rarely crevasse channels are observed | Progradation from crevasse channel into floodplain | Alcotas and Eslida Formations (Mjøs *et al.*, 1993; Brierley, 1991; Gersib & McCabe, 1981) |
| Floodplain fines (F.f) | Fl, Fr, Fsm | Sheet geometry with many kilometres in lateral dimensions. Ponds (in carbonate sediments) may also appear | Sediments of overbank sheet flow | Alcotas and Eslida Formations (Reinfelds & Nanson, 1993; Bentham *et al.*, 1993; Willis & Behrensmeyer, 1994) |

**Table 2.** Summary of fluvial styles of the Permian–Triassic alluvial deposits, southeast Iberian Basin and possible analogues.

| Fluvial style | Unit (from the study area) | Sediment type | Sinuosity (estimated) | General examples |
|---|---|---|---|---|
| Shallow gravel braided | Boniches Formation except lowest part | Gravel and minor sand | Low (< 1.2) | Middleton & Trujillo (1984), Dawson & Bryant (1987), Ramos & Sopeña (1983) |
| Sandy low-sinuosity braided isolated in the floodplain | Most of the Alcotas Formation and lower part of the Eslida Formation | Sand and fines | Low (< 1.2) | Nanson & Croke (1992), Olsen (1988), Rust & Legun (1983) |
| Sandy stable braided (shallow) | Cañizar Formation and part of the upper half of the Eslida Formation | Sand | Low–intermediate (< 1.2–1.5) | Blodgett & Stanley (1980), Crowley (1983), Miall (1976) |
| Sandy stable braided (deep) | Cañizar Formation and part of the upper half of the Eslida Formation | Sand | Low–intermediate (< 1.2–1.5) | Haszeldine (1983), Cant & Walker (1978), Rust & Legun (1983) |
| Sheetflood distal braided | Alcotas Formation and upper part of Eslida Formation | Sand and minor fines | Low (< 1.2) | Cotter & Graham (1991), Olsen (1989) |
| Sandy meandering | Some isolated bodies in Alcotas and Eslida Formations | Mainly sand and minor fines | High (> 1.5) | Alexander (1992), Plint (1983), Link (1984), Nami & Leeder (1978) |
| Anastomosed | Middle part of the Eslida Formation | Sand and fines | Low to high (< 1.2 to > 1.5) | Smith (1983), Eberth & Miall (1991), Smith & Smith (1980) |

it was an interior basin. Only the top of the Eslida Formation and the overlying Marines Formation (beyond the scope of this study, Fig. 2) were influenced by eustatic changes (López-Gómez & Arche, 1992, 1993a). The rate of subsidence and its changes represent a purely tectonic mechanism related to the crustal dynamics of the area. Subsidence depended upon the tectonic regime (extensional, strike-slip) and its pulses caused local and regional changes in slope, depocentre migration, avulsion control and the creation and destruction of local base levels (Salas & Casas, 1993; Arche & López-Gómez, 1996).

The Iberian Basin consists of a series of half-grabens. Sediments originated from source areas in the escarpment margin of the footwall block only during deposition of the lower part of the Boniches Formation. The hinged margin of the footwall block was a source only during deposition of the lower part of the Boniches Formation. The hinged margin of the hangingwall block did not contribute appreciably to sedimentation in the basin (López-Gómez & Arche, 1997) in contrast to other basins (Frostick & Reid, 1987). This probably was owing to the early development of a central high in the hangingwall block, which had acted as a drainage divide between the Iberian and Ebro Basins since Late Permian times (Fig. 9). Another important feature of the extensional tectonics of this time is the presence of transfer faults and relay zones (Gibbs, 1987) that linked several sub-basins and created local highs

across them. The main source of sediment was located in the Iberian Massif, to the north-west of the basin. The source area, which probably was very extensive and contained igneous and low-grade metamorphic rocks, had a feeder point in the northwest corner of the basin, and sediments were transported to the southeast by axially orientated rivers for most of the period under consideration (López-Gómez & Arche, 1993a; Arche & López-Gómez, 1996).

Rates of subsidence and their temporal variations control fluvial stratigraphy, if climate is more or less unchanged over the period considered, because it creates accommodation and the local and regional slope. The subsidence history of the Iberian Basin can be studied in detail, and in some way quantified using the backstripping method (Watts & Ryan, 1976; Steckler & Watts, 1978). Only a few attempts have been made to apply this method in the Iberian Range (Alvaro, 1987; Sánchez-Moya *et al.*; 1992; Salas & Casas, 1993; van Wees, 1994; Arche & López-Gómez, 1996) and it has never been used to relate subsidence rates to fluvial style. Two subsidence curves are presented here (Fig. 8), based on original sections in the southeast Iberian Range. They show the classic concave-upwards shape of extensional basins, with four episodes, three of rifting and one of thermal subsidence for the Permian–Triassic period (compare with van Wees, 1994; Goggin *et al.*, 1997; Ismail-Zadeh *et al.*, 1997).

The first episode, of Early Permian age (270–260 Ma), is represented only in the Cañete curve and was of limited extent in the study area. This episode led to the deposition of the Tabarreña Formation (Fig. 2), a slope breccia representing a phase of slow subsidence but moderate to high local slopes during the embryonic stage of rifting. Other isolated basins of related age are found in the north-west Iberian Basin. They were clearly underfilled and some of them evolved into lakes (i.e. Ermita Formation; Ramos, 1979). This episode is coeval with a magmatic and thermal event (Fig. 7), which probably caused doming in some parts of the Iberian Basin, resulting in increased slopes.

The second episode of rifting, of Late Permian age (256–247 Ma) was a more extensive and intense event (Fig. 8, Cañete and Chovar-Eslida curves). Normal faulting led to half-graben formation and, in the early stages, alluvial fans, confined to the more active south-west margin of the basin, were built by channelized and sheet flow (Boniches Formation, lower part). Palaeocurrents were transverse to the axis of the basin and the catchment, which cut directly across the fault escarpment, provided coarse sediments of local origin. Limited backfaulting on the basin boundary fault produced onlapping fan sequences on the hangingwall, hinged margin of the basin, and backstepping accompanied by clear vertical stacking of fan sequences. After the initial pulse, extension progressed steadily. Newly created accommodation led to a reorganization of the basin geometry, a much wider fluvial network, parallel to the axis of the basin, a change in the main source area to the northwest (López-Gómez & Arche, 1997) and a switch from alluvial fans to axially fed gravelly braided river deposits. The enlargement of the basin is also demonstrated by the vertical evolution from proximal to distal fluvial facies, as the feeder point receded with time.

A very important feature of the geometry of the basin during this period is the creation of a high shoulder of Palaeozoic basement along the south-west margin. Because of the changes in mass distribution related to extension, high basin shoulders imply a deep level of necking in the lithosphere (Braun & Beaumont, 1989; Kooi *et al.*, 1992). Thus, not only should the rate of extension be taken into account when relating fluvial style to tectonics, but also the depth of lithospheric necking.

Continental extension led to migration of the depocentre to the north-east (Fig. 4), the creation of hangingwall grabens and a much wider basin with greatly increased accommodation (Alcotas Formation). The predominant fluvial style of this period consisted of unstable, isolated channels of ephemeral to semi-permanent isolated character, and substantial overbank deposition. Axial rivers

migrated into new locations nearly parallel to the footwall scarp because of the high rate of subsidence (Tularosa type basin, Mack *et al.*, 1997). The feeder point migrated further to the north-west, and assuming that the avulsion frequency remained constant under stable climatic conditions, channels became more isolated with time as the basin became broader.

The local slope orientation changed frequently owing to subtle changes in regional subsidence and the channels had no time to migrate laterally or to amalgamate, as their position shifted frequently. The northwest and southeast sub-basins had a moderate to high subsidence rate (Leeder *et al.*, 1996; Leeder, 1997), leading to a situation of near equilibrium between sediment supply and subsidence rate. Brief periods of reduced sedimentation (underfilling) permitted carbonate and siliciclastic sediment to be deposited.

This extensional pulse had a contemporaneous igneous event, marked by the emplacement of episyenitic dykes in the nearby Central System (Fig. 7; González-Casado *et al.*, 1996).

The third period of extension (Late Permian–Early Triassic, 245–237 Ma) is a more complex period. It started with very slow, generalized subsidence, resulting from movement on both basin boundary faults (Fig. 8), which led to a graben configuration and slow creation of accommodation, leading to overfilling of the basin. The source area receded to the north-west and the drainage pattern was purely axial. These tectonic conditions led to a period of extensive reworking of the sandy braided river deposits of the Cañizar Formation, resulting in complex amalgamated channel forms and the total erosion of overbank sediments (Fig. 10). The situation changed completely in the southeast part of the study area during the early Anisian (Fig. 8, Chovar–Eslida curve). Most of the basin became stable following deposition of the Cañizar Formation, accompanied by development of deep weathering profiles. A segment of the Ateca-Maestrazgo basin boundary fault, however, split into at least three, normally faulted overlapping segments with relay ramps between segments (Fig. 11A). Along-strike temporal variations in displacement were prominent, leading to the semi-ellipsoidal shape of the basin, substantial differential subsidence and asymmetric distribution of the active channel belt (near the fault plane) and overbank deposits (away from the fault plane). The Eslida Formation was deposited during this period. It consists of a series of mainly upward-coarsening and -thickening cycles (Figs 3 & 10) onlapping the basement as a result of differential extension of the faults along strike (Fig. 11B & C). The basin was fed axially from the north-west along a relay ramp in the region of Teruel (Fig. 1) and sourced

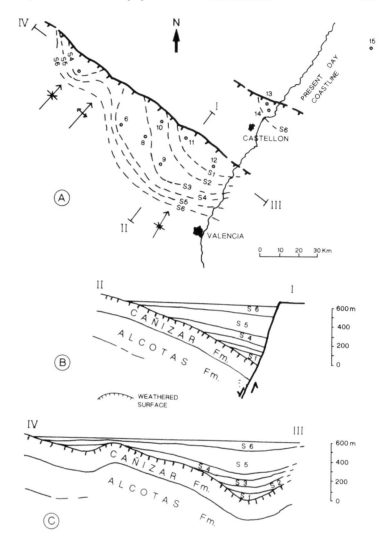

**Fig. 11.** (A) Synsedimentary faults and lateral extension of sequences (S1–S6) of the Eslida Formation. (B) Cross-section of the Eslida–Teruel basin (not to scale). (C) Longitudinal section of the same basin (not to scale). Note in (B) and (C) how the sequences onlap the basement, expanding with time. The top of the Cañizar Formation is a weathered surface throughout the study area.

by regional uplift of the Ateca High. Each individual pulse created accommodation, which increased withdrawal rate, leading to the formation of ribbon channels, which were at times vertically stacked, and overbank deposits (Fig. 6D). As the available accommodation diminished during quiescent periods, channels tended to form more complex amalgamated, multilateral–multistorey sand bodies. The vertical aggradation of cycles ended with cycle 5 (Fig. 10), interpreted to record cessation of movement along the fault and subsequent amalgamation and lateral migration of the channels.

Newly formed channels commonly reoccupy previous channels as differential compaction rates between overbank and channel deposits and subsequent erosion of pre-existing channel bodies accommodate sediment supply. Nested ribbon bodies are common in the lower part of sequences 1–5.

Sequence 6 (Fig. 3) is different from sequences 1–5 in that it consists of upward-fining successions containing a high percentage of overbank deposits. It is interpreted as the result of a subdued rate of subsidence and a rise in base level, related to a major sea-level rise that led to unstable conditions in the lower reaches of the fluvial system. Under these conditions, a substantial percentage of the fluvial sediment was trapped in floodplains (Mackey & Bridge, 1995; Allison *et al.*, 1997) and active channels underwent frequent lateral migration and/or avulsion as a result of subtle changes in local slope.

The third episode of extension was contemporaneous with the emplacement of monzogranitic dykes under an extensional regime in the nearby Central System (Fig. 7) (González-Casado *et al.*, 1996).

The fourth period of extension (Middle Triassic to Early Jurassic, 237–208 Ma) corresponds to the thermal subsidence interval (Fig. 8) and coastal and shallow-marine carbonate and evaporitic sediments onlapped the fluvial sediments and the Palaeozoic basement during this period.

## CONCLUSIONS

**1** The fluvial styles of the Permian–Triassic continental fill of the Iberian Basin were controlled mainly by variable subsidence rates, sediment supply and changes in source area, all of which were controlled by tectonics.

**2** The Iberian Basin experienced three pulses of synrift extension and a phase of thermal subsidence during the Permian–Triassic interval. Each period produced characteristic fluvial deposits.

**3** As climate was more or less stable during the period under consideration, changes in subsidence rate were the main control on fluvial style and show important temporal and spatial variations, simultaneously creating accommodation in some areas and uplift or quiescece in others.

**4** The Late Permian basins had half-graben geometries and were filled by alluvial fans fed by basin-transverse flow and evolved into braided river systems fed by longitudinal flow.

**5** Increased rates of subsidence led to further migration of the depocentres to the north-east, and evolution from a braided river network into ephemeral, unstable incised channels with substantial overbank deposition.

**6** The Early Triassic Epoch was a period of very low subsidence rates, which led to reduced accommodation and overfilling of the basin, repeated erosive events, amalgamation of sandy braided river deposits and almost complete erosion of overbank deposits.

**7** At the end of Early Triassic times, the southeast part of the basin was reactivated along one segment of the north-east basin boundary-fault system. Six tectonic pulses created accommodation leading to the formation of ribbon channel bodies and overbank deposits. As the accommodation diminished during periods of less frequent faulting, however, channels tended to form amalgamated multilateral–multistorey sand bodies.

## ACKNOWLEDGEMENTS

We thank Norman Smith, Peter DeCelles and Graham Evans for comments on an early version of this manuscript. We also thank Modesto Escudero and Carlos Sánchez for their help with illustrations, Mariano Marzo and Emilio Ramos for calculation of the subsidence curves, Sierd Cloetingh for comments during the research period, and Ana Burton for the English revision. This is a contribution to Project PB 95-0084 financed by the DIGICYT, Ministerio de Educación y Cultura, Spain.

## REFERENCES

ALEXANDER, J. (1992) Nature and origin of a laterally extensive alluvial sandstone body in the Middle Scalby Formation. *J. geol. Soc. London*, **149**, 431–441.

ALEXANDER, J. & LEEDER, M.R. (1987) Active tectonic control on alluvial architecture. In: *Recent Developments in Fluvial Sedimentology* (Eds Ethridge, F.G., Flores, R.M. & Harvey, M.D.), Spec. Publ. Soc. econ. Paleont. Miner., Tulsa, **39**, 243–252.

ALLEN, J.R.L. (1978) Studies in fluviatile sedimentation: an exploratory quantitative model for the architecture of avulsion controlled suites. *Sediment. Geol.*, **21**, 129–147.

ALLEN, J.R.L. (1980) Sand waves: a model of origin and internal structure. *Sediment. Geol.*, **26**, 281–328.

ALLEN, J.R.L. (1983) Gravel overpassing on humpback bars supplied with mixed sediment: examples from the Lower Old Red Sandstone, southern Britain. *Sedimentology*, **30**, 285–294.

ALLEN, J.R.L. (1984) *Sedimentary Structures: their Character and Physical Basis*. Developments in Sedimentology, Vol. 30, Elsevier, Amsterdam, 663 pp.

ALLISON, M.A., KUEHL, S.A., MARTIN, T.C. & HASSAN, A. (1997) Importance of flood-plain sedimentation for river sediment budgets and terrigenous input into the oceans: Insights from the Brahmaputra–Jamuna River. *Geology*, **26**, 175–178.

ALVARO, M. (1987) La subsidencia tectónica en la Cordillera Ibérica durante el Mesozoico. *Geogaceta*, **3**, 34–37.

ARCHE, A. & LÓPEZ-GÓMEZ, J. (1996) Origin of the Permian–Triassic Iberian Basin, central-eastern Spain. *Tectonophysics*, **266**, 443–464.

ARTHAUD, F. & MATTE, F. (1977) Late Palaeozoic strike-slip faulting in southern Europe and northern Africa as result of a right lateral shear zone between the Appalachians and the Urals. *Geol. Soc Am. Bull.*, **88**, 1305–1320.

ASHLEY, G.M. (1990) Classification of large-scale subaqueous bedforms: a new look at an old problem. *J. sediment. Petrol.*, **60**, 160–172.

AURELL, M., MELÉNDEZ, A., SAN ROMÁN, J., *et al.* (1992) Tectónica sinsedimentaria distensiva en el límite Triásico-Jurásico en la Cordillera Ibérica. *Actas III Congreso Geológico de España, Salamanca*, **1**, 50–54.

BARRON, E. & FAWCETT, P. (1995) The climate of Pangea: a review of climate model simulations of the Permian. In: *The Permian of Northern Pangea 1, Palaeogeography, Palaeoclimates, Stratigraphy* (Eds Scholle, P.A., Peryt, T.M. & Ulmer-Scholle, D.S.), pp. 37–52. Springer-Verlag, Berlin.

BENTHAM, P., TALLIN, P. & BURBANK, D. (1993) Braided stream and flood-plain deposition in a rapidly aggrading basin: the Scanilla Formation, Spanish Pyrenees. In: *Braided Rivers* (Eds Best, J.L. & Bristow, C.S.), Spec. Publ. geol. Soc. London, No. 75, pp. 177–194. Geological Society of London, Bath.

BLODGETT, R.H. & STANLEY, K.O. (1980) Stratification, bedforms and discharge relations of the Platte braided river system, Nebraska. *J. sediment. Petrol.*, **50**, 139–148.

BLUCK, B.J. (1976) Sedimentation in some Scottish rivers of low sinuosity. *Trans. R. Soc. Edinburgh, Earth Sci.*, **69**, 425–456.

BOULOUARD, CH. & VIALLARD, P. (1982) Réduction ou lacune du Trias inférieur sur la bordure mediterrannenne de la Chaine Ibérique: Arguments palynologiques. *C.R. Acad. Sci. Paris*, **295**, 803–808.

BRAUN, J. & BEAUMONT, C. (1989) A physical explanation on the relation between flank uplifts and breakup unconformity at rifted continental margins. *Geology*, **17**, 760–764.

BRIDGE, J.S. (1993) Description and interpretation of fluvial deposits: a critical perspective. *Sedimentology*, **40**, 801–810.

BRIDGE, J.S. & MACKEY, S.D. (1993) A revised alluvial stratigraphy model. In: *Alluvial Sedimentation* (Eds Marzo, M. & Puigdefábregas, C.), Spec. Publs int. Assoc. Sediment., No. 17, 319–336.

BRIERLEY, G.J. (1991) Floodplain sedimentology of the Squamish River, British Columbia: relevance of element analysis. *Sedimenology*, **38**, 735–750.

CANT, D.J. & WALKER, R.G. (1978) Fluvial processes and facies sequences in the sandy braided South Saskatchewan River, Canada. *Sedimentology*, **25**, 625–648.

CAPOTE, R. & GONZÁLEZ-LODEIRO, F. (1983) La estructura herciniana de los afloramientos palaeozoicos de la Cordillera Ibérica. In: *Libro Jubilar a José María Rios* (Ed. Instituto Geológico y Minero), Vol. **1**, pp. 513–529. Ministerio de Industria, Madrid.

COLLINSON, J.D. & THOMPSON, D.B. (1982) *Sedimentary Structures*. Allen & Unwin, London, 194 pp.

COTTER, E. & GRAHAM, J.R. (1991) Coastal plain sedimentation in the Late Devonian of southern Ireland: hummocky cross-stratification in fluvial deposits. *Sediment. Geol.*, **72**, 201–224.

CROWLEY, K.D. (1983) Large-scale bed configurations (macro-forms), Platte River Basin, Colorado and Nebraska: primary structures and formative processes. *Geol. Soc. Am. Bull.*, **94**, 117–133.

CROWLEY, T.J., HYDE, W.T. & SHORT, D.A. (1989) Seasonal cycle variations on the supercontinent of Pangaea. *Geology*, **17**, 457–460.

DAÑOBEITIA, J.J., ARGUEDAS, M., GALLART, J., BANDA, E. & MAKRIS, J. (1992) Deep crustal configuration of the Valencia Trough and its Iberian and Balearic borders from extensive refraction and wide-angle reflection seismic profiling. *Tectonophysics*, **302**, 27–35.

DAWSON, M.R. & BRYANT, I.D. (1987) Three-dimensional facies geometry in Pleistocene outwash sediments, Worcestershire, UK. In: *Recent Developments in Fluvial Sedimentology* (Eds Ethridge, F.G., Flores, R.M. & Harvey, M.D.), Spec. Publ. Soc. econ. Paleont. Miner., Tulsa, **31**, 131–136.

DECELLES, P.G., GRAY M.B., RIDWAY, K.D., *et al.* (1991) Controls on synorogenic alluvial-fan architecture, Beartooth conglomerate (Palaeocene), Wyoming and Montana. *Sedimentology*, **38**, 567–590.

DOBLAS, M., OYARZUN, R., SOPEÑA, A., *et al.* (1994) Variscan–Late Variscan–early Alpine progressive extensional collapse of central Spain. *Geodinám. Acta*, **7**, 1–14.

DOUBINGER, J., LÓPEZ-GÓMEZ, J. & ARCHE, A. (1990) Pollen and spores from the Permian and Triassic sediments of the southeastern Iberian Ranges, Cueva de Hierro (Cuenca) to Chelva-Manzanera (Valencia–Teruel) region, Spain. *Rev. Palaeobot. Palynol.*, **66**, 25–45.

DREYER, T. (1993) Quantified fluvial architecture in ephemeral stream deposits of the Esplugafreda Formation (Palaeocene), Tremp–Graus Basin, northern Spain. In: *Alluvial Sedimentation* (Eds Marzo, M. & Puigdefábregas, C.), Spec. Publs int Ass. Sediment., No. 17, 337–362. Blackwell Scientific Publications, Oxford.

EBERTH, D.A. & MIALL, A.D. (1991) Stratigraphy, sedimentology and evolution of a vertebrate bearing, braided to anastomosed fluvial system, Cutler Formation (Permian–Pennsylvanian), north-central New Mexico. *Sediment. Geol.*, **72**, 225–252.

FARRELL, K.M. (1987) Sedimentology and facies architecture of overbank deposits of the Mississippi River, False River region, Louisiana. In: *Recent Developments in Fluvial Sedimentology* (Eds Ethridge, F.G., Flores, R.M. & Harvey, M.D.), Spec. Publ. Soc econ. Paleont. Miner., Tulsa, **39**, 111–120.

FRIEND, P.F. (1983) Towards the field classification of alluvial architecture or sequence. In: *Modern and Ancient Fluvial Systems* (Eds Collinson, J.D. & Lewin, J.L.), Spec. Publs int. Ass. Sediment., No. 6, 345–354.

FRIEND, P.F., SLATER, M.J. & WILLIAMS, R.C. (1979) Vertical and lateral buildings of river sandstone bodies, Ebro Basin, Spain. *J. geol. Soc. London*, **136**, 39–46.

FROSTICK, L. & REID, I. (1987) A new look at rifts. *Geol. Today*, **3**, 122–133.

FROSTICK, L. & REID, I. (1989) Climatic versus tectonic controls of fan sequences: lessons from the Dead Sea, Israel. *J. geol. Soc. London*, **146**, 527–538.

GALLOWAY, W.E. (1977) Catahoula Formation of the Texas coastal plain: depositional systems, composition, structural development, ground-water flow, history, and uranium distribution. *Tex. Univ. Bur. econ. Geol. Rep. Invest.*, **87**, 59 pp.

GERSIB, G.A. & MCCABE, P.J. (1981) Continental coal-bearing sediments of the Port Hood Formation (Carboniferous), Cape Linzee, Nova Scotia, Canada. In: *Recent and Ancient Nonmarine Depositional Environments: Model for Exploration* (Eds Ethridge, F.G. & Flores, R.M.), Spec. Publ. Soc. econ. Paleont. Miner., Tulsa, **31**, 95–108.

GIBBS, A. (1987) Development of extension and mixed mode sedimentary basins. In *Continental Extensional Tectonics* (Eds Coward, M.P., Dewey, J.F. & Hancock, P.L.) Spec. Publ. geol. Soc. London, No. 28, pp. 19–33. Blackwell Scientific Publications, Oxford.

GOGGIN, V., JAQUIN, T. & GAULIER, J.M. (1997) Three-dimensional accommodation analysis of the Triassic in the Paris Basin. *Tectonophysics*, **282**, 205–222.

GONZÁLEZ-CASADO, J.M., CABALLERO, J.M., CASQUET, C., GALINDO, G. & TORNOS, F. (1996) Palaeostress and geotectonic interpretation of the Alpine cycle onset in the Sierra de Guadarrama (eastern Iberia Central System) based on evidence from epysyenites. *Tectonophysics*, **262**, 213–229.

GUIMERÁ, J. & ÁLVARO, M. (1990) Structure et evolution de la compression alpine dans la Chaine Iberique et la Chaine Cotière Catalane (Espagne). *Bull. Soc. geol. Fr.*, **6**, 339–348.

HARLAND, W.B., ARMSTRONG, R.L., COX, A.V., CRAIG, L.E., SMITH, A.G. & SMITH, D.G. (1990) *A Geologic Time Scale, 1989*. Cambridge University Press, Cambridge, 263 pp.

HASZELDINE, R.S. (1983) Descending tabular cross-bed sets and bounding surfaces from a fluvial channel in the Upper Carboniferous coalfield of Northeast England. In: *Modern and Ancient Fluvial Systems* (Eds Collinson, J.D. & Lewin, J.), Spec. Publs int. Ass. Sediment., No. 6, 449–456. Blackwell Scientific Publications, Oxford.

HIRST, J.P.P. (1991) Variations in alluvial architecture across the Oligo-Miocene Huesca fluvial system, Ebro Basin, Spain. In: *The Three-dimensional Facies Architecture of Terrigenous Clastic Sediments, and its Implications for Hydrocarbon Discovery and Recovery* (Eds Miall, A.D. & Tyler, N.), pp. 111–121. Concepts in Sedimentology and Paleontology, Vol. 3, Society of Economic Paleontologists and Mineralogists, Tulsa, OK.

HORNE, J.C. & FERN, J.C. (1976) *Carboniferous Depositional Environments in the Pocahontas Basin, Eastern Kentucky and Southern West Virginia*. Guidebook, Department of Geology, University of South Carolina, Columbia, 62 pp.

ISMAIL-ZADEH, A.T., KOSTYUCHENKO, S.L. & NAIMARK, B.M. (1997) The Timon–Pechora Basin, NE Russia: tectonic subsidence analysis and a model of formation mechanism. *Tectonophysics*, **283**, 205–218.

JORDAN, D.W. & PRIOR, W.A. (1992) Hierarchical levels of heterogeneity in a Mississippi river meander belt and application to reservoir systems. *Bull. Am. Assoc. petrol. Geol.*, **76**, 1601–1624.

KOOI, H., CLOETINGH, S. & BURRUS, J. (1992) Lithospheric necking and regional isostasy at extensional basins. 1. *J. Geophys. Res.*, **97**, 17553–17572.

KRAUS, M.J. & BOWN, T.M. (1988) Pedofacies analysis; a new approach to reconstructing ancient fluvial sequences. *Geol. Soc. Am. Spec. Pap.*, **216**, 143–152.

LAGO, M., POCOVI, A., ZACHMAM, D., ARRANZ, E. & CARLS, P. (1991) Comparación preliminar de las manifestaciones magmáticas, calcoalcalinas, stephaniense-pérmicas de la Cadena Ibérica. *Cuad. Lab. Xeol. Laxe.*, **16**, 95–107.

LAGO, M., ÁLVARO, J., ARRANZ, E., POCOVI, A. & VAQUER, R. (1992) Condiciones de emplazamiento, Petrología y Geoquímica de las riolitas calcoalcalinas estefanienses-pérmicas en las Cadenas Ibéricas. *Cuad. Lab. Xeol. Laxe.*, **17**, 187–198.

LEEDER, M.R. (1978) A quantitative stratigraphic model for alluvium, with special reference to channel deposit density and interconnectedness. In: *Fluvial Sedimentology* (Ed. Miall, A.D.), Mem. Can. Soc. petrol. Geol., Calgary, **5**, 587–596.

LEEDER, M.R. (1997) Sedimentary basins: tectonic recorders of sediment discharge from drainage catchments. *Earth Surf. Process. Landf.*, **22**, 229–237.

LEEDER, M.R. GREG, H.M., PEAKAL, J. & SALYARDS, S.L. (1996) First quantitative test of alluvial stratigraphic models: southern Rio Grande rift, New Mexico. *Geology*, **24**(1), 87–90.

LINK, M.H. (1984) Fluvial facies of the Miocene Ridge Route Formation, Ridge Basin, California. *Sediment. Geol.*, **38**, 263–286.

LONG, D.G.F. (1978) Proterozoic stream deposits: some problems of recognition and interpretation of ancient sandy fluvial systems. In: *Fluvial Sedimentology* (Ed. Miall, A.D.), Mem. Can. Soc. petrol. Geol., Calgary, **5**, 313–342.

LÓPEZ-GÓMEZ, J. & ARCHE, A. (1992) Palaeogeographical significance of the Röt (Anisian, Triassic) Facies (Marines Clays, Muds and Marls Fm.) in the Iberian Ranges, eastern Spain. *Palaeogeogr. Palaeoclimatol. Palaeoecol.*, **91**, 347–361.

LÓPEZ-GÓMEZ, J. & ARCHE, A. (1993a) Sequence stratigraphic analysis and palaeogeographic interpretation of the Buntsandstein and Muschelkalk facies (Permo-Triassic) in the SE Iberian Range, E. Spain. *Palaeogeogr. Palaeoclimatol. Palaeoecol.*, **103**, 179–201.

LÓPEZ-GÓMEZ, J. & ARCHE, A. (1993b) Architecture of the Cañizar fluvial sheet sandstones. In: *Alluvial Sedimentation* (Eds Marzo, M. & Puigdefábregas, C.), Spec. Publs Int. Ass. Sediment. No. 17, pp. 363–381. Blackwell Scientific Publications, Oxford.

LÓPEZ-GÓMEZ, J. & ARCHE, A. (1997) The Upper Permian Boniches Conglomerates Formation: evolution from alluvial fan to fluvial system environments and accompanying tectonic and climatic controls in the southeast Iberian Ranges, central Spain. *Sediment. Geol.*, **114**, 267–294.

LÓPEZ-GÓMEZ, J., MAS, J.R. & ARCHE, A. (1993) The evolution of the Middle Triassic (Muschelkalk) carbonate ramp in the SE Iberian Ranges, Eastern Spain: sequence stratigraphy, dolomitization processes and dynamic controls. *Sediment Geol.*, **87**, 165–193.

MACK, G.H., LOVE, D.W. & SEAGER, W.R. (1997) Spillover models for axial rivers in regions of continental extension: the Rio Mimbres and Rio Grande in the Southern Rio Grande Rift, USA. *Sedimentology*, **44**, 637–652.

MACKEY, S.D. & BRIDGE, J.S. (1995) Three-dimensional model of alluvial stratigraphy: theory and applications. *J. sediment. Res.*, **65**, 7–31.

MARZO, M. & ANADÓN, P. (1977) Evolución y características sedimentológicas de las facies fluviales basales del Buntsandstein de Olesa de Montserrat (provincia de Barcelona). *Cuad. Geol. Iber.*, **4**, 211–222.

MCKEE, E.D. & WEIR, G.W. (1953) Terminology for stratification and cross-stratification in sedimentary rocks. *Geol. Soc. Am. Bull.*, **64**, 381–389.

MIALL, A.D. (1976) Palaeocurrent and palaeohydrologic analysis of some vertical profiles through a Cretaceous braided stream deposit, Banks Island, Arctic Canada. *Sedimentology*, **23**, 459–484.

MIALL, A.D. (1985) Architectural-element analysis: a new method of facies analysis applied to fluvial deposits. *Earth Sci. Rev.*, **22**, 261–308.

MIALL, A.D. (1988a) Reservoir heterogeneities in fluvial sandstones: lesson from outcrop studies. *Bull. Am. Assoc. petrol. Geol.*, **72**, 682–697.

MIALL, A.D. (1988b) Facies architecture in clastic sedimentary basins. In: *New Perspectives in Basin Analysis* (Eds Kleinspehn, K. & Paola, C.), pp. 67–81. Springer-Verlag, Berlin.

MIALL, A.D. (1989) Architectural elements and bounding surfaces in channelized clastic deposits: notes on comparisons between fluvial and turbidite systems. In: *Sedimentary Facies in the Active Plate Margin* (Eds Taira, A. & Masuda, F.), pp. 3–15. Terra Scientific, Tokyo.

MIALL, A.D. (1991) Sedimentology of a sequence boundary within the nonmarine Torrivio Member, Gallup Sandstone (Cretaceous), San Juan Basin, New Mexico. In: *The Three-dimensional Facies Architecture of Terrigenous Clastic Sediments and its Implications for Hydrocarbon Discovery and Recovery* (Eds Miall, A.D. & Tyler, N.), pp. 224–232. Concepts in Sedimentology and Paleontology, Vol. 3, Society of Economic Paleontologists and Mineralogists, Tulsa, OK.

MIALL, A.D. (1996) *The Geology of Fluvial Deposits. Sedimentary Facies, Basin Analysis, and Petroleum Geology*. Springer-Verlag, New York, 582 pp.

MIALL, A.D. & GIBLING, M.R. (1978) The Siluro-Devonian clastic wedge of Somerset Island, Arctic Canada, and some regional palaeogeographic implications. *Sediment. Geol.*, **21**, 85–127.

MIDDLETON, G.V. & TRUJILLO, A. (1984) Sedimentology and depositional setting of the upper Proterozoic Scanlan Conglomerate, Central Arizona. In: *Sedimentology of Gravel and Conglomerates* (Eds Koster, E.H. & Steel, R.), Mem. Can. Soc. petrol. Geol., Calgary, **10**, 189–201.

MJØS, R., WALDERHAUG, O. & PRESTHOLM, E. (1993) Crevasse splay sandstone geometries in the Middle Jurassic Ravenscar Group of Yorkshire, UK. In: *Alluvial Sedimentation* (Eds Marzo, M. & Puigdefábregas, C.), Spec. Publs int. Ass. Sediment., No. 17, pp. 167–184. Blackwell Scientific Publications, Oxford.

NAMI, M. & LEEDER, M. (1978) Changing channel morphology and magnitude in the Scalby Formation (M. Jurassic) of Yorkshire, England. In: *Fluvial Sedimentology* (Ed. Miall, A.D.). Mem. Can. Soc. petrol. Geol., Calgary, **5**, 431–440.

NANSON, G.C. & CROKE, J.C. (1992) A genetic classification of floodplain, *Geomorphology*, **4**, 459–486.

NEMEC, W. & POSTMA, G. (1993) Quaternary alluvial fans in southwestern Crete: sedimentation processes and geomorphic evolution. In: *Alluvial Sedimentation* (Eds Marzo, M. & Puigdefábregas, C.), Spec. Publs int. Ass. Sediment., No. 17, 235–276. Blackwell Scientific Publications, Oxford.

OLSEN, H. (1988) The architecture of a sandy braided–meandering river system: an example from the Lower Triassic Solling Formation (M. Buntsandstein) in W-Germany. *Geol. Rundsch.*, **77**, 797–814.

OLSEN, H. (1989) Sandstone-body structures and ephemeral stream processes in the Dinosaur Canyon Member, Moenave Formation (Lower Jurassic), Utah, USA. *Sediment. Geol.*, **61**, 207–221.

ORE, H.T. (1964) Some criteria for recognition of braided stream deposits. *Wyo. Contrib. Geol.*, **3**, 1–14.

OUCHI, S. (1985) Response of alluvial rivers to slow active tectonic movement. *Geol. Soc. Am. Bull.*, **96**, 504–515.

PICARD, M.D. & HIGH, L.R. (1973) *Sedimentary Structures of Ephemeral Streams*. Elsevier, Amsterdam, 223 pp.

PLINT, A.G. (1983) Sandy fluvial point-bar sediments from the middle Eocene of Dorset, England. In: *Modern and Ancient Fluvial Systems* (Eds Collinson, J.D. & Lewin, J.), Spec. Publs int. Ass. Sediment., No. 6, pp. 355–368. Blackwell Scientific Publications, Oxford.

PUIGDEFÁBREGAS, C. & VAN VLIET, A. (1978) Meandering stream deposits from the Tertiary of the southern Pyrenees. In: *Fluvial Sedimentology* (Ed. Miall, A.D.), Mem. Can. Soc. petrol. Geol., Calgary, **5**, 469–485.

RAMOS, A. (1979) Estratigrafía y palaeogeografía del Pérmico y Triásico al oeste de Molina de Aragón (provincia de Guadalajara). *Semin. Estratigrafía, Ser. Monogr.*, **6**, 313 pp.

RAMOS, A. & SOPEÑA, A. (1983) Gravel bars in low-sinuosity streams (Permian and Triassic, central Spain). In: *Modern and Ancient Fluvial Systems* (Eds Collinson, J.D. & Lewin, J.) Spec. Publs int. Ass. Sediment., No. 6, pp. 301–312. Blackwell Scientific Publications, Oxford.

REINFELDS, I. & NANSON, G. (1993) Formation of braided river floodplains, Waimakariri River, New Zeland. *Sedimentology*, **40**, 1113–1127.

ROCA, E., GUIMERÁ, J. & SALAS, R. (1994) Mesozoic extensional tectonics in the southeastern Iberian Chain. *Geol. Mag.*, **131**, 155–168.

RØE, S.-L. & HERMANSEN, M. (1993) Processes and products of large, late Precambrian sandy rivers in northern Norway. In: *Alluvial Sedimentation* (Eds Marzo, M. & Puigdefábregas, C.), Spec. Publs int. Ass. Sediment., No. 17, pp. 151–166. Blackwell Scientific Publications, Oxford.

RUST, B.R. (1984) Proximal braidplain deposits in the Middle Devonian Malbaie Formation of eastern Gaspé, Quebec, Canada. *Sedimentology*, **31**, 675–695.

RUST, B.R. & LEGUN, A.S. (1983) Modern anastomosing-fluvial deposits in arid central Australia, and a Carboniferous analogue in New Brunswick, Canada. In: *Modern and Ancient Fluvial Systems* (Eds Collinson, J.D. & Lewin, J.), Spec. Publs int. Ass. Sediment., No. 6, pp. 385–392. Blackwell Scientific Publications, Oxford.

RUST, B.R., GIBLING, M.R. & LEGUN, A.S. (1984) Coal deposition in an anastomosed fluvial system: the Pensylvanian Cumberland Group south of Joggins, Nova Scotia, Canada. In: *Sedimentology of Coal and Coal-bearing Sequences* (Eds Rahmani, R.A. & Flores, R.M.), Spec. Publs int. Ass. Sediment., No. 7, pp. 105–120. Blackwell Scientific Publications, Oxford.

RUST, B.R. & NANSON, G.C. (1989) Bedload transport of mud as pedogenic aggregates in modern and ancient rivers. *Sedimentology*, **36**, 291–306.

SALAS, R. & CASAS, A. (1993) Mesozoic extensional tectonics, stratigraphy and crustal evolution during the Alpine Cycle of the eastern Iberian Basin. *Tectonophysics*, **228**, 33–55.

SÁNCHEZ-MOYA, Y., SOPEÑA, A., MUÑOZ, A. & RAMOS, A. (1992) Consideraciones teóricas sobre el análisis de la subsidencia: aplicaciones a un caso real en el borde de la Cuenca triásica Ibérica. *Rev. Soc. Geol. España*, **5**, 21–40.

SCOTESE, C.R. & LANGFORD, R.P. (1995) Pangea and the palaeogeography of the Permian. In: *The Permian of Northern Pangea 1, Palaeogeography, Palaeoclimates and Stratigraphy* (Eds Scholle, P.A., Peryt, T.M. & Ulmer-Scholle, D.S.), pp. 3–19. Springer-Verlag, Berlin.

SCHUMM, S.A. (1963) A tentative classification of alluvial river channels. *U.S. geol. Surv. Circ.*, **477**, 13.

SCHUMM, S.A. (1981) Evolution and response of the fluvial system, sedimentological implications. In: *Recent and Ancient Nonmarine Depositional Environments: Models for Exploration* (Eds Ethridge, F.G. & Flores, R.M.), Spec. Publ. Soc. econ. Paleont. Miner., Tulsa, **31**, 19–29.

SCHUMM, S.A. (1993) River response to baselevel change: implications for sequence stratigraphy. *J. Geol.*, **101**, 279–294.

SHEPHERD, R.G. (1987) Lateral accretion surfaces in ephemeral-stream point bars, Rio Puerco, New Mexico. In: *Recent Developments in Fluvial Sedimentology* (Eds Ethridge, F.G., Flores, R.M. & Harvey, M.D.), Spec. Publ. Soc. econ. Paleont. Miner., Tulsa, **39**, 93–98.

SMITH, D.G. (1983) Anastomosed fluvial deposits: modern examples from western Canada. In: *Modern and Ancient Fluvial Systems* (Eds Collinson, J.D. & Lewin, J.), Spec. Publs int. Ass. Sediment., No. 6, pp. 155–165. Blackwell Scientific Publications, Oxford.

SMITH, D.G. & SMITH, N.D. (1980) Sedimentation in anastomosed river systems: examples from alluvial valleys near Banff, Alberta. *J. sediment. Petrol.*, **50**, 157–164.

SOEGAARD, K. (1990) Fan-delta and braid-delta systems in Pennsylvanian Sandia Formation, Taos Trough, northern New Mexico: depositional and tectonic implications. *Geol. Soc. Am. Bull.*, **102**, 1325–1343.

SOPEÑA, A., LÓPEZ-GÓMEZ, J., ARCHE, A., *et al.* (1988) Permian and Triassic rift basins of the Iberian Peninsula. In: *Triassic–Jurassic Rifting* (Ed. Manspeizer, W.) Vol. 2, pp. 757–786. Elsevier, Amsterdam.

STECKLER, M.S. & WATTS, A.B. (1978) Subsidence of the Atlantic type continental margin of New York. *Earth planet. Sci. Lett.*, **41**, 1–13.

VAN WEES, J.D. (1994) *Tectonic modelling of basin deformation and inversion dynamics*. PhD thesis, Vrije Universiteit, Amsterdam, 164 pp.

VIRGILI, C., SOPEÑA, A., RAMOS, A., ARCHE, A. & HERNÁNDO, S. (1983) El relleno post-hercínico y el comienzo de la sedimentación mesozoica de la Cordillera Ibérica. In: *Libro Jubilar José María Ríos* (Ed. Fontboté, J.M.), Vol. 2, pp. 25–36. IGME, Madrid.

WATTS, A.B. & RYAN, W.B. (1976) Flexure of the lithosphere at continental margin basins. *Tectonophysics*, **36**, 25–44.

WILLIS, B.J. & BEHRENSMEYER, A.K. (1994) Architecture of Miocene overbank deposits in northern Pakistan. *J. sediment. Res., Sect. B*, **64**, 60–67.

*Spec. Publs int. Ass. Sediment.* (1999) **28**, 305–313

# Drainage evolution in active mountain belts: extrapolation backwards from present-day Himalayan river patterns

P. F. FRIEND*, N. E. JONES† *and* S. J. VINCENT‡

*\*Department of Earth Sciences, University of Cambridge, Downing Street, Cambridge CB2 3EQ, UK;*
*†B.P. Exploration, Dyce, Aberdeen AB21 7PB, UK; and*
*‡Cambridge Arctic Shelf Programme, University of Cambridge, West Building, Huntingdon Road,*
*Cambridge CB3 0DJ, UK*

## ABSTRACT

An understanding of how drainage patterns respond to active tectonics can provide insight into past deformational events within mountain belts. The Himalayan arc mountain belt is taken as an example because it is still strongly active. It also is old enough (55 Ma) and of sufficient extent that aspects of ongoing drainage modification in the outer Subhimalayan and Lower Himalayan zones can be compared with those internal parts of the mountain belt where drainage patterns have become fixed by gorge formation.

The drainage patterns of the outer lithotectonic zones have produced 0–8 km of exhumation over the last 15 Ma and provide many examples of deflection and gorge erosion during young episodes of thrusting. In contrast, the drainage systems of the Higher Himalayan zone have been evolving for longer, having produced 8–25 km of exhumation over the 55 Ma since the Himalaya began to rise. Generalized river gradients vary strongly in different parts of the belt corresponding to different amounts of rock uplift and this strongly influences the behaviour of the rivers.

## INTRODUCTION

Geologists have long used the record of fluvial sediments within depositional basins to provide information on the evolution of neighbouring active mountain belts, and their river systems. This has involved analysing features such as sedimentary facies, alluvial architecture and provenance. Less use, however, has been made of information on the erosional portions of the fluvial systems. This paper offers a brief review of the types of information on past drainage evolution that may be derived from present-day river patterns within active mountain belts. This is achieved through a study of the Himalayas, where the relative youth and continuing strong tectonic activity of the belt make it particularly suitable for many investigations of mountain building and its implications (Raymo *et al.*, 1988; Raymo & Ruddiman, 1992; Einsele *et al.*, 1996).

## TECTONIC EVOLUTION OF THE HIMALAYAS

### Convergence

Continental fragment reconstruction (Smith *et al.*, 1994), based primarily on ocean-floor magnetic anomalies, shows that greater India rifted from southern Africa 160 Ma (Mid–Late Jurassic) and from Antarctica 140 Ma (Early Cretaceous). Ninety million years ago (early Late Cretaceous), greater India began to converge rapidly with southern Eurasia. Fifty-five million years ago (early Early Eocene), a general convergence rate of about 20 cm yr$^{-1}$ dropped to some 5 cm yr$^{-1}$, based on palaeomagnetic studies (Klootwijk *et al.*, 1992). An overall convergence at about this rate is continuing today.

## Collision

The decrease in convergence rate at *c.* 55 Ma may have coincided in time with the initial collision of Indian and Eurasian continental crust. The termination of marine sedimentation along the Suture Zone between the two continents, in Ladakh, northern India, is biostratigraphically consistent (Searle *et al.*, 1997) and provides some support for this initial collision date.

Initial collision marked the onset of prolonged and continued deformation. Episodes of crustal strain have occurred at different times, not only along the main Himalayan arc but also along the suture. Similar, collision-related structures are also widely and discretely distributed further north and east across Eurasia (Dewey *et al.*, 1989).

## Himalayan uplift, structure and erosion

England & Molnar (1990) clarified the concepts involved in considering mountain uplift. Relative to the geoid or some other external datum, 'surface uplift' involves movement of the erosional surface (topography) of the crust and must be distinguished from 'uplift of rocks', which refers to movement of rock. These are both different from 'exhumation', which refers to movement of rocks in the crust relative to the erosional surface.

Between the suture zone in the north and the Indo-Gangetic alluvial plains in the south, the main Himalayan

arc can be divided into four major lithotectonic zones (Gansser, 1964, p. 243) (Fig. 1). The zones (from north to south), using Gansser's terminology, are listed below, along with the boundary faults or fault zones that are described briefly later in this paper:
**1** Tibetan Himalayas — South Tibetan Detachment (STD);
**2** Higher Himalayas (see Fig. 1) — Main Central Thrust (MCT);
**3** Lower Himalayas — Main Boundary Thrust (MBT);
**4** Subhimalayas — Main Frontal Thrust (MFT);
     Indo-Gangetic alluvium (Quaternary).

It seems reasonable to suggest that mountain uplift and compressional deformation began in the main Himalaya soon after initial continent–continent collision, *c.* 55 Ma (Early Eocene). Many detailed mineral studies have now been carried out in the Himalaya, allowing the reconstruction of the thermobarometric history of rock samples collected at the surface. Recent reviews by Einsele *et al.* (1996) and Searle (1996) have estimated the net amount of material removed by erosion. This exhumation can be generalized as follows: Tibetan Himalayas, 5–9 km; Higher Himalayas, 12–25 km; Lower Himalayas, 2–8 km.

Estimates of the uplift of rocks can be made, if the surface uplift can be assessed. Present-day erosional relief varies between about 6 km in parts of the Higher Himalayas and 1 km, or less, in parts of the Lower Himalayas and Subhimalayas.

It seems likely that the most important processes of

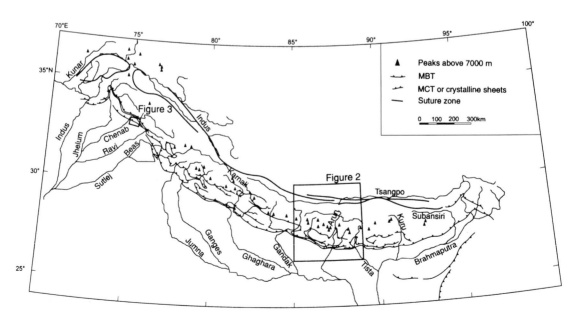

**Fig. 1.** Outline of the main drainage systems and morphotectonic features of the Himalaya. Simplified from Seeber & Gornitz (1983).

local erosion in the Himalaya have been various forms of downslope mass-movement under gravity, such as rock falls and landslips. The general exhumation of the range, however, has depended largely on the ability of the fluvial systems to transport material out of the range to its surrounding basins.

## PRESENT-DAY DRAINAGE PATTERNS

Four different river-system types can be distinguished in the Himalaya, based on their stream order (as defined by Strahler, 1952), catchment size and the lithotectonic zone from which their main catchment is formed. The Nepal section of the Himalaya is used as an example (Fig. 2).

**1** *Syntaxial river systems* are the two largest (Indus and Tsangpo–Brahmaputra) with catchment areas of 200 000 and 160 000 km$^2$ respectively, within the mountain belt. They originate from the southern margin of the Tibetan Plateau and flow west and east respectively, longitudinal to the mountain belt along the Indus–Tsangpo suture zone. They enter the Indo-Gangetic foreland basin at opposite ends of the Himalayan arc, via relatively short reaches of transverse flow through the mountain trend syntaxes, and have been called the 'syntaxial rivers' (Friend, 1998), as compared with the 'main-arc rivers', which are the broadly transverse, southerly directed types described next.

**2** *High-order, Tibetan Himalayan rivers* include the Sutlej, Karnali and Arun rivers, all of which originate in the Tibetan Himalaya. The sizes of their catchment areas vary from 20 000 to 55 000 km$^2$. The main east–west headwater tributaries coalesce behind the highest peaks of the Higher Himalayas and continue southward, transverse to the mountain belt, in deeply incised gorges through the Higher and Lower Himalayas. In the Subhimalaya, these rivers often show reaches of longitudinal flow, where they have been deflected around linear ridges associated with the MBT and the MFT.

**3** *Intermediate-order, Higher Himalayan rivers* originate from the southern margin of the Higher Himalaya and include the Tamur, Tista and tributaries of the Kosi rivers. Catchment areas vary from 4000 to 14 000 km$^2$. They often show transverse flow, although like the high-order rivers they show reaches of longitudinal flow hindward of the MBT and linear ridges associated with the imbricate thrusts of the Subhimalaya and MFT.

**4** *Low-order, Lower Himalayan and Subhimalayan rivers* originate from within the Lower Himalayas and the Subhimalaya. They have small catchments, often 100s of km$^2$, and never more than 3500 km$^2$, and generally flow transverse to the trend of the Himalayan arc, although

locally they may flow longitudinally. These rivers flow on to become the foothills-fed rivers of the plains (Sinha & Friend, 1994).

## EVIDENCE FOR THE HORIZONTAL (MAP-VIEW) MODIFICATION OF DRAINAGE PATTERNS

Maps and satellite data are the most important sources of information on drainage patterns. In this section, we consider situations in which the map-view patterns of river systems appear to have become modified. We start with the assumption that the river systems that formed when the Himalayan mountain belt first appeared, some 55 Ma, would have been regular, dendritic systems, that grew by headward erosion of the tributaries (Howard, 1967; Rodriguez-Iturbe & Rinaldo, 1997), and would have been 'consequent' on the northward and southward slopes of the belt. Any irregular features in the present-day systems must have formed during their evolution.

By comparing the river-system types and patterns described above with geological maps, the drainage of the Himalayan arc can be seen to be strongly influenced by three of the boundaries between lithotectonic zones that are listed above.

**1** *The STD (South Tibetan Detachment)* is a zone of normal faulting that formed in response to the gravity-collapse of the high Himalayas about 18 Ma (Burchfiel *et al.*, 1992; Einsele *et al.*, 1996). The high-order Tibetan Himalayan rivers have been deflected behind the high peaks of this zone to form long reaches of longitudinal drainage (Fig. 2). They then cross the Higher Himalaya through steep gorges, joining broad tracts of intermediate-order transverse rivers that flow southwards in the hangingwall of the MCT (Main Central Thrust).

**2** *The MBT (Main Boundary Thrust)* has brought Lower Himalayan zone material, in its hangingwall, over Subhimalayan zone sediments. In many cases the high- and intermediate-order rivers are deflected to run longitudinally on the hinterland (northward) side of the MBT. In Nepal (Fig. 2) the MBT and the MCT crop out very close to each other and their effect is more difficult to distinguish.

**3** *The MFT (Main Frontal Thrust)* has brought Subhimalayan zone sediments over or against the Quaternary sediments of the alluvial plains. Low-order rivers are both deflected longitudinally by the MFT and by a number of other structures within the Subhimalayas (Fig. 2). Numerous low-order rivers also appear to be consequent on the uplift of the hangingwall of the MFT and its associated structures. These small rivers often

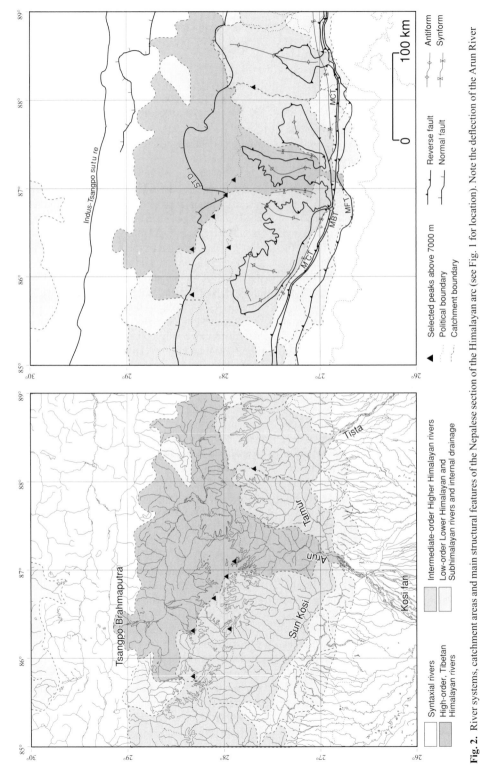

**Fig. 2.** River systems, catchment areas and main structural features of the Nepalese section of the Himalayan arc (see Fig. 1 for location). Note the deflection of the Arun River north of the highest Himalayan peaks, and the similar deflection of the Sun Kosi, Tamur and other intermediate-order Higher Himalayan rivers north of the MCT (Main Central Thrust) and MBT (Main Boundary Thrust). Low-order Lower Himalayan and Subhimalayan rivers are generally either deflected behind the MFT (Main Frontal Thrust), or are sourced from this feature. Catchment areas are delimited only in regions above 150 m a.s.l. The positions of geological structures are taken from Schelling (1992) and Burchfiel *et al.* (1992).

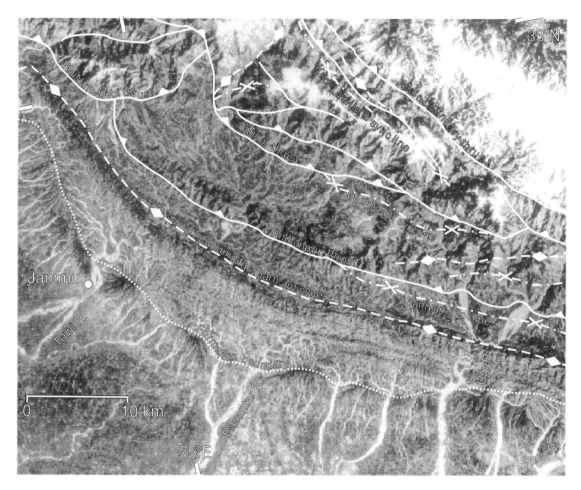

**Fig. 3.** Interpreted Landsat image of the Subhimalayan zone between the Chenab and Ravi rivers, north-west India (see Fig. 1 for location). Note the deflection of rivers behind successive structures and the marked changes in channel form across these structures. Further discussion is included in the text. Geological structures are from Raiverman *et al.* (1990), and the Landsat image was provided courtesy of B.P. Exploration.

show a very regular spacing along an individual thrust sheet (Talling *et al.*, 1997).

Figure 3 is part of a Landsat image of the Subhimalayan mountain front between the Chenab and Ravi rivers, with data from a detailed geological map (Raiverman *et al.*, 1990) superimposed. Outward from the Main Boundary Thrust, a number of thrusts and folds have been mapped, and the deflections of the Tawi and Ujh rivers behind the Kishanpur thrust and Suruim Mastgarh anticline are clear. Note also the changes in channel form across these boundaries. Other deflections are apparent as small rivers approach the 'Formation boundary', which is mapped at the edge of the alluvial plains to the south-west.

## EVIDENCE FOR THE MODIFICATION OF DRAINAGE PATTERNS IN THEIR VERTICAL POSITIONS

The deep gorges of the Himalaya have long been taken as evidence of the antecedence of the rivers (Wager, 1937; Holmes, 1965), i.e. that they have cut down to their present levels after establishing their drainage pattern at a higher level. Wager (1937) argued for the dominant role of downward cutting on the evidence that some gorges have been cut in bed-rock particularly resistant to river erosion in situations where less resistant bedrock could have been selected preferentially if general headward erosion from tributaries was occurring. When once a river has incised a gorge, plan-view deflection or modification is likely to become inhibited.

Profiling of the elevations of river beds has shown steps in the gradients, apparently corresponding to the surface emergence of young faults. This is an important potential tool for recognizing structures produced by young deformation, but may be difficult to distinguish in some cases from the erosional response to lithological contrasts in the bedrock.

Seeber & Gornitz (1983) compared the long-profiles of most of the main-arc rivers of the Himalaya (our high-order rivers) and found a distinct step in the region of the MCT (Main Central Thrust). They suggested that this zone of steep gradient corresponds to uplift above thick-skinned thrust ramps in this part of the orogen. The effect of these differences in gradient on river behaviour are likely to be considerable. High-gradient rivers are likely to incise strongly and resist deflection, whereas the low-gradient rivers of much of the Tibetan Himalaya and the Lower and Subhimalaya (Seeber & Gornitz, 1983) will be more prone to deflection.

The amounts of exhumation that have occurred in different zones of the orogen must have been, at least partly, a response to differences in rock and surface uplift, resulting from the differences in tectonic evolution that are represented in the different rocks and structures. The contrast in precipitation between the southern and northern flanks of the present mountain belt, although clearly an orographic response to the present mountain topography, however, is likely to have grown with it and always had some influence. So, the penetration of the high-order transverse-longitudinal rivers, draining the Tibetan Himalaya southwards between the peaks of the Higher Himalaya is likely to have been a tectonic phenomenon, influenced but not controlled by climate.

## EVIDENCE FROM THE
## FORELAND-BASIN FILL

Aspects of the Late Quaternary and present-day mega-fans of the Gandak and Kosi rivers (Fig. 2) have been described by Mohindra *et al.* (1992), Singh *et al.* (1993) and by Sinha & Friend (1994). The amounts of water and sediment carried on to and down these megafans by the mountain-fed rivers greatly exceed those of the intervening foothills and plains-fed rivers, and the elevation of the megafans above the intervening interfan areas provides evidence of the greater rates of net sediment accumulation. The spacing of the alluvial-plains major rivers, and their separation by areas of foothills- and plains-fed rivers

is a direct result of the deflection of earlier rivers by the growing structures and topography of the Subhimalaya and Lower Himalaya.

Growing knowledge of the stratigraphy of the Siwalik Group, which has been filling the foreland basin over the last 20 Ma, has been reviewed by Burbank *et al.* (1996a). The distribution of conglomerates in this stratigraphy appears to be particularly relevant to the reconstruction of the drainage within the mountain belt. Two types of conglomerate can be distinguished.

**1** Foothills-fed conglomerates, which form the culmination of a general coarsening upwards trend in most of the Upper Siwalik sediments cropping out in the Subhimalaya. These conglomerates seem to have been derived from the hanging walls of active thrusts in the Sub- and Lower Himalaya, starting about 5 Ma, but continuing since.

**2** Mountain-fed conglomerates, containing clasts from all the Himalayan lithotectonic zones and therefore regarded as the deposits of ancestral high-order, main-arc rivers. In the cases of the Indus and Beas rivers (Meigs *et al.*, 1995), distinct laterally restricted bodies of conglomerate were formed during Middle Siwalik times, dated from between 8 and 7 Ma. These bodies provide the first record in the uplifted sediments of the margin of the foreland basin of the ancestral high-order, main-arc rivers and show that the particular rivers have not experienced significant lateral movement since.

## GENERAL CONCLUSIONS ON THE
## EVOLUTION OF DRAINAGE
## THROUGH TIME

The Himalayan mountain system was initiated about 55 Ma and its drainage system began to form at that time. Subsequent evolution has involved exhumation of the mountain belt, involving erosion through the crust to the extent of as much as 25 km. Later drainage patterns incorporate any modifications of earlier patterns that have occurred during this erosion. The analysis of present-day patterns can provide a basis for extrapolation backwards towards the earlier patterns and demonstrates the processes that caused the patterns to change.

In the Subhimalaya and Lower Himalaya, rivers have been deflected by anticlinal ridges growing at the surface in response to thrust-fault movement below the surface, as clearly illustrated in the case of the Beas and Sutlej rivers (Jones & Vincent, 1990; Vincent, 1993; Gupta, 1997). Similar cases also have been investigated from the western USA (Burbank *et al.*, 1996b, c) and New Zealand (Jackson *et al.*, 1996). In all cases the deflection of the

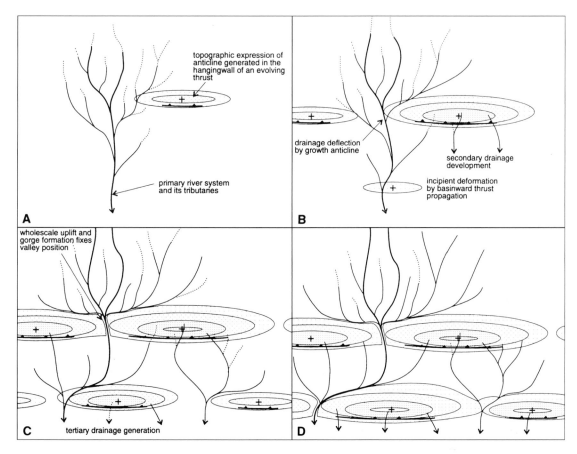

**Fig. 4.** Conceptual model illustrating drainage response to active structures within a foreland-propagating fold-and-thrust system. Note the deflection of river channels to form longitudinal reaches behind active structures, gorge formation where stream power is sufficiently large to enable valley incision to keep pace with or outpace uplift, and the generation of secondary drainage upon evolving structures. From Vincent (1993), based on ideas presented by Jones & Vincent (1990).

rivers has been a result of tectonic growth of anticlinal ridges along their plunging axes, in response to tip-line growth of the thrust faults active in their cores (Fig. 4). River deflection by active structures is most marked in the outermost fold and thrust structures, where differential uplift often exceeds the erosive potential of the low-gradient portions of the fluvial systems.

Mountain belts typically grow by the foreland propagation of thrusts and folds. This will result in the generation of new topographic barriers basinward of existing structures and will cause the further deflection of the downstream portions of river systems, as illustrated on Fig. 4. At the same time, gorges form at the points where lateral propagation of thrust tip-lines coalesce and the rivers are powerful enough to maintain flow (Fig. 4).

Although out-of-sequence thrusting will complicate this process, it can be assumed that, in the Lower

Himalayas and Subhimalayas, where thin-skinned thrusting is dominant, longitudinal deflections of river courses reflect the river pattern response to increasingly recent tectonic activity in areas closest to the mountain front. Thus, in the case of the Tawi and Ujh rivers (Fig. 3), two phases of thrust development appear to have caused deflections. Tributaries of the two rivers have been deflected behind the Kishanpur thrust before coalescing to pass through its topographic barrier. These rivers and others sourced on the southward side of the barrier have then been deflected by the Suruin Mastgarh anticline, which is the surface expression of a backthrust exposed further west. The deflected rivers flow longitudinally, parallel to this structure and then, below their confluence, they again turn south, and pass across the anticline, presumably because their combined flow was sufficient to maintain the river in the face of the rising structure. We note, in particular, the large channel complexes produced by the Ujh river upstream of the anticline. Other rivers, such as the Basantar, have their catchments on the basin-ward portions of this second anticlinal structure and form the next generation of river patterns. Interestingly, a further set of minor streams has developed at the outermost margin and a further drainage divide is apparent along the line mapped as 'Formation boundary' (Raiverman *et al.*, 1990; Fig. 3). Whether this boundary marks a rising structure or a lithological contrast will require further field investigation.

The Subhimalayas and Lower Himalayas appear to have lost very variable amounts of material, up to 8 km in thickness, by erosion of their surface. These zones have been propagating outwards by fold and thrust growth, over the last 10–15 Ma, and the deflection of the river systems by the growth of anticlinal ridges at the surface, as just described, provides evidence of the processes. Local gorges have formed where the rivers have been powerful enough to maintain their flow in spite of local tectonic surface uplift.

In the Higher Himalayas, high river gradients and gorge formation preclude deflection by presently active structures. Instead, given the 12–25 km of exhumation and the high relief of this zone, it seems inevitable that the patterns reflect dominant vertical incision. The Tibetan Himalayas show evidence of lateral modification of drainage, north of the South Tibetan Detachment, but it is unclear whether this is a result of active tectonic movement along this line, or as is more likely to be the case, both are the consequence of this also being the region containing the highest peaks within the Himalayan range.

## REFERENCES

BURBANK, D.W. BECK, R.A. & MULDER, T. (1996a) The Himalayan foreland basin. In: *The Tectonic Evolution of Asia* (Eds Yin, A. & Harrison, T.M.), pp. 149–188. Cambridge University Press, Cambridge.

BURBANK, D.W., LELAND, J., FIELDING, E., *et al.* (1996b) Bedrock incision, rock uplift and threshold hillslopes in the northwestern Himalayas. *Nature*, **379**, 505–510.

BURBANK, D.W. MEIGS, A. & BROZOVIC, N. (1996c) Interactions of growing folds and coeval depositional systems. *Basin Res.*, **8**, 199–223.

BURCHFIEL, B.T., CHEN, Z., HODGES, K.V., *et al.* (1992) The South Tibetan detachment system, Himalayan orogen: extension contemporaneous with and parallel to shortening in a collisional mountain belt. *Geol. Soc. Am. Spec. Pap.*, **269**, 41 pp.

DEWEY, J.F., CANDE, S. & PITMAN, III, W.C. (1989) Tectonic evolution of the India/Eurasia Collision Zone. *Eclogae geol. Helv.*, **82**, 717–734.

EINSELE, G., RATSCHBACHER, L. & WETZEL, A. (1996) The Himalaya–Bengal Fan Denudation–Accumulation System during the past 20 Ma. *J. Geol.*, **104**, 163–184.

ENGLAND, P. & MOLNAR, P. (1990) Surface uplift, uplift of rocks, and exhumation of rocks. *Geology*, **18**, 1173–1177.

FRIEND, P.F. (1998) General form and age of the denudation system of the Himalaya, **GGF**, **120**, 231–236.

GANSSER, A. (1964) *Geology of the Himalayas*. Wiley Interscience, London, 289 pp.

GUPTA, S. (1997) Himalayan drainage patterns and the origin of fluvial megafans in the Ganges foreland basin. *Geology*, **25**, 11–14.

HOLMES, A. (1965). *Principles of Physical Geology*, 2nd edn. Thomas Nelson, London, 1288 pp.

HOWARD, A.D. (1967) Drainage analysis in geologic interpretation: a summation. *Bull. Am. Assoc. petrol. Geol.*, **51**, 2246–2249.

JACKSON, J., NORRIS, R. & YOUNGSON, J. (1996) The structural evolution of active fault and fold systems in central Otago, New Zealand: evidence revealed by drainge patterns. *J. Struct. Geol.*, **18**(2), 17–234.

JONES, N.E. & VINCENT, S.J. (1990) Late stage fluvial systems in mountain belts and foreland basins. *Poster and Abstracts, 13th International Sedimentological Congress*, 26–31 August, Nottingham, p. 113.

KLOOTWIJK, C., GEE, J.S., PEIRCE, J.W., SMITH, G.M. & McFADDEN, P.L. (1992) An early India–Asia contact: paleomagnetic constraints from Ninetyeast Ridge, ODP Leg 121. *Geology*, **20**, 395–398.

MEIGS, A., BURBANK, D.W. & BECK, R.A. (1995) Middle–late Miocene (> 10 Ma) formation of the Main Boundary Thrust in the western Himalaya. *Geology*, **23**, 423–426.

MOHINDRA, R., PARKASH, B. & PRASAD, J. (1992) Historical geomorphology and pedology of the Gandak megafan, middle Gangetic plains, India. *Earth Surf. Process. Landf.*, **17**, 643–662.

RAIVERMAN, V., GANJU, J.L., RAM, J., JOSHI, S., KUNTE, S.V. & MUKHERJEA, A. (1990) *Geological Map of Himalayan Foothills*. Published by the Director, K.D. Malaviya, Institute of Petroleum Exploration, Oil and Natural Gas Corporation, Dehra Dun, India.

RAYMO, M.E. & RUDDIMAN, W.F. (1992) Tectonic forcing of late Cenozoic climate. *Nature*, **359**, 117–122.

RAYMO, M.E., RUDDIMAN, W.F. & FROELICH, P.N. (1988) Influence of late Cenozoic mountain building on ocean geochemical cycles. *Geology*, **15**, 649–653.

RODRÍGUEZ-ITURBE, I. & RINALDO, A. (1997) *Fractal River Basins. Chance and Self-Organisation*. Cambridge University Press, Cambridge, 547 pp.

SCHELLING, D. (1992) The tectonostratigraphy and structure of the Eastern Nepal Himalaya. *Tectonics*, **11**, 925–943.

SEARLE, M.P. (1996) Cooling history, erosion, exhumation and kinematics of the Himalaya–Karakoram–Tibet orogenic belt. In: *The Tectonic Evolution of Asia* (Eds Yin, A. & Harrison, T.M.), pp. 110–137. Cambridge University Press, Cambridge.

SEARLE, M.P., CORFIELD, R.I. STEPHENSON, B. & McCARRON, J. (1997) Structure of the North Indian continental margin in the Ladakh-Zanskar Himalayas: implications for the timing of obduction of the Spontang ophiolite, India–Asia collision and deformation events in the Himalaya. *Geol. Mag.*, **134**, 297–316.

SEEBER, L. & GORNITZ, V. (1983) River profiles along the Himalayan arc as indicators of active tectonics. *Tectonophysics*, **92**, 335–367.

SINGH, H., PARKASH, B. & GOHAIN, K. (1993) Facies analysis of the Kosi megafan deposits. *Sediment. Geol.*, **85**, 87–113.

SINHA, R. & FRIEND, P.F. (1994) River systems and their sediment flux, Indo-Gangetic plains, Northern Bihar, India. *Sedimentology*, **41**, 825–845.

SMITH, A.G., SMITH, D.G. & FUNNELL, B.M. (1994) *Atlas of Mesozoic and Cenozoic Coastlines*. Cambridge University Press, Cambridge, 99 pp.

STRAHLER, A.N. (1952) Hypsometric (area–altitude) analysis of erosional topography. *Geol. Soc. Am. Bull.*, **63**, 1117–1142.

TALLING, P., STEWART, M.D., STARK, C.P., GUPTA, S. & VINCENT, S.J. (1997). Regular spacing of drainage outlets from linear fault-blocks. *Basin Res.*, **9**, 275–302.

VINCENT, S.J. (1993) *Fluvial palaeovalleys in mountain belts: an example from the south central Pyrenees*. Unpublished PhD thesis, University of Liverpool, 407 pp.

WAGER, L.R. (1937) The Arun river drainage pattern and the rise of the Himalaya. *Geogr. J.*, **89**, 239–249.

*Spec. Publs int. Ass. Sediment.* (1990) **28**, 315–329

# Controls on the sedimentology of the November 1996 jökulhlaup deposits, Skeiðarársandur, Iceland

A. J. RUSSELL* *and* Ó. KNUDSEN†

*\*Department of Earth Sciences, Keele University, Keele, Staffordshire ST5 5BG, UK*
*(Email: a.j.russell@keele.ac.uk); and*
*†Klettur Consulting Engineers, Bíldshöfða 12, IS 112 Reykjavík, Iceland*
*(Email: oskar@vedur.is)*

## ABSTRACT

This paper examines controls on the sedimentology of deposits associated with the spectacular November 1996 jökulhlaup (glacial outburst flood, from the Icelandic 'jokul' (glacial) and 'hlaup' (outburst) pronounced 'yow-koul-lauwp') on the proximal zone of the Skeiðarársandur, Iceland, the world's largest active glacial outwash plain. A volcanic eruption beneath the Vatnajökull ice-cap began on 30 September 1996. On exiting the glacier, jökulhlaup flows coalesced into four main outlets. Flood flows initially included large debris lobes that surged from the glacier margin at velocities of up to 6 m s$^{-1}$. The main flood, however, was dominated by watery, turbulent runoff. Sections through ice-proximal fans in both the Gígjukvísl and Súla catchments showed distinctive upward-coarsening sequences indicative of rising-limb deposition. These deposits ranged from horizontally bedded sands, gravels, boulders and debris-rich ice blocks indicative of fluvial deposition, to matrix-supported units suggestive of hyperconcentrated and debris-flow deposition. These proximal, non-backwater-affected deposits were characterized by the simultaneous deposition of ice blocks. Matrix-supported sedimentary successions within a large ice-embayment contain large rip-up clasts, deposited rapidly from a high-energy hyperconcentrated flow into a slackwater zone.

Successions within the large Gígjukvísl proglacial outwash fan indicated waning-stage deposition from a hyperconcentrated flow. Late waning-stage ice-block grounding initiated rapid river-bed scour, resulting in the deposition of structureless gravel units indicative of 'traction carpet' conditions. Deposition into local backwater areas formed deltaic deposits containing sandy foreset beds up to 5 m in height. Successions downstream of a large backwater zone were dominated by trough and planar cross-stratified sands and gravels, deposited in the lee of both individual and clustered icebergs. In places, sections are capped by sandy, climbing-ripple sequences, indicating rapid waning-stage sedimentation. Sedimentary successions on the Gígjukvísl fan reflect both prolonged flow and continuously high sediment supply, even on the waning stage. In contrast, smaller outlets elsewhere experienced flows of shorter duration, where sediment supply was exhausted prior to flow cessation. The sedimentology of the 1996 jökulhlaup deposits was controlled by:

1 the high sediment concentration of flows exiting the glacier;
2 waning-stage sediment reworking;
3 the influence of large-scale channel geometry;
4 ice-block distribution;
5 flow duration.
The results of this study will allow better interpretation of the sedimentary impact of single high-magnitude jökulhlaups within both active glaciofluvial systems and the sedimentary record of former proglacial areas.

## AIMS AND RATIONALE

This paper seeks to account for the sedimentary structures associated with the November 1996 jökulhlaup, Skeiðarársandur, Iceland, by relating sedimentary characteristics to flow conditions observed and documented during the jökulhlaup (Snorrason *et al.*, 1997; Russell *et al.*, in press a; M.T. Guðmundsson, personal communication). This study aims to provide a better understanding of the controls on the characteristics and spatial variability of sedimentary successions deposited by a single, volcanically induced, high-magnitude glacier-outburst flood or 'jökulhlaup'.

Proglacial fluvial sedimentary successions can provide important information about the glacial systems which feed them (Maizels, 1979, 1993b, 1995; Sharp, 1988). Better knowledge of the controls on the sedimentary record of active proglacial outwash deposits is essential for improving our understanding of ancient glacial meltwater systems. Over recent years, increasing attention has focused on the role that jökulhlaups play in creating distinctive sedimentary successions in present-day proglacial environments (Maizels, 1989a,b, 1991, 1993a, 1995; Maizels & Russell, 1992; Russell, 1993; Russell & Marren, 1999). Owing to the paucity of data before, during and after single high-magnitude floods in active meltwater systems, most models of flood sedimentation have relied upon the dual approach of interpreting the sedimentary record and reconstructing palaeoflow parameters (Maizels, 1989a,b, 1991, 1993a). Indeed, the identification of individual flood units has, in many studies, been based upon interpretation of the sedimentary record, rather than on knowledge of the actual sedimentary succession associated with an individual jökulhlaup. So far, despite numerous historical accounts, hydrological studies and detailed reconstructions of jökulhlaups in Iceland, little specific information is available concerning the sedimentary record of a single 'observed' high-magnitude jökulhlaup where there is good pre- and post-jökulhlaup geomorphological data.

Jökulhlaups are known to have drained across the Skeiðarársandur, the world's largest active proglacial outwash plain, on a regular basis at least over the last few centuries (Churski, 1973; Thorarinsson, 1974; Björnsson, 1988, 1992; Guðmundsson *et al.*, 1995), and various aspects of their impact have been noted by Klimek (1972, 1973), Galon (1973) Nummedal *et al.* (1987), Boothroyd & Nummedal (1978), Maizels (1991, 1993b) and Russell & Marren (1999). Despite these previous studies, little specific information exists concerning the geomorphological or sedimentary impact of a single high-magnitude jökulhlaup. The November 1996 jökulhlaup therefore provides us with an excellent opportunity to examine the controls on the characteristics and spatial variability of deposits associated with a single high-magnitude event within the Skeiðarársandur.

## THE NOVEMBER 1996 JÖKULHLAUP

A volcanic eruption beneath the Vatnajökull ice-cap began on 30 September 1996 (Guðmundsson *et al.*,

1997). Over the next month, 3.8 km$^3$ of meltwater travelled subglacially into the Grimsvötn caldera subglacial lake until it reached a critical level for drainage (Björnsson, 1997). The jökulhlaup began on the most easterly outlet river, the Skeiðará, at 0730 hours on 5 November and reached a peak discharge of 45 000 m$^3$ s$^{-1}$ within 14 h, making this the shortest rising limb for any jökulhlaup recorded from the Grimsvötn caldera (Björnsson, 1997). The jökulhlaup also burst from single-conduit outlets and ice-margin-parallel crevasses up to 2 km in length (Snorrason *et al.*, 1997; Russell *et al.*, 1997a,b) (Fig. 1). Flood flows initially included large debris lobes that surged from the glacier margin at velocities up to 6 m s$^{-1}$. Flows began in the Gígjukvísl, further west (Fig. 1) at around 1015 hours on 5 November and reached the Súla bridge, farthest to the west, at 1613 hours on 5 November (Snorrason *et al.*, 1997) (Fig. 1). Flow in the Súla had the shortest duration and was observed from aerial photographs to have waned considerably by 1200 hours on 6 November, about 32 h later. The jökulhlaup had its longest duration (about 48 h) in the Gígjukvísl, with flows persisting well into 7 November (Snorrason *et al.*, 1997) (Fig. 1). The Skeiðeraa flow lasted 33 h. Specific flood powers are estimated to have been as high as 40 000 W m$^{-2}$ at the main Gígjukvísl outlet, decreasing to 8000–16 000 W m$^{-2}$ at the moraine constriction on the Gígjukvísl and 400–900 W m$^{-2}$ a further 2 km downstream (Russell *et al.*, 1997a,b). After the passage of the initial, debris-rich flood wave, flow was dominated by more watery, turbulent runoff. At flood peak, 60–100 × 10$^6$ m$^3$ of water was stored temporarily in a backwater lake, upstream of the Gígjukvísl moraine constriction (Russell *et al.*, 1997a,b) (Fig. 1).

## METHODS

Pre-jökulhlaup geomorphological mapping and sedimentological studies were undertaken in July, August and October 1996. Post-jökulhlaup sedimentological work was carried out during November 1996 and between May and August 1997. Bulk sediment samples of approximately 10 kg were taken to the Icelandic Roads Authority for sieving up to 50 mm (very coarse cobble). We used aerial photographs taken pre-flood (1991, 1992), during the flood (1200 hours, 6 November 1996) and after the flood (12 August 1997). Additional information was acquired from oblique photographs and video coverage taken during and after the jökulhlaup.

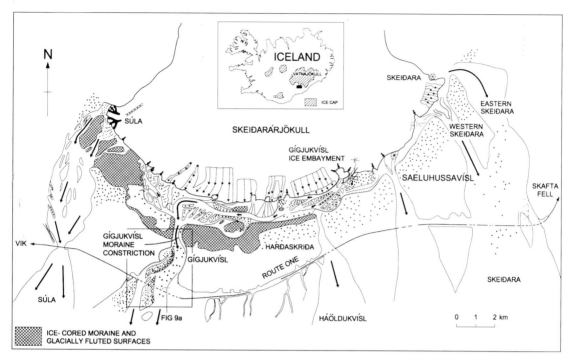

**Fig. 1.** Location map showing the location of the field area within Iceland and the main jökulhlaup outlets and channels examined. Note the different types of jökulhlaup outlets including: conduits, fountains, crevasses and large ice embayments.

## VERTICAL SEDIMENTARY SECTIONS

Vertical sedimentary sections are described and interpreted in a variety of locations across the proximal area of the Skeiðarársandur (Fig. 2). Section selection reflects both availability, local representativeness and the presence of pre- and post-jökulhlaup geomorphological data, allowing November 1996 jökulhlaup deposits to be located. Figure 2 illustrates some of the main sedimentary characteristics described below and shows how the sections discussed relate to the glacier margin, main jökulhlaup routeways and proglacial topography.

### Dissected outwash fans on the Súla and proximal Gígjukvísl

Distinctive ice-contact fans and depositional sheets deposited by the jökulhlaup in the Súla and proximal Gígjukvísl were dissected by waning-stage flows, revealing a range of inversely graded, normally graded and inverse–normally graded successions (Figs 2A & B & 3A–C). Laterally continuous, crudely bedded, upward-coarsening successions consist of openwork and sandy matrix-supported cobbles and gravels (Figs 2B & 3A). Cobble and pebble clasts are commonly held within a sandy matrix, giving a bimodal size distribution (Fig. 3A). Matrix content increases towards the top of these successions (Fig. 3A). At one location, the base of an inversely graded unit consists of well-sorted, openwork gravel layers (Fig. 3A). Ice blocks, up to 6 m in width, are commonly found wholly or partially buried within the fan sediment (Fig. 3B). Deposits are crudely bedded, with clasts occasionally showing *a*-axis parallel imbrication and clustering. The largest clasts on the fan surfaces show imbrication, although there is no surface-armoured layer or lag deposit. Imbrication clusters are composed occasionally of both ice and bedrock clasts. Thin sheets of frozen river-bed gravels are found deposited against obstacles, forming cluster bedforms. Some fan surfaces coalesce, suggesting possible overlap or interfingering of successions from different jökulhlaup outlets.

The sediments within these fans are interpreted as having been deposited during the rising stage of the jökulhlaup from a sediment-rich flow. Upward-coarsening successions indicate a progressive increase in coarse-sediment transport capacity on the rising-flow stage as sheets of material were deposited rapidly (Maizels, 1991, 1993a). The absence of well-defined stratification and the matrix-supported nature of much of

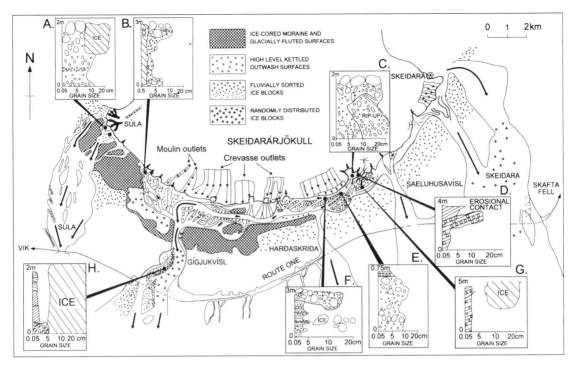

**Fig. 2.** Map of the proximal area of the Skeiðarársandur showing the location of sedimentary sections discussed in the text and their main characteristics (A–H). The relationship between each sedimentary succession and wholly or partly buried ice blocks is indicated.

these successions suggests deposition from hypercon-centrated flows (Costa, 1988; Maizels, 1989a,b; Kehew, 1993). This interpretation is supported by observations and video footage of the initiation of the jökulhlaup, when sediment-rich lobate flow-fronts were observed advanc-ing from the river channel (Snorrason *et al.*, 1997). Openwork, occasionally inversely graded, units of sedi-ment of up to cobble size are compatible with deposition from a highly concentrated, coarse-grained, traction layer, where non-Newtonian intertial forces were oper-ating to keep clasts suspended (Costa, 1988; Maizels, 1989a,b; Todd, 1989). Sedimentation was also rapid enough to bury grounded ice blocks completely (cf. Klimek, 1972) on the Skeiðarársandur. Jökulhlaup dura-tion in the Súla river, as a whole, was only of the order of 8–12 h at this location (Snorrason *et al.*, 1997), but could have been much shorter for these smaller tributaries. The ice-contact fans were rapidly incised by flows on the late-rising or waning stage, possibly the result of sediment exhaustion. Icebergs buried by earlier rising-stage flows were exhumed, resulting in the creation of distinctive ice-block obstacle marks (Russell, 1993).

## Sedimentation within an ice embayment

Vertical sedimentary sections were exposed in an ice-contact deposit, contained within a spectacular embay-ment, created by progressive ice-tunnel roof collapse during the jökulhlaup (Russell *et al.*, 1997a,b) (Figs 1, 2C & 4A). The embayment consists of two successively occupied jökulhlaup tunnel outlets draining into the Gígjukvísl system (Figs 1 & 4A). Deposits within the eastern outlet are raised some 10 m above those of the western outlet and are arranged as a staircase of ter-races sloping gently away from the embayment mouth. Bedding within the terraces consists of uniform, large-scale cross-strata, dipping at about 15° away from the embayment mouth. Terrace deposition is in the lee of an iceberg 25 m wide, which occupies the embayment mouth. Sections were found at the mouth of, and near to, the embayment backwall (Figs 2C & D & 4A–C).

The embayment mouth section consists of a matrix-supported unit containing sediment ranging from sand grains to boulders, deposited together with a 3-m-diameter rip-up clast composed of stratified sands and gravels (Figs 2C & 4C). The surrounding sand matrix-supported unit is almost completely massive, containing occasional signs of clast imbrication and clustering (Fig. 4C). Numerous clasts have their *a*-axes oriented vertically (Fig. 4B). The backwall section coarsens upwards, displays better sorting than found at the embay-ment mouth and is dominated by large-scale cross-stratification dipping towards the embayment backwall (Figs 2D & 4B). Cross-strata are composed of relatively well-sorted sands and gravels with occasional cobble-size strata (Figs 2D & 4B). Low-angle cross-strata are trun-cated by a series of higher angle strata in the region nearest to the embayment backwall (Fig. 4B).

The near-structureless and matrix-supported nature of the embayment-mouth succession suggests rapid deposi-tion from a high-energy hyperconcentrated flow capable of transporting large rip-up clasts (Lord & Kehew, 1987; Kehew & Lord, 1987; Costa, 1988; Maizels, 1989a,b; Kehew, 1993). Rip-up clasts are commonly found in depositional environments where subglacial material is eroded by meltwater (Knudsen, 1995). The cross-stratified backwall section represents the progressive deposition of sheets of sediment during the advance of a channel macroform, which appears to be graded to a former water level sloping away from the embayment mouth. Increased dip angles of cross-strata near the embayment backwall suggest reworking, either by water exiting a smaller meltwater tunnel outlet or by a circu-latory flow along the embayment margin transverse to the main flow direction.

It is clear from observations during the jökulhlaup that flows persisted in the western embayment through the entire waning stage at least until 7 November, whereas the upper surfaces of the two eastern embayment sedi-ment sections had emerged above the flood waters by 6 November (M.T. Guðmundsson, personal communica-tion). The fact that flows from the western embayment were clearly dominant suggests that water in the east-ern embayment was hydraulically ponded, resulting in slackwater conditions. This interpretation is supported by the fact that the eastern-embayment-mouth section reflects deposition under higher energy flows than those at the back of the eastern embayment. Such slackwater conditions, resulting in the formation of 'eddy' bars are well documented for high-magnitude floods (Baker & Kochel, 1988a,b; O'Connor, 1993). Observations suggest that the slackwater conditions must have prevailed during

**Fig. 3.** (*opposite*) (A) Openwork units at the base of the inverse-normally graded succession. Ice axe is 0.75 m in length. (B) Ice blocks buried within ice-contact proglacial fan, exposed by waning-stage incision (Fig. 2A). Ice axe is 30 cm across. (C) Upward-coarsening unit in ice-contact jökulhlaup fan on tributary of the Súla river, western margin of Skeiðarájökull (Fig. 2B). Note the increasing matrix content towards the section surface and the complete matrix-supported nature of the upper part of the section. Ice axe is 0.75 m long.

**Fig. 4.** (A) Oblique aerial photograph of the main outlet of the Gígjukvísl, viewed from south-east to north-west, showing the ice embayment and the flow expansion bar picked out by the patterns of ice-block deposition. The photograph was taken during the jökulhlaup waning-stage on 7 November 1996. Note waning-stage flows exiting western embayment. Sedimentary sections in (B) and (C) are at the mouth and to the rear of the eastern embayment. View spans 2 km at ice margin. (B) Distal section, near eastern embayment backwall, showing cross-stratification with distal increase in strata dip angle (Fig. 2D). (C) Proximal section in eastern embayment showing stratified rip-up clast within a massive, poorly sorted matrix-supported unit (Fig. 2C). Note 6-cm-long lighter for scale.

the period between the formation of the western embayment, after *c*. 1600 hours on 5 November and the morning of 6 November, when the slackwater conditions had subsided.

### Shallow sections on a large proglacial outwash fan

Several sections 1–3 m high were examined on the surface of a large 1.5 × 1 km outwash fan that was deposited in front of the double-embayment outlet of the Gígjukvísl (Russell *et al.*, 1997a,b) (Figs 2E & F, 4A & 5A).

A kettle-hole wall exposure, 700 m from the double embayment, revealed horizontally bedded, matrix-supported units, which display very poor sorting (Fig. 5A & B). These matrix-rich units are very densely compacted in comparison with other jökulhlaup sediments. Massive

units of 0.2–0.5 m thickness are capped by thinner units, displaying signs of small-scale channelization, marked by stringers of pebbles (Fig. 5B). These upper units have a lower matrix content and contain occasional clast-supported pebble clusters. Bulk grain-size analysis shows the deposits to be very poorly sorted and to contain 2–3% of silt size material or finer (Fig. 6A & B). Similarly massive, matrix-supported deposits are found in the proximal areas of the western Skeiðará channel, both *in situ*, with 2–3% silt-size material or finer, and as reworked slabs containing over 12% silt-size material and finer (Fig. 6C & D).

These deposits are interpreted as the products of hyperconcentrated flow pulses, which account for the matrix-rich units, subsequently capped by deposits associated with more fluidal, turbulent activity (Nemec & Steel,

**Fig. 5.** (A) Oblique aerial photograph of the main Gígjukvísl jökulhlaup fan, viewed from south to north, showing the location of kettle-hole and tributary outlet sections. The photograph was taken during the jökulhlaup waning stage on 7 November 1996. Note waning-stage flow obliquely across the tributary outlet section and the flow around numerous stranded ice blocks. The channel is 1 km wide at this location. Flow is from right to left. (B) Small section showing massive, matrix-supported, polymodal deposit. Note weak channelization, picked out by pebble stringers set within a dense matrix and openwork lenses at higher levels within the deposit (Fig. 2E). Note compass-clinometer for scale.

**Fig. 6.** Cumulative grain-size curves for the coarse-sediment fraction of sieved bulk samples. (A) & (B) Kettle-hole sections (Figs 2E & 5B); (C) & (D) Skeiðará units; (E) well-sorted gravelly sands at bottom of tributary section (Figs 2F & 7A–C); (F) massive clast-rich coarse gravel unit at surface of tributary section (Figs 2F & 7A–C); (G) waning-stage silty sands deposited on bar top within Gígjukvísl backwater zone.

**Fig. 7.** (A) Section, viewed from south to north, cut by flows exiting a nearby subglacial tunnel through sediment deposited by flow from the main ice embayment. Note the lateral continuity of sedimentary units, sedimentation around large ice blocks and the jeep for scale. Flow direction is from far right to middle left, with localized late-waning-stage flows towards the viewer. (B) Erosional contact between lower well-sorted sands and upper clast-rich unit, illustrating their contrasting characteristics. Lower well-sorted sand units contained clear ice blocks and occasional gravelly openwork lenses. Note knife for scale. (C) Trough eroded into lower unit and filled with cobble-sized material displaying crude cross-stratification. Erosional contact separates lower horizontally bedded gravels from a coarser grained, cross-stratified boulder infill. This structure may be the infill of an ice-block scour hollow after the re-entrainment of the ice block.

1984). Klimek (1972) attributed apparently similar sedimentary sequences on the Skeiðarársandur to deposition by a shallow rapid water flow. The fact that these sediments were deposited around stranded icebergs suggest deposition from waning-stage flows, which were observed to display signs of large-scale turbulence in the form of standing waves and flow separation around obstacles (M.T. Guðmundsson, personal communication). This suggests that with the exception of the western Skeiðará channel sediments, these deposits are unlikely to be the product of debris flows. It is possible, however, that the deposits of the lobate jökulhlaup flow fronts reflect high enough sediment concentrations to produce debris-flow conditions.

A section 200 m long and 4 m high, found more distally on the proglacial fan, was excavated by waning stage flows exiting a nearby conduit (Figs 2F & 5A). This section consists of a generally upward-coarsening sequence of sands and gravels with occasional cobble- and boulder-size clasts (Figs 2F & 7A). Deposition took place around large icebergs, up to 8 m in height. The lower part of the section contains well-sorted, horizontally bedded sands and gravels with lenses of openwork gravel interbedded with sand laminae (Fig. 7B). A block of clear ice 0.5 m in diameter was found deposited within this unit. The lower, well-sorted unit is capped by several cobble-rich units 0.3–0.7 m thick containing material of up to boulder size (Fig. 7B). The base of cobble-rich units is bounded by a sharp erosional contact mainly at a depth of 0.5 m, but locally increasing to approximately 2.5 m where it defines a channel-shaped trough 4–6 m in width (Fig. 7C). The upper units pinch out and interfinger with each other

**Fig. 8.** Large-scale sandy cross-strata deposited into slackwater, hydraulically ponded by major jökulhlaup outflow across the axis of the main proglacial trench. Foreset beds support ripple-drift cross-stratification. This type of deposit is probably superimposed upon a coarser-grained outwash surface, deposited by earlier rising-stage flows. A shallow trough-shaped erosional contact reflects occasional reworking of foreset fronts by flows transverse to the foreset-dip direction.

along the section. The upper units are massive, showing few signs of imbrication and clustering, where the clasts are supported by a matrix of poorly sorted coarse sand and fine gravel (Fig. 7B). At one location, ill-defined, large-scale cross-strata fill a trough-shaped hollow with sediment of up to boulder size (Fig. 7C). The coarse-grained unit is capped in places by up to 5 cm of horizontally bedded sands.

The lower unit is interpreted as reflecting deposition of laterally extensive, sheet-like sand and gravel waves on the waning-flow stage, sufficiently rapidly to allow the complete burial of small ice blocks and the partial burial of large ice blocks. It is clear from bulk grain-size analysis that all of this material could have been reworked from the very poorly sorted, massive, matrix-supported units exposed more proximally in the kettle-hole described earlier (Fig. 6A, B & E). Similarly, the upper massive coarse-grained unit also could be derived from the reworking of more proximal fan sediments (Fig. 6A, B & F). The massive nature of the upper coarse-grained units, together with their limited thickness, suggest deposition by instantaneous freezing of a coarse-grained clast-rich dispersion or 'traction carpet', where grains are supported by a dispersive pressure or inertia generated through grain collisions (Bagnold, 1954; Costa, 1988; Todd, 1989; Maizels, 1989a). The stacked nature of these coarse-grained traction carpet units suggests several pulses of bedload transport, initiated either by rapid scouring around grounded ice blocks (Russell, 1993) or by bed-armouring processes (Gomez, 1983). The section is located 200 m downstream of a group of 10 large icebergs with well-developed obstacle marks. It is suggested that rapid scour around these newly grounded icebergs generated

the sediment pulses responsible for the bedload pulses. The marked erosional contact, defining a channel-fill, may have been created by scour around an ice block, stranded temporarily on the waning-flow stage. After the ice block was re-entrained, the scour hollow filled rapidly with coarse-grained sediment. These processes are described by Russell (1993), who observed ice-block transport and grounding processes during a jökulhlaup in Greenland.

### Sediments deposited within backwater zones

Backwater conditions observed in various parts of the Skeiðarársandur were created mainly where ice-cored moraine ridges or blocks of older sandur sediment acted as a barrier to the free flow of the jökulhlaup from the glacier margin (Russell *et al.*, 1997a,b) (Figs 2 & 4A). Deposition into relatively calm bodies of temporarily stored water resulted in the formation of deltaic deposits with relief of up to 6 m. These deposits commonly consist of well-sorted sands containing scattered gravel clasts. Sandy foreset beds up to 5 m in height were deposited into backwater conditions upstream of the main Gígjukvísl embayment outlet (Figs 2A & 8). Individual beds contain climbing-ripple sequences deposited as sediment was being transported down the 10° inclined foreset slopes (Fig. 8). In places, horizontally laminated sands were deposited as bottom-sets in front of advancing coarser grained delta fronts. Upstream of the Gígjukvísl moraine constriction, high-relief bars were generated as waning-stage erosion incised higher stage, backwater-graded deposits. Backwater sediments appear to represent deposition from suspension, indicating the overwhelming

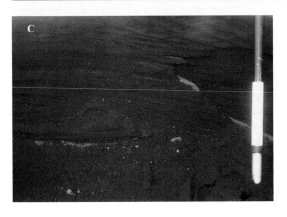

**Fig. 9.** (A) Aerial photograph (location in Fig. 1) of the Gígjukvísl downstream of the moraine constriction, viewed from south to north, showing rapid flow expansion between the moraine gap and the former bridge location. Photograph was taken at noon on 6 November on the waning jökulhlaup stage. Note large-scale bar pattern, reflected by the distribution of stranded ice blocks. Ice blocks occasionally form dense clusters, which take on a longitudinal pattern, separated by chute channels. Flow direction is from north to south, with standing waves clearly visible within the main channel near the eastern river bank. Copyright retained by Landmaelingar Islands. (B) Trough cross-stratified sands deposited in the lee of clusters of stranded ice blocks. (C) Ripple-drift cross-lamination showing rapid river bed aggradation on the waning flow stage.

dominance of sand-size sediment in transport during the main flood flow. Waning-stage fine sediment drapes across bar surfaces in the Gígjukvísl slackwater zone are composed of 45% silt-size material or finer (Fig. 6G).

### Gígjukvísl downstream of moraine constriction

Numerous small (< 2 m high) sedimentary sections were observed in the Gígjukvísl channel, downstream of the moraine constriction, immediately after the jökulhlaup on 8 November 1996. Channel-scale morphology is dominated by a large expansion bar of erosional origin (Figs 1, 2 & 9A). Numerous icebergs are deposited on this bar surface, forming locally very dense clusters (Fig. 9A). In places, the largest icebergs were linked by rafts of crushed ice, which appear to mark the local water surface during the waning-flow stage (Fig. 9A). Aerial photographs taken at noon on 6 November showed large sections of the highest bar surface exposed early on the waning-flow stage (Fig. 9A).

Sections are dominated by trough and planar cross-stratified sands and gravels deposited in the lee of both individual and clustered icebergs (Scholz *et al.*, 1988). Individual cross-strata sets are separated by numerous erosional contacts (Figs 2H & 9B & C). In places, sections are capped by sandy climbing-ripple sequences, indicating rapid waning-stage sedimentation. Cross-strata generally dip in a down-channel direction, diagonally across the main bar surface, but occasionally show backset bedding in the immediate wake of ice blocks (Gustavson, 1974) (Fig. 9B & C). These sedimentary structures reflect complex flow patterns around numerous, closely spaced ice blocks, inhibiting the formation of classic obstacle marks (Karcz, 1968; Richardson, 1968).

### DISCUSSION AND CONCLUSIONS

From the description and interpretation of individual sedimentary successions, several factors can be identified as major controls on the sedimentary characteristics of the 1996 jökulhlaup deposits.

### High sediment concentration of flows exiting the glacier

High sediment concentrations observed, monitored (Snorrason *et al.*, 1997) and inferred from geomorphological and hydraulic evidence, as the November 1996 jökulhlaup exited the glacier, are likely to have had a major influence on the sedimentary characteristics of ice-contact proglacial outwash fans. Poorly stratified, occa-

sionally massive, successions, comprising the bulk of the fan-like ice-contact jökulhlaup deposits (Figs 2A, B & 3C), are also compatible with rising-stage hyperconcentrated flow conditions. At the Gígjukvísl ice embayment, however, waning-stage flows were hyperconcentrated, because of continued reworking of subglacial sediment as conduit routing changed (Figs 2C & D, 5B & 7B). Although hyperconcentrated jökulhlaup flows have been inferred from the sedimentary record of other Icelandic outwash plains (Jonsson, 1982; Maizels, 1989a,b, 1991), this is the first time that the deposits of non-Newtonian flows have been identified on the Skeiðarársandur (Klimek, 1972; Galon, 1973; Boothroyd & Nummedal, 1978; Maizels, 1991).

### Waning-stage sediment reworking

This study emphasizes the importance of within-jökulhlaup sediment reworking for the creation of distinctive sedimentary units and landforms. Many rising-stage ice-contact proglacial fans were deeply incised during the later stages of the jökulhlaup, probably on the waning-flow stage (Fig. 3B). In some locations, such as the fan in front of the Gígjukvísl ice embayment, deposition occurred throughout the waning stage, punctuated by localized bed scour around stranded ice blocks (Figs 4A & 5A). The sudden re-entrainment of large amounts of sediment during these scour events generated clast-rich, hyperconcentrated, grain dispersions, depositing distinctive massive, poorly sorted units (Fig. 7B). The presence of an armoured surface layer over some proglacial fan surfaces suggests waning-stage winnowing, which may account for some of the grain-size differences observed between individual units (Figs 2F & 7A–C). Caution therefore must be exercised when using laterally extensive erosional contacts and bulk grain-size characteristics to identify individual jökulhlaups in the sedimentary record.

### Channel geometry and jökulhlaup sedimentation

Channel geometry played a crucial role in determining both the morphological and internal sedimentary characteristics of the November 1996 jökulhlaup deposits, similar to many other flood studies (Baker, 1973, 1984; Baker & Kochel 1988a,b; Russell & Marren, 1999). Backwater conditions prevailed both within the ice-walled Gígjukvísl outlet channel and within proglacial areas impounded by ice-cored moraine ridges and older outwash surfaces (Fig. 2). Major backwater ponding was noted for several proximal channel locations within the first few hours of the jökulhlaup, before the flood eroded various flow constrictions (M.T. Guðmundsson, personal communication). The backwater effects acted to reduce the transport of coarse-grained sediment from the glacier margin. As local backwater controls were removed during the jökulhlaup, there was a greater potential for earlier rising-stage sediments to be reworked and transported further down the fluvial system. Several cycles of storage and release can be identified for the Gígjukvísl channel system during the November 1996 jökulhlaup, as various backwater lakes were created and destroyed. Large-scale depositional morphology is strongly controlled by both the pre-existing geometry of the flood routeway and the creation of new channel geometries in the much-widened proglacial channels and the more complex newly created ice embayment.

### Ice blocks

Ice blocks created both distinctive bedforms and influenced the pattern and characteristics of sedimentation. Numerous ice blocks, ranging from 1 to 25 m in height, were deposited simultaneously with sediment within rising-stage deposits (Figs 2A & B & 3B). Some of the largest ice blocks were partially or completely exhumed by waning-stage flows, creating distinctive obstacle marks, whereas smaller, 2–3 m high, ice blocks remained buried and were able to melt out more slowly, producing kettle-holes. The depth of kettle-holes, resulting from the melting of completely buried ice blocks provides a minimum thickness of a jökulhlaup deposit. This may be a very useful tool for the reconstruction of the depositional impacts of previous jökulhlaups, both on the Skeiðarársandur and elsewhere. Ice blocks grounded at high-flow stage, controlled local patterns of early-waning stage deposition, by acting as a focus for deposition of finer-grained sediment (Fig. 9A). The largest ice blocks released during this jökulhlaup were deposited on the proglacial fan in front of the main Gígjukvísl outlet, forming elongate clusters, extending for distances of up to 1 km (Figs 1, 2 & 4A). Major flow separation around such clusters governed patterns of waning-stage erosion and deposition (Figs 4A & 5A). Scour around ice blocks deposited on the waning-flow stage provided localized pulses of coarse sediment, which are represented as individual sedimentary units (Figs 2F & 7B & C).

### Flow duration

The duration of the November 1996 jökulhlaup flows varied considerably among different outlets along the glacier margin (Snorrason *et al.*, 1997). Such variation is crucial in explaining variations in the preservation potential of both rising- and waning-stage successions along

the glacier margin. Short-duration flows on the western side of the Skeiðarárjökull resulted in the preservation of generally fine-grained rising-stage successions with a local supply of coarse sediment (Figs 2A & B & 3C). Waning-stage flows at these locations, however, have been mainly erosive, suggesting an exhaustion of sediment supply (Fig. 3B). Larger conduit outlets on the eastern side of Skeiðarárjökull show similar signs of waning-stage reworking of rising-stage sediments. In contrast, sedimentation from the Gígjukvísl outlets appears to have taken place both on the rising- and falling-flow stages, suggesting a more continuous subglacial sediment supply. The occurrence of rip-up clasts of turfs, till and glaciofluvial sediment across proximal areas of the sandur may provide a guide as to the amount of subglacial erosional activity and tunnel enlargement taking place through quarrying of subglacial sediment. The absence of rip-up clasts may suggest that flows were very short-lived, reflecting the initial passage of a sediment-laden flood wave through the glacier, prior to any enlargement of tunnels by removal of subglacial sediment. Rip-up clasts deposited at the mouth of the eastern embayment indicate active tunnel erosion at flood peak or even during waning flow. The use of rip-up clasts to indicate subglacial processes, however, may be complicated by the need to distinguish sediment-rich river icings (Collinson, 1971) from rip-up clasts (Krainer & Poscher, 1990; Knudsen, 1995) within the sedimentary record.

### Controls on the sedimentary characteristics of a high-magnitude jökulhlaup

Multiple jökulhlaup outlets and routeways, together with variable flow duration and sediment supply, produced a much wider range of flow rheologies, depositional settings and sedimentary characteristics than attributed previously to a single Icelandic jökulhlaup. A simple model is presented, illustrating idealized sedimentary characteristics for proglacial areas, with and without confining proglacial moraines (Fig. 10A–F). Jökulhlaup flows unconfined by proglacial moraine ridges and high-level palaeo-outwash surfaces will deposit fans comprising upward-coarsening successions, containing numerous completely and partially buried ice blocks and rip-up clasts (Fig. 10A & B). Deposited ice blocks will also show clear radial down-fan fining (Fig. 10). If flows wane rapidly, or sediment supply becomes limited during the jökulhlaup, fan incision may exhume buried ice blocks and rip-up clasts (Fig. 10A). Where flows and sediment supply persist on the waning stage, however, fan incision will be inhibited and rising stage deposits will be capped by better sorted waning stage deposits (Fig. 10B). The

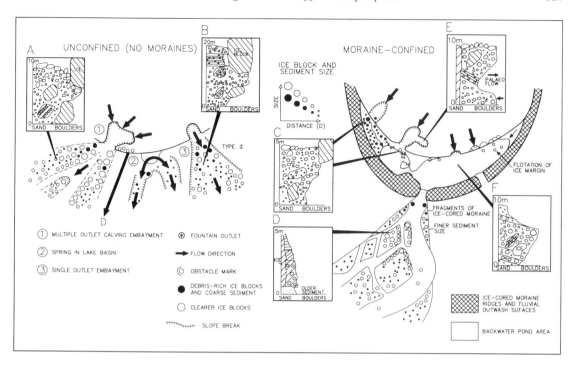

**Fig. 10.** Model showing how vertical sedimentary characteristics might vary depending on whether jökulhlaup waters were confined by moraine ridges (C–F) or were free to spread out from the ice-margin (A and B). Note increased spatial variability of expected sedimentary successions, associated with the moraine-confined conditions, where backwater conditions act as a major control on sedimentation at different times during the flood. Expected unconfined jökulhlaup successions are divided into locations where waning-stage sediment flux decreases (A) and where sediment flux remains high, even on the waning-stage flow (B). In the moraine-confined scenario, backwater conditions result in much finer grained deposits downstream of the backwater zone (D). Sedimentary successions associated with ice-contact proglacial fans also may vary depending upon degree of backwater ponding or whether deposition is into shallow fast flows (C) or into sluggish deep flows (F). Successions within the ice-embayment are expected to reflect a 180° change in flow direction from flow out of an embayment to slackwater flow into the embayment from a new tunnel outlet.

presence of proglacial moraine ridges and higher level abandoned outwash surfaces generates major backwater ponding during the jökulhlaup, resulting in turn, in a much wider range of sedimentary successions than for unconfined jökulhlaup flow (compare Fig. 10A & B with 10C–F). Sedimentation upstream of the backwater zone produces fans with similar sedimentary successions to those in Fig. 10A. Ice-margin parallel flow at major outlets, however, may result in fans with a shape that will be more strongly controlled by proglacial topography (Fig. 10C). Deposition into the backwater lake will take the form of either deltaic or fan deposits, depending upon whether sediments are graded to backwater lake levels (Fig. 10F). Non-backwater-graded fans will contain fewer buried ice blocks, owing to their increased tendency to float in greater water depths. Patterns of ice-block deposition within the backwater-influenced zone are also likely to be more random, owing to lower levels of fluvial action. The backwater zone will act as a sedi-

ment trap, reducing downstream grain sizes. Deposition downstream of the backwater zone is, therefore, more likely to be dominated by cross-stratified sands and gravels, controlled by complex flow patterns around stranded ice blocks (Fig. 10D). The backwater zone may increase the supply of ice blocks when jökulhlaup depths are sufficient to float sections of the glacier margin. The development of distinctive embayments into the ice margin provides suitable locations for high-energy, ice-contact deposition, resulting in sedimentary successions with complex palaeoflow patterns. This simple model may assist in the interpretation of the sedimentary record of the Skeiðarársandur and other active or relict proglacial areas. As volcanic activity in the Vatnajökull area is likely to increase over the next few decades (Larsen & Guðmundsson, 1997), further high-magnitude jökulhlaups are likely. This study therefore may help predict the sedimentary impact of jökulhlaups within active and relict proglacial regions world-wide.

## ACKNOWLEDGEMENTS

Field-work during the summer of 1996 was carried out as part of the Keele University Iceland Expedition of 1996. Research visits immediately before and after the jökulhlaup were funded by a NERC (National Environmental Research Council) grant awarded to A.J.R. (GR3/10960) in October 1996. Field-work during 1997 was funded by grants awarded to Ó.K. by the Icelandic Roads Authority and the Research Council of Iceland. Dr Judith Maizels, Philip Marren, Dolf de Jong and Helen Fay are thanked for valuable assistance and discussion in the field. Dr Magnus T. Guðmundsson provided us with invaluable discussion and observations, made at intervals during the jökulhlaup. Within Keele University's Department of Earth Sciences, Richard Burgess is thanked for photographic reproduction and Peter Greatbach, David Wilde and Terry Doyle are thanked for their cartography. We gratefully acknowledge the constructive criticism of our reviewers, Drs Basil Gomez and Frank Magillian.

## REFERENCES

BAGNOLD, R.A. (1954) Experiments on a gravity-free dispersion of large solid spheres in a Newtonian fluid under shear. *Proc. R. Soc. London Ser. A*, **255**, 49–63.

BAKER, V.R. (1973) Paleohydrology and sedimentology of lake Missoula flooding in Eastern Washington. *Geol. Soc. Am. Spec. Pap.*, **144**, 1–79 pp.

BAKER, V.R. (1984) Flood sedimentation in bedrock fluvial systems. In: *Sedimentology of Gravels and Conglomerates* (Eds Koster, E.H. & Steel, R.J.), Mem. Can. Soc. petrol. Geol., Calgary, **10**, 87–98.

BAKER, V.R. & KOCHEL, R.C. (1988a) Paleoflood analysis using slackwater deposits. In: *Flood Geomorphology* (Eds Baker, V.R., Kochel, R.C. & Patton, P.C.), pp. 357–376. John Wiley & Sons, Chichester.

BAKER, V.R. & KOCHEL, R.C. (1988b) Flood sedimentation in bedrock fluvial systems. In: *Flood Geomorphology* (Eds Baker, V.R., Kochel, R.C. & Patton, P.C.), pp. 123–137. John Wiley & Sons, Chichester.

BJÖRNSSON, H. (1988) *Hydrology of Ice Caps in Volcanic Regions*. Societas Scientarium Islandica, University of Iceland, Reykyavik, 139 pp.

BJÖRNSSON, H. (1992) Jökulhlaups in Iceland: prediction, characteristics and simulation. *Ann. Glaciol.*, **16**, 95–106.

BJÖRNSSON, H. (1997) Grímsvatnahlaup Fyrr og Nú. In: *Vatnajökull: Gos og hlaup* (Ed. Haraldsson, H.), pp. 61–77. Vegagerðin, Reykjavik.

BOOTHROYD, J.C. & NUMMEDAL, D. (1978) Proglacial braided outwash: a model for humid alluvial-fan deposits. In: *Fluvial Sedimentology* (Ed. Miall, A.D.), Mem. Can. Soc. petrol. Geol., Calgary, **5**, 641–668.

COLLINSON, J.D. (1971) Bedforms of the Tana River, Norway. *Geogr. Annaler*, **52A**, 31–56.

COSTA, J.E. (1988) Rheologic, geomorphic, and sedimentologic differentiation of water floods, hyperconcentrated flows, and debris flows. In: *Flood Geomorphology* (Eds Baker, V.R., Kochel, K.C. & Patton, P.C.), pp. 113–122. John Wiley & Sons, Chichester.

CHURSKI, Z. (1973) Hydrographic features of the proglacial area of Skeiðarárjökull. *Geogr. Pol.*, **26**, 209–254.

GALON, R. (1973) Geomorphological and geological analysis of the proglacial area of Skeiðarárjökull: central section. *Geogr. Pol.*, **26**, 15–56.

GOMEZ, B. (1983) Temporal variations in particle size distribution of surficial bed material: the effect of progressive bed armouring. *Geogr. Annaler*, **65A**(3–4), 183–192.

GUÐMUNDSSON, M.T., BJÖRNSSON, H. & PALLSSON, F. (1995) Changes in jökulhlaup sizes in Grimsvötn, Vatnajökull, Iceland, 1934–91, deduced from *in situ* measurements of subglacial lake volume. *J. Glaciol.*, **41**, 263–272.

GUÐMUNDSSON, M.T., SIGMUNDSSON, F. & BJÖRNSSON, H. (1997) Ice–volcano interaction of the 1996 Gjálp subglacial eruption, Vatnajökull, Iceland. *Nature*, **389**, 954–957.

GUSTAVSON, T.C. (1974) Sedimentation on gravel outwash fans Malaspina glacial foreland, Alaska. *J. sediment. Petrol.*, **44**, 374–389.

JONSSON, J. (1982) Notes on the Katla volcanoglacial debris flows. *Jökull*, **32**, 61–68.

KARCZ, I. (1968) Fluviatile obstacle marks from the wadis of the Negev. *J. sediment. Petrol.*, **38**, 1060–1072.

KEHEW, A.E. (1993) Glacial-lake outburst erosion of the Grand Valley, Michigan, and impacts on glacial lakes in the Lake Michigan basin. *Quat. Res.*, **39**, 36–44.

KEHEW, A.E. & LORD, M.L. (1987) Glacial outbursts along the mid-continental margins of the Laurentide Icesheet. In: *Catastrophic Flooding* (Eds Mayer, L. & Nash, D.), pp. 95–121. The Binghampton Symposium in Geomorphology, International Series, vol. 18. Allen & Unwin, Boston, London.

KLIMEK, K. (1972) Present day fluvial processes and relief of the Skeiðarársandur plain (Iceland). *Pol. Akad. Nauk Inst. Geogr.*, **94**, 129–139.

KLIMEK, K. (1973) Geomorphological and geological analysis of the proglacial area of the Skeiðarárjökull: extreme eastern and western sections. *Geogr. Pol.*, **26**, 89–113.

KNUDSEN, Ó. (1995) Concertina eskers, Brúarjökull, Iceland: an indicator of surge-type glacier behaviour. *Quat. Sci. Rev.*, **14**, 487–493.

KRAINER, K. & POSCHER, G. (1992) Ice-rich, redeposited diamicton blocks and associated structures in Quaternary outwash sediments of the Inn valley near Innsbruck, Austria. *Geogr. Annaler*, **72A**, 249–254.

LARSEN, G. & GUÐMUNDSSON, M.T. (1997) Gos Í Eldstöðvum undir Vatnajökli eftir 1200 AD. In: *Vatnajökull: Gos og hlaup 1996* (Ed. Haraldsson, H.), pp. 23–35. Vegagerðin, Reykjavik.

LORD, M.L. & KEHEW, A.E. (1987) Sedimentology and palaeohydrology of glacial-lake outburst deposits in southeastern Saskatchewan and northwestern North Dakota. *Geol. Soc. Am. Bull.*, **99**, 663–673.

MAIZELS, J.K. (1979) Proglacial aggradation and changes in braided channel patterns during a period of glacier advance: an Alpine example. *Geogr. Annaler*, **61**, 87–101.

MAIZELS, J.K. (1989a) Sedimentology, paleoflow dynamics and flood history of jökulhlaup deposits: paleohydrology of Holocene sediment sequences in southern Iceland sandur deposits. *J. sediment. Petrol.*, **59**, 204–223.

MAIZELS, J.K. (1989b) Sedimentology and palaeohydrology of Holocene flood deposits in front of a jökulhlaup glacier, South Iceland. In: *Floods, Hydrological, Sedimentological and Geomorphological Implications: an Overview* (Eds Bevan, K. & Carling, P.), pp. 239–253. John Wiley & Sons, Chichester.

MAIZELS, J.K. (1991) Origin and evolution of Holocene sandurs in areas of jökulhlaup drainage, south Iceland. In: *Environmental Change in Iceland: Past and Present* (Eds Maizels, J.K. & Caseldine, C.), pp. 267–300. Kluwer, Dortecht.

MAIZELS, J.K. (1993a) Lithofacies variations within sandur deposits: the role of runoff regime, flow dynamics and sediment supply characteristics. *Sediment. Geol.*, **85**, 299–325.

MAIZELS, J.K. (1993b) Quantitative regime modelling of fluvial depositional sequence: application to Holocene stratigraphy of humid-glacial braid-plains (Icelandic sandurs). In: *Characterisation of Fluvial and Aeolian Reservoirs* (Eds North, C.P. & Prosser, D.J.), Spec. Publ. geol. Soc. London, No. 73, pp. 53–78. Geological Society of London, Bath.

MAIZELS, J.K. (1995) Sediments and landforms of modern proglacial terrestrial environments. In: *Modern Glacial Environments: Processes, Dynamics and Sediments* (Ed. Menzies, J.), pp. 365–416. Butterworth-Heinemann, Oxford.

MAIZELS, J.K. & RUSSELL, A.J. (1992) Quaternary perspectives on jökulhlaup prediction. In: *Applications of Quaternary Research* (Ed. Gray, J.M.), pp. 133–153. Quaternary Proceedings vol. 2, Quaternary Research Association, Cambridge.

NEMEC, W. & STEEL, R.J. (1984) Alluvial and coastal conglomerates: their significant features and some comments on gravelly mass flow deposits. In: *Sedimentology of Gravels and Conglomerates* (Eds Koster, E.H. & Steel, R.J.), Mem. Can. Soc. petrol. Geol., Calgary, **10**, 1–31.

NUMMEDAL, D., HINE, A.C. & BOOTHROYD, J.C. (1987) Holocene evolution of the south-central coast of Iceland. In: *Glaciated Coasts* (Eds Fitzgerald, D.M. & Rosen, P.S.), pp. 115–150. Academic Press, London.

O'CONNOR, J. (1993) Hydrology, hydraulics, and geomorphology of the Bonneville Flood. *Spec. Pap. geol. Soc. Am.*, **274**, 1–83 pp.

RICHARDSON, P.D. (1968) The generation of scour marks near obstacles. *J. sedim. Petrol.*, **38**, 965–970.

RUSSELL, A.J. (1993) Obstacle marks produced by flows around stranded ice blocks during a jökulhlaup in West Greenland. *Sedimentology*, **40**, 1091–1113.

RUSSELL, A.J. & MARREN, P.M. (1999) Proglacial fluvial sedimentary sequences in Greenland and Iceland: a case study from active proglacial environments subject to jökulhlaups. In: *The Description and Analysis of Quaternary Stratigraphic Field Sections* (Eds Jones, A.P., Tucker, M.E. & Hart, J.K.), 171–208. Technical Guide 7, Quaternary Research Association, London.

RUSSELL, A.J., KNUDSEN, Ó., MAIZELS, J.K. & MARREN, P.M. (1997a) Controls on the geomorphological impact of the November 1996 jökulhlaup, Skeiðarársandur, Iceland. *Supplementi di Geogfia Fisica e Dinamica Quaternaria*, III, *Abstracts of the Fourth International Conference on Geomorphology*, pp. 335–336.

RUSSELL, A.J., KNUDSEN, Ó., MAIZELS, J.K. & MARREN, P.M. (1997b) Controls on the sedimentology of November 1996 jökulhlaup deposits, Skeiðarársandur, Iceland. In: *Abstracts of the Sixth International Conference on Fluvial Sedimentology* (Ed. Rogers, J.), 22–26 September, Cape Town, South Africa, p. 178.

RUSSELL, A.J., KNUDSEN, Ó., MAIZELS, J.K. & MARREN, P.M. (in press) Channel cross-sectional area changes and peak discharge calculations on the Gígjukvísl during the November 1996 jökulhlaup, Skeiðarársandur, Iceland. *Jökull*.

SCHOLZ, H., SCHREINER, B. & FUNK, H. (1988) Der einfluss von Gletscherlaufen auf die schmelzwasserablagerungen des Russell-Gletschers bei Söndre Strömfjord (WestGrönland). *Z. Gletscherkde. Glazialgeol.*, **24**, 55–74.

SHARP, M. (1988) Surging glaciers: geomorphic effects. *Prog. phys. Geog.*, **12**, 533–539.

SNORRASON, Á., JÓNSSON, P., PÁLSSON, S., *et al.* (1997) Hlaupið Á Skeiðarársandi Haustið 1996: Útbreiðsla, rennsli og aurburður. In: *Vatnajökull: Gos og hlaup* (Ed. Haraldsson, H.), pp. 79–137. Vegagerðin, Reykjavik.

THORARINSSON, S. (1974) *Vötnin Strið: Saga Skeiðarárhlaupa og Grimsvatnagosa*. Bókaútgáfa Menningarsjods, Reykjavik, 254 pp.

TODD, S.P. (1989) Stream-driven, high-density gravelly traction carpets: possible deposits in the Trabeg Conglomerate Formation, SW Ireland and some theoretical considerations of their origin. *Sedimentology*, **36**, 513–530.

# Alluvial Facies and Architecture

*Spec. Publs int. Ass. Sediment.* (1999) **28**, 333–346

# The influence of aggradation rate on braided alluvial architecture: field study and physical scale-modelling of the Ashburton River gravels, Canterbury Plains, New Zealand

P. J. ASHWORTH\*, J. L. BEST†, J. PEAKALL\*,† *and* J. A. LORSONG‡

\**School of Geography, University of Leeds, Leeds, West Yorkshire LS2 9JT, UK*
*(Email: P.Ashworth@geography.leeds.ac.uk);*
†*School of Earth Sciences, University of Leeds, Leeds, West Yorkshire LS2 9JT, UK; and*
‡*ARCO, El-Djazair Company, 2300 West Plano Parkway, Plano, Texas 7507, USA*

## ABSTRACT

Theoretical process-based models of braided alluvial architecture suggest that aggradation rate is a primary control on the geometry, stacking and heterogeneity of sedimentary deposits. This hypothesis is tested at the scale of the channel and bar using a combined field and flume modelling study, which quantifies the impact of a change in aggradation rate on the frequency of occurrence and geometry of the key depositional units that dominate coarse-grained, braided alluvial architecture. Aggradation of a 1 : 50 scale model of the braided Ashburton River, New Zealand, produces realistic alluvial architecture that closely corresponds to 7 km of logged field prototype outcrop. A twofold change in aggradation rate in the flume model and an order-of-magnitude change in the field outcrop, have *no* influence on the geometry and vertical distribution of fine- and coarse-grained depositional niches. Braided alluvial architecture at the channel scale therefore is determined by the local 'instantaneous' aggradation rate, related to individual flood events, rather than the long-term, regional aggradation rate.

## INTRODUCTION

Braided rivers comprise the most dynamic and intrinsically variable fluvial channel pattern and typically produce a highly complex, three-dimensional alluvial architecture within the subsurface (Bridge, 1985, 1993). Previous work has shown that the preservation and architectural geometry of braided-river alluvium is dependent on the interplay between a range of allocyclic and autocyclic variables. These variables include the magnitude and rate of channel scour, deposition and migration (e.g. Paola & Borgman, 1991; Webb, 1994; Best & Ashworth, 1997), downstream changes in grain size and sediment flux (e.g. Paola, 1988; Paola *et al.*, 1992; Miall, 1993), the frequency and extent of anabranch and channel belt avulsion (e.g. Mackey & Bridge, 1995; Heller & Paola, 1996), base- and/or sea-level change (e.g. Lawrence *et al.*, 1990; Salter, 1993; Leeder & Stewart, 1996) and the rate of basin subsidence and aggradation (e.g. Jervey, 1988; Bentham *et al.*, 1993; Willis, 1993). Assumptions concerning the effect of these variables on sediment preservation form the basis of several alluvial architecture

models, which seek to predict sandbody interconnectedness and alluvial heterogeneity (e.g. Bridge & Leeder, 1979; Webb, 1994; Mackey & Bridge, 1995; Heller & Paola, 1996).

Recent alluvial-architecture models have suggested that one of the key controls on the geometry of braided river alluvium is the rate of channel and basin-wide aggradation (Mackey & Bridge, 1995; Heller & Paola, 1996), which in turn is intimately linked to the frequency of channel avulsion (Törnqvist, 1994; Bryant *et al.*, 1995). In the simple case where it is assumed that there is a constant average avulsion frequency, the density and connectedness of channel sandbodies are inversely proportional to the basin-wide aggradation rate (Bridge & Leeder, 1979). In this case, low aggradation rates produce well-connected, sheet-like sand bodies, because less sediment is deposited between avulsion events. Conversely, high aggradation rates preserve isolated, ribbon-like channel bodies, separated by intervening fine-grained floodplain deposits. Currently, there are few data sets to test these fundamental ideas (cf. Heller & Paola, 1996, p. 298; Leeder *et al.*, 1996). In particular, the relationship between a changing aggradation rate and the geometry and spatial distribution of the principal alluvial architectural elements is poorly understood (Willis & Behrensmeyer, 1994; Törnqvist *et al.*, 1996).

Several theoretical models (e.g. Bridge & Leeder, 1979; Mackey & Bridge, 1995; Heller & Paola, 1996) simulate the preservation of alluvial architecture at the channel-belt scale, where the internal resolution is of the order of hundreds of metres (600 m in the case of Bridge & Leeder (1979)). Such resolution makes it difficult to compare model output with many field interpretations of coarse-grained braided alluvial architecture, which focus on deposition at the scale of the individual channel or bar (e.g. Miall, 1977; Steel & Thompson, 1983; Brierley, 1991) and limits the use of models in hydrocarbon reservoir characterization and hydrogeological problems, which require resolution of the order of 10–50 m (Tyler *et al.*, 1994; Webb & Anderson, 1996). Webb (1994, 1995) has developed a three-dimensional alluvial architecture model for braided rivers that simulates deposition at the architectural element scale by coupling facies types to channel Froude number, and uses a two-dimensional random-walk model to produce a set of channel topographies and sedimentary packages that are stacked through time. As Webb (1994, p. 229) states, however, the model 'sidesteps the issue of aggradation rate' because it back-calculates the aggradation rate from theoretical and known bedform-set thicknesses (cf. Paola & Borgman, 1991). Hence, there remains a pressing need for comprehensive data sets that can quantify the influence of aggradation rate on the geometry and frequency of individual architectural elements at the channel scale.

Research using physical models has shown that it is possible to simulate the dynamics of braided rivers (e.g. Schumm *et al.*, 1987; Ashmore, 1991a; Warburton, 1996) and isolate some of the key controls on braidplain evolution, such as bar creation (Hoey & Sutherland, 1991; Germanoski & Schumm, 1993; Ashworth, 1996), confluence kinetics (Klaassen, 1990; Ashmore, 1993), sediment supply (Schumm & Khan, 1972), tectonic uplift and tilting (Ouchi, 1985; Schumm, 1986; Jin & Schumm, 1987), base-level change (Schumm, 1993; Koss *et al.*, 1994; Wood *et al.*, 1994) and channel avulsion (Leddy *et al.*, 1993; Bryant *et al.*, 1995). More recently, Ashworth *et al.* (1994), Peakall (1995) and Peakall *et al.* (1996) have shown that it is possible to use Froude scale-modelling techniques to simulate sediment sorting and braidplain deposition and then, through aggradation, to produce a realistic three-dimensional alluvial architecture. This technique of scale modelling offers a powerful method for isolating the dominant controls on fluvial

deposition (Peakall *et al.*, 1996) and hence may be used to quantify the influence of aggradation rate on the channel-scale alluvial architecture of braided rivers.

This paper describes the simulation of alluvial architecture in a flume-scale model of the Ashburton River, Canterbury Plains, South Island, New Zealand, at two different imposed aggradation rates. The alluvial architecture simulated in the flume is compared with quantitative measurements of braided-river alluvium deposited by the Ashburton River over the past 35 000 yr, which is now exposed along an extensive erosional coastal cliffline. Dating of a suite of stacked sand bodies in the Ashburton exposure indicates that aggradation rate has changed over time and allows the field site to be used as an independent test of the flume results to examine the response of braided alluvial architecture to a change in aggradation rate. The aims of this paper are twofold:

**1** to quantify and compare the braided-alluvial architecture preserved in the Ashburton flume model and field exposure;

**2** to examine the influence of a change in aggradation rate on the geometry and frequency of occurrence of key depositional niches that dominate the channel-scale alluvial architecture of coarse-grained braided-river deposits.

## FIELD PROTOTYPE DESCRIPTION AND DATA ACQUISITION

The Canterbury Plains, on the east coast of South Island, New Zealand, extend over an area 70 km wide and 185 km long, from Timaru north-eastward to the Waipara River (Fig. 1), and were formed by coalescence of a series of broad alluvial fans, sourced from glacially fed rivers that emerged from the Southern Alps during the Quaternary (Brown *et al.*, 1988; Suggate, 1990; Bal, 1996). Deposition of much of the 700-m-thick unconsolidated Canterbury gravels is attributed to sedimentation during glacial episodes rather than interglacials (Soons & Gullentops, 1973), although deglaciation from *c.* 18 ka caused rapid deposition of fluvial gravels (Suggate, 1965). Sea-level rise, associated with deglaciation, reached its present highstand position at 6.5 ka (Gibb, 1986), but intense coastal erosion of up to 1 m $yr^{-1}$ (Gibb, 1978) has resulted in rapid retreat of the coastal cliff-line south of the Banks Peninsula (Figs 1 & 2A). In response to this coastal retrogradation, the braided rivers of the Canterbury Plains have incised by up to 6 mm $yr^{-1}$ (Leckie, 1994), providing excellent exposure of Quaternary outwash and fluvial gravels as a continuous, freshly weathered, vertical, 70 km long cliff, up to 30 m high (Fig. 2A & B). Optically stimulated luminescence (OSL) dating of 14 stacked sandstone bodies in the cliff

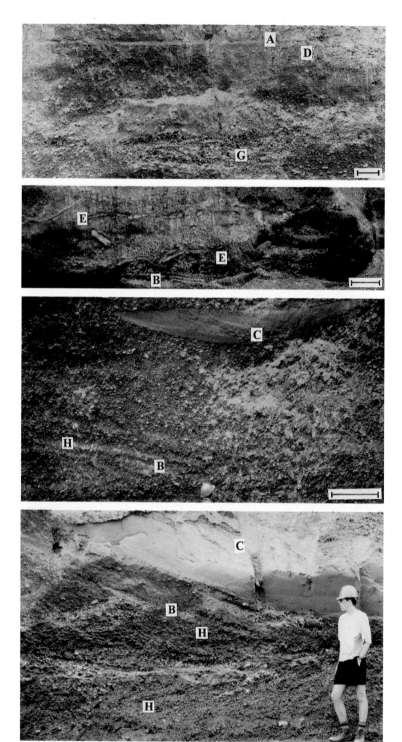

**Plate 1.** Examples of fine and coarse-grained depositional niches from the Ashburton field site (this page) and 1:50 Ashburton scale model (overleaf). Labels superimposed on the photographs correspond to depositional niches described in the text. Scale bar is 1 m in the field. Palaeoflow is towards the camera in the field and away from the camera in the flume. In the flume, coloured sand was mixed with the feeder sand to provide time horizons of flume deposition and to help distinguish internal structure. White silica flour is used to simulate fine-grained deposition and represents silt–fine-sand in the field. More details on the scaling principles and experimental protocol are given in the text. (*Continued overleaf*)

[*Facing p. 334*]

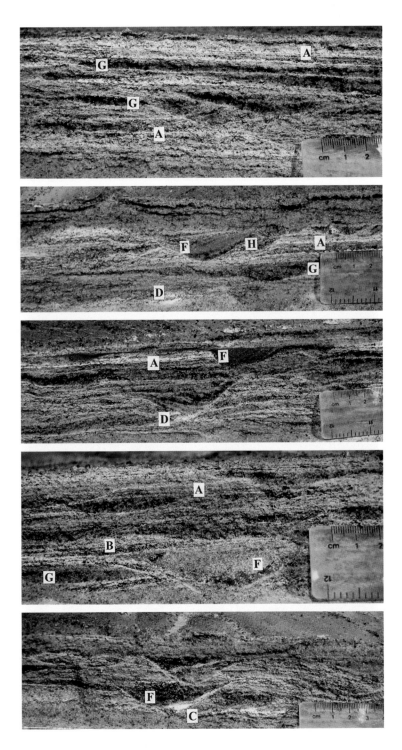

**Plate 1.** *Continued*

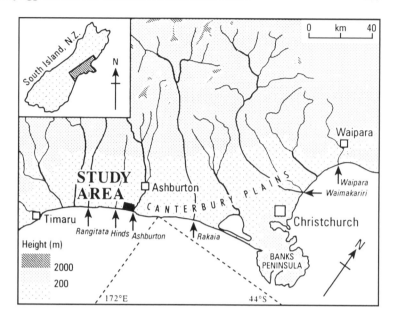

**Fig. 1.** Location of the Ashburton River, Canterbury Plains and the outcrop study site.

**Fig. 2.** Views of (A) the continuous cliff exposure of Quaternary outwash and fluvial gravels south of the Banks Peninsula (see Fig. 1) and (B) the 17-m-high cliffs that were logged near the mouth of the present Ashburton River. Two to 4 m of loess overlies the Ashburton gravels (see pale colour in (B)). Palaeoflow direction is out of the cliffs.

exposure at the modern Ashburton River mouth (Fig. 1) shows that the gravels range in age from approximately 35 to 8 ka from bottom to top of the cliff, with four distinct periods of fluvial and glacial deposition, each associated with a different aggradation rate, which ranges from 0.32 to 2.85 mm yr$^{-1}$ (Ashworth *et al.*, 1997).

Seven kilometres of outcrop were selected west of the Ashburton River mouth (Fig. 1). Cliff height ranged from 16.8 m to 13.5 m within the 7 km of section, this variation in height producing fewer measurements of deposit width and thickness in the top 3 m of section (see field-outcrop results later). Mean palaeoflow direction is perpendicular to the cliff face. The exposure reveals heterogeneous gravels (termed coarse-grained deposits), with inter-bedded sand bodies and clays (termed fine-grained deposits), which commonly possess sharp erosion surfaces.

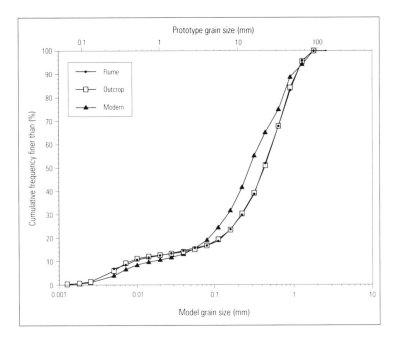

**Fig. 3.** Grain-size distributions for the modern Ashburton River and Ashburton outcrop together with the grain-size mix for the Ashburton flume model. A 1 : 50 length scale is used for modelling the prototype grain size. Note the close correspondence between the Ashburton outcrop (prototype) and model grain-size distributions.

Median ($D_{50}$) and 90th percentile ($D_{90}$) grain-size of the gravels are 21 mm (coarse pebbles) and 54 mm (very coarse pebbles), respectively (Fig. 3). Mean palaeochannel depth, estimated using fine-grained channel fills, is $0.87 \pm 0.06$ m ($n = 59$). The 7 km of cliff were photographed using a 35 mm camera with 50 mm lens from a fixed distance, in order to maintain a constant scale, and the overlapping (approximately 50%) photographs joined into a montage. Detailed logging of the outcrop was then undertaken and the key sedimentary units, erosion surfaces and internal sedimentary structures were superimposed on the photomontages. Over 3500 measurements were taken of the width, thickness and position (lateral and vertical) of every fine-grained sedimentary niche more than 0.2 m wide along the 7-km-long exposure, together with 115 measurements of coarse-grained niche geometry (see later for classification details). A correction was applied to all height and thickness measurements to account for distortion in the photographic image caused by a non-vertical alignment of the camera and cliff-face (see Wizevich, 1991). Attention focused on the measurement of the fine-grained depositional niches, because these were well defined in both outcrop and flume experiments (see later) and therefore provided an unambiguous record of the temporal and spatial change in alluvial architecture. The coarse-grained niches are valuable indicators of palaeochannel depth, but their lateral continuity was often ill-defined and difficult to interpret in these heterogeneous gravels (see discussion in Smith, 1990).

The close proximity of the logged outcrop to the Ashburton River permitted comparison of ancient and modern alluvial sediments and provided morphological and hydraulic information for the scaled flume model. Bulk samples of recent bar sedimentation had an average $D_{50}$ and $D_{90}$ grain size of 14 mm (medium pebbles) and 48 mm (very coarse pebbles) respectively (Fig. 3), which is slightly finer than sediments in the outcrop (compare curves in Fig. 3), because it represents the distal grain-size of the long profile of the modern river (the outcrop is truncated by coastal erosion). Twenty-seven measurements of the bankfull width and depth of all sizes of modern anabranches in the Ashburton River yielded a mean channel width:depth ratio of 31. The average bed slope of the South Ashburton is 0.0132, falling to 0.006 at the river mouth. A hydrograph from a 'characteristic' annual flood on 12 October 1990 was used to produce a scaled discharge hydrograph in the flume model.

## FLUME MODEL AND SCALING DETAILS

A 1 : 50 physical scale model of the Ashburton River was run in a tilting flume 5.50 m long and 3.65 m wide. Dry sediment was fed into a feeder channel 0.40 m wide and 2.30 m long, and thoroughly mixed with the recirculating discharge before entering a 400-mm-wide and

20-mm-deep straight channel dredged down the centre of a bed of flume sediment 60 mm thick. Bedload was trapped at the flume tail, but water and suspended sediment were recirculated. Water discharge was varied by a computer-controlled gate valve, which could be programmed to allow the modelling of flood hydrographs. The flume simulates aggradation through computer-controlled raising of an inner cradle nested within the inlet channel. Uplift of the sediment source is equivalent to basin (flume) subsidence. Two, overhead, time-lapse video cameras monitored channel change and the surface-velocity distribution (using floats), and a moveable bridge that spans the flume width permitted point gauging ($\pm 0.1$ mm) of bed topography, flow depth and water-surface slope. More details on the flume design and operation are given in Ashworth *et al.* (1994) and Peakall *et al.* (1996).

The Ashburton River flume model was scaled using prototype information from both the modern and ancient (outcrop) depositional environment, so that it aims to ensure geometric, kinematic (motion) and dynamic (force) similarity between model and prototype. The Ashburton model was not a scaled version of a specific reach, but instead replicated the process and form relationship in a braided river that has similar characteristics to the modern and ancient Ashburton. This type of modelling has been termed 'generic Froude scale modelling' by Church (personal communication, cited in Ashmore (1991a,b)), Ashworth *et al.* (1994), Peakall *et al.* (1996) and Warburton & Davies (1998). The key variables that must be satisfied in order to ensure model and prototype similarity are defined and discussed in Peakall *et al.* (1996), but in order to maintain hydraulic similarity the model flow must be fully rough and turbulent (flow Reynolds number, *Re*, > 2000), viscous forces minimal (grain Reynolds number, *Re**, > 15) and Froude numbers (*Fr*) close to 1 (i.e. on the border between sub- and super-critical flow). In the current experiments, a 1 : 50 length scale was chosen, which produced measured mean values of *Re* = 1702, *Re** = 27 and *Fr* = 0.92. The *Re* and *Re** values are in the transitional range between smooth and fully rough, turbulent flow but are similar to those found in several other scale models of braided rivers (cf. Peakall *et al.*, 1996, pp. 228–232). No bedforms were present in the Ashburton scale model, suggesting similarity in hydraulic and sediment transport conditions between model and prototype.

The Ashburton outcrop grain-size distribution was scaled at 1 : 50 and was reproduced by mixing different grades of washed sands and silts to give a flume sediment with $D_{50} = 0.42$ mm (medium sand) and $D_{90} = 1.07$ mm (very coarse sand) (Fig. 3). Use of inert, white, silica flour (ranging from 1 to 5 $\mu$m) to simulate the finest 7% of the model grain-size distribution (see Fig. 3) permitted the modelling and easy recognition of the fine-grained deposition niches that represent silt to fine sand in the prototype. Bed slope was set at 0.01 and the model was run for 25 h at a constant discharge to allow development of a fully braided channel planform. Twenty-three flood hydrographs were then simulated, at an aggradation rate of 0.5 mm h$^{-1}$, followed by a further 23 identical hydrographs at an aggradation rate of 1 mm h$^{-1}$. This produced approximately 70 mm of total aggradation. Each flood hydrograph ran for 120 min, with the discharge and sediment-feed rate changed at intervals of 30 s and 5 min respectively. Maximum discharge was 3.3 L s$^{-1}$ (field equivalent = 58 m$^3$ s$^{-1}$) and the maximum sediment-feed rate, lagged after the discharge peak, was 0.0187 kg s$^{-1}$. As the model hydrograph represents the prototype annual flood (and creates overbank conditions in the model), then the model aggradation rates are approximately equivalent to field aggradation rates of 5 and 25 mm yr$^{-1}$ (i.e. 0.1 and 0.5 mm multiplied by the length-scale per year). Although these aggradation rates are approximately an order of magnitude greater than rates reported from some alluvial basins (e.g. Smith, 1986; Törnqvist, 1993; Törnqvist *et al.*, 1996), they are within the range of accumulation rates ($10^0$–$10^1$ m 1000 yr$^{-1}$) given by Sadler (1981) for fluvial environments over short time periods ($10^2$–$10^3$ yr).

Coloured sand ($D_{50} = 0.37$ mm) (medium sand) was mixed with the feeder sediment on five occasions to provide time-horizons in the preserved alluvial sediment. On completion of the 46 runs, the flume was allowed to air-dry and then sectioned using a sharp metal plate. Thirty-three, 2-m-wide cross-sections were sectioned at 3-cm intervals down the centre of the flume and colour photographed using a large-format (5 × 4 cm) camera. Photomontages were then produced for each section and key sedimentary niches identified and measured utilizing the same method used for the field-outcrop images. Over 1500 fine and coarse-grained niches were measured in the flume sections.

## CLASSIFICATION OF FIELD AND FLUME BRAIDED ALLUVIAL ARCHITECTURE

Examination of the field and flume alluvial architecture enables identification of the key fine- and coarse-grained depositional niches (Plate 1, facing p. 334). This classification is based on the grain size, sorting, internal sedimentary structure and contacts with under-/overlying sediments. These fine- and coarse-grained facies bear many similarities to depositional niches defined in past studies of gravel-bed braided alluvium (e.g. Miall, 1977; Ramos & Sopeña, 1983; Steel & Thompson, 1983; Ramos *et al.*, 1986; Smith, 1990; Bentham *et al.*, 1993; Huggenberger, 1993; Siegenthaler & Huggenberger, 1993) and in past flume work (Ashworth *et al.*, 1994; Peakall *et al.*, 1996). The following depositional niches have been identified (Plate 1).

### Fine-grained sediment (sand, silt and clay in field; silt and fine-sand equivalents in flume)

**A** *Bar-top fines*: laterally extensive, thin, fine-grained sheets of sand/silt deposited on bar tops or low-angle bar margins during peak and waning flow. These fines are often eroded into discontinuous horizons in the field, but comprise a high percentage of the fine-grained deposits (see later). The bar-top fines are similar to the sandstone splays of Bentham *et al.* (1993), although they are smaller, representing deposition on individual bars, rather than across the entire braidplain.

**B** *Bar margins*: thin drapes on either steep or low-angle bar margins (both foresets and lateral-accretion surfaces) reflecting either fine-grained deposition in regions of separated flow in the lee side of steep bar margins or falling-stage drapes on low-angle accretion surfaces.

**C** *Channel fills*: rare plugs of fine-grained sediment in abandoned channels. These may form small plugs in either the initial or final stages of abandonment.

**D** *Erosional remnants*: these are undifferentiated, fine-grained remnants, which represent areas of isolated deposition (e.g. backwaters, floodplain depressions) or partly eroded fine-grained regions, which cannot be matched with their primary source. These are analogous to the mudstone facies of Ramos *et al.* (1986).

**E** *Clay intraclasts* (*field only*): clay clasts, decimetre in scale, concentrated in two distinct horizons (see later). These are generated by the reworking of fine-grained bar-top fines and abandoned channel fills but also represent bank slumping and the reworking of pockets of fine-grained deposition in floodplain topographic depressions. Angular clasts are common and may indicate minimal reworking and travel distance (cf. Smith, 1972), with some internal folding in the clay clasts suggesting deformation by ice, freeze–thaw activity and the effects of compaction.

### Coarse-grained sediment (gravel in field; coarse sand in flume)

**F** *Channel fills*: complete channel fills, often with arcuate basal erosion surfaces, which may have a massive or structured (e.g. bar margin) fill. Channel fills may exhibit distinct grain-size sorting, both vertically and laterally, with internal lateral accretion surfaces and basal lags. Morphology and internal structure are similar to the accretionary style C described by Smith (1990).

**G** *Bar cores*: coarse-grained, laterally extensive, thin gravel sheets which lack internal structure and which were generated from deposition at the centre of low-angle bars. Bar-core deposits are often overlain by bar-top fines and may interdigitate laterally with bar margins. Bar cores frequently comprise clean openwork gravels, which correspond to the sheet conglomerates of Ramos & Sopeña (1983) and Ramos *et al.* (1986), the lithofacies A of Steel & Thompson (1983) and gravel couplets of Siegenthaler & Huggenberger (1993). Mean length and dip of the openwork gravels are 1.33 m and 6.6° respectively ($n = 30$).

**H** *Bar margins*: accretion surfaces or slipfaces, which are well size-sorted as a result of either event-related deposition or avalanche sorting over the slipface of steeper bar margins. This facies type corresponds to the lateral accretion and cross-stratified conglomerates of Ramos & Sopeña (1983) and Ramos *et al.* (1986) and the pool-fill and foreset deposition of Huggenberger (1993).

### ALLUVIAL ARCHITECTURE: DEPOSITIONAL NICHE GEOMETRY AND FREQUENCY OF OCCURRENCE

#### Ashburton field outcrop

Table 1 shows that the fine-grained depositional niches, the bar-top fines and channel fills are dominated by sand deposition whereas the bar margins and erosional remnants are dominated by clays. On average, the width and thickness of the fine-grained niches are greatest for the channel fills and decrease in the order: channel fills > bar margin > bar-top fines >> erosional remnants. The distributions for all niches, except the channel fills, are log-normally distributed (Fig. 4A), as previously observed in the field (Geehan *et al.*, 1986) and flume (Ashworth & Best, 1994). As channel-fill width and thickness distributions are derived from far fewer data points than the other fine-grained niches (Table 1), it is difficult to derive full and reliable distributions. It is noteworthy, however, that the distribution of channel-fill thickness is less negatively skewed than those for the other fine-grained niches (Fig. 4A) and bears a similarity to the distributions of channel depth for the Ohau River (Mosley, 1982) and Niger River (Salter, 1993), which show a similar skewness (the peaks of the channel thickness and depth distributions for the

**Table 1.** Summary statistics of depositional-niche geometries and percentage occurrence from field and flume studies. Width and thickness statistics for the fine-grained deposits are for combined sand and clay data. Flume data are scaled-up to the Ashburton field prototype and are shown for the two aggradation rates. Data in italics are for the higher flume aggradation rate (1 mm h$^{-1}$). NA, no data available, SE, standard error of the mean.

| Depositional niche | Sand (%) | Clay (%) | Width (m) Minimum | Maximum | Mean | SE | Thickness (m) Minimum | Maximum | Mean | SE | Number and percentage occurrence |
|---|---|---|---|---|---|---|---|---|---|---|---|
| **Field** | | | | | | | | | | | |
| **Fine-grained:** | | | | | | | | | | | |
| bar-top fines | 85.5 | 14.5 | 0.28 | 9.89 | 1.87 | 0.041 | 0.06 | 2.86 | 0.31 | 0.006 | 900 (27.7) |
| bar margins | 12.0 | 88.0 | 0.17 | 9.55 | 2.04 | 0.069 | 0.06 | 1.58 | 0.87 | 0.424 | 473 (14.5) |
| channel fills | 98.3 | 1.7 | 0.11 | 17.74 | 4.67 | 0.424 | 0.13 | 2.66 | 0.87 | 0.059 | 57 (1.8) |
| erosional remnants | 33.3 | 66.7 | 0.06 | 2.39 | 0.44 | 0.007 | 0.06 | 2.90 | 0.19 | 0.003 | 1819 (56.0) |
| **Coarse-grained:** | | | | | | | | | | | |
| channel fills | NA | NA | 3.52 | 15.80 | 6.79 | 0.568 | 0.65 | 2.26 | 1.25 | 0.072 | 24 (20.9) |
| bar core | NA | NA | 2.50 | 17.28 | 8.55 | 0.409 | 0.46 | 4.12 | 1.48 | 0.074 | 74 (64.3) |
| bar margins | NA | NA | 0.91 | 6.70 | 4.13 | 0.455 | 0.64 | 3.27 | 1.53 | 0.141 | 17 (14.8) |
| **Flume** | | | | | | | | | | | |
| **Fine-grained:** | | | | | | | | | | | |
| bar-top fines | NA | NA | *0.24* | *16.1* | *2.43* | *0.151* | *0.03* | *0.24* | *0.055* | *0.003* | *270 (48.1)* |
| | | | 0.18 | 18.0 | 2.23 | 0.135 | 0.03 | 0.48 | 0.060 | 0.003 | 353 (39.9) |
| bar margins | NA | NA | *0.03* | *15.1* | *0.86* | *0.097* | *0.03* | *0.18* | *0.040* | *0.002* | *177 (31.6)* |
| | | | 0.06 | 3.45 | 0.73 | 0.035 | 0.03 | 0.57 | 0.045 | 0.003 | 285 (32.3) |
| channel fills | NA | NA | *0.07* | *0.07* | *0.072* | *0.363* | *0.03* | *0.03* | *0.031* | *0.000* | *2 (0.35)* |
| | | | 0.08 | 0.08 | 0.080 | 0.635 | 0.03 | 0.03 | 0.031 | 0.030 | 2 (0.23) |
| erosional remnants | NA | NA | *0.09* | *5.02* | *0.81* | *0.066* | *0.03* | *0.30* | *0.080* | *0.006* | *112 (20.0)* |
| | | | 0.12 | 9.38 | 0.76 | 0.048 | 0.03 | 0.24 | 0.072 | 0.003 | 243 (27.5) |
| **Coarse-grained:** | | | | | | | | | | | |
| channel fills | NA | NA | *0.66* | *7.75* | *3.30* | *0.30* | *0.03* | *1.27* | *0.518* | *0.04* | *43 (67.2)* |
| | | | 2.05 | 9.50 | 4.92 | 0.46 | 0.24 | 1.21 | 0.741 | 0.05 | 22 (84.6) |
| bar core | NA | NA | *3.51* | *12.1* | *7.47* | *1.78* | *0.24* | *0.30* | *0.265* | *0.014* | *21 (32.8)* |
| | | | 1.69 | 9.92 | 5.36 | 0.53 | 0.03 | 0.55 | 0.272 | 0.026 | 4 (15.4) |

Ashburton, Ohau and Niger Rivers are at 23, 23 and 22% of the data range respectively).

Coarse-grained niches are generally larger than their fine-grained counterparts (e.g. for channel fills, coarse-grained channel mean thickness = 1.25 m, compared with 0.87 m for fine-grained), possibly indicating the smaller original size of fine-grained units (i.e. fine-grained channel fills may represent only part of a channel-fill sequence), but also highlighting the lower preservation potential of fine-grained sediments (e.g. bar-top fines are 22% of the width of coarse bar cores). Erosional remnants dominate the fine-grained alluvial architecture (56%), again highlighting the vulnerability of sands and silts to reworking in a dynamic braided-river environment. Unambiguous, fine-grained channel fills are rarely preserved (1.8%), similar to the flume observations described by Ashworth & Best (1994) and in the Ashburton model (see below).

Figure 5 shows the change in fine-grained niche width, thickness and frequency of occurrence with height in the Ashburton gravels using all the geometric data from the 7 km of logged outcrop. These data are aggregated from the entire length of outcrop and are expressed as height above the outcrop (beach) base. Although these data are not expressed relative to individual erosion surfaces within the sections, which are often difficult to trace laterally more than 0.5 km, the Ashburton gravels show no lateral change in grain-size or channel dimensions over this 7-km section. Vertical frequency distributions of the coarse-grained niches are not presented owing to the sparse data base ($n = 115$) obtained in the present study over this length of outcrop. Examination of the coarse-grained niches and the lateral variation in geometry over distances > 30 km is the current focus of ongoing fieldwork. The peaks in some distributions in the fine-grained niches are related to sparse data in the 0.5 m height

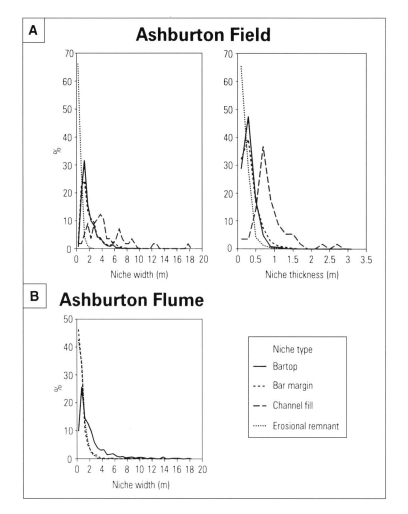

**Fig. 4.** (A) Width and thickness distributions of fine-grained (sand and clay) depositional niches from the Ashburton cliff gravels and (B) widths of fine-grained depositional niches from the Ashburton flume model. Widths have been scaled up from the flume model to their field equivalent using the 1 : 50 scaling ratio.

classes: this is clearly evident in the lower number of observations at the top of the cliff (16.8 m), which reflects changes in cliff height along the outcrop (see field prototype description) and for the channel fills, where the low sample number ($n = 57$) may affect the percentage occurrence, although it should be noted that mean width does not change with height in the section. The outcrop data show that there is no systematic change in width *or* frequency of occurrence of bar-top fines or bar-margin deposits with height in the Ashburton outcrop, a trend also displayed in the graph for the amalgamation of all niches (Fig. 5). This observation is supported by statistical analysis, which shows no significant difference in niche width and percentage occurrence with height at the 0.05 level. The only exceptions to this pattern are the erosional remnants and clay niches (which are both very

similar, because 67% of the erosional remnants are clay), which show a significant (at the 0.05 level) decrease in niche width with height up the cliff. The vertical distribution of erosional remnants and clays, however, are strongly affected by the switch between periods of different glacial and fluvial activity during the Late Quaternary (see below) and have two peaks that mark periods of probable braidplain abandonment (marked as A on Fig. 5).

Figure 5 also shows the change in fine-grained niche thickness with height in the Ashburton outcrop. Statistical analysis shows that in all cases except the channel fills, niche thickness displays a slight, but significant (at the 0.05 level), decrease up-section. This progressive, but small, decrease in niche thickness may be related to a change in both sediment supply and grain-size after the last glacial maximum, marking the gradual transition

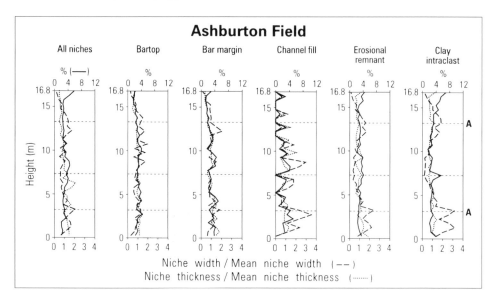

**Fig. 5.** Percentage occurrence and mean width and thickness of different depositional niches with height in the 7-km logged exposure of Ashburton gravels (0 m = base of cliff-line). Each plot shows both the relative frequency of occurrence of each niche, expressed as a percentage of the niche total (upper abscissa scale, plotted as a solid line) and the width and thickness of each niche, expressed as a proportion of the overall mean for that niche category (lower abscissa scale, plotted as dashed lines). Vertical bins of 0.5 m are used for each variable. Values of zero for the niche width and thickness indicate no data for that 0.5 m bin. The boundaries between units of different aggradation rate (from Ashworth *et al.*, 1997, fig. 3a) are shown as dotted horizontal lines and two distinct horizons of clay deposition are indicated by the label A (see text for explanation).

from glacial to fluvial conditions, which may reduce the supply of fine-grained sediment. It should be noted, however, that channel geometry does not show any systematic change over time (as expressed by the width and thickness of channel fills, Fig. 5), so the change in niche thickness cannot be explained by a reduction in discharge or the presence of smaller anabranch channels. The salient point to note from these data is that there is no response of niche thickness to an order-of-magnitude change in aggradation rate, which is related to periods of glacial and fluvial deposition during the Late Quaternary evolution of the Canterbury Plains (Brown *et al.*, 1988; Suggate, 1990). Aggradation rates, quantified using the OSL technique from 14 samples in these cliff sections, range from 0.32 to 2.85 mm yr$^{-1}$ (Ashworth *et al.*, 1997), representing four different periods of sedimentation (see lines on Fig. 5) over a period from approximately 35 to 8 ka. Although the OSL dates should be treated with caution because application of OSL dating to fluvial sediments is only in its infancy (Gibbons, 1997; Stokes, in press), the field results suggest that there is no discernible difference in either niche width, thickness or frequency of occurrence during the periods of different aggradation of the Ashburton gravels.

### Ashburton flume model

This discussion refers to the niche geometries at the two different flume aggradation rates; the role of changing aggradation rate is described separately later. Similar to the field, it is often difficult to delineate all the coarse-grained niches, as erosion surfaces may be very subtle. Additionally, the thickness of some fine-grained units in the flume model are below the resolution of measurement (set at 0.5 mm, i.e. 0.025 m field equivalent in Table 1) and these results therefore must be treated with caution. Nonetheless, the flume generates a large geometric data base to compare with the field prototype. The fine-grained niche widths are log-normally distributed (Fig. 4B) and, together with the frequency of occurrence, decrease in the order: bar-top fines > bar margins > erosional remnants > channel fills (Table 1). The lower frequency of occurrence of fine-grained erosional remnants in the flume (mean = 23%) compared with the field (56%) is probably explained by the difficulty in measuring small, isolated pockets of reworked sediments in the flume. This reduction in the percentage occurrence of erosional remnants accounts for the increase in the frequency of bar-top fines (mean = 44%, Table 1), which, on average, are 32% of the width of coarse-grained bar cores (similar to the field). The flume coarse-grained niches are wider and thicker than the fine-grained units, as also observed in the field (Table 1).

## FLUME–FIELD COMPARISON

The scaled-up geometries of the depositional niches show a good agreement between field and flume (Table 1), suggesting that on average the model-scaling procedures are correct and the flume data base is a realistic representation of the field. For example, mean channel depths obtained from the modern Ashburton River (0.59 m), the coarse- and fine-grained fills of channels in the Quaternary gravels (1.25 and 0.87 m respectively), the active channels in the flume (0.54 m), and the preserved coarse-grained channel fills in the flume sections (0.59 m) all fall within a narrow range, illustrating that channel morphology and depositional geometry are similar between modern, ancient and flume model. Mean widths of all fine- and coarse-grained niches in the flume and field are very closely matched (e.g. bar-top fines in field = 1.87 m, in flume = 2.43 m). The thicknesses of coarse-grained niches in the flume model and field outcrop are also in close agreement, but the fine-grained units are an order of magnitude thinner in the flume. Several factors may contribute towards this discrepancy. First, the fine-grained sediments in the field possess a degree of cohesion, especially the clays, which may undergo less reworking than the fines in the flume and hence produce thicker fines in the field. Second, although the hydrograph time was scaled in the flume, the interhydrograph time is shorter than in the field (i.e. floods do not run back-to-back in the field). This probably reduces the time available for settling of fines in abandoned channels and scours compared with the field prototype. Third, the flume may have a higher frequency of 'local' (i.e. anabranch) avulsion than the field, partly because channels that reach the flume sidewalls are sometimes deflected back into the model braidplain and also because aggradation rates are greater in the flume than field (see earlier) and hence will be associated with a higher avulsion frequency (Bryant *et al.*, 1995; Heller & Paola, 1996). This higher rate of channel mobility in the flume may lead to thinner fine-grained deposits because the channels will migrate and avulse, allowing less time for the deposition of fines, but maintaining a more constant thickness of the coarser grained deposits (Table 1).

## EFFECT OF CHANGING AGGRADATION RATE ON ALLUVIAL ARCHITECTURE

The data base of alluvial architecture in the flume and field can be used to test whether a change in aggradation rate has any effect on the frequency of occurrence and geometry of depositional niches within gravelly braided alluvium. Aggradation rate was doubled midway through the flume simulation and varied over an order of magnitude in the field with four periods of different sedimentation rate (Ashworth *et al.*, 1997). Table 1 shows no significant change in the fine- or coarse-grained niche geometry or frequency of occurrence between the lower and higher aggradation rate in the flume. For example, bar-top fines account for 40 and 48% of all fine-grained deposition and have mean widths of 2.23 m and 2.43 m at the lower and higher aggradation rate respectively. A similar pattern is seen in the field, where, despite a difference in sediment supply rate and depositional environment (i.e. fluvial or glacial), the vertical distribution and geometry of bar-top fines and bar-margin fines is invariant (Fig. 5). The only exception is the fine-grained erosional remnants and clay intraclasts, which show two broad peaks related to major periods of braidplain abandonment and subsequent floodplain reworking. Both the Ashburton field and flume data bases therefore suggest that channel-scale braided alluvial architecture is independent of aggradation rate. Although the thickness of all sedimentary niches, except channel fills, progressively decreases towards the top of the section in the field (see discussion above), this rate of decrease is minor and, most importantly, does not show any relationship to changing aggradation rate.

Theoretical models of braided alluvial architecture have shown that for rivers with constant floodplain and channel morphology, the preservation potential and geometry of sedimentary units is determined by (i) the basin-wide aggradation rate, (ii) regional and local avulsion frequency and (iii) the 'combing' depth of the channels (e.g. Bridge & Leeder, 1979; Paola & Borgman, 1991; Mackey & Bridge, 1995; Heller & Paola, 1996). Although these models have been developed at the scale of the entire floodplain, their arguments can be applied equally to the preservation of deposits *within* the channel belt that have been studied here. The Ashburton gravels and flume model suggest that the 'local' alluvial architecture (i.e. at the scale of the channel or bar) is independent of the long-term, aggradation rate. The geometry and stacking relationships of fine- and coarse-grained alluvial units are determined by the interrelationship between the 'instantaneous' aggradation rate and the channel 'combing depth' (intrinsic scour up to five times the mean channel depth, see Best & Ashworth (1997)). 'Local' alluvial architecture is determined by individual 'cut and fill' events, which operate in a stepwise manner over time as channels migrate and avulse and floodplain sedimentation responds to changes in sediment supply and the frequency and magnitude of floods. In the field, significant changes in sedimentation style occur only after a major 'regional' avulsion and channel-belt abandonment (e.g. the clay deposition at two horizons in the Ashburton outcrop). As the flume model cannot simulate major floodplain abandonment and there was little opportunity for rapid channel incision and sediment reworking because floods were run sequentially, the alluvial architecture was invariant with changing 'local' aggradation rate.

Hence, even though aggradation rates in the flume varied by 100%, and in the field varied by an order of magnitude, little response was recorded in the alluvial architecture. This result contradicts studies of basin infilling and braided alluvial architecture in the field (e.g. Willis, 1993; Törnqvist, 1994) and predictions from theoretical models of avulsion and braidplain sedimentation (e.g. Mackey & Bridge, 1995; Heller & Paola, 1996), which show a strong coupling between alluvial architecture and aggradation rate. It should be emphasized, however, that these studies were concerned with channel-belt deposits and there may be a case for treating the relationship between 'local' (i.e. channel/bar scale) and 'regional' (i.e. basin-wide scale) sedimentation rate as two separate temporal and spatial controls on braided-alluvial architecture (cf. Bryant *et al.*, 1995, p. 368). The flume results also suggest that theoretical models that operate at the depositional-element scale (e.g. Webb, 1994; Webb & Anderson, 1996) may require modification, as they assume that set thickness (i.e. sedimentary niche thickness) is a direct surrogate for aggradation rate. The range in aggradation rate calculated for the Ashburton field outcrop (0.32 to 2.85 mm yr$^{-1}$, Ashworth *et al.* (1997)) is similar to that recorded for other fluvial systems in the Holocene, such as the Rhine–Meuse and Mississippi deltas (*c.* 1.5 mm yr$^{-1}$, Törnqvist (1993) and Törnqvist *et al.* (1996)) and Canadian anastomosing rivers (1.5–6 mm yr$^{-1}$, Smith (1986)). This suggests that the 'regional' aggradation rate may have to vary by several orders of magnitude in braided rivers before it triggers a significant change in sedimentation style at the architectural-element scale.

# CONCLUSIONS

This research has provided the first results from scaled flume studies designed to investigate the influence of aggradation rate on alluvial architecture at the scale of the bar and channel in gravel-bed braided rivers. This work demonstrates the clear capability of combined field and flume studies to yield key insights into the preservation of braided alluvium. Three principal conclusions may be drawn from this work.

**1** Flume-scale modelling presents a powerful method by which to simulate alluvial aggradation and to investigate the dominant allo- and autocyclic controls on sediment preservation. Aggradation of a 1 : 50 scale model of the braided Ashburton River produces realistic internal architecture, grain-size sorting and erosion surfaces that can be compared to the field prototype. Although some flume–field discrepancies are inherent within the modelling, and techniques for more realistically scaling time and the effects of cohesion require development, the flume model yields information not available from other sources.

**2** For most depositional niches there is a good agreement between the frequency of occurrence and width of fine- and coarse-grained depositional niches preserved in the flume and field.

**3** The Ashburton gravels were deposited during four different aggradation rates that varied over an order of magnitude, whereas the Ashburton flume model simulated deposition at two aggradation rates that varied by 100%. Large data sets of field and flume alluvial architecture suggest that the width, thickness and frequency of occurrence of fine- and coarse-grained deposition in braided-river alluvium are independent of a change in aggradation rate of the magnitude recorded in the field and flume. Internal alluvial architecture, on the scale of the channel fill, is thus determined by the 'instantaneous' aggradation rate, related to individual flood events, rather than the mean regional aggradation rate. The thicknesses of sedimentary niches in the field show no clear trend with changing aggradation rate and marginally decrease up-section—possibly in response to a change in sediment supply and grain size.

This combined field and flume study suggests several issues that require further research. Firstly, there is a need for more information on the relationship between aggradation rate and both the type and frequency of channel avulsion (cf. Heller & Paola, 1996). In particular, data are required on the frequency of avulsion in braided rivers over time periods of $10^1$ to $10^3$ yr (cf. Mackey & Bridge,

1995) at the anabranch scale. Second, there is a need for more work on the impact of a changing aggradation rate on the three-dimensional architecture of braided-river alluvium. This work should encompass a wide range of aggradation rates and should aim to quantify the depth of intrinsic erosion generated by the river channel, its areal significance and its relation to regional avulsion. Third, there must be consideration of the relationship between alluvial architecture and down-basin changes in avulsion history, sedimentation rate, long-profile incision/aggradation and terrace formation over periods of $10^2$ to $10^4$ yr. Finally, there is a need to combine field, flume and theoretical geological information to create hybrid alluvial-architecture models (Webb & Anderson, 1996) that operate across a range of temporal and spatial scales. Until such robust and comprehensive field and flume geological data sets are obtained, all alluvial-architecture models can perhaps be treated only as working hypotheses (Heller & Paola, 1996).

## ACKNOWLEDGEMENTS

This research was funded by the Natural Environment Research Council (GR9/01640), ARCO Oil plc (USA and UK) and a University of Leeds Academic Development Fund grant. We are grateful to James Parr and other ARCO colleagues for their support of this work and to Angus Jackson and Steve Mardon who kindly helped with some of the flume measurements and data input. Dr Mark Bateman at the Centre for Dryland Research, University of Sheffield, processed the OSL samples and staff at the Canterbury Regional Council provided useful background information on the hydrology and geology of the Canterbury Plains. Don and Pauline Wilson are thanked for their logistical advice and great hospitality during the field work. The Royal Society kindly provided travel grants for PJA and JLB to attend the 6th International Conference on Fluvial Sedimentology in Cape Town, South Africa to present this paper. Finally, we are indebted to Peter Ashmore and Dru Germanoski, who provided thought-provoking and thorough reviews of this paper, which helped to sharpen our interpretation of the data and to significantly improve the scope of this work.

## REFERENCES

ASHMORE, P.E. (1991a) How do gravel-bed rivers braid? *Can. J. Earth Sci.*, **28**, 326–341.

ASHMORE, P.E. (1991b) Channel morphology and bed load pulses in braided, gravel-bed streams. *Geogr. Annaler*, **68**, 361–371.

ASHMORE, P.E. (1993) Anabranch confluence kinetics and sedimentation processes in gravel-braided streams. In: *Braided Rivers* (Eds Best, J.L. & Bristow, C.S.), Spec. Publ. geol. Soc. London, No. 75, pp. 129–146. Geological Society of London, Bath.

ASHWORTH, P.J. (1996) Mid-channel bar growth and its relationship to local flow strength and direction. *Earth Surf. Process. Landf.*, **21**, 103–123.

ASHWORTH, P.J. & BEST, J.L. (1994) *The Scale Modelling of Braided Rivers of the Ivishak Formation, Prudhoe Bay II: Shale Geometries and Response to Different Aggradation Rates.* Final Project Report, British Petroleum, Sunbury, Middlesex, August 1994, 247 pp.

ASHWORTH, P.J., BEST, J.L., LEDDY, J. & GEEHAN, G.W. (1994) The physical modelling of braided rivers and deposition of fine-grained sediment. In: *Process Models and Theoretical Geomorphology* (Ed. Kirkby, M.J.), pp. 115–139. John Wiley & Sons, Chichester.

ASHWORTH, P.J., BEST, J.L. & PEAKALL, J. (1997) *Aggradational Controls on the Three Dimensional Architecture of Braided Alluvium: a Combined Field and Flume Modelling Approach.* Final Report to the Natural Environment Research Council, Swindon, grant GR9/01640, April 1997, 9 pp.

BAL, A.A. (1996) Valley fills and coastal cliffs buried beneath an alluvial plain: evidence from variation of permeabilities in gravel aquifers, Canterbury Plains, New Zealand. *J. Hydrol. (NZ)*, **35**, 1–27.

BENTHAM, P.A., TALLING, P.J. & BURBANK, D.W. (1993) Braided stream and flood-plain deposition in a rapidly aggrading basin: the Escanilla Formation, Spanish Pyrenees. In: *Braided Rivers* (Eds Best, J.L. & Bristow, C.S.), Spec. Publ. geol. Soc. London, No. 75, pp. 177–194. Geological Society of London, Bath.

BEST, J.L. & ASHWORTH, P.J. (1997) Scour in large braided rivers and the recognition of sequence stratigraphic boundaries. *Nature*, **387**, 275–277.

BRIDGE, J.S. (1985) Paleochannel patterns inferred from alluvial deposits: a critical evaluation. *J. sediment Petrol.*, **55**, 579–589.

BRIDGE, J.S. (1993) Description and interpretation of fluvial deposits: a critical perspective. *Sedimentology*, **40**, 801–810.

BRIDGE, J.S. & LEEDER, M.R. (1979) A simulation model of alluvial stratigraphy. *Sedimentology*, **26**, 617–644.

BRIERLEY, G.J. (1991) Bar sedimentology of the Squamish River, British Columbia: definition and application of morphostratigraphic units. *J. sediment Petrol.*, **61**, 211–225.

BROWN, L.J., WILSON, D.D., MOAR, N.T. & MILDENHALL, D.C. (1988) Stratigraphy of the late Quaternary deposits of the northern Canterbury Plains, New Zealand. *N.Z. J. Geol. Geophys.*, **31**, 305–335.

BRYANT, M., FALK, P. & PAOLA, C. (1995) Experimental study of avulsion frequency and rate of deposition. *Geology*, **23**, 365–368.

GEEHAN, G.W., LAWTON, T.F., SAKURAI, S., *et al.* (1986) Geological prediction of shale continuity, Prudhoe Bay Field. In: *Reservoir Characterization* (Eds Lake, L.W. & Carroll, H.B.), pp. 63–82. Academic Press, New York.

GERMANOSKI, D. & SCHUMM, S.A. (1993) Changes in braided river morphology resulting from aggradation and degradation. *J. Geol.*, **101**, 451–466.

GIBB, J.G. (1978) Rates of coastal erosion and accretion in New Zealand. *N.Z. J. Mar. Freshwater Res.*, **12**, 429–456.

GIBB, J.G. (1986) A New Zealand regional Holocene eustatic sea-level curve and its application to determination of vertical tectonic movements. *R. Soc. N.Z. Bull.*, **24**, 377–395.

GIBBONS, A. (1997) Doubts over spectacular dates. *Science*, **278**, 220–222.

HELLER, P.L. & PAOLA, C. (1996) Downstream changes in alluvial architecture: an exploration of controls on channel-stacking patterns. *J. sediment Res.*, **66**, 297–306.

HOEY, T.B. & SUTHERLAND, A.J. (1991) Channel morphology and bedload pulses in braided rivers: a laboratory study. *Earth Surf. Process. Landf.*, **16**, 447–462.

HUGGENBERGER, P. (1993) Radar facies: recognition of facies patterns and heterogeneity estimation (Pleistocene Rhine gravel, N.E. Switzerland). In: *Braided Rivers* (Eds Best, J.L. & Bristow, C.S.), Spec. Publ. geol. Soc. London, No. 75, pp. 163–176. Geological Society of London, Bath.

JERVEY, M.T. (1988) Quantitative geological modeling of siliclastic rock sequences and their seismic expression. In: *Sea-level Changes — an Integrated Approach* (Eds Wilgus, C.K., Hastings, B.S., Kendall, C.G. St, Posamentier, H.W., Ross, C.A. & Van Wagoner, J.C.), Spec. Publ. Soc. econ. Paleont. Miner., Tulsa, **42**, 47–69.

JIN, D. & SCHUMM, S.A. (1987) A new technique for modelling river morphology. In: *International Geomorphology, 1986 Part I* (Ed. Gardiner, V.), pp. 681–690. John Wiley & Sons, Chichester.

KLAASSEN, G.J. (1990) On the scaling of braided sand-bed rivers. In: *Movable Bed Physical Models* (Ed. Shen, H.W.), pp. 59–71. Kluwer Academic, Dordrecht.

KOSS, J.E., ETHRIDGE, F.G. & SCHUMM, S.A. (1994) An experimental study of the effects of base-level change on fluvial, coastal plain and shelf systems. *J. sediment. Res.*, **B64**, 90–98.

LAWRENCE, D.T., DOYLE, M. & AIGNER, T. (1990) Stratigraphic simulation of sedimentary basins—concepts and calibrations. *Bull. Am. Assoc. petrol. Geol.*, **74**, 273–295.

LECKIE, D.A. (1994) Canterbury Plains, New Zealand — implications for sequence stratigraphic models. *Bull. Am. Assoc. Petrol. Geol.*, **78**, 1240–1256.

LEDDY, J.O., ASHWORTH, P.J. & BEST, J.L. (1993) Mechanisms of anabranch avulsion within gravel-bed braided rivers: observations from a physical scaled model. In: *Braided Rivers* (Eds Best, J.L. & Bristow, C.S.), Spec. Publ. geol. Soc. London, No. 75, pp. 119–127. Geological Society of London, Bath.

LEEDER, M.R. & STEWART, M. (1996) Fluvial incision and sequence stratigraphy: alluvial responses to relative sea-level fall and their detection in the geological record. In: *Sequence Stratigraphy in British Geology* (Eds Hesselbo, S.P. & Parkinson, D.N.), Spec. Publ. geol. Soc. London, No. 103, pp. 25–39. Geological Society of London, Bath.

LEEDER, M.R., MACK, G.H., PEAKALL, J. & SALYARDS, S.L. (1996) First quantitative test of alluvial stratigraphic models: Southern Rio Grande rift, New Mexico. *Geology*, **24**, 87–90.

MACKEY, S.D. & BRIDGE, J.S. (1995) Three-dimensional model of alluvial stratigraphy: theory and application. *J. sediment. Res.*, **B65**, 7–31.

MIALL, A.D. (1977) A review of the braided-river depositional environment. *Earth Sci. Rev.*, **13**, 1–62.

MIALL, A.D. (1993) The architecture of fluvial–deltaic sequences in the Upper Mesaverde Group (Upper Cretaceous), Book Cliffs, Utah. In: *Braided Rivers* (Eds Best, J.L. & Bristow, C.S.), Spec. Publ. geol. Soc. London, No. 75, pp. 305–333. Geological Society of London, Bath.

MOSLEY, M.P. (1982) Analysis of the effect of changing discharge on channel morphology and instream uses in a braided river, Ohau River, New Zealand. *Wat. Resour. Res.*, **18**, 800–812.

OUCHI, S. (1985) Response of alluvial rivers to slow active tectonic movement. *Bull. geol. Soc. Am.*, **96**, 504–515.

PAOLA, C. (1988) Subsidence and gravel transport in alluvial basins. In: *New Perspectives in Basin Analysis* (Eds Kleinspehn, K.L. & Paola, C.), pp. 231–243. Springer-Verlag, New York.

PAOLA, C. & BORGMAN, L. (1991) Reconstructing random topography from preserved stratification. *Sedimentology*, **38**, 553–565.

PAOLA, C., HELLER, P.L. & ANGEVINE, C.L. (1992) The large-scale dynamics of grain-size variation in alluvial basins, I, theory. *Basin Res.*, **4**, 73–90.

PEAKALL, J. (1995) *The influences of lateral ground-tilting on channel morphology and alluvial architecture.* PhD dissertation, University of Leeds, 333 pp.

PEAKALL, J., ASHWORTH, P.J. & BEST, J.L. (1996) Physical modelling in fluvial geomorphology: principles, applications and unresolved issues. In: *The Scientific Nature of Geomorphology* (Eds Rhoads, B.L. & Thorn, C.E.), pp. 221–253. John Wiley & Sons, Chichester.

RAMOS, A. & SOPEÑA, A. (1983) Gravel bars in low-sinuosity streams (Permian and Triassic, central Spain). In: *Modern and Ancient Fluvial Systems* (Eds Collinson, J.D. & Lewin, J.), Spec. Publs int. Ass. Sediment., No. 6, pp. 301–313. Blackwell Scientific Publications, Oxford.

RAMOS, A., SOPEÑA, A. & PEREZ-ARLUCEA, M. (1986) Evolution of Buntsandstein fluvial sedimentation in the northwest Iberian ranges (Central Spain). *J. sediment. Petrol.*, **56**, 862–875.

SADLER, P.M. (1981) Sediment accumulation rates and the completeness of stratigraphic sections. *J. Geol.*, **89**, 569–584.

SALTER, T. (1993) Fluvial scour and incision; models for the development of realistic reservoir geometries. In: *Characterisation of Fluvial and Aeolian Reservoirs* (Eds North, C.P. & Prosser, D.J.), Spec. Publ. geol. Soc. London, No. 73, pp. 33–51. Geological Society of London, Bath.

SCHUMM, S.A. (1986) Alluvial river response to active tectonics. In: *Active Tectonics* (Ed. Wallace, R.E.), pp. 80–94. National Academy Press, Washington, DC.

SCHUMM, S.A. (1993) River response to baselevel change: implications for sequence stratigraphy. *J. Geol.*, **101**, 279–294.

SCHUMM, S.A. & KHAN, H.R. (1972) Experimental study of channel patterns. *Bull. geol. Soc. Am.*, **83**, 1755–1770.

SCHUMM, S.A., MOSLEY, M.P. & WEAVER, W.E. (1987) *Experimental Fluvial Geomorphology.* John Wiley & Sons, New York, 413 pp.

SIEGENTHALER, C. & HUGGENBERGER, P. (1993) Pleistocene Rhine gravel: deposits of a braided river system with dominant pool preservation. In: *Braided Rivers* (Eds Best, J.L. & Bristow, C.S.), Spec. Publ. geol. Soc. London, No. 75, pp. 147–162. Geological Society of London, Bath.

SMITH, D.G. (1986) Anastomosing river deposits, sedimentation rates and basin subsidence, Magdalena River, northwestern Colombia, South America. *Sediment. Geol.*, **46**, 177–196.

SMITH, N.D. (1972) Flume experiments on the durability of mud clasts. *J. sediment. Petrol.*, **42**, 378–383.

SMITH, S.A. (1990) The sedimentology and accretionary style of an ancient gravel-bed stream: the Budleigh Salterton Pebble Beds (Lower Triassic), southwest England. *Sediment. Geol.*, **67**, 199–219.

SOONS, J.M. & GULLENTOPS, F.W. (1973) Glacial advances in the Rakaia Valley, New Zealand. *N.Z. J. Geol. Geophys.*, **16**, 425–438.

STEEL, R.J. & THOMPSON, D.B. (1983) Structures and textures in Triassic braided stream conglomerates ('Bunter' Pebble Beds) in the Sherwood Sandstone Group, North Staffordshire, England. *Sedimentology*, **30**, 341–367.

STOKES, S. (in press) Luminescence dating applications in geomorphological research. *Earth Surf. Process. Landf.*

SUGGATE, R.P. (1965) Late Pleistocene geology of the northern part of the South Island, New Zealand. *N.Z. geol. Surv. Bull.*, **77**, 1–91.

SUGGATE, R.P. (1990) Late Pliocene and Quaternary glaciations of New Zealand. *Quat. Sci. Rev.*, **9**, 175–197.

TÖRNQVIST, T.E. (1993) *Fluvial Sedimentary Geology and Chronology of the Holocene Rhine–Meuse Delta*, the Netherlands. Nederlandse Geografische Studies, 169 pp. Universiteit Utrecht, the Netherlands.

TÖRNQVIST, T.E. (1994) Middle and late Holocene avulsion history of the River Rhine (Rhine–Meuse delta, Netherlands). *Geology*, **22**, 711–714.

TÖRNQVIST, T.E., KIDDER, T.R., AUTIN, W.J., *et al.* (1996) A revised chronology for Mississippi River subdeltas. *Science*, **273**, 1693–1696.

TYLER, K., HENRIQUEZ, A. & SVANES, T. (1994) Modeling heterogeneities in fluvial domains: a review of the influence on production profiles. In: *Stochastic Modeling and Geostatistics* (Eds Yarus, J.M. & Chambers, R.L.), pp. 77–89. Computers in Applied Geology, Vol. 3, American Association of Petroleum Geologists, Tulsa, OK.

WARBURTON, J. (1996) (Ed.) Hydraulic modelling of braided gravel-bed rivers. *J. Hydrol. (NZ)*, **35**, 153–303.

WARBURTON, J. & DAVIES, T.R.H. (1998) The use of hydraulic models in management of braided gravel-bed rivers. In: *Gravel-bed Rivers in the Environment* (Eds Klingeman, P.C., Beschta, R.L., Komar, P.D. & Bradley, J.B.). Water Research Publications, Highlands Ranch, CO.

WEBB, E.K. (1994) Simulating the three-dimensional distribution of sediment units in braided-stream deposits. *J. sediment. Res.*, **B64**, 219–231.

WEBB, E.K. (1995) Simulation of braided-channel topology and topography. *Wat. Resour. Res.*, **31**, 2603–2611.

WEBB, E.K. & ANDERSON, M.P. (1996) Simulation of preferential flow in three-dimensional, heterogeneous conductivity fields with realistic internal architecture. *Wat. Resour. Res.*, **32**, 533–545.

WILLIS, B.J. (1993) Evolution of Miocene fluvial systems in the Himalayan foredeep through a two kilometer-thick succession in northern Pakistan. *Sediment. Geol.*, **88**, 77–121.

WILLIS, B.J. & BEHRENSMEYER, A.K. (1994) Architecture of Miocene overbank deposits in Northern Pakistan. *J. sediment Res.*, **B64**, 60–67.

WIZEVICH, M.C. (1991) Photomosaics of outcrops: useful photographic techniques. In: *The Three-Dimensional Facies Architecture of Terrigenous Clastic Sediments and its Implications for Hydrocarbon Discovery and Recovery* (Ed. Miall, A.D. & Tyler, N.), pp. 22–24. Concepts in Sedimentology and Paleontology, Vol. 3, Society of Economic Paleontologists and Mineralogists, Tulsa, OK.

WOOD, L.J., KOSS, J.E. & ETHRIDGE, F.G. (1994) Simulating unconformity development and unconformable stratigraphic relationships through physical experiments. In: *Unconformity-related Hydrocarbons in Sedimentary Sequences; Guidebook for Petroleum Exploration and Exploitation in Clastic and Carbonate Sediments* (Eds Dolson, J.C., Hendricks, M.L. & Wescott, W.A.), pp. 23–34. Rocky Mountain Association of Geologists, Denver, CO.

*Spec. Publs int. Ass. Sediment.* (1999) **28**, 347–362

# Sedimentary facies from ground-penetrating radar surveys of the modern, upper Burdekin River of north Queensland, Australia: consequences of extreme discharge fluctuations

C. R. FIELDING\*, J. ALEXANDER† *and* R. McDONALD‡

\**Department of Earth Sciences, University of Queensland, Queensland 4072, Australia;*
†*School of Environmental Sciences, University of East Anglia, Norwich NR4 7TJ, UK; and*
‡*Terradat Geophysics, P.O. Box 319, Cardiff CF1 3UJ, UK*

## ABSTRACT

Ground-penetrating radar (GPR), mapping and sediment-sampling techniques were used to investigate a reach of the upper Burdekin River, Australia after successive discharge events (1994–1997). The river is moderately sinuous, partly bedrock-controlled and has incised a channel 5–25 m deep and 300–400 m wide. It drains a large, mainly subhumid to semi-arid catchment that receives erratic summer rainfall, related to the passage of tropical cyclones across north-east Australia. Larger, short-duration discharge events cause major changes to the channel, bars, banks and bed but, because of their short duration (days to weeks), equilibrium channel form does not develop, channel-forming discharge is difficult to define and the deposit characteristics depend on the sequence of individual events. Discharge events that cause large-scale change to the deposits have a recurrence interval of about 8.5 yr. The extreme discharge variability of the Burdekin is reflected in its facies assemblage. A variety of barforms, including braid, lateral, transverse and point bars, are exposed on the river bed for much of any year and these are covered by dunes of varying geometry, plane beds, gravelly antidunes and gravel patches, with significant, stable arborescent vegetation (*Melaleuca argentea*), mainly in flow-parallel, linear groves. Successive GPR surveys, before and after the March 1997 (17 808 m$^3$ s$^{-1}$) channel-modifying event, demonstrate up to 6 m of bed erosion and deposition. The GPR data show distinct facies zonation relative to position on or in the major complex bars. Lower bar deposits are well-sorted, coarse- to very coarse-grained, gravelly sands and upper bar and bank-top deposits are principally silts to very fine-grained sands, with some interbedded coarser grained beds close to the active channel. Above the basal erosion surface, large-scale (to about 3 m) cross-bedded facies pass up-dip of major bedding surfaces and in some places vertically up into smaller scale cross-bedded facies. In places and at times the very large-scale cross-strata are truncated, recording reworking during the falling stage of major discharge events or during smaller events. Planar-stratified and cross-bedded facies are preserved above truncation surfaces. Lesser discharge events cause minor modification of topography, and deposit sediments on the bed and lower bar surfaces. The upper bar deposits are formed in major discharge events and deposits of individual events drape topography and thicken into swales. The facies assemblage and architecture is distinctive of this style of river and may be useful in identifying the deposits of variable-discharge rivers in the rock record.

## INTRODUCTION

The sediments of modern, tropical, variable-discharge rivers are poorly known, allowing virtually no realistic analogues for such deposits in the rock record. Such facies are nonetheless likely to be well represented in the stratigraphical record. We have studied the geomorphology, hydrology, water chemistry, surface and subsurface sediments and river-bed vegetation of the modern upper

Burdekin River in north Queensland, Australia (Fielding & Alexander, 1996; Alexander & Fielding, 1997; Fielding *et al.*, 1997). This paper describes the surface alluvial facies characteristics of the Burdekin from maps and observations of channel behaviour and deposits, and subsurface architecture from interpretation of ground-penetrating radar (GPR) data. The combination of GPR

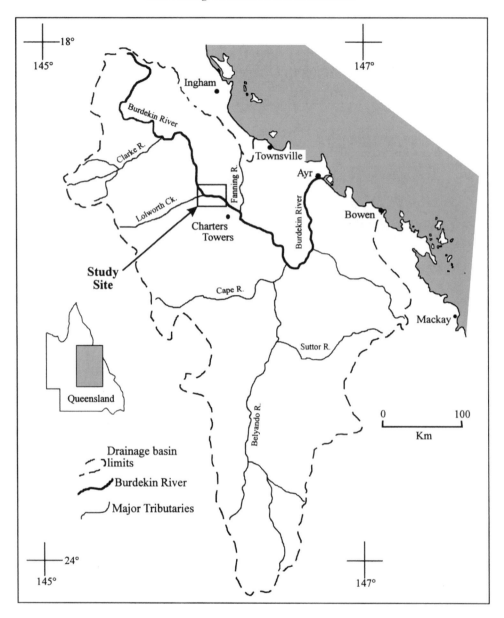

**Fig. 1.** Map showing the location of the Burdekin River drainage basin in north-east Australia.

data with new understanding of the bed- and barform processes, gained from examining a larger area over a longer period, has resulted in radically improved understanding of the system compared with that presented by Fielding & Alexander (1996). This analysis demonstrates that the character of the Burdekin River deposits strongly reflects the extreme discharge variability and may provide a geological 'fingerprint' for use in the rock record.

## THE BURDEKIN RIVER

The Burdekin River is one of the larger streams draining the Australian continent, yet until recently had received little attention from sedimentologists (see references in Fielding & Alexander, 1996). The sedimentology of the upper part of the river, the focus of the present study,

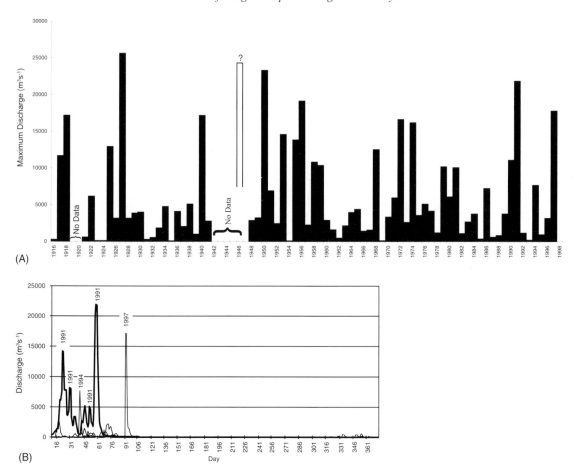

**Fig. 2.** (A) Annual maximum discharge series of the Burdekin River at Charters Towers, measured from Queensland Department of Primary Industries, Water Resources Commission, gauging stations at Charters Towers Weir and Sellheim. Note that although no gauging data were recorded at Charters Towers between 1941 and 1947, photographs and other information record a very large flood in March 1946, which was recorded at the Home Hill gauging station (on the Burdekin Delta). Note also the context of the 1997 flow event, which is the only flow event > 15 000 m³ s⁻¹ to have occurred since 1991. (B) Discharge for the period 1991–1997 recorded at the Sellheim gauging station near Charters Towers. The horizontal axis is days through the year, and the records for the 7 yr are superimposed.

has not been fully documented previously. The Burdekin River drains an area of 129 500 km² in central and northern Queensland (Fig. 1). The Burdekin has the largest mean annual runoff of any river on the east coast of Queensland ($9.8 \times 10^9$ m³ yr⁻¹), but experiences extreme variability ($2 \times 10^8$ to $2.9 \times 10^{10}$ m³ yr⁻¹), that is both highly seasonal and erratic (Fig. 2; Burdekin Project Committee, 1977; Belperio & Johnson, 1985; Pringle, 1991). Although the headwaters lie within tropical rainforested highlands, much of the drainage basin experiences annual rainfall of less than 700 mm.

More than 90% of the annual discharge occurs consist-

ently between January and April (Fig. 2B) and is related to precipitation from tropical cyclones (Pringle, 1991). The maximum discharge recorded for the upper Burdekin at Charters Towers is 25 659 m³ s⁻¹, on 12 February 1927, and for the lower river on the delta at Home Hill is 40 393 m³ s⁻¹, on 4 March 1946. Event hydrographs (e.g. Fig. 3) rise very rapidly and many events arrive at the study site as a bore. The flood waves have velocities of up to 4.3 m s⁻¹. Very high discharge lasts only a few hours to a few days and drops exponentially over periods of several days to a few weeks (Figs 2B & 3; see also Fielding & Alexander, 1996).

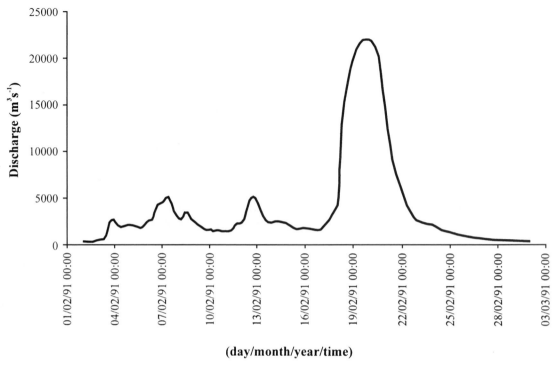

**Fig. 3.** Flood hydrograph for February 1991 overbank flow-event, based on gauging measurements at Sellheim, taken at 15-min intervals. Note the precipitous rising limb and the more gradual falling limb.

The modern upper Burdekin River is straight to moderately sinuous (Figs 4 & 5A & B). The channel typically is incised between 5 and 25 m into its own alluvium, and locally into Cenozoic basalt or Palaeozoic bedrock. Over the study area, the river has a mean gradient of 0.00058, although the longitudinal profile shows discrete steps at bedrock constrictions and deep, locally bedrock-floored, perennial water-holes are common at channel confluences (data from Burdekin Project Committee, 1977). Some straight stretches of the river are probably bedrock-controlled and some bends obviously are related to topographic obstacles. Nonetheless, the presence of point bars and ridge-and-swale topography indicates that parts of the river are undergoing meander migration. In the study area north-west of Charters Towers (Figs 1 & 4), the channel bed typically is 300–400 m wide and the width/depth ratio varies between 10 and 30.

During low-flow conditions (9 months of the year or more), the river bed is largely exposed, with only a narrow, 'misfit', perennial channel carrying a modest flow (typically $< 1$–$20$ m$^3$ s$^{-1}$). The dry river bed can be divided into areas of sand and gravel flats and bars of varying geometry and size (including point, lateral and braid bar forms). In places, these features combine to form sand and gravel sheets up to 3 km long. As expected, the bed generally slopes noticeably toward the outer bank of bends and is generally flat or gently sloping across and/or downstream on straight reaches. A notable feature of this and other watercourses in subhumid parts of Queensland is the abundance of mature, arborescent vegetation in the river bed, in the case of the Burdekin River mainly paperbark trees (*Melaleuca argentea*; see Fielding *et al.*, 1997).

Our observations (outlined below) show that the channel morphology is strongly controlled by major flow-events. The deposits record features of those events and, to a lesser extent, those of intermediate-size flow-events. *Channel-forming discharge* and *bankfull discharge* normally are considered to be the same, as most studies have concentrated on channels in more equitable conditions and assume the channel is at or near equilibrium with the discharge behaviour (large events last long enough and/or are frequent enough to establish channel form). Such concepts are problematic, however, in erratic-discharge regimes such as the Burdekin, where large flow-events are common, highly variable in magnitude and of very short duration. At the study site we have not been able to establish rating curves for large-discharge events, as we have insufficient data and cannot define bankfull on that basis. Furthermore, similar discharges may reach different heights in the same reach as bed topography and within-channel vegetation vary considerably from event to event. On the time-scale of our observations, the events that caused major impacts on bar, bed and bank conditions were those of 1991 (peak 21 900 $m^3 s^{-1}$) and March 1997 (17 808 $m^3 s^{-1}$). The 7694 $m^3 s^{-1}$ event in 1994 changed bedform distribution and locally modified banks and bars, but made no major modifications. The event was of very short duration and it is likely that if discharge had been maintained for longer, there would have been significant bed change. Using bed observations with preconceived ideas about channel behaviour would lead to the conclusion that the channel-forming discharge lies between 7694 and 17 808 $m^3 s^{-1}$. Such events have return periods (based on historical records) of 3.5–8.5 yr. This approach of defining channel-forming discharge is clearly misleading, as none of the discharge events reach equilibrium with the channel form.

## METHODS

A representative reach of the river, north-west of Charters Towers, was selected for study (Fig. 4). During times of minimal flow (July 1993, August 1994, August 1995, January–February 1996, August 1996 and December 1997), the surface of the river bed and banks was surveyed using a Geodimeter Total Station, and surface features, such as sediment type, bedform distribution and vegetation characteristics, were mapped on to aerial photographs and our own detailed topographic maps. Grain size, fabric and composition of surface-sediment samples were analysed and size, shape, orientation and lithology of gravel-grade sediment were measured *in situ*.

The falling stage of a flow that peaked at 7694 $m^3 s^{-1}$ was observed in March 1994. The falling stage of a flow that peaked at 3200 $m^3 s^{-1}$ and the rise and fall of another smaller flow event were observed in January 1996, and the early falling stage of an event that peaked at 17 808 $m^3 s^{-1}$ was observed in March 1997 (Fig. 2B). Video records were also obtained of the 1991 event, taken from a helicopter about 1 day after the peak flow. Surface-flow velocities were estimated by timing floating debris travelling between two sections. Impeller flow meters were impractical because of the lack of bridges and the high velocity (locally well in excess of 6 m $s^{-1}$).

The subsurface character of deposits was investigated by trenching and augering and by using GPR. Three

curved reaches were selected for GPR surveys on the basis of their likely sedimentary record and ease of access (Fig. 4). Two surveys were carried out, the first in August 1996 after a 3-yr period of little flow and the second, in December 1997, to examine the changes caused by a major discharge event in March 1997. A Sensors & Software PulseEKKO IV system with optional 50, 100 and 200 MHz antennae was used for the first survey (Fig. 5C) and a Mala Geosciences RAMAC/GPR with 100, 200 and 400 MHz antennae was used for the second survey (Fig. 5D). The reason for changing systems for the return visit was that the PulseEKKO system proved to be prone to failure in the hot, dusty working environment and much of the resulting data was degraded by system noise.

The selection of antenna frequency for each survey line was a compromise between achieving sufficient depth of penetration and optimizing resolution. Interpreted bedrock depths were found to be very shallow along most of the survey lines, which necessitated use of the 200 Mhz antenna for high resolution, but at several locations, notably where thicker alluvium was encountered, it proved necessary to use the 100 MHz antenna to improve the signal/noise ratio at bedrock time-depths.

The airwave and direct/ground wave, usually found at the top of a radar section, are absent from some of the GPR lines acquired from the first survey, because they were removed during attempts to reduce horizontal coherent system noise. Depth-conversion velocities of 0.12 m $ns^{-1}$ (unsaturated sediments) and 0.06 m $ns^{-1}$ (saturated sediments) have been derived from isolated CMP tests, standard tables and processing trials to establish near-horizontal water tables in the vicinity of the river channel. Time-depths to the water table vary laterally, as a result of surface topographic changes and true water-elevation trends, which prevents reliable use of a depth conversion scale on the *y*-axis of radar sections. Conversion factors of 1 ns = 0.06 m in dry sand/gravel and 1 ns = 0.03 m in wet sand/gravel were used to convert two-way time to depth.

A traverse line, surveyed in 1996 using both 100 MHz and 200 MHz antennas (line 101: Fig. 4), was resurveyed in 1997 using a 200 MHz antenna. Data acquired in 1996 using the 200 MHz antenna show the major reflectors clearly, but otherwise data quality is poor as a result of prevailing system noise. In order to illustrate changes in surface and subsurface character between the 1996 and 1997 surveys, all three data sets are shown in Fig. 6 (see foldout), aligned so as to be directly comparable.

The 1996 survey recorded approximately 9 km of line data across Dalrymple, Brigalow and Big Bends (Fig. 4), to achieve a grid coverage of the major areas of interest.

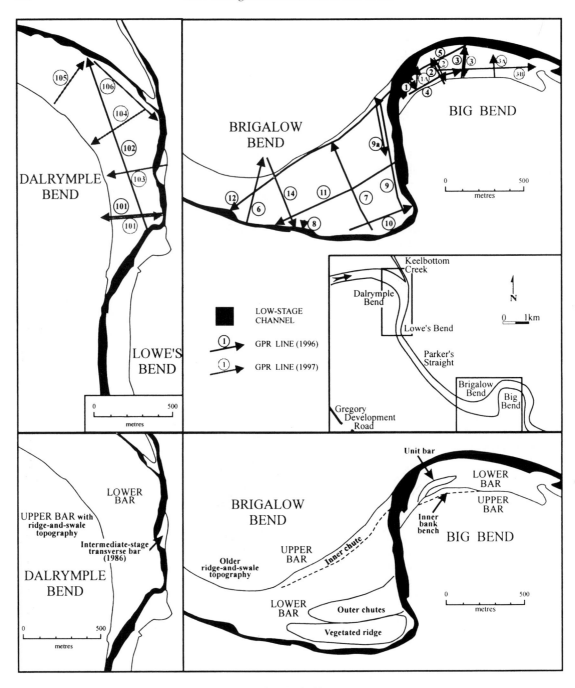

**Fig. 4.** The study sites and the distribution of GPR surveys referred to in this paper.

**Fig. 5.** Representative views of the upper Burdekin River and its sedimentary facies. (A) Aerial view of Dalrymple Bend, looking downstream, taken in August 1994. Note the restriction of flow to the misfit low-stage channel around the outer edge of the bend and the confluence of Keelbottom Creek in the centre of the view. (B) Aerial view of Brigalow Bend, looking upstream, taken in August 1994. Note the eccentric planform of the point bar and the presence of arborescent vegetation on the bar surface. (C) View of GPR operations during the 1996 survey. The antennae were rigged on to a sled which is towed by the lead person. Fibre-optic cables connect the antennae to the control system mounted in the wheelbarrow. (D) The 1997 GPR survey, showing the RAMAC/GPR system operated by one person, crossing the inner bank bench on Big Bend. (E) View across the Dalrymple point bar in December 1997, taken from the vegetated upper bar and looking towards the low-stage channel. Note the occurrence of gravel patches in the upper part of the lower bar (by figures), and the large dunes that cross the lower part of the lower bar in the distance. Flow is from left to right. (F) Front and slip-face of flat-topped transverse bar at upstream end of Big Bend, with waterlogged lee trough colonized by *Melaleuca argentea* saplings.

**Fig. 5.** (G) View of lower part of Big Bend point bar following modest flow-event in January 1996, showing washed-out dunes and drain-off features on surface. (H) Gravel antidune at the upstream end of Brigalow Bend: trench highlights the reversed asymmetry of the bedform, with the shorter, steeper limb facing upstream. (I) Amalgamated beds of small-scale cross-bedded sand and gravel exposed in a cutbank exposure on the lower point bar at Dalrymple Bend (close to GPR line 101: see text for details). (J) Cutbank section through the outer edge of the upper bar at Big Bend, close to the inner bank bench. The section, about 2 m thick, comprises alternating sharp-based beds of loose, coarse to very coarse sand and fine gravel, and more cohesive silt to very fine sand beds, some of which show pedogenic modification.

Although data quality was generally excellent over the lower bars and river bed, silt to very fine sand on the upper vegetated bar and ridge-and-swale areas caused signal attenuation, to the extent that no significant penetration was achieved in these areas (e.g. Fig. 6B). The December 1997 survey was focused on Dalrymple and Big Bends, with the principal aims of resurveying selected 1996 lines to determine effects of the March 1997 event and surveying certain specific barforms and vegetational features (Fig. 4).

Data processing involved the following steps.

**1** Dewow: this process removes low-frequency noise, which occurs as a result of saturation of the recording-instrument electronics by the large-amplitude air and direct waves.

**2** Set time zero: this aligns the first reflected arrival with the time of zero nanoseconds.

**3** Remove background: this subtracts a background trace from each trace in the profile, effectively removing horizontal banding, present in data as a result of either external or internal noise and which is more evident in areas where data amplitudes are lower.

**4** Gain: application of gain amplifies the reflected signals. For the present data, a programmed automatic gain control was applied using a window of 20 ns.

**5** Filtering: unwanted frequencies were removed using a trapezoidal bandpass filter designed to remove noise frequencies either side of the centre frequency of the antenna.

**6** Depth conversion: elevation statics were applied to the data to correct for surface topography.

The software used for processing was GRADIX by Interpex Ltd. Lines were corrected for topography using survey data acquired using a Geodimeter Total Station.

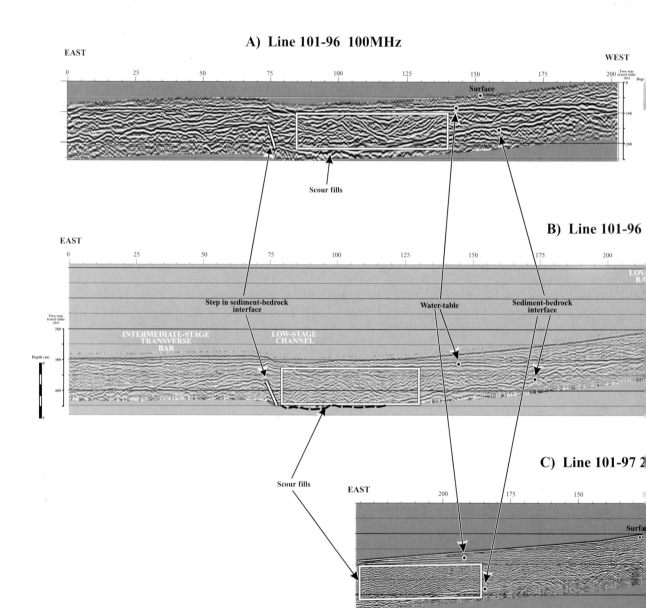

**Fig. 6.** Ground-penetrating radar (GPR) records of line 101 at Dalrymple Bend (see Fig. 4 for location): (A) 1996 survey (PulseEKKO IV, 100 MHz antenna); (B) 1996 survey with 200 MHz antenna; (C) 1997 survey (RAMAC/GPR, 200 MHz antenna). The three records are fully processed, corrected for topography, reproduced at the same scale and aligned relative to each other. The principal surface and subsurface features are annotated (see text for details). In both sections, flow direction is into the plane of the paper. Note the principal changes between the 1996 and 1997 surveys: rearrangement of the trough-shaped reflectors beneath the low-stage channel, interpreted as scour fills, and the formation of a low-relief, ?incipient unit bar at 10–70 m on the 1997 record.

# LINE 101
# Dalrymple Bend

**FLOW DIRECTION
INTO PLANE OF PAPER**

**200MHz**

WEST

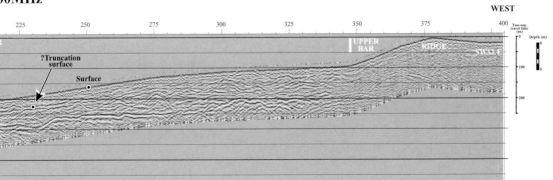

?Truncation surface

Surface

UPPER BAR

RIDGE

SWALE

Two-way travel time (ns)

Depth (m)

**100MHz**

Truncation surface

WEST

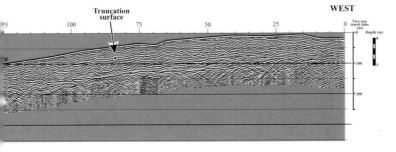

Two-way travel time (ns)

Depth (m)

Interpretation follows the principles of seismic-stratigraphic analysis as recommended by Gawthorpe *et al.* (1993) and the terminology used to describe reflection geometry is after Huggenberger (1993) and van Heteren *et al.* (1998). Ground-truthing was provided by cutbank exposures, trenches and auger holes.

## RESULTS

All GPR lines show a persistent, high-amplitude event that from its local intersection with the surface can be identified as the sediment–bedrock unconformity. This irregular surface lies within a few metres of the surface across the study sites. The water table also can be identified clearly in places, where it is not apparently coincident with the sediment–bedrock interface. Within the alluvial record are well-defined features that can be interpreted confidently as sedimentary structures (mainly cross-bedding on a variety of scales) and their bounding stratal surfaces. The cross-sectional geometry of bedsets on the various bar surfaces and of the alluvial deposit as a whole, is clearly visible on GPR lines. Point diffractions are common and are interpreted as the result of buried trees/large wood debris, boulders or irregularities in the bedrock surface, depending on the context of the feature.

### Dalrymple Bend

Dalrymple Bend has the surface form of a point bar (Figs 4 & 5A), with the perennial channel situated close to the outer bank. The geomorphology is complicated locally by the confluence of Keelbottom Creek and by an area of bedrock exposure near the southern (downstream) exit of the bend. A gently sloping sand- and gravel-covered lower bar rises to a vegetated upper bar, which passes inward into a succession of ridges and swales with an amplitude of approximately 5 m and a wavelength of 100–150 m.

Although the surface distribution of bedforms varies from year to year, a general pattern has been recognized. Within the low-flow channel, small (0.05–0.20 m amplitude) dunes remain active throughout most years, and these modify falling to intermediate stage, flat-topped transverse bars (0.4–1.0 m amplitude). The lowermost part of the lower bar surface is a gently sloping plane bed, with a few, mainly small and in many cases washed-out dunes and local development of gravelly antidunes (Alexander & Fielding, 1997). At certain times, parts of this surface are covered in a thin layer of mud. Upslope, the surface is covered in dunes, which vary in size both spatially and with successive events and include large,

slightly sinuously crested bedforms with slip-faces up 3 m in height (Fig. 5E). The uppermost part of the lower bar is covered by a field of smaller dunes and large gravel patches (Fig. 5E). A sharp break of slope defines the foot of the upper, vegetated bar. The dune fields noted above were particularly well defined in December 1997 (Fig. 5E), suggesting that they are related directly to major flow-events.

Figure 4 shows the location of GPR lines 101 and 102 (1996 survey) across Dalrymple Bend. Line 101 is a dip line to examine the structure of the lower bar and low-stage channel in a direction perpendicular to flow, and was repeated during the 1997 survey, whereas line 102 was intended to provide a section across the lower bar in a direction parallel to flow. Line 101 (1996) shows a gross internal structure that varies with position on the point bar (Fig. 6A & B, see foldout). Beneath the thalweg, the GPR record shows a series of oblique-tangential and trough-shaped reflections. These are interpreted as deep (up to 3 m), narrow (up to 40 m) scours, that appear to have a complex, cross-bedded fill and incise almost down to the bedrock surface. Laterally, this facies passes up-dip (towards the inside of the bend) into a different radar facies, characterized by smaller oblique-tangential and hummocky reflections, interpreted as smaller-scale cross-bedding, with discrete hyperbolas attributed to buried boulders or tree trunks. Although there is some interference from the water table, the scour fills and their correlative strata appear to terminate locally below a surface-subparallel reflection, which is interpreted as a truncation or bounding surface. This surface can be traced tentatively up-dip across the lower bar. Above this surface, the radar facies comprises poorly developed, oblique-tangential and discontinuous, subparallel reflections. This is interpreted to indicate structure dominated by planar bedding and low-angle cross-bedding, locally with a downslope-dipping and downlapping geometry.

Line 101 (1997) shows a similar internal architecture (Fig. 6C), but the position and geometry of the scour fills has changed somewhat and where visible, the truncation surface has changed position and geometry, suggesting reorganization of the alluvium during a major flow-event between August 1996 and December 1997. The only flow capable of effecting such changes was the March 1997 event. Furthermore, a low-relief barform on the surface of the upper part of the lower bar, noted during the 1997 survey (Fig. 6C), also can be attributed to sediment deposition during the March 1997 event and may represent the inception of a unit bar.

Line 102 provides a section 1100 m+ long across Dalrymple Bend, in a direction broadly parallel to flow. Two representative portions of the line are reproduced as

## A Line 102: 225-350 m

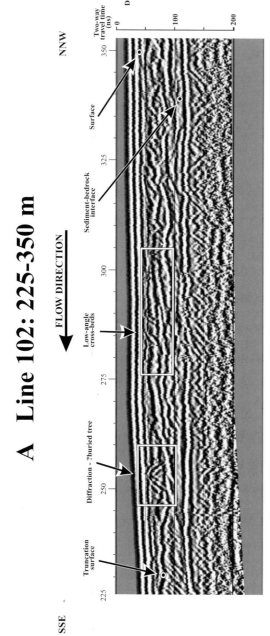

## B Line 102: 675-800m

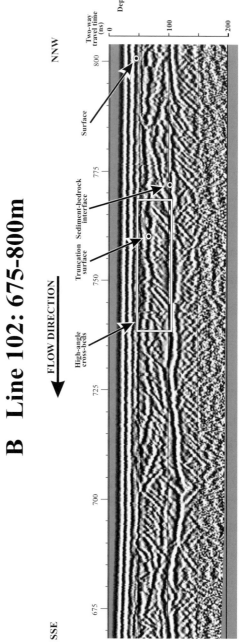

**Fig. 7.** The GPR records of parts of line 102 at Dalrymple Bend (1996 survey: PulseEKKO IV system with 100 MHz antenna: see Fig. 4 for location). (A) 255–350 m, near the downstream end of the bend. (B) 670–800 m, near the upstream end of the bar. The two records are fully processed, corrected for topography and reproduced at the same scale. The principal surface and subsurface features are annotated (see text for details). In both illustrations, flow is from right to left.

Fig. 7. In Fig. 7A, on the downstream portion of the lower bar, the radar facies consists principally of groups of oblique-parallel reflections, bounded by more persistent, parallel–subparallel continuous reflections. These reflections are interpreted as planar bedding and low-angle cross-bedding, with minor, high-angle cross-bedding, defined by slightly irregular bounding surfaces. Less common, asymmetrical, trough-shaped reflectors in the upper part of the fill may be interpreted as oblique sections through small trough cross-beds. Figure 7B shows another portion of line 102 (675–800 m), towards the upstream end of Dalrymple Bend. In this section, the principal radar facies is oblique-parallel and oblique-tangential reflections in the lower part, overlain by less well-developed, parallel–subparallel, continuous reflection character above a bounding (truncation) surface. The former is interpreted as mostly flow-parallel sections through cross-beds typically 2 m thick, some of which may be traced on the GPR record for up to 100 m. Trough forms, developed locally (e.g. between 740 and 770 m), are interpreted as more oblique sections through the same trough cross-sets, or filled scour hollows. The data appear to indicate up-section change in bed character across a laterally persistent bounding surface, which occurs one-half to two-thirds of the way up the sedimentary section.

Dalrymple provides the thickest sedimentary record in the study area, up to 6 m. Lines 101 and 102 show a consistent horizontal and vertical arrangement of radar facies. With the exception of the deep, narrow scour fills, seen beneath the thalweg in line 101 and the low-angle cross-beds, all bedding features imaged by the GPR can be related directly to features visible on the surface or in cutbank sections near the lines (e.g. Fig. 5I). The low-angle cross-beds are problematic: their thickness (typically about 2 m) suggests that they are unlikely to be related to the gravelly antidunes described from the same area by Alexander & Fielding (1997). They might, however, record the migration of low-amplitude unit bars across the point-bar surface or the arrival of a sediment wave (as sediment transport lags discharge in this system). The high-angle cross-beds seen in line 102 are of a similar scale to those seen in flow-transverse section in line 101 and are interpreted as part of the same suite of features. The change in structure within these sets is evident by comparing the 1996 and 1997 surveys of line 101, and can be interpreted as arising from reorganization of the river bed during the bankfull flow-event of March 1997.

We suggest that during rising stage and at peak discharge of major flow-events, sediment moves in bedforms or sheets mobile down to the basal erosion surface.

Large, three-dimensional dunes, which start to develop during peak discharge and continue to migrate during early falling stage (because of bedform lag effects), give rise to the large-scale cross-bedding noted in deeper parts of the channel, with sediment waves depositing low-angle cross-beds and dunes forming high-angle cross-bedding in shallower parts. Truncation surfaces may record fluctuations in scour depth (e.g. in migrating dune troughs) during the falling stage of major flow events (e.g. March 1997, as in the 1997 record), or may reflect the depth to which lesser flow-events are able to rework sediment (1–1.5 m) as in 1996 record. In the latter scenario, the smaller scale structures above the truncation surface can be interpreted as arising from such events. The close relationship between the low-angle cross-beds and truncation surface evident from Fig. 7A suggests that the low-angle cross-beds are formed during the early falling stage of major events.

## Brigalow Bend

Brigalow Bend has the form of a point bar (Fig. 4). The bend comprises an outer low-stage channel, an adjacent, elevated, vegetated ridge, a large area of gently sloping sand and gravel crossed by a series of flat-floored chute channels and an incised chute adjacent to the inner bend, with an adjacent zone of ridge-and-swale topography at the upstream end. Several groves of mature, live *Melaleuca argentea* trees occur on the lower bar surface, together with isolated remnants of a floodplain vegetational community on ridges. The existence of these high areas of floodplain vegetation within the area of the bend, together with the ridge-and-swale distribution, suggest that the meander expanded across the area with the point bar building up to allow bar-top conditions to develop extensively, recent high-discharge events having then caused significant erosion of the bar surface.

The surface of the lower bar is covered mainly with coarse to very coarse sand with minor gravel and local discrete gravel accumulations (mainly adjacent to obstacles such as tree groves). The dominant surface bedforms are small dunes with plane beds in the outer chutes and gravel antidunes (Fig. 5H) at the upstream end of the bar (Alexander & Fielding, 1997). The vegetated ridge surface is covered by silt and very fine sand, with natural exposures indicating that cross-bedded, coarser gravelly sand dominates lower in the section. Silt and very fine sand also dominates the surface of the inner and outer banks. The low-stage channel is similar to that at Dalrymple. Bedrock crops out locally, mainly in the low-stage channel (see Fielding & Alexander (1996) for a more detailed account of Brigalow Bend).

A grid of GPR lines was acquired during the 1996 survey (Fig. 4). Many of these lines did not produce interpretable data, however, owing to the shallow bedrock interface (< 2 m over much of the site), locally steep topography and/or the silt to very fine sand at or near the surface causing attenuation of the signal. The modest thickness of alluvium across the middle (low topography) part of the bar agrees with the suggestion that the chute–channel system has been developing through much or all of the 20th century (Newman-Sutherland, 1995) and this may be an example of chute cutoff in progress.

Figure 8 (see foldout) shows part of line 9 (Fig. 4), which crosses the downstream end of Brigalow Bend in a direction roughly perpendicular to flow. The sediment–bedrock interface is less well defined across much of this line, but the subsurface sediment character of the vegetated ridge and outer chutes is clearly visible. Beneath the ridge, a predominantly parallel-continuous radar facies with some oblique-tangential reflections is disrupted locally by prominent diffractions (Fig. 8). This is interpreted to record the mainly flat-stratified and cross-stratified nature of the coarse sand sediment below the surface, penetrated in places by deep roots from eucalypts and other trees growing on the ridge. Radar facies below the chutes are also mainly parallel-continuous or -discontinuous with some oblique-tangential reflections, again recording the preservation of flat- and cross-stratified coarse, gravelly sands. An oblique-parallel facies is also evident on the northern side of a low, gravel-covered rise within the outer chutes (Fig. 8), and is interpreted as the result of local lateral accretion.

Other GPR lines over Brigalow Bend show sediment patterns that, although less well defined and considerably thinner, are consistent with those found on Dalrymple. *Melaleuca argentea* trees, found living on the lower bar surface, probably survive through having their roots and lower trunks anchored in bedrock.

## Big Bend

Big Bend is a horseshoe-shaped meander, partly constrained by bedrock, which includes a complex bar with the form of a point bar at its upstream end (Fig. 4). Exposed bedrock 'islands', which provide anchor points for definition of the bedrock surface, occur adjacent to the low-stage channel and define the inner bank at the upstream end of the bend. A sloping lower bar surface, covered in sand and gravel, rises irregularly to a vegetated, upper bar that is covered in silt and very fine sand. Intermediate-stage transverse bars and high-stage dunes have been recorded on the lower part of the lower bar (e.g. Fig. 5F). A train of small gravel antidunes (wavelength about 9 m) also has been recorded on this surface.

Oblique, bank-attached unit bars have been observed on the inner bend and an inner bank bench (thought to have developed by growth of unit bars) forms the uppermost part of the lower bar along parts of the southern bank.

The surface distribution of bedforms is similar to that at Dalrymple (although on a smaller scale), with some modifications. Plane beds and small (amplitude 0.05–0.3 m), often washed-out, dunes (Fig. 5G), together with the transverse bars (and remnants of large dunes) occupy the lowest parts of the lower bar, passing upward into fields of dunes that are up to 3 m high at the upstream end of the bar, becoming smaller downstream. Locally and during some years, the lower bar surface is covered by a thin layer (< 0.15 m) of mud, which cracks extensively during the dry season. Morphological changes to the upstream end of Big Bend point bar were observed in each successive field season from 1993 to 1997 and in 1997 it was evident, from the presence of newly exposed, 2-m-high bedrock knolls, that the March 1997 flow-event had removed significant volumes of sediment from the bar. Repeat GPR surveys of 1996 lines 1, 3 and 5 (Fig. 4) support this interpretation.

Line 2 (1996 survey), a dip-line running from the upper bar/lower bar boundary down to the low-stage channel (Fig. 9, see foldout), shows the steep, vegetated slope of the upper bar flattening abruptly on to the inner bank bench, which then slopes down on to the lower part of the lower bar. In the upper bar (0–40 m: Fig. 9), GPR reflectors appear continuous and planar, parallel to the surface: a nearby exposure (Fig. 5J) confirms this geometry and indicates that these planar, inclined strata comprise alternating, sharp-based, very coarse to coarse-grained, partly cross-bedded sand beds and organic-debris-rich mud to very fine-grained, ripple cross-laminated sand layers, some of which are pedogenically altered. Two samples of detrital wood from one of these layers yielded conventional radiocarbon ages of 140 ± 50 yr and modern (Fielding & Alexander, 1996). The surface-parallel, planar strata drape a mound across the inner bank bench, and become less well defined down-dip off the bench. This change in character may result from a change in sediment character, with sand becoming more dominant, or possibly could result from the presence of the mud layers in the shallow subsurface affecting the signal. A mud layer was exposed locally along the outer edge of the bench. The deeper subsurface of the bench and of the lower bar downslope from the bench has discontinuous, inclined reflectors (oblique-tangential), that are interpreted as a combination of point diffractions and some trough cross-bedding. Some of these features appear to be truncated by surface-parallel radar reflectors at the outer edge of the inner bank bench.

Lower on the point bar (60–160 m: Fig. 9), a low-relief, unit bar passes down on to the plane-bedded, lowest part

# Line 9: 0-500 m

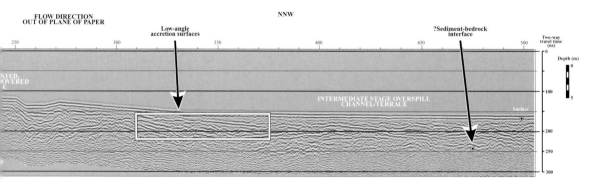

**Fig. 8.** The GPR record of part of line 9 at Brigalow Bend (1996 survey: PulseEKKO IV system with 200 MHz antenna: see Fig. 4 for location). The record is fully processed and corrected for topography. The principal surface and subsurface features are annotated (see text for details). Flow direction is out of the plane of the paper.

2

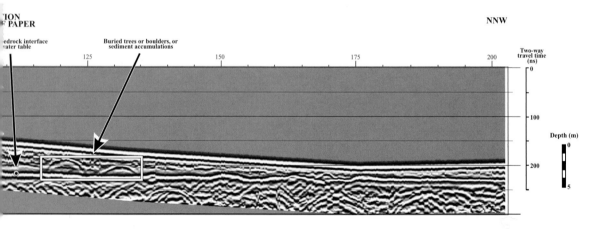

**Fig. 9.** The GPR record of line 2 at Big Bend (1996 survey: PulseEKKO IV system with 100 MHz antenna: see Fig. 4 for location). The record is fully processed and corrected for topography. The principal surface and subsurface features are annotated (see text for details). Flow direction is out of the plane of the paper.

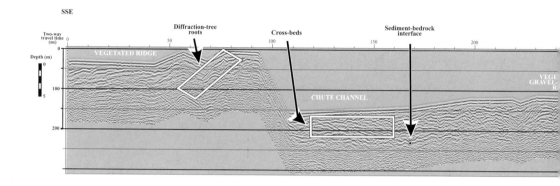

Line

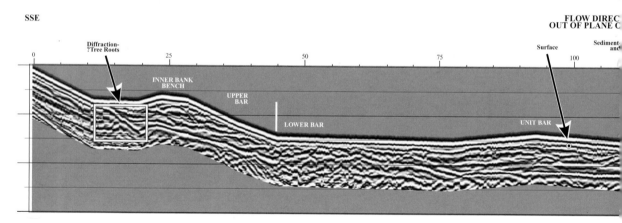

of the exposed river bed. The near-surface structure appears to be dominated by surface-parallel reflectors that drape the unit bar and offlap/downlap towards the low-stage channel. Beneath the draping strata the unit bar shows a similar reflection character to that below the inner bank bench, with a few, mainly discontinuous, flat and inclined reflectors, separated by zones of few reflections. Prominent hummocky reflections (?diffractions) at ~ 120 m and 130 m may represent buried boulders or trees, or sediment accumulations.

Line 2 provides a useful contrast to line 101 at Dalrymple, illustrating a situation where only a thin sediment cover is preserved. At the lower end of line 2, the low-stage channel is bedrock-floored and hence the sediment–bedrock interface can be identified reliably. Because of its minimal sediment cover, no deep scours, such as were found at Dalrymple, were observed at Big Bend (or Brigalow Bend). No clear distinction between large-scale cross-bedded and smaller scale structured units was evident from GPR data, although the sedimentary record in line 2 can be divided clearly into a lower unit, showing a combination of discontinuous, flat and inclined reflectors with point diffractions, and an upper layer, characterized by surface-parallel reflections. The character of the uppermost part of the record may result from the presence of mud layers in the shallow subsurface; up to three such layers were encountered in auger holes drilled nearby in 1995 (Fielding *et al.*, 1997).

Big Bend is considerably smaller than Dalrymple. The GPR lines over Big Bend depict a complex cross-sectional morphology, with an inner bank bench and unit bars that are not as evident from Dalrymple GPR lines. Big Bend lacks any well-developed ridge-and-swale topography, but it is tempting to speculate whether the inner bank bench and unit bar noted at Big Bend are related to the incipient formation of such topography. Both features are clearly composite, being deposited and modified over several flood cycles.

## FACIES ASSEMBLAGE

The examples illustrated and described above are representative of the extensive GPR surveys conducted over the study reach. These data are combined with our extensive sedimentological data base to construct a model for the three-dimensional distribution of lithofacies in the upper Burdekin River. The modern Burdekin River is not undergoing net aggradation and allowances for this fact must be made when comparing with the rock record.

Lower bar deposits are composed principally of coarse to very coarse, gravelly sand with local gravel accumulations (Fig. 10). In areas of significant sediment thickness close to the low-stage channel (Fig. 10A), deep, narrow trough scours are formed during the rising stage of major flow-events and filled by angle-of-repose cross-bedding, that often has a complex internal geometry and where individual sets can be traced in a current-parallel direction for up to 100 m. Set-size decreases up the slope of the lower point bar. In some parts, low-angle cross-bedding is common and may be the result of the migration of low-amplitude, unit bars or sediment waves across the point-bar surface.

Intervals deposited from individual major flow-events are likely to pass laterally from the large-scale cross-bedded sands and gravels into smaller scale cross-bedded strata, as shown by line 101 (Fig. 6). In most comparable ancient fluvial deposits, however, a composite record is more likely to be preserved, where some sediments are deposited from major flow-events, but others are deposited from or reworked during sequences of more modest flow-events, as seen in the 1996 survey of line 101 (Fig. 6A & B). In both situations, large-scale cross-bedded strata are truncated by a flat to scalloped bounding surface, which is overlain by amalgamated, mainly planar-bedded or cross-bedded sand/gravel (Fig. 10A). The deposits may include discontinuous beds of mud and the *in situ* remains of tree saplings, which became established in the river bed during periods of modest flow (Fielding *et al.*, 1997). Antidune deposits also may be more abundant in the deposits of intermediate-scale flows, as a result of lateral confinement of such flows by the channel forms developed by the larger discharge events. Planar cross-bedding could develop from transverse, low–intermediate-stage bars, but these may have low preservation potential (Fig. 10A).

Higher up the slope of the point bar (Fig. 10B), strata are dominated by intermediate-scale cross-bedded gravelly sands, with the potential for formation of unit bars and inner bank benches internally composed of such strata. These strata may be truncated by planar-bedded, surface-concordant bedsets, that drape and offlap pre-existing topography formed by unit bars and inner bank benches. Upper bar deposits comprise alternating beds of coarse to very coarse gravelly sand (recording high flow-events) and pedogenically modified silt to very fine-grained sands, which are surface concordant and drape pre-existing topography. Some of these layers may be preserved within the upper part of the stratigraphy of inner bank benches, where developed. Slump-scar exposures at Dalrymple Bend indicate that the silt to very fine sand deposits that blanket ridge-and-swale topography overlie the facies described and illustrated above.

The facies assemblage found in the upper Burdekin River is composed principally of coarse to very coarse

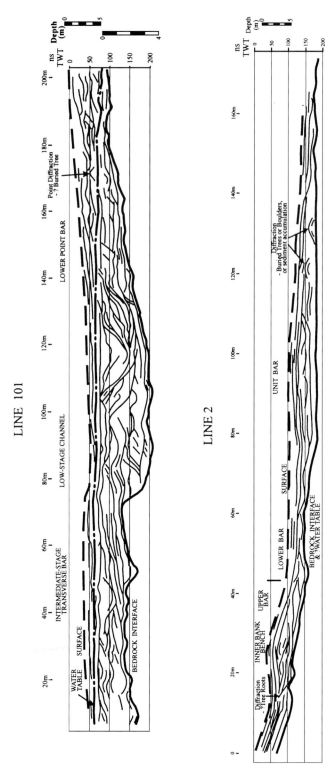

**Fig. 10.** Line drawings from (A) GPR line 101 at Dalrymple, and (B) line 2 at Big Bend, to illustrate the cross-sectional fabric of sedimentary facies in the alluvium of the upper Burdekin River, in an orientation transverse to flow direction. Line 101 illustrates the architecture expected of the thalweg area of the channel, passing up on to the lower slope of the point bar. The outer bank is immediately to the left of the 0 m mark. Line 2 shows the middle and upper portions of the point bar, including a unit bar and inner bank bench, the latter connected to the upper, vegetated bar by a steep slope on which the line begins. See text for details of sedimentary facies.

gravelly sands, with a variety of sedimentary structures, zoned according to their geomorphological position within the river, and lesser amounts of fine sand and silt. The vertical sequence fines up abruptly at the top of the lower bar deposit, via interbedding of coarse, gravelly sand and very fine sand/silt, into the fine-grained sediments of the upper bar and bank. Such vertical successions have been recorded from modern and ancient examples of meandering and other rivers and are not diagnostic of any particular fluvial style. Furthermore, although the interpreted, cross-sectional architecture illustrated in Fig. 10 shows some morphological similarities with other studies of point-bar deposits (e.g. Willis, 1989, 1993), the internal radar structure of the Burdekin point bars is quite different from other GPR images of meandering rivers (Bridge *et al.*, 1995; Leclerc & Hickin, 1997). In particular, although dipping reflectors recording lateral accretion are preserved within the alluvium, their low angle of inclination and irregularity would make identification in the rock record difficult. The possibility therefore exists that deposits of rivers such as the Burdekin, preserved in the ancient record, might be misinterpreted as braided river facies.

An additional feature of the Burdekin River is that the discharge characteristics efficiently segregate the sediment by size. The consequence of this is that the lower bars are composed of coarse to very coarse sand and so no ripples are found on their surfaces. This may represent a significant difference from the deposits of large channels with less variable discharge. The vertical change in sedimentary structure therefore will tend to be more pronounced in the deposits of highly variable discharge channels, with no ripple cross-lamination preserved in lower deposits and an abrupt upward transition into interbedded coarse and fine sand with ripples at the margin of the vegetated upper bar.

The differences noted above can be attributed to the considerable contrast in discharge regime between the mainly modest-sized, perennially flowing streams of the temperate Northern Hemisphere mentioned above and the flashy discharge regime of the Burdekin River. We submit that the facies assemblage and internal architecture documented herein are a reflection of the extreme variations in discharge experienced by the Burdekin River and suggest that similar facies assemblages, preserved in the rock record, might record the deposits of large rivers that experienced comparable discharge variability.

## SUMMARY AND CONCLUSIONS

An investigation of the sedimentary facies characteristics of the upper Burdekin River in north Queensland, Australia, has been carried out as part of a broader, multidisciplinary study of that river. The river is one of a class of streams draining subhumid to semi-arid parts of tropical north-east Australia, where discharge can increase by two to four orders of magnitude in less than 24 h, with major flow-events of 15 000+ $m^3 s^{-1}$ occurring on a regular basis (8.5 yr statistical recurrence). The geomorphology, sediment character and bedform distribution near Charters Towers were mapped and two GPR surveys, designed to investigate the subsurface character of the alluvium, were conducted. Work was focused on three curved sections of the river, named Dalrymple, Brigalow and Big Bends.

The three bends all showed the characteristics of complex point bars, with a (sparsely to non-vegetated) lower bar covered with coarse- to very coarse-grained, gravelly sand and a variety of dune, plane bed and antidune bedforms, passing upslope into a vegetated upper bar covered with silt to very fine-grained sand and, at Dalrymple Bend, with well-developed ridge-and-swale topography. The distribution of sedimentary structures, as interpreted from radar facies, shows a strong zonation relative to position on the point bar. Adjacent to the thalweg, deep, narrow scours are filled by large-scale cross-bedding, which passes up-dip into smaller scale cross-bedding. Comparison between 1996 and 1997 surveys indicates that the bankfull event of March 1997 was responsible for mobilizing sediment down to the bedrock interface and redepositing a similar body of material. Both surveys record a variably developed truncation surface in the middle part of the sedimentary section, separating large-scale from overlying smaller scale cross-bedded facies. In the 1996 survey the overlying strata are interpreted as the result of reworking of the top of the 1991 major-event deposits by successive, more modest flow-events prior to March 1997, whereas in the 1997 survey much, if not all, of the section may owe its bedding character to the March 1997 event.

Unit bars were observed locally on the lower bar. These are composed of trough cross-bedded gravelly sands, truncated by a surface-parallel, concordant radar facies, which is interpreted as the result of surface reworking. An inner bank bench at Big Bend has a similar internal structure to the unit bar preserved in the same area. The upper bar is draped by a surface-parallel, concordant radar facies that corresponds to interlayered silt/very fine sands and coarser sand beds and which drapes down across the inner bank bench at Big Bend. Each coarse-grained/fine-grained couplet was deposited during a major discharge event and some finer grained layers may have resulted from more moderate discharge events. Deposition of suspension sediment would be promoted by flow separation adjacent to the inner bank at Big Bend during the peak and falling stages of flow events (such features have been noted in several locations and at different flow conditions during the study).

Given the capacity of the Burdekin River to scour to bedrock, at least locally, during bankfull flows, it follows that up to 6 m of coarse- to very coarse-grained, gravelly sand may be mobilized during major discharge events and are deposited during the falling stage to show the lateral and vertical changes in sedimentary structure described above. Taking into account the recurrence interval for major flow-events in the Burdekin, however, it seems likely that the stratigraphical record of such a river (even if it were situated in a subsiding sedimentary basin) would consist of several, truncated major-event deposits (including large-scale cross-bedded strata in certain parts of the sediment body), with strongly amalgamated intervals of smaller scale cross-bedded strata representing periods of modest flow in the river. Were the Burdekin to be filled and to avulse, its fill in flow-transverse section would be a lens-shaped body, up to 25 m thick and 200–500 m wide (more if the river were allowed to meander freely), flanked by interfluve deposits consisting of pedogenically modified silts and very fine-grained sands. The internal structure of the channel body would be as described herein.

We believe that this account of the facies assemblage and internal facies architecture of the Burdekin River will be useful as an analogue for recognition of deposits of extreme-discharge rivers in the rock record. The deposits and GPR profiles described herein could readily by mistaken for the deposits of braided or other non-meandering streams, without exposure on a scale comparable to that of the channel fill (i.e. > 200 m in a direction normal to flow direction). Clearly, there is a need for caution in the diagnosis of channel style from the rock record, but equally there is a similar need for more data on the sediments of extreme-discharge rivers world-wide.

## ACKNOWLEDGEMENTS

This work was made possible by financial support from the Australian Research Council (A39531816) to CRF and grants from the Nuffield Foundation and Royal Society to JA. Jo Hamilton acted as a field assistant during the 1996 GPR survey. Scott Brownlaw and Peter Jorgensen drafted illustrations. Anne and Peter Finlay of James Cook University's Fletcherview station are thanked for their hospitality. Reviewers H. Jol and J. Diemer are thanked for their constructive criticism of the submitted manuscript.

## REFERENCES

ALEXANDER, J. & FIELDING, C.R. (1997) Gravel antidunes in the tropical Burdekin River, Queensland, Australia. *Sedimentology*, **44**, 327–337.

BELPERIO, A.P. & JOHNSON, D.P. (1985) Postglacial sedimentation, Burdekin delta and Townsville coastal plain. In: *1985 Field Conference, Mackay–Collinsville–Townsville Region* (Eds Johnson, D.P. & Stephens, A.W.), pp. 62–68. Geological Society of Australia (Queensland Division), Brisbane.

BRIDGE, J.S., ALEXANDER, J., COLLIER, R.E.Ll., GAWTHORPE, R.L. & JARVIS, J. (1995) Ground-penetrating radar and coring used to study the large-scale structure of point-bar deposits in three dimensions. *Sedimentology*, **42**, 839–852.

BURDEKIN PROJECT COMMITTEE (1977) *Resources and Potential of the Burdekin River Basin, Queensland*. Australian Government Publishing Service, Canberra, 195 pp.

FIELDING, C.R. & ALEXANDER, J. (1996) Sedimentology of the upper Burdekin river of north Queensland, Australia — an example of a tropical, variable discharge river. *Terra Nova*, **8**, 447–457.

FIELDING, C.R., ALEXANDER, J. & NEWMAN-SUTHERLAND, E.N. (1997) Preservation of *in situ*, arborescent vegetation and fluvial bar construction in the Burdekin River of north Queensland, Australia. *Palaeogeogr. Palaeoclimatol. Palaeoecol.*, **135**, 123–144.

GAWTHORPE, R.L., COLLIER, R.E.Ll., ALEXANDER, J., BRIDGE, J.S. & LEEDER, M.R. (1993) Ground penetrating radar: application to sandbody geometry and heterogeneity studies. In: *Characterisation of Fluvial and Aeolian Reservoirs* (Eds North, C.P. & Prosser, D.J.), Spec. Publ. geol. Soc. London, No. 73, pp. 421–432. Geological Society of London, Bath.

HUGGENBERGER, P. (1993) Radar facies: recognition of facies patterns and heterogeneities within Pleistocene Rhine gravels, NE Switzerland. In: *Braided Rivers* (Eds Best, J.L. & Bristow, C.S.), Spec. Publ. geol. Soc. London, No. 75, pp. 163–176. Geological Society of London, Bath.

LECLERC, R.F. & HICKIN, E.J. (1997) The internal structure of scrolled floodplain deposits based on ground penetrating radar, North Thompson River, British Columbia. *Geomorphology*, **21**, 17–38.

NEWMAN-SUTHERLAND, E. (1995) *Ecology of bottomland flora, upper Burdekin River, north Queensland*. Unpublished thesis, University of Queensland, 157 pp.

PRINGLE, A.W. (1991) Fluvial sediment supply to the north-east Queensland coast, Australia. *Austr. geogr. Stud.*, **29**, 114–138.

VAN HETEREN, S., FITZGERALD, D.M., MCKINLAY, P.A. & BUYNEVICH, I.V. (1998) Radar facies of paraglacial barrier systems: coastal New England, USA. *Sedimentology*, **45**, 181–200.

WILLIS, B.J. (1989) Palaeochannel reconstructions from point bar deposits; a three-dimensional perspective. *Sedimentology*, **36**, 757–766.

WILLIS, B.J. (1993) Interpretation of bedding geometry within ancient point-bar deposits. In: *Alluvial Sedimentation* (Eds Marzo, M. & Puigdefabregas, C.), Spec. Publs int. Ass. Sediment., No. 17, pp. 101–114. Blackwell Scientific Publications, Oxford.

*Spec. Publs int. Ass. Sediment.* (1999) **28**, 363–379

# Meander bend reconstruction from an Upper Mississippian muddy point bar at Possum Hollow, West Virginia, USA

B. R. TURNER* *and* K. A. ERIKSSON†

*\*Department of Geological Sciences, University of Durham, Durham DH1 3LE, UK; and*
*†Department of Geological Sciences, Virginia Polytechnic and State University, Blacksburg,*
*Virginia, 24061-0420, USA*

## ABSTRACT

The terrestrial to shallow-marine Hinton Formation (Upper Mississippian) exposed at Possum Hollow, West Virginia, contains laterally accreted point-bar deposits, which overlie a vertic palaeosol and are capped by marine transgressive deposits. These form part of an unconformably bounded 3–5 Myr sequence, controlled by a third-order eustatic cycle and/or tectonic loading, in which the palaeosol and point bar are interpreted as the up-dip equivalent of a fourth-order base-level rise expressed in down-dip areas by tidal estuarine successions. The scoured base of the point bar is overlain by a channel lag conglomerate that passes sharply upwards into centimetre to decimetre thick, sandstone–mudstone couplets forming individual lateral accretion units. These couplets differ from those in tidal channels in that the mud-rich part is thicker, their thickness and frequency is less regular, and there is less separation of sand and clay. Dip variations along individual accretionary surfaces define bounding surface discontinuities, which enclose flood-generated couplets of broadly similar shape. The bounding surface discontinuities allow the point bar to be divided into six genetic packages, which are attributed to normal variations in flood discharge and channel bend migration, possibly linked to changes in meander bend dimensions. Meander bend reconstruction, based on interpretation of genetic packaging of strata, suggests that the point bar and channel bend evolved through time as follows:

1 the initial channel had a relatively low width:depth ratio and low transverse bar slopes;
2 the overall channel width increased, point bar platforms developed and meander bend curvature decreased;
3 short-term decrease in channel width:depth and increased suspension load;
4 increased width:depth and lower radius of meander bend curvature;
5 erosion and increased energy levels (?major flood) or relocation of the point bar into a higher energy part of the meander bend;
6 decreased width:depth and increased mud content consistent with abandonment.

## INTRODUCTION

The sediment load carried by a channel influences channel morphology and depositional style (Schumm, 1963, 1972). Coarse bedload and suspension-load channels represent end members of a spectrum of possible channel types (Galloway, 1981). Most studies have been concerned with bedload and mixed-load channel types, whereas relatively little attention has been paid to mud-dominated, fine-grained, suspension-load channels for cases where they are tidally influenced (Jackson, 1978; Thomas *et al.*, 1987; Rahmani, 1988; Smith, 1988; Maskaske & Nap, 1995). Laboratory and theoretical–empirical studies of point bars (e.g. Jackson, 1975;

Bridge & Jarvis, 1976; Willis, 1993) also have been concerned primarily with bars composed of fine–medium non-cohesive sand (cf. Schumm & Khan, 1972; Hickin, 1983). The problem is compounded by the fact that epsilon cross-stratification, common in ancient point bars, is difficult to detect in modern point bars. Where the internal architecture of modern point bars has been examined directly (Jackson, 1978; Nanson, 1980) or indirectly (Bridge *et al.*, 1995), however, features resembling epsilon cross-stratification are evident. Moreover, most investigations of modern point bars have been concerned, of necessity, with low-stage deposits above the water

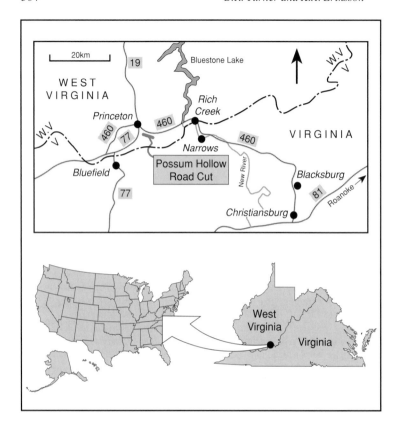

**Fig. 1.** Location of muddy point bar deposit exposed in road-cut, 11 km east of Princeton at the intersection with Possum Hollow road, on US Highway 460.

table (cf. Bridge *et al.*, 1995). As a result, modelling of point bars in the rock record, using modern analogues, is based on a limited data set, and there is a need for more robust predictive models of lateral growth and migration of bars (Jackson, 1978; Hickin, 1983; Willis, 1989, 1993), especially fine-grained fluvial systems.

A fine-grained muddy point bar deposit, in the middle part of the Upper Mississippian Hinton Formation, is exposed in a road-cut along US Highway 460 in Mercer County, southern West Virginia (Fig. 1). The road-cut allows for detailed study of vertical and lateral facies architecture, based on measured vertical sections and photomosaics of the face, which has a local structural dip of between 5° and 12° to the west. In this paper we describe spatial changes in the point-bar deposit, and interpret these changes in terms of point-bar growth, changes in channel dimensions and the response of the fluvial system to shifts in base level.

## GEOLOGICAL BACKGROUND

In late Mississippian times, south-west Virginia and southern West Virginia lay on the eastern margin of the NE–SW trending Appalachian foreland basin. In this basin, rates of sedimentation closely balanced rates of subsidence, thereby maintaining shallow water conditions throughout most of its depositional history (Colton, 1970). Predominantly terrigenous clastic rocks occur within the study area, supplied by the Appalachian orogenic belt bordering the eastern margin of the basin. Marine conditions dominated late Mississippian times, but with a change eastwards from shallow marine into more proximal nearshore and marginal marine, coastal and non-marine environments. These non-marine environments became more important during later Mississippian and early Pennsylvanian times, when sedimentation rates exceeded basin subsidence and forced retreat of the shallow Appalachian sea, concomitant with progradation of shallow coastal and non-marine depositional environments on to the adjacent shelf. Regionally traced unconformities and the stacking pattern of estuarine–marine units throughout the Upper Mississippian section define 3–5-Myr-cycle base-level changes, which provide a framework for sequence stratigraphic analysis of the Upper Mississippian succession (Fig. 2) (Miller & Eriksson, 1997). Ages for the Upper Mississippian section in West Virginia is based on biostratigraphical correlations with

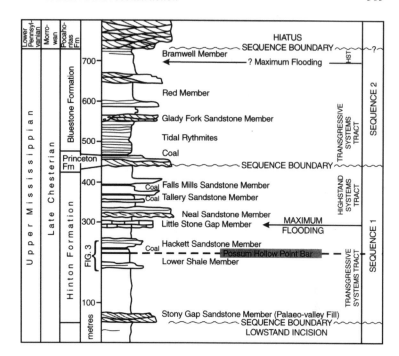

**Fig. 2.** Sequence stratigraphic interpretation of the Late Mississippian succession in south-west Virginia and southern West Virginia, showing the location of the Possum Hollow point-bar section, which forms part of a transgressive systems tract (details in text).

European sections using marine fossils, plant megafossils and sensitive high resolution microprobe (SHRIMP) dating of zircons (Dan Miller, personal communication, 1998).

Along the eastern margin of the Appalachian Basin, Upper Mississippian strata crop out in a NE–SW belt of folded rocks defining the structural front of the folded Appalachian Mountains. These rocks are represented by the following formations, in ascending order: (i) Greenbrier Limestone, (ii) Bluefield, (iii) Hinton, (iv) Princeton and (v) Bluestone. The Hinton Formation, hosting the point bar, is of late Chesterian age (Fig. 2). Although floras are rare, and comprise mostly *Stigmaria stellata* and *Sphenopteris elegans*, they correlate with Namurian A floras of western Europe (Pfefferkorn & Gillespie, 1981). Lithologically, the Hinton Formation consists of up to 400 m of interbedded shale, siltstone, sandstone and limestone with rare impure coal and underclay (Englund *et al.*, 1981; Hoare, 1993). The sediments were deposited in alternating coastal alluvial plain and nearshore marine environments. In contrast, Sunborg *et al.* (1990) described the Hinton Formation as being dominantly terrestrial in origin, despite the presence of marine faunas at various levels in the succession (Hoare, 1993). Stratigraphically, the Possum Hollow road-cut section comprises part of the lower Hinton Formation (the Lower Shale Member of Reger (1926); the Middle Red Member of Wilpolt & Marden (1959); the Adria Member of Thomas (1959) ), one of the thickest and most conspicuous red-bed successions in the region (Fig. 2). The Hackett Sandstone Member of Reger (1926) that overlies the study interval provides a useful stratigraphical marker (Figs 2 & 3).

Despite the predominance of terrestrial red-bed facies, the Upper Mississippian section in Mercer County includes scattered estuarine and marine units that provide evidence for recurring marine incursions. Estuarine tidal rhythmites have been recognized in the Stony Gap, Neal, Tallery and Falls Mills Sandstone Members of the Hinton Formation; and the Possum Hollow point bar is bounded above and below by thin marine shales (Fig. 3). Intercalation of terrestrial and marine facies is consistent with repeated relative sea-level change upon an inferred coastal setting in which fluvial systems drained westwards towards a nearby shoreline. The Possum Hollow section lies within a 200-m-thick succession of red, coastal alluvial plain sandstones and mudstones. The mudstones contain lacustrine–playa lake carbonates and palaeosols, whereas the interbedded sandstones include lenticular channel sandbodies (> 5 m thick), thin (< 3 m thick) tabular crevasse-splay sandstones and rare muddy point bars, of which Possum Hollow is the best example. The incompletely exposed section beneath the Possum Hollow point bar contains four closely spaced palaeosols (Fig. 3). The two lower palaeosols show only weakly developed pedogenic features, whereas the upper one (palaeosol 4) contains some of the most compelling

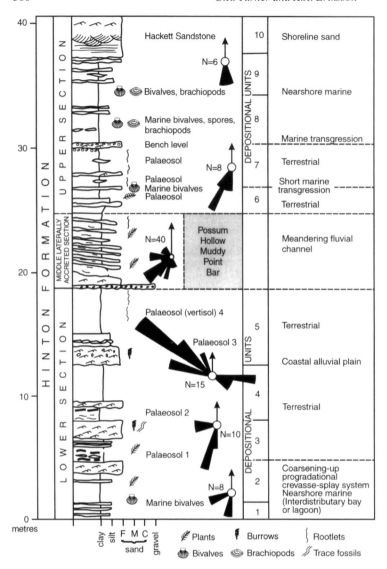

**Fig. 3.** Measured section and environmental interpretation of the Upper Mississippian Hinton Formation exposed in the Possum Hollow road-cut.

evidence for soil-forming processes. It consists of a massive, blocky, red-maroon, locally mottled and root-penetrated mudstone containing:

**1** numerous polished and faintly grooved arcuate, shiny, clay-coated slickenside surfaces;

**2** drab haloed root traces;

**3** sandy reduced zone at the top with large vertical rootlet structures;

**4** large intersecting slickensides forming parallelipeds.

These features are characteristic of modern vertisols, which are dominated by shrink–swell processes in response to seasonal wetting and drying, and the high expandable clay content. As vertic palaeosols tend to

form on relatively flat topography (Caudill *et al.*, 1996) their presence beneath the point-bar deposit (Fig. 3) suggests that the alluvial plain was flat and low-lying, and relatively well-drained. Palaeosol 3 (Fig. 3) is a calcareous nodular mudstone, which is interpreted to be the lower nodular horizon of the vertic palaeosol 4 above, or the C horizon of a calcic vertic palaeosol, with the overlying palaeosol 4 corresponding to the middle horizon. The upper horizon is missing, possibly as a result of erosion by the point bar above (Fig. 3). Evidence of this is seen in the presence of dark grey, pedogenic calcareous nodules in the conglomerate along the base of the point bar. The nodules are mostly irregularly shaped, poor to moderately

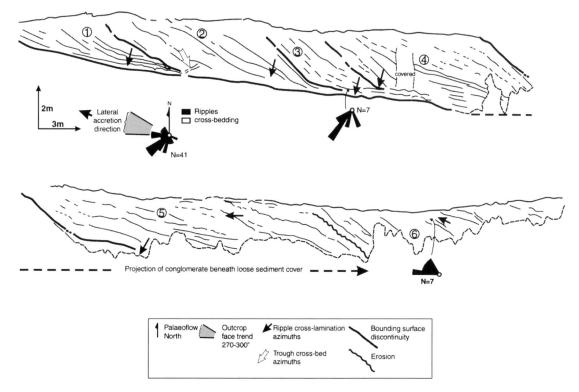

**Fig. 4.** Lateral profile of Possum Hollow point bar, drawn from photomosaics, showing genetic packages, bounding-surface discontinuities and palaeocurrent directions. Each arrow represents the vector mean of between three and six ripple or cross-bed foreset azimuths.

rounded, with internal cracks and/or growth discontinuities filled with clays and carbonate. In thin-section, some show internal layering, whereas others have a simpler, homogeneous, turbate microcrystalline texture. A few of these homogeneous types have concave-shaped marginal cracks, often extending to the surface of the nodule, suggesting an impact origin.

## POSSUM HOLLOW SUCCESSION

The succession at Possum Hollow comprises a lower marine-influenced section, a middle laterally accreted section of inclined heterolithic strata and an upper more marine influenced section (Fig. 3).

### Middle laterally accreted section

This deposit is 3–4 m thick and can be traced laterally for over 90 m along the road-cut (Fig. 1). It is not present on the north side of the road-cut. At its western end, the succession dips into the subsurface, but there is no evidence

at the surface of an abandoned channel. The section overlies a red-maroon, palaeosol with vertic affinities, and is capped by a marine transgressive sequence (Fig. 3). It comprises an erosion surface overlain by a conglomerate facies, overlain in turn by a laterally accreted facies of inclined heterolithic strata (IHS of Thomas *et al.*, 1987) (Figs 4 & 5). Palaeocurrent directions are generally into the outcrop face or at an acute angle to it.

### Conglomerate facies

*Description.* The conglomerate is 5–30 cm thick and overlies a slightly irregular horizontal to subhorizontal scour surface cut into the underlying vertic palaeosol (Fig. 6). It has a fine sandy matrix enclosing granules and small pebbles of dark grey, rounded, pedogenic calcareous nodules, minor red mudstone and siltstone intraclasts, iron-oxide-rich argillaceous (?sesquioxide) nodules, blackened rolled vertebrate bones, plant material, feldspar grains and granules, and small bivalves. The conglomerate fines at the very top, where it passes sharply into sandstone or silty mudstone containing local lenses

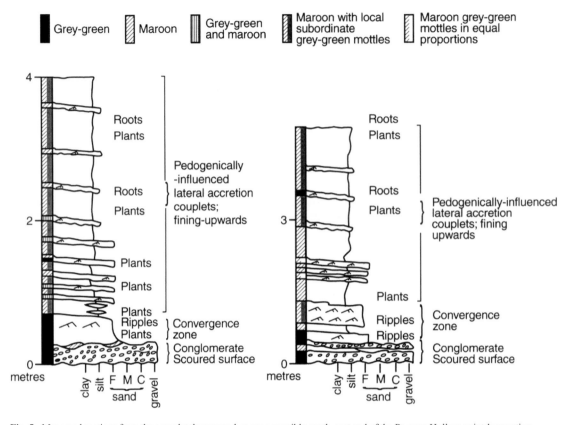

**Fig. 5.** Measured sections from the completely exposed, more accessible south-east end of the Possum Hollow point-bar section showing lateral-accretion sandstone–mudstone couplets (inclined heterolithic strata) and upward-fining trend.

Lateral accretion
couplets

Convergence zone

Conglomerate

Vertic palaeosol

**Fig. 6.** Conglomerate at the base of the point bar, overlying scoured vertic red palaeosol and overlain by a sand-rich zone and sandstone–mudstone lateral-accretion couplets.

of fine-grained conglomerate (Fig. 5). The conglomerate is structureless, except for occasional 2–3-cm-amplitude current ripples in the finer grained upper part, which have steep face azimuths to the south-west (209°). Rare internal scours occur in places along with locally developed reddish diffuse mottles and lenses (Fig. 5).

*Interpretation.* There is no evidence of deep incision of the immediately underlying deposit beneath the conglomerate, which represents a channel floor lag reworked mainly from a nearby calcretic palaeosol, possibly the now eroded upper horizon of the immediately underlying vertic palaeosol (Fig. 3). Interbedding of the conglomerate

**Fig. 7.** Lateral-accretion couplets in lower part of the Possum Hollow point-bar section, just above the basal conglomerate. Note the sharp base to the sandstone part of the sand–mud couplet and gradational relationship between them.

with the basal, thickened sandy part of the overlying lateral accretion deposits, which themselves may contain streaks and lenses of fine conglomerate, indicate that the two facies were deposited contemporaneously. Intraformational conglomerates of this type are common in the lower reaches of river systems lacking a coarse bedload component, specially in mature, low-gradient, coastal alluvial plains (Fisk, 1944; Stewart, 1983). Discharge variation generated internal scours, and local traction currents, developed during deposition of the conglomerate, formed waning-flow ripple bedforms, now preserved in the finer parts of the conglomerate.

*Lateral accretion facies*

*Description.* This facies is composed of sandstone and mudstone beds organized into centimetre to decimetre thick sandstone–mudstone couplets, which form individual lateral accretion units up to 35 cm thick. These lateral accretion units are better developed in the lower part of the facies (Fig. 7) and dip to the west-north-west (270°–300°) (Fig. 4). Individual beds making up the couplets sometimes show complex lateral and vertical relationships. The sandstone beds are up to 7.5 cm thick and fine- to very fine-grained, often approaching silt grade in size. They are greenish-grey or maroon-red sandstones, with maroon and grey mottles. Greenish-grey colours dominate, however, particularly in the lower part of the facies. Although the sandstones usually have sharp bases, there is no evidence of significant erosion (Fig. 5). They have gradational tops, fining upwards over a few millimetres into the mudstone-dominated part of the couplet. Internally, the sandstones contain small-scale current ripple cross-lamination and carbonaceous plant fragments (Fig. 5).

The mudstone beds are blocky, maroon-red silty mudstones with subordinate, irregular, and often diffuse greenish-grey and fawn mottles. Specks of black carbonaceous plant material occur in the centre of some mottles. The mudstones, which attain thicknesses of up to 25 cm, are typically structureless, but have a variable silt content and contain silty streaks and lenses. Sedimentation-tube grain-size analysis of two mudstone samples show that they contain up to 30% clay with 30% silt, 30% fine sand and 10% fine–medium sand (200–425 μm). They also contain plant material, pale greenish-grey, shallow penetrative root structures and a few, small, hard calcareous and ferruginous-rich nodules < 2 cm in diameter. The sand- and mud-rich couplets generally persist laterally from the highest to the lowest part of the point bar, although definition in the upper more mudstone-rich part is sometimes indistinct (Fig. 4). No body fossils or burrows are found in the lateral accretion facies.

The mudstone part of the couplets shows a systematic upward increase in thickness at the expense of sandstone throughout the point-bar deposit, producing an upward-fining trend. The ratio of sandstone to mudstone in the lower part of deposit is about 1 : 1 but it decreases in the upper part to 1 : 3 (Fig. 5). Up-dip, the mudstone component merges into the mudstone-dominated deposits above, with no preserved evidence of any bar-top topography. As a result, the upper surface of the point bar is not easy to distinguish from the overlying overbank succession. The lateral accretion facies fines upwards, thus:

**1** throughout the entire facies;

**2** within individual coarse to fine sandstone–mudstone couplets;

**3** within individual sandstone beds of the couplets.

Despite this, no significant variation in grain size occurs along lateral accretion surfaces. In addition there is an overall upward increase in mottles, rootlets and small pedogenic-type calcareous nodules, concomitant with the increase in mudstone.

Dips on individual lateral accretion surfaces vary from 4 to 26°, with an average of 14° (Fig. 4). The steepest dip recorded from ancient fluvial lateral accretion deposits to date is 29° for a mudstone-dominated inclined heterolithic strata (IHS) set described by Visser (1986) from the Cretaceous Judith River Formation in Alberta, but in modern settings dips may be as high as 36° (Thomas *et al.*, 1987). Such steep dips are typical of inclined heterolithic strata sets with a high mudstone content, although the dip itself commonly changes when lateral accretion surfaces are traced in the down-dip direction. At Possum Hollow, variations in dip along lateral-accretion surfaces define a number of profile geometries, which tend to occur in genetically related packages of broadly similar shape, separated by bounding surface discontinuities (Fig. 4).

Thicker amalgamated sandstone beds, locally interbedded with conglomerate, occur where lateral accretion couplets flatten-out downwards (Figs 5 & 6). These thicker amalgamated sandstone beds are typically horizontal to subhorizontal and they extend laterally for 3–4 m before terminating against the conglomerate or lateral-accretion bounding surface discontinuity of the next package down-dip. These basal sandstones are ripple cross-laminated with steep-face azimuths indicating flow towards the south-west (204°), subperpendicular to the direction of point-bar accretion (Fig. 4). Rare, small trough cross-beds, < 15 cm thick, occur within the basal sandstone beds (Fig. 4). Some foreset azimuths indicate flow to the south-east (120°), opposite the lateral accretion direction.

The upper finer grained part of the lateral accretion deposit consists of red mudstone, silty mudstone and siltstone containing a few red, and less commonly greenish-grey, fine-grained sandstone interbeds. The mudstone, silty mudstone and siltstone beds show complex relationships and typically are blocky and jointed with locally developed slickenside surfaces. They contain diffuse, greenish-grey mottles and rare rootlet structures, but apart from some faint traces of ripple cross-lamination and irregular lamination in the siltstones they are structureless. At one level, the mudstones show evidence of deformation, mobilization and intrusion into the overlying

siltstone in a complex and irregular manner, reminiscent of flame structures. The sandstone interbeds are up to 18 cm thick, and contain current ripple cross-lamination and rare vertical to subvertical greenish-grey root structures with a maroon sandy fill. Ripples in the lower part of one of the thicker sandstone beds are up to 0.6 cm in preserved amplitude and 7.5 cm in preserved wavelength, whereas in the upper, slightly coarser part of the same bed they are up to 2 cm in amplitude and 25 cm in wavelength. Steep-face ripple azimuths indicate flow towards the west (270°) and south-west (214°), subparallel and slightly oblique to the dip direction of the lateral accretion units (Fig. 4). A 2-cm-thick deformed zone with convolute laminations occurs at the base of this sandstone, which rests sharply, but non-erosively, on the sandstone below. Most of these sandstone beds represent the poorly defined lower part of lateral-accretion couplets. Ripple azimuths in the lower and upper parts of the laterally accreted facies indicate a consistent downstream component of flow to the southwest, but with a spread of 200° (Fig. 4). This spread reflects the change in palaeocurrent directions at the western end of the outcrop and cross-bed azimuths directed up the point-bar slope.

*Interpretation.* The laterally accreted facies is interpreted to have been deposited in a small, laterally migrating, relatively shallow, meandering channel occupying part of a seasonally dry coastal alluvial plain (Englund *et al.*, 1981; Cecil *et al.*, 1985; Donaldson *et al.*, 1985). The argillaceous-dominated lateral accretion bedding (IHS) was deposited on the inside bend of the channel as a point bar, contemporaneous with deposition of the conglomerate. Only about 10% sand coarser than fine sand (> 200 µm) grade was transported by the channel system at this time; most of the sediment load (60%) was silt and clay carried in suspension.

The laterally accreted sandstone–mudstone couplets formed during individual flood events (Smith, 1988). Each repeat couplet records rising-flood-stage erosion and subsequent falling-stage deposition of silt and clay. Following flood events, the point bar surface was exposed and subjected to desiccation and pedogenic processes. Evidence of weak pedogenesis is seen in the presence of drab mottles, often associated with root traces, rootlets, small calcareous nodules and abundant plant material in the mudstones. X-ray diffraction analysis of mudstone samples from two pedogenically influenced sandstone–mudstone couplets shows that illite is the dominant clay mineral (69%, 68%) with kaolinite (21%, 18%) and chlorite (14%, 10%) present in lesser amounts. Some 18% and 15% of expandable mixed-layer clays are associated with the illite. The vertic affinities of the palaeosols in the local succession (Fig. 3) suggests that they originally contained significant amounts of expandable clays, which underwent extensive illitization during burial diagenesis (Caudill *et al.*, 1996); a process that also may have occurred during diagenesis of the laterally accreted mudstone. The general lack of bedding and lamination in these mudstones may likewise reflect pedogenesis and/or the expansion and wetting of clay (cf. Woodyer *et al.*, 1979). Pedogenic modification of the point-bar surface implies a considerable fluctuation in discharge and water level in the channel, and supports a dry climate. Pedogenesis and desiccation of the mud help to resist erosion by the next flood event (Woodyer *et al.*, 1979), and the growth of rooted vegetation on exposed point-bar surfaces increases its tensile and shear strength (Bridge, 1985). Although drying and cracking of mud may develop after exposure for only a few hours (Levey, 1978), there is no evidence of desiccation cracks in the mudstones. This possibly reflects the slow rate of deposition, which may provide the opportunity for formational waters to cause the cracks to swell, shut and close prior to significant filling, especially in the presence of swelling clays, as documented for vertisols (Blodgett, 1985; Caudill *et al.*, 1992). Deformed mudstones showing evidence of deformation and mobilization are attributed to the swelling and contraction of sediments rich in swelling clays (Taylor & Woodyer, 1978) or possibly as the end product of wet–dry cycles in vertic soils (Caudill *et al.*, 1992).

Inclined, alternating sand–mud couplets have been described from modern meandering river point bars, but the sand–mud couplets are generally much thinner than those described here (Nanson, 1980; Gibson & Hickin, 1997). Basal sand–mud flood deposits in modern floodplains seldom attain 50 cm couplet$^{-1}$ during a single flood event. This suggests that the couplets may be composite units, deposited by more than one flood event, as indicated by the local internal scour surfaces, or that they were deposited in a meander that occupied a non-tidal backwater. This latter scenario may occur as a result of downstream control by a local base level (E.J. Hickin, personal communication, 1998). Sand–mud couplets in meandering rivers are generally thought to be more typical of tidally influenced channels (Thomas *et al.*, 1987: Smith, 1988; Maskaske & Nap, 1995). The Possum Hollow couplets, however, differ from those in tidally influenced channels in that:

**1** the mud-rich part of the couplet is generally thicker;
**2** mud is preserved in the lower part of the point bar (cf. Rahmani, 1988);
**3** the thickness and frequency of the couplets is less regular;

**4** the mudstone is blocky and unlaminated, and contains silty streaks and lenses unlike the well-defined sand–clay separation and cyclic bundling typical of tidal environments.

The lack of deep incision and dominance of current ripples in the sands and silts probably reflects low-energy conditions, rather than the fine grain size of the bedload. The mudstones, for example, contain up to 30% of sand grains in the range 100–425 μm and the proportion is likely to be much higher in the coarser sandy part of the couplet. These grain sizes fall well within the stability field for dune and ripple bedforms (Middleton & Southard, 1984). The lack of mud intraclasts at the base of sandstone beds, however, emphasizes that the critical shear velocity for erosion of the mud layers was not exceeded. Nevertheless, the locally thickened, more gently dipping amalgamated basal sand-rich beds probably reflect deposition from bedload during high-stage flow under slightly higher energy conditions on the lower part of the point bar, where curve-crested dunes are the most common type of bedform (Harms & Fahnestock, 1965; Smith, 1971; Cant, 1978; Crowley, 1983; Bridge & Jarvis, 1982). Although normally oblique to the local channel direction, some dunes at Possum Hollow migrated up the distal base of the point-bar slope, where flow separation is more strongly developed (Bridge & Leeder, 1976).

## GENETIC PACKAGES

Point bars are complex, dynamically active, three-dimensional sediment bodies, the shape and location of which constantly change in response to channel migration. The interactive controls on bar movement and geometry, which are numerous and variable (Willis, 1993), have been assessed in this study on the basis of the lateral variability of the bar, as expressed by discontinuity bounded genetic packaging of strata seen in outcrop (Fig. 4).

Six genetically related lateral accretion packages, separated by bounding surface discontinuities, have been recognized in the Possum Hollow point bar (Fig. 4). These discontinuities, which are equivalent to the third-order bounding surfaces of Miall (1996), are defined by changes in dip, and are not usually accompanied by significant erosion, except in one example between packages 5 and 6 (Fig. 4). Such packages have been attributed to a variety of causes, including longer term periodicity of major flood events, high- or low-stage flow modification and discontinuous bar growth, and variable discharge over time (Thomas *et al.*, 1987; Bridge *et al.*, 1995). We have used profile geometries of lateral accretion surfaces,

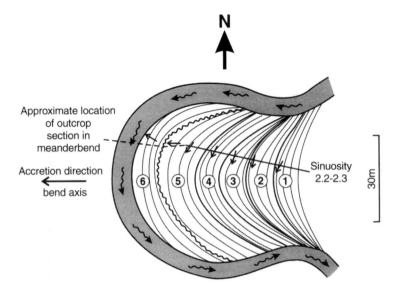

**Fig. 8.** Possum Hollow meander bend reconstruction, showing inferred location of genetic packages relative to the channel-bend axis as the point bar evolved.

within these genetically related packages, in an attempt to provide a more detailed analysis of the temporal and spatial evolution of the point bar, and the initiation and growth of the meander bend. The genetic packages at Possum Hollow are considered to represent the lateral, possible step-wise growth elements of a single point bar during the active life of the channel (Thomas *et al.*, 1987; Willis, 1993). Evidence in support of this is the similar dip direction of lateral accretion couplets, and similar palaeocurrents in packages 1–4 with minor changes in 5 and 6 (Fig. 8). Each package is made up of between five and ten lateral accretion couplets (Fig. 5), each couplet representing a single flood event (cf. Bridge & Diemer, 1983; Smith, 1988). Thus, it is tempting to relate individual packages to longer term flood cycles with recurrence intervals of decades or more, except for the lack of significant erosion surfaces normally associated with such high-magnitude events (Bridge *et al.*, 1995). Thomas *et al.* (1987), for example, related erosion surfaces between inclined heterolithic strata sets to major flood events of unspecified frequency. Bridge & Diemer (1983), on the other hand, considered each lateral accretion couplet (their lateral accretion bedsets) to have been deposited by a single major flood event. Major and minor discontinuities in point bars were attributed to exceptional and normal flood events by Elliott (1976). The general lack of significant erosion, associated with all but one of the discontinuities at Possum Hollow, and the similar palaeocurrent directions and grain sizes across discontinuity surfaces between packages 1 and 4, argues against flood events of significantly higher than normal magnitude and periodicity for their growth. At geomorphological time-scales, most work is done by rivers during flood events of modest magnitude, but relatively high frequency, not by extreme floods (Hickin, 1983). We interpret the discontinuities to be primarily a response to normal variations in flow intensity during different episodes of channel migration and deposition, which may, in turn, be linked to changes in channel-bend dimensions.

### Genetic package 1

This package records the initiation and early development of the point bar. It has a simple overall geometry comprising straight to slightly concave-up, gently dipping, regularly spaced, lateral accretion surfaces (Fig. 4). These flatten off sharply towards the bottom, where they pass into tangentially based horizontal to subhorizontal beds, that merge down-dip into a locally thickened, amalgamated basal sand-rich zone.

Straight to slightly concave-up lateral accretion geometries are typical of point bars composed of alternating sand–mud couplets and abundant suspension load (Thomas *et al.*, 1987). The generally low dips and basal flattening probably reflect a decrease in the suspension load and/or mud, producing relatively low transverse bar slopes independent of W : D ratios, which remained low (Table 1). The down-dip transition, from sand–mud couplets into more gently dipping basal sand-rich zones has been attributed to high-energy conditions on the lower part of mixed-load point bars (Thomas *et al.*, 1987).

**Table 1.** Possum Hollow palaeomeander-bend reconstructed morphology and hydrology for genetic packages 1–3 (see text for details). $W_b$ is bankfull channel width; $D_{max}$ is maximum channel depth; $D_m$ is bankfull mean depth; $W:D$ is width to depth ratio; $I_m$ is meander wavelength (Williams, 1987); $R_c$ is meander loop radius of curvature (Williams, 1987); $Si$ is sinuosity (Williams, 1987); $Q$ is average discharge (Osterkamp & Hedman, 1982); $Q_b$ is bankfull discharge (Williams, 1984); $Q_2$ is discharge with 2-yr recurrence interval (Osterkamp & Hedman, 1982); $Q_5$ is discharge with 5-yr recurrence interval (Osterkamp & Hedman, 1982); $Q_{2.33}$ is mean annual peak flow (Schumm, 1972; Williams, 1987); $V$ is mean flow velocity (Lacey, 1934).

| Parameter | Equation | Standard error (log units) | Genetic package 1 | Genetic package 2 | Genetic package 3 |
|---|---|---|---|---|---|
| $W_b$* | Two-thirds relationship | | 15 m | 19 m | 13 m |
| $D_{max}$† | Maximum vertical thickness of lateral accretion units plus 10% | | 3.0 m | 3.6 m | 2.3 m |
| $D_m$ | $D_m = 0.12W_b^{0.69}$ | 0.287 | 1.5 m | 1.7 m | 1.4 m |
| $D_m$‡ | Between one-half and a third of $D_{max}$ | | 1.0–1.5 m | 1.2–1.8 m | 0.8–1.2 m |
| $W:D$ | $W_b/D_{max}$ | | 5.0 | 5.3 | 5.6 |
| $L_m$ | $L_m = 7.5W_b^{1.12}$ | 0.219 | 198 m | 258 m | 169 m |
| $R_c$ | $R_c = 1.5W_b^{1.12}$ | 0.182 | 33 m | 42 m | 32 m |
| $R_c$§ | $R \approx 2.5W_b$ | | 37 m | 47 m | 32 m |
| $Si$ | $Si = 3.5(W_b/D_{max})^{-0.27}$ | 0.061 | 2.2 | 2.3 | 2.3 |
| $Q$ | $Q = 0.027W_b^{1.71}$ | 0.31 | 3 m³⁻¹ | 5 m³⁻¹ | 2 m³⁻¹ |
| $Q$¶ | $Q = 0.029W_b^{1.28}D_{max}^{1.10}$ | 0.36 | 3 m³⁻¹ | 5 m³⁻¹ | 2 m³⁻¹ |
| $Q_b$ | $Q_b = 0.011L_m^{1.54}$ | 0.171 | 38 m³⁻¹ | 57 m³⁻¹ | 30 m³⁻¹ |
| $Q_2$ | $Q_2 = 1.9W_b^{1.22}$ | 0.41 | 52 m³⁻¹ | 70 m³⁻¹ | 43 m³⁻¹ |
| $Q_5$ | $Q_5 = 5.8W_b^{1.10}$ | 0.42 | 114 m³⁻¹ | 148 m³⁻¹ | 97 m³⁻¹ |
| $Q_{2.33}$ | $Q_{2.33} = 2.66W_b^{0.90}D_{max}^{0.68}$ | 0.182 | 64 m³⁻¹ | 90 m³⁻¹ | 47 m³⁻¹ |
| $V$ | $V = 11D^{0.67}S^{0.33}$ | Standard deviation $c.$ 30% | 0.70–0.81 m s⁻¹ | 0.72–0.85 m s⁻¹ | 0.66–0.78 m s⁻¹ |

* Coefficient of 0.85 (see text and Allen, 1970). † 10% correction factor (Bridge & Diemer, 1983). ‡ See text and Bridge & Diemer (1983). § Leopold *et al*. (1964). ¶ Schumm (1972).

### Genetic package 2

The lateral accretion bounding discontinuity to this package, and immediately succeeding lateral accretion surfaces, flatten-off locally at two distinct levels along the accretionary profile, separated by a more steeply dipping part (15°–17°) (Fig. 4). This feature becomes less obvious in the down-dip direction, where profiles begin to exhibit a crude sigmoidal shape, but with the down-dip end once again characterized by a locally flattened and thickened sand-rich zone at the toe of the package. The lateral accretion surfaces are longer and laterally more persistent than those in laterally adjacent genetic packages 1 and 3, possibly reflecting a temporary change in overall channel width and/or transverse bar slope. Estimate bankfull channel widths at this time show a significant increase (Table 1), which may have influenced the bar slope (Bridge, 1985).

The local flattening of the lateral accretion profile at two distinct levels without significant erosion divides this package into three component parts, which may relate to the development of point-bar platforms of lower, middle and upper levels, typical of modern suspended-load streams (Taylor & Woodyer, 1978). In terms of this analogue, the basal sand-rich zones with rare cross-beds, may represent the more sandy, poorly vegetated low bench, which thins up-sequence as intervening mud beds become more frequent, and grades into the interbedded, suspension-load dominated, sand and mud of the middle and high benches, producing an overall upward-fining trend. The steep local increase in dip in this package may then be equivalent to the more steeply dipping beds of the middle bench (Taylor & Woodyer, 1978, Fig. 7). Rapid bank erosion also has been linked to point-bar platforms, which form as a result of the relatively more rapid accretion of the lower part of the point bar (Ikeda, 1989). As there is no evidence of strong erosional activity associated with this genetic package, however, rapid bank erosion is unlikely. The down-dip change towards more sigmoidally shaped profiles indicates that the point-bar platforms were relatively short-lived features and were replaced by smoother profile geometries. This may represent a response to short-term intrapackage decrease in meander bend curvature, prior to the emplacement

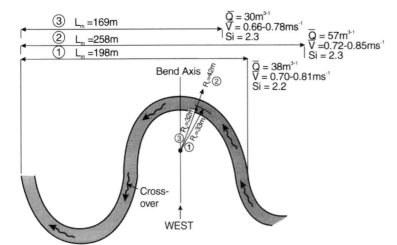

**Fig. 9.** Possum Hollow meander bend, showing variations in bend wavelength ($L_m$), radius of curvature ($R_c$), sinuosity ($Si$), bankfull discharge ($Q$) and velocity ($V$) during deposition of genetic packages 1–3 (see Table 1 and text for details).

of package 3. In modern meandering-river systems, more gently curved meanders typically produce more sigmoidally shaped point-bar lateral accretion profiles (Thomas *et al.*, 1987).

### Genetic package 3

The bounding surface discontinuity between genetic packages 2 and 3 marks a distinct increase in dip of the lateral accretion surfaces, which have a relatively simple profile geometry (Fig. 4). Although lateral accretion surfaces are generally poorly defined in this package, there appears to be a sand-rich zone at the down-dip toe. The narrow width of this package suggests that it represents a relatively short-term change in lateral point-bar growth compared with other genetic packages. The steeper dip of accretionary surfaces above discontinuities in sandy lateral accretion sequences are commonly attributed to major discharge fluctuations (Willis, 1993). Increased suspension load and decrease in the channel W : D ratio also may account for increased dips (Thomas *et al.*, 1987). This is not supported by the palaeochannel data (Table 1), however, which show increased W : D ratios at this time. The poor definition of lateral accretion surfaces, especially in the upper part of the package, reflects the increased mud content. Variations in meander bend wavelength ($L_m$) between packages 1 and 3 (Table 1, Fig. 9) serve to emphasize fluctuations in channel discharge (Willis, 1993).

### Genetic package 4

This package is marked by a sudden decrease in dip and the development of laterally extensive, very low angle,

subhorizontal accretion surfaces that steepen slightly towards the top. They also increase noticeably in dip at the very down-dip end of the package, just before the next, more steeply dipping bounding surface discontinuity, where the conglomerate and scoured based of the point bar begin to disappear beneath loose scree cover (Fig. 4).

Temporal changes in the ratio of channel width to depth, particularly mean depth, can have an important influence on bedding inclination (Willis, 1993). The very low regular dips may relate to the increased width to depth ratio of the channel at this time, and the reduced radius of curvature of the meander bend. Simultaneous tightening of the bend and increasing the form ratio (W/D) of the channel are normally short-term events, however, as rivers develop increasingly asymmetric cross-sections as migration tightens the bend. An alternative possibility, therefore, is variation in mud content and deposition from suspension fallout. Thus, the slight increase in dip in the upper part of lateral accretion units could be explained by increased mud content and more rapid deposition from suspension fallout, or it may be equivalent to the inner accretionary bank of Bluck (1971).

### Genetic package 5

The lateral accretion surfaces of this incompletely exposed package show consistently shallow dips of 2–3°, which appear to parallel the lower flattened part of the bounding surface discontinuity; the upper steeper part is onlapped by the lateral accretion surfaces. This is the widest package in the point bar, and the first to show the presence, at the better exposed down-dip end, of a thicker than normal, sand-rich accretionary unit (Fig. 4).

The shallow dip of the lateral accretion couplets and onlap relationships with the bounding surface discontinuity suggests the possibility of erosional reworking of the upper part of the point bar, during a major flood event, or falling stage, prior to the onset of a new phase of point-bar accretion within a relatively wide, but still shallow, channel. The thick, down-dip sand-rich unit and associated change in palaeocurrent direction (Fig. 4), are probably a function of sudden increased energy levels (bed shear stress and flow velocity). Although this supports the idea of a major flood event, it could relate to a shift of the meander bend itself, relocating the point bar into a higher energy part of the bend, now orientated more perpendicular to the channel trend, and parallel to the axis (Fig. 8).

## Genetic package 6

Only the upper part of this package is seen, although the lateral accretion surfaces are locally well defined. The sand-rich relatively thick bounding discontinuity to this package shows a slight increase in dip and a sharp, irregular lower scoured surface, disconformable with package 5 (Fig. 4). Local relief of up to 4 cm occurs on the scoured base of the sandstone, which is structured internally by ripple cross-lamination. The succeeding lateral accretion couplets are conformable with this bounding discontinuity and consistent in dip, except towards the end of the outcrop where they show an increase in dip (Fig. 4). The full extent of the surfaces and evidence of additional genetic packages, however, are obscured by vegetation and the structural dip carrying it into the subsurface.

The erosively based bounding surface discontinuity, and change in palaeoflow direction consistent with that during the late-stage emplacement of package 5 (Fig. 4), suggests a higher than normal flood event. Major disconformities like this, associated with erosion and changes in inclination of lateral accretion couplets, have been attributed to flood events with recurrence intervals of decades to centuries in modern river bends (Bridge *et al.*, 1995). Flood events of this order of magnitude also may lead to adjustments of the channel in terms of its dimensions (Schumm & Lichty, 1963) and/or position, relocating the point bar closer and more parallel relative to the bend axis, as evidenced by the consistent palaeocurrents (Fig. 8). An alternative explanation is to be found in the work of Hickin (1974) on meandering growth patterns in the Beatton River. He found a critical value for the ratio of radius of channel curvature ($r$) to channel width ($W_m$), which resulted from a change in the direction of lateral bend erosion. This change represents a response to fluctuations in flow resistance as a result of changes in channel curvature and is accompanied by a change in the direction

of lateral accretion. Thus, the erosive bounding surface discontinuity could conceivably record changes in bend curvature as the meander bend evolved. Estimates of ($r$)/$W_m$ for genetic packages 1–3 yield values of 2.2, 2.2 and 2.4, but could not be obtained for this incompletely exposed package.

The increase in dip of the lateral accretion couplets at the end of the outcrop may signify a decrease in the width to depth ratio of the channel, and/or increased mud content (Thomas *et al.*, 1987), consistent with the initiation of an abandonment phase. Increased inclination of bedding into abandoned channel fills is a characteristic feature of deposition on migrating point bars (Willis, 1989). Certainly, there is no evidence at outcrop or in the subsurface that the point bar continues further west. Moreover, the presence, immediately west of the road-cut, of a valley feature, strongly suggests the dominant presence of mudstone.

## MEANDER BEND RECONSTRUCTION

Although low-stage bedforms, such as ripples, are not regarded as the most reliable indicators of mean channel direction (cf. Edwards *et al.*, 1983), in low-energy systems such as Possum Hollow, they can prove extremely useful in delineating local channel orientation (Harms *et al.*, 1963; Schwartz, 1978). Furthermore, under steady-state conditions, sediment transport on point bars is considered to be primarily in the downstream direction (Allen, 1970; Bridge, 1978; Bridge & Jarvis, 1976). The relatively low spread of readings in individual genetic packages (Fig. 4) supports this view, especially as most of these were measured in the more accessible lower part of the point bar, where palaeocurrent indicators are generally more reliable (Coleman, 1969; Bridge, 1985). Small mud-rich point bars, such as Possum Hollow, are typical of suspension-load dominated, low-gradient channels with a high W : D ratio and high sinuosity (Schumm, 1972; Jackson, 1978). In high sinuosity meander bends, orthogonal sections should show palaeocurrents essentially perpendicular to the direction of lateral accretion (Fig. 8). With increasing distance around the bend from the bend axis (upstream and downstream), the palaeocurrents should show an increasingly oblique trend relative to the axis, eventually lying more or less parallel to the lateral accretion direction close to the entrance/exit of the bend, but in opposite flow directions (Fig. 8). Based on this idealized model the relationship between lateral accretion direction and palaeocurrents at Possum Hollow suggests that the outcrop during development of packages 1–4 was orientated at a relatively low angle (*c.* 30°) to, and just upstream of the bend axis (Fig. 8). The location of the outcrop section close to the bend axis provides for more accurate estimates of palaeochannel parameters and meander bend geometries. Packages 1–4 suggest essentially unidirectional growth stages along a continuous erosional axis. Following this, a change in palaeocurrent direction occurred on either side of the discontinuity separating packages 5 and 6, when the section more closely paralleled the channel axis (Fig. 8). The reasons for this change are discussed below.

Palaeochannel parameters have been estimated from the exposed cross-sectional area of the lateral accretion deposits measured in the field and on photomosaics. Measurements have been confined to completely exposed packages, in which the lateral accretion deposits have not been subjected to significant bounding surface erosion, and where the lateral accretion surfaces can be traced without break from the uneroded top of the point bar to the conglomerate base. This approach limits the analysis to genetic packages 1–3 (Figs 4 & 8). The maximum depth of the channel, neglecting the effects of compaction, can be represented by the maximum vertical thickness of lateral accretion surfaces with the steepest transverse slope (Table 1) (Bridge & Diemer, 1983), from the top of the lateral accretion deposits to the scoured base of the channel fill. The maximum bankfull depth will have been about 10% more than this value and the mean bankfull depth between one-half and one-third of this value (Bridge, 1978). Bankfull channel width can be estimated from the length of lateral accretion surfaces multiplied by a value in the range 0.70–0.95 (Allen, 1970) to obtain the full channel width at the bend apex. It should be stressed that both bankfull depth and width in modern channels vary along the channel, attaining a maximum value at the bend apex and a minimum value at the cross-over (Williams, 1987). As the Possum Hollow cross-section is thought to be located just upstream of the bend apex, a coefficient of 0.85 has been used to estimate bankfull channel width (cf. Allen, 1970; Cotter, 1971).

Estimates of channel parameters, bend geometry and palaeoflow characteristics have been based, wherever possible, on the empirical equations of Williams (1987). These are more reliable than earlier Holocene methodologies because of their more extensive data base. Other empirical equations have been used to estimate discharge (average, bankfull, and 2- and 5-yr flood discharges), slope and velocity. Average channel discharge, for example, was estimated according to the method of Osterkamp & Hedman (1982), who established a set of equations for different categories of bed and bank material. Grain-size analysis of the Possum Hollow lateral-accretion couplets suggests a medium-silt–clay channel

bed (31–60% silt–clay) according to the classification of Osterkamp & Hedmen (1982) (see also Williams, 1987, table 4). Details of the estimates and methods are given in Table 1. Although the empirical approach towards palaeohydraulic analysis has its limitations (see discussion in Williams, 1987), it does provide rough estimates of meander bend characteristics, which can be used to constrain models for meander bend reconstruction and point-bar evolution. Values for W : D ratios of < 10 and sinuosities of > 2.0 (Table 1 & Fig. 9) are consistent with the values of Schumm (1963) for single suspended-load channels in which bedload makes up < 3% of total load.

## CONCLUSIONS

A muddy point bar in Upper Mississippian coastal alluvial plain strata of southern West Virginia serves as an example of the influence of base-level changes on the spatial and temporal development of fine-grained fluvial point bars. The point bar lies within a thick, predominantly terrestrial section, which is interpreted to represent the transgressive systems tract of a third-order sequence. Within this systems tract, fourth-order rises in base level are manifested by an upward progression from palaeosols to aggradational fluvial deposits to thin marine shales.

The point bar consists of a laterally accreted section of inclined heterolithic strata comprising centimetre to decimetre thick sandstone–mudstone couplets. These were deposited during single flood events within a small, shallow, meandering channel occupying a relatively flat, well-drained coastal alluvial plain. The channel was characterized by large discharge fluctuations, which periodically exposed the point-bar surface to desiccation and pedogenic modification. The point-bar profile has been divided into six disconformity-bounded genetic packages made up of five to ten lateral-accretion sandstone–mudstone couplets. These packages represent step-wise growth elements of the point bar during the active life of the meander bend and are attributed to normal variations in flow intensity, linked to migration and changes in meander bend dimensions.

Reconstructions of meander bend dimensions and palaeoflow characteristics indicate that the Possum Hollow section was located just upstream of the bend axis and aligned parallel to subparallel to its axial trend. Channel W : D ratios were < 10 with sinuosities of > 2.0. Interpretation of the geometry and internal structure of the genetic packages making up the internal macroform architecture of the point-bar sand body, suggests that the point bar and meander bend evolved sequentially as follows:

1 low suspension load and/or mud, producing low transverse bar slopes independent of W : D ratios, but with higher energy conditions on the lower point bar;

2 development of short-lived point-bar platforms, possibly reflecting a decrease in meander bend curvature;

3 steeper accretionary profiles resulting from a short-term increase in suspension load;

4 increased W : D ratios and reduced curvature of the meander bend, accompanied by increased deposition of mud on the upper point bar;

5 erosional reworking of the upper point bar as a result of a major flood or shift of meander bend, relocating the point bar into a higher energy part of the bend;

6 fluctuations in flow as a result of flood or change in channel curvature and direction of lateral accretion, followed by abandonment.

## ACKNOWLEDGEMENTS

The project was initiated and carried out mainly by the first author during his sabbatical leave at Virginia Polytechnic and State University (VPSU) in 1995. We are indebted to Durham University Research Fund for financial support (BRT) and to the Department of Geological Sciences at VPSU for their hospitality and support during the course of the project. We thank Dan Miller for his support in the field, and for his critical and helpful review of earlier versions of the manuscript. Reviews by Ted Hickin and Brian Willis greatly improved the manuscript.

## REFERENCES

ALLEN, J.R.L. (1970) *Physical Processes of Sedimentation.* Allen & Unwin, London, 248 pp.

BLODGETT, R.H. (1985) Paleovertisols — their utility in reconstructing ancient fluvial floodplain sequences. *Abstracts, Third International Conference on Fluvial Sedimentology,* Fort Collins, CO, p. 10.

BLUCK, B.J. (1971) Sedimentation in the meandering River Endrick. *Scott. J. Geol.,* 7, 93–138.

BRIDGE, J.A. (1978) Palaeohydraulic interpretation using mathematical models of contemporary flow and sedimentation in meandering channels. In: *Fluvial Sedimentology* (Ed. Miall, A.D.), Mem. Can. Soc. petrol. Geol., Calgary, 5, 723–742.

BRIDGE, J.A. (1985) Paleochannel patterns inferred from alluvial deposits: a critical evaluation. *J. sediment. Petrol.,* 55, 579–589.

BRIDGE, J.A. & DIEMER, J.A. (1983) Quantitative interpretation of an evolving ancient river system. *Sedimentology,* 30, 599–623.

BRIDGE, J.S. & JARVIS, J. (1976) Flow and sedimentary processes in the meandering River South Esk, Glen Clova, Scotland. *Earth Surf. Process.,* 1, 303–336.

BRIDGE, J.S. & JARVIS, J. (1982) The dynamics of a river bend: a study in flow and sedimentary processes. *Sedimentology,* 29, 499–542.

BRIDGE, J.A. & LEEDER, M.R. (1976) Sedimentary model for intertidal mudflat channels with examples from the Solway Firth, Scotland. *Sedimentology,* 23, 533–552.

BRIDGE, J.S., ALEXANDER, J., COLLIER, R.E.L., GAWTHORPE, R.L. & JARVIS, J. (1995) Ground penetrating radar and coring used to study the large-scale structure of point bar deposits in three dimensions. *Sedimentology,* 42, 839–852.

CANT, D.J. (1978) Development of a facies model for sandy braided river sedimentation: comparison of the South Saskatchewan River and Battery Point Formation. In: *Fluvial Sedimentology* (Ed. Miall, A.D.), Mem. Can. Soc. petrol. Geol., Calgary, 5, 627–640.

CAUDILL, M.R., MORA, C.I., TOBIN, K.J. & DRIESE, S.G. (1992) Preliminary interpretations of paleosols associated with Late Mississippian marginal marine deposits, Pennington Formation, Monterey, T.N. In: *Paleosols, Paleoweathering Surfaces, and Sequence Boundaries* (Eds Driese, S.G., Mora, C.I. & Walker, K.R.), pp. 57–78. Studies in Geology, Vol. 21, University of Tennessee Department of Geological Sciences, Knoxville, TN.

CAUDILL, M.R., DRIESE, S.G. & MORA, C.I. (1996) Preservation of a paleo-vertisol and an estimate of Late Mississippian paleo-precipitation. *J. sediment. Res.,* A66, 58–70.

CECIL, C.B., STANTON, R.W., NEUZIL, S.G., DULONG, F.T., RUPPERT, L.F. & PIERCE, B.S. (1985) Palaeoclimatic controls on Late Palaeozoic sedimentation and peat formation in the central Appalachian Basin (U.S.A.). *Int. J. Coal Geol.,* 5, 195–230.

COLEMAN, J.M. (1969) Brahmaputra River: channel processes and sedimentation. *Sediment. Geol.,* 3, 129–239.

COLTON, G.W. (1970) The Appalachian basin — its depositional sequences and their geological relationships. In: *Studies of Appalachian Geology* (Eds Fisher, G.W., Pettijohn, F.J., Reed, J.C. & Weaver, K.N.), pp. 5–48. Interscience Publishers, New York.

COTTER, E. (1971) Paleoflow characteristics of a Late Cretaceous river in Utah from analysis of sedimentary structures in the Ferron Sandstone. *J. sediment. Petrol.,* 41, 129–138.

CROWLEY, K.D. (1983) Large-scale bed configurations (macroforms), Platte River basin, Colorado and Nebraska: primary structures and formative processes. *Geol. Soc. Am. Bull.,* 94, 117–133.

DONALDSON, A.C., RENTON, J.J. & PRESLEY, M.W. (1985) Pennsylvanian deposystems and palaeoclimates of the Appalachians. *Int. J. Coal Geol.,* 5, 167–193.

EDWARDS, M.B., ERIKSSON, K.A. & KEIR, R.S. (1983) Paleochannel geometry and flow patterns determined from exhumed Permian point bars in north central Texas. *J. sediment. Petrol.,* 53, 1261–1270.

ELLIOTT. T. (1976) The morphology, magnitude and regime of a Carboniferous fluvial-distributary channel. *J. Sediment. Petrol.,* 46, 70–76.

ENGLUND, K.J., HENRY, T.W. & CECIL, C.B. (1981) Upper Mississippian and Lower Pennsylvanian depositional environments, southwestern Virginia and southern West Virginia. *Geological Society of America, 81, Cincinnati, Field Trip Guidebook No. 4. Mississippian–Pennsylvanian Boundary in the Central Part of the Appalachian Basin. Part 1: Southwestern Virginia–Southern West Virginia,* pp. 171–194. Geological Society of America, Boulder, CO.

FISK, H.N. (1944) *Geological Investigation of the Alluvial Valley of the Lower Mississippi River*. Mississippi River Commission, Vicksburg, MS, 78 pp.

GALLOWAY, W.E. (1981) Depositional architecture of Cenozoic Gulf Coastal plain fluvial systems. In: *Recent and Ancient Non-marine Depositional Environments: Models for Exploration* (Eds Etheridge, F.G. & Flores, R.M.), Spec. Publ. Soc. econ. Paleont. Miner., Tulsa, **31**, 127–155.

GIBSON, J.W. & HICKIN, E.H. (1997) Inter- and supratidal sedimentology of a fjord-head estuary, south-western British Columbia. *Sedimentology*, **44**, 1031–1052.

HARMS, J.C. & FAHNESTOCK, R.K. (1965) Stratification, bedforms and flow phenomena (with an example from the Rio Grande). In: *Primary Sedimentary Structures and their Hydrodynamic Interpretation* (Ed. Middleton, G.V.), Spec. Publ. Soc. econ. Paleont. Miner., Tulsa, **12**, 84–112.

HARMS, J.C., MACKENZIE, D.B. & MCCUBBIN, D.G. (1963) Stratification in modern sands of the Red River, Louisiana. *J. Geol.*, **71**, 566–580.

HICKIN, E.J. (1974) Development of meanders in natural river channels. *Am. J. Sci.*, **274**, 414–442.

HICKIN, E.J. (1983) River channel changes: retrospect and prospect. In: *Modern and Ancient Fluvial Systems* (Eds Collinson, J.D. & Lewin, J.), Spec. Publs int. Ass. Sediment., No. 6, pp. 61–84. Blackwell Scientific Publications, Oxford.

HOARE, R.D. (1993) Mississippian (Chesterian) bivalves from the Pennsylvanian stratotype area in West Virginia and Virginia. *J. Palaeontol.*, **67**, 374–396.

IKEDA, H. (1989) Sedimentary controls on channel migration and origin of point bars in sand-bedded meandering rivers. In: *River Meandering* (Eds Ikeda, S. & Parker, G.). *Am. geophys. Union, Water Resour. Monogr.*, **12**, 51–68.

JACKSON, R.G. (1975) Velocity–bedform–textural patterns of meander bends in the Lower Wabash River of Illinois and Indiana. *Geol. Soc. Am. Bull.*, **86**, 1511–1522.

JACKSON, R.G. (1978) Preliminary evaluation of lithofacies models for meandering alluvial streams. In: *Fluvial Sedimentology* (Ed. Miall, A.D.), Mem. Can. Soc. petrol. Geol., Calgary, **5**, 543–576.

LACEY, G. (1934) Uniform flow in alluvial rivers and canals. *Min. Proc. Inst. civ. Eng.*, **237**, 421–453.

LEOPOLD, L.B., WOLMAN, M.G. & MILLER, J.P. (1964) *Fluvial Processess in Geomorphology*. Freeman, San Francisco, 522 pp.

LEVEY, R.A. (1978) Bedform distribution and internal stratification of coarse-grained point bars, Upper Congaree River, S.C. In: *Fluvial Sedimentology* (Ed. Miall, A.D.), Mem. Can. Soc. petrol. Geol., Calgary, **5**, 105–128.

MASKASKE, B. & NAP, R.L. (1995) A transition from a braided to a meandering channel facies, showing inclined heterolithic stratification (Late Weichselian, central Netherlands). *Geol. Mijnbouw*, **74**, 13–20.

MIALL, A.D. (1996) *The Geology of Fluvial Deposits*. Springer-Verlag, Berlin, 582 pp.

MIDDLETON, G.V. & SOUTHARD, J.B. (1984) *Mechanics of Sediment Motion*. Soc. econ. Paleont. and Miner., Tulsa, Short Course 3, 401 pp.

MILLER, D.J. & ERIKSSON, K.A. (1997) Sequence-stratigraphic framework of Upper Mississippian strata in the Central Appalachians: evolution of fluvial style in response to changes in accommodation. *Abstracts, 6th International Conference on Fluvial Sedimentology*, Cape Town, 22–26 September, p. 139.

NANSON, G.C. (1980) Point bar and floodplain formation of the meandering Beatton River, northeastern British Columbia, Canada. *Sedimentology*, **27**, 3–29.

OSTERKAMP, W.R. & HEDMAN, E.R. (1982) Perennial-streamflow characteristics related to channel geometry and sediment in Missouri River basin. *U.S. geol. Surv. Prof. Pap.*, **1242**, 1–37.

PFEFFERKORN, H.W. & GILLISPIE, W.H. (1981) Biostratigraphic significance of plant megafossils near the Mississippian–Pennsylvanian boundary in southern West Virginia and southwest Virginia. *Geological Society of America 81, Cincinnati, Field Trip Guidebook No. 4. Mississippian–Pennsylvanian Boundary in the Central Part of the Appalachian Basin. Part 1: Southwestern Virginia–Southern West Virginia*, pp. 159–164. Geological Society of America, Boulder, CO.

RAHMANI, R.A. (1988) Estuarine tidal channel and nearshore sedimentation of a Late Cretaceous epicontinental sea, Drumheller, Alberta, Canada. In: *Tide-influenced Sedimentary Environments and Facies* (Eds de Boer, P.L., van Gelder, A. & Nio, S.D.), pp. 433–471. Reidel, Dordrecht.

REGER, D.B. (1926) *Mercer, Monroe and Summers Counties: County Report*. West Virginia Geological and Economic Survey, Morgantown, 963 pp.

SCHUMM, S.A. (1963) A tentative classification of alluvial river channels. *U.S. geol. Surv. Circ.*, 477, 10 pp.

SCHUMM, S.A. (1972) Fluvial palaeochannels. In: *Recognition of Ancient Sedimentary Environments* (Eds Rigby, J.K. & Hamblin, W.K.), Spec. Publ. Soc. econ. Paleont. Miner., Tulsa, **16**, 98–107.

SUHUMM, S.A. & KHAN, H.R. (1972) Experimental study of channel patterns. *Geol. Soc. Am. Bull.*, **83**, 1755–1770.

SCHUMM, S.A. & LICHTY, R.W. (1963) Channel widening and flood plain construction along the Cimarron River in southwestern Kansas. *U.S. geol. Surv. Prof. Pap.*, **352-D**, 71–88.

SCHWARTZ, D.E. (1978) Hydrology and current orientation analysis of a braided-to-meandering transition: the Red River in Oklahoma and Texas, U.S.A. In: *Fluvial Sedimentology* (Ed. Miall, A.D.), Mem. Can. Soc. petrol. Geol., Calgary, **5**, 105–127.

SMITH, D.G. (1988) Modern point bar deposits analogous to the Athabasca Oil Sands, Alberta, Canada. In: *Tide-influenced Sedimentary Environments and Facies* (Eds de Boer, P.L., van Gelder, A. & Nio, S.D.), pp. 417–432. Reidel, Dordrecht.

SMITH, N.D. (1971) Transverse bars and braiding in the Lower Platte River, Nebraska. *Geol. Soc. Am. Bull.*, **82**, 3407–3420.

STEWART, D.J. (1983) Possible suspended-load channel deposits from the Wealden Group (Lower Cretaceous) of southern England. In: *Modern and Ancient Fluvial Systems* (Eds Collinson, J.D. & Lewin, J.), Spec. Publs int. Ass. Sediment., No. 6, pp. 369–384.

SUNBORG, F.A., BENNINGTON, J.B., WIZEVICH, M.C. & BAMBACH, R.K. (1990) Upper Carboniferous (Namurian) amphibian trackways from the Bluefield Formation, West Virginia, USA. *Ichnos*, **1**, 111–124.

TAYLOR, G. & WOODYER, K.D. (1978) Bank deposition in suspended-load streams. In: *Fluvial Sedimentology* (Ed. Miall, A.D.), Mem. Can. Soc. petrol. Geol., Calgary, **5**, 257–275.

THOMAS, R.G., SMITH, D.G., WOOD, J.M., VISSER, J., CALVERLEY-RANGE, E.A. & KOSTER, E.H. (1987) Inclined heterolithic stratification — terminology, description, interpretation and significance. *Sediment. Geol.*, **53**, 123–179.

THOMAS, W.A. (1959) *Upper Mississippian stratigraphy of southwestern Virginia, southern West Virginia and eastern*

*Kentucky*. Unpublished PhD thesis, Virginia Polytechnic Institute, Blacksburg, VA, 322 pp.

VISSER, J. (1986) *Sedimentology and taphonomy of a Styracosaurus bone bed in the Late Cretaceous Judith River Formation, Dinosaur Provincial Park, Alberta*. Unpublished MSc thesis, University of Calgary, Calgary, 150 pp.

WILLIAMS, G.P. (1984) Paleohydrologic equations for rivers. In: *Developments and Applications of Geomorphology* (Eds Costa, J.E. & Fleisher, P.J.), pp. 343–367. Springer-Verlag, Berlin.

WILLIAMS, G.P. (1987) Paleofluvial estimates from dimensions of former channels and meanders. In: *Flood Geomorphology* (Eds Baker, V.R., Kochel, R.C. & Patton, P.C.), pp. 321–334. Wiley Interscience, New York.

WILLIS, B.J. (1989) Palaeochannel reconstruction from point bar deposits: a three-dimensional perspective. *Sedimentology*, **36**, 757–766.

WILLIS, B.J. (1993) Interpretation of bedding geometry within ancient point-bar deposits. In: *Alluvial Sedimentation* (Eds Marzo, M. & Puigdefabregas, C.), Spec. Publs int. Ass. Sediment., No. 17, pp. 101–114. Blackwell Scientific Publications, Oxford.

WILPOLT, R.H. & MARDEN, D.W. (1959) Geology and oil and gas possibilities of Upper Mississippian rocks of southwestern Virginia, southern West Virginia, and eastern Kentucky. *U.S. geol. Surv. Bull.*, **1072-K**, 587–656.

WOODYER, K.D., TAYLOR, G. & CROOK, K.A.W. (1979) Depositional processes along a very low gradient suspended load stream: the Barwon River, New South Wales. *Sediment. Geol.*, **22**, 97–120.

*Spec. Publs int. Ass. Sediment.* (1999) **28**, 381–392

# Palaeohydrological parameters of a Proterozoic braided fluvial system (Wilgerivier Formation, Waterberg Group, South Africa) compared with a Phanerozoic example

M. VAN DER NEUT* *and* P. G. ERIKSSON†

*\*Council for Geoscience, Private Bag X112, Pretoria 0001, South Africa; and*
*†Department of Geology, University of Pretoria, Pretoria 0002, South Africa (Email: pat@scientia.up.ac.za)*

## ABSTRACT

Palaeohydrological data for mean and bankfull channel depth and width, mean annual discharge and mean annual bankfull discharge, palaeoslope, drainage area and principal stream length are estimated for braided, bedload rivers of the *c.* 1800 Ma Wilgerivier Formation red beds in South Africa. The age of these deposits implies deposition within a pre-vegetative continental setting, where soil formation was probably limited, where weathering was aggressive and erosion rapid, and there is no evidence for significant suspended-load influence on the Wilgerivier fluvial style. These data are compared with the same hydrological variables calculated previously for the Middle Triassic Molteno Formation (predominantly braided bedload channels, with subordinate suspended load and floodplain deposits). The drainage area and principal stream length of the Molteno braided channels were significantly higher, and it thus follows that Molteno channel dimensions (and their bankfull equivalents) as well as discharge were higher than for individual braid-channels of the Wilgerivier Formation. Bankfull discharge values estimated for the latter exhibit a greater range than those derived from Molteno Formation cross-bed set thicknesses. Palaeoslope values estimated for the Precambrian Wilgerivier braid-channels lie between those generally found for alluvial fans (> 0.026) and for rivers (< 0.007), including the Molteno channels (mean = 0.0058). These higher palaeoslopes, combined with shorter principal stream lengths, indicate that the Wilgerivier source areas were much closer than those inferred for the Molteno Formation. The margins of the Wilgerivier graben basin would have provided smaller, fault-bounded drainage areas with steep slopes. In the Precambrian, at about 1800 Ma, weathering was still relatively aggressive, providing much sandy detritus close to source terrains, and the generally semi-arid conditions combined with lack of vegetation meant that rapid aeolian and aqueous erosion would have removed fine material. Gravity flow deposits on the high Wilgerivier palaeoslopes were thus uncommon and sandy material was transported by braided channel systems on high-gradient floodplains, most likely as a result of intermittent torrential rainstorms. The different palaeoclimate inferred for the Wilgerivier Formation, combined with unique weathering–erosional regimes of the Precambrian, most probably were responsible for the significant differences in estimated palaeohydraulic parameters when compared with the Phanerozoic Molteno braid-channels.

## INTRODUCTION

The approximately 1800 Ma Wilgerivier Formation forms part of the Waterberg Group, one of the earliest red bed successions of South Africa (Twist & Cheney, 1986). The Waterberg Group comprises fault-related continental sedimentary deposits (Callaghan *et al.*, 1991) laid down within the so-called Main Waterberg basin and a smaller Middelburg basin, located to the south-east of the main depository (Fig. 1). The Wilgerivier Formation is the only stratigraphical unit within the Middelburg basin, the present study area, and unconformably overlies rocks of the

2100 Ma (Burger & Walraven, 1980) Loskop Formation. It is, in turn, succeeded by Mesozoic deposits of the Karoo Supergroup. Horizontally stratified and cross-bedded sandstones predominate within the Middelburg basin, with subordinate conglomerate beds, particularly in the western portion thereof. These strongly red-coloured rocks dip gently towards the centre of the basin at approximately 10°, on average. Maximum dips up to 70° occur on the north-eastern margin of the basin, reflecting post-depositional inversion tectonics. The Wilgerivier

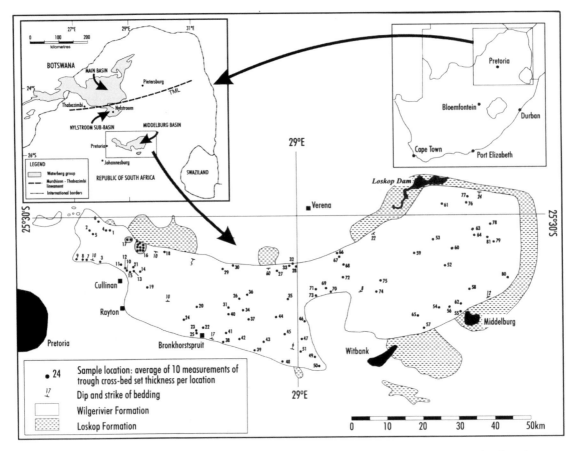

**Fig. 1.** Locality map showing the Main Waterberg basin and smaller Middelburg basin, South Africa (inset, top left). The Nylstroom sub-basin of the Main Waterberg basin represents a fault-bounded protobasin, within which the lowermost two formations of that depository were laid down. Note 81 numbered locations within the Middelburg basin, where an average of 10 trough cross-bed set thicknesses was measured at each locality (main map). Note also dip of Wilgerivier Formation beds towards the centre of the preserved Middelburg basin.

conglomerates tend to define the bases of upward-fining sandstone successions, capped by thin mudrocks (Eriksson & Vos, 1979; Van der Neut et al., 1991). Sandstone and conglomerate beds exhibit lenticular geometry, with lateral extents up to 850 m and thicknesses mostly from 5 to 15 m; contacts between the two rock types are commonly gradational.

Previous workers consider the Waterberg Group in the Main basin to reflect predominantly alluvial-fluvial braid-plain deposition within a graben setting, with subordinate aeolian and lacustrine or littoral palaeoenvironments (e.g. Vos & Eriksson, 1977; Eriksson & Vos, 1979; Tankard et al., 1982; Callaghan, 1987; Callaghan et al., 1991). Little work has been done within the Middelburg basin, apart from that by Van der Neut et al. (1991), who propose a distal alluvial-fan–braidplain palaeoenvironmental

setting and a graben tectonic framework. A braided-stream model is thus considered to be generally applicable to the Wilgerivier Formation (Van der Neut & Eriksson, 1997).

The aim of this paper is to derive a variety of hydrological parameters operating during the deposition of the Proterozoic fluvial Wilgerivier Formation. Very few such studies have been undertaken on Precambrian rocks, an exception being that by Els (1990) on the Middelvlei gold placer in the Late Archaean Witwatersrand Supergroup. Palaeohydraulic investigations of Phanerozoic to Recent fluvial deposits are more common. For example, Turner (1978, 1980) investigated the relationship between the sedimentary lithofacies and various palaeohydrological parameters of the Molteno Formation, Karoo Supergroup in South Africa. Zawada (1994) studied the magnitude of

Cenozoic palaeofloods of the Buffels River, South Africa, in order to obtain a better understanding of the modern river's hydrology and how these data can aid future development, such as the construction of dams.

Palaeohydrological data from the Wilgerivier Formation are compared in this paper with Phanerozoic data from the Molteno Formation, taken from Turner (1980). Both the Wilgerivier and Molteno Formations reflect braided-stream deposition within a setting influenced by active tectonic uplift of the source areas (e.g. Turner, 1975; Van der Neut *et al.*, 1991), and in both studies palaeohydraulic data were collected at a number of sites across the preserved deposits. In addition, both formations are characterized by sandy bedload fluvial deposits with subordinate pebbly sandstones and conglomerates. In contrast, the inferred palaeoclimate in the drainage area of the Molteno braided channels was cool and humid (e.g. Cairncross *et al.*, 1995), whereas the Waterberg rocks in both graben basins reflect semi-arid conditions (e.g. Callaghan *et al.*, 1991). Unique atmospheric and weathering–erosional conditions pertaining in the Precambrian (e.g. Eriksson *et al.*, 1998) would most likely have influenced the Wilgerivier channel systems and a comparison with a broadly similar Phanerozoic example is thus potentially interesting.

## PALAEOHYDROLOGY: METHODS

Various authors have proposed different formulae for the estimation of hydrological parameters in ancient river systems, largely making use of grain-size analyses, clast sizes (in conglomerates/gravels) and set thicknesses measured from cross-bed structures (e.g. Ethridge & Schumm, 1978). It is important to view the results obtained by such means with due caution, as the calculations are effectively only estimates of hydrological conditions during deposition of the fluvial sediments. Standard errors in their application are often significant and vary depending on the formulae utilized and the number of data points used to derive the equations (see Williams (1984) for a review of the problem). In this paper, we will utilize different formulae in order to minimize the implicit errors and also to provide an estimation of the range of palaeohydraulic parameters being investigated. Channel-fill structures containing pebbles are limited in the present study area and recourse had thus to be made to using methods based on cross-bedding, a procedure also used by Turner (1980). All of these calculations are in one way or another related to the set thickness of preserved cross-beds. The latter are used to obtain a mean water depth ($d_m$), in metres, by applying Allen's (1968) formula:

$$h = 0.086 \, (d_m)^{1.19} \qquad (1)$$

where $h$ is the mean set thickness of trough cross-beds in metres, and $d_m$ is the mean water depth over the sedimentary structure, in metres. This formula was used by Miall (1976) also, who indicates that it provides only 50% accuracy, as water depth above a flow-form of given height may vary up to this amount (Turner, 1980). In spite of this severe limitation, Miall was still of the opinion that the formula was effective for relative comparison of water depth. For the Wilgerivier Formation, trough cross-bed set thicknesses were determined for 81 localities within the study area (Fig. 1), with an average of 10 measurements at each site, for a total of 810 set thicknesses (see Table 2). This data set is considerably larger than the 137 cross-beds measured by Turner (1980), or the number of data points used to derive most of the palaeohydraulic formulae (Williams, 1984). In addition, preserved sandy channel-fill structures were also recorded at these same localities and these data are given in Table 1. They provide a possible means of evaluating the accuracy of calculated channel depths and widths from the various formulae discussed here.

The ratio between channel width and depth ($F$) may be estimated by:

$$F = 225 \, M^{-1.08} \quad \text{(Schumm, 1968a)} \qquad (2)$$

where $M$ is the sediment load variable, or the percentage of silt and clay in the channel perimeter. According to Schumm (1968a,b), coarse bedload streams typically have $M$ values of less than 5%. The palaeostreams of the Wilgerivier Formation generally fall in this category, as they have low matrix contents (Van der Neut *et al.*, 1991), and $M$ therefore is assumed to be 5% in this study. Substituting this value in the above equation gives a channel width to depth ratio of $F = 45$. This result enables estimation of the width of the channel, $w$ (in metres), by calculating:

$$w = 45 \, d_m \quad \text{(Schumm, 1968a)} \qquad (3)$$

From this, it is then possible to estimate the average daily discharge (also commonly called the mean annual discharge), $Q_m$ (m$^3$ s$^{-1}$), as

$$Q_m = v \, A \qquad (4)$$

where $A$ is the mean cross-sectional surface area of the channel (m$^2$), approximated by multiplying the mean water depth ($d_m$) by channel width ($w$). Although Turner (1980) used this formula to derive an approximation of mean annual discharge, it actually provides an estimate of the maximum instantaneous water discharge (Williams, 1984). Turner (1980) and Miall (1976) both assumed a velocity of 0.75 m s$^{-1}$ when using this formula, as

**Table 1.** Width and depth of preserved sandy channel-fill structures present in the Wilgerivier Formation.

| Locality number (see Fig. 1) | Width (m) | Depth (m) | Locality number (see Fig. 1) | Width (m) | Depth (m) |
|---|---|---|---|---|---|
| 2 | 10 | 0.4 | 18 | 3 | 0.1 |
| 2 | 24 | 0.9 | 20 | 24 | 0.8 |
| 3 | 3.6 | 0.2 | 22 | 3 | 0.12 |
| 4 | 2 | 0.25 | 22 | 7 | 0.3 |
| 6 | 12 | 0.2 | 22 | 2 | 0.12 |
| 10 | 5 | 0.25 | 22 | 6 | 0.4 |
| 10 | 6.5 | 0.5 | 23 | 2.5 | 0.18 |
| 12 | 11 | 0.45 | 25 | 10 | 0.45 |
| 12 | 12 | 0.5 | 25 | 28 | 1.4 |
| 12 | 9 | 0.25 | 33 | 5 | 0.36 |
| 12 | 42 | 2 | 33 | 14 | 0.7 |
| 12 | 0.7 | 0.05 | 40 | 10 | 0.23 |
| 12 | 8 | 0.4 | 40 | 22 | 0.8 |
| 15 | 8 | 0.4 | 40 | 3 | 0.15 |
| 15 | 12 | 0.4 | 40 | 11 | 0.35 |
| 15 | 5 | 0.25 | 40 | 11 | 0.25 |
| 15 | 28 | 0.7 | 41 | 3.5 | 0.2 |
| 15 | 23 | 0.7 | 41 | 3 | 0.5 |
| 15 | 19 | 0.5 | 41 | 4.5 | 0.6 |
| 16 | 20 | 1.1 | 55 | 8.2 | 0.26 |
| 16 | 46 | 0.9 | 55 | 6.5 | 0.3 |
| 16 | 24 | 0.5 | 55 | 13 | 0.3 |
| 16 | 17 | 0.35 | 59 | 10 | 0.9 |
| 16 | 6 | 0.3 | 60 | 1.6 | 0.15 |
| 16 | 6 | 0.2 | 60 | 1.6 | 0.25 |
| 16 | 21 | 0.4 | 61 | 2.4 | 0.12 |
| 16 | 23 | 0.6 | 63 | 5 | 0.2 |
| 16 | 10 | 0.4 | 63 | 3.5 | 0.2 |
| 16 | 11 | 0.3 | 63 | 3 | 0.2 |
| 16 | 10 | 0.5 | 63 | 1.2 | 0.12 |
| 16 | 4.5 | 0.3 | 63 | 3.8 | 0.15 |
| 16 | 4 | 0.25 | 72 | 1.3 | 0.12 |
| 16 | 8 | 0.35 | 75 | 10 | 0.18 |
| 16 | 30 | 3 | 75 | 12 | 0.4 |
| 17 | 9 | 0.3 | 80 | 8.8 | 0.35 |
| 17 | 4 | 0.12 | 80 | 9 | 0.7 |
| 17 | 4.5 | 0.1 | | | |

large-scale dunes migrate under velocities ranging from 0.5–1 m s$^{-1}$. Velocities obtained from measuring pebble axes cannot be used here, as there would naturally be large differences between the velocities of sand grains and large clasts.

The mean bankfull channel depth ($d_b$) (metres) can be calculated from:

$$d_b = 0.6\, M^{0.34}\, Q_m^{0.29} \quad \text{(Schumm, 1969)} \qquad (5)$$

Calculation of bankfull channel width, $w_b$ (m), from

$$w_b = 8.9\, d_b^{1.40} \quad \text{(Leeder, 1973)} \qquad (6)$$

enables a more realistic estimation of $Q_m$ (average daily discharge) than eqn (4), when applied to the formula of Osterkamp & Hedman (1982):

$$Q_m = 0.027\, w_b^{1.71} \qquad (7)$$

Comparison of the two estimates of $Q_m$, as determined above, will necessarily reveal large differences (Table 2), as eqn (4) reflects maximum instantaneous discharge and eqn (7) the average daily discharge. A check of relative accuracy by comparing the two $Q_m$ estimates is thus not viable.

According to Schumm (1972), the gradient or slope of a river is one of the fundamental parameters controlling channel morphology. Two of his formulae were used to estimate stream palaeoslope ($S$) (m m$^{-1}$), using parameters already determined above:

$$S = 60\, M^{-0.38}\, Q_m^{-0.32} \quad \text{(Schumm, 1968a)} \qquad (8a)$$
$$S = 30\, (F^{0.95}/w^{0.98}) \quad \text{(Schumm, 1972)} \qquad (8b)$$

As only the second estimation of $Q_m$ (eqn 7) may sensibly be applied to eqn (8a), two different estimations of $S$ (Table 2) can be obtained, and these provide an

**Table 2.** Palaeohydraulic data estimated for the Wilgerivier Formation.

| Locality (see Fig. 1) | $h$ (m) | $d_m$ (m) | $w$ (m) | $Q_m$ (m³ s⁻¹) | $d_b$ (m) | $w_b$ (m) | $Q_m(1)$ (m³ s⁻¹) | $s$ (m m⁻¹) | $s(1)$ (m m⁻¹) | $Q_b$ (m³ s⁻¹) | $Q_b(1)$ (m³ s⁻¹) | $A_d$ (km²) | $A_d(1)$ (km²) | $L$ (km) | $L(1)$ (km) |
|---|---|---|---|---|---|---|---|---|---|---|---|---|---|---|---|
| 1 | 0.18 | 1.86 | 83.71 | 116.79 | 4.12 | 64.70 | 33.73 | 0.010559 | 0.018751 | 964.70 | 1132.99 | 9532.08 | 11811.42 | 341.70 | 388.61 |
| 2 | 0.45 | 4.02 | 180.79 | 544.75 | 6.45 | 120.89 | 98.24 | 0.007499 | 0.010160 | 3206.49 | 3491.08 | 47283.10 | 52959.74 | 893.21 | 956.08 |
| 3 | 0.15 | 1.60 | 71.82 | 85.96 | 3.77 | 57.13 | 27.26 | 0.011302 | 0.021182 | 759.62 | 905.69 | 6930.98 | 8762.73 | 282.23 | 324.87 |
| 4 | 0.103 | 1.16 | 52.37 | 45.70 | 3.14 | 44.20 | 17.58 | 0.013005 | 0.027236 | 464.08 | 570.79 | 3593.02 | 4734.82 | 190.28 | 224.55 |
| 5 | 0.43 | 3.87 | 174.01 | 504.68 | 6.30 | 117.20 | 93.16 | 0.007628 | 0.010474 | 3020.98 | 3301.49 | 43671.42 | 49160.25 | 851.62 | 914.32 |
| 6 | 0.1648 | 1.73 | 77.73 | 100.69 | 3.95 | 60.92 | 30.43 | 0.010912 | 0.019890 | 859.34 | 1016.64 | 8169.97 | 10222.54 | 311.50 | 356.34 |
| 7 | 0.1451 | 1.55 | 69.84 | 81.30 | 3.71 | 55.85 | 26.23 | 0.011443 | 0.21658 | 727.26 | 869.49 | 6540.11 | 8298.88 | 272.57 | 314.44 |
| 8 | 0.15 | 1.60 | 71.82 | 85.96 | 3.77 | 57.13 | 27.26 | 0.011302 | 0.021182 | 759.62 | 905.69 | 6930.98 | 8762.73 | 282.23 | 324.87 |
| 9 | 0.109 | 1.22 | 54.92 | 50.26 | 3.23 | 45.94 | 18.78 | 0.012733 | 0.026224 | 499.84 | 611.90 | 3966.77 | 5194.81 | 201.93 | 237.39 |
| 10 | 0.353 | 3.28 | 147.42 | 362.23 | 5.73 | 102.44 | 74.00 | 0.008211 | 0.011951 | 2332.47 | 2590.99 | 30933.00 | 35586.63 | 692.44 | 753.18 |
| 11 | 0.136 | 1.47 | 66.14 | 72.91 | 3.60 | 53.43 | 24.32 | 0.011724 | 0.022616 | 668.06 | 803.01 | 5840.12 | 7463.76 | 254.67 | 295.05 |
| 12 | 0.344 | 3.21 | 144.26 | 346.85 | 5.66 | 100.65 | 71.81 | 0.008290 | 0.012160 | 2254.83 | 2510.09 | 29567.75 | 34112.99 | 673.94 | 734.31 |
| 13 | 0.25 | 2.45 | 110.32 | 202.85 | 4.84 | 80.95 | 49.48 | 0.009340 | 0.015053 | 1483.90 | 1696.07 | 16925.50 | 20226.75 | 482.23 | 536.64 |
| 14 | 0.246 | 2.42 | 108.84 | 197.42 | 4.80 | 80.06 | 48.56 | 0.009396 | 0.015216 | 1452.86 | 1662.81 | 16455.02 | 19699.50 | 474.14 | 528.20 |
| 15 | 0.3 | 2.86 | 128.59 | 275.58 | 5.29 | 91.67 | 61.21 | 0.008725 | 0.013325 | 1884.52 | 2121.74 | 23277.38 | 27263.95 | 583.83 | 641.92 |
| 16 | 0.377 | 3.46 | 155.80 | 404.58 | 5.91 | 107.14 | 79.91 | 0.008012 | 0.011437 | 2542.51 | 2808.99 | 34701.70 | 39633.83 | 741.89 | 803.47 |
| 17 | 0.3 | 2.86 | 128.59 | 275.58 | 5.29 | 91.67 | 61.21 | 0.008725 | 0.013325 | 1884.52 | 2121.74 | 23277.38 | 27263.95 | 583.83 | 641.92 |
| 18 | 0.22 | 2.20 | 99.08 | 163.63 | 4.55 | 74.19 | 42.62 | 0.009796 | 0.016396 | 1254.96 | 1449.64 | 13556.56 | 16406.45 | 421.73 | 473.30 |
| 19 | 0.417 | 3.77 | 169.58 | 479.30 | 6.21 | 114.77 | 89.88 | 0.007716 | 0.010691 | 2901.83 | 3179.33 | 41389.96 | 46750.01 | 824.64 | 887.15 |
| 20 | 0.291 | 2.79 | 125.34 | 261.83 | 5.21 | 89.79 | 59.07 | 0.008825 | 0.013599 | 1810.76 | 2043.83 | 22070.59 | 25937.43 | 565.48 | 622.99 |
| 21 | 0.326 | 3.06 | 137.89 | 316.89 | 5.51 | 97.02 | 67.44 | 0.008458 | 0.012605 | 2101.44 | 2349.76 | 26916.79 | 31239.07 | 637.00 | 696.54 |
| 22 | 0.3 | 2.86 | 128.59 | 275.58 | 5.29 | 91.67 | 61.21 | 0.008725 | 0.013325 | 1884.52 | 2121.74 | 23277.38 | 27263.95 | 583.83 | 641.92 |
| 23 | 0.36 | 3.33 | 149.88 | 374.39 | 5.78 | 103.82 | 75.72 | 0.008151 | 0.011796 | 2393.29 | 2654.23 | 32013.04 | 36749.51 | 706.84 | 767.85 |
| 24 | 0.2 | 2.03 | 91.46 | 139.41 | 4.34 | 69.52 | 38.14 | 0.010151 | 0.017475 | 1107.57 | 1289.51 | 11459.41 | 14035.63 | 381.62 | 430.99 |
| 25 | 0.222 | 2.22 | 99.84 | 166.14 | 4.57 | 74.65 | 43.08 | 0.009763 | 0.016297 | 1269.94 | 1465.84 | 13752.38 | 16651.38 | 425.75 | 477.53 |
| 26 | 0.28 | 2.70 | 121.34 | 245.41 | 5.12 | 87.46 | 56.47 | 0.008953 | 0.013954 | 1721.56 | 1949.36 | 20633.10 | 24351.33 | 543.08 | 599.85 |
| 27 | 0.238 | 2.35 | 105.85 | 186.75 | 4.73 | 78.28 | 46.72 | 0.009513 | 0.015556 | 1391.24 | 1596.64 | 15531.13 | 18661.36 | 457.98 | 511.32 |
| 28 | 0.35 | 3.25 | 146.37 | 357.08 | 5.70 | 101.84 | 73.27 | 0.008237 | 0.012020 | 2306.52 | 2563.97 | 30474.99 | 35092.72 | 686.27 | 746.89 |
| 29 | 0.325 | 3.06 | 137.53 | 315.26 | 5.50 | 96.82 | 67.20 | 0.008468 | 0.012631 | 2093.00 | 2340.91 | 26772.64 | 31082.31 | 634.95 | 694.44 |
| 30 | 0.338 | 3.16 | 142.14 | 336.74 | 5.61 | 99.45 | 70.35 | 0.008345 | 0.012304 | 2203.42 | 2456.43 | 28672.27 | 33144.09 | 661.61 | 721.72 |
| 31 | 0.17 | 1.77 | 79.78 | 106.09 | 4.01 | 62.22 | 31.55 | 0.010786 | 0.019481 | 895.06 | 1056.18 | 8625.83 | 10756.04 | 321.82 | 367.38 |
| 32 | 0.425 | 3.83 | 172.31 | 494.85 | 6.27 | 116.27 | 91.90 | 0.007661 | 0.010556 | 2975.02 | 3254.41 | 42787.74 | 48227.66 | 841.24 | 903.87 |
| 33 | 0.12 | 1.32 | 59.54 | 59.08 | 3.38 | 49.06 | 21.01 | 0.012284 | 0.024591 | 566.97 | 688.59 | 4692.63 | 6080.58 | 223.35 | 260.91 |
| 34 | 0.21 | 2.12 | 95.29 | 151.32 | 4.45 | 71.87 | 40.37 | 0.009968 | 0.016914 | 1180.72 | 1369.14 | 12479.50 | 15203.03 | 401.65 | 452.15 |
| 35 | 0.175 | 1.82 | 81.75 | 111.39 | 4.07 | 63.46 | 32.63 | 0.010670 | 0.019107 | 929.72 | 1094.46 | 9074.12 | 11278.92 | 331.75 | 378.00 |
| 36 | 0.15 | 1.60 | 71.82 | 85.96 | 3.77 | 57.13 | 27.26 | 0.011302 | 0.021182 | 759.62 | 905.69 | 6930.98 | 8762.73 | 282.23 | 324.87 |
| 37 | 0.256 | 2.50 | 112.54 | 211.10 | 4.90 | 82.27 | 50.87 | 0.009257 | 0.014816 | 1530.76 | 1746.20 | 17641.83 | 21027.75 | 494.37 | 549.29 |
| 38 | 0.1525 | 1.62 | 72.82 | 88.38 | 3.80 | 57.78 | 27.79 | 0.011233 | 0.020949 | 776.26 | 924.26 | 7134.14 | 9003.15 | 287.17 | 330.19 |
| 39 | 0.325 | 3.06 | 137.53 | 315.26 | 5.50 | 96.82 | 67.20 | 0.008468 | 0.012631 | 2093.00 | 2340.91 | 26772.64 | 31082.31 | 634.95 | 694.44 |
| 40 | 0.1971 | 2.01 | 90.34 | 136.03 | 4.31 | 68.83 | 37.49 | 0.010207 | 0.017647 | 1086.57 | 1266.58 | 11170.57 | 13703.90 | 375.81 | 424.85 |
| 41 | 0.34 | 3.17 | 142.85 | 340.10 | 5.62 | 99.85 | 70.83 | 0.008327 | 0.012255 | 2220.52 | 2474.29 | 28969.45 | 33465.85 | 665.72 | 725.92 |
| 42 | 0.23 | 2.29 | 102.86 | 176.32 | 4.65 | 76.47 | 44.89 | 0.009635 | 0.015916 | 1330.26 | 1530.98 | 14630.19 | 17645.22 | 441.85 | 494.43 |
| 43 | 0.163 | 1.71 | 77.01 | 98.85 | 3.93 | 60.46 | 30.04 | 0.010957 | 0.020037 | 847.06 | 1003.02 | 8014.64 | 10040.34 | 307.93 | 352.51 |
| 44 | 0.45 | 4.02 | 180.79 | 544.75 | 6.45 | 120.89 | 98.24 | 0.007499 | 0.010160 | 3206.49 | 3491.08 | 47283.10 | 52959.74 | 893.21 | 956.08 |

*(continued on p. 386)*

**Table 2.** *(continued)*

| Locality (see Fig. 1) | $h$ (m) | $d_m$ (m) | $w$ (m) | $Q_m$ (m³ s⁻¹) | $d_b$ (m) | $w_b$ (m) | $Q_m(1)$ (m³ s⁻¹) | $s$ (m m⁻¹) | $s(1)$ (m m⁻¹) | $Q_b$ (m³ s⁻¹) | $Q_b(1)$ (m³ s⁻¹) | $A_d$ (km²) | $A_d(1)$ (km²) | $L$ (km) | $L(1)$ (km) |
|---|---|---|---|---|---|---|---|---|---|---|---|---|---|---|---|
| 45 | 0.26 | 2.53 | 114.02 | 216.67 | 4.93 | 83.15 | 51.80 | 0.009204 | 0.014663 | 1562.19 | 1779.77 | 18126.43 | 21568.45 | 502.48 | 557.72 |
| 46 | 0.1283 | 1.40 | 62.98 | 66.11 | 3.50 | 51.35 | 22.72 | 0.011981 | 0.023515 | 618.92 | 747.53 | 5274.50 | 6784.35 | 239.57 | 278.63 |
| 47 | 0.1825 | 1.88 | 84.68 | 119.52 | 4.15 | 65.31 | 34.27 | 0.010504 | 0.018579 | 982.30 | 1152.35 | 9764.67 | 12081.24 | 346.68 | 393.91 |
| 48 | 0.11 | 1.23 | 55.34 | 51.04 | 3.24 | 46.23 | 18.98 | 0.012690 | 0.026064 | 505.86 | 618.80 | 4030.60 | 5273.08 | 203.87 | 239.53 |
| 49 | 0.1 | 1.14 | 51.08 | 43.49 | 3.10 | 43.32 | 16.99 | 0.013150 | 0.027779 | 446.44 | 550.44 | 3412.11 | 4511.10 | 184.48 | 218.12 |
| 50 | 0.4 | 3.64 | 163.75 | 446.92 | 6.09 | 111.56 | 85.62 | 0.007836 | 0.010993 | 2747.74 | 3020.90 | 38485.90 | 43669.82 | 789.42 | 851.60 |
| 51 | 0.14 | 1.51 | 67.77 | 76.55 | 3.65 | 54.50 | 25.15 | 0.011597 | 0.022182 | 693.94 | 832.11 | 6143.63 | 7826.60 | 262.53 | 303.58 |
| 52 | 0.138 | 1.49 | 66.96 | 74.72 | 3.62 | 53.97 | 24.73 | 0.011660 | 0.022397 | 680.97 | 817.53 | 5991.06 | 7644.34 | 258.60 | 299.32 |
| 53 | 0.163 | 1.71 | 77.01 | 98.85 | 3.93 | 60.46 | 30.04 | 0.010957 | 0.020037 | 847.06 | 1003.02 | 8014.64 | 10040.34 | 307.93 | 352.51 |
| 54 | 0.144 | 1.54 | 69.40 | 80.26 | 3.70 | 55.56 | 25.99 | 0.011476 | 0.021768 | 720.04 | 861.40 | 6453.70 | 8196.11 | 270.40 | 312.10 |
| 55 | 0.188 | 1.93 | 86.82 | 125.64 | 4.21 | 66.64 | 35.48 | 0.010389 | 0.018213 | 1021.28 | 1195.15 | 10284.79 | 12683.16 | 357.64 | 405.57 |
| 56 | 0.117 | 1.30 | 58.28 | 56.62 | 3.34 | 48.22 | 20.40 | 0.012401 | 0.025011 | 548.46 | 667.51 | 4489.50 | 5833.64 | 217.49 | 254.50 |
| 57 | 0.121 | 1.33 | 59.95 | 59.91 | 3.40 | 49.34 | 21.22 | 0.012246 | 0.024455 | 573.17 | 695.64 | 4761.19 | 6163.78 | 225.30 | 263.05 |
| 58 | 0.199 | 2.02 | 91.07 | 138.24 | 4.33 | 69.28 | 37.91 | 0.010170 | 0.017534 | 1100.32 | 1281.59 | 11359.46 | 13920.89 | 379.61 | 428.87 |
| 59 | 0.783 | 6.40 | 287.95 | 1381.91 | 8.44 | 176.42 | 187.48 | 0.006098 | 0.007015 | 6627.55 | 6892.74 | 124490.83 | 131176.45 | 1596.65 | 1647.56 |
| 60 | 0.25 | 2.45 | 110.32 | 202.85 | 4.84 | 80.95 | 49.48 | 0.009340 | 0.015053 | 1483.90 | 1696.07 | 16925.50 | 20226.75 | 482.23 | 536.64 |
| 61 | 0.117 | 1.30 | 58.62 | 56.62 | 3.34 | 48.22 | 20.40 | 0.012401 | 0.025011 | 548.46 | 667.51 | 4489.50 | 5833.64 | 217.49 | 254.50 |
| 62 | 0.1867 | 1.92 | 86.32 | 124.18 | 4.20 | 66.33 | 35.19 | 0.010415 | 0.018298 | 1012.04 | 1185.00 | 10160.81 | 12539.86 | 355.05 | 402.81 |
| 63 | 0.1464 | 1.56 | 70.37 | 82.52 | 3.73 | 56.19 | 26.50 | 0.011405 | 0.021529 | 735.81 | 879.07 | 6642.86 | 8420.98 | 275.13 | 317.21 |
| 64 | 0.1819 | 1.88 | 84.45 | 118.87 | 4.15 | 65.16 | 34.14 | 0.010517 | 0.018620 | 978.07 | 1147.70 | 9708.63 | 12016.26 | 345.48 | 392.64 |
| 65 | 0.15 | 1.60 | 71.82 | 85.96 | 3.77 | 57.13 | 27.26 | 0.011302 | 0.021182 | 759.62 | 905.69 | 6930.98 | 8762.73 | 282.23 | 324.87 |
| 66 | 0.18 | 1.86 | 83.71 | 116.79 | 4.12 | 64.70 | 33.73 | 0.010559 | 0.018751 | 964.70 | 1132.99 | 9532.08 | 11811.42 | 341.70 | 388.61 |
| 67 | 0.0833 | 0.97 | 43.81 | 31.99 | 2.83 | 38.24 | 13.72 | 0.014078 | 0.031390 | 351.35 | 439.80 | 2479.29 | 3344.53 | 152.31 | 182.28 |
| 68 | 0.195 | 1.99 | 89.53 | 133.60 | 4.29 | 68.33 | 37.03 | 0.010248 | 0.017774 | 1071.42 | 1250.03 | 10963.39 | 13465.62 | 371.62 | 420.40 |
| 69 | 0.13 | 1.42 | 63.68 | 67.59 | 3.52 | 51.81 | 23.07 | 0.011923 | 0.023309 | 629.69 | 759.72 | 5397.26 | 6932.17 | 242.90 | 282.26 |
| 70 | 0.1667 | 1.74 | 78.48 | 102.65 | 3.97 | 61.39 | 30.84 | 0.010866 | 0.019738 | 872.35 | 1031.06 | 8335.31 | 10416.25 | 315.27 | 360.38 |
| 71 | 0.1375 | 1.48 | 66.75 | 74.27 | 3.62 | 53.83 | 24.63 | 0.011676 | 0.022451 | 677.74 | 813.90 | 5953.17 | 7599.04 | 257.62 | 298.25 |
| 72 | 0.365 | 3.37 | 151.63 | 383.17 | 5.82 | 104.80 | 76.95 | 0.008109 | 0.011687 | 2436.96 | 2699.58 | 32794.18 | 37589.02 | 717.14 | 778.33 |
| 73 | 0.42 | 3.79 | 170.61 | 485.11 | 6.23 | 115.34 | 90.64 | 0.007695 | 0.010640 | 2929.22 | 3207.45 | 41911.80 | 47302.03 | 830.86 | 893.42 |
| 74 | 0.15 | 1.60 | 71.82 | 85.96 | 3.77 | 57.13 | 27.26 | 0.011302 | 0.021182 | 759.62 | 905.69 | 6930.98 | 8762.73 | 282.23 | 324.87 |
| 75 | 0.14 | 1.51 | 67.77 | 76.55 | 3.65 | 54.50 | 25.15 | 0.011597 | 0.022182 | 693.94 | 832.11 | 6143.63 | 7826.60 | 262.53 | 303.58 |
| 76 | 0.155 | 1.64 | 73.82 | 90.83 | 3.83 | 58.42 | 28.33 | 0.011165 | 0.020723 | 792.98 | 942.91 | 7339.80 | 9246.10 | 292.10 | 335.51 |
| 77 | 0.2056 | 2.08 | 93.61 | 146.03 | 4.40 | 70.84 | 39.39 | 0.010047 | 0.017155 | 1148.40 | 1333.99 | 12026.08 | 14684.90 | 392.83 | 442.84 |
| 78 | 0.1522 | 1.62 | 72.70 | 88.09 | 3.80 | 57.70 | 27.73 | 0.011241 | 0.020977 | 774.26 | 922.03 | 7109.63 | 8974.17 | 286.57 | 329.55 |
| 79 | 0.132 | 1.43 | 64.52 | 69.39 | 3.55 | 52.37 | 23.50 | 0.011853 | 0.023067 | 642.74 | 774.46 | 5546.89 | 7112.08 | 246.92 | 286.63 |
| 80 | 0.1223 | 1.34 | 60.50 | 61.00 | 3.42 | 49.70 | 21.48 | 0.012198 | 0.024281 | 581.26 | 704.83 | 4850.95 | 6272.59 | 227.84 | 265.83 |
| 81 | 0.1105 | 1.23 | 55.55 | 51.43 | 3.25 | 46.37 | 19.09 | 0.012669 | 0.025985 | 508.87 | 622.25 | 4062.67 | 5312.39 | 204.84 | 240.60 |
| Average | 0.23 | 2.23 | 100.57 | 198.32 | 4.50 | 74.19 | 45.44 | 0.01 | 0.02 | 1378.19 | 1564.18 | 16760.77 | 19624.53 | 440.42 | 489.29 |
| Standard deviation | 0.12 | 0.94 | 42.25 | 190.90 | 1.05 | 24.81 | 27.76 | 0.00 | 0.01 | 967.64 | 1015.29 | 17005.64 | 18250.84 | 237.56 | 244.96 |

$h$ is mean set thickness ($n = 10$ per locality) of trough cross-beds; $d_m$ is mean water depth (eqn 1); $w$ is channel width (eqn 3); $Q_m$ is maximum instantaneous water discharge (eqn 4); $d_b$ is mean bankfull channel depth (eqn 5); $w_b$ is bankfull channel width (eqn 6); $Q_m(1)$ is average daily discharge (eqn 7); $s$ is stream palaeoslope (eqn 8a, using $Q_m(1)$ values); $s(1)$ is stream palaeoslope (eqn 8b); $Q_b$ is bankfull water discharge (eqn 9, using $s$ values); $Q_b(1)$ is bankfull water discharge (eqn 9, using $s(1)$ values); $A_d$ is drainage area (eqn 10, using $Q_b$ values); $A_d(1)$ is drainage area (eqn 10, using $Q_b(1)$ values); $L$ is principal stream length (eqn 11, using $A_d$ values); $L(1)$ is principal stream length (eqn 11, using $A_d(1)$ values).

approximate range of possible palaeoslope values for the Wilgerivier braided channels. These two $S$ values, in turn, enable two estimates of the bankfull water discharge ($Q_b$) ($m^3 s^{-1}$) to be made, using:

$$Q_b = 4.0 \, A_b^{1.21} \, S^{0.28} \text{ (Williams, 1978)} \qquad (9)$$

where $A_b = d_b \times w_b$.

Leopold *et al.* (1964) provide an estimation of the probable drainage area for river systems by the formula:

$$Q_b = A_d^b \qquad (10)$$

where $A_d$ is the drainage area ($km^2$), with the value of $b$ ranging from 0.65 to 0.8. An average value of $b = 0.75$ is proposed by Leopold *et al.* (1964), which was also used by Turner (1980) as well as in this paper. According to Leopold *et al.* (1964), the principal stream length bears a stable and constant relationship to the drainage area and is approximated by the equation:

$$L = 1.4 \, A_d^{0.6} \qquad (11)$$

where $L$ is the stream length (km). With estimation of both $A_d$ and $L$, the two $Q_b$ values obtained from eqn (9) enable two determinations each for these two palaeohydrological parameters (Table 2). In determining all of the palaeohydraulic parameters for the Wilgerivier Formation, it must be stressed that they apply only to individual braided channels within the whole braidplain, and the same restriction obviously also applies to Turner's (1980) data. In view of the large number of trough cross-beds measured in the present study, the palaeohydraulic data should, however, provide a reasonable sample of different individual channels in the Wilgerivier Formation.

## HYDROLOGICAL DATA

Table 2 provides an overview of the various hydrological parameters (as defined above) estimated in this study for the Wilgerivier Formation (including mean and one standard deviation). Equivalent palaeohydrological data (mean only) for the Mesozoic Molteno Formation are taken from Turner (1980) and are shown in Table 3. Figure 2 provides a binary plot of mean annual bankfull discharge ($Q_b$) and drainage area ($A_d$) estimated for the Palaeoproterozoic Wilgerivier Formation, and for the Mesozoic Molteno Formation by Turner (1980). In Fig. 3, $Q_b$ and palaeoslope ($S$) estimates for the Wilgerivier and Molteno Formations are plotted against each other.

Examining Table 2, it is evident that all except one hydrological parameter estimated for the Wilgerivier Formation are lower than their equivalent values derived from the Molteno (Table 3) river systems. The exception is $S$, the palaeoslope, which is significantly higher for the Wilgerivier Formation. Although bankfull discharge values ($Q_b$) are, on average, greater for the Molteno streams, the range of those estimated for individual Wilgerivier Formation channels is larger (Table 2 & Fig. 3). The widely accepted direct causative relationship between drainage area and discharge (e.g. Miall, 1996) can explain the almost straight-line trend observed in both Precambrian and Phanerozoic examples, illustrated in Fig. 2.

Comparing widths and depths of preserved sandy channel-fill structures in the Wilgerivier Formation (Table 1) with calculated values ($w_m$ and $d_m$ and their bankfull equivalents, $w_b$ and $d_b$, in Table 2) indicates large and significant differences; dimensions measured in the field are noticeably smaller. This, however, probably reflects an inability to recognize, and thus also measure, channel-fill structures in the Wilgerivier Formation greater than about 50 m (maximum recorded width = 46 m; $n = 73$) width in the field, rather than excessive errors in the relevant palaeohydraulic formulae. Outcrops of the exposed Wilgerivier fluvial deposits seldom exceed several tens of metres in width and a few metres in height.

**Table 3.** Palaeohydraulic data estimated for the Molteno Formation (from Turner, 1980).

| Locality | $h$ (m) | $n$ | $d_m$ (m) | $d_b$ (m) | $w_b$ (m) | $Q_m$ ($m^3 s^{-1}$) | $Q_b$ ($m^3 s^{-1}$) | $S$ (m m$^{-1}$) | $A_d$ (km$^2$) | $L$ (km) |
|---|---|---|---|---|---|---|---|---|---|---|
| A | 0.63 | 13 | 5.3 | 6.5 | 239 | 950 | 3300 | 0.0047 | 49 000 | 900 |
| B | 0.69 | 29 | 5.7 | 6.8 | 257 | 1098 | 3650 | 0.0049 | 56 000 | 990 |
| C | 0.61 | 38 | 5.2 | 6.4 | 234 | 912 | 3200 | 0.0053 | 47 000 | 880 |
| D | 0.58 | 27 | 5.0 | 6.3 | 225 | 843 | 3000 | 0.0055 | 43 000 | 850 |
| E | 0.54 | 21 | 4.7 | 6.0 | 212 | 747 | 2650 | 0.0058 | 37 000 | 770 |
| F | 0.44 | 9 | 3.9 | 5.4 | 176 | 514 | 2150 | 0.0070 | 27 700 | 650 |
| Mean | 0.56 | (Total = 137) | 4.8 | 6.1 | 215 | 774 | 2800 | 0.0058 | 39 000 | 800 |

Locality is given by Turner (1980); $h$ is mean trough cross-bed set thickness, per locality; $n$ is number of trough cross-bed set thicknesses measured; $d_m$ is mean channel depth; $d_b$ is mean bankfull channel depth; $w_b$ is bankfull channel width; $Q_m$ is approximation of mean annual discharge (derived from eqn 4 in text); $Q_b$ is bankfull mean annual discharge; $S$ is mean channel slope; $A_d$ is drainage area; $L$ is principal stream length.

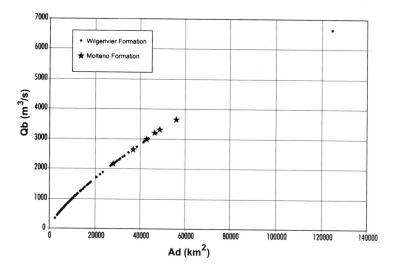

**Fig. 2.** Binary plot of mean annual bankfull discharge ($Q_b$) and drainage area ($A_d$) values estimated for the Wilgerivier (data from Table 2; columns headed $Q_b$ and $A_d$, not from $Q_b(1)$ and $A_d(1)$) and Molteno (data from Turner, 1980) Formations.

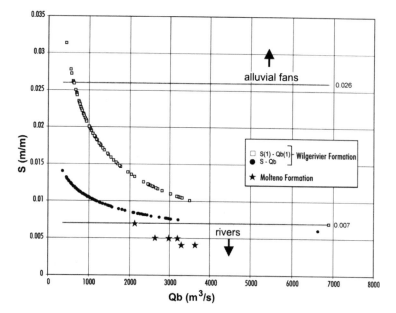

**Fig. 3.** Binary plot of palaeoslope ($S$) and mean annual bankfull discharge ($Q_b$) values estimated for the Wilgerivier (data from Table 2; columns headed $S$ and $Q_b$ for lower curve; columns headed $S(1)$ and $Q_b(1)$ for upper curve) and Molteno (data from Turner, 1980) Formations. Note that eqns (8a), (8b) and (9) (see text) enable estimation of two values each for both $S$ and $Q_b$, thus providing the two Wilgerivier curves in this figure, and a range of estimated values. Note also the maximum gradient of 0.007 associated with modern rivers and that modern alluvial fans generally have slopes in excess of 0.026 m m$^{-1}$ (Blair & McPherson, 1994); the Wilgerivier palaeoslope estimates lie between these two values.

The generally accepted downstream increase in river channel discharge ($Q_m$ and $Q_b$) values often can be related directly to measured palaeocurrent trends (e.g. Turner, 1980). For the Wilgerivier Formation, more than 5000 recorded directions (first author's unpublished data) support a consistent easterly flowing fluvial system for most of the preserved basin and throughout its preserved stratigraphical thickness (up to 5500 m in the east of the basin; Eriksson & Vos, 1979). Unlike the Molteno Formation, where Turner (1980) demonstrated an increase in average $Q_m$ values at six sampling stations

in a down-palaeocurrent direction, examination of the west-to-east changes in $Q_m$ for the Wilgerivier Formation in the study area indicates a random pattern of increasing and decreasing discharge estimates. The Molteno Formation has a maximum preserved thickness of 460 m (Turner, 1975) and much of the northern half of the formation is < 60 m thick (Eriksson, 1984). Turner (1980) determined palaeohydraulic parameters for only one member in this much thinner formation and this provides a reasonable explanation of the downstream increase in discharge values observed in his study. For the much thicker Wilgerivier Formation and for the significantly larger data base of trough cross-bed set thicknesses used to estimate palaeohydraulic conditions for individual braided channels, downstream changes in discharge values would tend to be masked and confused by stacking of a large number of fluvial deposits within the preserved stratigraphy. As these fluvially deposited beds dip inwards within the Middelburg basin (Fig. 1), cross-bed measurements from west to east would naturally tend to record deposition within increasingly higher stratigraphical intervals.

## DISCUSSION

Both the Molteno and Wilgerivier Formations are inferred to reflect predominantly braided-river deposition (e.g. Turner, 1980; Van der Neut *et al.*, 1991). The Precambrian Wilgerivier Formation was laid down with an absence of terrestrial vegetation, and this, combined with the significant mechanical weathering processes generally inferred for the Precambrian period for most palaeoclimatic settings (e.g. Condie, 1997), and with rapid chemical breakdown also interpreted for much of early Precambrian time (e.g. Corcoran *et al.*, 1998), would have provided large quantities of detritus. Without the stabilizing effect of land vegetation, rivers with generally low bank stabilities would easily have removed the abundant bedload material (e.g. Mueller & Corcoran, 1998). For these reasons, a braided-river pattern is normally accepted for pre-vegetative times (e.g. Schumm, 1968b; Cotter, 1978; Long, 1978; Eriksson *et al.*, 1998). As a caveat, it should be emphasized that recognition of fluvial deposits and even more so of fluvial style in the rock record is problematic and relies on an accumulation of different lines of evidence (Jackson, 1978; Miall, 1996).

In general terms, braided-channel systems are favoured by rapid and large fluctuations in river discharge (i.e. $Q_m$ and $Q_b$ variables), a large available supply of sandy and pebbly bedload sediment and easily erodible banks (Cant, 1982). Evidence for a braided, bedload stream interpretation for the Wilgerivier Formation includes the predominance of cross-bedded (mainly planar) medium- to coarse-grained sheet sandstones, subordinate imbricate conglomerate beds, pebbly and more common sandy channel-fill structures, minor desiccated mudrock partings and consistently orientated palaeocurrent trends exhibiting a small variation in direction (Van der Neut *et al.*, 1991). In addition to a lack of vegetation for these approximately 1800 Ma fluvial deposits, the Wilgerivier sediments would also most probably have

formed in a continental environment lacking (to a large degree) in soil-forming biota (e.g. Eriksson *et al.*, 1998). The combined absence of vegetation and extensive soil development would have inhibited the formation of anastomosing channel systems during the Precambrian (e.g. Cant, 1982) and the scarcity of mudrocks preserved in the Wilgerivier Formation mitigate against a meandering model. The interpretation of a braided, bedload fluvial style for the Wilgerivier Formation, as adopted in the present comparative hydrological study, thus seems reasonable.

In contrast to the Precambrian palaeoenvironment, deposition of the Molteno Formation during the Middle Triassic took place within a continental setting where both vegetation and soil formation were well established (e.g. Visser, 1984). Turner (1975, 1980, 1983, 1986) has provided evidence for a perennial braided stream origin for the Molteno sediments. This formation essentially comprises three bedload-dominated fluvial wedges derived from tectonically uplifted source terrains. Extensive sheets of pebbly, poorly sorted and upward-fining sandstones are often interbedded with finer and more mature sandstones (Smith *et al.*, 1993), suggesting the more proximal parts of such a braided-river system (Turner, 1986), with reworking of sand flats during exposure of bedforms at low-flow stages (Eriksson, 1984). A braided, bedload fluvial style for the Molteno can be ascribed to a combination of high source-area relief (Turner, 1986), generally cool-temperate climatic conditions (Anderson, 1976) and sparsely vegetated interfluves (Smith *et al.*, 1993). Detailed analysis of the relationship between plant remains and sedimentary facies within the Molteno Formation supports the views of previous workers: a predominant braidplain environment, with subordinate mixed-load meandering channels and more distal lacustrine to marshy floodplain settings with restricted coal development (Cairncross *et al.*, 1995).

A strong measure of tectonic control on Molteno braided fluvial sedimentation is generally accepted (e.g. Turner, 1980; Smith *et al.*, 1993), with source uplift and braidplain palaeoslopes inferred to have been at a maximum for such river systems. Similarly, the Waterberg Group is thought to have been deposited within an active tectonic setting, in two grabens, preserved partially as the Main and Middelburg basins (Callaghan *et al.*, 1991; Van der Neut *et al.*, 1991). Graben basins typically have a relatively large number of alluvial systems draining the margins, with the lower gradient hangingwall dip slopes tending to form larger systems from more extensive catchment areas than those developed along the steeper footwall sides (e.g. Leeder & Gawthorpe, 1987). Analogously, in the East African rift system, opposite-facing border faults lead generally to segmented basin margins and thus, concomitantly, to a number of alluvial catchment areas of varying, although overall limited size (e.g. Frostick *et al.*, 1986). In contrast, a long linear provenance is reconstructed for the Molteno Formation, with an inferred lateral extent of approximately 700 km (Turner, 1980).

The much larger catchment area postulated for the braided streams in the Molteno Formation by Turner (1980) is reflected in the significantly higher drainage area ($A_d$) and principal stream length ($L$) values estimated for these channels (Table 3). It is thus also logical that the individual Molteno braided channels were, on average, deeper ($d_m$ and $d_b$ values) and wider ($w_b$ values) than those of the Wilgerivier Formation, and that they were characterized by generally larger set thicknesses in preserved trough cross-beds (Tables 2 & 3). Although both Molteno and Wilgerivier Formations were deposited on extensive braidplains, that preserved in the Molteno is at least an order larger (in length and width) than that found in the Wilgerivier Formation (Turner, 1980; Van der Neut *et al.*, 1991). The direct dependence of discharge values on the size of the drainage area, applicable to river systems in general, is also reflected in the palaeohydraulic estimates for both the Wilgerivier and Molteno Formations, as illustrated in the binary plot of $Q_b$ against $A_d$ (Fig. 2). The greater average discharge values for the Molteno channels, compared with values estimated for individual braid channels in the Wilgerivier example are also evident in this figure.

With the exception of the palaeoslope data estimated for the two formations, the palaeohydraulic parameters determined here, and their differences, thus may be ascribed logically to the larger catchment or drainage area for the Molteno braided channels. The palaeoslopes ($S$ values in Table 3) estimated for the Molteno Formation are thought to be a reflection of conditions of maximum source-area uplift (Turner, 1980). The significantly higher palaeoslopes calculated for the Wilgerivier Formation (Table 2 & Fig. 3) therefore must be regarded as meaningful. Schumm & Khan (1972) demonstrated that increase in valley floor slope favours development of braided fluvial styles at the expense of meandering rivers. As the Molteno Formation bears evidence of subordinate mixed-load channels and floodplain deposits, it is logical that the average palaeoslope calculated for the Wilgerivier Formation, which lacks such evidence, is higher than that for the Molteno. Blair & McPherson (1994) point out that there is a distinct break in nature in the longitudinal slopes of fluvial distributary systems, between those found on alluvial fans (slopes ranging

from 0.026 to 0.466) and those characteristic of fluvial systems (maximum slopes of approximately 0.007). The Molteno Formation $S$ values (Table 3) have an average slope of 0.0058 and a maximum of 0.007, which are in accord with the inferred tectonic setting for these braided channels. Palaeoslopes estimated for the Wilgerivier Formation fall almost entirely into the gap discussed above (Fig. 3), indicating braided channels different in character to either modern fans or bedload fluvial systems. A possible solution lies in their Precambrian age.

As discussed previously, the Precambrian was characterized by aggressive weathering, largely as a result of the different composition of the atmosphere (e.g. Corcoran *et al.*, 1998), and by very rapid rates of erosion in the absence of land vegetation and soil biota (e.g. Eriksson *et al.*, 1998). The Middelburg basin lies essentially within the eastern part of the pre-existing, Early Proterozoic Pretoria Group basin, characterized largely by mudrocks and mature quartzose sandstones (e.g. Eriksson & Reczko, 1995). Along the steeply sloping margins of the Middelburg graben, large quantities of sandy detritus would thus most likely have formed by these combined conditions. A semi-arid palaeoclimate is inferred for the Waterberg sediments in general, owing to the presence of preserved alluvial and aeolian deposits (e.g. Callaghan *et al.*, 1991), and for the Wilgerivier Formation in particular (e.g. Van der Neut *et al.*, 1991). Archaean and Early Proterozoic climates up till about 2.3 Ga appear to have been warm and moist, but increasing proportions of evaporites and red beds in the Early Proterozoic rock record after 2.3 Ga point to an increase in arid to semi-arid settings (Condie, 1997). Such palaeoclimatic conditions would have led to the removal of fine, muddy detritus by deflation. Intermittent torrential rainfall is surmised to have reworked the sandy detritus close to the relatively small Wilgerivier catchment areas and transported this sediment in braided, bedload channels.

Rapid Precambrian weathering would preferentially have formed sand grains rather than pebbles and boulders from Pretoria Group quartzite source rocks, thereby inhibiting formation of alluvial fans close to the provenance areas. Instead, the sandy material was rapidly resedimented by high-gradient braided channels, draining limited, fault-controlled drainage areas in the Wilgerivier Formation. A unique fluvial style, in terms of palaeoslope, may thus have applied during parts of the Precambrian, when such suitable conditions were present for the formation of Wilgerivier-type braided channels. In contrast, $S$ values estimated for the Molteno Formation braided channels are compatible with maximum palaeoslope values for modern rivers. A much larger catchment area led to higher discharge values and larger individual channel dimensions for the Molteno, and palaeoclimate was most likely cool and temperate.

## CONCLUSIONS

In this paper, empirical data are presented for various hydrological variables for the braided-stream deposits in the approximately 1800 Ga Wilgerivier Formation, Waterberg Group, South Africa. Although it must be emphasized that the calculated parameters are not necessarily very accurate, owing to factors inherent in the formulae used, the data are nevertheless suitable for comparative purposes with data sets derived from other fluvial deposits and systems. The hydrological data from the Wilgerivier Formation reflect a braided-river setting *sensu stricto*, with no evidence of significant suspended load or floodplain influences, and in view of their age, land vegetation was absent and soil-forming biota were not important. A comparison is made here with similar hydrological data from the Mesozoic Molteno Formation of South Africa. The Molteno rivers appear to have been predominantly braided, and formed when both vegetation and soil formation were common. In addition, the Molteno differed from the Wilgerivier channels in that subordinate suspended load and floodplain settings were present. Both Wilgerivier and Molteno braided, bedload channel systems formed within a setting characterized by active tectonism in provenance areas.

A much larger drainage area estimated for the Molteno Formation braided channels can explain most of the differences in the palaeohydraulic parameters determined, except for the palaeoslopes. Palaeoslopes calculated for the Molteno Formation (Table 3) are close to the maximum known from modern rivers (0.007), but those estimated for the Wilgerivier Formation (0.01–0.02; Table 2) fall approximately between this maximum and values found in modern fans (> 0.026) (Fig. 3). It is inferred that the faulted margins of the Wilgerivier graben provided a number of smaller drainage areas and that the combination of lack of vegetation and soil biota, as well as aggressive weathering and high erosion rates typical for Precambrian times promoted formation of sandy detritus close to source areas. High-gradient braided, bedload fluvial systems developed within this framework, and intermittent torrential storms are thought to have played an important role on these braided floodplains. In contrast, the Phanerozoic Molteno Formation braidplain had a large, single provenance terrain with a cool temperate palaeoclimate, and this, combined with active tectonic uplift, led to high discharge rates on palaeoslopes close to the maximum found for modern rivers.

## ACKNOWLEDGEMENTS

The first author acknowledges the Director of the Council for Geoscience, South Africa, for permission to publish this manuscript. The second author is grateful to the Foundation for Research Development and the University of Pretoria for research funding. We wish also to acknowledge numerous fruitful discussions held with many geological colleagues and the insight gained from previous workers on the subject matters covered in this paper. Constructive reviews by Michael Wizewich and Brian Turner proved most useful and we also appreciate, most sincerely, the insight and encouragement of Norm Smith.

## REFERENCES

ALLEN, J.R.L. (1968) *Current Ripples.* North Holland, Amsterdam.

ANDERSON, H.M. (1976) *A revision of the genus Dicrodium from the Molteno Formation.* Unpublished PhD thesis, University of the Witwatersrand.

BLAIR, T.C. & McPHERSON, J.G. (1994) Alluvial fans and their natural distinction from rivers based on morphology, hydraulic processes, sedimentary processes, and facies assemblages. *J. sediment. Res.*, **A64**, 450–489.

BURGER, A.J. & WALRAVEN, F. (1980) Summary of age determinations carried out during the period April 1978 to March 1979. *Ann. geol. Surv. S. Afr.*, **14**, 109–118.

CAIRNCROSS, B., ANDERSON, J.M. & ANDERSON, H.M. (1995) Palaeoecology of the Triassic Molteno Formation, Karoo Basin, South Africa — sedimentological and palaeontological evidence. *S. Afr. J. Geol.*, **98**(4), 452–478.

CALLAGHAN, C.C. (1987) *The geology of the Waterberg Group in the southern portion of the Waterberg Basin.* Unpublished MSc thesis, University of Pretoria.

CALLAGHAN, C.C., ERIKSSON, P.G. & SNYMAN, C.P. (1991) The sedimentology of the Waterberg Group in the Transvaal, South Africa: an overview. *J. Afr. Earth Sci.*, **13**, 121–139.

CANT, D.J. (1982) Fluvial facies models and their application. In: *Sandstone Depositional Environments* (Eds Scholle, P.A. & Spearing, D.), Mem. Am. Assoc. petrol. Geol., Tulsa, **31**, 115–137.

CONDIE, K.C. (1997) *Plate Tectonics and Crustal Evolution*, 4th edn. Butterworth-Heinman, Oxford.

CORCORAN, P.L., MUELLER, W.U. & CHOWN, E.H. (1998) Climatic and tectonic influences on fan deltas and wave- to tide-controlled shoreface deposits: evidence from the Archean Keskarrah Formation, Slave Province, Canada. *Sediment. Geol.*, **120** (1–4), 125–152.

COTTER, E. (1978) The evolution of fluvial style, with special reference to the central Appalachian Paleozoic. In: *Fluvial*

*Sedimentology* (Ed. Miall, A.D.), Mem. Can. Soc. petrol. Geol., Calgary, **5**, 361–383.

ELS, B.G. (1990) Determination of some palaeohydraulic parameters for a fluvial Witwatersrand succession. *S. Afr. J. Geol.*, **93**(3), 531–537.

ERIKSSON, K.A. & VOS, R.G. (1979) A fluvial fan depositional model for Middle Proterozoic red beds from the Waterberg Group, South Africa. *Precambrian Res.*, **9**, 169–188.

ERIKSSON, P.G. (1984) A palaeoenvironmental analysis of the Molteno Formation in the Natal Drakensberg. *Trans. geol. Soc. S. Afr.*, **87**, 237–244.

ERIKSSON, P.G. & RECZKO, B.F.F. (1995) The sedimentary and tectonic setting of the Transvaal Supergroup floor rocks to the Bushveld Complex. *J. Afr. Earth Sci.*, **21**, 487–504.

ERIKSSON, P.G., CONDIE, K.C., TIRSGAARD, H., *et al.* (1998) Precambrian clastic sedimentation systems. *Sediment. Geol.*, **120**(1–4), 5–53.

ETHRIDGE, F.G. & SCHUMM, S.A. (1978) Reconstructing paleochannel morphologic and flow characteristics: methodology, limitations, and assessment. In: *Fluvial Sedimentology* (Ed. Miall, A.D.), Mem. Can. Soc. petrol. Geol., Calgary, **5**, 703–721.

FROSTICK, L.E., RENAUT, R.W. & REID, I. (Eds) (1986) *Sedimentation in the African Rifts*. Spec. Publ. geol. Soc. London, No. 25, 382 pp. Blackwell Scientific Publications, Oxford.

JACKSON, R.G., II (1978) Preliminary evaluation of lithofacies models for meandering alluvial streams. In: *Fluvial Sedimentology* (Ed. Miall, A.D.), Mem. Can. Soc. petrol. Geol., Calgary, **5**, 543–576.

LEEDER, M.R. (1973) Fluviatile fining-upwards cycles and the magnitude of palaeochannels. *Geol. Mag.*, **110**, 265–276.

LEEDER, M.R. & GAWTHORPE, R.L. (1987) Sedimentary models for extensional tilt-block/half-graben basins. In: *Continental Extension Tectonics* (Eds Coward, M.P., Dewey, J.F. & Hancock, P.L.), Spec. Publ. geol. Soc., London, No. 28, pp. 139–152. Blackwell Scientific Publications, Oxford.

LEOPOLD, L.B., WOLMAN, G.M. & MILLER, J.P. (1964) *Fluvial Processes in River Geomorphology*. Freeman, San Francisco.

LONG, D.G.F. (1978) Proterozoic stream deposits: some problems of recognition and interpretation of ancient sandy fluvial systems. In: *Fluvial Sedimentology* (Ed. Miall, A.D.), Mem. Can. Soc. petrol. Geol., Calgary, **5**, 313–342.

MIALL, A.D. (1976) Palaeocurrent and palaeohydrologic analysis of some vertical profiles through a Cretaceous braided stream deposit, Banks Island, Arctic Canada. *Sedimentology*, **23**, 459–483.

MIALL, A.D. (1996) *The Geology of Fluvial Deposits: Sedimentary Facies, Basin Analysis and Petroleum Geology*. Springer-Verlag, Heidelberg.

MUELLER, W.U. & CORCORAN, P.L. (1998) Characteristics of pre-vegetational, late-orogenic basins: examples from the Archean Superior Province, Canada. *Sediment. Geol.*, **120**(1–4), 177–203.

OSTERKAMP, W.R. & HEDMAN, E.R. (1982) Perennial-streamflow characteristics related to channel geometry and sediment in Missouri River basin. *Prof. Pap. U.S. geol. Surv.*, **1242**, 37 pp.

SCHUMM, S.A. (1968a) River adjustment to altered hydrologic regimen — Murrumbidgee River and paleochannels, Australia. *Prof. Pap. U.S. geol. Surv.*, **598**, 65 pp.

SCHUMM, S.A. (1968b) Speculations concerning paleohydro-

logic controls of terrestrial sedimentation. *Bull. Geol. Soc. Am.*, **79**, 1573–1588.

SCHUMM, S.A. (1969) River metamorphosis. *Proc. Am. Soc. civ. Eng., J. Hydraul. Div.*, **HY1**, 255–273.

SCHUMM, S.A. (1972) Fluvial paleochannels. In: *Recognition of Ancient Sedimentary Environments* (Eds Rigby, J.K. & Hamblin, W.K.), Spec. Publ. Soc. econ. Paleont. Miner., Tulsa, **16**, 98–107.

SCHUMM, S.A. & KHAN, H.R. (1972) Experimental study of channel patterns. *Bull. Geol. Soc. Am.*, **83**, 1755–1770.

SMITH, R.M.H., ERIKSSON, P.G. & BOTHA, W.J. (1993) A review of the stratigraphy and sedimentary environments of the Karoo-aged basins of Southern Africa. *J. Afr. Earth Sci.*, **16**, 143–169.

TANKARD, A.J., JACKSON, M.P.A., ERIKSSON, K.A., HOBDAY, D.K., HUNTER, D.R. & MINTER, W.E.L. (1982) *Crustal Evolution of Southern Africa*. Springer-Verlag, New York.

TURNER, B.R. (1975) *The stratigraphy and sedimentary history of the Molteno Formation in the main Karoo Basin of South Africa and Lesotho*. Unpublished PhD thesis, University of the Witwatersrand.

TURNER, B.R. (1978) Palaeohydraulics of clast transport during deposition of the Upper Triassic Molteno Formation in the main Karoo Basin of South Africa. *S. Afr. J. Sci.*, **74**, 171–173.

TURNER, B.R. (1980) Palaeohydraulics of an upper braided river system in the main Karoo Basin, South Africa. *Trans. geol. Soc. S. Afr.*, **83**, 425–431.

TURNER, B.R. (1983) Braidplain deposition of the Upper Triassic Molteno Formation in the main Karoo (Gondwana) Basin, South Africa. *Sedimentology*, **30**, 77–89.

TURNER, B.R. (1986) Tectonic and climatic controls on continental depositional facies in the Karoo Basin of northern Natal, South Africa. *Sediment. Geol.*, **46**, 231–257.

TWIST, D. & CHENEY, E.S. (1986) Evidence for the transition to an oxygen-rich atmosphere in the Rooiberg group, South Africa — a note. *Precambrian Res.*, **33**, 255–264.

VAN DER NEUT, M. & ERIKSSON, P.G. (1997) A sedimentological investigation of the predominantly fluvial Wilgerivier Formation, Waterberg Group, South Africa. *Abstracts, 6th International Conference on Fluvial Sedimentology*, Cape Town, 22–26 September, p. 218.

VAN DER NEUT, M., ERIKSSON, P.G. & CALLAGHAN, C.C. (1991) Distal alluvial fan sediments in early Proterozoic red beds of the Wilgerivier Formation, Waterberg Group, South Africa. *J. Afr. Earth Sci.*, **12**, 537–547.

VISSER, J.N.J. (1984) A review of the Stormberg Group and Drakensberg volcanics in Southern Africa. *Palaeont. Afr.*, **25**, 5–27.

VOS, R.G. & ERIKSSON, K.A. (1977) An embayment model for tidal and wave-swash deposits occurring within a fluvially dominated Proterozoic sequence in South Africa. *Sediment. Geol.*, **18**, 161–173.

WILLIAMS, G.P. (1978) Bank-full discharge of rivers. *Water Resour. Res.*, **14**, 1141–1154.

WILLIAMS, G.P. (1984) Paleohydrologic equations for rivers. In: *Developments and Applications of Geomorphology* (Eds Costa, J.E. & Fleisher, P.J.), pp. 343–367. Springer-Verlag, Berlin.

ZAWADA, P.K. (1994). Palaeoflood hydrology of the Buffels River, Laingsburg, South Africa: was the 1981 flood the largest? *S. Afr. J. Geol.*, **97**, 21–32.

*Spec. Publs int. Ass. Sediment.* (1999) **28**, 393–407

# Sand- and mud-dominated alluvial-fan deposits of the Miocene Seto Porcelain Clay Formation, Japan

K. NAKAYAMA

*Department of Geoscience, Shimane University, Matsue 690-8504, Japan (Email: nakayama@riko.shimane-u.ac.jp)*

## ABSTRACT

The Late Miocene Seto Porcelain Clay Formation, best exposed in the northern part of Seto City in Japan, was deposited over a period of 1 Myr around 9 Ma. It is distributed sporadically in small collapse basins less than 30 km$^2$ in area. The formation is underlain mainly by Cretaceous granite, as well as Palaeozoic-Mesozoic pelagic formations and Early to Middle Miocene shallow-marine formations (22–15 Ma). After deposition of the marine formations, the granite was strongly weathered and decomposed *in situ*, forming sand- and mud-size material. Many plant fossils occur within the Seto Porcelain Clay Formation, which suggest a warm climate with a low annual temperature range.

Three facies associations (FA1–FA3) are recognized. Facies association 1 consists mainly of poorly sorted gravel beds and cross-stratified sandy gravel beds, with associated massive mud beds. Gravel clasts were derived from nearby uplifted basement blocks. Sand beds commonly occur in channels, and mud beds contain rare plant roots. Facies association 2 is dominated by cross-stratified coarse- to very coarse-grained sand beds, with mud beds that commonly contain plant roots, and is subdivided into FA2a and FA2b on the basis of sand-bed architecture. Sand beds in FA2a commonly occur in channels, whereas those in FA2b have a sheet-like geometry. Plant roots are more abundant in the mud beds of FA2b than in FA2a. Facies association 3 comprises mainly plant-root- and plant-debris-dominated mud and lignite beds, with subordinate massive muddy sand beds.

Palaeocurrents occur in two radiating domains. In each domain, facies associations are ordered FA1, FA2a, FA2b and FA3 away from the palaeopiedmont line. Palaeocurrents indicate that two alluvial fan systems existed, and the arrangement of facies associations implies lateral facies change. The fans have radial lengths of 1.5 km and 2.0 km. Slope gradient in FA2 is estimated to be about 0.1° to 0.2°, based on channel depth and grain size in sand beds. The combination of voluminous sand supply from decomposed granite, together with collapse of small basement blocks and mud entrapment by vegetation, enabled the formation of the sand- and mud-dominated alluvial-fan systems.

## INTRODUCTION

Alluvial fans commonly are defined by geomorphological character, rather than by a characteristic fluvial style. Many studies have distinguished alluvial-fan facies in the ancient record. The most distinctive facies components have been considered to be debris-flow deposits, although not all the depositional systems termed fans necessarily contain debris flows (Miall, 1978; Rust, 1978). Geomorphological characters, such as radiating distributaries and cone-shaped architecture are key criteria for recognition of alluvial fans. It is usually difficult, however, to recognize these geomorphological characteristics in ancient records, owing to limited exposure. The Late Miocene Seto Porcelain Clay Formation (Matsuzawa *et al.*, 1960)

is exceptionally well exposed in the northern part of Seto City in central Japan, because it is quarried as a raw material for pottery. The formation consists mainly of unlithified mud and sand, and yields many plant fossils. Numerous studies have described its clay mineralogy and palaeobotany, but sedimentological studies are few. Nakayama (1991, 1993, 1994) provided some descriptions of sedimentary facies, and reported the possibility of an alluvial fan. Those studies, however, did not establish the cause of fan development. In this paper, sand- and mud-dominated alluvial fans are recognized in the formation, based on facies analysis, bed architecture and palaeocurrent directions of three-dimensional exposures.

The factors controlling the development of the fan systems also are discussed.

## GEOLOGICAL OUTLINE

The Japanese island-arc system is divided geologically into northeast Japan and southwest Japan by the Itoigawa–Shizuoka Tectonic Line (Fig. 1). Three late Cenozoic fluvio-lacustrine sedimentary basins occur in the eastern part of southwest Japan. These are the Tokai, Kobiwako and Osaka Groups, from east to west (Ikebe, 1957). The basins are intramassif basins, located between fore-arc (Pacific Ocean side) and back-arc (Japan Sea side) settings. Each basin covers an area of 200–1000 km², and results from the collapse of multiple small basement blocks less than 80 km² in area (Nakayama, 1996).

The Seto Porcelain Clay Formation lies in the lowermost part of the Tokai Group. Nakayama (1994) considered that the formation was deposited around 9 Ma, based on stratigraphical correlation. The Seto Porcelain Clay Formation is distributed sporadically, occurring in small

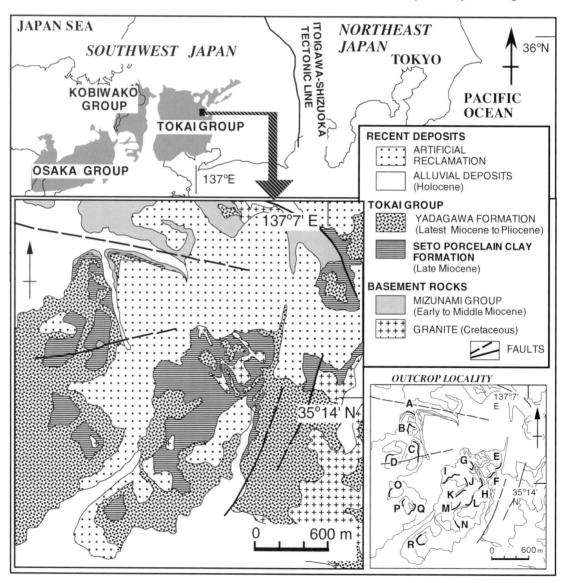

**Fig. 1.** Geological map of the study area. The bottom right inset map shows the location of exposures.

**Table 1.** Facies associations recognized in the Seto Porcelain Clay Formation.

| Facies association | Dominant lithofacies type* | Minor lithofacies type* | Architectural element† | Distribution (locality number) | Interpretation |
|---|---|---|---|---|---|
| FA1 | Gci, Gcm, Gt, Sm, Fsm | Gmm, Gmg, Gh, St, Sr | SG, GB, CH | A, B, E, G > C, F, H, I, K, Q | Debris-flow-dominated fluvial system |
| FA2a | St, Ss, Fsm, Sm, Sh | Gt, Sp, Sl, Fr | CH, SB, SG | C, D, E, F, H, J, L, O > A, B, G, I, K, P | Channelized braided (or low sinuosity) system |
| FA2b | St, Fsm, Sm, Sh, Fr | Sp, Sl, C | SB, SG‡, LS | C, D, F, H, I, J, K, O > E, L, P, Q | Sheet-flow-dominated braided (or low sinuosity) system |
| FA3 | Fr, C, Sm, Fsm | Fl, Fm, St | FF, SG‡, LS | M, N, Q, R > O, P | Floodplain and shallow stagnant water system |

\* Classification from Miall (1978) and Miall (1996). † Classification from Miall (1985) and Miall (1996). ‡ SG composed of Sm.

collapse basins less than 30 km² in area. These small basins correspond to single or several small collapsed basement blocks. Plant assemblages in the formation indicate a warm climate with a low annual temperature range (Miki, 1963). Mineralogical evidence suggests derivation of the sediments from decomposed granite and argillized volcanic ash (Tanemura, 1964; Fujii, 1968).

The study area lies within a single collapse basin. Basement rocks are Cretaceous granite and Early to Middle Miocene shallow-marine formations of the Mizunami Group (Fig. 1). The granite is strongly weathered, and is physically disintegrated and chemically decomposed throughout the study area. The overlying Mizunami Group is composed of tuffaceous mudstone and sandstone beds, along with conglomerate beds containing very angular to angular Palaeozoic-Mesozoic shale and chert cobbles and boulders. The Palaeozoic-Mesozoic pelagic formations that are the source of these cobbles and boulders are exposed 1 km to the north of the study area.

An unconformity lies between the Seto Porcelain Clay Formation and the basement rocks. The formation is about 20–40 m thick, and is flat-lying to slightly southwestward dipping. It is unconformably overlain by the Yadagawa Formation, in which fluvial gravel beds predominate. A few normal faults are recognized or inferred in the study area. The two main trends of these faults are WNW–ESE and NNE–SSW, although another WSW–ENE fault trend has also been recognized (Fig. 1). All faults except the WSW–ENE fault have been active several times since the Early Miocene. According to the displacements of the Mizunami and Tokai Groups, they were active just before initial deposition of the Seto Porcelain Clay Formation, but did not move during deposition of the formation. The single WSW–ENE fault was active only after deposition of the formation (Nakayama, 1994).

# FACIES ASSOCIATIONS

Three representative facies associations (FA1–FA3; Table 1) have been distinguished, based on grain size, facies assemblages and facies architecture. Stratigraphical sections and sketches have been described from 18 major exposures in the area, and numerous other small exposures have been examined. Sketches of three major exposures are shown in Fig. 2. Each facies association shows an obvious upward-fining succession. The facies associations, their spatial and stratigraphical variations and interpretation of their depositional systems are given below.

## Facies association 1

### Description

Facies association 1 comprises mainly poorly sorted boulder beds, cross-stratified pebble to cobble beds, massive muddy sand beds and massive mud beds. The boulder beds are clast-supported, and some show inverse grading. Boulders are very angular to subangular, and were derived from the Palaeozoic-Mesozoic formations to the north (Fig. 3A). The pebble to cobble beds are sand matrix-supported, and are trough cross-stratified. Pebbles to cobbles are angular to subrounded (Fig. 3B), and also originate from the Palaeozoic-Mesozoic formations. Massive disorganized muddy sand beds consist mostly of very coarse sand to very fine pebble-size quartz grains. Massive mud beds include some sporadic very coarse sand to very fine pebble-size quartz grains. As a result, some massive mud beds fall in the 'muddy sand' category of the traditional sand–silt–clay triangular diagram. Differences in sorting between the muddy sand and the sandy mud are apparent, however, in cumulative

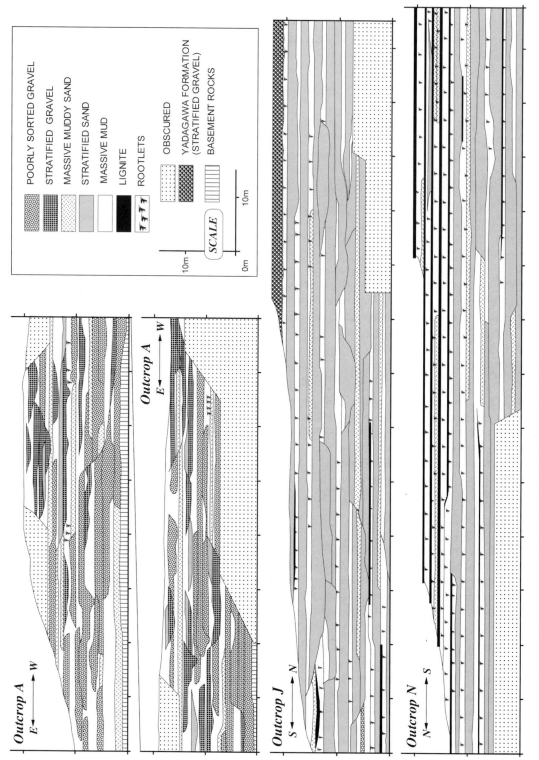

**Fig. 2.** Sketches of representative exposures of the Seto Porcelain Clay Formation, located in Fig. 1. Exposures A, J and N correspond to proximal, medial and distal fan settings, respectively. In these exposures, beds are horizontal, because the outcrop direction is approximately parallel to the strike (beds generally dip south–westward, but dip southward around exposure A). Facies association 1 predominates in exposure A. Exposure J consists of FA2a (mainly lower half) and FA2b (mainly upper half). Exposure N is FA2b in the lower half, and FA3 in the rest of the section. Typical columnar sections of each facies association (FA1–FA3) are given in Fig. 10. Massive muddy sand beds deposited by sediment gravity flow are intercalated in all facies associations.

**Fig. 3.** Gravel beds in FA1. Numbers on the scale bar are 10 cm intevals. (A) Clast-supported boulder bed at exposure A. The clasts are very angular to subangular chert and pelagite, and were derived from the Palaeozoic–Mesozoic formations cropping out 1 km to the north of the study area. (B) Trough cross-stratified gravel bed at exposure C; note angular to subrounded clasts.

grain-size curves. Grain-size distribution is described in a later section.

Boulder beds, massive sand beds and massive mud beds show sheet-like geometries without apparent basal erosion. Sharp margins of the boulder beds and the massive sand beds suggest abrupt termination, whereas those of the massive mud beds usually thin out more gradually. Many cross-stratified pebble to cobble beds crop out as channels or sheet geometries, with erosional surfaces. Channel shapes observed within single exposures (less than 150 m in lateral length) are usually narrow and deep, with an approximate width/thickness ratio (W/T) of 10.

Upward-fining successions in FA1 are dominated by two types: firstly from boulder bed to massive sand or mud bed; and secondly from a cross-stratified gravel bed to massive sand or mud bed (Fig. 2, exposure A). In places, successions comprising both massive boulder bed and cross-stratified gravel bed and massive sand or mud beds are observed, but these are uncommon. Thicknesses of the upward-fining successions vary from 1.0 to 4.0 m. The lower half to two-thirds of each succession is usually gravel. Palaeocurrent directions within each succession are very consistent, as described later.

*Interpretation*

The poorly sorted boulder beds are considered to be the deposits of debris flows, as suggested by inverse grading, clast shape, lack of erosional surfaces, and abrupt lateral terminations of beds. The cross-stratified pebble to cobble beds originate from the bedload of gravelly rivers. From the very high proportion of angular clasts in these gravel beds, it is likely that the basement which supplied the clasts was exposed in the immediate vicinity. The

abundance of muddy material, disorganized sand grains, and lack of basal erosion in the massive sand beds imply deposition from sediment gravity flow. From the above, FA1 is interpreted as the deposits of debris flow and related flow-dominated fluvial systems.

**Facies association 2**

*Description*

Facies association 2 is characterized by the dominance of trough cross-stratified sand beds. Massive mud beds, massive muddy sand beds and plane-laminated sand beds also are observed. Rarely, lignite beds, 0.5 m thick, are recognized. The trough cross-stratified sand beds are composed mainly of very coarse sand to very fine pebble-size quartz and feldspar grains. Occasionally boulder-size quartz megacrystals and tree trunks also are found. Two types of massive mud bed are recognized. Although both typically include quartz grains, the two types are distinguished by the presence of rare or abundant rootlets and plant fragments, respectively. The massive muddy sand beds of FA2 superficially resemble those of FA1. Some muddy sand beds in FA2, however, contain rootlets and plant fragments. Plane-laminated sand beds are composed of medium to very coarse sand grains.

Both sheet and channel geometries are seen in the trough cross-stratified sand beds and the massive mud beds. Boulder-size quartz crystals and logs in the cross-stratified sand bed are interpreted as channel base lags (Fig. 4A). Channel shape usually is narrow and deep (W/D < 15). The massive muddy sand sheet beds end abruptly in places (Fig. 2, exposure J).

Upward-fining successions from trough cross-stratified

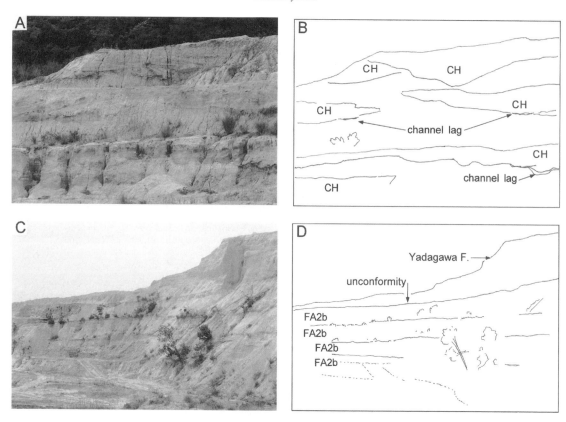

**Fig. 4.** Outcrop photographs and sketches of FA2. (A) Facies association 2a, channelized cross-stratified sand and channel-lag deposits at exposure L. Height of the exposure is about 7 m. (B) Sketch of exposure L. (C) Sheet geometry in FA2b in exposure I. Vegetated horizons mark the boundaries of the facies associations. Four accumulations of FA2b can be seen. Total thickness of the four accumulations is about 9 m. (D) Sketch of exposure I.

sand beds to a massive mud bed predominate in FA2. Massive muddy sand beds and plane-laminated sand beds are intercalated either above or below the trough cross-stratified sand beds. Successions commonly range between 1.5 and 2.5 m in thickness, and channels vary from 0.7 to 4.0 m in thickness. Trough cross-stratified sand comprises half to two-thirds of each succession.

Facies association 2 is subdivided into FA2a and FA2b, based on the dominant lithofacies and bed geometry. Facies association 2a is comparatively poor in rootlets and plant fragments, and channel geometry is commonly observed (Fig. 4A & B). Facies association 2b is characterized by beds including rootlets and plant fragments, and sheet geometry predominates (Fig. 4C & D). Palaeocurrent directions obtained from trough cross-stratifications in each succession are consistent. Multiple channels are recognized at several stratigraphical horizons (e.g. middle–upper horizon of exposure J in Fig. 2).

*Interpretation*

The dominance of trough cross-stratified sand beds and palaeocurrent consistency in each succession imply that FA2 was deposited from the bedload of a sandy braided or a low-sinuosity fluvial system. Two other characteristics are remarkable: one is the presence of muddy sand beds, which can be interpreted as the deposits of sediment gravity flows; the other is the channel geometry of FA2a, which is especially narrow and deep. Width/depth ratio of normal braided systems are more than 40 (Schumm, 1963). These points will be discussed in the section on depositional processes.

**Facies association 3**

*Description*

Facies association 3 is composed mainly of massive mud beds, lignite beds and associated massive muddy sand

**Fig. 5.** Facies association 3 exposure photographs. (A) Rootlet-dominated mud beds and lignite beds (horizontal black line) at exposure M. Two cycles of FA3 are seen. Height of the photograph is 2.5 m. (B) Rootlet-dominated massive sandy mud beds at exposure N. Horizontal dark grey bed in the upper middle of the photograph grades laterally into a lignite bed. Bar scale is 3 m. (C) Rootlet-dominated massive sand bed at exposure Q. Height of the photograph is 30 cm.

beds (Fig. 5A, upper horizon of exposure N in Fig. 2). The massive mud beds commonly contain rootlets (Fig. 5B), which range from 1 to 80 cm in length, and from 0.5 to 10 cm in diameter. Massive mud beds without rootlets also occur. Lignite beds contain tree trunks, stems and branches. The largest tree trunk observed was more than 3 m long, and 25 cm in diameter. Plant leaves, however, are very rare in the lignite beds. Massive muddy sand beds also are common, similar to those of FA2, either with or without rootlets and plant fragments (Fig. 5C).

Beds of FA3 show sheet geometries without erosional surfaces. Abrupt termination of massive muddy sand beds is also observed (Fig. 2, exposure N), as in the muddy sand beds of FA1 and FA2. Lignite beds and massive mud beds thin out laterally.

An upward succession from massive mud beds to lignite beds predominates, with massive sand beds intercalated at any horizon. These successions range from 0.5 to 2.0 m in thickness. The ratio of massive mud to lignite is roughly 1 : 1, or greater.

*Interpretation*

The lignite and massive mud beds represent deposition

from suspension. The presence of rootlets suggests that the mud beds were exposed to subaerial conditions. Exposure may have been temporary, however, as indicated by the small size of the rootlets, and the lack of palaeosols. The absence of leaves was discussed by Nakayama (1993). He concluded that many stems, branches, tree trunks and leaves were carried into temporary lakes, but the leaves sank and decomposed during periods of draining.

Facies association 3 is considered to represent floodplain deposits, with temporary shallow lakes and marshes. The muddy sand beds in FA3 are unusual, when compared with typical deposits of modern floodplains (e.g. Umitsu, 1994). They are considered to be sediment gravity flow deposits.

### Spatial and stratigraphical variation of facies associations

Figure 6 shows the distribution of the facies associations, based on observations made at 18 major exposures and numerous minor exposures, and from descriptions of 78 drill cores (Otsuka *et al.*, 1968). Facies association 1 occurs chiefly in the northern (northwestern) part of the study area, whereas FA3 is limited to the southern

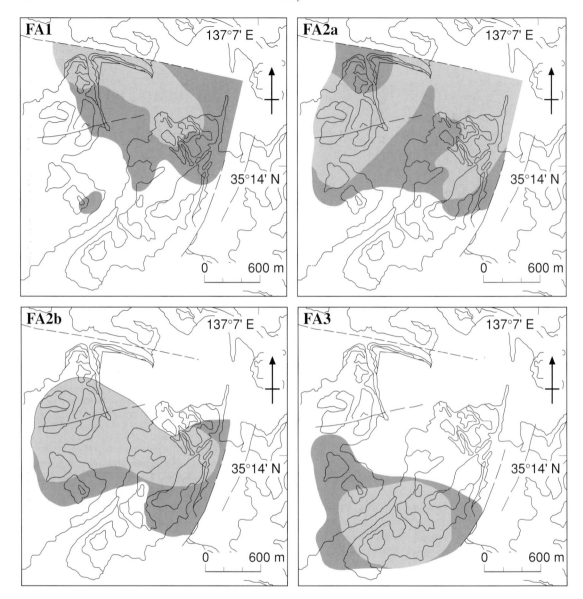

**Fig. 6.** Spatial and stratigraphical variation of facies associations. Dark shading outlines areas where three or more cycles are recognized, and the light shaded area outlines the occurrence of one or two cycles. Thin solid lines are the boundaries of the geological map units and faults shown in Fig. 1.

part. Although FA2 occurs over almost the entire basin, FA2a is found mainly in the central and northern areas, and FA2b is chiefly in the central area.

Stratigraphically, FA1 tends to be best developed in the lower and middle horizons, whereas FA3 occurs in the upper horizon. Facies association 1 and FA2 occasionally unconformably overlie the basement rocks, whereas FA3 is never in contact with the basement. The difference in

stratigraphical distribution between FA2a and FA2b is small, and is not considered significant. Facies association 2a generally overlies FA1 in the northern part of the study area, and its base is commonly confined to channels eroded into FA1. This relationship is observed at exposures C, G and F. In the southern part of the study area, upward transitions from FA2 to FA3 are recognized with non-erosional horizontal surfaces (Fig. 2, exposure J).

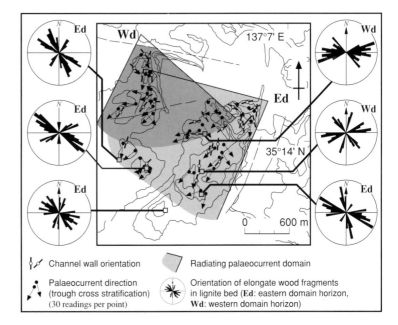

**Fig. 7.** Spatial distribution of palaeocurrent directions measured from trough cross-stratified sand in FA2 and FA3. Two overlapping radiating domains can be identified. Orientations of elongate wood fragments in lignite beds also are shown. These orientations tend to be arranged perpendicular to their respective palaeocurrent domain direction.

Facies associations are thus laterally distributed in the order FA1, FA2a, FA2b and FA3 towards the south and a stratigraphical arrangement of FA1, FA2 and FA3 is recognized (see Fig. 2). The spatial separation of the facies association is clearer than the stratigraphical arrangement.

## PALAEOCURRENT ANALYSIS

Palaeocurrents were determined by measurement of channel-wall orientations and trough cross-stratification directions. In Fig. 7, individual channel-wall orientations are shown, whereas palaeocurrent directions are shown as the mean directions of 30 troughs measured from each bed. Variance at each point is within a range of 10°. Multiple sites measured within single beds at large exposures such as C and I also show consistent directions.

Palaeocurrents occur in two radiating domains. In the western domain, currents trend from north to south, whereas in the eastern domain the trend is from northeast to southwest. According to stratigraphical relationships between the palaeocurrent trends, the eastern domain prograded westward over the western domain. Fragments of the Palaeozoic–Mesozoic formation are found in areas and horizons that correspond with the distribution of the western-domain beds.

Elongated wood fragments in the lignite beds show significant orientation. The long axes of 50 fragments at each point are roughly perpendicular to the local palaeocurrent direction.

## GRAIN COMPOSITION AND SIZE

Grain compositions and grain size were measured in the five main lithofacies of the Seto Porcelain Clay Formation, in the basement granite, and in the Palaeozoic-Mesozoic formations exposed in the immediate vicinity. Weathering of the granite occurred during the period between the end of deposition of the Miocene Mizunami Group (15 Ma) and the initial deposition of the Seto Porcelain Clay Formation (about 10 Ma).

Grain compositions of the two gravelly lithofacies were measured directly in outcrop, utilizing a standard point-counting method (more than 200 points). Samples from the sand and mud lithofacies and weathered granite were fixed with adhesive (P-Resin), and then cut as thin-sections for microscope measurement. Samples from fresh granite and the Palaeozoic-Mesozoic formations were cut as thin-sections without adhesive. More than 200 points were counted within these thin sections.

Grain-size distributions of the sand and mud lithofacies and decomposed granite were analysed with hand-crushed samples, using standard sieving and SK Laser Micron Sizer (Shinsei PRO-7000). Disintegrated granite was measured by micrometer from thin-sections at the same time as the grain compositions.

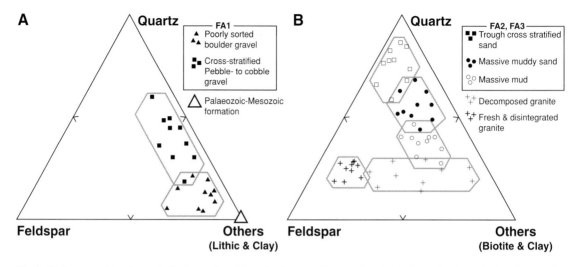

**Fig. 8.** Grain compositions in quartz–feldspar–others triangular diagrams. (A) Samples of the Palaeozoic-Mesozoic formations, and FA1 gravel facies. 'Others' are mainly lithic fragments and clay. (B) Samples of weathered granite, sand and mud facies from FA2 and FA3. 'Others' are dominated by biotite and clay.

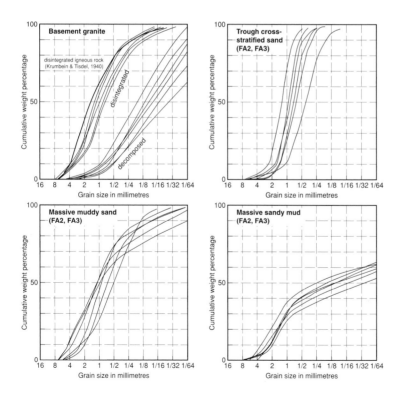

**Fig. 9.** Grain-size distribution. The Seto Porcelain Clay Formation samples are from FA2 and FA3. Plant fragments were removed before measurement. Granite samples were obtained within the study area shown in Fig. 1.

Grain compositions of the sediments and their source materials are illustrated in Fig. 8. The Palaeozoic-Mesozoic pelagite source essentially lacks quartz and feldspar of the size found in the sediments of the Seto Porcelain Clay Formation, and so plots at the 'Others' apex (Fig. 8A). Mechanically disintegrated granites plot in a confined area around $c$. $Q_{20-30}F_{50-70}O_{10}$ (Fig. 8B). In contrast, chemically weathered granites trend across the diagram at almost constant Q towards the Q–O junction, reflecting *in situ* conversion of feldspar to clay (Fig. 8B). Poorly sorted boulder clasts of FA1 plot closer to the O apex at low Q and F, and FA1 cross-stratified pebble to cobble gravels trend toward the Q apex at low F (Fig. 8A). These trends suggest that the FA1 sediments were not derived solely from weathered granites, but that there was considerable input from the surrounding Palaeozoic-Mesozoic formations. In contrast, trough cross-stratified sands, massive muddy sands and massive muds from FA2 and FA3 plot in three well-defined groups, which trend from near the Q apex towards the field of weathering granite with increasing sorting. The areal distribution and volume of FA2 and FA3 facies are greater than FA1, and grain compositions clearly indicate that they were derived chiefly from deeply weathered granite.

Grain-size distributions of the trough cross-stratified sand, massive muddy sand and massive sandy mud from FA2 and FA3, and also of weathered granite, are shown in Fig. 9. Cumulative curves for disintegrated granite from the study area closely correspond to the curve of disintegrated igneous rock from Krumbein & Tisdel (1940). These curves conform to Rosin's Law, indicating that the grain-size distribution and sorting of the disintegrated granite is similar to that of crushed rock. The cumulative curves of all Seto Porcelain Clay Formation samples are finer than those of the disintegrated granite. This implies that the weathered granite could not supply clasts larger than 8 mm in diameter to the Seto Porcelain Clay Formation. Rare boulder-size quartz megacrystals, probably derived from geodes in the granite (Mizota *et al.*, 1998), are the exception.

## DEPOSITIONAL PROCESSES

Interpretation of each facies association and the other sedimentological data mentioned above can be synthesized to identify the depositional processes of the two alluvial-fan systems. The spatial distribution of facies association are arranged as FA1, FA2a, FA2b and FA3 in each radiating domain, as deduced from palaeocurrents. Facies association 1 (debris-flow-dominated fluvial deposits) represents proximal-fan deposits. Similarly,

FA2 corresponds to medial-fan deposits and FA3 corresponds to distal-fan deposits. Faults in the study area were active immediately prior to the deposition of the Seto Porcelain Clay Formation (Nakayama, 1994), and are likely to have formed palaeopiedmont fronts. The depositional system of the formation is illustrated in Fig. 10.

The existence of an alluvial-fan system is supported by the presence of massive muddy sand beds in all facies associations. Generally, such massive muddy sand of sediment gravity flows would not develop in sandy braided or low-sinuosity systems, or in floodplain and shallow stagnant-water systems. Facies association 2 and FA3 developed within 2 km of the basin periphery, as represented by the inferred palaeopiedmont line. Over this short distance, sediment gravity flows could easily reach the medial- and distal-fan environments in which FA2 and FA3 accumulated. An alluvial-fan system is further supported by the unusually narrow and deep channels described from FA1 and FA2a. Width/depth ratios of less than 15 are seen in both FA1 and FA2. Such narrow and deep channels could develop as incised channels on the proximal fan. The arrangement of the confined-channel facies association of FA2a and the unconfined-channel facies association of FA2b is consistent with the growth direction of an alluvial-fan system.

The western (-domain) fan supplied fragments from both granite and the Palaeozoic-Mesozoic formations, whereas the eastern (-domain) fan supplied mainly weathered granitic fragments. Facies association 1 is poorly developed in the eastern fan, and clasts of the Palaeozoic-Mesozoic formations are rarely found within it. The fragments supplied reflect the neighbouring basement rocks within a few kilometres of the inferred palaeopiedmont line. From stratigraphical evidence, the western fan prograded only during the early depositional stages of the formation, whereas the eastern fan grew throughout the entire depositional period. Some parts of the western fan were thus covered by deposits of the eastern fan.

In general, the depositional successions of prograding alluvial-fan systems commonly show an upward-coarsening trend. Such an upward-coarsening trend, however, is not evident in the Seto Porcelain Clay Formation. There are at least two reasons for this. Firstly, the system of the western fan supplied coarser clasts than the eastern fan, and also was covered by deposits of the eastern fan. Secondly, the drainage area of the eastern fan comprised mainly weathered granite, which was poor in gravel-size clasts. Progradation of the eastern fan thus consisted mainly of sandy material, even in the proximal areas.

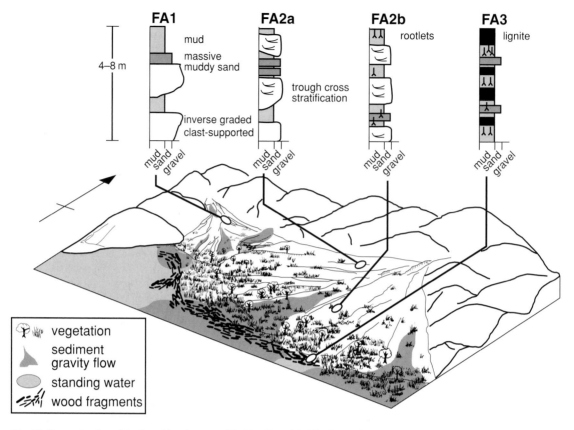

**Fig. 10.** Reconstruction of the depositional system of the Seto Porcelain Clay Formation and its facies associations. The illustration represents approximately the same as the study area shown in Fig. 1.

Woody fragments in the lignite beds of FA3 are orientated perpendicular to local palaeocurrents. By comparison with modern systems, Nakayama (1993) found that elongated wood fragments were orientated parallel to the palaeoshoreline of temporary lakes by wind action (Fig. 10).

## ESTIMATION OF FAN MORPHOLOGY

The spatial extent of the two fan systems can be estimated, based on the distribution of facies associations. The radial lengths of the western and the eastern fans are about 1.5 km and 2.0 km, respectively (Fig. 7). The formation is flat-lying to slightly south-westward dipping, so it is very difficult to estimate palaeoslope gradient of the fans, based on outcrop observation and drill-core data. Palaeoslope gradient, however, can be estimated using hydraulic relations. The palaeoslope of FA2b is estimated here, and is considered to represent the middle to

middle–distal part of the fan. To determine palaeoslope, estimates of water depth and grain size are necessary. The thickness of trough cross-stratified sand beds in FA2b can be utilized as a measure of water depth. Measured thicknesses range between 0.4 and 3.0 m, but generally are in the narrower range of 0.9–1.8 m. Mean grain diameters of these sands range from 0.4 to 1.3 mm (medium to very coarse sand) (Fig. 7). The average of the mean diameters is 0.80 mm (coarse sand), which is the value used in this evaluation.

Minimum slope gradient is considered to be the slope at which grains can be entrained in fluid flow. This slope condition can be calculated using the following equation:

$$\gamma g D S_e = 0.163 d^{1.213} \qquad (1)$$

where, $\gamma$ is water density (1000 kg m$^{-3}$), $g$ is gravitational acceleration of 9.8 m s$^{-2}$, $D$ is the water depth in metres, $S_e$ is the slope, and $d$ is the grain diameter in millimetres. The equation was obtained by combining the equation of DuBoys (1879) relating shear stress and slope gradient,

**Table 2.** Estimation of hydraulic palaeoslope gradient of FA2b.

| D<br>Water<br>depth<br>(m) | d<br>Grain<br>diameter<br>(mm) | $S_e$<br>Slope gradient<br>to entrain clasts* | | $S_c$<br>Critical slope for Fr = 1† | | | |
| --- | --- | --- | --- | --- | --- | --- | --- |
| | | | | n = 0.012 | | n = 0.022 | |
| | | H/L | Degrees | H/L | Degrees | H/L | Degrees |
| 0.4 | 0.80 | 0.0000317 | 0.00182 | 0.0019 | 0.11 | 0.0064 | 0.37 |
| 0.9 | 0.80 | 0.0000141 | 0.00081 | 0.0015 | 0.08 | 0.0049 | 0.28 |
| 1.8 | 0.80 | 0.0000070 | 0.00040 | 0.0012 | 0.07 | 0.0039 | 0.22 |
| 3.0 | 0.80 | 0.0000042 | 0.00024 | 0.0010 | 0.06 | 0.0033 | 0.19 |

* Slope gradient to entrain clasts is calculated by eqn (1).
† Critical slope gradient for Froude number 1 is by eqn (2). Manning numbers (n) used are 0.012 and 0.022.

and the equation of Costa (1983) relating shear stress and grain diameter. The calculations for a range of cross-stratified sand beds of FA2b is shown in Table 2. The results indicate slopes between 0.00081° and 0.00040° for water depths ranging between 0.9 and 1.8 m.

Maximum slope gradient can be estimated as the slope necessary to attain critical flow at the Froude number (Fr) of 1, using the following equation modified from Blair & McPherson (1994):

$$S_c = n^2 g / \sqrt[3]{D} \qquad (2)$$

in which, g and D are as in eqn (1), $S_c$ is the slope for critical flow and n is the Manning roughness coefficient. Manning roughness varies depending on the bottom materials, and values of 0.012 and 0.022 for an open natural channel with a sand bottom are used here (Chow, 1959; Japan Society of Civil Engineers, 1985). The trough cross-stratified sand is formed by three-dimensional dunes under subcritical flow (Fr < 1). Consequently, the palaeoslope of FA2b could never reach $S_c$. The $S_c$ values estimated here are less than 0.28° for the cases of water depths of 0.9 and 1.8 m (Table 2). Three-dimensional dunes are formed under comparatively rapid flow within subcritical flow. From the above, the palaeoslope gradient of FA2b is thought to be about 0.1–0.2°.

# SAND- AND MUD-DOMINATED FANS: PRIMARY FACTORS AND PROPERTIES

Numerous alluvial fans have been recognized at piedmont zones formed by faults (e.g. Longwell, 1930). The fan systems of the Seto Porcelain Clay Formation are related to palaeopiedmont lines, which are inferred from the basin-margin faults. The fan systems in this study, especially the eastern fan, are predominately sand and mud and are poor in gravel clasts. This results from the

presence of strongly weathered granite in the drainage area, which could supply only sand and mud. The basin-margin faults and the deeply weathered granite source hinterland are the primary factors controlling the sand- and mud-dominated fan system of the Seto Porcelain Clay Formation.

The sand- and mud-dominated fan systems have a radial length of less than 2 km, and low slope gradient of about 0.1–0.2° in the medial to medial–distal part of the fan, which was less than 1.5 km from the fan apex. A low slope gradient is thus characteristic of these fans and the low proportion of debris-flow deposits is significant. In contrast, muddy sand deposited from sediment gravity flows is widespread, even in distal parts of these fans. As the fan systems are small, many muddy sands may have been debouched directly from the weathered-granite drainage area. Extensive development of sandy sediment gravity flow deposits is also a peculiar property of these fan systems.

Rootlets dominate in FA2b and FA3 and are recognized even in FA2a and FA1. Although this implies that vegetation covered the entire fans at times, the greater incidence of rootlets in FA2b and FA3 suggests that vegetation cover was more significant or persistent in the medial and distal parts of the fans. This cover may have enhanced mud accumulation by entrapment of silt and clay.

Modern analogues and corresponding ancient records of similar fan systems are rare, although Nichols (1987) and Hirst (1991) described sandy fan or sandy fan-related systems. Genetically, the willow and spruce fans of the Northwest Territories, Canada (Legget et al., 1966) are possible modern analogues of the fan systems described here. The climatic conditions of the Canadian fans, however, are quite different from those of this study. The Canadian analogues are composed of fine-grained material because their drainage area is underlain solely by

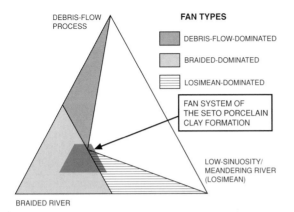

**Fig. 11.** Position of the Seto Porcelain Clay Formation fans on the triangular alluvial-fan classification diagram of Stanistreet & McCarthy (1993). Probable position is shown as a shaded polygon, based on areal proportions of debris flow, braided, and losimean deposits in the 18 major exposures. Facies association 1 is identifiable as debris-flow deposits and comprises 12% of the exposures. As obscured intervals are more common in the bases of sections, however, where FA1 is likely to be the best developed, this value is regarded as the lower boundary of the polygon. The upper limit is an estimate based on the maximum possible occurrence of debris-flow deposits from the areal extent of obscured exposure. Differentiation between the braided and losimean apices is more difficult, however. Although FA3 is regarded as the deposits of losimean systems and defines the left edge of the polygon, FA2 cannot be assigned simply to either apex. Much of FA2 clearly comprises braided-river deposits (multiple channels in FA2a and dominant sheet geometry in FA2b), but some losimean elements also may occur. The ratio between braided and losimean deposits is thus partially qualitative.

mudrocks. The primary role of drainage-area clast-size in controlling the dominant grains of the fan is a feature common to this study and that of Legget *et al.* (1966).

McPherson & Blair (1993) and Blair & McPherson (1994) give strict criteria for the recognition of alluvial fans, which the fan system in this study does not fully conform to. Domination of bedload sand (cross-stratified sand) and suspended mud, the small amount of gravelly debris flow deposits, and very low slope gradient are outside the criteria for the alluvial fan proposed by McPherson & Blair (1993) and Blair & McPherson (1994). Stanistreet & McCarthy (1993), however, define three fan types: debris-flow dominated, braided dominated, and losimean (low-sinuosity meandering) dominated. The fan system in this paper conforms to the braided-dominated fan type of Stanistreet & McCarthy (1993). The fan defined by McPherson & Blair (1993) is considered by Miall (1996) to correspond to the debris-flow-dominated fan of Stanistreet & McCarthy (1993). In

detail, the fan systems described in this paper also have components of debris-flow-dominated and losimean (low-sinuosity meandering) -dominated fans (Fig. 11). Debris-flow deposits are recognized in FA1 and narrow and deep low-sinuosity channels and possible low-gradient slopes are also recognized.

## CONCLUSIONS

Sand- and mud-dominated alluvial-fan systems have been identified in the ancient record of the Late Miocene Seto Porcelain Clay Formation in Japan. Three facies associations are recognized in this formation. Facies association 1 comprises mainly poorly sorted gravel beds and cross-stratified sandy gravel beds with associated massive mud beds; FA2 consists mainly of cross-stratified sand beds with mud beds containing rootlets; FA3 is composed primarily of plant-root- and plant-debris-dominated mud beds and lignite beds. Facies association 2 is subdivided into FA2a, dominated by channel geometry, and FA2b, which is dominated by sheet geometry. Palaeocurrents occur in two radiating domains, which define two fan systems. In each domain, facies associations are arranged in order FA1, FA2a, FA2b and FA3, which correspond with proximal (apex) to distal fan segments. Facies association 2 is the dominant facies association. Massive muddy sand beds, deposited from sediment gravity flows, are recognized in all facies associations. The morphology of the two fans suggests a radial length of less than 2 km, with palaeoslope gradients of about 0.1–0.2°. The drainage areas of the two fan systems, especially the eastern fan system, consist mainly of deeply weathered granite, which supplies sand- and mud-size material rather than gravel clasts. This is the primary cause of sand and mud domination. Vegetation cover also may play a significant role in the arrangement of the facies associations. Rootlets are most common in the medial and distal fan area, corresponding to a greater incidence of mud. This suggests that the vegetation present in these areas promoted mud entrapment.

## ACKNOWLEDGEMENTS

I would like thank Dr B.P. Roser for his critical reading and helpful comments on the manuscript. Two reviewers M. Goedhart and I. Stanistreet and editor N.D. Smith are thanked for their constructive comments on the submitted manuscript. I also thank Mr K. Kubo for discussion in the field. This study was supported by the Japanese Ministry of Education (grant-in-aid) 05740317, 10440142.

# REFERENCES

BLAIR, T.C. & MCPHERSON, J.G. (1994) Alluvial fans and their natural distinction from rivers based on morphology, hydraulic processes, sedimentary processes, and facies assemblages. *J. sediment Res.*, **A64**, 450–489.

CHOW, V.T. (1959) *Open Channel Hydraulics*. McGraw-Hill, New York.

COSTA, J.E. (1983) Paleohydraulic reconstruction of flash-flood peaks from boulder deposits in the Colorado Front Range. *Geol. Soc. Amer. Bull.*; **94**, 986–1004.

DUBOYS, M.P. (1879) Etudes du régime et l'action exercée par les eaux sur un lit a fond de graviers indefiniment affouiable. *Ann. Ponts Chaussées*, **18**, 141–195.

FUJII, N. (1968) Genesis of the fireclay deposits in Tajimi-Toki district, Gifu Prefecture, central Japan. *Rep. geol. Surv. Jpn*, **230**, 1–56.

HIRST, J.P.P. (1991) Variations in alluvial architecture across the Oligo-Miocene Huesca fluvial system, Enbo basin, Spain. In: *Three-dimensional Facies Architecture of Terrigenous Clastic Sediments, and its Implications for Hydrocarbon Discovery and Recovery* (Eds Miall, A.D. & Tyler, N.), pp. 111–121. Concepts in Sedimentology and Paleontology, Vol. **3**, Society of Economic Paleontologists and Mineralogists, Tulsa, OK.

IKEBE, N. (1957) Cenozoic sedimentary basin in Japan, with special reference to the Miocene sedimentary basin. *Cenozoic Res. (Shinseidai no Kenkyu)*, **24–25**, 1–10. (In Japanese with English abstract.)

JAPAN SOCIETY OF CIVIL ENGINEERS (1985) *Formulae of Hydraulics*. Giho-do, Tokyo. (In Japanese.)

KRUMBEIN, W.C. & TISDEL, F.W. (1940) Size distributions of source rocks of sediments. *Am. J. Sci.*, **238**, 296–305.

LEGGETT, R.F., BROWN, R.J.E. & JOHNSON, G.H. (1966) Alluvial fan formation near Aklavik, Northwest Territories, Canada. *Geol. Soc. Am. Bull.*, **77**, 15–30.

LONGWELL, C.R. (1930) Faulted fans west of Sheep Range, southern Nevada. *Am. J. Sci.*, **20**, 1–13.

MATSUZAWA, I., KATO, R., KUWAHARA, T., KIMURA, T., UEMURA, T. & TUZUKI, Y. (1960) *Geology of Southern Areas of Sanageyama, with Geological Map*. Aichi Prefecture, Aichi. (In Japanese.)

MCPHERSON, J.G. & BLAIR, T.C. (1993) Alluvial fans. Fluvial or not? *Keynote Address, 5th International Conference on Fluvial Sedimentology, Brisbane*, pp. K33–K41.

MIALL, A.D. (1978) Lithofacies types and vertical profile models in braided river deposits: a summary. In: *Fluvial Sedimentology* (Ed. Miall, A.D.), Mem. Can. Soc. petrol. Geol., Calgary, **5**, 597–904.

MIALL, A.D. (1985) Architectural-element analysis: a new method of facies analysis applied to fluvial deposits. *Earth Sci. Rev.*, **22**, 261–308.

MIALL, A.D. (1996) *The Geology of Fluvial Deposits, Sedimentary Facies, Basin Analysis and Petroleum Geology*. Springer-Verlag, Berlin.

MIKI, S. (1963) Further study of plant remains in *Pinus trifolia* beds, Central Hondo, Japan. *Chigaku Kenkyu, Spec. Vol.*, 80–93. (In Japanese.)

MIZOTA, C., FAURE, K., NAKAYAMA, K. & ZENG, N. (1998) Origin of boulder-sized euhedral quartz in the Seto Porcelain Clay Formation, central Japan. *Geochem. J.*, **32**, 59–63.

NAKAYAMA, K. (1991) Depositional process of the Neogene Seto Procelain Clay Formation in the northern part of Seto City, central Japan. *J. geol. Soc. Jpn*, **97**, 945–958. (In Japanese with English abstract.)

NAKAYAMA, K. (1993) Depositional process of lignite bed in the Upper Miocene Seto Porcelain Clay Formation, Central Japan. *Internat. Project Paleolimol. Late Cenozoic Climate News Lett.*, **7**, 147.

NAKAYAMA, K. (1994) Stratigraphy of the Upper Cenozoic Tokai Group around the east coast of Ise Bay, Central Japan. *J. Geosci. Osaka City Univ.*, **37**, 77–143.

NAKAYAMA, K. (1996) Depositional processes for fluvial sediments in an intra-arc basin: an example from the upper Cenozoic Tokai Group in Japan. *Sediment. Geol.*, **101**, 193–211.

NICHOLS, G. (1987) Structural controls on fluvial distributary systems – Luna system, northern Spain. In: *Recent Developments in Fluvial Sedimentology* (Eds Ethridge, F.G., Flores, R.M. & Harvey, M.D.), Spec. Publ. Soc. econ. Paleont. Miner., Tulsa, **39**, 269–277.

OTSUKA, T., KONDO, Y., SASAKI, M., TAKEDA, Y. & SHIMOSAKA, Y. (1968) *Quartz Sands and Fireclay in the Vicinity of Seto in Central Japan*. Aichi Prefecture, Aichi, 43 pp. (In Japanese.)

RUST, B.R. (1978) A classification of alluvial channel systems. In: *Fluvial Sedimentology* (Ed. Miall, A.D.), Mem. Can. Soc. petrol. Geol., Calgary, **5**, 187–198.

SCHUMM, S.A. (1963) Sinuosity of alluvial rivers on the Great Plains. *Geol. Soc. Am. Bull.*, **74**, 1089–1100.

STANISTREET, I.G. & MCCARTHY, T.S. (1993) The Okavango fan and the classification of subaerial systems. *Sediment. Geol.*, **85**, 115–133.

TANEMURA, M. (1964) Geological and mineralogical studies of clay and silica sand deposits in Seto district, Aichi Prefecture. *Rep. geol. Surv. Jpn*, **203**, 1–42. (In Japanese.)

UMITSU, M. (1994) *Late Quaternary Environment and Landform Evolution of Riverine Coastal Lowlands*. Kokin Shoin, Tokyo. (In Japanese.)

*Spec. Publs int. Ass. Sediment.* (1999) **28**, 409–434

# Sedimentology of the Gwembe Coal Formation (Permian), Lower Karoo Group, mid-Zambezi Valley, southern Zambia

I. A. NYAMBE*

*Department of Geology, University of Ottawa, Ottawa-Carleton Geoscience Centre, Ottawa,
Ontario KIN 6N5, Canada*

## ABSTRACT

The Lower Permian Gwembe Coal Formation has a maximum thickness of 280 m. It is underlain and overlain by the Siankondobo Sandstone and Madumabisa Mudstone formations, respectively. The three formations constitute the Lower Karoo Group. The Gwembe Coal Formation represents fluvial deposition of sandstones, siltstones and mudstones in channels and on the floodplains. Fourteen lithofacies in the formation are grouped into four facies associations, namely Maamba Sandstone, Coal, Mudrock and Sandstone A facies. The Maamba Sandstone facies association is probably a high-sinuosity meandering stream deposit. The coal facies association includes the Interseam Sandstone and the coal lithofacies. The productive coals (Main Seam), with thicknesses from 5 to 12 m, were deposited in shallow swampy areas of the floodplain. Accumulation of organic deposits in the swamps was interrupted by deposition of channel, crevasse channel and splay, levee (Interseam Sandstone) and overbank fine deposits. Rootlets in underlying sandstones indicate an *in situ* origin for the Main Seam Coal. The Sandstone A facies association contains features consistent with a change in fluvial style from a proximal braided system to high-sinuosity meandering stream system. The mudrock facies association is mainly overbank fine deposits with abundant siderite concretionary beds that were precipitated diagenetically.

## INTRODUCTION

Major economically important coalfields occur in the mid-Zambezi Valley of southern Zambia (Fig. 1). The discovery of this coal in the mid-1960s resulted in the first substantial surface and subsurface investigation of the coal-bearing sequence. This sequence is now known as the Gwembe Coal Formation, a Permian fluvial sequence within the continental Karoo Supergroup. Because of its economic significance, this sequence has received more attention than any of the other formations in the supergroup. Following the studies related to coal exploration in the late 1960s, however, essentially no further geological work was carried out until the formation was re-examined by Nyambe (1993) as part of a doctoral study that encompassed the pre-Karoo Sinakumbe Group and the Karoo Supergroup in the Zambezi Valley (Fig. 1).

A thorough, up-to-date understanding of the sedimentological context of these Permian coal deposits is seen as having considerable potential for guiding continued exploration and exploitation in the mid-Zambezi Valley Basin. The objectives of this paper, therefore, are to provide a comprehensive account of the character, extent, geometry and interrelationships of the sedimentary facies of the Gwembe Coal Formation, and to interpret the depositional setting, making use of the significant recent advances in the understanding of facies models in clastic sedimentology.

## METHODS AND GENERAL FEATURES

The Gwembe Coal Formation reaches the surface in an extensive outcrop belt near the northwestern margin of the mid-Zambezi Valley rift basin. The formation rests disconformably and locally unconformably on rocks of the Permo-Carboniferous Siankondobo Sandstone Formation and is conformably overlain by the Permian Madumabisa Mudstone Formation. These three formations constitute the Upper Palaeozoic Lower Karoo

* Present address: Department of Geology, School of Mines, University of Zambia, Box 32379, Lusaka, Zambia (Email: inyambe@mines.unza.zm)

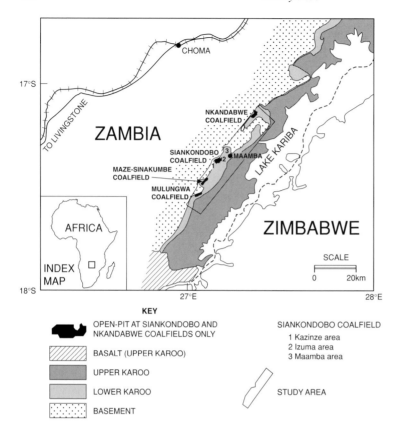

**Fig. 1.** General geology of the mid-Zambezi Valley Basin, southern Zambia, coalfield localities and location of the study area.

Group (Fig. 2) in the area. Field data for this study were collected primarily in 1990 and 1991 at various locales in the mid-Zambezi Valley (Fig. 1). A number of formal and informal stratigraphical names (Fig. 3) have been applied to portions of this coal-bearing formation, and because they have been widely used during exploitation are also used in this paper.

The formation generally is recessive because of its predominant mudrock (mudstone and siltstone) constituents. Mining operations, however, such as the now abandoned open pit of the Nkandabwe coalfield (Fig. 1), and the active Kazinze and Izuma open pits of the Maamba Coalfield (Fig. 1), provide superb surface exposures that facilitate lateral and vertical profiling, using photomosaics, and which help particularly in the understanding of lateral facies relationships. Drill cores and logs related to coal exploration also provide an important source of data. Palaeocurrent directions were measured mainly from unidirectional sedimentary structures; axes of trough cross-strata were considered the most reliable indicators (cf. Potter & Pettijohn, 1963). Colours used in rock descriptions correspond largely to the Munsell colour chart (Geological Society of America, 1980).

## LITHOFACIES DESCRIPTION

The Gwembe Coal Formation consists mainly of carbonaceous and silty mudstones and siltstones (mudrocks) with subordinate interbedded coal seams and sandstones (Fig. 3). Fourteen lithofacies, distinguished on the basis of lithology, grain size and sedimentary structures, are described in Table 1, and are grouped and interpreted in terms of depositional relationships into four facies associations.

For convenient cross-reference, the headings used for the lithofacies in Table 1 incorporate the stratigraphical occurrence of each within the formation. These are described briefly below.

### Maamba Sandstone Member

The Maamba Sandstone is composed mainly of units of buff, coarse-grained to pebbly, slightly feldspathic and micaceous sandstone. In many places it consists of a basal conglomerate that grades upwards through carbonaceous, fine-grained sandstone and siltstone into carbonaceous

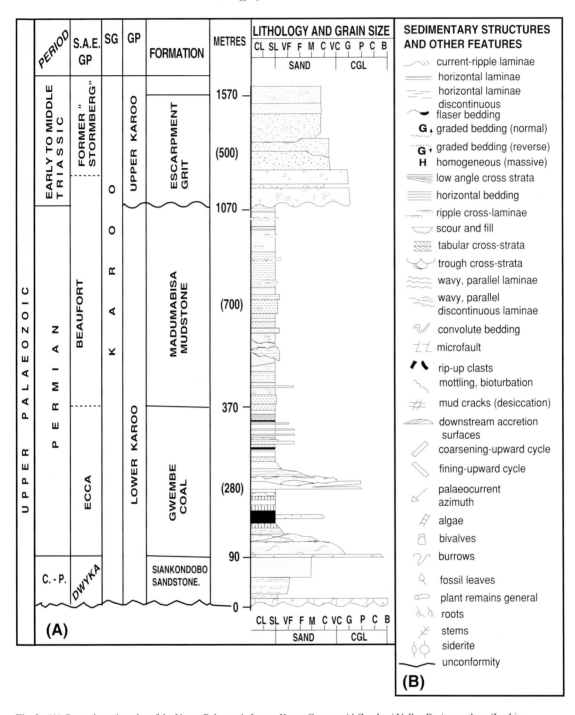

**Fig. 2.** (A) General stratigraphy of the Upper Palaeozoic Lower Karoo Group, mid-Zambezi Valley Basin, southern Zambia. Abbreviations: CL, clay; SL, silt; VF, very fine-grained; F, fine-grained; M, medium-grained; C, coarse-grained; VC, very coarse-grained; G, granules; P, pebbles; C, cobbles; B, boulders; CGL, conglomerate; GP, group; S.A.E. GP, South African equivalent group; C.-P., Carboniferous to Permian; SG, supergroup. (B) Key to sedimentary structures and other features for all subsequent stratigraphical sections. For key to lithology see Fig. 3.

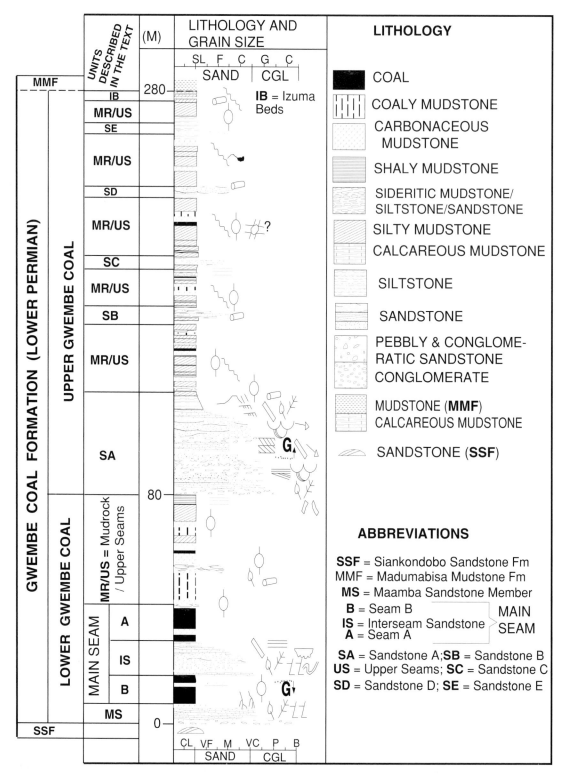

**Fig. 3.** Generalized stratigraphical column of Gwembe Coal Formation, mid-Zambezi Valley Basin, southern Zambia (not to scale). See Fig. 2 for key to sedimentary structures, other features and abbreviations.

**Table 1.** Summary of characteristics and interpretation of Gwembe Coal Formation lithofacies.

| Stratigraphical interval | Lithofacies | Colour | Lithology | Bedding and sedimentary structures/features | Thickness/geometry/ nature of bounding surfaces | Mineralogy and other components | Depositional environment |
|---|---|---|---|---|---|---|---|
| Maamba Sandstone Member | Conglomerate (SC$_{p-c}$) | Whitish grey | Matrix: fine to conglomeratic sand; coarse fraction (subangular to subrounded clasts): granules to cobbles up to 20 cm in diameter, poorly to moderately sorted, matrix-supported conglomerate with minor pebbly sandstone | Massive, some show crude stratification; low-angle stratified pebbly sandstone; cross-bedding, clasts may define bedding | Irregular bases; upward-fining sequences; conglomerate units, generally 2–3 m thick; pebbly sandstone (0.5 to 2 m thick) fining-upwards into the upper sandstone; units up 17 m thick including the overlying S$_{vf-vc}$ lithofacies | Rock fragments (vein quartz, quartzite, sandstone of Siankondobo Sandstone, schists); polymictic | Fluvial: point bars of meandering river channel system |
| | Very fine- to very coarse-grained sandstone (S$_{vf-vc}$) | Whitish grey to dark grey | Very fine to very coarse sand, moderately well sorted; matrix- to framework-supported carbonaceous sandstone with thin coaly, carbonaceous mudstones/coal layers | Horizontal bedding (Sh), massive (Sm), trough cross-bedding (St); soft sediment deformation, current scours, ripple lamination, pinch and swell, flaser bedding; plant stem impressions up to 1 m; tree stem impressions up to 2 m | Tabular and amalgamated; irregular bases and tops; upward-fining units; minor small-scale channels; beds alternate, commonly forming cyclic sequence of sandstone, carbonaceous sandstone, siltstone, mudstone and coaly mudstone, muddy coal and coal | Grains: quartz (predominant), feldspar, rock fragments, muscovite. Matrix: limonite, sericite and chlorite. Heavy minerals: zircon, epidote, sphene, metallic minerals, tourmaline, rutile, kaolinite, illite | |
| Main Seam (Coal and Interseam Sandstone) | Coal (C) | Black | Coal that may contain argillaceous laminae mainly comprising carbonaceous and coaly mudstone with minor silty intercalations and lenses of mudstone, siltstone and sandstone | Thinly (< 1 cm) to very thickly bedded units (over 1 m), beds from < 1 up to 50 cm are common (Fig. 5a). Massive and horizontal lamination | Tabular: generally irregular bases and flat tops with overlying sandstones, up to 12 m thick. Upper seams usually thin (c. 50 cm) with abundant vitrain laminae | See Table 2 | Alluvial: swamp |

*(continued on p. 414)*

**Table 1.** (*cont'd*)

| Stratigraphical interval | Lithofacies | Colour | Lithology | Bedding and sedimentary structures/features | Thickness/geometry/nature of bounding surfaces | Mineralogy and other components | Depositional environment |
|---|---|---|---|---|---|---|---|
| | Sheet-type interseam sandstone (S$_{sis}$) | Shades of grey to whitish grey (cream white) and weathers to yellowish brown/brownish black | Very fine sand to pebbles (1 cm maximum across), moderately well sorted, generally framework-supported sandstone. | Massive; locally normal or reverse grading; horizontal to low-angle stratification; root traces (Fig. 5b) | Tabular (sheets) up to 50 cm thick (Fig. 5b). Sharp and erosional contacts | Subangular to subrounded grains: quartz (predominant), feldspar, muscovite, rock fragments of siltstone, mudstone and coal; grain contacts are planar to concavo-convex, no preferred grain orientation; in laminated sandstones, laminae defined by composition and grain-size differences. Clay minerals: kaolinite, illite, and smectite | Alluvial: crevasse splay deposits |
| | Lenticular-type interseam sandstone (S$_{lis}$) | | Carbonaceous conglomeratic lags. Plant fragments and stem impressions over 1 m in length | Massive, or graded (normal or reverse); scour and fill lenses draped by streaks and stringers of carbonaceous matter. Dewatering structures, slumps and folded sandstones (Sd) (Fig. 5d) | Irregular beds or amalgamated beds; erosional basal contacts; beds from 0.5 to 1.5 m thick (Fig. 5c) in units up to 4 m containing lensoids up to 50 cm long and 15 cm thick, which become smaller upward in unit; thin beds grade into channel-form bodies up to 10 m wide and less than 2 m deep (Fig. 5c) | | Alluvial: crevasse channel deposits |
| Mudrocks | Coaly mudstone (M$_{cl}$) | Greyish black to black | Abundant coal fragments in M$_{ca}$; intercalations of coal, carbonaceous, sideritic and M$_{sl}$ interlayers and fragments are common | Predominantly laminated with minor massive units. Plant stem impressions and leaf fragments form bright aggregates; cream–white siderite spherulites, 2–3 mm across are characteristic | Generally tabular beds with sharp contacts. Beds 1–10 cm thick forming units over 7 m; horizontal lamination is defined by alternating dull and bright laminae (clarain and vitrain, respectively) | Coaly (60%) and woody fragments, exinous, organic debris; quartz, feldspar, mica, mudrock intraclasts, siderite, kaolinite; bright vitrain (< 1 mm) made of fossil leaves; massive blocks contain abundant comminuted vitrain; calcite and pyrite present | Alluvial: floodplain and swamp |
| | Carbonaceous mudstone (M$_{ca}$) | Dark grey/ black; weathers to reddish, yellowish, and orange–brown colours | Carbonaceous mud and clay, with scattered detrital grains; highly carbonaceous mudstone | Massive but commonly contains irregular, usually discontinuous horizontal laminae and stringers of organic matter, locally separated by thin, light grey, silty mudstone intercalations; vitrinite | Commonly gradational contacts; occurs as thinly to very thickly bedded units over 2 m thick, or as thin intercalations within other lithofacies such as M$_{sl}$; units can be up to 17 m thick characterized by | Disseminated or large clasts of organic matter (50%), quartz, feldspar, siderite; detrital grains can be up to 25% in the silty intercalations, but usually less than 15% in the M$_{ca}$; grains are mostly silt-sized, rarely exceed 0.3 mm in | Alluvial: floodplain and swamp |

| Facies | Colour | Lithology | Sedimentary structures | | Composition | Interpretation |
|---|---|---|---|---|---|---|
| | | | forms discontinuous horizontal and irregular bright laminae and stringers; normal grading noted at thin-section scale | intercalations of light-coloured silty mudstone (Fig. 6a), bright black $M_{cl}$ and yellowish mottled sideritic mudstone; coal beds also present | diameter (maximum 0.5 mm) and are mainly quartz; subangular intraclasts of $M_{ca}$ are present with coaly laminae bent around them; some coaly fragments display poorly preserved cell-like structures | Alluvial: levees, channel and floodplain, distal crevasse splays |
| Silty mudstone ($M_{sl}$) | Shades of grey, commonly light olive grey; weathers commonly to yellowish and orange-brown | Silty (slightly carbonaceous) mudstone, intraclasts common; local poorly sorted mud-clast breccias and conglomerates | Very thinly to very thickly bedded units (cm to 10+ m), massive (predominant) and laminated; lenticular and flaser bedding; small-scale cross-lamination; small-scale sigmoidal lamination; bioturbation (Fig. 6b); rootlet casts (Fig. 6b). Laminated $M_{sl}$ form units up to 1.5 m thick with beds < 10 cm thick; commonly horizontally laminated (locally resembling varves, Fig. 6c) | Blocky, tabular; sharp and gradational contacts; units over 10 m thick; upward-fining units consist of basal mudstone breccia overlain by mud-clast conglomerate that grades upwards into dark $M_{ca}$ with scattered intraclasts of $M_{sl}$; locally the units form cycles (repeated sequences) (Fig. 6a) | Detrital silt-size grains (up to 50%): quartz (predominant), feldspar, muscovite and mudrock intraclasts, carbonaceous matter (5–10%) dispersed in the clay–carbonate matrix (30–60%); siderite, calcite, dolomite? tourmaline, epidote; lamination resulting from differences in composition (Fig. 6c) | |
| Sideritic mudstone/siltstone/ sandstone ($MSS_{sd}$) | Light olive grey, blue-grey to purple-grey, weathers to purple, pink and red, rusty brown 'ironstone' | Mud to medium sand, diagenetic concentration of siderite; commonly mudstone, but sideritic bodies also occur in sandstone and siltstone | Massive; concretionary beds consist of individual concretions that coalesced laterally to form impersistent but extensive irregular bodies up to 15 m long; tabular beds consist of alternating, thin, irregular coaly laminae and thicker siderite-rich (mottled) laminae | Concretionary (Fig. 6d) and tabular (sheet — Fig. 7a) forms; concretionary form generally irregular sharp bases and sharp convex-upward tops; impersistent, irregular concretionary beds 0 to 1.4 m thick within $M_{cl}$ and $M_{ca}$ lithofacies; up to 2.2 m thick in $M_{sl}$; possible root tubules; tabular beds up to 50 cm thick commonly associated with $M_{cl}$ and $M_{ca}$ (Fig. 7a) | Spherulites of siderite (up to 95%) in matrix of clay, quartz (up to 0.06 mm across), locally up to 20%, feldspar, muscovite; siderite altered to iron oxide (30–95%); cell-like structures (probably plant cells) and lenses of organic matter; matrix (5–60%) as interparticle fillings; *Microcodium* present; kaolinite | Originally alluvial, diagenetically altered by siderite cementation |

*(continued on p. 416)*

**Table 1.** (*cont'd*)

| Stratigraphical interval | Lithofacies | Colour | Lithology | Bedding and sedimentary structures/features | Thickness/geometry/ nature of bounding surfaces | Mineralogy and other components | Depositional environment |
|---|---|---|---|---|---|---|---|
| Sandstone A | Microconglomerate to very coarse-grained sandstone (CS$_{g-vc}$) | Whitish grey (cream) | Granules and pebbles (up to 95%); little sandy matrix, moderately to well sorted; minor intercalations of medium- to coarse-grained sandstone; framework-supported | Thinly to very thickly bedded units ranging from 30 cm to over 2 m; trough (St – up to 2.5 m wide) and planar (Sp) cross-beds; massive (Sm) or graded (normal and reverse); horizontal bedding; scour and fill; low-angle and tangential scoop-shaped cross-bedding (Sl) over 10 m wide; erosional truncation planes and minor parting lineation | Erosional sharp basal contacts; amalgamated beds; beds over 2 m thick in units up to 25 m including overlying S$_{vfc}$ and M$_{sm}$ lithofacies (Fig. 7b); locally upward-fining units | Grains: quartz (> 90%), feldspar, rock fragments, mica, rutile; matrix (2–4%): quartz, chlorite, sericite with chalcedony as cement; heavy minerals: zircon, epidote, tourmaline and rutile | Fluvial: various dune forms in braided streams |
| | Very fine- to coarse-grained sandstone (S$_{vfc}$) | Greenish grey (shades of grey) (Fig. 10d) | Very fine to coarse sand; moderately well sorted matrix- to framework-supported | Irregular, amalgamated beds are commonly massive (Sm); trough cross-bedding (St); ripple cross-laminated sandstone (Sr), horizontally stratified sandstone (Sh), and planar bedded mudrock (mainly siltstone), with fine lamination and very small ripples (Fl), in decreasing order of abundance; erosional scours with intraclasts (Se) are present locally; fossil plant stems and leaf impressions are common; S$_{vfc}$ is gradational into mudrocks | Fining-upward units with a change in grain size from CS$_{g-vc}$ at the base to very fine-grained sandstone (Fig. 7c) at the top accompanied by a change in size of sedimentary structures from large trough cross-beds at the base to small-scale trough cross-beds at the top; units range from 10 cm to 5 m thick; contacts commonly gradational, locally sharp | Quartz, feldspar, mica | |

| | Colour | Lithology | Sedimentary structures | Geometry | Composition | Interpretation |
|---|---|---|---|---|---|---|
| Siltstone (Sl) | Light to dark grey | Carbonaceous siltstone and subordinate mudstone; moderately well sorted | Massive and small-scale cross-lamination; siltstone and mudstone commonly occur as minor intercalations in the $CS_{g-vc}$ and $S_{vf-c}$ | Poorly defined units (i.e. poor exposure); occupy top portions of upward-fining units of Sandstone A | Quartz, feldspar, mica; plant debris abundant | |
| Other Sandstones — Sandstones B, C, D and E (Sst B, C, D and E) | Sst B — blue- to purple-grey; Sst C — light grey to greenish grey; Sst D — white with dark beds; Sst E — whitish to pink | Sst B — fine-grained, siderite-rich, muddy siltstone to a coarse-grained, muddy sandstone; Sst C — medium- to fine-grained, non-ferruginous and sideritic sandstone; Sst D — medium-grained sandstone; Sst E — coarse-grained, pebbly arkose; sandstones are poorly to moderately well sorted; matrix- to framework-supported | Massive (Sm), horizontal and small-scale cross-lamination common in most sandstones; Sst C — small tree trunk impressions (up to 20 cm long) and fragments of siltstone and fine sandstone occur; Sst D — small- to medium-scale cross-bedding with a lag conglomerate; conglomerate lenses (up to 60 cm) locally; occasional dark siltstone beds and carbonaceous partings, and interbedded flaggy mudstones and silty mudstones; Sst E — low-angle cross-bedding | Mainly wedge-shaped, lenticular beds; units up to 10 m; upward-fining units common. Sst E — units up to 5 m thick, overlying fine- to medium-grained sandstone | Sst B — siderite; Sst C — generally K-feldspar predominant; Sst D — quartz, feldspar, rock fragments (siltstone); Sst E — feldspar (over 25%), quartz and some biotite; heavy minerals in Sst C, D and E include zircon, sphene, epidote, garnet, amphibole, pyroxene; others are calcite, kaolinite and illite | Fluvial: crevasse channel and splay from avulsion of the main channel on the flood-plain |
| Izuma Beds | Light to dark grey | Similar to ($M_{ca}$) and ($M_{sl}$) lithofacies above | Occur as alternating beds of carbonaceous mudstone and silty mudstone/siltstone | Similar to ($M_{ca}$) and ($M_{sl}$) lithofacies above; unit up to 20 m thick | Similar to ($M_{ca}$) and ($M_{sl}$) lithofacies above | |

**Fig. 4.** Kazinze open pit, exposing the Maamba Sandstone Member on the floor, overlain by Main Seam (B), Interseam Sandstone (IS), Main Seam (A) and mainly carbonaceous mudstone (CM). Scale top right corner of photograph.

and coaly mudstones, and some coal. The Maamba Sandstone is divided into two broad lithofacies, a lower conglomerate and an upper sandstone. The conglomerate is well exposed at the Izuma waterfall, Siankondobo area (Fig. 1), where it is up 6 m thick and shows characteristics of a channelized body that fines upward from conglomerate, at the base, into low-angle stratified pebbly sandstone, which is in turn overlain by massive, carbonaceous, coarse-grained sandstone, and eventually into coal. At the Izuma waterfall, the conglomerate consists of scattered pebbles and cobbles in a pebbly sandstone matrix. The very fine- to very coarse-grained sandstone lithofacies comprises mainly alternating sandstone and mudstone with thin, coaly mudstones and coal seams commonly terminated by wash-outs (Money *et al.*, 1974). Macerated plant fragments, plant impressions and vitrain shards also are characteristic features of the Maamba Sandstone.

## Coal and Interseam Sandstones

The Main Seam is comprised almost entirely of coal, containing only subordinate intercalated mudstone (Fig. 4). The seam consists of massive durain with a characteristic dull metallic lustre and conchoidal fracture (Fig. 4). Argillaceous laminae, specks of siderite (roses and spherulites) oxidized to limonite, marcasite 'flowers', barite on joint and bedding planes, and layers of pyrite nodules occur locally in the Main Seam (Money *et al.*, 1968). The seam grades laterally and vertically into more carbonaceous mudstone that contains impersistent closely spaced vitrain laminae. The Upper Seams are

impersistent, thin coals. Those in the upper part of the Maamba Sandstone Member (Fig. 3) are similar in composition to the Main Seam; those in the upper Gwembe Coal Formation (Fig. 3) are slightly different in composition. One lithofacies, namely coal, has been defined to encompass the Main Seam and Upper Seams of the Gwembe Coal Formation of the Lower Karoo Group.

From microscopic features, the Main Seam consists of two characteristic coal types, banded (Fig. 5A) and non-banded (massive). The massive type is composed largely of semifusinite associated with other macerals in minor amounts; the banded type (Fig. 5A) consists of alternating thick semifusinite layers and thin vitrinite layers. Laminae and intercalations of other macerals (e.g. liptinite) are present locally in association with the major ones. The two types are consistent with field observations in that the microscopically banded type is usually associated with the finely banded coals (lamina thickness from 1 mm to slightly more than 1 cm) and the non-banded type is associated with the more coarsely banded (bed thickness over 1 cm) but internally massive coals.

Upper Seams are laminated with numerous vitrain laminae up to 1 cm thick, separated by duller, fine-grained clarain–fusain laminae and locally thin, muddy partings of similar thickness. The Upper Seams consist mainly of mineral matter alternating with vitrinite laminae.

The Zambian coal shows that all three maceral groups are present, with the inertinite (largely semifusinite) accounting for 84%, vitrinite 12% and liptinite 4% (Table 2). Similar results were reported by Alpern (1968; cited in Money & Drysdall, 1975), namely 95% inertinite, 3% vitrinite and 2% exinite (liptinite). The ash content is relatively high, with an average of 22%. Fixed carbon is 54.7% (61.2%, this study), volatiles 18.6% (22%, this study), moisture 2.4% and combustible sulphur 1.5% (1.3%, this study) and the coals have a calorific value of 5933 kcal kg$^{-1}$ (Money & Drysdall, 1975). Total sulphur, mainly from pyrites, is much higher (up to 3.24%) in the Kazinze open pit than in the Mulungwa and Nkandabwe areas (0.77% and 1.1%, respectively).

The Interseam Sandstone (Figs 4 & 5B–D) is a very fine- to very coarse-grained sandstone with subordinate conglomerate lags, and intercalated siltstone and mudstone. The sandstone varies in thickness (up to 4 m thick), with well-developed cross-bedding and upward-fining sequences, usually with basal pebbly lag conglomerates. Plant fragments and stem impressions over 1 m in length, and fragments of siltstone, mudstone and coal also are present. Two lithofacies (Table 1) are recognized based on geometry, namely sheet-type interseam (Fig. 5B) and lenticular-type interseam (Fig. 5C) sandstones.

**Fig. 5.** Features of the Coal and Interseam Sandstone lithofacies. (A) Horizontally laminated coal, with abundant cleats. Pen is 14 cm long. (B) Sheet sandstone lithofacies. Very coarse-grained, generally massive sandstone becoming faintly stratified (horizontal to low-angle) towards top. Notice root traces (black arrow). Kazinze open pit. Pencil is 16 cm long. (C) Lenticular sandstone lithofacies in channel-form (*c.* 10 m long and 1.5 m thick) sandstone body, in Kazinze open pit, behind person (arrow). (D) Deformed and contorted, highly carbonaceous sandstone of the lenticular sandstone lithofacies. Mulungwa area. Knife is 9.5 cm long.

**Table 2.** Characteristics of coal from the Main Seam, Maamba Mine, Kazinze open pit.

| | | |
|---|---|---|
| Coal maceral petrography | Intervals | Petrographic intervals present, indicating vertical changes in composition of seam |
| | Lithotypes | Durain greatly predominates in the Main Seam. Various amounts of vitrain, clarain and fusain |
| | Amounts | Inertinite always more than 80%; vitrinite 12% or less; lipinite 4% or less |
| | Dominant types | Inertinite occurs as semifusinite, micrinite, inertodetrinite and fusinite; vitrinite mainly as collinite; and lipinite as liptite |
| | Preservation | Good; preserved cellular structures in semifusinite and fusinite |
| | Mineral matter | Siderite (rosettes and spherulites) altered to limonite and other iron oxides; clay minerals and carbonates (calcite and ankerite); syngenetic pyrite mainly finely disseminated and as patches, clots, blebs and aggregates |
| Proximate analysis of productive coals | Fixed Carbon | From 58% to 65% |
| | Ash | From 13% to 22% |
| | Volatile matter | From 17% to 25% |
| | Rank | Low volatile bituminous |
| Ultimate analysis of productive coals | Carbon | From 68% to 75% |
| | Hydrogen | From 2.5% to 4.0% |
| | Nitrogen | From 1.4% to 1.6% |
| | Sulphur | From 0.35% to 3.25% |
| | Ash | From 13% to 22% |
| | Oxygen (by difference) | From 4.4% to 6.3% |

**Fig. 6.** Features of the mudrock lithofacies. (A) Laterally persistent, thinly to thickly bedded, massive, light grey to dark grey silty mudstone and siltstone (light coloured). Carbonaceous mudstone lithofacies intercalated with coaly mudstone lithofacies at the base. Beds offset by faults. Kazinze open pit. Person for scale (arrow). (B) Polished slab of silty mudstone strongly bioturbated towards top (up arrow). Light grey alteration is the result of weathering. Notice area of black coloration (arrow) that fades downwards and many dark grey traces, probably representing rootlets or burrows. Kazinze open pit. Scale in centimetres. (C) Polished slab showing horizontally laminated silty mudstone. Mulungwa area. Scale in centimetres. (D) Discontinuous concretionary beds in coaly to carbonaceous mudstone lithofacies, Siankondobo area. Hammer is 34 cm long (arrow).

## Mudrocks

Mudrock (Figs 6 & 7A) forms about 70% of the Gwembe Coal Formation (Fig. 3). It is generally black to dark grey and very rich in organic matter (Fig. 6A–D). Money *et al.* (1974) recognized four types of mudstone in the formation, i.e. coaly mudstone, carbonaceous mudstone, sideritic mudstone and dark grey mudstone, with gradational to sharp contacts. The coaly and carbonaceous mudstone units occur mainly in the lower Gwembe Coal Formation and the grey and silty mudstone members in the upper Gwembe Coal Formation. In fresh exposures the distinction between sideritic mudstone and other mudstones can be difficult. In weathered exposures, however, the sideritic units are easy to distinguish because siderite is oxidized to iron oxides. In this study, three mudstone lithofacies, namely coaly, carbonaceous and

silty (slightly carbonaceous) mudstones, have been distinguished on the basis of different carbon content (Table 1). The three lithofacies are usually massive, but where mica flakes are abundant, their alignment results in shaly and fissile mudstone. A fourth lithofacies, defined here on the basis of high siderite content, also includes siltstone and sandstone and is termed sideritic mudstone, siltstone and sandstone (Table 1). The siderite-rich lithofacies occurs in very thinly to very thickly bedded alternating units (cycles) (Figs 6D & 7A). Seventeen cyclic units in a 12-m sequence, involving bright laminated coal, coaly mudstone and carbonaceous mudstone were recorded by Tavener-Smith (1960). Carbonaceous matter occurs mainly as fine disseminations, stringers, streaks and impersistent laminae of vitrain as well as plant fragments. In silty mudstones, light grey elongate irregular intraclasts, which increase in proportion upwards, are common, whereas in

**Fig. 7.** Features of mudrock and Sandstone A lithofacies. (A) Cyclic alternation of tabular sideritic lithofacies (see T) and coaly and carbonaceous mudstones. Kazinze River, Siankondobo area. Scale 2.1 m (arrow). (B) Photograph showing lithofacies of Sandstone A sharply overlying carbonaceous and coaly mudstone lithofacies, Maze–Sinakumbe area. Notice amalgamated beds and the channel-like structures towards top. Scale is 2.1 m (arrow). (C) Polished core slab showing normal grading — from conglomeratic sandstone at base to medium-grained sandstone at the top. Mulungwa area. Scale in centimetres.

places, light grey elongate irregular intraclasts in greyish black matrix are brecciated and generally aligned parallel to bedding. Figure 6B shows bioturbation and a local black coloration that fades downwards (top left of photograph), and contains many dark grey traces, possibly rootlets or burrows. Films associated with marcasite occur locally (Money *et al.*, 1974). Pyrite 'flowers' occur on joint and bedding planes. Mudstone pellet breccia and conglomerate units overlain by dark carbonaceous mudstone with scattered pellets of paler mudstone define upward-fining units and cycles in the mudstone (Money *et al.*, 1974). Symmetrical sand and silt lenses (1 cm in length), small-scale symmetrical loading folds and flame structures are apparent in the mudstones (Money *et al.*, 1974).

## Sandstone A

Sandstone A is well developed throughout the study area; the other sandstones (B–E) are minor and have not been located in surface exposures in their type locality (Maamba Mine Licence area) during the present field-work. Because of this, only Sandstone A is fully covered in this section, and the other sandstones are described under the general heading 'Lithofacies of other sand-

stones and the Izuma Beds'. Eight sections were studied. The sandstone varies in thickness from 0 to over 30 m. Dips of the sandstone beds are generally less than 12°. In areas between streams and rivers, the sandstone presumably is continous along strike, as is evident from some of the drill cores. Wedging out of these sandstone bodies, however, also is common.

Sandstone A (Fig. 7B & C) is composed mainly of medium- to very coarse-grained sandstone, pebbly sandstone, quartz microconglomerate (granule conglomerate) and subordinate siltstone and mudstone. The coarser siliciclastic facies are more abundant at the base, and grade upwards into very fine-grained clastic facies (Fig. 7B). Sandstone A is almost invariably cross-bedded, with foreset laminae commonly defined by concentrations of granules and pebbles in basal sections of units or by discontinuous stringers and intercalations of mainly carbonaceous and silty material towards the top, and in the finer interbeds within the lithofacies. Three lithofacies are recognized, namely microconglomerate to very coarse-grained sandstone, very fine- to coarse-grained sandstone, and siltstone (Table 1). In the Sinakumbe area (Fig. 1), Sandstone A is exposed (Fig. 7B), overlying laminated carbonaceous, coaly and silty mudstone. The

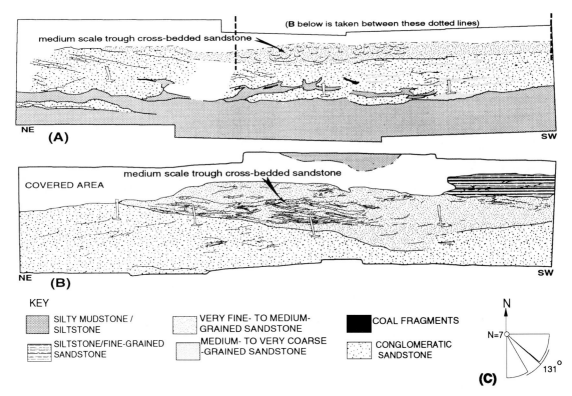

**Fig. 8.** (A) Upward-fining unit (point bar), from basal conglomeratic sandstone (with abundant intercalations (locally contorted) and stringers of sandstone and siltstone/mudstone) to very fine-grained sandstone and siltstone then into silty mudstone. Note coal fragments. Sandstone A facies association of the Gwembe Coal Formation, Mulungwa area, mid-Zambezi Valley, southern Zambia. (B) Represents enlargement of the upper right part of (A). (C) Palaeocurrent measured from trough cross-beds indicate direction to the south-east. Hammer outlines are 34 cm long.

basal part of the section starts with an erosional sharp contact, followed by 15 m of amalgamated, massive microconglomerate. This is in turn overlain by channel-like bodies (Fig. 7B). In the Mulungwa area (Fig. 1), 15-m-thick upward-fining units of Sandstone A with interbeds of finer grained sandstone are exposed.

Two photomosaic profiles were made of Sandstone A, one in the Mulungwa area (Fig. 8), the other in the north-western part of the Siankondobo area (Fig. 9). In the Mulungwa area the mosaic was taken in the top 2–3 m of a 15-m upward-fining unit of Sandstone A. The base of the microconglomerate that starts the upward-fining unit is poorly defined, because isolated discontinuous layers of microconglomerate are, in places, enclosed within fine to coarse-grained sandstone, and together they have been contorted. To the northeast (about 10 m) of the profile, however, a bounding surface is overlain by large-scale trough cross-beds (3.2 m wide and 2 m thick) that become crudely stratified to massive laterally to the south-west. These microconglomerate layers locally contain

intercalations of distorted finer sediments and coal fragments. The trough cross-beds and the equivalent massive bed are succeeded by medium- to small-scale trough cross-beds (Fig. 8), which grade upward into mudstone. The lithofacies commonly contain intraclasts, mainly of coal and coaly fragments, and fossil plant stem impressions. Measurement of trough cross-bed orientations (seven readings) showed a palaeocurrent direction to the south-east (Fig. 8). Figure 9 represents a 171-m-long slope profile from the Siankondobo area. The profile shows downstream accretion surfaces with a downcurrent change in grain size from granular (microconglomerate), to fine- to medium-grained sandstone. Sedimentary structures are obscured by weathering, but in places horizontal bedding parallel to bounding surfaces is present, as well as crude trough cross-bed sets. Orientations of the ribs and furrows (144 readings) exposed on bedding surfaces within the vicinity of the profile indicated an average palaeocurrent direction of 121° (southeast) with a spread from northeast to southwest (Fig. 9).

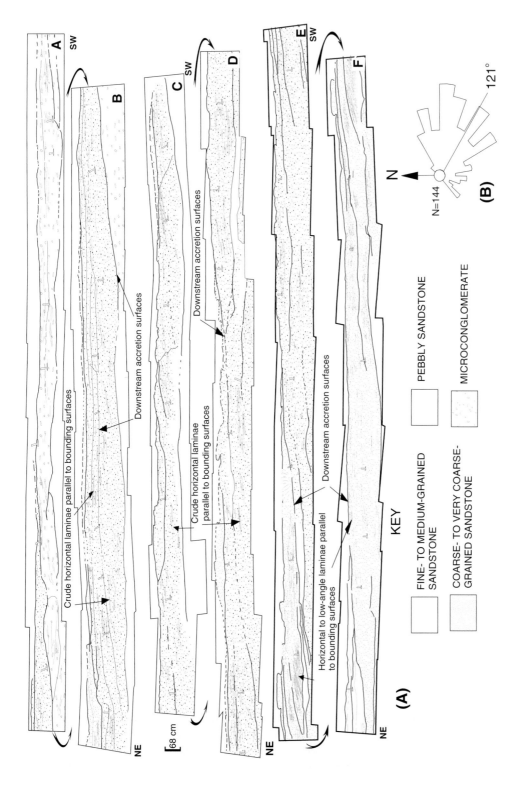

**Fig. 9.** (A) Downstream accretion surfaces with crude stratification and grain size reduction in Sandstone A facies association of the Gwembe Coal Formation, mid-Zambezi Valley Basin. Towards end and beginning of profiles A and B respectively, is a vertical sequence of microconglomerate overlain by crudely stratified pebbly sandstone. Total distance (profiles A, B, gap in between B and C, C to F) is 171 m. Scale left of profile C is 68 cm or hammer outlines are 34 cm long. (B) Shows the general palaeocurrent direction (readings taken from the bedding surfaces exposing rib and furrow (trough)) within the vicinity of the profiles (A–F)) to the southeast.

The microconglomerate lithofacies grades upwards into very fine- to coarse-grained sandstones lithofacies that become increasingly carbonaceous (Fig. 7C).

## Other sandstones and the Izuma Beds

Four other sandstones were described in the Siankondobo area (Radosevic *et al.*, 1968), which include sandstones B, C, D and E (Fig. 3). In the Mulungwa area (Fig. 1), sandstones B, C and D have been recognized, but only muddy feldspathic, medium-grained sandstone units up to 1 m thick, grading into siltstones, are recognized in the Nkandabwe area (Fig. 1). The sandstones are 5–7 m thick (maximum 11 m for sandstone D in the Mulungwa area) and are separated by carbonaceous and silty mudstone. Four lithofacies are defined as lithofacies of sandstones B, C, D and E. The fifth is a lithofacies of the Izuma Beds, which consist of a carbonaceous and a silty mudstone lithofacies. The lithofacies occur as alternating beds of carbonaceous mudstone and silty mudstone/siltstone.

## FACIES ASSOCIATIONS

A summary of the characteristics and interpretation of the Gwembe Formation lithofacies is give in Table 1. Stratigraphically, the facies associations alternate between coarse-grained facies (sandstone) and fine-grained facies (mudstone) (Figs 10 & 11). These alternations indicate rapid changes in current flow and give rise to cyclic deposits (Figs 3, 4, 6A, 7A & 10). Whereas the lithofacies within the Maamba Sandstone and Sandstone A are mainly coarse grained and not closely related, the finer grained lithofacies are closely related. For example, coaly mudstone is associated with coal, and carbonaceous mudstone occupies an intermediate position between the underlying coaly mudstone and the overlying grey silty mudstone. In general, a progressive upward decrease in carbonaceous matter from the Main Seam is observed as the mudstone becomes greyer and siltier (Figs 3, 4, 6A & 12). In Fig. 11, the lateral and vertical variation of the Interseam Sandstone, coal and mudrock bodies indicate the widespread occurrence of these, with the variable geometry reflecting local changes in depositional environment. The contacts between the lithofacies commonly are gradational, with subordinate sharp contacts particularly between coaly mudstone and the other mudstone lithofacies, and between sideritic lithofacies and the host rock type. These finer grained lithofacies are cyclic in nature. Because the coal lithofacies of the Main Seam are interpreted as formed *in situ*, whereas the mudstones (coaly, carbonaceous and silty) were transported, the coal

lithofacies has been grouped with the intervening sand-stones (Interseam Sandstone lithofacies) as the Coal facies association (CFA), and the mudstones (coaly, carbonaceous and silty), sideritic lithofacies, the lithofacies of the 'Other Sandstones' and Izuma Beds lithofacies are grouped as the Mudrock facies association (MFA). The conglomerate and the very fine- to very coarse-grained sandstone lithofacies of the Maamba Sandstone Member are grouped as the Maamba Sandstone facies association (MSFA). The three lithofacies of Sandstone A (micro-conglomerate to very coarse-grained sandstone, very fine- to coarse-grained sandstone and siltstone) are grouped as Sandstone A facies association (SAFA) (Fig. 10).

## Maamba Sandstone facies association

The MSFA comprises conglomerate and sandstone inter-bedded with subordinate coaly, carbonaceous and silty mudstones, and coal. The coarseness and sedimentary structures (mainly cross-stratification) indicate that the conglomerate and sandstone lithofacies of the Maamba Sandstone can be interpreted as high-energy fluvial deposits. The upward-fining sequences are interpreted as point bars of a meandering-river channel system. The Izuma Waterfall conglomerate–sandstone indicates a former channel with a palaeoflow approximately to the south-east. This also is consistent with the palaeocurrent readings (one station), which indicated sediment input from the south-west (Fig. 12).

The MSFA forms an upward-fining sequence (*c*. 18 m thick) that grades into the Main Seam. The thickness, geometry and sedimentary structures suggest that the assemblage represents a channelized sandstone body deposited on an alluvial plain by a high-sinuosity stream. Money *et al.* (1968) suggested that coarse-grained basal sediments of the Maamba Sandstone were probably formed by slumping in an unstable environment along the delta front, and that such instability was in part a result of episodic isostatic readjustment after the retreat of the Dwyka ice sheet.

## Coal facies association

The lenticular sandstone lithofacies (CFA) of the Interseam Sandstone is interpreted mainly as crevasse-channel deposits, and the sheet-sandstone lithofacies (CFA) as crevasse-splay deposits (Figs 10 & 11). The thickness of the upward-fining units (< 2 m) of the len-ticular sandstone lithofacies and the paucity of low-angle stratification characteristic in many fluvial point bars indicate that these are minor channels, probably crevasse channels. The low-angle stratification in the

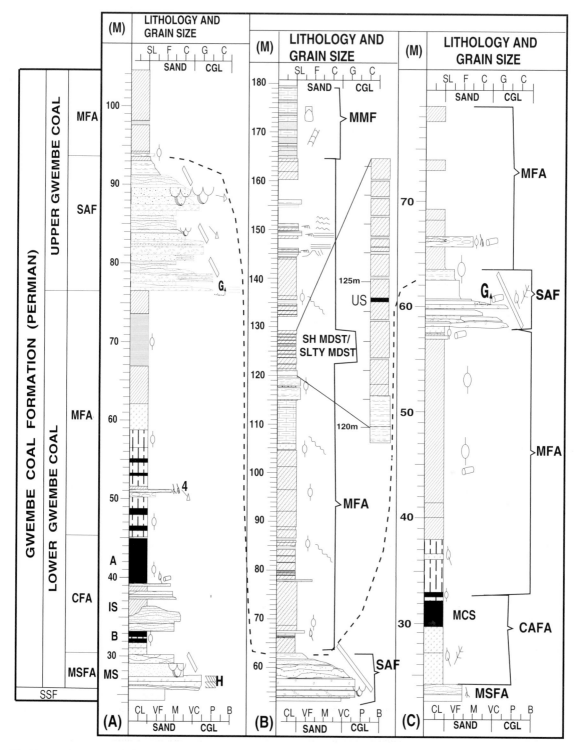

**Fig. 10.** Graphic logs showing the facies associations in the Gwembe Coal Formation, from northwest to northeast starting with (A) Mulungwa River, through (B) Zhimu River to (C) Kasika River, mid-Zambezi Valley Basin, southern Zambia. Note that SAF is thicker in the Mulungwa River (north-west) in contrast to Kasika River (north-east). For key to lithofacies, sedimentary structures and features, see Figs 2 & 3. Abbreviations: MSFA, Maamba Sandstone facies association; CFA, Coal facies association; MFA, Mudrock facies association; SAFA, Sandstone A facies association; MCS, Main Coal Seam; SH MDST/SLTY MDST, shaly and silty mudstone.

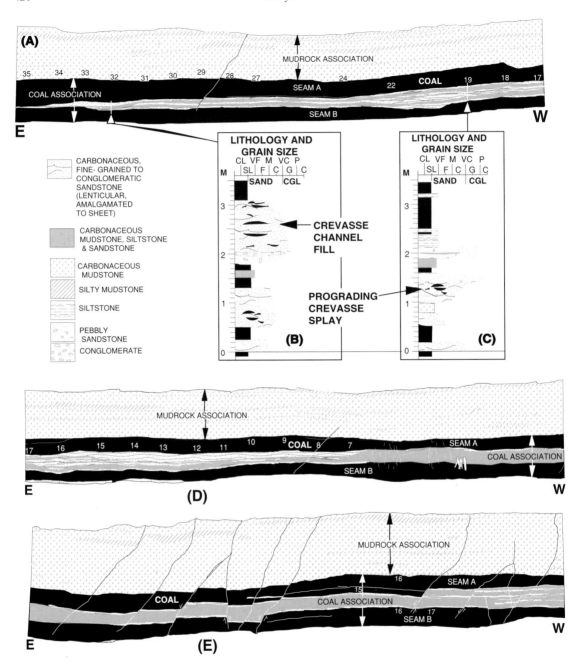

**Fig. 11.** Diagram from photomosaic showing lateral variation in the coal and mudrock facies associations of the Gwembe Coal Formation, Kazinze open pit. Note lateral pinchout of siltstone and sandstone beds. Profile continues to west in profiles (D) and (E). (B) and (C) are representative vertical profiles and interpretations of the Interseam Sandstone at the sampling points indicated (numbers 7–35). Profiles showing lateral variation in the coal and mudrock facies associations of the Gwembe Coal Formation, Kazinze open pit (continues to the east from (D) to (A) and to the west in (E) ). Note post-Lower Karoo Group faulting. Outlines of human figures for scale.

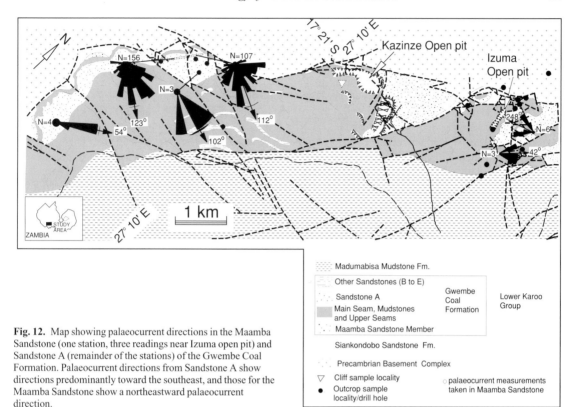

**Fig. 12.** Map showing palaeocurrent directions in the Maamba Sandstone (one station, three readings near Izuma open pit) and Sandstone A (remainder of the stations) of the Gwembe Coal Formation. Palaeocurrent directions from Sandstone A show directions predominantly toward the southeast, and those for the Maamba Sandstone show a northeastward palaeocurrent direction.

sheet-sandstone lithofacies probably represents crevasse-splay deposits (cf. Rust, 1978) and the rootlet structures (therefore a palaeosol) in this lithofacies indicate an *in situ* origin of at least the Main Seam. In Fig. 11, some coal beds grade laterally into coaly mudstone and then into silty mudstone, and some lenticular sandstone bodies grade into channel-form bodies. These lateral changes indicate local variations in sediment supply. The Interseam Sandstone shows variable thicknesses (*c.* 4 m) and consists of generally upward-fining bodies with subordinate upward-coarsening units that divide the Main Seam into four seams (A, B, C and D). These upward-fining units fill channels that are generally shallow (*c.* 2 m) and wide (10 to 20 m). Channel filling was not continuous, as indicated by carbonaceous/coal-draped surfaces and small, cross-cutting channels within the sandstone body.

The coal lithofacies formed by accumulation of peat in relatively shallow swamps on an alluvial floodplain. Peat accumulation was interrupted periodically by introduction of crevasse deposits (Interseam Sandstone). Extensive calcite lenses possibly resulted from evaporation in small pans or open-water areas (cf. Falcon, 1989). Iron sulphide minerals (e.g. pyrite), such as those found in the

coal lithofacies, are known to develop in reducing environments, usually in humic-rich waters (Berner, 1971).

The variation in microlithotype (largely fusite to clarain), maceral and mineral content of the mid-Zambezi Valley coal indicates that the coals formed under variable conditions. These variations may represent periods of reworking, local flooding or rising groundwater levels, and/or changing geochemical conditions. Such factors likely varied between areas, depending on the proximity to river channels and valley flanks, the degree of subsidence and compaction, and the length of time under specific conditions. Studies of microlithotype distribution suggest that the variation in microlithotype is an indication of the time spent in a particular palaeoenvironment or under certain geochemical conditions (Hagelskamp & Snyman, 1988; Hagelskamp *et al.*, 1988). These studies have shown that:

**1** a vitrite (vitrinite) layer indicates plant matter accumulating in relatively quiet, reducing, water-logged conditions;

**2** a semifusinite layer (inertite) indicates plant material accumulating *in situ* in exposed semi-oxidizing to oxidizing conditions;

**3** an inertodetrite layer (inertite) indicates either decomposition of cellulose-rich plant matter (herbaceous forms) *in situ*, in oxidizing and semi-aquatic conditions to form fine, granular gel-like material, or the removal of a variety of plant material from its original site of accumulation and partial gelification, followed by re-deposition in a waterborne detrital form.

The common occurrence of quartz and clay minerals, associated with abundant inertodetrites, is considered to be indicative of relatively high-energy, freshwater flood deposits, whereas clays, which appear to have flocculated in coal-forming environments, may indicate the existence of specific and different geochemical conditions in standing peat-swamp waters (Postma, 1983). The abundance of pyrite within the Interseam Sandstone and Coal seams of the Gwembe Coal Formation suggests reducing conditions; its formation is favoured by the presence of dissolved sulphide. Freshwater contains some sulphate, which in the reduced zone will inevitably result in the formation of some sulphide together with siderite. Postma (1983) concluded that it formed through the reduction of ferric oxyhydroxides by organic matter. The reduction of ferric oxyhydroxides by organic matter provides a source of both $Fe^{2+}$ and carbonate for siderite formation. This initially was demonstrated theoretically by Berner (1964, 1971) who showed that siderite is not stable in marine environments, because it is only stable at extremely low dissolved sulphide activities, lower than in seawater. In Late Pleistocene and Recent deposits of northern Lake Tanganyika (East Africa rift), Baltzer (1991) indicated that pyrite is associated with layers in which sulphur is a direct function of iron and that siderite is typical of layers in which sulphur occurs in minor amounts and is distributed independently from iron. However, although siderite is common in ancient nonmarine sediments, it has not been observed precipitating today (Kelts & Hsu, 1978).

**Mudrock facies association**

Sediments of the MFA generally are massive (Fig. 6a), with minor lamination (Fig. 6c). The coaly mudstone lithofacies is interpreted as overbank fines that settled from suspension on the floodplain. Massive, fine-grained sediments have long been interpreted as suspension deposits resulting from waning currents (Collinson & Thompson, 1989). Abundant intraformational fragments, commonly elongate parallel to lamination, suggest current activity. The silty mudstone clasts in the coaly mudstone lithofacies resemble the diamictite texture in some Karoo sequences in South America (N. Eyles, personal communication, 1992). If this is true for the Gwembe

Coal Formation, it indicates that short-lived periods of glaciation still occurred from time to time in the mid-Zambezi Valley. Another possible interpretation is that the clasts were blown in by wind. The carbonaceous mudstone lithofacies also are thought to represent overbank fines deposited from suspension during floods on the floodplain and in swamps. The normal grading observed in some beds suggests deposition by waning flow during falling flood stage. Laminae bent around small grains suggest that these are drop-grains blown into the area by wind.

The silty mudstone lithofacies is interpreted as mainly overbank fines deposited in levees and in upward-fining units in channels. The mud-clast breccias and conglomerates in the upward-fining cycles suggest reworking of the mudstones during flood events, and deposition by waning currents. The coarser varieties of the lithofacies (siltstone) are interpreted as distal crevasse splays and channel deposits. Rootlets in some beds indicate subaerial exposure and that plants flourished on the floodplain.

The sideritic mudstone, siltstone and sandstone lithofacies indicate siderite cementation beneath a high water table during diagenesis. The concretions are characteristic of waterlogged soils that are neutral to alkaline, occurring where the surface of the floodplain is continuously saturated (Retallack, 1988). This study suggests that where extensive areas of the floodplain were saturated, the siderite developed a tabular form and where the floodplain was locally saturated, it developed a concretionary form. Postma (1983) confirmed that siderite is dominant in freshwater sediments with pore waters containing little chloride or dissolved sulphide, and that are supersaturated with ferrous iron. Although the root traces suggest subaerial exposure, grey colour and high organic content indicate accumulation in a reducing environment. The occurrence of siderite is consistent with an alluvial floodplain origin for the Gwembe Coal Formation. *Microcodium* in the sideritic lithofacies was long thought to be algal, inorganic bacterial or actinomycete in origin, but is now considered to represent calcified mycorhizae, a symbiotic association between soil fungi and cortical cells of roots of higher plants (Klappa, 1978). The presence of *Microcodium* further supports the floodplain origin for the Gwembe Formation.

The 'other sandstone' lithofacies are interpreted as crevasse channel and splay deposits resulting from avulsion of the main channel on the floodplain. Money & Drysdall (1975) indicated that the succeeding Gwembe sandstones (B, C, D and E) marked the periodic re-establishment of a complex river system. The localized occurrence and thinness (< 10 m thick) of these lithofacies, however, suggest deposition in crevasse channels and splays during episodic flooding.

The mudrocks of the mid-Zambezi Valley are interpreted as overbank fines that accreted vertically from suspension during episodic flooding of the floodplain. The coarser, horizontally laminated to small-scale cross-laminated sediments (silty mudstone, siltstone and very fine-grained sandstone) are interpreted as distal deposits of crevasse channels, splays and levees. Silt-dominated overbank fines were likely deposited in splays and channels, with peat accumulating in the swamps to form coaly and carbonaceous mudstones, depending on the ratio of mud to organic matter. Away from the swamps, non-carbonaceous silt accumulated to form the silty mudstones. Pedogenesis, particularly in the mudrock assemblage, is expressed by the occurrence of root tubules in the sideritic concretionary facies and by mottling of the mudrock. Mottling in the massive mudstone, particularly in the silty mudstone lithofacies, is interpreted as resulting from seasonally oscillating water tables, and may have developed during pedogenesis either as a result of fluctuating Eh–pH conditions or through redistribution of iron oxide/hydroxide particles (Buurman, 1980; Platt & Keller, 1992).

The colour of sediments is generally considered to be an early diagenetic phenomenon controlled by geochemical conditions (Eh) and biological activity (Downing & Squirrell, 1965; Walker, 1967; Thompson, 1970; McBride, 1974; McPherson, 1980; Myrow, 1990). Redox reactions result in green-grey to red coloration. As the $Fe^{+3}/Fe^{+2}$ ratio in a rock decreases, the resulting colours can range from red through purple to grey (McBride, 1974). Yellow coloration is imparted by limonite, brown results from goethite, and red and reddish mottling from hematite (ferric *c.* $Fe_2O_3$) (Myrow, 1990). Yellowish brown/brown colours are thus alteration products. The laterally persistent brick red (blood-red) siderite beds indicate that the mudrock assemblage was deposited on a broad alluvial plain. In summary, crevasse channel and splay deposits, swamp deposits and high-sinuosity meandering channels and the cyclic nature of lithofacies indicate that the Gwembe Coal Formation was deposited on an alluvial floodplain.

## Sandstone A facies association

The SAFA (Figs 3, 8, 9 & 10) marked a re-establishment of more active fluvial conditions on the floodplain. The microconglomerate to very coarse sandstone lithofacies of Sandstone A suggests deposition in a high-energy environment. Reverse grading indicates progradation of channels during floods and normal grading represents settling of grains during waning current flow. Large trough cross-beds are interpreted as dune bedforms of the lower

flow regime, whereas planar cross-beds likely resulted from migration of linguoid and transverse bars (Miall, 1977; Bluck, 1979). Palaeocurrent plots for the SAFA in the Siankondobo area indicate sediment transport mainly from the north-west (Fig. 12). One station (six readings) in the north-eastern part of the map (Fig. 12) for Sandstone A showed a source area to the north-east and can be explained as belonging to a separate minor Izuma basin (Izuma open pit area) that is separated from the Kazinze open pit area by a major fault. The opposing asymmetry in these basins led to opposing sediment source areas.

Thicker Sandstone A units in the north-western part of the Siankondobo area and the northern part of the Maze–Sinakumbe area (Fig. 7B) reflect active subsidence with a lateral transition from dominantly low-sinuosity channel deposits to dominantly high-sinuosity channel deposits. Money & Drysdall (1975) interpreted Sandstone A as a lateral accretion deposit of a meandering channel on the floodplain. The very coarse-grained nature, the large-scale cross-bedding (trough and planar) and the paucity of mudrocks in the microconglomerate to very coarse sandstone lithofacies of Sandstone A, however, indicate that the sandstone probably was deposited in a proximal braided stream system at the margin of the basin. Rust (1978) pointed out that gravel in braided systems in distinguished from that in meandering systems by the dominance of framework-support and that sand in braided systems is distinguished by the presence of suspension-deposited intraclasts as opposed to bedload. The framework-supported grains of the microconglomerate lithofacies and coaly intraclasts (likely to have been moved in suspension) in the lithofacies suggest a braided influence. Presence of downstream accretion surfaces, common in braided streams, further supports the braided-stream interpretation. These streams probably fed a high-sinuosity river running parallel to the basin axis, which produced lateral accretion bodies of the meandering type. B. Rust (personal communication, 1990) compared Sandstone A to the Triassic Hawkesbury Sandstone of Australia, which is interpreted as a braided river system deposit.

## GWEMBE COAL FORMATION AS A FLOODPLAIN DEPOSITIONAL SYSTEM

At the end of deposition of the Siankondobo Sandstone Formation, during early post-glacial conditions, a broad floodplain was established in the mid-Zambezi Valley on which a high-sinuosity meandering river system appears to have developed (Fig. 13). Deposits of the river channel system constitute the Maamba Sandstone facies association at the base of the Gwembe Coal Formation and the

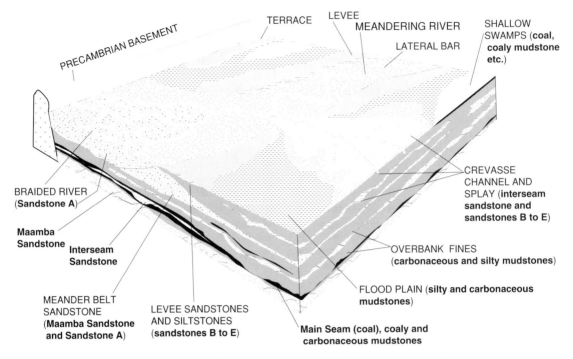

**Fig. 13.** Depositional model of the Gwembe Coal Formation, mid-Zambezi Valley Basin, southern Zambia.

overlying lithofacies are deposits in various subenvironments on the floodplain. The water level in freely meandering river channels is commonly above that of the surrounding floodplain, confined there (except in flood) by its levees (Walker & Cant, 1984). The presence of such a high water table also would explain the formation of siderite (see lithofacies interpretation). Therefore, a high water table would have been present across the floodplain, where swamps were established. Peat accumulated over a broad expanse, where subsidence or flooding did not drown the vegetation, leading to accumulation of the coal lithofacies. The swamps received sand incursions as crevasse channels and splays (Interseam Sandstone), which resulted in lateral splitting of the coal seams and introduced clay into parts of the swamp farther from the channel. The mudrock assemblage was deposited mainly as overbank fines during floods.

Tectonic uplift of the source area provided the clastic sediment for the braided systems, resulting in deposition of Sandstone A and a subsequent change to freely meandering river channels. In braided channels, the water level generally is below that of adjacent areas, and therefore peat accumulation is unlikely to be thick or widespread, because vegetated areas are not extensive and the water table is too low most of the time (Rust *et al.*, 1987).

Therefore the coal seams above Sandstone A are thin and of poor quality. Lateral movement of the active channel system would have eroded the peat by undermining the clastic sediment beneath it. On proximal braidplains, represented by the pebbly components of Sandstone A, the coal seams are remarkably thin, and coal and coaly fragments in the pebbly sandstone indicate reworking of the coal. The change from a braided system to a meandering system meant that active channels became, by definition, significantly lower than the floodplain. The lower water table would have resulted in effects such as the abundant red/brown coloration of the mudstone and sandstone above Sandstone A in the upper part of the Gwembe Coal Formation. The colour changes and mottling indicate subaerial exposure coupled with strong bioturbation. Hubert & Reed (1978) observed that red hematitic coloration is imparted to sediments under oxidizing conditions by the early post-depositional alteration of iron-bearing minerals. This would explain the predominant red coloration in the formation.

Kaolinite is the most abundant clay mineral in the coal of the Gwembe Coal Formation. Previous studies have shown that kaolinite is abundant in bituminous coals (O'Gorman & Walker, 1971) and its proposed origin may include:

**1** incorporation by normal sedimentary processes such as overbank flooding of the swamp, and reworking of sedimentary material within the swamp (Mackowsky, 1968; Spears, 1987);

**2** diagenetic alteration of pre-existing aluminosilicates within the peat (Davis *et al.*, 1984), possibly by including amorphous, plant-associated inorganic material (Renton & Cecil, 1979);

**3** alteration of volcanic ash-fall material, usually present in discrete layers known as tonsteins (Price & Duff, 1969; Spears & Kanaris-Sotiriou, 1979; Dewison, 1989).

Price & Duff (1969) indicated that diagenetic recrystallization of aluminosilicate material to form kaolinite is known to occur to a greater extent in coal than in associated mudstones owing to more suitable conditions (greater acidity and lower $K^+$ activity) for kaolinite formation in the coal-forming environment (Garrels & Christ, 1965). In the mid-Zambezi Valley Basin, the kaolinite is more likely to have originated as a result of processes (1) and (2) because of the lack of tonsteins and the commonly associated mineral, anatase. Kaolinite shows better developed crystallinity, indicated by its uniform XRD peaks in the coals, than in the Interseam Sandstone and mudstone, where other clays are present. Patterson & Murray (1984) noted a similar pattern, and concluded that kaolinite formed in the presence of lesser amounts of other clay minerals tends to develop greater crystallinity.

The Gwembe Coal Formation (Fig. 3) shows cyclicity, mainly as alternating silty mudstone and carbonaceous/sideritic/coaly mudstone or coal. Several mechanisms for control of cyclicity have been proposed (Beerbower, 1964; Miall, 1980). Murchison & Westoll (1968), divided these into three groups.

**1** Periodic relative movement of land and sea-levels such as:

(a) repeated subsidence of the area of sedimentation followed by sedimentary accumulation to or above sea-level, either simply, or with oscillations;

(b) eustatic movements controlled from outside the basin by glaciation (glacio-eustatic change) or as a result of changes in the ocean basins owing to increases or decreases in volume of oceanic spreading ridges (tectono-eustatic change);

(c) repeated uplift and erosion of the continental hinterland.

**2** Repeated climatic changes (e.g. glaciation–deglaciation).

**3** Cyclic effects from steady change, such as:

(a) isostatic adjustments;

(b) adjustment to tectonic stress;

(c) trapping and releasing of sediments from heavily vegetated swamps;

(d) effects of compaction of sediments;

(e) effects related to the continuous action of sedimentary processes in rivers and deltas.

The operation of these mechanisms is a subject of current debate and research, but generally it is accepted that sedimentary cycles are related to eustatic changes of sea-level (i.e. as a result of glacio- and tectono-eustatic changes), local tectonics and rates of sedimentation (Plint *et al.*, 1992). Allen (1964) attributed cyclicity in the Lower Old Red Sandstone of the Anglo-Welsh Basin to three mechanisms:

**1** autocyclic, in which each river wandered across a portion of the floodplain under conditions of steady subsidence and sediment supply;

**2** autocyclic, influenced by allocyclic sea-level changes in which rivers in unison alternately eroded and aggraded the floodplain under steady subsidence and sediment supply conditions;

**3** a combination of allocyclic and autocyclic in which bursts of tectonic activity in the source area led to increased sediment supply through rejuvenated streams that eroded and aggraded the floodplain with no variation in the base level and rate of subsidence of the basin.

Tavener-Smith (1960) suggested that the Gwembe coal-bearing sequence was of lacustrine origin. Money & Drysdall (1975) argued for a floodplain origin for the Gwembe Coal Formation. As an analogue for the mid-Zambezi Valley coal deposits, they used the present-day Niger delta in which peat accumulates in linear back-swamps. The comparable Gwembe coal swamp was envisaged by Money & Drysdall (1975) as a south-westerly sloping environment, vegetated probably by small plants and shrubs, with few trees, comparable to present-day vegetation of the Upper Zambezi floodplain between Mongu and Senanga, western Zambia. Semi-permanent to permanent water in places (e.g. ox-bow lakes) meant that peat accumulated in both reducing (anaerobic) and oxidizing (aerobic) environments. Peat accumulation and development was controlled by palaeorelief and size of the floodplain. Distribution of the ash content was controlled by palaeogeography, in particular position of the main river channel, such that away from active streams thicker coal developed. Money & Drysdall (1975) suggested that an abundance of inertinite in mid-Zambezi Valley coal indicates a floodplain origin, which allowed strong aerobic decomposition of plant residues at the peat surface as a result of subaerial exposure; in contrast, the anaerobic conditions prevalent in lakes would result in the deposition of syngenetic sulphide, resulting in unsystematic distribution of sulphur values in the Main Seem.

Differences in the composition and character of the coal seams of Northern and Southern hemispheres (e.g.

absence of tree trunks in Gondwana coals) are attributed to generic conditions by several researchers arguing for *in situ* (autochthonous) origin for the northern coals and drift origin (allochthonous) for the southern coals (Lamplugh, 1907; Steart, 1919; Lightfoot, 1929). Recent studies (Watson, 1958; Plumstead, 1966; Money & Drysdall, 1975) argue for an *in situ* origin of the southern coal based on extent and thickness (up to 15 m in places). The present study has documented rootlets in the sheet sandstones below the B Seam in the Kazinze open pit (Fig. 5B), which supports *in situ* origin of the Main Seam Coal. The differences can be explained by the type of vegetation, which in turn would have been controlled by climate. Plumstead (1966) indicated that Northern Hemisphere coals were mainly from evergreen vegetation in nearly stagnant waters of extensive coastal swamps under humid tropical conditions, whereas Gondwana coals were from decidous plants that grew in floodplain and lacustrine swamps, in a comparatively cool seasonal climate during and following retreat of the Dwyka ice age. This conclusion was echoed by Money & Drysdall (1975), who indicated that the base of the Main Seam is well defined, and the top is interbedded with coaly and carbonaceous mudstone, suggesting that the greater basal part of the Main Seam is of *in situ* origin, but that redistribution of plant material within the floodplain resulted in an allochthonous origin for the upper part. The sedimentology, palaeogeography and maceral characteristics of the Main Seam all suggest an *in situ* origin (Money & Drysdall, 1975; this study).

In summary, the lower and upper members in the Gwembe Coal Formation represent two upward-fining cycles. Most likely both allocyclic and autocyclic processes were involved in their formation. Bursts of tectonic activity in the source area would have led to increased sediment supply through rejuvenated streams that eroded and aggraded the floodplain with no variation in the base level and rate of subsidence of the basin. The first cycle, which starts with the Maamba Sandstone facies association, began with a probable tectonic pulse that resulted in increased sediment supply through rejuvenated streams or as a result of isostatic rebound following glacial retreat. A tectonic pulse probably started the second cycle, i.e. that produced in Sandstone A. For the most part, however, the lower cycle consists of alternating units of silty mudstone and carbonaceous/coaly mudstone that progressively become silty towards the top, and these are best explained by periodic flooding. The second cycle is dominated by thick silty mudstone (slightly carbonaceous) with intercalations of carbonaceous/coaly mudstones and upper coal seams alternating with sandstones A–E. These features can be explained best by (i) fluvial —

high-sinuosity migration and avulsion — and (ii) floodplain flooding and vertical accretion. These processes are autocyclic (Allen, 1982) and not influenced by external forces such as changes in sea-level or tectonism (Allen, 1964; Beerbower, 1964).

## CONCLUSIONS

Fourteen lithofacies of the Gwembe Coal Formation can be grouped into four facies associations: Maamba Sandstone, Coal, Mudrock and Sandstone A. The Maamba Sandstone facies association is interpreted as meandering stream deposits. The Coal facies association accumulated in swamps on the floodplain that were interrupted periodically by crevasse channels and splays that deposited the Interseam Sandstone. The Mudrock facies association consists of overbank fines, whereas the 'lithofacies of the other sandstones (B, C, D and E,)' are attributed to crevasse channels and splays, and siderite to diagenetic precipitation. The Sandstone A facies association (FAFA) represents proximal braided stream deposits that grade distally into meandering-river-system deposits. The characteristics of the lithofacies and facies associations support the deposition of the Gwembe Coal Formation on the floodplain.

## ACKNOWLEDGEMENTS

This work was part of a PhD project conducted under the guidance of the late Professor B.R. Rust (University of Ottawa) and funded by his NSERC grant. The author is thankful to the late Professor Rust for his contribution in the supervision, financial support and encouragement during the field work of the project. The author also wishes to thank O. Dixon and A. Donaldson for taking over the guidance and financial support of the project. O. Dixon is thanked for tireless encouragement and suggestions that shaped the manuscript. Anayawa Nyambe is thanked for typing the manuscript. I also would like to thank the two reviewers P. Eriksson and M. Wizevich for the critical comments that led to improvement of the manuscript. N.D. Smith is thanked for providing final suggestions that further improved the manuscript.

## REFERENCES

ALLEN, J.R.L. (1964) Studies in fluviatile sedimentation: six cyclothems from the Lower Old Red Sandstone, Anglo-Welsh basin. *Sedimentology*, **3**, 163–198.

ALLEN, P.A. (1982) Cyclicity of Devonian fluvial sedimentation,

Cunningsburgh Peninsula, SE Shetland. *J. geol. Soc. London*, **139**, 49–58.

BALTZER, F. (1991) Late Pleistocene and Recent detrital sedimentation in the deep parts of northern Lake Tanganyika (East African Rift). In: *Lacustrine Facies Analysis* (Eds Anadón, P., Cabrera, L. & Kelts, K.), Spec. Publs int. Ass. Sediments., No. 13, pp. 147–173. Blackwell Scientific Publications, Oxford.

BEERBOWER, R.J. (1964) Cyclothems and cyclic depositional mechanism in alluvial plain sedimentation. In: *Symposium on Cyclic Sedimentation* (Ed. Merriam, D.F.). *Kans. State geol. Surv. Bull.*, **169**, 31–42.

BERNER, R.A. (1964) Stability fields of iron minerals in anaerobic marine sediments. *J. Geol.*, **72**, 826–834.

BERNER, R.A. (1971) *Principles of Chemical Sedimentology*. McGraw-Hill, New York, 240 pp.

BLUCK, B.J. (1979) Structure of coarse grained stream alluvium. *Trans. R. Soc. Edinburgh*, **70**, 181–221.

BUURMAN, P. (1980) Palaeosols in the Reading Beds (Paleocene) of Alum Bay, Isle of Wight, U.K. *Sedimentology*, **27**, 593–606.

COLLINSON, J.D. & THOMPSON, D.B. (1989) *Sedimentary Structures*, 2nd edn. Allen and Unwin, London, 207 pp.

DAVIS, A., RUSSEL, S.J., RIMMER, S.M. & YAEKEL, J.D. (1984) Some genetic implications of silica and alumino-silicates in coal. *Int. J. Coal Geol.*, **3**, 293–314.

DEWISON, G.M. (1989) Dispersed kaolinite in the Barnsley Seam coal (U.K.): evidence for a volcanic origin. *Int. J. Coal Geol.*, **11**, 291–304.

DOWNING, R.A. & SQUIRRELL, H.C. (1965) On the red and green beds in the Upper Coal Measures of the eastern part of the South Wales Coalfield. *Bull. geol. Surv. G.B.*, **23**, 45–56.

FALCON, R.M.S. (1989) Macro- and micro-factors affecting coal-seam quality and distribution in southern Africa with particular reference to the No. 2 seam, Witbank coalfield, South Africa. *Int. J. Coal Geol.*, **12**, 681–731.

GARRELS, R.M. & CHRIST, C.L. (1965) *Solutions, Minerals and Equilibria*. Harper and Row, New York, 450 pp.

GEOLOGICAL SOCIETY OF AMERICA (1980) *Rock Color Chart*. Geological Society of America, Boulder, CO, 10 pp.

HAGELSKAMP, H.H.B. & SNYMAN, C.P. (1988) On the origin of low-reflecting inertinites in coals from the Highveld Coalfield, South Africa. *Fuel*, **67**, 307–314.

HAGELSKAMP, H.H.B., ERICKSSON, P.G. & SNYMAN, C.P. (1988) The effect of depositional environment on coal distribution and quality parameters in a portion of the Highveld Coalfield, South Africa. *Int. J. Coal Geol.*, **10**, 51–78.

HUBERT, J.F. & REED, A.A. (1978) Red-bed diagenesis in the East Berlin Formation, Newark Group, Connecticut Valley. *J. sediment. Petrol.*, **48**, 175–184.

KELTS, K. & HSÜ, K.J. (1978) Freshwater carbonate sedimentation. In: *Lakes – Chemistry, Geology, and Physics* (Ed. Lerman, A.), pp. 295–323. New York, Springer-Verlag.

KLAPPA, C.F. (1978) Biolithogenesis of *Microcodium*; elucidation. *Sedimentology*, **25**, 489–522.

LAMPLUGH, G.W. (1907) The geology of the Zambesi Basin around the Batoka Gorge (Rhodesia). *Q.J. geol. Soc. London*, **63**, 162–216.

LIGHTFOOT, B. (1929) The geology of the central part of the Wankie Coalfield. *Geol. Surv. S. Rhodesia Bull.*, **15**, 62.

MACKOWSKY, M.-TH. (1968) Mineral matter in coal. In: *Coal and Coal-bearing Strata* (Murchison, D.G. & Westoll, T.S.), pp. 105–123. Oliver and Boyd, Edinburgh.

McBRIDE, E.F. 1974. Significance of color in red, green purple, olive, brown and gray beds of Difunta Group, northeastern Mexico. *J. sediment. Petrol.*, **44**, 760–773.

McPHERSON, J.G. (1980) Genesis of variegated redbeds in the fluvial Aztec Siltstone (Late Devonian), Southern Victoria Land, Antarctica. *Sedimentary Geol.*, **27**, 119–142.

MIALL, A.D. (1977) A review of the braided-river depositional environment. *Earth Sci. Rev.*, **13**, 1–62.

MIALL, A.D. (1980) Cyclicity and the facies model concept in fluvial deposits. *Bull. Can. Petrol. Geol.*, **28**, 59–80.

MONEY, N.J. & DRYSDALL, A.R. (1975) The geology, classification, paleogeography and origin of the mid-Zambezi coal deposits of Zambia. *International Union of Geological Sciences Commission on Stratigraphy, Subcommittee Gondwana Stratigraphy, Palaeontology, Gondwana Symposium, Proceedings Paper No. 3 (Gondwana Geology)*, pp. 249–270.

MONEY, N.J., DENMAN, P.D. & RADOSEVIC, B. (1968) Sedimentology of the Lower Karoo rocks of the Siankondobo and Mulungwa areas of the mid-Zambezi Valley. *Zambia geol. Surv. Rec.*, **11**, 17–27.

MONEY, N.J., DENMAN, P.D. & DRYSDALL, A.R. (1974) *Geology of the Coalfields of the Mid-Zambezi Valley, Zambia*. Unpublished Bulletin 6, Geological Survey of Zambia, Lusaka, 175 pp.

MURCHISON, D. & WESTOLL, T.S. (1968) *Coal and Coal-Bearing Strata*. Oliver and Boyd, Edinburgh, 379 pp.

MYROW, M.P. (1990) A new graph for understanding colors of mudrocks and shales. *J. Geol. Educ.*, **38**, 16.

NYAMBE, I.A. (1993) Sedimentology, tectonic framework and economic potential of the Sinakumbe Group (?Ordovician to Devonian) and Karoo Supergroup (Permo-Carboniferous to Lower Jurassic) in the mid-Zambezi Valley Basin, southern Zambia. Unpublished PhD thesis, University of Ottawa, 425 pp.

O'GORMAN, J.V. & WALKER, P.L., JR. (1971) Mineral matter characteristics of some American coals. *Fuel*, **50**, 135–151.

PATTERSON, S.H. & MURRAY, H.H. (1984) Kaolin, refractory clay, ball clay and halloysite in North America, Hawaii and Caribbean region. *U.S. geol. Surv. Prof. Pap.*, **1306**, 1–56.

PLATT, N.H. & KELLER, B. (1992) Distal alluvial deposits in a foreland basin setting: the Lower Freshwater Molasse (Lower Miocene), Switzerland: sedimentology, architecture and paleosols. *Sedimentology*, **39**, 545–565.

PLINT, A.G., EYLES, H., EYLES, H.C. & WALKER, R.G. (1992) Control of sea level change. In: *Facies Models* (Eds Walker, R.G. & James, N.P.), pp. 15–25. Geological Association of Canada, St John's, Newfoundland.

PLUMSTEAD, E.P. (1966) The story of South Africa's coal. *Optima*, **16** (12), 387–402.

POSTMA, D. (1983) Pyrite and siderite oxidation in swamp sediments. *J. Soil Sci.*, **34**, 163–182.

POTTER, P.E. & PETTIJOHN, F.J. (1963) *Palaeocurrents and Basin Analysis*. Springer-Verlag, New York, 296 pp.

PRICE, N.B. & DUFF, P. McL. D. (1969) Mineralogy and chemistry of tonsteins from Carboniferous sequences in Great Britain. *Sedimentology*, **13**, 45–69.

RADOSEVIC, B., MONEY, N.J. & DENMAN, P.D. (1968) Petrography of the Lower Karoo sandstones in the Siankondobo area of the mid-Zambezi Valley. *Zambia geol. Surv. Rec.*, **11**, 9–15.

RENTON, J.J. & CECIL, C.B. (1979) The origins of mineral matter in coal. In: *Carboniferous Coal Guidebook B-37-1* (Eds Donaldson, A., Presley, M.W. & Renton, J.J.), pp. 206–233. Morgantown, W. Virginia Geological Survey.

RETALLACK, G.J. (1988) Field recognition of paleosols. In: *Paleosols and Weathering Through Geologic Time: Principles and Applications* (Eds Reinhardt, J. & Siegles, W.R.), *Geol. Soc. Am. Spec. Pap.*, **216**, 1–20.

RUST, B.R. (1978) Depositional models for braided alluvium. In: *Fluvial Sedimentology* (Ed. Miall, A.D.), Mem. Can. Soc. petrol. Geol., Calgary, **5**, 605–625.

RUST, B.R., GIBLING, M.R., BEST, M.A., DILLES, S.J. & MASSON, A.G. (1987) A sedimentological overview of the coal-bearing Morien Group (Pennsylvanian), Sydney Basin, Nova Scotia, Canada. *Can. J. Earth Sci.*, **24**, 1869–1885.

SPEARS, D.A. (1987) Mineral matter in coals, with special reference to the Pennie Coalfields. In: *Geology Society Symposium on Coal and Coal-Bearing Strata* (Ed. Scott, A.C.), Egham, Surrey, 1986, pp. 171–185.

SPEARS, D.A. & KANARIS-SOTIRIOU, R. (1979) A geochemical and mineralogical investigation of some British and other European tonsteins. *Sedimentology*, **26**, 407–425.

STEART, F.A. (1919) Some notes on the geology of the north-western portion of the Natal coalfield. *Trans. geol. Soc. S. Afr.*, **22**, 90–111.

TAVENER-SMITH, R. (1960) The Karroo System and coal resources of the Gwembe District, south-west section. *N. Rhodesia Geol. Surv. Bull.*, **4**, 1–84.

THOMPSON, A.M. (1970) Geochemistry of color genesis in red-bed sequence, Juniata and Bald Eagle Formations, Pennsylvania. *J. sediment. Petrol.*, **40**, 599–615.

WALKER, R.G. & CANT, D.J. (1984) Sandy fluvial systems. In: *Facies Models*, 2nd edn (Ed. Walker, R.G.), pp. 71–89. Geoscience Canada Reprint Series No. 1, Geological Society of Canada, Waterloo, Ontario.

WALKER, T.R. (1967) Formation of red beds in modern and ancient deserts. *Geol. Soc. Am. Bull.*, **78**, 353–368.

WATSON, R.L.A. (1958) The origin of Wankie Coal. *Trans. Geol. Soc. S. Afr.*, **61**, 167–181.

*Spec. Publs int. Ass. Sediment.* (1999) **28**, 435–448

# Sedimentology of the Section Peak Formation (Jurassic), northern Victoria Land, Antarctica

R. CASNEDI *and* A. DI GIULIO

*Dipartimento di Scienze della Terra, Via Ferrata 1, 27 100 Pavia, Italy*
*(Email: rcasnedi@unipv.it and digiulio@unipv.it)*

## ABSTRACT

Models of Gondwana reconstruction agree in linking South Africa and Antarctica in Devonian to Triassic times, with an overall slow accumulation of shallow-water and alluvial deposits on a craton. In the Transantarctic Mountains, this sedimentary context, with the Gondwanide uplift, is well represented up to the Triassic by the rocks of the Beacon Supergroup; later similar alluvial deposition is recorded in northern Victoria Land (Priestley and Campbell Valleys) by the 'Beacon-like' sandstones of the Section Peak Formation, which interfingers with and is capped by basalt flows of the Ferrar Group. In this study, sedimentological observations of these sandstones, previously ascribed to the upper part of the Beacon Supergroup, are described.

The Section Peak Formation shows variously stratified alluvial facies associations with aeolian trough cross-bedding. Several sections have been measured and studied, leading to interpretations of the vertical evolution of the deposits, the lateral variations of the sedimentary environments and the provenance framework of clastic sediments and their palaeogeographical interpretation.

A sandy braided-stream environment is envisaged, composed of channels with dunes and foreset macroforms associated with both slip-faces and lee sides with descending dunes and rare plane beds. The abundance of thin sandstone beds with aeolian sedimentary structures indicates common aeolian reworking. Thicker sandstone bodies can be interpreted as the deposits of relatively persistent fluvial channels.

An aeolian environment, different from that described above, has been observed locally in the upper part of the area studied; large-scale, cross-bedded, fine-grained sandstones show foresets close to the angle of repose.

The palaeogeographical evolution, from Triassic to Middle Jurassic, indicates a low-gradient slope with meandering channels, grading into a more complex river system with braided streams (Section Peak Formation). Sandstone petrology proves that it drained mostly high-grade granulite-facies metamorphic and plutonic rocks of the basement (Wilson Terrane), but also intermittently reworked volcanic debris. These were eroded from coeval basaltic lava flows, erupted in the Section Peak basin during the initiation of lithospheric stretching, which preceded the Gondwana break-up.

## INTRODUCTION AND GEOLOGICAL OUTLINE

Studies of the rocks of the Section Peak Formation by Gair *et al.* (1965) and Collinson *et al.* (1986), in the Priestley and Rennick Nevé area, are here extended to the Terra Nova Bay area, where the formation, along with the Kirkpatrick Basalt cover, forms cliff faces in the Prince Albert Mountains and Deep Freeze Range. Previously ascribed to the Late Triassic–Early Jurassic because of the flora content (Tessensohn & Madler, 1987), we suggest a younger age (Middle Jurassic), at least for the upper part of the formation, on the basis of relationships with dated Ferrar volcanic rocks.

The flat-lying Section Peak Formation rests on basement rocks deformed during the Cambro-Ordovician Ross Orogeny. Three terranes form this basement in northern Victoria Land (Fig. 1): the Wilson, the Bowers and the Robertson Bay terranes. The Section Peak rocks rest on the Priestley Formation (Cambrian) and the Granite Harbour Intrusives (Ordovician; Carmignani *et al.*, 1988), which form the youngest rocks of the Wilson Terrane.

The unconformity between rocks of the Section Peak Formation and the basement is one of the most

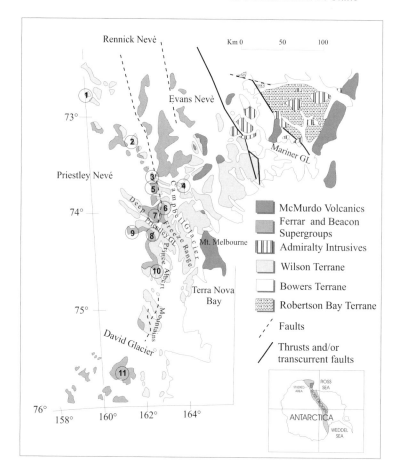

**Fig. 1.** Location map of the main outcrops of the Section Peak Formation with the numbers related to the section descriptions. Geological map after Carmignani *et al.* (1988). 1, Roberts Butte; 2, Section Peak; 3, Vantage Hills; 4, Vulcan Hills; 5, Archambault Ridge; 6, Mount Adamson; 7, Shafer Peak; 8, Timber Peak–Mount New Zealand; 9, Skinner Ridge; 10, Mount Nansen; 11, Mount Bowen.

spectacular features in the Terra Nova Bay area (Fig. 2), and is represented by an almost flat erosion surface. The Sub-Beacon erosion surface occurs throughout the Transantarctic Mountains and was first named the Kukri Peneplain by Gunn & Warren (1962). The erosion surface is post-Ordovician, being pre-Devonian in southern Victoria Land (Barrett, 1991), whereas in northern Victoria Land the erosion surface could have been exposed as recently as the Late Palaeozoic (Maya erosion surface).

In northern Victoria Land, rocks of the upper portion of the Beacon Supergroup belong to the Victoria Group (see Barrett (1991) for a general review), which includes the Section Peak Formation and the Takrouna Formation (Walker, 1983; Collinson *et al.*, 1986). The Permian Takrouna Formation occurs north of the study area and rests on the Wilson Terrane and partly on the Bowers Terrane.

The present work deals with the Section Peak Formation; its age (previously given as Triassic) indicates

as possible correlatives the upper Lashly Formation in southern Victoria Land and the Falla Formation in the Beardmore Glacier area of the Central Transantarctic Mountains (Barrett *et al.*, 1986). Elliot *et al.* (1986a) and Elliot (1996), on the basis of field data collected in the Mesa Range and in the Beardmore Glacier, proposed changes in nomenclature by introducing the Exposure Hill Formation and the Hanson Formation, which could roughly represent in part a correlative of the sections described in the present work.

According to K.J. Woolfe (personal communication), the youngest (Middle Jurassic) sandstones studied by us could be related directly to the Ferrar Group, in a 'Beacon-like' facies still unknown in the Ferrar, owing to the location of the base of the Ferrar Group at the beginning of the basic flows. The studied sections therefore could not belong to the Beacon Supergroup, as believed previously (Collinson *et al.*, 1986, Casnedi *et al.*, 1994, Di Giulio *et al.*, 1997a,b), but at least for the upper part,

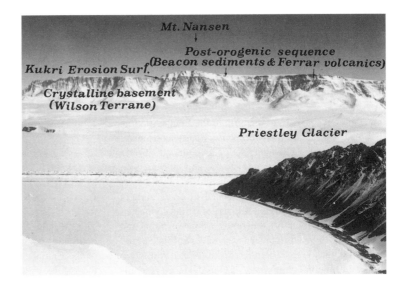

**Fig. 2.** General view from the lower Priestley Glacier of Mount Nansen (2737 m; Prince Albert Mountains) with the angular unconformity (locally non-conformity) between the deformed basement (Granite Harbour Intrusives) and the Section Peak Formation. It is interlayered with and covered by the Ferrar volcanics at the top of the range with the ice-cap. The sub-Section Peak unconformity surface, possibly corresponding to the Kukry erosion surface, is highly weathered and dips at 2–5° westwards.

including the lava flows, to the Ferrar Group. This problem is still open, as in the Beardmore area (Elliot, 1996).

The Section Peak Formation consists of sandstones of various grain sizes, mostly deposited in a braided-stream environment. An important feature of the formation is the common interbedding of basic igneous rocks. They generally are interpreted as doleritic intrusions, clearly younger than the sedimentary country rocks. New data, derived from the petrographic analyses of sandstones lying on some of these igneous layers, have led us to interpret them as basaltic extrusions, and consequently to review the relationships between sandstone deposition and Ferrar volcanism. This interpretation is based mainly on:

**1** the absence of thermal metamorphism above the igneous layers;

**2** the distribution of basic volcanic clasts in the sandstones.

The deposition of the Section Peak Formation and the volcanism of the Ferrar are strictly connected with the Jurassic tholeiitic magmatism at about 175–180 Ma (Elliot *et al.*, 1986b; Elliot, 1992, 1996; Heimann *et al.*, 1994), comprising pyroclastic rocks and basalts, which temporally coincided with the fragmentation of Gondwana. The structural environment in which this magmatism took place is still not well defined, because different geochemical characteristics have been reported over a large area of Antarctica (Transantarctic Mountains and Weddell Province). Most explanations invoke rifting and extension related to Gondwana fragmentation.

## MAIN PHASES OF GONDWANA BREAK-UP, RELATED TO UPPERMOST BEACON SEDIMENTATION AND FERRAR MAGMATISM

The separation of the Gondwana plates possibly began in the latest Precambrian or earliest Palaeozoic. The Pan-African event was responsible for crustal weakening along a line later followed by rifting that separated East and West Gondwanaland.

As far as the initial relative motion of the plates was concerned, some reconstructions report a continuity of the edge of the shield between East and West Gondwanaland and the trend of the Early Palaeozoic fold belt, whereas in other interpretations the two features are offset by some 1000 km. In Antarctica, the intrusion of Late Ordovician to Devonian dolerites in the Shackleton Range, associated with basement thermal reactivation, gives evidence of probable Early Palaeozoic fragmentation.

The break-up took place in different phases; the first, between 200 and 160 Ma, was coeval with flows and intrusions of the Ferrar Supergroup. Separation of Antarctica from South Africa followed a line along the eastern flank of the Mozambique Ridge, formerly connected with Dronning Maud Land, Antarctica. The initial movement is believed to have occurred along a left-lateral shear zone, allowing Dronning Maud Land to shift in a north-easterly direction (Cox, 1992). A second stage of break-up may have begun as long ago as 160 Ma, with the separation of Madagascar and East Antarctica, probably as part of the same plate, from Africa.

Relative motions of the Antarctic, African and South American plates originated at a triple junction, in existence since the Early Palaeozoic, with a counter-clockwise rotation of Antarctica relative to Africa and of Africa to South America. This second phase of fragmentation, which caused the birth of the South Atlantic, began about 130 Ma, was preceded by intense volcanism.

The next stages of fragmentation are dated at 120 Ma (separation of Antarctica from India) and 80 Ma (from Australia). Basalts found in northern Victoria Land at Litell Rocks (120–90 Ma) are related by some authors (Elliott & Foland, 1986) to the phase of rifting connected with the separation of Antarctica from Australia and Tasmania. Graben formation and shoulder uplift occurred in the Rennick area (where the Section Peak Formation crops out). Although no direct-dating evidence is available, this phase of rifting may have coincided with the initial break-up, leading to the above-mentioned separation (Roland & Tessensohn, 1987). The final isolation of Antarctica (30 Ma) followed the opening of the Drake Passage.

The above-mentioned first phase of fragmentation (200–160 Ma) is proven by the voluminous tholeiitic basalts and related mafic intrusive rocks that are preserved along 3500 km of the Transantarctic Mountains. These rocks consist of thick flood flows of the Kirkpatrick Basalt (Kyle *et al.*, 1981; Brotzu *et al.*, 1988) and thick sills and dykes of the Ferrar Dolerite (Elliot, 1992). According to Schmidt & Rowley (1986) and Woolfe & Barrett (1995), this igneous activity was caused by a continental rift during the Jurassic. Beacon Supergroup sedimentation terminated during or immediately prior to the initiation of the igneous activity (Woolfe & Barrett, 1995). The Beacon Supergroup and correlated units are a Devonian–Jurassic, mostly alluvial, succession that covers a great part of Antarctica, with the best outcrops in Dronning Maud Land and along the Transantarctic Mountains (Barrett, 1991).

## DESCRIPTION OF SECTIONS WITH FIELD DATA, RADIOMETRIC DATA AND FOSSIL CONTENT

The Section Peak Formation is best exposed along both sides of the Priestley Glacier, on the western side of the Campbell Glacier and on scattered exposures in the Priestley–Rennick Nevè area (Figs 2 & 3). Before our survey, the area had been investigated mostly by Collinson *et al.* (1986) and their data have been utilized in addition to ours. The descriptions of the northern outcrops (from Roberts Butte to Vantage Hills) therefore are taken from Collinson *et al.* (1986).

**1** Roberts Butte: an 18-m-thick trough cross-bedded, medium-grained sandstone crops out at the top of an 800-m-high cliff. This sequence overlies an irregular surface of granite and is capped by a diabase sill (Collinson *et al.*, 1986).

**2** Section Peak (in the Lichen Hills): a 42-m-thick sandstone sequence is intercalated with two diabase sills. The lower part (25 m) is coarse-grained sandstone with large-scale trough cross-bedding with pebbles up to 3 cm

**Fig. 3.** View of the Section Peak Formation represented by the thin, horizontal, light strip of sandstone on the left-hand side of the photograph. It is interlayered between the Priestley Formation at the bottom and the Kirkpatrick Basalt at the top. The sandstone is cut sharply to the right by the ice-cap connected with the polar plateau. The basement is made up of the Priestley Formation (Cambrian black turbidites) largely intruded by granites (light colour on the right). Both are represented in the sandstone clasts but with a strong prevalence of granite-derived quartz and feldspar. In the foreground, glaciers flow to the western side of the Priestley Glacier; in the background is the Eisenhower Range with Mount Baxter (2430 m) on the right.

in diameter. Coalified fossil wood is abundant. The middle part, between the two sills, consists of 2 m of medium-grained sandstone. The upper part is a 15-m-thick medium- to fine-grained sandstone. This part is dominated by large-scale trough cross-beds with a few sets of planar cross-beds at the base in fine-grained sandstone. Blocks of black shale with coalified wood occur within the diabase. This section lies between granite at the base and diabase at the top (Collinson *et al.*, 1986).

This section was first reported by Gair *et al.* (1965) who assigned the basic rock at the top to a lava flow of the Kirkpatrick Basalt. Collinson *et al.* (1986) interpreted it as a diabase sill. Gair *et al.* (1965) reported microfloras of Late Triassic–Early Jurassic age, but Norris (1965) thought that they were slightly younger (Early Jurassic).

**3 Vantage Hills:** a 27-m-thick sandstone section overlies 300 m of basic igneous rock, which covers the granitic basement. Another basic layer caps the top of the sandstone. Coarse- to medium-grained sandstone shows large-scale trough cross-bedding. Planar tabular sets 1–1.5 m thick also occur. Four upward-fining cycles, 4–12 m thick, are separated by channels containing large sandstone and siltstone clasts.

**4 Vulcan Hills:** a sequence of sandstone, of about 80 m estimated thickness, crops out between two basic igneous layers. Only the middle part has been described and sampled. Coarse-grained sandstone shows high-angle cross-bedding, with channels filled with pebbly sandstone showing basal erosion surfaces. Thin beds of plant-bearing black mudstone are intercalated with light-coloured sandstone, some containing abundant fossil wood debris, and volcaniclastic rocks (Tessensohn & Madler, 1987). These authors ascribe a Rhaetian age to the flora.

**5 Archambault Ridge:** this is the thickest (80 m) and the most differentiated, as far as facies are concerned, of the sections measured. The lower part, 28 m thick, is made up of coarse- to medium-grained cross-bedded sandstone (Fig. 4) with some basic igneous interbeds. No thermal metamorphism is displayed by the covering sediments. The upper part, 52 m thick, is composed of finer grained sandstone. It shows large-scale cross-bedding (Fig. 5), with foresets close to the angle of repose. The section overlies granite and is capped by a basic igneous rock (probably the Kirkpatrick Basalt).

**6 Mount Adamson:** this section, 30 m thick, is very similar in lithology and facies to the lower part of the Archambault Ridge section. The igneous interbeds are abundant and thick. No thermal metamorphism has been observed in the sediments overlying these interbeds.

**7 Shafer Peak:** owing to the steepness and altitude of the cliff (Fig. 6), the only observations were made by helicopter. The section is several tens of metres thick

**Fig. 4.** Large-scale cross-bedding with scour deposits in the lower part of the Mount Archambault section. Channel bar in braided-river deposits.

and is composed of two sandstone intervals (each some 30 m thick) separated by black igneous rocks. Both intervals have erosive bases, the upper on the black igneous rock, the lower on granite or metasediments.

**8 Timber Peak–Mount New Zealand:** the section is separated into three parts by two igneous interbeds. The lowest section is 35 m thick and displays, at the base, sandstone with large-scale bidirectional cross-bedding, resembling the herringbone structure of a possible tidal environment. Conglomerates occur as thin beds associated with erosion surfaces and form channel lags. The middle part, 12 m thick, is medium to fine grained; the highest, 25 m thick, is finer grained. It contains manganese nodules and displays planar cross-stratification. Basic igneous lavas occur at the base and at the top of the section.

A flora was described by Ricker (1964), who found silicified logs up to 4.5 m in length and 20–30 cm in diameter, some in growth position. Samples were recovered by Gair *et al.* (1965) and discussed together with the flora of Section Peak, which is slightly younger. Norris (1965) was inclined to refer the Timber Peak flora to the Late Triassic.

**9 Skinner Ridge–Mount Mackintosh:** the clastic sequence is intercalated between two thick levels of basic igneous rocks. The lower part of the section contains horizontally bedded or cross-stratified gravels, which fine upwards into cross-bedded gravelly sand. Imbrication is common. A sandy facies follows, with planar and trough cross-stratification. This lower part of the section, 25 m thick, ends with a basaltic flow, 0.5 m thick. The upper part, 10 m thick, is made up of medium- to fine-grained sandstone with shale and thin coal interbeds. A well-preserved piece of tree trunk has been collected from

**Fig. 5.** (A) Cross-bedding formed as a result of the lateral and downstream advance of a channel bar; foresets are high-angle and south-dipping. Sets of gently dipping sandstone occur, with climbing ripple lamination and diagenetic black manganiferous nodules in the upper right. Finer sand and silt with horizontal bedding and small ripples occur at the base. Braided-river environment in the lower part of the Mount Archambault section. (B) Coarse-grained sandstone with channel lags of a braided-river environment at the bottom, passing upwards into finer sandstones with sedimentary structures (dunes) of aeolian origin, possibly as a result of a sharp transition from wet to arid conditions. The aeolian deposit reworks the underlying flood-generated beds. Pebbles in the lower part of the aeolian deposit could represent a deflation surface. Middle part of the Mount Archambault section. (C) Large-scale aeolian dunes with foresets dipping at up to 30°; cosets are 2–10 m in thickness. Upper part of the Mount Archambault section.

**Fig. 6.** The south-western wall of Shafer Peak (3600 m) facing the Priestley Glacier. The Section Peak Formation crops out at two levels, separated by an approximately 300-m-thick black layer of Ferrar basalt. The lower level rests on the Priestley Formation (on the left) or on the Granite Harbour Intrusives (right side). In the upper level the sandstones are closely interbedded with Ferrar volcanics and finally capped by the Kirkpatrick Basalt, which extends northwards to the Mesa Range.

the detritus at the base of the sequence (Fig. 7). It has been recognized as Gymnosperm wood of post-Carnian age (G. Brambilla, palaeobotanist of Pavia University, personal communication).

**10** Mount Nansen: this is the most spectacular outcrop of the Section Peak Formation, forming, with the Kirkpatrick Basalt at the top, a cliff on the Eisenhower Range, covering a large slope cut in the Granite Harbour Intrusives facing the lower Priestley Glacier (Fig. 2). The section sampled on the Thern Promontory is some 15 km from Mount Nansen. The section, 70 m thick, differs from the others in the abundance of coarse-grained material. Conglomerate beds have erosive bases with sharp lateral facies changes. Coarse and gravelly sandstones associated with medium-grained sandstones are present only in the middle part of the section (20 m thick). The section is covered by a thick dolerite sill, studied in detail by Brotzu *et al.* (1988) for petrographic analyses and by Lanza & Zanella (1993) for palaeomagnetic determinations.

**11** At Mount Bowen, about 20 m of sandstones crop out intercalated with two basic igneous layers (Fig. 8). High-angle, cross-bedded, coarse-grained sandstones are intercalated with finer beds. Conglomeratic sandstones with erosive bases are interbedded or overlain by ripple cross-laminated siltstone.

## FACIES ANALYSIS AND DEPOSITIONAL ENVIRONMENTS

The wide range of sedimentary features, the large spacing between sections studied and outcrops observed, and the absence of marker beds make it difficult to correlate the outcrops. The coarsest facies seem to be best developed to the southwest and an overall upward-fining trend was observed. A number of different facies can be described.

**1** Conglomeratic facies. Coarse-grained beds, 0.5–1 m thick, massive and clast-supported, are made up mainly of well-rounded clasts up to 5 cm in diameter and rarely larger. Imbrication is common. The mutually erosive sedimentary bodies are combined into major pebbly sandstone sheets with virtually no evidence of overbank fines. Sharp lateral facies changes of beds of different grain size are common. This facies is found mainly in the Mount Nansen (Thern Promontory) area.

**2** Coarse- to medium-grained sandstone facies with channel lags. This facies comprises a wide range of grain sizes, from pebbly sandstone to thin beds of fine sand. Trough cross-bedding is common, with sets up to 2 m thick; planar cross-bedding is less frequent. The coarse-grained sets show epsilon cross-bedding, dipping at right angles to the palaeocurrent direction, determined from ripple cross-stratification.

**3** Medium- to fine-grained sandstone facies with large-scale cross-bedding. Set thickness is around 1 m, and coset thickness reaches 10 m. First-order bounding surfaces are extensive and of very low inclination; they bound composite units of cross-bedded sets, which are in turn defined by second-order bounding surfaces. This facies has been observed in the upper part of the Archambault section.

The association of the conglomeratic facies with the interbedded, laterally impersistent coarse sandstone facies could indicate a proximal environment, such as a large

**Fig. 7.** Fragment of wood collected in the detritus at the base of the Skinner Ridge (A and B), with transverse thin-sections. Yearly growth rings, up to 1.5 mm wide, are recognizable; their number shows a possible age of 60–80 yr and a diameter of more than 20 cm. The characteristics of the wood: simple tracheids (C and D), uniserial rays up to 15 cells high (average of six), uniserial (90%) and biserial (10%) dotting, demonstrate that the wood is from a conifer of post-Carnian age (i.e. latest Triassic or younger; G. Brambilla, Pavia University, personal communication). Lower left scale bar = 1 mm. Lower right scale bar = 0.25 mm.

**Fig. 8.** Mount Bowen (1874 m) with the sandstone (about 20 m thick) of the Section Peak Formation cross-cut by dykes of Ferrar dolerite near the base of the formation. In this case the intrusion is younger than the sandy deposits and the sandstone lacks any volcanic debris.

alluvial fan dominated by braided-stream processes. High-gradient fluvial systems can carry highly concentrated suspensions of coarse-grained sediment. Rapid downcurrent facies changes from structureless pebbly sandstones into sandstone with traction features were observed in the Mount Nansen section. Individual flood units vary from coarse-grained, highly lenticular units, bounded by erosional surfaces to laterally extensive beds of graded sandstones. Massive conglomerates represent longitudinal bar deposits and the horizontally bedded sandstones lenses are bar-top facies.

The association of medium- to fine-grained sandstone facies with pebbly sandstone and fine sandstone is interpreted as a braided alluvial system, recognized earlier by Collinson *et al.* (1986). Channels are filled with the coarsest material; levee deposits are represented by the sharp lateral transitions, with sands gently sloping from the river bank into flood basins away from the channel. Typical overbank deposits, formed by fine-grained material, are less well developed. Lateral migration of channel bars produces steeper concave sides and gentle convex sides with foreset laminae (Fig. 5). Overall, the beds are characterized by an increasingly tabular geometry. This facies association could represent the downcurrent facies of the flood-dominated, coarse-grained units already described. In a downcurrent direction, textural segregation, produced by transport, gives way to progressively better sorted and finer grained units.

At the scale of the outcrops, the overall proximal-to-distal trend of facies distribution cannot be established easily on the basis of textural characteristics. Although

the downcurrent transformations of flood-generated flows can be inferred from facies changes in the same direction, these changes take place over relatively great distances and cannot be detected over the large spacing between the outcrops. Analogous environments associated with flood basalts and volcaniclastic dispersal were described by Eriksson & Simpson (1993) in Australia and Haughton (1993) in Scotland. Probable aeolian sedimentary structures (mostly sand dunes) have been observed in the fine-grained layers. These could signify aeolian reworking of some beds in a fluvial environment with dry episodes.

Medium- to fine-grained sandstone facies with large-scale cross-bedding may have an aeolian origin. Many properties can indicate an aeolian environment: presence of large-scale cross-bedding, with foresets close to the angle of repose, high-index ripples, fine- to medium-grained well-sorted sands and high degrees of sphericity and roundness. The sets are 1 m thick and are arranged in cosets, which can exceed 10 m. They may have been formed by migrating transverse or barchan dunes. First-order bounding surfaces are extensive and bound composite units of cross-bedded sets, which are in turn defined by second-order bounding surfaces. The outcrops in the upper part of the Mount Archambault section are mostly parallel with the palaeowind: the cross-stratification suggests winds from the north-west. This vertical facies evolution, from true alluvial to aeolian, involves a change of climate from humid to dry conditions. Climate could be the main factor controlling sedimentation patterns and facies distribution in the system considered. The same characteristics have been described for the Clarens Formation (formerly Cave Sandstone; Beukes, 1970), which represents the uppermost stratigraphical unit of Karoo sedimentation in South Africa (Tankard *et al.*, 1982). It is interlayered with and covered by the Karoo volcanics (Cox, 1992), correlative with the Ferrar. It is very interesting to compare the scattered outcrops of the Section Peak Formation with the spectacular and continuous exposures of the Clarens Formation at the Golden Gate National Park in South Africa (Smith *et al.*, 1997) and in the Drakensberg, which were connected to Victoria Land before the Gondwana break-up.

## PROVENANCE OF SANDSTONES AND PALAEOTECTONIC INTERPRETATION

The sandstone composition of the Section Peak Formation has been studied by conventional point-counting techniques, performed on 76 samples collected in the Mount Archambault, Mount Adamson, Mount New Zealand, Skinner Ridge and Mount Nansen sections

(Fig. 9); in addition, compositional data reported by Collinson *et al.* (1986) from the Roberts Butte, Section Peak and Vantage Hill sections have been considered. This data set provides the available petrographic constraints for discussing the provenance of the Section Peak sandstones and the related palaeotectonic setting.

Sandstone composition results are highly variable from sample to sample within the same stratigraphical section; sandstone compositions range from subarkoses and arkoses to lithic arenites (Fig. 10). The bulk of the detritus generally is composed of monocrystalline quartz grains, mainly with undulose extinction patterns and sometimes containing mineral inclusions, such as rutile needles, zircons, biotite and muscovite; polycrystalline quartz grains, both coarse grained and fine grained (single crystals > and < 0.062 mm respectively) occur in minor amounts. Chert is present only occasionally, in the volcanic-bearing samples, suggesting its origin from devitrified volcanic glass. Both K-feldspar and plagioclase occur in highly variable relative amounts (P/F ratio from 0 to 0.8 with an average value of 0.3), possibly as a result of changing plagioclase input from volcanic rocks and selective

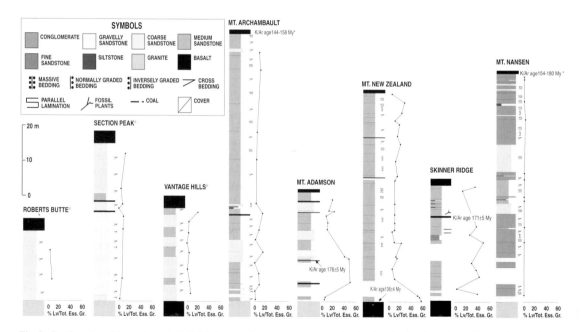

**Fig. 9.** Stratigraphy of the sections studied: lithology, sedimentary structures, distribution of volcanic grains and K/Ar data. Only sections sampled systematically are reported (including the first three, described by Collinson *et al.* (1986)). A clear correlation exists between the percentage of volcanic clasts in the sandstone and the occurrence of basaltic bodies; it proves the partial reworking of basaltic lava flows into the stream deposits. K/Ar ages of basalts are reported from Brotzu *et al.* (1988) and Di Giulio *et al.* (1997). The Middle Jurassic age is demonstrated for most of the samples, the younger ages being explained as the result of Ar loss. The interaction of the Section Peak sandstones with the lava flows indicates coeval sedimentation. They correlate with ages reported in Heimann *et al.* (1994) for the Central Transantarctic Mountains and Victoria Land. These authors report an eruptive activity for the Kirkpatrick Basalt within a short interval of less than about 1 Myr at 176.6 ± 1.8 Ma).

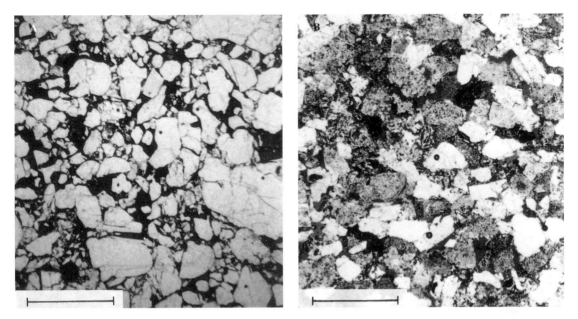

**Fig. 10.** (A) Contrasting composition of sandstones occurring in the Section Peak Formation; plane polarized light thin-section photomicrograph of a subarkose sample. (B) Thin-section photomicrograph of a lithic arenite sample made up mostly of volcanic grains; these types of sandstone represent the two end-members of the compositional range covered by the Section Peak sandstone. Scale bars = 1 mm.

dissolution during diagenesis. Among the lithic grains, metamorphic rock fragments generally are present, but in quite small amounts. In contrast, fragments of basic, deeply chloritized volcanic rock, frequently with fluidal and microlithic textures, occur in highly variable amounts from sample to sample. They constitute the most common grain type in some samples and cause the highly variable composition of the Section Peak sandstones.

This compositional framework is interpreted to be the result of mixing of sand-size grains from two completely different sources, which interplayed in the Section Peak basin. This also is clearly shown by the wide scattering of samples within the provenance-oriented quartz–feldspar–lithic grains (QFL) diagram (Fig. 11), where Section Peak sandstones cover an elongated area lying between two compositional end-members: a relatively stable continental basement source and an undissected volcanic source (Di Giulio *et al.*, 1997).

The lithology of the continental basement source has been investigated by the study of the extinction pattern and the mineral inclusions of quartz grains (Di Giulio *et al.*, 1999). Consistently, these different provenance approaches suggest that the basement-derived grains of the Section Peak sandstones were sourced from a mainly granitic continental basement, including high-grade granulite-facies metamorphic rocks; these features

fit well with that of the Wilson Terrane, which forms the substratum of the Section Peak Formation–Ferrar Group in northern Victoria Land. This conclusion is confirmed by the chemical composition of detrital garnets contained in the Section Peak sandstones, which closely resembles the composition of the garnets from granulite-facies rocks of the Wilson Terrane (Di Giulio *et al.*, 1999).

In contrast, the origin of volcanic grains in the sandstones is constrained by their stratigraphical distribution within the stratigraphical sections sampled (Fig. 9). In most cases it was clear that there is a greater abundance of volcanic grains in the Section Peak samples close to the basaltic rocks at the base or interbedded with the clastic sequences. Their occurrence rapidly decreases away from the basalts and is nearly zero where the basalts are absent. This fact, together with the lack of contact metamorphism in the sandstones resting directly on the basalt, even when they rest on thick volcanic layers, as at the base of some stratigraphical sequences (e.g. in the Mount New Zealand section), supports two fundamental conclusions about the origin of such grains.

**1** It confirms that at least part of the basaltic rocks associated with the Section Peak sediments was produced by subaerial volcanic flows, as supposed initially by Gair *et al.* (1965), Sturm & Carryer (1968) and Crowder (1968). Sturm & Carryer (1968) also reported basalt

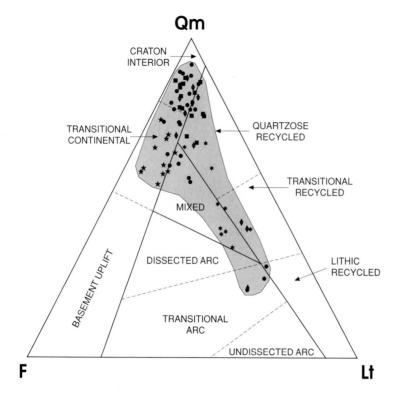

**Fig. 11.** Provenance-oriented Qm–F–Lt plot showing the composition of the Section Peak sandstone in the framework of provenance fields, after Dickinson (1985). The sandstones are widely scattered in an elongated area resting between the stable continental block and the undissected arc end-members. Symbols: circles, Mount New Zealand section; squares, Mount Archambault section; stars, Mount Nansen section; rhombs, Mount Adamson section; asterisks, Skinner Ridge section.

clasts and volcanic bombs present in some of the inter-bedded sediments.

**2** It indicates that the volcanic grains included in the sandstones were produced by erosion of coeval lava flows by the braided streams flowing in the Section Peak basin. Therefore, the undissected volcanic source that fed the volcanic grains to the Section Peak sandstones appears to have been an anorogenic one, and not a subduction-related volcanic arc, as is considered in current prove-nance models (e.g. Dickinson, 1985); indeed, it is not surprising that an undissected volcanic source produces detritus rich in rock fragments whatever its geodynamic setting.

The stratigraphical relationships between the Section Peak clastic sediments and the basaltic lava flows provide a tool for refining the poorly constrained age of the Section Peak Formation through the radiometric dating of interbedded flows. The result is a Mid-Jurassic age, which agrees with the post-Carnian age suggested for the piece of conifer trunk collected in the Skinner Ridge section (Fig. 7).

The resulting palaeogeographical picture is one of a peneplaned continental basement undergoing the begin-ning of tectonic stretching, coupled with continental basaltic volcanism, which preceded the beginning of

Gondwana break-up in northern Victoria Land (Fig. 12). In this framework, the Section Peak Formation records the persistence of 'Beacon-like' sedimentary envir-onments during the beginning of the Ferrar volcanic activity.

## CONCLUSIONS

The new sedimentological and petrographical analyses carried out in the Section Peak Formation, cropping out in northern Victoria Land, throw light on some aspects of the palaeoenvironmental evolution and on the relation-ship with the magmatism related to the Gondwana break-up in that area.

A braided fluvial environment is envisaged for most of the outcrops, in agreement with pre-existing inter-pretations, but in one section the sedimentary structures, coupled with the grain size and textures, demonstrate an evolution to aeolian conditions. This passage entails climatic changes, from humid to dry, for an area (Victoria Land) which was connected to South Africa in Jurassic times. It is not surprising, therefore, that there is a correspondence of environments between the Antarctic sequence and the upper Karoo sequence. Moreover,

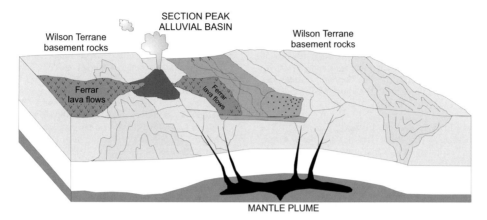

**Fig. 12.** Sketch depicting the inferred relationship between sedimentation and volcanism related to an early phase of rifting in the Section Peak basin. In fact the sub-Beacon peneplain has been found in the Rennick area (Rennick Graben) at different elevations. They are related by Roland & Tessensohn (1987) to a phase of rifting, possibly connected with the initial separation of Antarctica from Australia, although no direct dating evidence is available. In our reconstruction, the early faulting is considered to be older and coeval with the Mid-Jurassic sandy sedimentation, indicating a fault-controlled basin. In other words, the faults might have originated during the Mid-Jurassic rifting and reactivated in younger times.

thin-section study and radiometric age determinations on some basaltic interbeds give the following results.

**1** Some basaltic interbeds are not dolerite sills, as believed previously, but are basaltic lava flows, interlayered with sandstones. In this respect it is not surprising to observe the abundance of volcanic detritus in the sandstones above basalt layers, its rapid decrease away from the basalts and its absence where the basalts do not occur (Fig. 9). Sills, of course, cannot provide volcanic detritus to the overlying sediments. Moreover, these overlying sandstones, even when they cover thick basaltic layers (e.g. in the Mount New Zealand section) do not show any evidence of contact metamorphism. This does not mean that all basalt interbeds are lava flows. In fact, some sills, obviously younger, also have been observed (Fig. 8). The volcanic clasts, produced by coeval flows, are mixed with arkosic components, derived from the granitic and metamorphic basement of the older Wilson Terrane.

**2** The presence of lava flows interlayered with sandstones clearly indicates that the two phases, sedimentary and magmatic, were contemporaneous. Owing to the difficulties of exact age determinations of the flora (the determinations do vary and a Triassic age can be extended to the Jurassic for the Section Peak Formation fossil content), new radiometric data obtained from these lava flows provide an additional tool for defining the age of the Section Peak Formation. They ascribe the volcanism to the Mid-Jurassic and this age therefore must be extended to the coeval fluvial–aeolian sedimentation. Owing to the presence of basaltic lava flows caused by extension and extrusion following the Gondwana break-

up, this phase could be included directly in the Ferrar Supergroup (K.J. Woolfe, personal communication).

**3** The Mid-Jurassic continental rifting, which preceded the Gondwana break-up, could have strongly modified the morphology of the basin, the site of the sandy sedimentation. During the late Palaeozoic–early Mesozoic, the basin was already continental and filled mostly with alluvial braided-river deposits, as depicted by previous authors. The subsequent formation of structural slopes, resulting from extensional tectonics, could have reactivated the efficiency of river transport. As a result, steep, deeply channelled rivers, with a wide range of palaeocurrent directions, transported coarse-grained sediments into a newly formed basin, probably a fault-controlled graben. In fact, the varying elevations of the Section Peak Formation outcrops proves the existence of phases of faulting, the earliest of which resulted from the Mid-Jurassic rifting. No direct proof of synsedimentary faulting has been found up to now, because the main faults are covered by the ice-cap.

## ACKNOWLEDGEMENTS

This research was funded by grants from the Italian Antarctic National Research Programme (PNRA-ENEA). We wish to thank the geological team and the technical staff, and particularly P.C. Pertusati, A. Rossi, F. Rossetti and F. Storti for the field collaboration. Many thanks to J. Isbell and K.J. Woolfe for their careful review, which greatly improved the original paper.

# REFERENCES

BARRETT, P.J. (1991) The Devonian to Jurassic Beacon Supergroup of the Transantarctic Mountains and correlatives in other parts of Antarctica. In: *The Geology of Antarctica* (Ed. Tingey, R.J.), pp. 120–152. *Oxford Monographs on Geology and Geophysics*, **17**, Oxford Scientific Publications, Oxford.

BARRETT, P.J., ELLIOT, D.H. & LINDSAY, J.F. (1986) The Beacon Supergroup (Devonian–Triassic) and Ferrar Group (Jurassic) in Beardmore Glacier area, Antarctica. *Antarct. Res. Ser.*, **36**, 339–428.

BEUKES, N.J. (1970) Stratigraphy and sedimentology of the Cave Sandstone Stage, Karoo System. *Proceedings and Papers, 2nd International Gondwana Symposium*, CSIR, Pretoria, 1970; pp. 321–342.

BROTZU, P., CAPALDI, G., CIVETTA, L., MELLUSO, L. & ORSI, G. (1988) Jurassic Ferrar Dolerites and Kirkpatrick Basalts in Northern Victoria Land (Antarctica): stratigraphy, geochronology and petrology. *Mem. Soc. Geol. Ital.*, **43**, 97–116.

CARMIGNANI, L., GHEZZO, C., GOSSO, G., *et al.* (1988) Geological map of the area between David and Mariner Glaciers (Victoria Land — Antarctica). *Mem. Soc. Geol. Ital.*, **33**.

CASNEDI, R., DI GIULIO, A. & ROSSI, A. (1994) The sandstones of the Beacon Supergroup near Terra Nova Bay (Northern Victoria Land, Antarctica): preliminary results of facies and petrographic analyses. *Terra Antartica*, **1**, 92–95.

COLLINSON, J.W., PENNINGTON, D.C. & KEMP, N.R. (1986) Stratigraphy and petrology of Permian and Triassic fluvial deposits in Northern Victoria Land, Antarctica. *Antarct. Res. Ser.*, **46**, 211–242.

COX, K.G. (1992) Karoo igneous activity, and the early stages of the break-up of Gondwanaland. In: *Magmatism and the Causes of Continental Break-up* (Eds Storey, B.C., Alabaster, T. & Pankhurst, R.J.), Spec. Publ. geol. Soc. London, No. 68, pp. 137–148. Geological Society of London, Bath.

CROWDER, D.C. (1968) Geology of part of North Victoria Land, Antarctica. *U.S. geol. Surv. Prof. Pap.*, **600d**, D95–D107.

DICKINSON, W.R. (1985) Interpreting provenance relations from detrital modes of sandstones. In: *Provenance of Arenites* (Ed. Zuffa, G.G.), pp. 333–361. Reidel Publ., Dordrect.

DI GIULIO, A., CASNEDI, R., CERIANI, A., ORTENZI, A. & ROSSI, A. (1997) Sandstone composition of the Section Peak Formation (Beacon Supergroup, Northern Victoria Land, Antarctica) and relations with the Ferrar Group Volcanics. In: *The Antarctic Region: Geological Evolution and Processes* (Ed. Ricci, C.A.), pp. 297–304. Terra Antartica Pub., Siena.

DI GIULIO, A., TRIBUZIO, R., CERIANI, A. & RICCARDI, M.P. (1999) Integrated analyses constraining the provenance of sandstones, a case study: the Section Peak Formation (Beacon Supergroup, Antarctica). *Sed. Geol.* **124**, 169–183.

ELLIOT, D.H. (1992) Jurassic magmatism and tectonism associated with Gondwanaland break-up: an Antarctic perspective. In: *Magmatism and the Causes of Continental Break-up* (Eds Storey, B.C., Alabaster, T. & Pankhurst, R.J.), Spec. Publ. geol. Soc. London, No. 68, pp. 165–184. Geological Society of London, Bath.

ELLIOT, D.H. (1996) The Hanson Formation: a new stratigraphical unit in the Transantarctic Mountains, Antarctica. *Antarct. Sci.*, **8** (4), 389–394.

ELLIOT, D.H. & FOLAND, K.A. (1986) K–Ar age determination of the Kirkpatrick Basalt, Mesa Range. In: *Geological Investigations in Northern Victoria Land* (Ed. Stump, E.). *Am. geophys. Union, Antarct. Res. Ser.*, **46**, 279–288.

ELLIOT, D.H., HABAN, M.A. & SIDERS, M. (1986a) The Exposure Hill Formation, Mesa Range. In: Geological Investigations in Northern Victoria Land (Ed. Stump, E.). *Am. geophys. Union, Antarct. Res. Ser.*, **46**, 267–278.

ELLIOT, D.H., SIDERS, M. & HABAN, M.A. (1986b) Jurassic tholeiites in the region of the upper Rennick Glacier, north Victoria Land. In: *Geological Investigations in Northern Victoria Land* (Ed. Stump, E.). *Am. geophys. Union, Antarct. Res. Ser.*, **46**, 249–265.

ERIKSSON, K.A. & SIMPSON, E.L. (1993) Siliciclastic braided-alluvial sediments intercalated within continental flood basalts in the Early to Middle Proterozoic Mount Isa Inlier, Australia. In: *Alluvial Sedimentation* (Eds Marzo, M. & Puigdefabregas, C.), Spec. Publs int. Ass. Sediment., No. 17, pp. 473–488. Blackwell Scientific Publications, Oxford.

GAIR, H.S., NORRIS, G. & RICKER, J. (1965) Early Mesozoic microfloras from Antarctica. *N. Z. J. Geol. Geophys.*, **8**, 231–235.

GUNN, B.M. & WARREN, G. (1962) Geology of Victoria Land between the Mawson and Mulock Glaciers, Antarctica. *N.Z. geol. Surv. Bull.*, **80**, 1–100.

HAUGHTON, P.D.W. (1993) Simultaneous dispersal of volcaniclastic and non-volcanic sediment in fluvial basins: examples from the Lower Old Red Sandstone, east-central Scotland. In: *Alluvial Sedimentation* (Eds Marzo, M. & Puigdefabregas, C.), Spec. Publs Int. Ass. Sediment., No. 17, pp. 451–471. Blackwell Scientific Publications, Oxford.

HEIMANN, A., FLEMING, T.H., ELLIOTT, D.H. & FOLAND, K.A. (1994) A short interval of Jurassic continental flood basalt volcanism in Antarctica as demonstrated by $^{40}Ar/^{39}Ar$ geochronology. *Earth Planet. Sci. Lett.*, **121**, 19–41.

KYLE, P.R., ELLIOTT, D.H. & SUTTER, J.F. (1981) Jurassic Ferrar Supergroup tholeiites from the Transantarctic Mountains, Antarctica, and their relationship with the initial fragmentation of Gondwana. In: *Gondwana Five* (Eds Cresswell, M. & Vella, P.), pp. 283–287. A.A. Balkema, Rotterdam.

LANZA, R. & ZANELLA, E. (1993) Palaeomagnetism of the Ferrar dolerite in the northern Prince Albert Mountains (Victoria Land, Antarctica). *Geophys. J. Int.*, **114**, 501–511.

NORRIS, G. (1965) Triassic and Jurassic miospores and acritarchs from the Beacon and Ferrar Groups, Victoria Land, Antarctica. *N. Z. J. Geol. Geophys.*, **8**, 236–277.

RICKER, J. (1964) Outline of the Geology between Mawson and Priestley Glaciers, Victoria Land. In: *Antarctic Geology, SCAR Proceedings 1963, IV, General Geology* (Ed. Adie, R.J.), pp. 265–275. North Holland Pub. Co., Amsterdam.

ROLAND, N.W. & TESSENSOHN, F. (1987) Rennick faulting — an early phase of Ross Sea rifting. *Geol. Jahrb.*, *Reihe B*, **66**, 203–229.

SCHMIDT, D.L. & ROWLEY, P.D. (1986) Continental rifting and transform faulting along the Jurassic Transantarctic rift, Antarctica. *Tectonics*, **5**, 279–291.

SMITH, R., TURNER, B., HANCOX, J. & GROENEWALD, G. (1997) Evolving fluvial landscapes in the main Karoo Basin. *Guidebook 6th International Conference on Fluvial Sedimentology* (Eds Rogers, J. & Cotter, E.), **9**, pp. 1–162. University of Cape Town, Cape Town, 22–26 September.

STURM, A. & CARRYER, S. (1968) Geology of the region between the Matusevich and Tucker Glaciers, North Victoria Land, Antarctica, *N. Z. J. Geol. Geophys*, **13**, 408–435.

TANKARD, A.J., JACKSON, M.P.A., ERIKSSON, K.A., HOBDAY,

D.K., HUNTER, D.R. & MINTER, W.E.L. (1982) *Crustal Evolution of Southern Africa*. Springer-Verlag, Berlin, 523 pp.

TESSENSOHN, F. & MADLER, K. (1987) Triassic plant fossils from North Victoria Land, Antarctica. *Geol. Jahrb., Reihe B*, **66**, 187–201.

WALKER, B.C. (1983) The Beacon Supergroup of Northern Victoria Land, Antarctica. In: *Antarctic Earth Sciences* (Eds Oliver, R.L., James, P.R. & Jago, J.B.), pp. 211–214. Australian Academy of Science, Canberra.

WOOLFE, K.J. & Barrett, P.J. (1995) Constraining the Devonian to Triassic tectonic evolution of the Ross Sea Sector. *Terra Antarct.*, **2** (1), 7–21.

*Spec. Publs int. Ass. Sediment.* (1999) **28**, 451–466

# Reconstruction of fluvial bars from the Proterozoic Mancheral Quartzite, Pranhita–Godavari Valley, India

TAPAN CHAKRABORTY

*Geological Studies Unit, Indian Statistical Institute, 203 B.T. Road, Calcutta 700 035, India*
*(Email: tapan@isical.ac.in)*

## ABSTRACT

Reconstruction of bars in ancient fluvial deposits is crucial in interpreting channel pattern and palaeohydraulics. Excellent exposures of the late Proterozoic alluvial deposits of the Mancheral Quartzite around Ramgundam (18°48′N, 79°25′E) provide an opportunity for such reconstruction. The Mancheral Quartzite is a pebbly, coarse-grained sandstone and conglomerate succession 30–76 m thick. Several braided fluvial facies and an alluvial-plain aeolian facies are well developed in the Mancheral Quartzite at Ramgundam. The planar cross-bedded bar facies overlie major erosional surfaces and is characterized by complex interlayering of planar cross-beds 30–120 cm thick and trough cross-beds 5–15 cm thick.

A number of low-angle, downcurrent-inclined cross-bed bounding surfaces indicate accretion of the bar lee surface. Thickening of the cross-bed sets in a down-palaeoflow direction, the presence of well-developed alternately coarse- and fine-grained foresets and counter-current ripple lamination at the toe of the cross-beds, denote flood-stage accretion of the bedforms into the adjacent pools. Lateral transformation of the larger planar cross-sets into cosets of smaller cross-beds, truncation by smaller trough cross-beds showing flow at high angle to those of the larger planar sets and thin muddy sandstones denote modification of the bar form during falling water stage.

Combined palaeocurrent data from all planar cross-beds show a mean south-westerly orientation, which is assumed to be the local channel direction. Palaeocurrent data from the bar deposits (both planar and trough cross-beds combined) indicate downcurrent, oblique or symmetric accretion of the bars with respect to the local channel direction and are inferred to document lateral or mid-channel braid bar deposits.

The thickness of the bar deposits suggests a shallow depth to the channels (~ 2.5 m). Discharge was characterized by rapid and pronounced flow fluctuation. Internal organization of the bars is intermediate in character between topographically differentiated large braid bars and simple linguoid dunes, which commonly characterize many shallow sand-bed braided channels.

## INTRODUCTION

Bars are the principal depositional elements within rivers. Although most rivers show the presence of lateral, upstream and downstream accretion of bars (Bristow & Best, 1993), different channel types are characterized by a dominant bar type (Bluck, 1979; Haszeldine, 1983a,b; Willis, 1993; Miall, 1994). Reconstruction of bars from ancient alluvium, therefore, serves as an important tool for palaeoenvironmental analysis and reconstruction of palaeochannel characteristics.

A variety of bars, differing in scale and morphology, has been documented from ancient alluvium (e.g. Cant & Walker, 1978; Bluck, 1980; Haszeldine, 1983b; Kirk 1983; Rust & Jones, 1987; Wizevich, 1992; Roe &

Hermansen, 1993; Willis, 1993; Miall, 1994). Such studies have become more popular with increasing use of the techniques of architectural element analysis (*sensu* Miall, 1985). Few of these studies, however, dwell on the details of the internal sedimentary structures that can provide important palaeohydraulic information other than the dominant migration direction of these features. This paper presents a detailed account of inferred bar deposits from the Proterozoic alluvial strata of the Mancheral Quartzite. It also presents estimates of palaeohydrological parameters and attempts to present a generalized model for the reconstruction of shallow braid bars from ancient fluvial deposits.

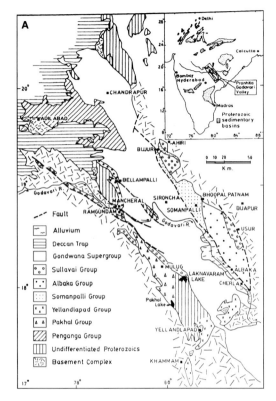

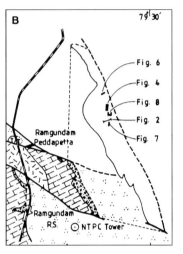

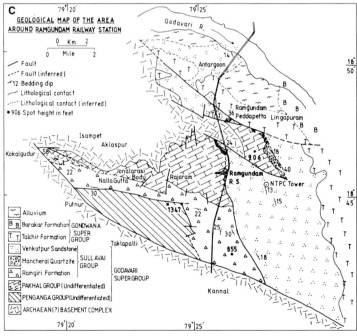

**Fig. 1.** (A) Generalized geological map of Pranhita–Godavari Valley. Inset shows Proterozoic sedimentary basins of India. (B) Location of sections 1 to 4 and vertical log (Fig. 2) through the bar succession in the enlarged outcrop map of Mancheral Quartzite at Ramgundam Gutta. (C) Geological map of the area around Ramgundam.

## GEOLOGICAL SETTING

The Pranhita–Godavari Valley Basin is one of the major Proterozoic basins of India, where two linear belts of Proterozoic rocks, flanking an axial belt of Permo-Jurassic Gondwana rocks are exposed (Fig. 1). The Sullavai Group is the youngest unit of the Proterozoic succession exposed in this basin (see Table 1). The Mancheral Quartzite, a constituent of the Sullavai Group, is very well exposed around Mancheral and Ramgundam in the south-western Proterozoic belt (Fig. 1). In the Ramgundam section, the Mancheral Quartzite unconformably overlies the rocks of the middle Proterozoic Pakhal Group and is overlain by the Venkatpur Sandstone (Table 1). In and around Mancheral, the fluvial strata of the Mancheral Quartzite erosively overlies the Ramgiri Formation of the Sullavai Group and gradationally passes upward through a transition zone of flat-bedded sabkha-playa sediments into the erg deposits of the Venkatpur Sandstone (Chakraborty, 1991a, 1994).

The Mancheral Quartzite comprises poorly sorted, coarse-grained pebbly sandstones and minor conglomerates. Typically, 1–6 m thick, salmon-red, fine- to very fine-grained sandstones occur at different stratigraphical levels and comprise a subordinate part of the succession. The sandstones and conglomerates of the Mancheral Quartzite are interpreted to have been deposited from high-gradient braided streams and the well-sorted, fine-grained sandstone interlayers are the product of fluvial–aeolian interactions in the distal part of the sandy braidplain (Chakraborty, 1991b; Chakraborty & Chaudhuri, 1993). The facies recognized within the Mancheral Quartzite are summarized in Table 2, and Fig. 2 shows the vertical log through a well-exposed section of the succession. Details of the coarse-grained planar cross-bedded sandstone facies (facies 6 in Table 2) are discussed below.

## RECONSTRUCTION OF BARS IN THE MANCHERAL QUARTZITE

Fluvial bars are defined as depositional highs within channels that might become partially exposed during the falling-river stage (Miall, 1981; Bridge, 1985; Roe & Hermansen, 1993). In modern braided rivers, definition of bars has been strongly influenced by channel dimensions. Both a simple bedform (Collinson, 1970; Smith, 1971) and a large periodic to quasi-periodic 'macroform' with complex erosional and depositional history (Bluck, 1976, 1979; Cant & Walker, 1978; Crowley, 1983) have

been designated as bars. The 'bars' within braided streams include a hierarchy of bed configurations, e.g. (in increasing scale) individual bedforms (two-dimensional or three-dimensional), small 'unit' bars, bar complexes or sandflats and mature vegetated islands (Walker & Cant, 1984).

In view of the morphological complexities involved and the difficulty of their recognition in the ancient record, Miall (1981) proposed a simple classification of fluvial bars consisting of three categories:

1 gravelly, planar or massive bedded bars;
2 sandy, simple foreset bars;
3 compound bars of sand or gravel.

Recognition of the first two categories may be easier, but recognition of compound bars requires reconstruction of the palaeomorphology of the bar and involves lateral tracing of sets or cosets of strata, identification of the different order of bounding surfaces and measurement of the palaeocurrent pattern of these structures (Bluck, 1976, 1979). The concept has been applied successfully to the analysis of several ancient fluvial deposits (e.g. Bluck, 1980; Allen, 1983; Haszeldine, 1983a,b; Kirk, 1983; Steel & Thompson, 1983; Wizevich, 1992). The 'bars' inferred in this study are mostly compound types as defined above.

### The bar succession

The inferred bar succession of the Mancheral Quartzite consists predominantly of cosets of planar cross-beds (facies 6, Fig. 3), with minor amounts of small trough cross-beds (facies 5) and laterally impersistent layers of muddy fine-grained sandstone (facies 7). The upward-fining succession has an overall upward-fining trend but may show local coarsening upwards (Fig. 2).

The succession at Ramgundam is organized into several channel-fill successions. Each channel-fill succession is marked at its base by a major erosional surface that is flat to concave-up and can be traced laterally for up to several hundred metres. The surfaces may show local relief of up to a few metres (Chakraborty, 1991a; see also Fig. 8 in Chakraborty & Chaudhuri, 1993). The dominantly planar cross-bedded succession, described below, is separated by a major erosion surface from the underlying trough cross-bedded sandstone of facies 4 (Fig. 2). Elsewhere in the Ramgundam Gutta (hill) the bar succession overlies the conglomeratic sandstone (facies 2) or may uncomformably overlie the Pakhal Limestone. Details of the primary structures, internal bounding surfaces, facies architecture and palaeocurrents of the bar succession are well exposed in several outcrop sections described below.

**Table 1.** Stratigraphical sequence of the southwestern Proterozoic belt, Pranhita-Godavari valley.

| Age | Supergroup/Group | Formation — Mancheral area | Formation — Ramgundam area | Broad lithology and sedimentary structures | Depositional environment |
|---|---|---|---|---|---|
| Permian – Early Cretaceous | Gondwana Supergroup | Talchir Formation — Venkatpur Sandstone (48 m+) | Talchir Formation — Venkatpur Sandstone | Red, fine- to medium-grained, well-sorted sandstone with metre-scale planar cross-beds and flat beds | Erg and erg-margin deposits (Chakraborty, 1991a) |
| Proterozoic | Godavari Supergroup | Mancheral Quartzite (76 m) | *Angular unconformity*<br>Mancheral Quartzite (23 m) ? / Ramgiri Formation (456 m+) ? ? ? | Mancheral Quartzite: purple to red, coarse-grained pebbly sandstone and conglomerate; upward-fining sequences; interlayered fine-grained sandstone units with adhesion structures | Braided fluvial with interlayered aeolian units (Chakraborty & Chaudhuri, 1993) |
| | | Ramgiri Formation (250 m+) | | Ramgiri Formation: red conglomerate and pebbly arkose; conglomerates reverse graded; sandstone cross-bedded; laterally extensive, sheet-like sedimentation units | Distal alluvial-fan–braided-river (Chakraborty, 1991b, 1994) |
| | Penganga Group | Sat Nala Shale / Chanda Limestone / Pranhita Sandstone | *Angular unconformity*<br>Mulug Subgroup<br>*Disconformity*<br>Mallampalli Subgroup<br>*Angular unconformity* | | |
| Archaean (?) | | Gneissic Basement Complex | Gneissic Basement Complex | | |

**Table 2.** Summary of lithofacies recognized in the Mancheral Quartzite.

| Facies number | Brief description | Palaeocurrent | Interpretation |
|---|---|---|---|
| 1 | Matrix- or clast-supported, pebble to boulder conglomerate; beds 11–140 cm thick and show upward-coarsening maximum particle size and upward-fining trend; MPS of individual beds varies from 5 to 84 cm; beds show MPS: bed-thickness positive correlation; few clast-supported beds show clast imbrication and crude horizontal stratification; pervasive hematitic cement locally present; grades upward into trough cross-bedded sandstone of facies 4 | Imbrication in few beds show south-westerly transport | Viscous to non-cohesive debris flow or hyperconcentrated flood flow deposit on alluvial fans |
| 2 | Poorly sorted pebbly sandstone to sandy conglomerate; sandy conglomerate massive; pebbly sandstone cross-bedded; 4–40 mm clasts dispersed throughout the cross-bedded units; cross-sets 8–60 cm thick; occasional sigmoidal foresets; soft-sediment deformation common; facies occurs near the base of the channel-fill sequences; interlayered with facies 3 and grades upward into facies 4 | Exposure mean direction varies from 251° to 296°; dispersion locally high | Rapid deposition from sediment-laden, high velocity flow in shallow channels |
| 3 | Medium- to coarse-grained sandstone with large trough cross-beds; set thickness > 45 cm; > 30 m wide in bedding plane exposures; occur interlayered with facies 2 | Consistent mean flow towards 291° | Deposition from large three-dimensional dunes in deeper parts of the channels |
| 4 | White to grey, fine to very coarse-grained sandstone with few pebbles; 2–15-cm-thick trough cross-beds ubiquitous; cosets of troughs form sheet-like sand bodies bounded by pebble-strewn flat erosional surfaces; sand bodies 30–81 cm (average 50 cm) thick and traceable in strike-parallel direction for 80–110 m; gradationally overlain by facies 7 or 8; at places facies 6 erosively overlies this facies unit | At Ramgundam exposure mean direction varies from 151° to 244°; overall mean 215°; at Mancheral mean direction 006° | Deposition from small three-dimensional dunes in shallow, wide channels occurring at the higher topographic levels/proximal floodplain of braided streams |
| 5 | Red to reddish brown coarse-grained sandstone locally with granules and small pebbles; 2–10-cm-thick trough cross-beds; cosets 5–25 cm thick; occur always as lenticular units interlayered with facies 6 and often fill shallow scours overlying large planar set/coset | Exposure mean varies from 118° to 275°; flow was usually at high angle to that of the enclosing planar cross-beds of facies 6 | Three-dimensional dunes in small, low-stage channels dissecting top of larger bedforms/bars in braided rivers |
| 6 | Deep brown to purple, coarse-grained sandstone locally with granules and small pebbles; planar cross-beds 10–130 cm thick; cosets 50–160 cm thick; planar sets tabular to lenticular and often evolve downcurrent into compound cross-beds; downcurrent-inclined set/coset boundaries; counter-current ripples and reactivation surfaces common; interlayered with facies 5 and 7 | Dispersion is locally high, overall flow consistently south-west; exposure mean direction from large planar cross-beds varies from 204° to 284° | Deposition from migrating braid bars that experienced rapid fluctuation of flow depth and velocity |
| 7 | Red to purple fine-grained muddy sandstone/mudstone; usually ripple laminated; interference ripple marks, shallow channels, pools and desiccation cracks common; lenticular units up to 15 cm thick; interlayered with or overlie facies 6 and 4 | Variable ripple orientation | Deposition in pools/sluggish channels on bar tops or higher topographic levels of braided streams |
| 8 | Typically salmon red, very fine to medium-grained, well-sorted, well-rounded sandstone; abundant adhesion structures; aeolian strata comprise c. 40% of the facies succession; 5–25 cm thick, lenticular aqueous units of massive/faintly cross-laminated sandstone; ferricrete layers at places; in Mancheral area polygonal salt-ridge structures common; gradationally overlie/laterally intertongue with facies 4 | Highly variable palaeowind direction measured from adhesion cross-laminae | Aeolian/aqueous deposits or weathering profile material in the highest topographic levels of the overbank areas of braided rivers |

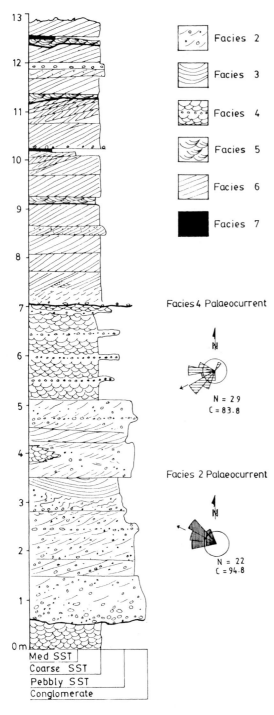

**Fig. 2.** Vertical log through a representative section of Mancheral Quartzite, Ramgundam Gutta. For location of the log see Fig. 1B. The vertically striped rose shows palaeocurrent data from facies 2 and rose with horizontal stripes shows data from facies 4. For all the rose diagrams N = number of data and C = consistency ratio.

## Hierarchy of bounding surfaces

The most prominent bounding surfaces observed within Mancheral Quartzite are those marking the base of the channel-fill successions. These are designated as first-order bounding surfaces. Sections 1, 2 and 3 below show these surfaces as basal erosion surfaces underlying bar successions. Within the overlying planar cross-bedded succession, most prominent bounding surfaces form the erosional bases of the larger, solitary or compound planar cross-sets. These surfaces are irregular to flat and are either subparallel or at a low-angle to the first-order bounding surfaces. These surfaces are designated as second-order bounding surfaces and are displayed most prominently in section 1 (Fig. 4, see foldout). It should be noted, however, that within the limits of outcrop available, these two surfaces have not been found to truncate each other. First-order surfaces are easily recognized, because they separate distinct facies successions and are laterally much more extensive than the second-order surfaces. For example in section 4 (see below), although the first-order surfaces are not exposed the erosional lower bounding surfaces of the larger planar cross-beds in this section facilitate easy recognition of the second-order bounding surfaces. Internal reactivation surfaces within the cross-sets or the intraset boundaries of compound cross-sets constitute the third-order bounding surfaces within the Mancheral bar successions. The shapes of the third-order surfaces are highly variable and have been found to change from convex-up to concave-up within the same set.

## Description of the outcrop sections

### Section 1 (Fig. 4)

The northern end of the section is marked by a triangular unit, comprising a 70-cm-thick, partly deformed planar cross-set (L) and an overlying thin unit of rippled, fine-grained sandstone (8–25 m, Fig. 4). This triangular unit is overlain to the north by a coset of planar cross-sets with set boundaries dipping to the north. To the south second-order bounding surfaces, overlying this deformed cross-bed, are inclined downcurrent. Combined the second-order bounding surfaces in this section define a form set geometry (*sensu* Anatase *et al.*, 1997). South of set L, three major downcurrent-inclined second-order surfaces accommodate three planar cross-sets (marked A, B and C in Fig. 4). These second-order surfaces are inclined 3–8° towards the south with respect to the first-order bounding surface at the base of the section. Each of the three sets thickens considerably (about 10 times in the case of set A) as they migrate over the downcurrent-

**Fig. 3.** Coset of planar cross-beds in a bar succession. Hammer handle is 34 cm long.

**Fig. 5.** Compound cross-beds at the downcurrent end of a large planar cross-bed, section 1, Ramgundam Gutta (set 'B', Fig. 4). Lower left-hand corner of the photograph shows upcurrent-dipping laminae at the southern margin of underlying set 'A' (Fig. 4). The hammer handle is 34 cm long.

dipping basal erosional surfaces. The basal erosional surfaces of sets A and B, in their downcurrent extremity, however, become flat to slightly concave-up, producing a scoop-like shape. Consequently, the overlying cross-sets have a lenticular set geometry. At the northern end, the lower bounding surface of set A (as well as an internal reactivation surface) descends stepwise as it erodes down into the underlying succession (between 15 and 27 m, Fig. 4). Note that finer grained, parallel lamina sets lap over the step-like reactivation surface at its upcurrent end and counter-current ripples are present in the thickest part of set A (between 27 and 30 m, Fig. 4). Inclination of the foreset laminae of the set diminishes in the downcurrent direction and ultimately evolves into concave-up, channel-form layers near its downcurrent termination (34–43 m, Fig. 4). The top of the channel-form layers have been dissected by several small trough cross-beds. The overlying set B also thickens as it migrates over a downcurrent-inclined surface. The upcurrent part of set B is dominated by avalanche foresets, with few intervals containing reactivation surfaces, whereas in the downcurrent part, it evolves into a compound set (Fig. 5). The reactivation surfaces in the upcurrent part of set B

are convex-up, whereas at the terminal part of the set, the third-order (intraset) bounding surfaces become very gentle, slightly convex-up and subparallel to the underlying concave-up (second-order) bounding surface (46–55 m, Fig. 4). As in set A, the downstream margin of set B is dissected by small trough cross-beds containing coarse-grained sandstone with dispersed granules. Palaeocurrent measurements from the small trough cross-beds indicate palaeoflow at a high angle to that of the large planar sets. Near its downcurrent termination, set B and overlying trough cross-beds are capped by a 15-cm-thick unit of rippled, muddy sandstone of facies 7 (52–62 m, Fig. 4). In the central part of the section (around 35 m, Fig. 4), smaller planar cross-sets with granules and pebbles overlie set B. Near the southern end of the section, set B is succeeded by another downcurrent-dipping second-order surface overlain by planar cross-beds (set C) that also thicken in the downcurrent direction.

*Section 2* (Fig. 6; exposed north of section 1)

A basal sheet conglomerate separates the planar cross-stratified succession from the underlying succession of

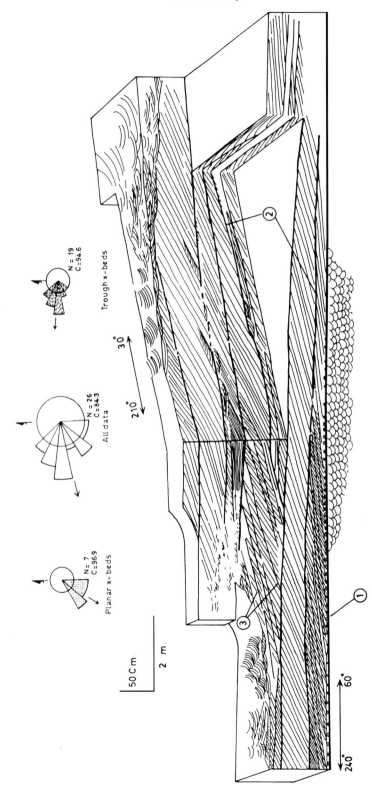

**Fig. 6.** Detailed outcrop diagram of B section 2, exposed north of section 1 (see Fig. 1B for location). Note different orders of bounding surfaces marked by circled numbers. The first-order bounding surface lined with lag conglomerate at the base of the section overlies a succession of smaller trough cross-beds of facies 4.

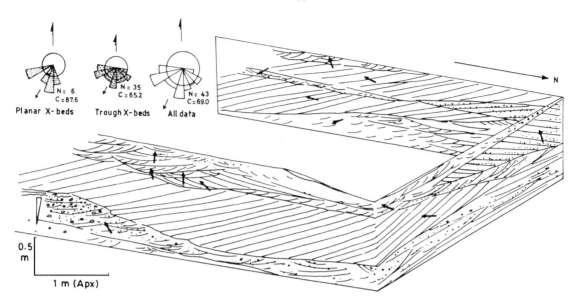

**Fig. 8.** Fence diagram showing the details of section 4. Note divergent accretion of large planar sets and locally upward-coarsening trend within small trough cross-beds (indicated by the inverted arrow, east wall). Palaeocurrent arrows are plotted with respect to the north direction shown in the top right-hand corner of the fence diagram.

facies 4. The section is dominated by superposed sets of large planar cross-beds with a persistent west-south-westerly flow. Planar sets are tabular to wedge-shaped and are characterized by alternating coarse- and fine-grained avalanche foresets with rare to common countercurrent ripples. In the downcurrent direction, some of the foresets become finer grained; their inclination decreases and some evolve into compound cross-beds. Some other planar sets pass laterally into ripple-laminated and small trough cross-bedded zones. Shallow channels (up to 30 cm deep) cut into the top of the planar sets and are filled with small trough cross-beds. Palaeoflow directions of these trough cross-beds are at high angles to the planar sets (Fig. 6).

### Section 3 (Fig. 7, see foldout; south of section 1)

Compared with sections 1 and 2, coarse-grained, small trough cross-bedded units are volumetrically more abundant in this section. Large planar cross-sets interbed with, or laterally pinch out within, trough cross-bedded units at different stratigraphical levels. The troughs show flow both to the south-west and east (Fig. 7). Several rippled, muddy sandstone units (up to 23 cm thick) occur in the sequence and often are associated with the low-angle planar cross-beds having convex-up foresets (near the southern margin of the section).

### Section 4 (Fig. 8; located between sections 1 and 3)

The section, about 50 m in the inferred downstream direction of section 1, contains three different orientations of outcrop faces and allows a three-dimensional reconstruction of the depositional architecture. Organization of facies in the two parallel north–south trending faces is essentially similar to those in other sections. Second-order bounding surfaces are generally irregular, but some of these surfaces are distinctly dipping to the south-west or south-east. Consequently, some of the planar cross-sets thicken in the downcurrent direction (north and west wall of the exposure). Cosets of smaller cross-sets with oblique palaeoflow directions, interbed with or erode down into the larger planar sets. Concentrations of granules and pebbles locally define upward-coarsening trends (southern end of eastern wall). The east–west trending section shows large planar sets migrating oblique to the streamwise direction and thickening away from the axial part of the exposure (Fig. 8).

### Interpretation of the sections

Planar cross-beds that dominate the bar succession probably were deposited by migrating two-dimensional dunes (Ashley, 1990). Bedforms were both simple and compound types, the latter with superposed smaller bedforms.

Similar planar cross-bedded successions are common in many modern and ancient braided fluvial deposits (Collinson, 1970; Smith, 1970; Banks, 1973; Bluck, 1976; Cant, 1978; Haszeldine, 1983a,b; Roe, 1987). Paucity of fine-grained sediments and low directional variability of the planar cross-beds both within and between the exposures (Figs 4, 6, 7 & 8) support deposition in low-sinuosity bedload streams (Willis, 1993). Transition of solitary planar sets downcurrent into compound cross-sets (Figs 4 & 6), presumably formed by the migration of smaller bedforms at falling stages and are reported to be common features of the braided Platte River (Blodgett & Stanley, 1980). Reactivation surfaces indicate frequent fluctuation of flow depth and velocity. Change in the shape of the reactivation surfaces from convex-up to concave-up (set B, Fig. 4) probably represents a progressive lowering of the flow stage (Jones & McCabe, 1980). Erosional truncation of the set or coset of planar cross-beds by lenticular, trough cross-bedded units (Figs 4, 6 & 8) is inferred to represent the divergent, shallow, low-stage channels that dissected the bar tops. Rippled fine sediments (Figs 4 & 7) probably record deposition in slough channels on emergent bar tops (cf. Bluck, 1976).

Many of the planar cross-beds of the sections described above show a remarkable tendency to thicken downcurrent as they migrate over the inclined bounding surfaces (sets A, B and C in Fig. 4; sets in the north and west wall of Fig. 8; and also a few sets in Figs 6 & 7). In order to achieve net deposition, migrating bedforms must climb in the downcurrent direction (Rubin & Hunter, 1982; Rubin, 1987). Consequently the 3–8° southward dips of the set bounding surfaces within the Mancheral bar succession must imply the presence of sloping primary depositional surfaces. In the overall context of the braided river origin of the Mancheral Quartzite, inclined surfaces in these sections (particularly in section 1, Fig. 4) are inferred to represent the lee slopes of downcurrent accreting fluvial bars. Straight-crested dunes grew larger as they migrated down the lee side of the bars into deeper water of adjacent pools (cf. Haszeldine, 1983a,b; Kirk, 1983; Willis 1993). One of the most remarkable features of section 1 is the downcurrent transition of a thick planar cross-bed (set A) into a concave-up trough-shaped lamina-set (Fig. 4). Similar features have been reported earlier from sandy and gravelly braided rivers and have been interpreted variously as channel-fill between two bars (Ramos *et al.*, 1986) or as pool-filling structures (Sigenthaler & Huggenberger, 1993). The second-order bounding surface at the base of the set defines a scoop-shaped depression with a steep, upcurrent margin and slightly concave-up downcurrent end. Set A filling the scour

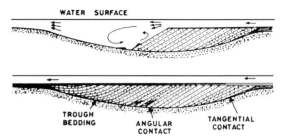

**Fig. 9.** Schematic diagram showing the changing foreset shape of a bedform as it fills a scoop-shaped scour. Note regressive ripples in the thickest, central part of the cross-set (after Jopling, 1965).

shows systematic variation in the foreset geometry. The shallow upcurrent part shows parallel laminae (18–23 m, set A, Fig. 4) and the thickest central part shows angular foresets and counter-current ripples. At the downcurrent end it forms trough-like laminae. The variable foreset geometry is best explained by invoking the depth-ratio concept (depth of water : depth of basin or scour being filled by a bedform (Jopling, 1965; Fig. 9)). At the upstream margin of the scour pool, the flow was shallow (high depth ratio) and formed upper regime plane beds. As the bedform prograded into deeper waters of the scour pool, increased bedform height (lower depth ratio) ensured formation of angular foresets. The increased bedform height also produced well-developed flow separation in the lee of the bedform and back-flow eddies formed regressive ripples at the toe of the foresets. As the bedform approached the other end of the scour, decreasing depth of the scour caused the foresets to become low-angle and tangential. Eventually it formed laminae subparallel to the upcurrent-dipping base of the scour, resulting in trough-shaped lamina (Jopling, 1965; Fig. 9).

The above interpretation supports the inferred existence of deeper-water pools in front of the depositional high, represented by the sequence of deformed cross-beds (set L) and overlying fine sandstone in Fig. 4. The second-order surface at the base of the overlying set B has a similar geometry. The bounding surface at the base of set C, although incompletely exposed, hints at a similar trend. These surfaces are inferred to represent the successive positions of the advancing pool in front of a bar. It is apparent from Fig. 4 that the scour pools migrated both laterally and vertically and the depth of the successive scours varied. Formset-like geometry (cf. Anastase *et al.*, 1997) of the onlapping second-order surfaces across the deformed cross-set (set L) in Fig. 4, probably indicate that the deformed cross-set and the overlying fines acted as a depositional high or core of the bar succession preserved in this section. The up-dip migrating planar sets to the north of the section probably denote upcurrent accretion of the bar core (Miall, 1985; Bristow, 1987; Bristow & Best, 1993). The scours, defined by the scoop-shaped to down-dipping second-order surfaces in the lee of the inferred bar core probably represent positions of migrating-pool structures. Similar bar–pool units have been reported from modern braided rivers as well as laboratory models (Ashmore, 1993; Bluck, 1976; Bristow *et al*., 1993; Clifford, 1993; Ferguson, 1993). Pools probably were formed at the flood stage and dune bedforms encroached upon them as the discharge fell (Clifford, 1993). Progradation of the bar probably was achieved through the migration of the scour pool and the accretion of the pool-filling cross-stratified units on to the bar lee. As the river stage fell, smaller superposed bedforms overtook the larger dunes, forming compound cross-beds and eventually shallow channels dissected the emergent bars. Less commonly, fine-grained sediments formed on top of the exposed bars, to be partly removed during a subsequent flood, when a new scour was formed in the lee of the existing depositional high. Local concentrations of coarser material, within the smaller trough cross-bedded units, probably reflect the winnowing action of shallow flows over emergent bars (Bluck, 1976). Two fine-grained sandstone units are preserved in section 1 (13–17 m and 52–62 m in Fig. 4). Assuming that the fines accumulated at the top of the bars, occurrence of these two units, roughly at the same stratigraphical level above the first-order bounding surface, indicates that positive relief of the bar (with respect to the base of the channel) was maintained over several flood and low-stage cycles.

The shapes of successive scoop-shaped second-order bounding surfaces in Fig. 4 have a remarkable similarity to the computer generated scalloped cross-bedding (fig. 17, Rubin, 1987). In the computer generated cross-bed, bounding surfaces cyclically scoop down into underlying deposits (as in the second-order surfaces in Fig. 4). Rubin (1987) postulated that such structures are produced by migration of bedforms undergoing large and rapid fluctuations of height. Such change in bedform height also has been reported from different depositional settings, including the Brahmaputra River (Coleman, 1969). When combined with downcurrent accretion, changing height of the fluvial bars in response to imposed flow condition is likely to produce features similar to those generated by computer simulation.

It should be noted that the average thicknesses of planar cross-beds in sections 2 and 3 are less than those in sections 1 and 4. In sections 2 and 3, lower bounding surfaces of the planar sets are subparallel to or inclined at very low-angle to the basal first-order bounding surfaces (Figs 6 & 7). These surfaces lack the well-developed concave-up scour morphology observed in section 1. Some features in these sections are consistent with their deposition on braid bars:

1 transformation of larger solitary planar sets into compound cross-beds;
2 erosional truncation by cosets of smaller trough sets;
3 low palaeocurrent dispersion of planar cross-beds as compared with small trough cross-beds.

Smaller bedform size and absence of deep scour pools, as observed in section 1, however, probably indicate that the cross-stratified successions were formed in shallow channels, probably smaller anabranches within the Mancheral braided streams. Downstream transition of solitary planar sets into cosets of small cross-beds and higher dispersion of the palaeocurrent azimuths of smaller trough cross-beds (sections 2, 3 and 4) probably represent converging flow at the lee of larger bedforms during the falling-river stages and is very similar to the type c sandstones described by Roe & Hermansen (1993) from the Precambrian fluvial deposits of Norway. These larger two-dimensional individual bedforms acted as emergent depositional highs during falling stages and were modified or dissected by low-stage shallow flows.

A variety of scours has been reported from modern rivers (see Slater, 1993). Two of the commonest type of scours in braided streams are riffle-pools and confluence scours (Keller & Melhorn, 1978; Ashmore, 1993). Flow convergence can be inferred from sections 3 and 4, but similar evidence is absent in section 1 (Fig. 4), which preserves the scour structures. Therefore, there is no first-hand evidence to relate these features to confluence scours. Correlation of these features with riffle-pool successions is problematic, because the relationship between riffles and braid bars is not very well understood (E. Wohl and N. Clifford, personal communications, 1994). Although it is not known whether the riffles evolve into braid bars the presence of an elongated pool flanking the braid bar is well documented from modern braided rivers (Ashmore, 1993, Ferguson, 1993). Close association with depositional highs, low height and evidence of periodic migration and infilling by unidirectional, downstream-accreting two-dimensional dunes suggest that the scour structures of Mancheral bar successions probably represent bar–pool units (cf. Ashmore, 1993). Development of strong eddy currents, during high flood at the lee of the channel bars, possibly provided the driving mechanism for the formation of these scours. Data from the two-dimensional outcrops of the Mancheral Quartzite, however, are inadequate for precise process-based interpretation of these scour structures.

## PALAEOCURRENTS AND BAR TYPES

Azimuths of planar cross-beds in each of the exposures show a very consistent mean towards the south-west (Figs 4, 6–8) and is inferred to reflect the mean channel direction (Rust, 1972; Bluck, 1976). The small trough cross-beds, within and between exposures, show much greater dispersion (up to 220°) as compared with that of the planar sets (maximum spread of about 100° about the mean). This palaeocurrent pattern is consistent with the origin of small trough cross-beds from shallow, divergent flows that dissected the bar top at the low-river stage (cf. Bluck, 1979).

Palaeocurrent patterns in section 1 indicate dominant flood-stage accretion of the bar to the south-west along the channels. The smaller cross-beds show a more westerly flow direction (Fig. 4). This is inferred to indicate deposition as a mid-channel bar that during falling stages accreted more towards the right (west) bank of the local channel (Bluck, 1976; Hazeldine, 1983b).

Disposition of large planar cross-beds in section 4 (Fig. 8) indicates divergent accretion of larger bedforms into the deeper channels on both flanks of a topographic high, as is typical of mid-channel braid bars (Bluck, 1980, pp. 33–37; Allen, 1983; Bristow, 1993). Erosively overlying troughs in this section show a higher spread, but a symmetrical orientation with respect to the inferred mean channel direction. This pattern also is consistent with the mid-channel bar interpretation of the section.

In section 3 (Fig. 7) some of the larger planar cross-beds initially thicken downcurrent and then pinch out laterally into cosets of smaller trough cross-beds. Whereas the planar cross-beds show consistent flow to the south-west, the trough axes show much greater variation, with stronger modes towards the east or south-west. This probably indicates flow convergence in the lee of the larger bedforms and preferential growth of the bar to the east (left bank of local channel) during low stage. Section 2 (Fig. 6) contains similar successions, but palaeocurrent patterns indicate preferential low-flow-stage accretion to the west (right bank of local channels).

Summarizing the data from the sections, it can be inferred that most of these bars developed as mid-channel depositional features with dominant flood-stage accretion in the downstream direction. During low stage, flow either converged at the lee of the larger two-dimensional bedforms or bars (section 3) or flowed obliquely over the bars, either towards the right (section 1 and 2) or left (section 4) bank of the stream. Low-stage accretion of the bars towards alternate banks suggests a transformation from mid-channel to lateral bars (Bluck, 1976; Haszeldine, 1983b).

## DEPTH OF THE PALAEOCHANNELS

Rippled, fine-grained sandstones in Fig. 4 are inferred to have formed at the top of the emergent bars. Two fine-grained sandstone units (at 13–17 m and 52–62 m in Fig. 4) are about 190 cm above the basal, first-order erosion surface (the inferred channel base). Therefore, one estimate of the relief of the preserved bar in this section is 190 cm. The heights of the bars provide the minimum estimate of the bankful channel depth (Allen, 1983). Accordingly the channels were at least 2 m deep.

As already discussed, the formset geometry of the second-order bounding surfaces in Fig. 4 (with a short stoss and a longer lee side) probably reflect the depositional topography of the palaeobar (cf. Anastase *et al.*, 1997). The highest elevation of the brink point of this topography (above 17 m) is 235 cm above the basal first-order bounding surface. This value provides an alternative estimate of the bar height and would imply a minimum channel depth of about 2.5 m for this section.

It is difficult to make similar estimates for the succession in Fig. 6, because of the lack of preserved depositional topography. The minimum height of the low-stage channels, presumed to be flowing over emergent bars, from the basal first-order erosion surface, however, is about 90 cm (south-western end of Fig. 6). Assuming this value to reflect the thickness of the emergent bar/ bedform in this section, the minimum estimate of the channel depth should be around 1 m. Lower estimates of the palaeochannel depth from the succession in Fig. 6 is consistent with the lower heights of the preserved cross-bed sets.

Flow depth also can be estimated using the relationship

$$H = 0.086d^{1.19} \quad \text{(Allen, 1968)} \tag{1}$$

where $H$ is the mean dune height and $d$ is the flow depth. The calculated value of flow depth ($d$) can be modified further by incorporating compensation for (i) compaction of sand as a result burial, $c$, and (ii) variation in flow depth between straight and sinuous reaches, $c'$. The modified depth is given by

$$d_m = cc'd \quad \text{(Khan, 1987)} \tag{2}$$

Values of $c$ range from 1.1 to 1.2 and those of $c'$ range from 0.585 to 1.0 (Khan, 1987, and references therein). The mean thicknesses of planar cross-beds were calculated for section 1 and section 2 and 117 measurements were collected throughout the Ramgundam area (Table 3). The flow depth calculated by eqns (1) and (2) for the succession preserved in section 1 (Fig. 4) ranges between 146 and 272 cm and that for section 2 varies between 97

**Table 3.** Estimation of the palaeoflow depth on the basis of thickness of facies 6 planar cross-beds.

| Section number/location | Number of data | Mean height of dunes ($H$) (cm) | Depth $d_m = cc' (H/0.086)^{1/1.19}$ (cm) | Remarks |
|---|---|---|---|---|
| Section 1 (Fig. 4); Ramgundam Gutta | 9 | 54.55 | 145.7–271.7 | Height of reconstructed bar 235 cm |
| Section 2 (Fig. 6); Ramgundam Gutta | 10 | 33.66 | 97.01–181.01 | Minimum height of reconstructed bar 90 cm |
| All data from facies 6; Ramgundam area | 117 | 31.65 | 92.20–143.28 | — |

and 181 cm. The average flow depth estimated for the Mancheral braided system (on the basis of facies 6 planar cross-bed thicknesses) ranges between 92 and 145 cm. The calculated values tally closely with the values estimated from the bar morphology and thus support the bar dimensions reconstructed on geometric grounds as discussed earlier.

## TOWARDS A MODEL FOR SHALLOW BRAIDED-RIVER BARS

Bars have been the focus of many sedimentological studies aimed at deciphering the distinctive signature of their depositional processes (e.g. Williams & Rust, 1969; Smith, 1971, 1974; Bluck, 1976, 1979; Cant, 1978; Brierley, 1991; Bristow 1993). Slater (1993) and Bristow *et al.* (1993) recently pointed out that other features of stream beds, notably scours of different types, having high preservation potential and sedimentological implications, have received little attention from sedimentologists. These scours include both quasi-regular riffle-pool sequences (Keller & Melhorn, 1978; Clifford, 1993) and more irregular confluence and constriction scours (Mosley, 1976; Best, 1986; Ashmore *et al.*, 1992). Recognition of scour pools within the Mancheral Quartzite underlines the importance of these features as an essential element of ancient braided-river deposits and provides an actualistic basis for constructing a braided-bar facies model.

The common features that characterize the bar successions in the Mancheral Quartzite are:
1 the presence of a succession of planar cross-beds above a major erosion surface and paucity of fine-grained sediments in the succession;
2 downcurrent thickening of some of the larger planar cross-beds as they migrated over downcurrent-dipping second-order bounding surfaces;
3 transition of the simple planar cross-beds into compound cross-beds or into cosets of smaller trough cross-beds in the downcurrent direction;

4 the presence of shallow channels that cut down into the top of the larger bedforms and are filled with small trough cross-beds — flow in these shallow channels was at a high angle to that forming the larger cross-sets.
5 low dispersion of palaeocurrents measured from large planar cross-beds both within and between exposures — in contrast, directions from the smaller trough cross-beds show higher dispersion.

These features collectively indicate the following inferences.
1 Bars dominantly are comprised of large two-dimensional dunes.
2 The presence of a 'depositional high' or bar and a deeper-water pool in its lee, comparable to the 'pool-bar' units described from modern braided rivers and scaled laboratory models (Ashmore, 1993). In comparatively deeper channels, these scour pools were well developed and in current-parallel sections reveal a 'spoon-shaped' geometry (cf. Ashmore, 1993, p. 131). In shallower channels, pools had a shallower depth and the lower bounding surfaces of the two-dimensional dunes (the inferred bar–pool interface) are either subparallel or subtend a small angle with the first-order bounding surfaces at the base of the sand body (the inferred channel base).
3 The bar successions in the Mancheral Quartzite were formed during rapidly fluctuating flow conditions. During low-flow stage, smaller bedforms developed initially in the lee of the larger bedforms (Blodgett & Stanley, 1980) but as water level fell further, bedforms were dissected by shallow divergent flows.
4 Mancheral river channels had a low sinuosity. Bars were mostly mid-channel type. During low-flow conditions, some of the bars accreted either to the left or right bank of the stream channels.
5 The Mancheral fluvial system was characterized by the contemporaneous existence of both shallow and deeper channels similar to the different order of coexisting channels in braided streams (Bristow & Best, 1993).

Bluck (1976, 1979) recognized multiple topographic levels within braid bars and termed them platform, supra-platform bar-head, bar-head lee and bar tail. A hierarchy of bounding surfaces, inferred to have developed as a result of the lateral migration of these topographically differentiated bars was used by Haszeldine (1983a,b) to reconstruct a 10-m-thick bar succession. A hierarchy of simple periodic to complex quasi-periodic bedforms is known to exist within braided channels and consequently the scale and internal organization of the bar succession is expected to vary considerably. The scale of bedforms considered to be a braid bar is probably strongly influenced by the scale of the channels in which they exist, varying from simple dunes to large channel islands (see Smith, 1978; Miall, 1981). The existence of pool–bar units, convergence of flow in the lee of the bars, dissection of bars or larger bedforms at low stage, coexistence of multiple orders of channels and the presence of mid-channel bars with flow on both sides of the depositional high, however, are typical of most braided rivers, independent of their scale or size of the bedload (Smith, 1970; Rust, 1972; Bluck, 1976; Cant, 1978; Bristow, 1987; Ashmore, 1993; Ferguson. 1993). Recognition of these features provides the key to the reconstruction of the bars in the Mancheral Quartzite. Simpler internal organization and the lower thickness of the bar succession in the Mancheral Quartzite appear to reflect the lack of multiple topographic levels and the simpler morphology of the bars. The bars of the Mancheral Quartzite are probably intermediate in character between large bars with multiple topographic levels and simple linguoid dunes that occur within braided streams.

In many respects, Mancheral bars are analogous to the bar sequences described by Roe & Hermansen (1993) from the Precambrian fluvial deposits of northern Norway. As they point out, unstable sandy banks and the absence of cohesive fines in the overbank areas (see Chakraborty & Chaudhuri (1993) for a description of overbank deposits of the Mancheral Quartzite) of these Precambrian rivers favoured channel widening, in response to increased flood discharge, resulting in a high width : depth ratio. Shallow channels resulted in low bar heights (Roe & Hermansen, 1993, their figs 4 & 8). The internal organization of these Norwegian bars, notably the downcurrent thickening of the larger cross-sets, up- or downcurrent transition of these larger sets into cosets of smaller cross-sets, erosional dissection of the top of the larger sets and low palaeocurrent variability, bears a striking resemblance to the Mancheral bar successions. The grain size of these Norwegian rocks, however, is much finer, which may account for the abundance of plane beds with parting lineation and sigmoidal cross-sets in the succession. The coarser grained deposits of the Mancheral Quartzite contain very few of these features.

## CONCLUSIONS

Well-exposed sections of the planar cross-bedded facies of the Mancheral Quartzite allow reconstruction of the shallow braid bars that characterized the precursor channels. The exposures reveal:

**1** sets of cross-beds with downcurrent-inclined bounding surfaces, which denote accretion of bedforms on the front of the depositional highs (bars) into the adjacent pools;

**2** scoop-shaped second-order bounding surfaces and overlying sets or cosets of planar cross-beds are inferred to be pool-filling successions;

**3** the bar successions were characterized by fluctuating flow, flow convergence in the lee of the bars/bedforms and dissection of the bar tops;

**4** depositional channels had a low sinuosity and bars were mostly of mid-channel types;

**5** different lines of evidence suggest maximum bar-forming depths of about 2.5 m.

## ACKNOWLEDGEMENTS

Part of the data presented here has been extracted from the author's PhD thesis, carried out under the supervision of Professor A.K. Chaudhuri of the Indian Statistical Institute. Very competent field assistance was provided by S.N. Das. Drafting and redrafting of the diagrams were patiently carried out by A.K. Das. I am grateful to all of them. Exchanges with Nicholas Clifford (University of Hull) and Ellen Wohl (Colorado State University) on the riffle-pool sequences were really instructive. I gratefully acknowledge the constructive reviews of Michael Wizevich and an anonymous reviewer and the editorial efforts of Norm Smith that greatly improved the manuscript.

## REFERENCES

ALLEN, J.R.L. (1968) *Current Ripples, Their Relation to Patterns of Water and Sediment Motion*. North Holland, Amsterdam, 433 pp.

ALLEN, J.R.L. (1983) Studies in fluviatile sedimentation: bars, bar-complexes and sandstone sheets (low-sinuosity braided streams) in Brownstones (L. Devonian), Welsh Borders. *Sediment. Geol.*, **33**, 237–293.

ANASTASE, A.S., DALRYMPLE, R.W., JAMES, N.P. & NELSON, C.S. (1997) Cross-stratified calcarenites from New Zealand: subaqueous dunes in a cool-water, Oligo-Miocene seaway. *Sedimentology*, **44**, 869–891.

ASHLEY, G.M. (1990) Classification of large-scale subaqueous bedforms: a new look at an old problem. *J. sedim. Petrol.*, **60**, 160–172.

ASHMORE, P. (1993) Anabranch confluence kinetics and sedimentation process in gravel-braided streams. In: *Braided Rivers* (Eds Best, J.L. & Bristow, C.S.), Spec. Publ. geol. Soc. London, No. 75, pp. 129–146. Geological Society of London, Bath.

ASHMORE, P.E., FERGUSON, R.I., PRESTGAARD, K.L., ASHWORTH, P.J. & PAOLA, C. (1992) Secondary flow in anabrach confluences of a braided gravel bed stream. *Earth Surface Process. Landf.*, **17**, 299–311.

BANKS, N.L. (1973) The origin and significance of some down-current dipping cross-stratified sets. *J. sedim. Petrol.*, **43**, 423–427.

BEST, J.L. (1986) The morphology of river channel confluences. *Progr. phys. Geogr.*, **10**, 157–174.

BLODGETT, R.H. & STANLEY, K.O. (1980) Stratification, bedforms and discharge relations of the Platte braided river system. *J. sedim. Petrol.*, **50**, 139–148.

BLUCK, B.J. (1976) Sedimentation in some Scottish rivers of low sinuosity. *Trans. R. Soc. Edinburgh*, **69**, 425–456.

BLUCK, B.J. (1979) Structure of coarse grained braided stream alluvium. *Trans. R. Soc. Edinburgh*, **70**, 181–221.

BLUCK, B.J. (1980) Structure, generation and preservation of upward fining, braided stream cycles in the Old Red Sandstone of Scotland. *Trans. R. Soc. Edinburgh*, **71**, 29–46.

BRIDGE, J.S. (1985c) Paleochannels inferred from the alluvial deposits: a critical evaluation. *J. sedim. Petrol.*, **55**, 579–589.

BRIERLEY, G.J. (1991) Bar sedimentology of the Squamish River, British Columbia: definition and application of morphostratigraphic units. *J. sediment. Petrol.*, **61**, 211–225.

BRISTOW, C.S. (1987) Brahmaputra River: channel migration and deposition. In: *Recent Developments in Fluvial Sedimentology* (Eds Ethridge, F.G., Flores, R.M. & Harvey, M.D.), Spec. Publ. Soc. econ. Paleont. Miner., Tulsa, **39**, 63–74.

BRISTOW, C.S. (1993) Sedimentary structures exposed in the bar tops in the Brahmaputra River, Bangladesh. In: *Braided Rivers* (Eds Best, J.L. & Bristow, C.S.), Spec. Publ. geol. Soc. London, No. 75, pp. 277–289. Geological Society of London, Bath.

BRISTOW, C.S. & BEST, J.L. (1993) Braided rivers: perspectives and problems. In: *Braided Rivers* (Eds Best, J.L. & Bristow, C.S.), Spec. Publ. geol. Soc. London, No. 75, pp. 1–11. Geological Society of London, Bath.

BRISTOW, C.S., BEST, J.L. & ROY, A.G. (1993) Morphology and facies models of channel confluences. In: *Alluvial Sedimentation* (Eds Marzo, M. & Puigdefabrigas, C.), Spec. Publs int. Ass. Sediment., No. 17, pp. 91–100. Blackwell Scientific Publications, Oxford.

CANT, D.J. (1978) Bedforms and bar types in South Saskatchewan River. *J. sedim. Petrol.*, **48**, 1321–1330.

CANT, D.J. & WALKER, R.G. (1978) Fluvial processes and facies sequences in the sandy braided South Saskatchewan River, Canada. *Sedimentology*, **25**, 625–648.

CHAKRABORTY, T. (1991a) Sedimentology of a Proterozoic erg: the Venkatpur Sandstone, Pranhita-Godavari Valley, south India. *Sedimentology*, **38**, 301–322.

CHAKRABORTY, T. (1991b) *Stratigraphy and sedimentation of the Proterozoic Sullavai Group in the south-central part of the Pranhita-Godavari Valley, Andhra Pradesh, India.* PhD thesis, Jadavpur University, Calcutta, 188 pp.

CHAKRABORTY, T. (1994) Stratigraphy of the Late Proterozoic Sullavai Group, Pranhita-Godavari Valley, Andhra Pradesh. *Indian J. Geol.*, **66**, 124–147.

CHAKRABORTY, T. & CHAUDHURI, A.K. (1993) Fluvial aeolian interactions in a Proterozoic alluvial plain: example from Mancheral Quartzite, Sullavai Group, Pranhita-Godavari Valley, India. In: *Dynamics and Environmental Context of Aeolian Sedimentary Systems* (Ed. Pye, K.), Spec. Publ. geol. Soc. London, No. 72, pp. 127–141. Geological Society of London, Bath.

CLIFFORD, N.J. (1993) Formation of riffle-pool sequences: field evidences for an autogenic process. *Sediment. Geol.*, **85**, 39–51.

COLEMAN, J.M. (1969) Brahmaputra River: channel processes and sedimentation. *Sediment. Geol.*, **3**, 129–239.

COLLINSON, J.D. (1970) Bedforms of the Tana river, Norway. *Geogr. Annaler.*, **52a**, 31–56.

CROWLEY, K.D. (1983) Large scale bed configurations (macroforms), Platte River basin, Colorado and Nebraska: Primary structures and formative processes. *Geol. Soc. Am. Bull.*, **94**, 117–133.

FERGUSON, R.I. (1993) Understanding braiding processes in gravel-bed rivers: progress and unsolved problems. In: *Braided Rivers* (Eds Best, J.L. & Bristow, C.S.), Spec. Publ. geol. Soc. London, No. 75, pp. 1–11. Geological Society of London, Bath.

HASZELDINE, R.S. (1983a) Descending tabular cross-bed sets and bounding surfaces from a fluvial channel, Upper Carboniferous coalfield of Northeast England. In: *Modern and Ancient Fluvial Systems* (Eds Collinson, J.D. & Lewin, J.), Spec. Publs int. Ass. Sediment., No. 6, pp. 449–456. Blackwell Scientific Publications, Oxford.

HASZELDINE, R.S. (1983b) Fluvial bars reconstructed from deep straight channel, Upper Carboniferous coalfield of northeast England. *J. sediment. Petrol.*, **53**, 1223–1247.

JONES, C.M. & MCCABE, P.J. (1980) Erosion surfaces within giant fluvial cross-beds of Carboniferous in Northern England. *J. sediment. Petrol.*, **50**, 613–620.

JOPLING, A.V. (1965) Hydraulic factors and shape of laminae. *J. sediment. Petrol.*, **35**, 777–791.

KELLER, E.A. & MELHORN, W.N. (1978) Rhythmic spacing and origin of pools and riffles. *Geol. Soc. Am. Bull.*, **89**, 723–730.

KIRK, M. (1983) Bar developments in a fluvial sandstone (Westphalian 'A'), Scotland. *Sedimentology*, **30**, 727–742.

KHAN, Z.A. (1987) Paleodrainage and paleochannel morphology of a Barakar river (Early Permian) in the Rajmahal Gondwana basin, Bihar, India. *Palaeogeogr. Palaeoclimatol. Palaeoecol.*, **58**, 235–247.

MIALL, A.D. (1981) *Analysis of Fluvial Depositional Systems*. American Association of Petroleum Geologists, Tulsa, Education Course Note Series No. 20, 75 pp.

MIALL, A.D. (1985) Architectural element analysis: a new method of facies analysis applied to fluvial deposits. *Earth Sci. Rev.*, **22**, 261–308.

MIALL, A.D. (1994) Reconstructing fluvial macroform architecture from two-dimensional outcrops: examples from Castlegate Sandstone, Book Cliffs, Utah. *J. sediment. Res.*, **B64**, 146–158.

MOSLEY, M.P. (1976) An experimental study of channel confluences. *J. Geol.*, **84**, 535–562.

RAMOS, A., SOPEÑA, A. & PÉREZ-ARLUCEA, M. (1986) Evolution of the Buntsandstein fluvial sedimentation in the northwest Iberian ranges (central Spain). *J. sediment. Petrol.*, **56**, 862–875.

ROE, S.L. (1987) Cross-strata and bedforms of probable transitional dunes to upper-stage plane-bed origin from a Late

Precambrian fluvial sandstone, northern Norway. *Sedimentology*, **34**, 89–101.

ROE, S.L. & HERMANSEN, M. (1993) Processes and products of large Late Precambrian sandy rivers. In: *Alluvial Sedimentation* (Eds Marzo, M. & Puigdefabrigas, C.), Spec. Publs int. Ass. Sediment., No. 17, pp. 151–166. Blackwell Scientific Publications, Oxford.

RUBIN, D. (1987) *Cross-bedding, Bedforms and Paleocurrent.* Society of Economic Paleontologists and Mineralogists, Tulsa, OK, Concepts in Sedimentology and Palaeontology, **V**, 187 pp.

RUBIN, D. & HUNTER, R. (1982) Bedform climbing in theory and nature. *Sedimentology*, **29**, 121–138.

RUST, B.R. (1972) Structure and processes in a braided river. *Sedimentology*, **18**, 221–245.

RUST, B.R. & JONES, B.G. (1987) The Hawkesbury Sandstone south of Sydney, Australia: Triassic analogue of a large braided river. *J. sediment. Petrol.*, **57**, 222–233.

SIGENTHALER, C. & HUGGENBERGER, P. (1993) Evidence of dominant pool preservation in Rhine gravels. In: *Braided Rivers* (Eds Best, J.L. & Bristow, C.S.), Spec. Publ. geol. Soc. London, No. 23, pp. 291–304. Geological Society of London, Bath.

SLATER, T. (1993) Fluvial scour and incision: model for their influence on the development of realistic reservoir geometry. In: *Characterisation of Fluvial and Aeolian Reservoirs* (Eds North, C.P. & Prosser, D.J.), Spec. Publ. geol. Soc. London, No. 73, pp. 33–51. Geological Society of London, Bath.

SMITH, N.D. (1970) Braided stream depositional environment: comparison of the Platte river with some Silurian clastic rocks, north-central Appalachian. *Geol. Soc. Am. Bull.*, **81**, 2993–3014.

SMITH, N.D. (1971) Transverse bars and braiding in the lower Platte river, Nebraska. *Geol. Soc. Am. Bull.*, **81**, 3407–3420.

SMITH, N.D. (1974) Sedimentology and bar formation in the upper Kicking Horse River, a braided outwash stream. *J. Geol.*, **82**, 205–224.

SMITH, N.D. (1978) Some comments on terminology for bars in shallow rivers. In: *Fluvial Sedimentology* (Ed. Miall, A.D.), Mem. Can. Soc. petrol. Geol., Calgary, **5**, 85–88.

STEEL, R.J. & THOMPSON, D.B. (1983) Structures and textures in Triassic braided stream conglomerate ('Bunter' Pebble Beds) in the Sherwood Sandstone Group, North Staffordshire, England. *Sedimentology*, **30**, 341–367.

WALKER, R.G. & CANT, D.J. (1984) Sandy fluvial systems. In: *Facies Models*, 2nd edn (Ed. Walker, R.G.). Geosci. Can. Reprint Ser. 1, 23–31.

WILLIAMS, G.E. & RUST, B.R. (1969) The sedimentology of a braided river. *J. sediment. Petrol.*, **39**, 649–679.

WILLIS, B. (1993) Ancient river system from the Himalayan foredeep, Chinji Village area, northern Pakistan. *Sediment. Geol.*, **8**, 1–76.

WIZEVICH, M.C. (1992) Sedimentology of the Pennsylvanian quartzose sandstones of the Lee Formation, central Appalachian Basin: fluvial interpretation based on lateral profile analysis. *Sediment. Geol.*, **78**, 1–47.

# Index

Page numbers in *italics* refer to charts and figures; page numbers in **bold** refer to tables.

Kazinze (Zambia)
  open pit, 410, *419*, 432
    coal characteristics, **419**
    facies associations, *426*
    mudstones, *420*
Keelbottom Creek (Australia), *353*, 355
Kimberley (Australia), 78
King River (Australia), avulsions, 175
Kirkpatrick Basalt (Antarctica), 435, 438, *441*
  lava flows, 439
Klein Hydroscan, Model 401, applications, 34
kolks, mechanisms, 4
Kootenay River (Canada)
  channel bottom profiles, 121, *122*
  channel fills, 121
  eddy accretions
    cross-sectional profiles, 121, *123*
    studies, 113–30
  floodplain, 116
  flow velocity, profiles, 121–3
  point bars, 119
  reaches, confined, 128
  separation-zone ridges, 124
  stratigraphy, *117*, *120*
  study area, 115–16
  vibracoring studies, *117*
Kosi River (Himalayas)
  catchment area, 307
  megafans, 310
kriging, applications, 45, 65
Kruger National Park (South Africa), 132
Kukri Peneplain (Antarctica), 436

lakes, evolution, *246*
LANDSAT images, *195*, *198*, *199*, *224*, *225*, *309*
Languedoc Basin (France), alluvial architecture, 278
Lashly Formation (Antarctica), 436
lateral migration, processes, 172
Liard River (Canada), reaches, unconfined, 128
lithofacies
  and eddy accretions
    interpretation, 116–19
    successions, 121
  Gwembe Coal Formation, 410–24
    characteristics, **413–17**
  Mancheral Quartzite, **455**
  point bars, 119
lithostratigraphical logs
  applications, 116
  eddy accretion studies, 116–19
Little River (US), 213
Liveringa (Australia), floods, *80*
Lobith (Netherlands)
  discharge rates
    frequency distributions, *66*
    measurements, 62
  sediment transport, **65**, *66*
log jams, and avulsions, 176

Long Island Sound (US), bedform studies, 40
Loskop Formation (South Africa), 381
Lower Himalayas, 306
  river systems, 307
    deflections, 310–12
Lower Karoo Group (Zambia)
  sedimentology, 409–34
  stratigraphy, *411*
Lower Mississippi Valley (LMV) (US)
  avulsion failure, 211–20
  avulsions, 264
  geomorphology, *213*
  gradients, 214
  meanders, 211
  study area, 213

Maamba Coalfield (Zambia)
  coal, characteristics, **405**, **419**
  exposure, *418*
  facies, 410
Maamba Sandstone (Zambia)
  deposition, 429–30
  facies associations, 424, 432
  palaeocurrents, *427*
  stratigraphy, 410–18
macroturbulence, generation, 54
Madumabisa Mudstone Formation (Zambia), 409
Main Waterberg basin (South Africa)
  sample area, *382*
  sedimentology, 381
Mala Geosciences RAMAC/GPR *see* RAMAC/GPR system
Mancheral Quartzite (India)
  bars
    characterization, 463
    planar cross-beds, *457*
    reconstruction, 451–66
    succession, 453
    types of, 462
  bedforms, foreset geometry, *460*
  bounding surfaces, 456, *458*
  composition, 453
  facies, 453
  geological setting, 453
  lithofacies, **455**
  outcrop sections, 456–9
    fence diagram, *459*
    interpretation, 459–61
  palaeochannels, depth, 462–3
  stratigraphy, **454**
  study areas, *452*
  vertical log, *456*
Manning's resistance coefficient, *140*
Margaret River (Australia), 78
  satellite image, *83*
Marines Formation (Spain)
  eustatic changes, 297
  fluvial styles, 293
mean velocity
  normalized, *7*
    contour plots, *8*, *10*

mean water depth, determination, 383
meander bend reconstruction
  genetic packages, 371–5
  morphology and hydrology, **373**
  and point bars, 363–79
  processes, 375–6
meanders
  confined
    eddy accretions in, 113–30
    floodplain-width vs. channel-width, *128*
  direction changes, 114–15
  Fitzroy River, *88*
  formation, 252
megafans, 310
Meghna River (Bangladesh), 44, 223
*Melaleuca argentea*, Burdekin River, 350, *353*, 357, 358
Melonhole Creek (Australia), 81
Meuse, River (Netherlands)
  avulsions, 264
  crevasse splays, 252, 262
  dunes, 30
*Microcodium* spp., occurrence, 428
mid-Zambezi Valley Basin, 431
  mudrocks, 429
  sedimentology, 409–34
  stratigraphy, *411–12*
  study area, 410
Middelburg basin (South Africa)
  location, 390
  sample area, *382*
  sedimentology, 381
Milk River (Canada)
  reaches
    confined, 128
    unconfined, 128
mining
  historical background, 161
  impacts, on estuarine sediments, 161–8
Mississippi River (US)
  aggradation, 212–13
  avulsion deposits, thickness, 264
  avulsion failure, 211–20
  avulsions, 175, 212–13
  crevasse channel, 212
  discharge rates, 218–19
  distributaries, 218
  eddy accretions, 113, 114, 124
  floods, 132
  meanders, 114–15, 211, *212*, 215
  navigation traffic, 159
  and sea-level lowering, 273
  study area, 213
  valley slope, 274
  *see also* Lower Mississippi Valley (LMV) (US)
Missouri River (US)
  aerial photographs, 181, *182*, *183*, *187*
  dams, 181
  flow regulation, 189–90
  sediment transport, 180
  study area, *180*